Probability and Random Processes

Probability and Random Processes

SECOND EDITION

G. R. GRIMMETT
Statistical Laboratory, University of Cambridge

D. R. STIRZAKER
Mathematical Institute, Oxford University

CLARENDON PRESS · OXFORD

Oxford University Press, Walton Street, Oxford OX2 6DP

Oxford New York
Athens Auckland Bangkok Bombay
Calcutta Cape Town Dar es Salaam Delhi
Florence Hong Kong Istanbul Karachi
Kuala Lumpur Madras Madrid Melbourne
Mexico City Nairobi Paris Singapore
Taipei Tokyo Toronto
and associated companies in
Berlin Ibadan

Oxford is a trade mark of Oxford University Press

Published in the United States by
Oxford University Press Inc., New York

First published 1982
Second edition 1992
Reprinted 1993, 1994, 1995

A catalogue record for this book is available from the British Library

Library of Congress Cataloging in Publication Data
(Data applied for)

ISBN 0 19 853665 8 (Pbk)

Printed in Hong Kong

Lastly, numbers are applicable even to such things as seem to be governed by no rule, I mean such as depend on chance: the quantity of probability and proportion of it in any two proposed cases being subject to calculation as much as anything else. Upon this depend the principles of game. We find sharpers know enough of this to cheat some men that would take it very ill to be thought bubbles; and one gamester exceeds another, as he has a greater sagacity and readiness in calculating his probability to win or lose in any particular case. To understand the theory of chance thoroughly, requires a great knowledge of numbers, and a pretty competent one of Algebra.

John Arbuthnot
An essay on the usefulness of mathematical learning
25 November 1700

Preface to the second edition

Our objectives in preparing this second edition have remained unaltered since the publication of the first. We hope that it attains some of the following targets:

to be suitable for mathematics undergraduates at all levels, and to be a useful reference book for graduate students and others with interests in probability and its applications;

to be a rigorous introduction to probability theory without burdening the reader with a great deal of measure theory;

to discuss random processes in some depth with many examples;

to include certain topics which are suitable for undergraduate courses but which are rarely taught;

to impart to the beginner some flavour of more advanced work.

We assume that the reader has certain fundamental mathematical skills, such as a familiarity with elementary set theory and analysis.

The better to achieve these targets, we have reworked much of the material in the first edition, and have made some substantial changes and additions. Principally,

(a) new and expanded sections in the early chapters provide more illustrative examples, and introduce more ideas early on,
(b) two new chapters provide fuller treatments of the simpler properties of martingales and diffusion processes,
(c) we have added exercises at the ends of almost all sections, together with many new problems at the ends of chapters.

In a companion volume entitled *Probability and random processes: problems and solutions* (Clarendon Press, Oxford, 1992) appear complete worked solutions to all exercises and problems of this second edition.

The basic layout of the book remains unchanged. Chapters 1–5 begin with the foundations of probability theory, move through the elementary properties of random variables, and finish with the weak law of large numbers and the central limit theorem; on route, the reader meets random walks, branching processes, and characteristic functions. This material is suitable for about two lecture courses at a moderately elementary level. The rest of the book is largely concerned with random processes. Chapter 6 deals with Markov chains, treating discrete-time chains in some detail (and including an easy proof of the ergodic theorem for chains with countably infinite state spaces) and continuous-time chains largely by example. Chapter 7 contains a general discussion of convergence, together with simple but rigorous accounts of the strong law of large numbers, and martingale convergence. Each of these two chapters could be used as a basis of a lecture course. Chapters 8–13 are more fragmented and provide suitable material for about five shorter lecture courses on stationary processes and ergodic theory, renewal processes, queues, martingales, and diffusions.

The ends of proofs and examples are indicated by the symbols ■ and ●.

We thank those who read and commented upon sections of this work. Brian Davies read an early draft, and Tim Brown read large chunks of the final draft of the first edition; both had many valuable ideas and suggestions about possible improvements. Sean Collins made many suggestions about our treatment of discrete-time Markov chains; Stephen Suen found too many errors for our liking. We are grateful also to Geoff Eagleson, Professor G. E. H. Reuter, David Green, and Bernard Silverman.

Of great value in the preparation of the second edition were the detailed criticisms of Michel Dekking, Frank den Hollander, and Torgny Lindvall. Certain sections of the current version owe much to Erwin Bolthausen and Colin McDiarmid.

Finally, we are both especially grateful to Dominic Welsh. He taught us both when we were undergraduates and graduate students, and later gave us the idea of writing this book. He must take sole responsibility for all serious mathematical errors, details of which may be addressed directly to him.

Bristol and Oxford G.R.G.
June 1991 D.R.S.

Contents

1 Events and their probabilities

1.1 Introduction

Much of our life is based on the belief that the future is largely unpredictable. For example, games of chance such as dice or roulette would have few adherents if their outcomes were known in advance. We express this belief in chance behaviour by the use of words such as 'random' or 'probability', and we seek, by way of gaming and other experience, to assign quantitative as well as qualitative meanings to such usages. Our acquaintance with statements about probability relies on a wealth of concepts, some more reasonable than others. A mathematical theory of probability will incorporate those concepts of chance which are expressed and implicit in common rational understanding. Such a theory will formalize these concepts as a collection of axioms, which should lead directly to conclusions in agreement with practical experimentation. This chapter contains the essential ingredients of this construction.

1.2 Events as sets

Many everyday statements take the form 'the chance (or probability) of A is p', where A is some event (such as 'the sun shining tomorrow', 'Cambridge winning the Boat Race', ...) and p is a number or adjective describing quantity (such as 'one-eighth', 'low', ...). The occurrence or non-occurrence of A depends upon the chain of circumstances involved. This chain is called an *experiment* or *trial*; the result of an experiment is called its *outcome*. In general, we cannot predict with certainty the outcome of an experiment in advance of its completion; we can only list the collection of possible outcomes.

(1) **Definition.** The set of all possible outcomes of an experiment is called the **sample space** and is denoted by Ω.

(2) **Example.** A coin is tossed. There are two possible outcomes, heads (denoted by H) and tails (denoted by T), so that

$$\Omega = \{H, T\}.$$

We may be interested in the possible occurrences of the following events:

(a) the outcome is a head;
(b) the outcome is either a head or a tail;

(c) the outcome is both a head and a tail (this seems very unlikely to occur);
(d) the outcome is not a head. ●

(3) **Example.** A die is thrown once. There are six possible outcomes depending on which of the numbers 1, 2, 3, 4, 5, or 6 is uppermost. Thus

$$\Omega = \{1, 2, 3, 4, 5, 6\}.$$

We may be interested in the following events:

(a) the outcome is the number 1;
(b) the outcome is an even number;
(c) the outcome is even but does not exceed 3;
(d) the outcome is not even. ●

We see immediately that each of the events of these examples can be specified as a subset A of the appropriate sample space Ω. In the first example they can be rewritten as

(a) $A = \{H\}$, (b) $A = \{H\} \cup \{T\}$,
(c) $A = \{H\} \cap \{T\}$, (d) $A = \{H\}^c$,

whilst those of the second example become

(a) $A = \{1\}$, (b) $A = \{2, 4, 6\}$,
(c) $A = \{2, 4, 6\} \cap \{1, 2, 3\}$, (d) $A = \{2, 4, 6\}^c$.

(The *complement* of a subset A of Ω is denoted here and subsequently by A^c; henceforth subsets of Ω containing a single member, such as $\{H\}$, will be written without the containing braces.)

Henceforth we think of *events* as subsets of the sample space Ω. Whenever A and B are events in which we are interested, then we can reasonably concern ourselves also with the events $A \cup B$, $A \cap B$, and A^c, representing 'A or B', 'A and B', and 'not A' respectively. Events A and B are called *disjoint* if their intersection is the empty set $\varnothing$; $\varnothing$ is called the *impossible event*. The set Ω is called the *certain event*, since some member of Ω will certainly occur.

Thus events are subsets of Ω, but need all the subsets of Ω be events? The answer is in the negative, but some of the reasons for this are too difficult to be discussed here. It suffices for us to think of the collection of events as a subcollection $\mathcal{F}$ of the set of all subsets of Ω. This subcollection should have certain properties in accordance with the earlier discussion:

(a) if $A, B \in \mathcal{F}$ then $A \cup B \in \mathcal{F}$ and $A \cap B \in \mathcal{F}$;
(b) if $A \in \mathcal{F}$ then $A^c \in \mathcal{F}$;
(c) the empty set $\varnothing$ belongs to $\mathcal{F}$.

Any collection $\mathcal{F}$ of subsets of Ω which satisfies these three conditions is called a *field*. It follows from the properties of a field $\mathscr{F}$ that

$$\text{if} \quad A_1, A_2, \ldots, A_n \in \mathscr{F} \quad \text{then} \quad \bigcup_{i=1}^{n} A_i \in \mathscr{F};$$

that is to say, $\mathscr{F}$ is closed under finite unions and hence under finite intersections also (see Problem (1.8.3)). This is fine when Ω is a finite set, but we require slightly more to deal with the common situation when Ω is infinite, as the following example indicates.

(4) **Example.** A coin is tossed repeatedly until the first head turns up; we are concerned with the number of tosses before this happens. The set of all possible outcomes is the set

$$\Omega = \{\omega_1, \omega_2, \omega_3, \ldots\}$$

where ω_i denotes the outcome when the first $i-1$ tosses are tails and the ith toss is a head. We may seek to assign a probability to the event A, that the first head occurs after an even number of tosses:

$$A = \omega_2 \cup \omega_4 \cup \omega_6 \cup \cdots .$$

This is an infinite countable union of members of Ω, and we require that such sets lie in $\mathscr{F}$ in order that we can discuss its probability. ●

Thus we also require that the collection of events be closed under the operation of taking countable unions. Any collection of subsets of Ω with these properties is called a *σ-field*.

(5) **Definition.** A collection $\mathscr{F}$ of subsets of Ω is called a **σ-field** if it satisfies the following conditions:

(a) $\varnothing \in \mathscr{F}$;

(b) if $A_1, A_2, \ldots \in \mathscr{F}$ then $\bigcup_{i=1}^{\infty} A_i \in \mathscr{F}$;

(c) if $A \in \mathscr{F}$ then $A^c \in \mathscr{F}$.

It follows from Problem (1.8.3) that σ-fields are closed under the operation of taking countable intersections. Here are some examples of σ-fields.

(6) **Example.** The smallest σ-field associated with Ω is the collection $\mathscr{F} = \{\varnothing, \Omega\}$. ●

(7) **Example.** If A is any subset of Ω then $\mathscr{F} = \{\varnothing, A, A^c, \Omega\}$ is a σ-field. ●

(8) **Example.** The *power set* of Ω, which is written $\{0, 1\}^\Omega$ and contains all subsets of Ω, is obviously a σ-field. For reasons beyond the scope of this book, it is often too large a collection for probabilities to be assigned reasonably to all its members. ●

To recapitulate, with any experiment we may associate a pair $(\Omega, \mathscr{F})$, where Ω is the set of all possible outcomes or *elementary events* and $\mathscr{F}$ is a σ-field of subsets of Ω which contains all the events in whose occurrences we may be interested; henceforth, to call a set A an *event* is equivalent to asserting that A belongs to the σ-field in question. We usually translate statements about combinations of events into set-theoretic jargon; for example, the event that both A and B occur is written as $A \cap B$. Table 1.1 is a translation chart.

Table 1.1

Typical notation	Set jargon	Probability jargon
Ω	Collection of objects	Sample space
ω	Member of Ω	Elementary event, outcome
A	Subset of Ω	Event that some outcome in A occurs
A^c	Complement of A	Event that no outcome in A occurs
$A \cap B$	Intersection	Both A and B
$A \cup B$	Union	Either A or B, or both
$A \setminus B$	Difference	A, but not B
$A \triangle B$	Symmetric difference	Either A or B, but not both
$A \subseteq B$	Inclusion	If A, then B
$\varnothing$	Empty set	Impossible event
Ω	Whole space	Certain event

Exercises

(9) Let $\{A_i: i \in I\}$ be a collection of sets. Prove that

$$\text{(a)} \left(\bigcup_i A_i\right)^c = \bigcap_i A_i^c, \qquad \text{(b)} \left(\bigcap_i A_i\right)^c = \bigcup_i A_i^c.$$

These are sometimes called 'de Morgan's laws'.

(10) Let A and B belong to some σ-field $\mathscr{F}$. Show that $\mathscr{F}$ contains the sets $A \cap B$, $A \setminus B$, and $A \triangle B$.

(11) A conventional knock-out tournament (such as that at Wimbledon) begins with 2^n competitors and has n rounds. There are no play-offs for the positions $2, 3, \ldots, 2^n - 1$, and the initial table of draws is specified. Give a concise description of the sample space of all possible outcomes.

(12) Let $\mathscr{F}$ be a σ-field of subsets of Ω and suppose that $B \in \mathscr{F}$. Show that $\mathscr{G} = \{A \cap B: A \in \mathscr{F}\}$ is a σ-field of subsets of B.

1.3 Probability

We wish to be able to discuss the likelihoods of the occurrences of events. Suppose that we repeat an experiment a large number N of times, keeping the initial conditions as equal as possible, and suppose that A is some event which may or may not occur on each repetition. Our experience of most scientific experimentation is that the proportion of times that A occurs settles down to some value as N becomes larger and larger; that is to say, writing

$N(A)$ for the number of occurrences of A in the N trials, $N(A)/N$ converges to a constant limit as N increases. We can think of the ultimate value of this ratio as being the probability $\mathbf{P}(A)$ that A occurs on any particular trial†; it may happen that the ratio does not behave in a coherent manner and our intuition fails us at this level, but we shall not discuss this here. Clearly, the ratio is a number between zero and one; if $A = \varnothing$ then $N(\varnothing) = 0$ and the ratio is 0, whilst if $A = \Omega$ then $N(\Omega) = N$ and the ratio is 1. Furthermore, suppose that A and B are two disjoint events, each of which may or may not occur at each trial. Then

$$N(A \cup B) = N(A) + N(B)$$

and so the ratio $N(A \cup B)/N$ is the sum of the two ratios $N(A)/N$ and $N(B)/N$. We now think of these ratios as representing the probabilities of the appropriate events. The above relations become

$$\mathbf{P}(A \cup B) = \mathbf{P}(A) + \mathbf{P}(B), \qquad \mathbf{P}(\varnothing) = 0, \qquad \mathbf{P}(\Omega) = 1.$$

This discussion suggests that the probability function $\mathbf{P}$ should be *finitely additive*, which is to say that

$$\text{if } A_1, A_2, \ldots, A_n \text{ are disjoint events, then } \mathbf{P}\left(\bigcup_{i=1}^{n} A_i\right) = \sum_{i=1}^{n} \mathbf{P}(A_i);$$

a glance at Example (1.2.4) suggests the more extensive property that $\mathbf{P}$ be *countably additive*, in that this should hold for countable collections $A_1, A_2, \ldots$ of disjoint events.

These relations are sufficient to specify the desirable properties of a probability function $\mathbf{P}$ applied to the set of events. Any such assignment of likelihoods to the members of $\mathscr{F}$ is called a *probability measure*.

(1) **Definition.** A **probability measure** $\mathbf{P}$ on $(\Omega, \mathscr{F})$ is a function $\mathbf{P}: \mathscr{F} \to [0, 1]$ satisfying

(a) $\mathbf{P}(\varnothing) = 0, \qquad \mathbf{P}(\Omega) = 1;$

(b) if $A_1, A_2, \ldots$ is a collection of disjoint members of $\mathscr{F}$, so that $A_i \cap A_j = \varnothing$ for all pairs i, j satisfying $i \neq j$, then

$$\mathbf{P}\left(\bigcup_{i=1}^{\infty} A_i\right) = \sum_{i=1}^{\infty} \mathbf{P}(A_i).$$

The triple $(\Omega, \mathscr{F}, \mathbf{P})$, comprising a set Ω, a σ-field $\mathscr{F}$ of subsets of Ω, and a probability measure $\mathbf{P}$ on $(\Omega, \mathscr{F})$, is called a **probability space**.

We can associate a probability space $(\Omega, \mathscr{F}, \mathbf{P})$ with any experiment, and all questions associated with the experiment can be reformulated in terms of this space. It may seem natural to ask for the numerical value of the

† This superficial discussion of probabilities is inadequate in many ways; questioning readers may care to discuss the philosophical and empirical aspects of the subject amongst themselves (see Appendix III).

probability $\mathbf{P}(A)$ of some event A. The answer to such questions must be contained in the description of the experiment in question. For example, the assertion that a *fair* coin is tossed once is equivalent to saying that heads and tails have an equal probability of occurring; actually, this is the *definition* of fairness.

(2) **Example.** A coin, possibly biased, is tossed once. We can take $\Omega = \{H, T\}$ and $\mathcal{F} = \{\varnothing, H, T, \Omega\}$, and a possible probability measure $\mathbf{P}: \mathcal{F} \to [0, 1]$ is given by

$$\mathbf{P}(\varnothing) = 0, \qquad \mathbf{P}(H) = p, \qquad \mathbf{P}(T) = 1 - p, \qquad \mathbf{P}(\Omega) = 1$$

where p is a fixed real number in the interval $[0, 1]$. If $p = \frac{1}{2}$, then we say that the coin is *fair*, or unbiased. ●

(3) **Example.** A die is thrown once. We can take $\Omega = \{1, 2, 3, 4, 5, 6\}$, $\mathcal{F} = \{0, 1\}^{\Omega}$, and the probability measure $\mathbf{P}$ given by

$$\mathbf{P}(A) = \sum_{i \in A} p_i \quad \text{for any } A \subseteq \Omega$$

where $p_1, p_2, \ldots, p_6$ are specified numbers from the interval $[0, 1]$ with unit sum. The probability that i turns up is p_i. The die is fair if $p_i = \frac{1}{6}$ for each i, in which case

$$\mathbf{P}(A) = \tfrac{1}{6}|A| \quad \text{for any } A \subseteq \Omega$$

where $|A|$ denotes the cardinality of A. ●

$(\Omega, \mathcal{F}, \mathbf{P})$ denotes a typical probability space. We now give some of its simple but important properties.

(4) **Lemma.**

(a) $\mathbf{P}(A^c) = 1 - \mathbf{P}(A)$
(b) *if* $B \supseteq A$ *then* $\mathbf{P}(B) = \mathbf{P}(A) + \mathbf{P}(B \setminus A) \geqslant \mathbf{P}(A)$
(c) $\mathbf{P}(A \cup B) = \mathbf{P}(A) + \mathbf{P}(B) - \mathbf{P}(A \cap B)$
(d) *more generally, if* $A_1, A_2, \ldots, A_n$ *are events then*

$$\mathbf{P}\left(\bigcup_{i=1}^{n} A_i\right) = \sum_i \mathbf{P}(A_i) - \sum_{i<j} \mathbf{P}(A_i \cap A_j) + \sum_{i<j<k} \mathbf{P}(A_i \cap A_j \cap A_k) - \cdots$$
$$+ (-1)^{n+1}\mathbf{P}(A_1 \cap A_2 \cap \cdots \cap A_n)$$

where, for example, $\sum_{i<j}$ *sums over all unordered pairs* (i, j) *for* $i \neq j$.

Proof.

(a) $A \cup A^c = \Omega$ and $A \cap A^c = \varnothing$, so $\mathbf{P}(A \cup A^c) = \mathbf{P}(A) + \mathbf{P}(A^c) = 1$.

(b) $B = A \cup (B \setminus A)$. However, this is the union of disjoint sets and so

$$\mathbf{P}(B) = \mathbf{P}(A) + \mathbf{P}(B \setminus A).$$

(c) $A \cup B = A \cup (B \setminus A)$, which is a disjoint union. So

$$\begin{aligned}\mathbf{P}(A \cup B) &= \mathbf{P}(A) + \mathbf{P}(B \setminus A) = \mathbf{P}(A) + \mathbf{P}(B \setminus (A \cap B)) \\ &= \mathbf{P}(A) + \mathbf{P}(B) - \mathbf{P}(A \cap B)\end{aligned}$$

by the result of (b).

(d) The proof is by induction on n, and is left as an *exercise* (see Exercise (1.3.9)). ■

In (4b) $B \setminus A$ denotes the set of members of B which are not in A. In order to write down the quantity $\mathbf{P}(B \setminus A)$, we require that $B \setminus A$ is in $\mathscr{F}$, the domain of $\mathbf{P}$; this is always true when A and B are in $\mathscr{F}$ and to prove this was part of Exercise (1.2.10). Notice that each proof proceeded by expressing an event in terms of disjoint unions and then applying $\mathbf{P}$. It is sometimes easier to calculate the probabilities of intersections of events rather than their unions; part (d) of the lemma is useful then, as we shall discover soon. The next property of $\mathbf{P}$ is more technical, and says that $\mathbf{P}$ is a *continuous* set function; this property is essentially equivalent to the condition that $\mathbf{P}$ is countably additive rather than just finitely additive (see Problem (1.8.16) also).

(5) **Lemma.** *Let $A_1, A_2, \ldots$ be an increasing sequence of events, so that $A_1 \subseteq A_2 \subseteq A_3 \subseteq \cdots$, and write A for their limit:*

$$A = \bigcup_{i=1}^{\infty} A_i = \lim A_i.$$

Then $\mathbf{P}(A) = \lim \mathbf{P}(A_i)$.

Similarly, if $B_1, B_2, \ldots$ is a decreasing sequence of events, so that $B_1 \supseteq B_2 \supseteq B_3 \supseteq \cdots$, then

$$B = \bigcap_{i=1}^{\infty} B_i = \lim B_i$$

satisfies $\mathbf{P}(B) = \lim \mathbf{P}(B_i)$.

Proof. $A = A_1 \cup (A_2 \setminus A_1) \cup (A_3 \setminus A_2) \cup \cdots$ is the union of a disjoint family of events. Thus, by Definition (1),

$$\begin{aligned}\mathbf{P}(A) &= \mathbf{P}(A_1) + \sum_{i=1}^{\infty} \mathbf{P}(A_{i+1} \setminus A_i) \\ &= \mathbf{P}(A_1) + \lim_{n\to\infty} \sum_{i=1}^{n-1} [\mathbf{P}(A_{i+1}) - \mathbf{P}(A_i)] \\ &= \lim_{n\to\infty} \mathbf{P}(A_n).\end{aligned}$$

To show the result for decreasing families of events, take complements and use the first part (*exercise*). ■

To recapitulate, statements concerning chance are implicitly related to experiments or trials, the outcomes of which are not entirely predictable. With any such experiment we can associate a probability space $(\Omega, \mathscr{F}, \mathbf{P})$ the properties of which are consistent with our shared and reasonable conceptions of the notion of chance.

Here is a final piece of jargon. An event A is called *null* if $\mathbf{P}(A) = 0$. Null events should not be confused with the impossible event $\varnothing$. Null events are happening all around us, even though they have zero probability; after all, what is the chance that a dart strikes any given point of the target at which it is thrown? That is, the impossible event is null, but null events need not be impossible.

Exercises

(6) Let A and B be events with probabilities $\mathbf{P}(A) = \frac{3}{4}$ and $\mathbf{P}(B) = \frac{1}{3}$. Show that $\frac{1}{12} \leqslant \mathbf{P}(A \cap B) \leqslant \frac{1}{3}$, and give examples to show that both extremes are possible. Find corresponding bounds for $\mathbf{P}(A \cup B)$.

(7) A fair coin is tossed repeatedly. Show that a head is bound to turn up sooner or later. Show similarly that any given finite sequence of heads and tails occurs eventually with probability one.

(8) Six cups and saucers come in pairs: there are two cups and saucers which are red, two white, and two with stars on. If the cups are placed randomly on to the saucers, find the probability that no cup is upon a saucer of the same pattern.

(9) Let $A_1, A_2, \ldots, A_n$ be events where $n \geqslant 2$, and prove that

$$\mathbf{P}\left(\bigcup_{i=1}^{n} A_i\right) = \sum_i \mathbf{P}(A_i) - \sum_{i<j} \mathbf{P}(A_i \cap A_j) + \sum_{i<j<k} \mathbf{P}(A_i \cap A_j \cap A_k) - \cdots + (-1)^{n+1}\mathbf{P}(A_1 \cap A_2 \cap \cdots \cap A_n).$$

In each packet of Corn Flakes may be found a plastic bust of one of the last five Vice-Chancellors of Cambridge University, the probability that any given packet contains any specific Vice-Chancellor being $\frac{1}{5}$, independently of all other packets. Show that the probability that each of the last three Vice-Chancellors is obtained in a bulk purchase of six packets is $1 - 3(\frac{4}{5})^6 + 3(\frac{3}{5})^6 - (\frac{2}{5})^6$.

1.4 Conditional probability: a fundamental lemma

Many statements about chance take the form 'if B occurs, then the probability of A is p', where B and A are events (such as 'it rains tomorrow' and 'the bus being on time' respectively) and p is a likelihood as before. To include this in our theory, we should return briefly to the discussion about proportions at the beginning of the previous section. An experiment is repeated N times, and on each occasion we observe the occurrences or non-occurrences of two events A and B. Now, suppose we only take an interest in those outcomes for which B occurs; all other experiments are disregarded. In this smaller collection of trials the proportion of times that A occurs is $N(A \cap B)/N(B)$, since B occurs at each of them. However,

$$\frac{N(A \cap B)}{N(B)} = \frac{N(A \cap B)/N}{N(B)/N}.$$

If we now think of these ratios as probabilities, we see that the probability that A occurs, given that B occurs, should be reasonably defined as $\mathbf{P}(A \cap B)/\mathbf{P}(B)$.

Probabilistic intuition leads to the same conclusion. Given that an event B occurs, it is the case that A occurs if and only if $A \cap B$ occurs. Thus the conditional probability of A given B should be proportional to $\mathbf{P}(A \cap B)$, which is to say that it equals $\alpha\mathbf{P}(A \cap B)$ for some constant α. However, the conditional probability of Ω given B must equal 1, and thus $\alpha\mathbf{P}(\Omega \cap B) = 1$, yielding $\alpha = \{\mathbf{P}(B)\}^{-1}$.

We formalize these notions as follows.

(1) **Definition.** If $\mathbf{P}(B) > 0$ then the **conditional probability** that A occurs given that B occurs is defined to be

$$\mathbf{P}(A \mid B) = \frac{\mathbf{P}(A \cap B)}{\mathbf{P}(B)}.$$

$\mathbf{P}(A \mid B)$, pronounced 'the probability of A given B', is our notation for this conditional probability. We sometimes speak of $\mathbf{P}(A \mid B)$ as 'the probability of A conditioned on B'.

(2) **Example.** Two fair dice are thrown. Given that the first shows 3, what is the probability that the total exceeds 6? The answer is obviously $\frac{1}{2}$, since the second must show 4, 5, or 6. However, let us labour the point. Clearly $\Omega = \{1, 2, 3, 4, 5, 6\}^2$, the set of all ordered pairs (i, j) for $i, j \in \{1, 2, \dots, 6\}$ (remember that $A \times B = \{(a, b): a \in A, b \in B\}$ and that $A \times A = A^2$), and we can take $\mathcal{F}$ to be the set of all subsets of Ω, with $\mathbf{P}(A) = |A|/36$ for any $A \subseteq \Omega$. Let B be the event that the first die shows 3 and A be the event that the total exceeds 6. Then

$$B = \{(3, b): 1 \leqslant b \leqslant 6\}, \qquad A = \{(a, b): a + b > 6\}$$

$$A \cap B = \{(3, 4), (3, 5), (3, 6)\}.$$

Thus

$$\mathbf{P}(A \mid B) = \frac{\mathbf{P}(A \cap B)}{\mathbf{P}(B)} = \frac{|A \cap B|}{|B|} = \frac{3}{6}. \qquad \bullet$$

(3) **Example.** A family has two children. What is the probability that both are boys, given that at least one is a boy? The older and younger child may each be male or female, so there are four possible combinations of sexes, which

we assume to be equally likely. Hence we can represent the sample space in the obvious way as

$$\Omega = \{GG, GB, BG, BB\}$$

where $\mathbf{P}(GG) = \mathbf{P}(GB) = \mathbf{P}(BG) = \mathbf{P}(BB) = \frac{1}{4}$. From the definition of conditional probability

$$\begin{aligned}\mathbf{P}(BB \mid \text{one boy at least}) &= \mathbf{P}(BB \mid GB \cup BG \cup BB)\\ &= \frac{\mathbf{P}(BB \cap (GB \cup BG \cup BB))}{\mathbf{P}(GB \cup BG \cup BB)}\\ &= \frac{\mathbf{P}(BB)}{\mathbf{P}(GB \cup BG \cup BB)} = \frac{1}{3}.\end{aligned}$$

A popular but incorrect answer to the question is $\frac{1}{2}$. This is the correct answer to another question: for a family with two children, what is the probability that both are boys given that the younger is a boy? In this case

$$\begin{aligned}\mathbf{P}(BB \mid \text{younger is a boy}) &= \mathbf{P}(BB \mid GB \cup BB)\\ &= \frac{\mathbf{P}(BB \cap (GB \cup BB))}{\mathbf{P}(GB \cup BB)} = \frac{\mathbf{P}(BB)}{\mathbf{P}(GB \cup BB)} = \frac{1}{2}.\end{aligned}$$

The usual dangerous argument contains the assertion

$$\mathbf{P}(BB \mid \text{one child is a boy}) = \mathbf{P}(\text{other child is a boy}).$$

Why is this meaningless? ●

The next lemma is crucially important in probability theory. A family $B_1, B_2, \ldots, B_n$ of events is called a *partition* of Ω if

$$B_i \cap B_j = \varnothing \quad \text{when} \quad i \neq j, \quad \text{and} \quad \bigcup_{i=1}^{n} B_i = \Omega.$$

Each elementary event $\omega \in \Omega$ belongs to exactly one set in a partition of Ω.

(4) **Lemma.** *For any events A and B*

$$\mathbf{P}(A) = \mathbf{P}(A \mid B)\mathbf{P}(B) + \mathbf{P}(A \mid B^c)\mathbf{P}(B^c).$$

More generally, let $B_1, B_2, \ldots, B_n$ be a partition of Ω. Then

$$\mathbf{P}(A) = \sum_{i=1}^{n} \mathbf{P}(A \mid B_i)\mathbf{P}(B_i).$$

Proof. $A = (A \cap B) \cup (A \cap B^c)$. This is a disjoint union and so

$$\begin{aligned}\mathbf{P}(A) &= \mathbf{P}(A \cap B) + \mathbf{P}(A \cap B^c)\\ &= \mathbf{P}(A \mid B)\mathbf{P}(B) + \mathbf{P}(A \mid B^c)\mathbf{P}(B^c).\end{aligned}$$

The second part is similar (see Problem (1.8.10)). ■

(5) **Example.** We are given two urns, each containing a collection of coloured balls. Urn I contains two white and three blue balls, whilst urn II contains three white and four blue balls. A ball is drawn at random from urn I and put into urn II, and then a ball is picked at random from urn II and examined. What is the probability that it is blue? We assume unless otherwise specified that a ball picked randomly from any urn is equally likely to be any of those present. The reader will be relieved to know that we no longer need to describe $(\Omega, \mathscr{F}, \mathbf{P})$ in detail; we are confident that we could do so if necessary. Clearly, the colour of the final ball depends on the colour of the ball picked from urn I. So let us 'condition' on this. Let A be the event that the final ball is blue and B be the event that the first one picked was blue. Then, by Lemma (4),

$$\mathbf{P}(A) = \mathbf{P}(A \mid B)\mathbf{P}(B) + \mathbf{P}(A \mid B^c)\mathbf{P}(B^c).$$

We can easily find all these probabilities:

$$\mathbf{P}(A \mid B) = \mathbf{P}(A \mid \text{urn II contains three white and five blue balls}) = \tfrac{5}{8}$$

$$\mathbf{P}(A \mid B^c) = \mathbf{P}(A \mid \text{urn II contains four white and four blue balls}) = \tfrac{1}{2}$$

$$\mathbf{P}(B) = \tfrac{3}{5}, \qquad \mathbf{P}(B^c) = \tfrac{2}{5}.$$

Hence

$$\mathbf{P}(A) = \tfrac{5}{8}\cdot\tfrac{3}{5} + \tfrac{1}{2}\cdot\tfrac{2}{5} = \tfrac{23}{40}.$$ ●

Unprepared readers may have been surprised by the sudden appearance of urns in this book. In the seventeenth and eighteenth centuries, lotteries often involved the drawing of slips from urns, and voting was often a matter of putting slips or balls into urns. In France today *aller aux urnes* is synonymous with voting. It was therefore not unnatural for the numerous Bernoullis and others to model births, marriages, deaths, fluids, gases, and so on, using urns containing balls of varied hue.

(6) **Example.** Only two factories manufacture zoggles. 20 per cent of the zoggles from factory I and 5 per cent from factory II are defective. Factory I produces twice as many zoggles as factory II each week. What is the probability that a zoggle, randomly chosen from a week's production, is satisfactory? Clearly this satisfaction depends on the factory of origin. Let A be the event that the chosen zoggle is satisfactory and let B be the event that it was made in factory I. Arguing as before,

$$\begin{aligned}\mathbf{P}(A) &= \mathbf{P}(A \mid B)\mathbf{P}(B) + \mathbf{P}(A \mid B^c)\mathbf{P}(B^c)\\ &= \tfrac{4}{5}\cdot\tfrac{2}{3} + \tfrac{19}{20}\cdot\tfrac{1}{3} = \tfrac{51}{60}.\end{aligned}$$

If the chosen zoggle is defective, what is the probability that it came from factory I? In our notation this is just $\mathbf{P}(B \mid A^c)$. However,

$$\mathbf{P}(B \mid A^c) = \frac{\mathbf{P}(B \cap A^c)}{\mathbf{P}(A^c)}$$

$$= \frac{\mathbf{P}(A^c \mid B)\mathbf{P}(B)}{\mathbf{P}(A^c)} = \frac{\frac{1}{5} \cdot \frac{2}{3}}{1 - 51/60} = \frac{8}{9}. \qquad \bullet$$

This section is terminated with a cautionary example. It is not untraditional to perpetuate errors of logic in calculating conditional probabilities. Lack of unambiguous definitions and notation has led astray many probabilists, including even Boole, who was credited by Russell with the discovery of pure mathematics and by others for some of the logical foundations of computing. The well-known 'prisoners' paradox' illustrates some of the dangers here.

(7) **Example. Prisoners' paradox.** In a dark country, three prisoners have been incarcerated without trial. Their warder tells them that the country's dictator has decided arbitrarily to free one of them and to shoot the other two, but he is not permitted to reveal to any prisoner the fate of that prisoner. Prisoner A knows therefore that his chance of survival is $\frac{1}{3}$. In order to gain information, he asks the warder to tell him in secret the name of some prisoner (but not himself) who will be killed, and the warder names prisoner B. What now is prisoner A's assessment of the chance that he will survive? Could it be $\frac{1}{2}$: after all, he knows now that the survivor will be either A or C, and he has no information about which? Could it be $\frac{1}{3}$: after all, according to the rules, at least one of B and C has to be killed, and thus the extra information cannot reasonably affect A's earlier calculation of the odds? What does the reader think about this? The resolution of the paradox lies in specifying the 'protocol' used by the warder in naming a doomed prisoner in the situation when either response (B or C) is possible. ●

Exercises

(8) Prove that $\mathbf{P}(A \mid B) = \mathbf{P}(B \mid A)\mathbf{P}(A)/\mathbf{P}(B)$ whenever $\mathbf{P}(A)\mathbf{P}(B) \neq 0$. Show that, if $\mathbf{P}(A \mid B) > \mathbf{P}(A)$, then $\mathbf{P}(B \mid A) > \mathbf{P}(B)$.

(9) For events $A_1, A_2, \ldots, A_n$, prove that

$$\mathbf{P}(A_1 \cap A_2 \cap \cdots \cap A_n) = \mathbf{P}(A_1)\mathbf{P}(A_2 \mid A_1)\mathbf{P}(A_3 \mid A_1 \cap A_2) \cdots \mathbf{P}(A_n \mid A_1 \cap A_2 \cap \cdots \cap A_{n-1}),$$

whenever $\mathbf{P}(A_1 \cap A_2 \cap \cdots \cap A_{n-1}) > 0$.

(10) A man possesses five coins, two of which are double-headed, one is double-tailed, and two are normal. He shuts his eyes, picks a coin at random, and tosses it. What is the probability that the lower face of the coin is a head?

He opens his eyes and sees that the coin is showing heads; what is the probability that the lower face is a head?

He shuts his eyes again, and tosses the coin again. What is the probability that the lower face is a head?

He opens his eyes and sees that the coin is showing heads; what is the probability that the lower face is a head?

He discards this coin, picks another at random, and tosses it. What is the probability that it shows heads?

(11) What do you think of the following 'proof' by Lewis Carroll that an urn cannot contain two balls of the same colour? Suppose that the urn contains two balls, each of which is either black or white; thus, in the obvious notation, $\mathbf{P}(BB) = \mathbf{P}(BW) = \mathbf{P}(WB) = \mathbf{P}(WW) = \frac{1}{4}$. We add a black ball, so that $\mathbf{P}(BBB) = \mathbf{P}(BBW) = \mathbf{P}(BWB) = \mathbf{P}(BWW) = \frac{1}{4}$. Next we pick a ball at random; the chance that the ball is black is (using conditional probabilities) $1 \cdot \frac{1}{4} + \frac{2}{3} \cdot \frac{1}{4} + \frac{2}{3} \cdot \frac{1}{4} + \frac{1}{3} \cdot \frac{1}{4} = \frac{2}{3}$. However, if there is probability $\frac{2}{3}$ that a ball, chosen randomly from three, is black, then there must be two black and one white, which is to say that originally there was one black and one white ball in the urn.

1.5 Independence

In general, the occurrence of some event B changes the probability that another event A occurs; the original probability $\mathbf{P}(A)$ is replaced by $\mathbf{P}(A \mid B)$. If this probability remains unchanged, that is to say $\mathbf{P}(A \mid B) = \mathbf{P}(A)$, then we call A and B 'independent'. This is well defined only if $\mathbf{P}(B) > 0$. Definition (1.4.1) of conditional probability leads us to the following.

(1) **Definition.** Events A and B are called **independent** if

$$\mathbf{P}(A \cap B) = \mathbf{P}(A)\mathbf{P}(B).$$

More generally, a family $\{A_i : i \in I\}$ is called **independent** if

$$\mathbf{P}\left(\bigcap_{i \in J} A_i\right) = \prod_{i \in J} \mathbf{P}(A_i)$$

for all finite subsets J of I.

A common student error is to say that A and B are independent if $A \cap B = \varnothing$; this is clearly false. If the family $\{A_i : i \in I\}$ has the property that

$$\mathbf{P}(A_i \cap A_j) = \mathbf{P}(A_i)\mathbf{P}(A_j) \quad \text{for all } i \neq j$$

then it is called *pairwise independent*. Pairwise-independent families are not necessarily independent, as the following example shows.

(2) **Example.** Suppose $\Omega = \{abc, acb, cab, cba, bca, bac, aaa, bbb, ccc\}$, and each of the nine elementary events in Ω occurs with equal probability $\frac{1}{9}$. Let A_k be the event that the kth letter is a. It is left as an *exercise* to show that $\{A_1, A_2, A_3\}$ is pairwise independent but not independent. ●

(3) **Example (1.4.6) revisited.** The events A and B of this example are clearly dependent because

$$\mathbf{P}(A \mid B) = \tfrac{4}{5}, \qquad \mathbf{P}(A) = \tfrac{51}{60}.$$ ●

(4) **Example.** Choose a card at random from a pack of 52 playing cards, each being picked with equal probability 1/52. We claim that the suit of the chosen card is independent of its rank. For example,

$$\mathbf{P}(\text{king}) = 4/52, \qquad \mathbf{P}(\text{king} \mid \text{spade}) = 1/13.$$

Alternatively,

$$\mathbf{P}(\text{spade king}) = 1/52$$
$$= \tfrac{1}{4} \cdot \tfrac{1}{13} = \mathbf{P}(\text{spade})\mathbf{P}(\text{king}).$$ ●

Let C be an event with $\mathbf{P}(C) > 0$. To the conditional probability measure $\mathbf{P}(\cdot \mid C)$ corresponds the idea of *conditional independence*. Two events A and B are called *conditionally independent given* C if

(5) $$\mathbf{P}(A \cap B \mid C) = \mathbf{P}(A \mid C)\mathbf{P}(B \mid C);$$

there is a natural extension to families of events.

Exercises

(6) Let A and B be independent events; show that A^c, B are independent, and deduce that A^c, B^c are independent.

(7) We roll a die n times. Let A_{ij} be the event that the ith and jth rolls produce the same number. Show that the events $\{A_{ij}: 1 \leqslant i < j \leqslant n\}$ are pairwise independent but not independent.

(8) A fair coin is tossed repeatedly. Show that the following two statements are equivalent:

(a) the outcomes of different tosses are independent,
(b) for any given finite sequence of heads and tails, the chance of this sequence occurring in the first m tosses is 2^{-m}, where m is the length of the sequence.

(9) Let $\Omega = \{1, 2, \dots, p\}$, $\mathcal{F}$ be the set of all subsets of Ω, and $\mathbf{P}(A) = |A|/p$ for all $A \in \mathcal{F}$. Suppose p is prime. Show that, if A and B are independent events, then at least one of A and B is either $\varnothing$ or Ω.

1.6 Completeness and product spaces

This section should be omitted at the first reading, but we shall require its contents later. It contains only a sketch of complete probability spaces and product spaces; the reader should look elsewhere for a more detailed treatment (see Billingsley 1986). We require the following result.

(1) **Lemma.** *If $\mathcal{F}$ and $\mathcal{G}$ are two σ-fields of subsets of Ω then $\mathcal{F} \cap \mathcal{G}$ is a σ-field also. More generally, if $\{\mathcal{F}_i: i \in I\}$ is a family of σ-fields of subsets of Ω then $\mathcal{I} = \bigcap_{i \in I} \mathcal{F}_i$ is a σ-field also.*

The proof is not difficult and is left as an *exercise*. Note that $\mathcal{F} \cup \mathcal{G}$ may not be a σ-field, although it may be extended to a unique smallest σ-field, written $\sigma(\mathcal{F} \cup \mathcal{G})$, as follows. Let $\{\mathcal{I}_i: i \in I\}$ be the collection of all σ-fields which contain both $\mathscr{F}$ and $\mathscr{G}$ as subsets; this collection is non-empty since it contains the set of all subsets of Ω. Then $\mathscr{I} = \bigcap_{i \in I} \mathscr{I}_i$ is the unique smallest σ-field which contains $\mathscr{F} \cup \mathscr{G}$.

(A) Completeness. Let $(\Omega, \mathscr{F}, \mathbf{P})$ be a probability space. Any event A which has zero probability, that is $\mathbf{P}(A) = 0$, is called *null*. It may seem reasonable to suppose that any subset B of a null set A will itself be null, but this may be without meaning since B may not be an event and thus $\mathbf{P}(B)$ may not be defined.

(2) **Definition.** A probability space $(\Omega, \mathscr{F}, \mathbf{P})$ is called **complete** if all subsets of null sets are events.

Any incomplete space can be completed. Let $\mathscr{N}$ be the collection of all subsets of null sets in $\mathscr{F}$ and let $\mathscr{G} = \sigma(\mathscr{F} \cup \mathscr{N})$ be the smallest σ-field which contains all sets in $\mathscr{F}$ and $\mathscr{N}$. It can be shown that the domain of $\mathbf{P}$ may be extended in an obvious way from $\mathscr{F}$ to $\mathscr{G}$; $(\Omega, \mathscr{G}, \mathbf{P})$ is the *completion* of $(\Omega, \mathscr{F}, \mathbf{P})$.

(B) Product spaces. The probability spaces discussed in this chapter have usually been constructed around the outcomes of one experiment, but instances occur naturally when we need to combine the outcomes of several independent experiments into one space (see (1.2.4) and (1.4.2)). How should we proceed in general?

Suppose two experiments have associated probability spaces $(\Omega_1, \mathscr{F}_1, \mathbf{P}_1)$ and $(\Omega_2, \mathscr{F}_2, \mathbf{P}_2)$ respectively. The sample space of the pair of experiments, considered jointly, is the collection $\Omega_1 \times \Omega_2 = \{(\omega_1, \omega_2): \omega_1 \in \Omega_1, \omega_2 \in \Omega_2\}$ of ordered pairs. The appropriate σ-field of events is more complicated to construct. Certainly it should contain all subsets of $\Omega_1 \times \Omega_2$ of the form $A_1 \times A_2 = \{(a_1, a_2): a_1 \in A_1, a_2 \in A_2\}$ where A_1 and A_2 are typical members of $\mathscr{F}_1$ and $\mathscr{F}_2$ respectively. However, the family of all such sets, $\mathscr{F}_1 \times \mathscr{F}_2 = \{A_1 \times A_2: A_1 \in \mathscr{F}_1, A_2 \in \mathscr{F}_2\}$, is not in general a σ-field. By the discussion after (1), there exists a unique smallest σ-field $\mathscr{G} = \sigma(\mathscr{F}_1 \times \mathscr{F}_2)$ of subsets of $\Omega_1 \times \Omega_2$ which contains $\mathscr{F}_1 \times \mathscr{F}_2$. All we require now is a suitable probability function on $(\Omega_1 \times \Omega_2, \mathscr{G})$. Let $\mathbf{P}_{12}: \mathscr{F}_1 \times \mathscr{F}_2 \to [0, 1]$ be given by

(3) $$\mathbf{P}_{12}(A_1 \times A_2) = \mathbf{P}_1(A_1)\mathbf{P}_2(A_2)$$

for any $A_1 \in \mathscr{F}_1$ and $A_2 \in \mathscr{F}_2$. It can be shown that the domain of $\mathbf{P}_{12}$ can be extended from $\mathscr{F}_1 \times \mathscr{F}_2$ to the whole of $\mathscr{G} = \sigma(\mathscr{F}_1 \times \mathscr{F}_2)$. The ensuing probability space $(\Omega_1 \times \Omega_2, \mathscr{G}, \mathbf{P}_{12})$ is called the *product space* of $(\Omega_1, \mathscr{F}_1, \mathbf{P}_1)$ and $(\Omega_2, \mathscr{F}_2, \mathbf{P}_2)$. Products of larger numbers of spaces are constructed similarly. The measure $\mathbf{P}_{12}$ is sometimes called the 'product measure' since its defining equation (3) assumed that the two experiments are independent. There are of course many other measures that can be applied to $(\Omega_1 \times \Omega_2, \mathscr{G})$.

In many simple cases this technical discussion is unnecessary. Suppose that Ω_1 and Ω_2 are finite, and that their σ-fields contain all their subsets; this is the case in (1.2.4) and (1.4.2). Then $\mathscr{G}$ contains all subsets of $\Omega_1 \times \Omega_2$.

1.7 Worked examples

Here are some more examples to illustrate the ideas of this chapter. The reader is now equipped to try his hand at a substantial number of those problems which exercised the pioneers in probability. These frequently involved experiments having equally likely outcomes, such as dealing whist hands, putting balls of various colours into urns and taking them out again, throwing dice, and so on. In many such instances, the reader will be pleasantly surprised to find that it is not necessary to write down $(\Omega, \mathscr{F}, \mathbf{P})$ explicitly, but only to think of Ω as being a collection $\{\omega_1, \omega_2, \dots, \omega_N\}$ of possibilities, each of which may occur with probability $1/N$. Thus, $\mathbf{P}(A) = |A|/N$ for any $A \subseteq \Omega$. The basic tools used in such problems are as follows.

(a) Combinatorics: remember that the number of permutations of n objects is $n!$ and that the number of ways of choosing r objects from n is $\binom{n}{r}$.

(b) Set theory: to obtain $\mathbf{P}(A)$ we can compute $\mathbf{P}(A^c) = 1 - \mathbf{P}(A)$ or we can partition A by conditioning on events B_i and use (1.4.4).

(c) Use of independence.

(1) **Example.** Consider a series of hands dealt at bridge. Let A be the event that in a given deal each player has one ace. Show that the probability that A occurs at least once in seven deals is approximately $\frac{1}{2}$.

Solution. The number of ways of dealing 52 cards into four equal hands is $52!/(13!)^4$. There are $4!$ ways of distributing the aces so that each hand holds one, and there are $48!/(12!)^4$ ways of dealing the remaining cards. Thus

$$\mathbf{P}(A) = \frac{4!\,48!/(12!)^4}{52!/(13!)^4} \simeq \frac{1}{10}.$$

Now let B_i be the event that A occurs for the first time on the ith deal. Clearly $B_i \cap B_j = \varnothing$, $i \neq j$. Thus

$$\begin{aligned}\mathbf{P}(A \text{ occurs in seven deals}) &= \mathbf{P}(B_1 \cup \cdots \cup B_7)\\ &= \sum_1^7 \mathbf{P}(B_i) \quad \text{using (1.3.1).}\end{aligned}$$

Since successive deals are independent, we have

$$\begin{aligned}\mathbf{P}(B_i) &= \mathbf{P}(A^c \text{ occurs on deal 1 and } A^c \text{ occurs on deal 2}\\ &\qquad \cdots \text{ and } A^c \text{ occurs on deal } i-1 \text{ and } A \text{ occurs on deal } i)\\ &= (\mathbf{P}(A^c))^{i-1}\mathbf{P}(A) \quad \text{using (1.5.1)}\\ &\simeq \left(1 - \frac{1}{10}\right)^{i-1} \frac{1}{10}.\end{aligned}$$

Thus

$$\mathbf{P}(A \text{ occurs in seven deals}) = \sum_{1}^{7} \mathbf{P}(B_i) \simeq \sum_{1}^{7} \left(\frac{9}{10}\right)^{i-1} \frac{1}{10} \simeq \frac{1}{2}.$$

Can you see an easier way of obtaining this answer? ●

(2) **Example.** There are two roads from A to B and two roads from B to C. Each of the four roads has probability p of being blocked by snow, independently of all the others. What is the probability that there is an open road from A to C?

Solution.

$$\begin{aligned}\mathbf{P}(\text{open road}) &= \mathbf{P}((\text{open road from A to B}) \cap (\text{open road from B to C}))\\ &= \mathbf{P}(\text{open road from A to B})\mathbf{P}(\text{open road from B to C})\end{aligned}$$

using the independence. However, p is the same for all roads; thus, using (1.3.4),

$$\begin{aligned}\mathbf{P}(\text{open road}) &= (1 - \mathbf{P}(\text{no road from A to B}))^2\\ &= (1 - \mathbf{P}((\text{first road blocked}) \cap (\text{second road blocked})))^2\\ &= (1 - \mathbf{P}(\text{first road blocked})\mathbf{P}(\text{second road blocked}))^2\end{aligned}$$

using the independence. Thus

(3) $$\mathbf{P}(\text{open road}) = (1 - p^2)^2.$$

Further suppose that there is also a direct road from A to C, which is independently blocked with probability p. Then by (1.4.4),

$$\begin{aligned}\mathbf{P}(\text{open road}) &= \mathbf{P}(\text{open road} \mid \text{direct road blocked}) \cdot p\\ &\quad + \mathbf{P}(\text{open road} \mid \text{direct road open}) \cdot (1 - p)\\ &= (1 - p^2)^2 \cdot p + 1 \cdot (1 - p)\end{aligned}$$

using (3). ●

(4) **Example. Symmetric random walk (or 'Gambler's ruin').** A man is saving up to buy a new Jaguar at a cost of N units of money. He starts with k $(0 < k < N)$ units and tries to win the remainder by the following gamble with his bank manager. He tosses a fair coin repeatedly; if it comes up heads then the manager pays him one unit, but if it comes up tails then he pays the manager one unit. He plays this game repeatedly until one of two events occurs: either he runs out of money and is bankrupted or he wins enough to buy the Jaguar. What is the probability that he is ultimately bankrupted?

Solution. This is one of many problems the solution to which proceeds by the construction of a linear difference equation subject to certain boundary conditions. Let A denote the event that he is eventually bankrupted, and let

B be the event that the first toss of the coin shows heads. By (1.4.4)

(5) $$\mathbf{P}_k(A) = \mathbf{P}_k(A \mid B)\mathbf{P}(B) + \mathbf{P}_k(A \mid B^c)\mathbf{P}(B^c),$$

where $\mathbf{P}_k$ denotes probabilities calculated relative to the starting point k. We want to find $\mathbf{P}_k(A)$. Consider $\mathbf{P}_k(A \mid B)$. If the first toss is a head then his capital increases to $k+1$ units and the game starts afresh from a different starting point. Thus $\mathbf{P}_k(A \mid B) = \mathbf{P}_{k+1}(A)$ and similarly $\mathbf{P}_k(A \mid B^c) = \mathbf{P}_{k-1}(A)$. So, writing $p_k = \mathbf{P}_k(A)$, (5) becomes

(6) $$p_k = \tfrac{1}{2}(p_{k+1} + p_{k-1}) \quad \text{if} \quad 0 < k < N,$$

which is a linear difference equation subject to the boundary conditions $p_0 = 1$, $p_N = 0$. The analytical solution to such equations is routine, and we shall return later to the general method of solution. In this case we can proceed directly. We put $b_k = p_k - p_{k-1}$ to obtain $b_k = b_{k-1}$ and hence $b_k = b_1$ for all k. Thus

$$p_k = b_1 + p_{k-1} = 2b_1 + p_{k-2} = \cdots = kb_1 + p_0$$

is the general solution to (6). The boundary conditions imply that $p_0 = 1$, $b_1 = -1/N$, giving

(7) $$\mathbf{P}_k(A) = 1 - k/N.$$

As the price of the Jaguar rises, that is as $N \to \infty$, ultimate bankruptcy becomes very likely. This is the problem of the 'symmetric random walk with two absorbing barriers' to which we shall return in more generality later. ●

Our experience of student calculations leads us to stress that probabilities lie between zero and one; any calculated probability which violates this must be incorrect.

(8) **Example. Testimony.** A court is investigating the possible occurrence of an unlikely event T. The reliability of two independent witnesses called Alf and Bob is known to the court: Alf tells the truth with probability α and Bob with probability β, and there is no collusion between the two of them. Let A and B be the events that Alf and Bob assert (respectively) that T occurred, and let $\tau = \mathbf{P}(T)$. What is the probability that T occurred given that both Alf and Bob declare that T occurred?

Solution. We are asked to calculate $\mathbf{P}(T \mid A \cap B)$, which is equal to $\mathbf{P}(T \cap A \cap B)/\mathbf{P}(A \cap B)$. Now $\mathbf{P}(T \cap A \cap B) = \mathbf{P}(A \cap B \mid T)\mathbf{P}(T)$ and

$$\mathbf{P}(A \cap B) = \mathbf{P}(A \cap B \mid T)\mathbf{P}(T) + \mathbf{P}(A \cap B \mid T^c)\mathbf{P}(T^c).$$

We have from the independence of the witnesses that A and B are conditionally independent given either T or T^c. Therefore

$$\mathbf{P}(A \cap B \mid T) = \mathbf{P}(A \mid T)\mathbf{P}(B \mid T) = \alpha\beta,$$

$$\mathbf{P}(A \cap B \mid T^c) = \mathbf{P}(A \mid T^c)\mathbf{P}(B \mid T^c) = (1-\alpha)(1-\beta),$$

so that

$$\mathbf{P}(T \mid A \cap B) = \frac{\alpha\beta\tau}{\alpha\beta\tau + (1-\alpha)(1-\beta)(1-\tau)}.$$

As an example, suppose that $\alpha = \beta = 9/10$ and $\tau = 10^{-3}$. Then $\mathbf{P}(T \mid A \cap B) = 81/1080$, which is somewhat small as a basis for a judicial conclusion. ●

(9) **Example. Zoggles revisited.** A new process for the production of zoggles is invented, and both factories of Example (1.4.6) install extra production lines using it. The new process is cheaper but produces fewer reliable zoggles, only 75 per cent of items produced in this new way being reliable.

Factory I fails to implement its new production line efficiently, and only 10 per cent of its output is made in this manner. Factory II does better: it produces 20 per cent of its output by the new technology, and now produces twice as many zoggles in all as factory I.

Is the new process beneficial to the consumer?

Solution. Both factories now produce a higher proportion of unreliable zoggles than before, and so it might seem at first sight that there is an increased proportion of unreliable zoggles on the market.

Let A be the event that a randomly chosen zoggle is satisfactory, B the event that it came from factory I, and C the event that it was made by the new method. Then

$$\begin{aligned}\mathbf{P}(A) &= \frac{1}{3}\mathbf{P}(A \mid B) + \frac{2}{3}\mathbf{P}(A \mid B^c)\\ &= \frac{1}{3}\left(\frac{1}{10}\mathbf{P}(A \mid B \cap C) + \frac{9}{10}\mathbf{P}(A \mid B \cap C^c)\right)\\ &\quad + \frac{2}{3}\left(\frac{1}{5}\mathbf{P}(A \mid B^c \cap C) + \frac{4}{5}\mathbf{P}(A \mid B^c \cap C^c)\right)\\ &= \frac{1}{3}\left(\frac{1}{10}\cdot\frac{3}{4} + \frac{9}{10}\cdot\frac{4}{5}\right) + \frac{2}{3}\left(\frac{1}{5}\cdot\frac{3}{4} + \frac{4}{5}\cdot\frac{19}{20}\right) = \frac{523}{600} > \frac{51}{60},\end{aligned}$$

so that the proportion of satisfactory zoggles has been increased. ●

(10) **Example. Simpson's paradox.** A doctor has performed clinical trials to determine the relative efficacies of two drugs, with the following results.

	Women		Men	
	Drug I	Drug II	Drug I	Drug II
Success	200	10	19	1000
Failure	1800	190	1	1000

Which drug is the better? Here are two conflicting responses.

1. Drug I was given to 2020 people, of whom 219 were cured. The success rate was 219/2020, which is much smaller than the corresponding figure, 1010/2200, for drug II. Therefore drug II is better than drug I.
2. Amongst women the success rates of the drugs are 1/10 and 1/20, and amongst men 19/20 and 1/2. Drug I wins in both cases.

This well-known statistical paradox may be reformulated in the following more general way. Given three events, A, B, C, it is possible to allocate probabilities such that

(11) $$\mathbf{P}(A \mid B \cap C) > \mathbf{P}(A \mid B^c \cap C) \quad \text{and} \quad \mathbf{P}(A \mid B \cap C^c) > \mathbf{P}(A \mid B^c \cap C^c)$$

but

(12) $$\mathbf{P}(A \mid B) < \mathbf{P}(A \mid B^c).$$

We may think of A as the event that treatment is successful, B as the event that drug I is given to a randomly chosen individual, and C as the event that this individual is female. The above inequalities imply that B is preferred to B^c when C occurs and when C^c occurs, but B^c is preferred to B overall.

Setting

$$a = \mathbf{P}(A \cap B \cap C), \qquad b = \mathbf{P}(A^c \cap B \cap C),$$
$$c = \mathbf{P}(A \cap B^c \cap C), \qquad d = \mathbf{P}(A^c \cap B^c \cap C),$$
$$e = \mathbf{P}(A \cap B \cap C^c), \qquad f = \mathbf{P}(A^c \cap B \cap C^c),$$
$$g = \mathbf{P}(A \cap B^c \cap C^c), \qquad h = \mathbf{P}(A^c \cap B^c \cap C^c),$$

and expanding (11)–(12), we arrive at the (equivalent) inequalities

(13) $$ad > bc, \qquad eh > fg, \qquad (a + e)(d + h) < (b + f)(c + g),$$

subject to the conditions $a, b, c, \dots, h \geq 0$ and $a + b + c + \cdots + h = 1$. Inequalities (13) are equivalent to the existence of two rectangles R_1 and R_2, as in Figure 1.1, satisfying

$$\text{area}(D_1) > \text{area}(D_2), \quad \text{area}(D_3) > \text{area}(D_4), \quad \text{area}(R_1) < \text{area}(R_2).$$

Many such rectangles may be found, by inspection, as for example those with $a = 3/30$, $b = 1/30$, $c = 8/30$, $d = 3/30$, $e = 3/30$, $f = 8/30$, $g = 1/30$, $h = 3/30$. Similar conclusions are valid for finer partitions $\{C_i : i \in I\}$ of the sample space, though the corresponding pictures are harder to draw. ●

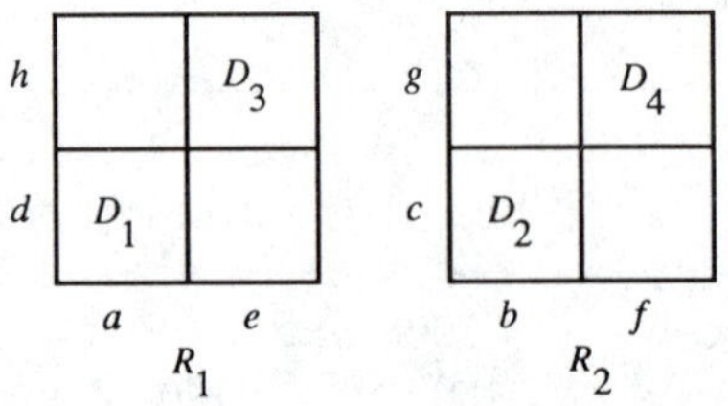

Fig. 1.1 Two unions of rectangles illustrating Simpson's paradox.

Exercises

(14) There are two roads from A to B and two roads from B to C. Each of the four roads is blocked by snow with probability p, independently of the others. Find the probability that there is an open road from A to B given that there is no open route from A to C.

If, in addition, there is a direct road from A to C, this road being blocked with probability p independently of the others, find the required conditional probability.

(15) Calculate the probability that a hand of 13 cards dealt from a normal pack of 52 contains exactly two kings and one ace. What is the probability that it contains exactly one ace given that it contains exactly two kings?

(16) A symmetric random walk takes place on the integers $0, 1, 2, \ldots, N$ with absorbing barriers at 0 and N, starting at k. Show that the probability that the walk is never absorbed is zero.

(17) The so called 'sure thing principle' asserts that if you prefer x to y given C, and also prefer x to y given C^c, then you surely prefer x to y. Agreed?

(18) A pack contains m cards, labelled $1, 2, \ldots, m$. The cards are dealt out in a random order, one by one. Given that the label of the kth card dealt is the largest of the first k cards dealt, what is the probability that it is also the largest in the pack?

1.8 Problems

1. A traditional fair die is thrown twice. What is the probability that

   (a) a six turns up exactly once?
   (b) both numbers are odd?
   (c) the sum of the scores is 4?
   (d) the sum of the scores is divisible by 3?

2. A fair coin is thrown repeatedly. What is the probability that on the nth throw

   (a) a head appears for the first time?
   (b) the numbers of heads and tails to date are equal?
   (c) exactly two heads have appeared altogether to date?
   (d) at least two heads have appeared to date?

3. Let $\mathscr{F}$ and $\mathscr{G}$ be σ-fields of subsets of Ω.

   (a) Use elementary set operations to show that $\mathscr{F}$ is closed under countable intersections; that is, if $A_1, A_2, \ldots$ are in $\mathscr{F}$, then so is $\bigcap_i A_i$.
   (b) Let $\mathscr{H} = \mathscr{F} \cap \mathscr{G}$ be the collection of subsets of Ω lying in both $\mathscr{F}$ and $\mathscr{G}$. Show that $\mathscr{H}$ is a σ-field.
   (c) Show that $\mathscr{F} \cup \mathscr{G}$, the collection of subsets of Ω lying in either $\mathscr{F}$ or $\mathscr{G}$, is not necessarily a σ-field.

4. Describe the underlying probability spaces for the following experiments:

   (a) a biased coin is tossed three times;
   (b) two balls are drawn without replacement from an urn which originally contained two ultramarine and two vermilion balls;
   (c) a biased coin is tossed repeatedly until a head turns up.

5. Show that the probability that *exactly* one of the events A and B occurs is $\mathbf{P}(A) + \mathbf{P}(B) - 2\mathbf{P}(A \cap B)$.

6. Prove that $\mathbf{P}(A \cup B \cup C) = 1 - \mathbf{P}(A^c \mid B^c \cap C^c)\mathbf{P}(B^c \mid C^c)\mathbf{P}(C^c)$.

7. (a) If A is independent of itself, show that $\mathbf{P}(A)$ is 0 or 1.
(b) If $\mathbf{P}(A)$ is 0 or 1, show that A is independent of all events B.

8. Let $\mathscr{F}$ be a σ-field of subsets of Ω, and suppose $\mathbf{P}: \mathscr{F} \to [0, 1]$ satisfies: (i) $\mathbf{P}(\Omega) = 1$, and (ii) $\mathbf{P}$ is additive (in that $\mathbf{P}(A \cup B) = \mathbf{P}(A) + \mathbf{P}(B)$ whenever $A \cap B = \varnothing$). Show that $\mathbf{P}(\varnothing) = 0$.

9. Suppose $(\Omega, \mathscr{F}, \mathbf{P})$ is a probability space and $B \in \mathscr{F}$ satisfies $\mathbf{P}(B) > 0$. Let $\mathbf{Q}: \mathscr{F} \to [0, 1]$ be defined by $\mathbf{Q}(A) = \mathbf{P}(A \mid B)$. Show that $(\Omega, \mathscr{F}, \mathbf{Q})$ is a probability space. If $C \in \mathscr{F}$ and $\mathbf{Q}(C) > 0$, show that $\mathbf{Q}(A \mid C) = \mathbf{P}(A \mid B \cap C)$; discuss.

10. Let $B_1, B_2, \ldots$ be a partition of the sample space Ω, each B_i having positive probability, and show that

$$\mathbf{P}(A) = \sum_{j=1}^{\infty} \mathbf{P}(A \mid B_j)\mathbf{P}(B_j).$$

11. Prove Boole's inequalities:

$$\text{(a) } \mathbf{P}\left(\bigcup_1^n A_i\right) \leqslant \sum_1^n \mathbf{P}(A_i), \qquad \text{(b) } \mathbf{P}\left(\bigcap_1^n A_i\right) \geqslant 1 - \sum_1^n \mathbf{P}(A_i^c).$$

12. Prove that

$$\mathbf{P}\left(\bigcap_1^n A_i\right) = \sum_i \mathbf{P}(A_i) - \sum_{i<j} \mathbf{P}(A_i \cup A_j) + \sum_{i<j<k} \mathbf{P}(A_i \cup A_j \cup A_k) - \cdots - (-1)^n \mathbf{P}(A_1 \cup A_2 \cup \cdots \cup A_n).$$

13. Let $A_1, A_2, \ldots, A_n$ be events, and let N_k be the event that exactly k of the A_i occur. Prove the result sometimes referred to as Waring's theorem:

$$\mathbf{P}(N_k) = \sum_{i=0}^{n-k} (-1)^i \binom{k+i}{k} S_{k+i}$$

where

$$S_j = \sum_{i_1 < i_2 < \cdots < i_j} \mathbf{P}(A_{i_1} \cap A_{i_2} \cap \cdots \cap A_{i_j}).$$

Use this result to find an expression for the probability that a purchase of six packets of Corn Flakes yields exactly three distinct busts (see Exercise (1.3.9)).

14. Prove Bayes's formula; that is, if $A_1, A_2, \ldots, A_n$ is a partition of Ω, each A_i having positive probability, then

$$\mathbf{P}(A_j \mid B) = \frac{\mathbf{P}(B \mid A_j)\mathbf{P}(A_j)}{\sum_1^n \mathbf{P}(B \mid A_i)\mathbf{P}(A_i)}.$$

15. A random number N of dice is thrown. A_i is the event that $N = i$, and $\mathbf{P}(A_i) = 2^{-i}$, $i \geqslant 1$. The sum of the scores is S. Find the probability that:

(a) $N = 2$ given $S = 4$;
(b) $S = 4$ given N is even;
(c) $N = 2$, given that $S = 4$ and the first die showed 1;
(d) the largest number shown by any die is r, where S is unknown.

16. Let $A_1, A_2, \ldots$ be a sequence of events. Define

$$B_n = \bigcup_{m=n}^{\infty} A_m, \qquad C_n = \bigcap_{m=n}^{\infty} A_m.$$

Clearly $C_n \subseteq A_n \subseteq B_n$. The sequences $\{B_n\}$ and $\{C_n\}$ are decreasing and increasing respectively with limits

$$\lim B_n = B = \bigcap_n B_n = \bigcap_n \bigcup_{m \geqslant n} A_m,$$

$$\lim C_n = C = \bigcup_n C_n = \bigcup_n \bigcap_{m \geqslant n} A_m;$$

B and C are called $\limsup_{n\to\infty} A_n$ and $\liminf_{n\to\infty} A_n$ respectively. Show that

(a) $B = \{\omega \in \Omega: \omega \in A_n$ for infinitely many values of $n\}$,
(b) $C = \{\omega \in \Omega: \omega \in A_n$ for all but finitely many values of $n\}$.

We say that the sequence $\{A_n\}$ converges to a limit $A = \lim A_n$ if B and C are the same set A. Suppose that $A_n \to A$ and show that

(c) A is an event, in that $A \in \mathscr{F}$,
(d) $\mathbf{P}(A_n) \to \mathbf{P}(A)$.

17. In Problem (16) above, show that B and C are independent whenever B_n and C_n are independent for all n. Deduce that if this holds and furthermore $A_n \to A$, then $\mathbf{P}(A)$ equals either zero or one.

18. Show that the assumption that $\mathbf{P}$ is *countably* additive is equivalent to the assumption that $\mathbf{P}$ is continuous. That is to say, show that if a function $\mathbf{P}: \mathscr{F} \to [0, 1]$ satisfies $\mathbf{P}(\varnothing) = 0$, $\mathbf{P}(\Omega) = 1$, and $\mathbf{P}(A \cup B) = \mathbf{P}(A) + \mathbf{P}(B)$ whenever $A, B \in \mathscr{F}$ and $A \cap B = \varnothing$, then $\mathbf{P}$ is countably additive (in the sense of satisfying (1.3.1b)) if and only if $\mathbf{P}$ is continuous (in the sense of (1.3.5)).

19. Anne, Betty, Chloë, and Daisy were all friends at school. Subsequently each of the $\binom{4}{2} = 6$ subpairs meet up; at each of the six meetings the pair involved quarrel with some fixed probability p, or become firm friends with probability $1 - p$. Quarrels take place independently of each other. In future, if any of the four hears a rumour, then she tells it to her firm friends only. If Anne hears a rumour, what is the probability that

(a) Daisy hears it?
(b) Daisy hears it if Anne and Betty have quarrelled?
(c) Daisy hears it if Betty and Chloë have quarrelled?
(d) Daisy hears it if she has quarrelled with Anne?

20. A biased coin is tossed repeatedly. Each time there is a probability p of a head turning up. Let p_n be the probability that an even number of heads has occurred after n tosses (zero is an even number). Show that $p_0 = 1$ and that $p_n = p(1 - p_{n-1}) + (1 - p)p_{n-1}$ if $n \geqslant 1$. Can you solve this difference equation?

21. A biased coin is tossed repeatedly. Find the probability that there is a run of r heads in a row before there is a run of s tails, where r and s are positive integers.

22. A bowl contains twenty cherries, exactly fifteen of which have had their stones removed. A greedy pig eats five whole cherries, picked at random, without remarking on the presence or absence of stones. Subsequently, a cherry is picked randomly from the remaining fifteen.

    (a) What is the probability that this cherry contains a stone?
    (b) Given that this cherry contains a stone, what is the probability that the pig consumed at least one stone?

23. The 'ménages' problem poses the following question. Some consider it to be desirable that men and women alternate when seated at a circular table. If n couples are seated randomly according to this rule, show that the probability that nobody sits next to his or her partner is

$$\frac{1}{n!}\sum_{k=0}^{n}(-1)^k\frac{2n}{2n-k}\binom{2n-k}{k}(n-k)!.$$

You may find it useful to show first that the number of ways of selecting k non-overlapping pairs of adjacent seats is $\binom{2n-k}{k}2n(2n-k)^{-1}$.

24. An urn contains b blue balls and r red balls. They are removed at random and not replaced. Show that the probability that the first red ball drawn is the $(k+1)$th ball drawn equals $\binom{r+b-k-1}{r-1}\Big/\binom{r+b}{b}$. Find the probability that the last ball drawn is red.

25. An urn contains a azure balls and c carmine balls, where $ac \neq 0$. Balls are removed at random and discarded until the first time that a ball (B, say) is removed having a different colour from its predecessor. The ball B is now replaced and the procedure restarted. This process continues until the last ball is drawn from the urn. Show that this last ball is equally likely to be azure or carmine.

26. A pack of four cards contains one spade, one club, and the two red aces. You deal two cards face downwards at random in front of a truthful friend. She inspects them and tells you that one of them is the ace of hearts. What is the chance that the other card is the ace of diamonds? Perhaps $\frac{1}{3}$?

    Suppose that your friend's protocol was

    (a) with no red ace, say 'no red ace',
    (b) with the ace of hearts, say 'ace of hearts',
    (c) with the ace of diamonds but not the ace of hearts, say 'ace of diamonds'.

    Show that the probability in question is $\frac{1}{3}$.

    Devise a possible protocol for your friend such that the probability in question is zero.

27. **Eddington's controversy.** Four witnesses, A, B, C, and D, at a trial each speak the truth with probability $\frac{1}{3}$ independently of each other. In their testimonies, A claimed that B denied that C declared that D lied. What is the (conditional) probability that D told the truth?

2 Random variables and their distributions

2.1 Random variables

We shall not always be interested in an experiment itself, but rather in some consequence of its random outcome. For example, many gamblers are more concerned with their losses than with the games which give rise to them. Such consequences, when real valued, may be thought of as functions which map Ω into the real line $\mathbb{R}$, and these functions are called 'random variables'.

(1) **Example.** A fair coin is tossed twice: $\Omega = \{HH, HT, TH, TT\}$. For $\omega \in \Omega$, let $X(\omega)$ be the number of heads, so that

$$X(HH) = 2, \qquad X(HT) = X(TH) = 1, \qquad X(TT) = 0.$$

Now suppose that a gambler wagers his fortune of £1 on the result of this experiment. He gambles cumulatively so that his fortune is doubled each time a head appears and is annihilated on the appearance of a tail. His subsequent fortune W is a random variable given by

$$W(HH) = 4, \qquad W(HT) = W(TH) = W(TT) = 0. \qquad \bullet$$

After the experiment is done and the outcome $\omega \in \Omega$ is known, a random variable $X: \Omega \to \mathbb{R}$ takes some value. In general this numerical value is more likely to lie in certain subsets of $\mathbb{R}$ than in certain others, depending on the probability space $(\Omega, \mathscr{F}, \mathbf{P})$ and the function X itself. We wish to be able to describe the distribution of the likelihoods of possible values of X. Example (1) above suggests that we might do this through the function $f: \mathbb{R} \to [0, 1]$ defined by

$$f(x) = \text{probability that } X \text{ is equal to } x,$$

but this turns out to be inappropriate in general. Rather, we use the *distribution function* $F: \mathbb{R} \to \mathbb{R}$ defined by

$$F(x) = \text{probability that } X \text{ does not exceed } x.$$

More rigorously, this is

(2) $$F(x) = \mathbf{P}(A(x))$$

where $A(x) \subseteq \Omega$ is given by

$$A(x) = \{\omega \in \Omega: X(\omega) \leqslant x\}.$$

However, $\mathbf{P}$ is a function on the collection $\mathscr{F}$ of events; we cannot discuss

$\mathbf{P}(A(x))$ unless $A(x)$ belongs to $\mathscr{F}$, and so we are led to the following definition.

(3) **Definition.** A **random variable** is a function $X: \Omega \to \mathbb{R}$ with the property that $\{\omega \in \Omega: X(\omega) \leqslant x\} \in \mathscr{F}$ for each $x \in \mathbb{R}$.

If the reader so desires, then he may pay no attention to the technical condition in the definition and think of random variables simply as functions mapping Ω into $\mathbb{R}$. (If $\{\omega \in \Omega: X(\omega) \leqslant x\} \in \mathscr{F}$ for each $x \in \mathbb{R}$ then we say that X is $\mathscr{F}$*-measurable*.) We shall always use upper-case letters, like X, Y, and Z, to represent generic random variables, whilst lower-case letters, like x, y, and z, will be used to represent possible numerical values of these variables. Do not confuse this notation in your written work.

Every random variable has a distribution function, given by (2); distribution functions are *very* important and useful.

(4) **Definition.** The **distribution function** of a random variable X is the function $F: \mathbb{R} \to [0, 1]$ given by

$$F(x) = \mathbf{P}(X \leqslant x).$$

This is the obvious abbreviation of equation (2). Events written as $\{\omega \in \Omega: X(\omega) \leqslant x\}$ are commonly abbreviated to $\{\omega: X(\omega) \leqslant x\}$ or $\{X \leqslant x\}$. We write F_X where it is necessary to emphasize the role of X.

(5) **Example (1) revisited.** The distribution function F_X of X is given by

$$F_X(x) = \begin{cases} 0 & \text{if} \quad x < 0 \\ \frac{1}{4} & \text{if} \quad 0 \leqslant x < 1 \\ \frac{3}{4} & \text{if} \quad 1 \leqslant x < 2 \\ 1 & \text{if} \quad x \geqslant 2, \end{cases}$$

and is sketched in Figure 2.1. The distribution function F_W of W is given by

$$F_W(x) = \begin{cases} 0 & \text{if} \quad x < 0 \\ \frac{3}{4} & \text{if} \quad 0 \leqslant x < 4 \\ 1 & \text{if} \quad x \geqslant 4, \end{cases}$$

and is sketched in Figure 2.2. This illustrates the important point that the distribution function of a random variable X tells us about the values taken

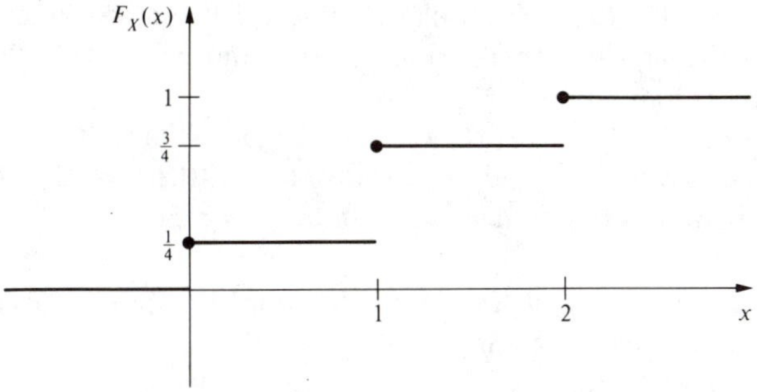

Fig. 2.1 The distribution function F_X of X.

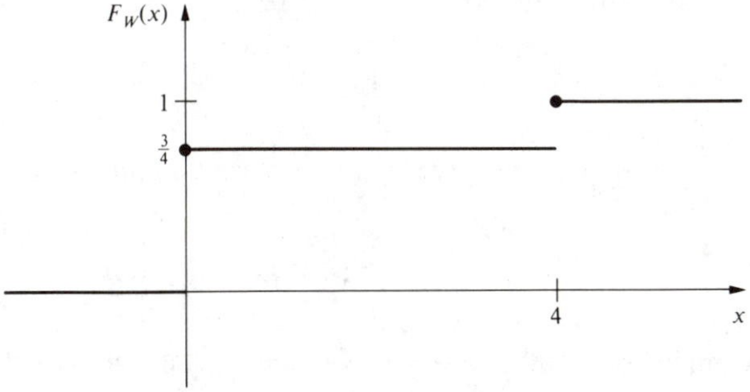

Fig. 2.2 The distribution function F_W of W.

by X and their relative likelihoods, rather than about the sample space and collection of events. ●

(6) **Lemma.** *A distribution function F has the following properties*:

(a) $\lim_{x \to -\infty} F(x) = 0, \qquad \lim_{x \to \infty} F(x) = 1,$

(b) *if* $x < y$ *then* $F(x) \leqslant F(y)$,

(c) F *is right-continuous (that is,* $F(x + h) \to F(x)$ *as* $h \downarrow 0$).

Proof.

(a) Let $B_n = \{\omega \in \Omega: X(\omega) \leqslant -n\} = \{X \leqslant -n\}$. Then $B_1, B_2, \ldots$ is a decreasing family of events with the empty set as limit. Thus, by (1.3.5), $\mathbf{P}(B_n) \to \mathbf{P}(\varnothing) = 0$. The other part is similar.

(b) Let $A(x) = \{X \leqslant x\}$, $A(x, y) = \{x < X \leqslant y\}$. Then $A(y) = A(x) \cup A(x, y)$ is a disjoint union, and so by (1.3.1),

$$\mathbf{P}(A(y)) = \mathbf{P}(A(x)) + \mathbf{P}(A(x, y))$$

giving

$$F(y) = F(x) + \mathbf{P}(x < X \leqslant y) \geqslant F(x).$$

(c) This is an *exercise*. Use (1.3.5). ■

Actually, this lemma characterizes distribution functions. That is to say, F is the distribution function of some random variable if and only if it satisfies (6a), (6b), and (6c).

For the time being we can forget all about probability spaces and concentrate on random variables and their distribution functions. The distribution function F of X contains a great deal of information about X.

(7) **Example. Constant variables.** The simplest random variable takes a constant value on the whole domain Ω.

Let $c \in \mathbb{R}$ and define $X: \Omega \to \mathbb{R}$ by

$$X(\omega) = c \quad \text{for all } \omega \in \Omega.$$

The distribution function $F(x) = \mathbf{P}(X \leqslant x)$ is the step function

$$F(x) = \begin{cases} 0 & x < c \\ 1 & x \geqslant c. \end{cases}$$

Slightly more generally, we call X *constant* (*almost surely*) if there exists $c \in \mathbb{R}$ such that

$$\mathbf{P}(X = c) = 1.$$

●

(8) **Example. Bernoulli variables.** Consider Example (1.3.2). Let $X: \Omega \to \mathbb{R}$ be given by

$$X(H) = 1, \qquad X(T) = 0.$$

Then X is the simplest non-trivial random variable, with two possible values, 0 and 1. Its distribution function $F(x) = \mathbf{P}(X \leqslant x)$ is

$$F(x) = \begin{cases} 0 & x < 0 \\ 1 - p & 0 \leqslant x < 1 \\ 1 & x \geqslant 1. \end{cases}$$

X is said to have the *Bernoulli distribution.*

(9) **Example. Indicator functions.** A particular class of Bernoulli variables is very useful in probability theory. Let A be an event and let $I_A: \Omega \to \mathbb{R}$ be the *indicator function* of A; that is,

$$I_A(\omega) = \begin{cases} 1 & \text{if } \omega \in A \\ 0 & \text{if } \omega \in A^c. \end{cases}$$

Then I_A is a Bernoulli random variable taking the values 1 and 0 with probabilities $\mathbf{P}(A)$ and $\mathbf{P}(A^c)$ respectively. The following is a useful identity. Suppose $\{B_i: i \in I\}$ is a family of disjoint events with $A \subseteq \bigcup_{i \in I} B_i$. Then

(10) $$I_A = \sum_i I_{A \cap B_i}.$$

●

(11) **Lemma.** *Let F be the distribution function of X. Then*

(a) $\mathbf{P}(X > x) = 1 - F(x)$,
(b) $\mathbf{P}(x < X \leqslant y) = F(y) - F(x)$,
(c) $\mathbf{P}(X = x) = F(x) - \lim_{y \uparrow x} F(y)$.

Proof. (a) and (b) are *exercises.*
(c) Let $B_n = \{x - 1/n < X \leqslant x\}$ and use the method of proof of Lemma (6). ■

Note one final piece of jargon for future use. A random variable X with distribution function F is said to have two 'tails' given by

$$T_1(x) = \mathbf{P}(X > x) = 1 - F(x)$$

$$T_2(x) = \mathbf{P}(X \leqslant -x) = F(-x)$$

where x is large and positive. We shall see later that the rates at which the T_i decay to zero as $x \to \infty$ have a substantial effect on the existence or non-existence of certain associated quantities called the 'moments' of the distribution.

Exercises

(12) Let X be a random variable on a given probability space, and let $a \in \mathbb{R}$. Show that

(i) aX is a random variable,
(ii) $X - X = 0$, the random variable taking the value 0 always, and $X + X = 2X$.

(13) A random variable X has distribution function F. What is the distribution function of $Y = aX + b$, where a and b are real constants?

(14) A fair coin is tossed n times. Show that, under reasonable assumptions, the probability of exactly k heads is $\binom{n}{k}(\frac{1}{2})^n$. What is the corresponding quantity when heads appears with probability p on each toss?

(15) Show that if F and G are distribution functions and $0 \leqslant \lambda \leqslant 1$ then $\lambda F + (1 - \lambda)G$ is a distribution function.

2.2 The law of averages

We may recall the discussion in Section 1.3 of repeated experimentation. In each of N repetitions of an experiment, we observe whether or not a given event A occurs, and we write $N(A)$ for the total number of occurrences of A. One possible philosophical underpinning of probability theory requires that the proportion $N(A)/N$ settles down as $N \to \infty$ to some limit interpretable as the 'probability of A'. Is our theory to date consistent with such a requirement?

With this question in mind, let us suppose that $A_1, A_2, \ldots$ is a sequence of independent events having equal probability $\mathbf{P}(A_i) = p$, where $0 < p < 1$; such an assumption requires of course the existence of a corresponding probability space $(\Omega, \mathcal{F}, \mathbf{P})$, but we do not plan to get bogged down in such matters here. We think of A_i as being the event 'that A occurs on the ith experiment'. We write $S_n = \sum_{i=1}^{n} I_{A_i}$, the sum of the indicator functions of $A_1, A_2, \ldots, A_n$; S_n is a random variable which counts the number of occurrences of A_i for $1 \leqslant i \leqslant n$ (certainly S_n is a function of Ω, since it is the sum of such functions, and it is left as an *exercise* to show that S_n is $\mathcal{F}$-measurable). The following result concerning the ratio $n^{-1}S_n$ was proved by James Bernoulli before 1692.

(1) **Theorem.** *It is the case that $n^{-1}S_n$ converges to p as $n \to \infty$ in the sense that, for all $\varepsilon > 0$,*

$$\mathbf{P}\left(p - \varepsilon \leqslant \frac{1}{n} S_n \leqslant p + \varepsilon\right) \to 1 \quad \textit{as} \quad n \to \infty.$$

There are certain technicalities involved in the study of the convergence of random variables (see Chapter 7), and this is the reason for the careful statement of the theorem. For the time being, we encourage the reader to interpret the theorem as asserting simply that the proportion $n^{-1}S_n$ of times that the events $A_1, A_2, \ldots, A_n$ occur converges as $n \to \infty$ to their common probability p. We shall see later how important it is to be careful when making such statements.

Interpreted in terms of tosses of a fair coin, the theorem implies that the proportion of heads is (with large probability) near to $\frac{1}{2}$. As a caveat regarding the difficulties inherent in studying the convergence of random variables, we remark that it is *not* true that, in a 'typical' sequence of tosses of a fair coin, heads outnumber tails about one-half of the time.

Proof. Suppose that we toss a coin repeatedly, and heads occurs on each toss with probability p. The random variable S_n has the same probability distribution as the number H_n of heads which occur during the first n tosses, which is to say that $\mathbf{P}(S_n = k) = \mathbf{P}(H_n = k)$ for all k. It follows that, for small positive values of ε,

$$\mathbf{P}\left(\frac{1}{n} S_n \geqslant p + \varepsilon\right) = \sum_{k \geqslant n(p+\varepsilon)} \mathbf{P}(H_n = k).$$

We have from Exercise (2.1.14) that

$$\mathbf{P}(H_n = k) = \binom{n}{k} p^k (1-p)^{n-k} \quad \text{for} \quad 0 \leqslant k \leqslant n,$$

and hence

(2)
$$\mathbf{P}\left(\frac{1}{n} S_n \geqslant p + \varepsilon\right) = \sum_{k=m}^{n} \binom{n}{k} p^k (1-p)^{n-k}$$

where $m = \lceil n(p+\varepsilon) \rceil$, the least integer not less than $n(p+\varepsilon)$. The following argument is standard in probability theory. Let $\lambda > 0$ and note that $e^{\lambda k} \geqslant \exp([\lambda n(p+\varepsilon)]$ if $k \geqslant m$. Writing $q = 1 - p$, we have that

$$\begin{aligned}\mathbf{P}\left(\frac{1}{n} S_n \geqslant p + \varepsilon\right) &\leqslant \sum_{k=m}^{n} \exp\{\lambda[k - n(p+\varepsilon)]\} \binom{n}{k} p^k q^{n-k} \\ &\leqslant e^{-\lambda n \varepsilon} \sum_{k=0}^{n} \binom{n}{k} (p\, e^{\lambda q})^k (q\, e^{-\lambda p})^{n-k} \\ &= e^{-\lambda n \varepsilon} (p\, e^{\lambda q} + q\, e^{-\lambda p})^n,\end{aligned}$$

by the binomial theorem. It is a simple *exercise* to show that $e^x \leqslant x + e^{x^2}$ for $x \in \mathbb{R}$. With the aid of this inequality, we obtain

(3) $$\mathbf{P}\left(\frac{1}{n} S_n \geqslant p + \varepsilon\right) \leqslant e^{-\lambda n \varepsilon}[p \exp(\lambda^2 q^2) + q \exp(\lambda^2 p^2)]^n \leqslant \exp(\lambda^2 n - \lambda n \varepsilon).$$

We can pick λ to minimize the right-hand side, and this value is $\lambda = \frac{1}{2}\varepsilon$, giving

(4) $$\mathbf{P}\left(\frac{1}{n} S_n \geqslant p + \varepsilon\right) \leqslant \exp(-n\varepsilon^2/4) \quad \text{for} \quad \varepsilon > 0,$$

an inequality known as 'Bernstein's inequality'. It follows immediately that $\mathbf{P}(n^{-1}S_n \geqslant p + \varepsilon) \to 0$ as $n \to \infty$. An exactly analogous argument shows that $\mathbf{P}(n^{-1}S_n \leqslant p - \varepsilon) \to 0$ as $n \to \infty$, and thus the theorem is proved. ■

Bernstein's inequality (4) is rather powerful, asserting that the chance that S_n exceeds its mean by a quantity of order n tends to zero *exponentially fast* as $n \to \infty$; such an inequality is known as a 'large-deviation estimate'. We may use the inequality to prove rather more than the conclusion of the theorem. Instead of estimating the chance that, for a specific value of n, S_n lies between $n(p - \varepsilon)$ and $n(p + \varepsilon)$, let us estimate the chance that this occurs *for all large n*. Writing $A_n = \{p - \varepsilon \leqslant n^{-1}S_n \leqslant p + \varepsilon\}$, we wish to estimate $\mathbf{P}(\bigcap_{n=m}^{\infty} A_n)$. Now the complement of this intersection is the event $\bigcup_{n=m}^{\infty} A_n^c$, and the probability of this union satisfies, by the inequalities of Boole and Bernstein,

(5) $$\mathbf{P}\left(\bigcup_{n=m}^{\infty} A_n^c\right) \leqslant \sum_{n=m}^{\infty} \mathbf{P}(A_n^c) \leqslant \sum_{n=m}^{\infty} 2\exp(-n\varepsilon^2/4) \to 0, \quad \text{as} \quad m \to \infty,$$

giving that, as required,

(6) $$\mathbf{P}\left(p - \varepsilon \leqslant \frac{1}{n} S_n \leqslant p + \varepsilon \text{ for all } n \geqslant m\right) \to 1 \quad \text{as} \quad m \to \infty.$$

Exercises

(7) You wish to ask each of a large number of people a question to which the answer 'yes' is embarrassing. The following procedure is proposed in order to determine the

embarrassed fraction of the population. As the question is asked, a coin is tossed out of sight of the questioner. If the answer would have been 'no' and the coin shows heads, then the answer 'yes' is given. Otherwise people respond truthfully. What do you think of this procedure?

(8) A coin is tossed repeatedly and heads turns up on each toss with probability p. Let H_n and T_n be the numbers of heads and tails in n tosses. Show that, for $\varepsilon > 0$,

$$\mathbf{P}\left(2p - 1 - \varepsilon \leqslant \frac{1}{n}(H_n - T_n) \leqslant 2p - 1 + \varepsilon\right) \to 1 \quad \text{as} \quad n \to \infty.$$

2.3 Discrete and continuous variables

Much of the study of random variables is devoted to distribution functions, characterized by (2.1.6). The general theory of distribution functions and their applications is quite difficult and abstract and is best omitted at this stage. It relies on a rigorous treatment of the construction of the Lebesgue–Stieltjes integral; this is sketched in Section 5.6. However, things become much easier if we are prepared to restrict our attention to certain subclasses of random variables specified by properties which make them tractable. We shall consider in depth the collection of 'discrete' random variables and the collection of 'continuous' random variables.

(1) **Definition.** The random variable X is called **discrete** if it takes values in some countable subset $\{x_1, x_2, \ldots\}$, only, of $\mathbb{R}$.

We shall see that the distribution function of a discrete variable has jump discontinuities at the values $x_1, x_2, \ldots$ and is constant in between; such a distribution is called *atomic*. This contrasts sharply with the other important class of distribution functions considered here.

(2) **Definition.** The random variable X is called **continuous** if its distribution function can be expressed as

$$F(x) = \int_{-\infty}^{x} f(u)\,du \qquad x \in \mathbb{R},$$

for some integrable function $f: \mathbb{R} \to [0, \infty)$.

The distribution function of a continuous random variable is certainly continuous (actually it is 'absolutely continuous'). For the moment we are concerned only with discrete variables and continuous variables. There is another sort of random variable, called 'singular', for a discussion of which the reader should look elsewhere. A common example of this phenomenon is based upon the Cantor ternary set (see Grimmett and Welsh 1986,

Billingsley 1986, or Kingman and Taylor 1966). Other variables are 'mixtures' of discrete, continuous, and singular variables. Note that the word 'continuous' is a misnomer when used in this regard: in describing X as continuous, we are referring to a property of its distribution function rather than of the random variable (function) X itself.

(3) **Example. Discrete variables.** The variables X and W of Example (2.1.1) take values in the sets $\{0, 1, 2\}$ and $\{0, 4\}$ respectively; they are both discrete. ●

(4) **Example. Continuous variables.** A straight rod is flung down at random on to a horizontal plane and the angle ω between the rod and true north is measured. The result is a number in $\Omega = [0, 2\pi)$. Never mind about $\mathcal{F}$ for the moment; we can suppose that $\mathcal{F}$ contains all nice subsets of Ω, such as the collection of open subintervals like (a, b), where $0 \leqslant a < b < 2\pi$. The implicit symmetry suggests the probability measure $\mathbf{P}$ which satisfies $\mathbf{P}((a, b)) = (b - a)/(2\pi)$; that is to say, the probability that the angle lies in some interval is directly proportional to the length of the interval. Here are two random variables X and Y:

$$X(\omega) = \omega$$
$$Y(\omega) = \omega^2.$$

Notice that Y is a function of X in that $Y = X^2$. The distribution functions of X and Y are

$$F_X(x) = \begin{cases} 0 & x \leqslant 0 \\ x/(2\pi) & 0 \leqslant x < 2\pi \\ 1 & x \geqslant 2\pi \end{cases}$$

$$F_Y(y) = \begin{cases} 0 & y \leqslant 0 \\ y^{\frac{1}{2}}/(2\pi) & 0 \leqslant y < 4\pi^2 \\ 1 & y \geqslant 4\pi^2. \end{cases}$$

To see this, let $0 \leqslant x < 2\pi$ and $0 \leqslant y < 4\pi^2$. Then

$$\begin{aligned} F_X(x) &= \mathbf{P}(\{\omega \in \Omega : 0 \leqslant X(\omega) \leqslant x\}) \\ &= \mathbf{P}(\{\omega \in \Omega : 0 \leqslant \omega \leqslant x\}) = x/(2\pi), \\ F_Y(y) &= \mathbf{P}(\{\omega : Y(\omega) \leqslant y\}) \\ &= \mathbf{P}(\{\omega : \omega^2 \leqslant y\}) = \mathbf{P}(\{\omega : 0 \leqslant \omega \leqslant y^{\frac{1}{2}}\}) = \mathbf{P}(X \leqslant y^{\frac{1}{2}}) \\ &= y^{\frac{1}{2}}/(2\pi). \end{aligned}$$

X and Y are continuous because

$$F_X(x) = \int_{-\infty}^{x} f_X(u)\,\mathrm{d}u, \qquad F_Y(y) = \int_{-\infty}^{y} f_Y(u)\,\mathrm{d}u$$

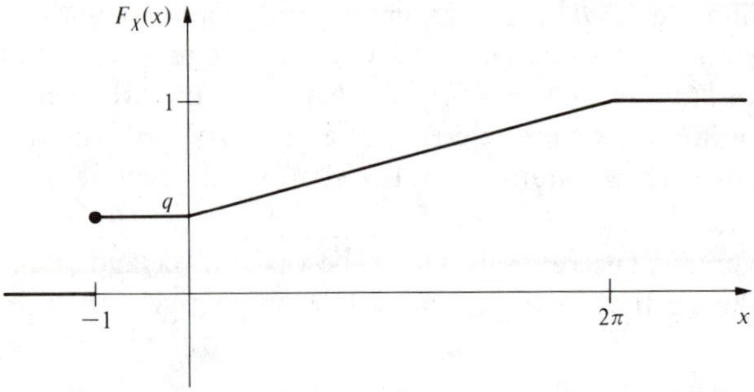

Fig. 2.3 The distribution function F_X of X in Example (5).

where

$$f_X(u) = \begin{cases} 1/(2\pi) & \text{if } 0 \leqslant u \leqslant 2\pi \\ 0 & \text{otherwise} \end{cases}$$

$$f_Y(u) = \begin{cases} u^{-\frac{1}{2}}/(4\pi) & \text{if } 0 \leqslant u \leqslant 4\pi^2 \\ 0 & \text{otherwise.} \end{cases}$$ ●

(5) **Example. A random variable which is neither continuous nor discrete.** A coin is tossed, and a head turns up with probability p $(= 1 - q)$. If a head turns up then a rod is flung on the ground and the angle measured as in Example (4). Then $\Omega = \{T\} \cup \{(H, x): 0 \leqslant x < 2\pi\}$, in the obvious notation. Let $X: \Omega \to \mathbb{R}$ be given by

$$X(T) = -1, \qquad X((H, x)) = x.$$

X takes values in $\{-1\} \cup [0, 2\pi)$ (see Figure 2.3 for a sketch of its distribution function). We say that X is continuous, except for a 'point mass (or *atom*) at -1'. ●

Exercises

(6) Let X be a random variable with distribution function F, and let $a = (a_m: -\infty < m < \infty)$ be a strictly increasing sequence of real numbers satisfying $a_{-m} \to -\infty$ and $a_m \to \infty$ as $m \to \infty$. Define $G(x) = \mathbf{P}(X \leqslant a_m)$ when $a_{m-1} \leqslant x < a_m$, so that G is the distribution function of a discrete random variable. How does the function G behave as the sequence a is chosen in such a way that $\sup_m |a_m - a_{m-1}|$ becomes smaller and smaller?

(7) Let X be a random variable and let $g: \mathbb{R} \to \mathbb{R}$ be continuous and strictly increasing. Show that $Y = g(X)$ is a random variable.

(8) Let X be a random variable with distribution function

$$\mathbf{P}(X \leqslant x\} = \begin{cases} 0 & \text{if } x \leqslant 0 \\ x & \text{if } 0 < x \leqslant 1 \\ 1 & \text{if } x > 1. \end{cases}$$

Let F be a distribution function which is continuous and strictly increasing. Show that $Y = F^{-1}(X)$ is a random variable having distribution function F. Is it necessary that F be continuous and/or strictly increasing?

2.4 Worked examples

(1) **Example. Darts.** A dart is flung at a circular target of radius 3. We can think of the hitting point as the outcome of a random experiment; we shall suppose for simplicity that the player is guaranteed to hit the target somewhere. Setting the centre of the target at the origin of $\mathbb{R}^2$, we see that the sample space of this experiment is

$$\Omega = \{(x, y): x^2 + y^2 < 9\}.$$

Never mind about the collection $\mathcal{F}$ of events. Let us suppose that, roughly speaking, the probability that the dart lands in some region A is proportional to the area $|A|$ of A. Thus

(2) $$\mathbf{P}(A) = |A|/(9\pi).$$

The scoring system is as follows. The target is partitioned by three concentric circles C_1, C_2, and C_3, centred at the origin with radii 1, 2, and 3. These circles divide the target into three annuli A_1, A_2, and A_3, where

$$A_k = \{(x, y): k - 1 \leqslant (x^2 + y^2)^{\frac{1}{2}} < k\}.$$

We suppose that the player scores an amount k if and only if the dart hits A_k. The resulting score X is the random variable given by

$$X(\omega) = k \quad \text{whenever} \quad \omega \in A_k.$$

What is its distribution function?

Solution. Clearly

$$\begin{aligned} \mathbf{P}(X = k) &= \mathbf{P}(A_k) \\ &= |A_k|/(9\pi) \\ &= (2k - 1)/9, \qquad \text{for } k = 1, 2, 3, \end{aligned}$$

and so the distribution function of X is given by

$$F_X(r) = \mathbf{P}(X \leqslant r) = \begin{cases} 0 & \text{if} \quad r < 1 \\ \lfloor r \rfloor^2/9 & \text{if} \quad 1 \leqslant r < 3 \\ 1 & \text{if} \quad r \geqslant 3 \end{cases}$$

where $\lfloor r \rfloor$ denotes the largest integer not larger than r (see Figure 2.4).

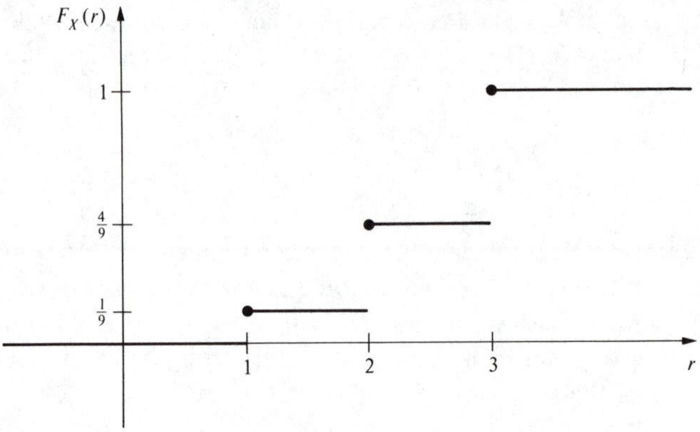

Fig. 2.4 The distribution function F_X of X. ●

(3) **Example. Continuation of (1).** Let us consider a revised method of scoring in which the player scores an amount equal to the distance between the hitting point ω and the centre of the target. This time the score Y is a random variable given by

$$Y(\omega) = (x^2 + y^2)^{\frac{1}{2}}, \quad \text{if} \quad \omega = (x, y).$$

What is the distribution function of Y?

Solution. For any real r let C_r denote the disc with centre $(0, 0)$ and radius r, that is

$$C_r = \{(x, y): x^2 + y^2 \leqslant r\}.$$

Then

$$F_Y(r) = \mathbf{P}(Y \leqslant r) = \mathbf{P}(C_r) = r^2/9 \quad \text{if} \quad 0 \leqslant r \leqslant 3.$$

This distribution function is sketched in Figure 2.5.

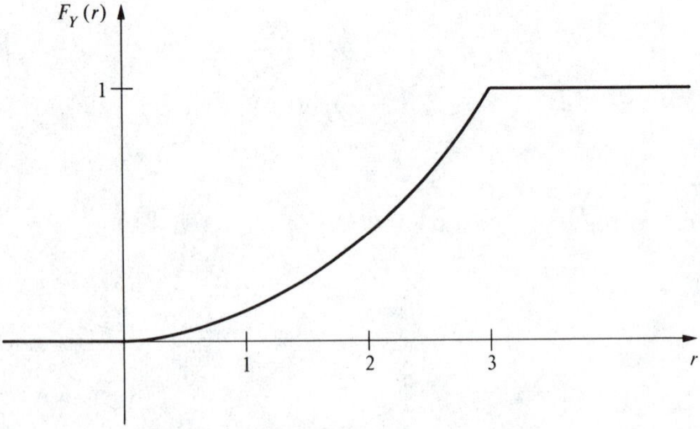

Fig. 2.5 The distribution function F_Y of Y. ●

(4) **Example. Continuation of (1).** Now suppose that the player fails to hit the target with fixed probability p; if he is successful then we suppose that the distribution of the hitting point is described by (2). His score is specified as follows. If he hits the target then he scores an amount equal to the distance between the hitting point and the centre; if he misses the target then he scores 4. What is the distribution function of his score Z?

Solution. Clearly Z takes values in $[0, 4]$. Use (1.4.4) to see that

$$\begin{aligned} F_Z(r) &= \mathbf{P}(Z \leqslant r) \\ &= \mathbf{P}(Z \leqslant r \mid \text{hits target})\mathbf{P}(\text{hits target}) \\ &\quad + \mathbf{P}(Z \leqslant r \mid \text{misses target})\mathbf{P}(\text{misses target}) \\ &= \begin{cases} 0 & \text{if } r < 0 \\ (1-p)F_Y(r) & \text{if } 0 \leqslant r < 4 \\ 1 & \text{if } r \geqslant 4, \end{cases} \end{aligned}$$

where F_Y is given in (3) (see Figure 2.6 for a sketch of F_Z).

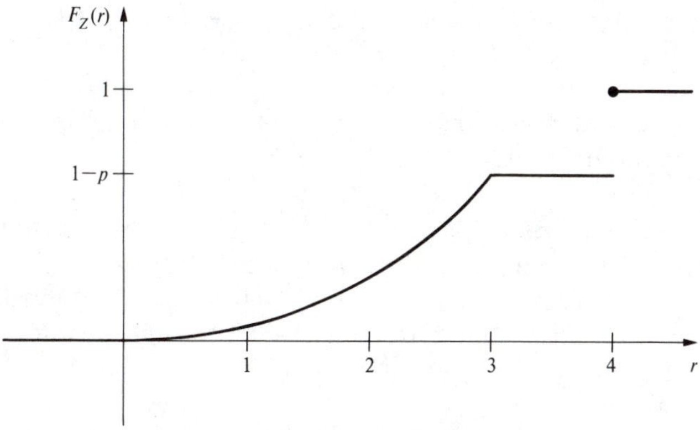

Fig. 2.6 The distribution function F_Z of Z. ●

Exercises

(5) Let X be a random variable with a continuous distribution function F. Find expressions for the distribution functions of the following random variables:

(a) X^2, (b) $\sqrt{X}$,
(c) $\sin X$, (d) $G^{-1}(X)$,
(e) $F(X)$, (f) $G^{-1}(F(X))$,

where G is a continuous and strictly increasing function.

(6) **Truncation.** Let X be a random variable with distribution function F, and let $a < b$. Sketch the distribution functions of the 'truncated' random variables Y and Z given by

$$Y = \begin{cases} a & \text{if } X < a \\ X & \text{if } a \leqslant X \leqslant b \\ b & \text{if } X > b \end{cases} \qquad Z = \begin{cases} X & \text{if } |X| \leqslant b \\ 0 & \text{if } |X| > b. \end{cases}$$

Indicate how these distribution functions behave as $a \to -\infty$, $b \to \infty$.

2.5 Random vectors

Suppose that X and Y are random variables on the probability space $(\Omega, \mathcal{F}, \mathbf{P})$. Their distribution functions, F_X and F_Y, contain information about their associated probabilities. But how may we encapsulate information about their properties *relative to each other*? The key is to think of X and Y as being the components of a 'random vector' (X, Y) taking values in $\mathbb{R}^2$, rather than being unrelated random variables each taking values in $\mathbb{R}$.

(1) **Example. Tontine** is a scheme wherein subscribers to a common fund each receive an annuity from the fund during his or her lifetime, this annuity increasing as the other subscribers die. When all the subscribers are dead, the fund passes to the French government (at least, this was the case in the first such scheme designed by Lorenzo Tonti around 1653). The performance of the fund depends on the lifetimes $L_1, L_2, \ldots, L_n$ of the subscribers (as well as on their wealths), and we may record these as a vector $(L_1, L_2, \ldots, L_n)$ of random variables. ●

(2) **Example. Darts.** A dart is flung at a conventional dartboard. The point of striking determines a distance R from the centre, an angle Θ with the upward vertical (measured clockwise, say), and a score S. With this experiment we may associate the random vector (R, Θ, S), and we note that S is a function of the pair (R, Θ). ●

(3) **Example. Coin tossing.** Suppose that we toss a coin n times, and set X_i equal to 0 or 1 depending on whether the ith toss results in a tail or a head. We think of the vector $\boldsymbol{X} = (X_1, X_2, \ldots, X_n)$ as describing the result of this composite experiment. The total number of heads is the sum of the entries in $\boldsymbol{X}$. ●

An individual random variable X has a distribution function F_X defined by $F_X(x) = \mathbf{P}(X \leqslant x)$ for $x \in \mathbb{R}$. The corresponding 'joint' distribution function of a random vector $(X_1, X_2, \ldots, X_n)$ is the quantity $\mathbf{P}(X_1 \leqslant x_1, X_2 \leqslant x_2, \ldots, X_n \leqslant x_n)$, a function of n real variables $x_1, x_2, \ldots, x_n$. In order to aid the notation, we introduce an ordering of vectors of real numbers: for vectors $\boldsymbol{x} = (x_1, x_2, \ldots, x_n)$ and $\boldsymbol{y} = (y_1, y_2, \ldots, y_n)$ we write $\boldsymbol{x} \leqslant \boldsymbol{y}$ if $x_i \leqslant y_i$ for each $i = 1, 2, \ldots, n$.

(4) **Definition.** The **joint distribution function** of a random vector $\mathbf{X} = (X_1, X_2, \ldots, X_n)$ on the probability space $(\Omega, \mathscr{F}, \mathbf{P})$ is the function $F_{\mathbf{X}}: \mathbb{R}^n \to [0, 1]$ given by $F_{\mathbf{X}}(\mathbf{x}) = \mathbf{P}(\mathbf{X} \leqslant \mathbf{x})$ for $\mathbf{x} \in \mathbb{R}^n$.

As before, the expression $\{\mathbf{X} \leqslant \mathbf{x}\}$ is an abbreviation for the event $\{\omega \in \Omega: \mathbf{X}(\omega) \leqslant \mathbf{x}\}$. Joint distribution functions have properties similar to those of ordinary distribution functions. For example, Lemma (2.1.6) becomes the following.

(5) **Lemma.** *The joint distribution function $F_{X,Y}$ of the random vector (X, Y) has the following properties*:

(a) $\lim_{x,y\to-\infty} F_{X,Y}(x, y) = 0$, $\lim_{x,y\to\infty} F_{X,Y}(x, y) = 1$,
(b) *if* $(x_1, y_1) \leqslant (x_2, y_2)$ *then* $F_{X,Y}(x_1, y_1) \leqslant F_{X,Y}(x_2, y_2)$,
(c) $F_{X,Y}$ *is continuous from above, in that*

$$F_{X,Y}(x + u, y + v) \to F_{X,Y}(x, y) \quad \textit{as} \quad u, v \downarrow 0.$$

We state this lemma for a random vector with only two components X and Y, but the corresponding result for n components is valid also. The proof of the lemma is left as an *exercise*. Rather more is true. It may be seen without great difficulty that

(6) $$\lim_{y\to\infty} F_{X,Y}(x, y) = F_X(x) \; (= \mathbf{P}(X \leqslant x))$$

and

(7) $$\lim_{x\to\infty} F_{X,Y}(x, y) = F_Y(y) \; (= \mathbf{P}(Y \leqslant y));$$

this more refined version of part (a) of the lemma tells us that we may recapture the individual distribution functions of X and Y from a knowledge of their joint distribution function. The converse is false: it is not generally possible to calculate $F_{X,Y}$ from knowledge of F_X and F_Y alone. The functions F_X and F_Y are called the 'marginal' distribution functions of $F_{X,Y}$.

(8) **Example.** A schoolteacher asks each member of his or her class to flip a fair coin twice and to record the outcomes. The diligent pupil D does this and records a pair (X_D, Y_D) of outcomes. The lazy pupil L flips the coin only once and writes down the result twice, recording thus a pair (X_L, Y_L) where $X_L = Y_L$. Clearly X_D, Y_D, X_L, and Y_L are random variables with the same distribution functions. However, the pairs (X_D, Y_D) and (X_L, Y_L) have different *joint* distribution functions. In particular, $\mathbf{P}(X_D = Y_D = \text{heads}) = \frac{1}{4}$ since only one of the four possible pairs of outcomes contains heads only, whereas $\mathbf{P}(X_L = Y_L = \text{heads}) = \frac{1}{2}$. ●

Once again there are two classes of random vectors which are particularly interesting: the 'discrete' and the 'continuous'.

(9) **Definition.** The random variables X and Y on the probability space $(\Omega, \mathscr{F}, \mathbf{P})$

are called **(jointly) discrete** if the vector (X, Y) takes values in some countable subset of $\mathbb{R}^2$ only.

(10) **Definition.** The random variables X and Y on the probability space $(\Omega, \mathscr{F}, \mathbf{P})$ are called **(jointly) continuous** if their joint distribution function can be expressed as

$$F_{X,Y}(x, y) = \int_{u=-\infty}^{x} \int_{v=-\infty}^{y} f(u, v)\, du\, dv \qquad x, y \in \mathbb{R},$$

for some integrable function $f: \mathbb{R}^2 \to [0, \infty)$.

We shall return to such questions in later chapters. Meanwhile here are two concrete examples.

(11) **Example. Three-sided coin.** We are provided with a special three-sided coin, each toss of which results in one of the possibilities H (heads), T (tails), E (edge), each having probability $\frac{1}{3}$. Let H_n, T_n, and E_n be the numbers of such outcomes in n tosses of the coin. The vector (H_n, T_n, E_n) is a vector of random variables satisfying $H_n + T_n + E_n = n$. If the outcomes of different tosses have no influence on each other, then it is not difficult to see that

$$\mathbf{P}\big((H_n, T_n, E_n) = (h, t, e)\big) = \frac{n!}{h!\,t!\,e!}\left(\frac{1}{3}\right)^n$$

for any triple (h, t, e) of non-negative integers having sum n. The random variables H_n, T_n, E_n are (jointly) discrete and are said to have (jointly) the **trinomial** distribution. ●

(12) **Example. Darts.** Returning to the flung dart of Example (2), let us assume that no region of the dartboard is preferred unduly over any other region of equal area. It may then be shown (see (2.4.3)) that

$$\mathbf{P}(R \leqslant r) = \frac{r^2}{\rho^2}, \quad \mathbf{P}(\Theta \leqslant \theta) = \frac{\theta}{2\pi}, \quad \text{for} \quad 0 \leqslant r \leqslant \rho,\ 0 \leqslant \theta \leqslant 2\pi,$$

where ρ is the radius of the board, and furthermore

$$\mathbf{P}(R \leqslant r, \Theta \leqslant \theta) = \mathbf{P}(R \leqslant r)\mathbf{P}(\Theta \leqslant \theta).$$

It follows that

$$F_{R,\Theta}(r, \theta) = \int_{u=0}^{r} \int_{v=0}^{\theta} f(u, v)\, du\, dv$$

where

$$f(u, v) = \frac{u}{\pi\rho^2}, \quad 0 \leqslant u \leqslant \rho,\ 0 \leqslant v \leqslant 2\pi.$$

The pair (R, Θ) is (jointly) continuous. ●

Exercises

(13) A fair coin is tossed twice. Let X be the number of heads, and let W be the indicator function of the event $\{X = 2\}$. Find $\mathbf{P}(X = x, W = w)$ for all appropriate values of x and w.

(14) Let X be a Bernoulli random variable, so that $\mathbf{P}(X = 0) = 1 - p$, $\mathbf{P}(X = 1) = p$. Let $Y = 1 - X$ and $Z = XY$. Find $\mathbf{P}(X = x, Y = y)$ and $\mathbf{P}(X = x, Z = z)$ for $x, y, z \in \{0, 1\}$.

(15) The random variables X and Y have joint distribution function

$$F_{X,Y}(x, y) = \begin{cases} 0 & \text{if } x < 0 \\ (1 - e^{-x})\left(\dfrac{1}{2} + \dfrac{1}{\pi}\tan^{-1} y\right) & \text{if } x \geqslant 0. \end{cases}$$

Show that X and Y are (jointly) continuously distributed.

(16) Let X and Y have joint distribution function F. Show that

$$\mathbf{P}(a < X \leqslant b, c < Y \leqslant d) = F(b, d) - F(a, d) - F(b, c) + F(a, c)$$

whenever $a < b$ and $c < d$.

2.6 Monte Carlo simulation

It is presumably the case that the physical shape of a coin is one of the major factors relevant to whether or not it will fall with heads uppermost. In principle the shape of the coin may be determined by direct examination, and hence we may arrive at an estimate for the chance of heads. Unfortunately, such a calculation would be rather complicated, and it is easier to estimate this chance by simulation, which is to say that we may toss the coin many times and record the proportion of successes. Similarly, roulette players are well advised to observe the behaviour of the wheel with care in advance of placing large bets, in order to discern its peculiarities (unfortunately, casinos are now wary of such observation, and change their wheels at regular intervals).

Here is a related question. Suppose that we know that our coin is fair (so that the chance of heads is $\frac{1}{2}$ on each toss), and we wish to know the chance that a sequence of 50 tosses contains a run of outcomes of the form $HTHHT$. In principle this probability may be calculated explicitly and exactly. If we require only an estimate of its value, then another possibility is to simulate the experiment: toss the coin $50N$ times for some N, divide the result into N runs of 50, and find the proportion of such runs which contain $HTHHT$.

It is not unusual for a specific calculation to be possible in principle but extremely difficult in practice, often owing to limitations on the operating speed or the size of the memory of a computer. Simulation can provide a way around such a problem. Here are some examples.

(1) **Example. Gambler's ruin revisited.** The gambler of (1.7.4) eventually won his Jaguar after a long period devoted to tossing coins, and he has now decided

to save up for a yacht. His bank manager has suggested that, in order to speed things up, the stake on each gamble should not remain constant but should vary as a certain prescribed function of the gambler's current fortune. The gambler would like to calculate the chance of winning the yacht in advance of embarking on the project, but he finds himself incapable of doing so.

Fortunately, he has kept a record of the extremely long sequence of heads and tails encountered in his successful play for the Jaguar. He calculates his sequence of hypothetical fortunes based on this information, until the point when this fortune reaches either zero or the price of the yacht. He then starts again, and continues to repeat the procedure until he has completed it a total of N times, say. He estimates the probability that he will actually win the yacht by the proportion of the N calculations which result in success. Can you see why this method will make him overconfident? He might do better to retoss the coins. ●

(2) **Example. A dam.** It is proposed to build a dam in order to regulate the water supply, and in particular to prevent seasonal flooding downstream. How high should the dam be? Dams are expensive to construct, and some compromise between cost and risk is necessary. It is decided to build a dam which is just high enough to ensure that the chance of a flood of some given extent within ten years is less than 10^{-2}, say. No one knows exactly how high such a dam need be, and a young probabilist proposes the following scheme. Through examination of existing records of rainfall and water demand we may arrive at an acceptable model for the pattern of supply and demand. This model includes, for example, estimates of the distributions of rainfall on successive days over long periods. With the aid of a computer, the 'real world' situation is simulated many times in order to study the likely consequences of building dams of various heights. In this way we may arrive at an accurate estimate of the height required. ●

(3) **Example. Integration.** Let $g: [0, 1] \to [0, 1]$ be a continuous but nowhere differentiable function. How may we calculate its integral $I = \int_0^1 g(x)\,dx$? The following experimental technique is known as the 'hit or miss Monte Carlo technique'.

Let (X, Y) be a random vector having the uniform distribution on the unit square. That is, we assume that $\mathbf{P}[(X, Y) \in A] = |A|$, the area of A, for any nice subset A of the unit square $[0, 1]^2$; we leave this assumption somewhat up in the air for the moment, and shall return to such matters in Chapter 4. We declare (X, Y) to be 'successful' if $Y \leqslant g(X)$. The chance that (X, Y) is successful equals I, the area under the curve $y = g(x)$. We now repeat this experiment a large number N of times, and calculate the proportion of times that the experiment is successful. Following the law of averages (2.2.1) we may use this value as an estimate of I.

Clearly it is desirable to know the accuracy of this estimate. This is a harder problem to which we shall return later. ●

Simulation is a dangerous game, and great caution is required in interpreting the results. There are two major reasons for this. First, a computer simulation is limited by the degree to which its so-called 'pseudo-random number generator' may be trusted. It has been said for example that the summon-according-to-birthday principle of conscription to the United States armed forces may have been marred by a pseudo-random number generator with a bias for some numbers over others. Secondly, in estimating a given quantity, one may in some circumstances have little or no idea how many repetitions are necessary in order to achieve an estimate within a specified accuracy.

We have made no remark about the methods by which computers calculate 'pseudo-random numbers'. Needless to say they do not flip coins, but rely instead on operations of sufficient numerical complexity that the outcome, although deterministic, is apparently unpredictable except by an exact repetition of the calculation.

These techniques were named in honour of Monte Carlo by Metropolis, von Neumann, and Ulam, while they were involved in the process of building bombs at Los Alamos in the 1940s.

2.7 Problems

1. Each toss of a coin results in a head with probability p. The coin is tossed until the first head appears. Let X be the total number of tosses. What is $\mathbf{P}(X > m)$? Find the distribution function of X.

2. (a) Show that any discrete random variable may be written as a linear combination of indicator variables.
   (b) Show that any random variable may be expressed as the limit of an increasing sequence of discrete random variables.
   (c) Show that the limit of any increasing convergent sequence of random variables is a random variable.

3. (a) Show that if X and Y are random variables on a probability space $(\Omega, \mathscr{F}, \mathbf{P})$, then so are $X + Y$, XY, and $\min\{X, Y\}$.
   (b) Show that the set of all random variables on a given probability space $(\Omega, \mathscr{F}, \mathbf{P})$ constitutes a vector space over the reals. If Ω is finite, write down a basis for this space.

4. Let X have distribution function

$$F(x) = \begin{cases} 0 & \text{if} \quad x < 0 \\ \frac{1}{2}x & \text{if} \quad 0 \leqslant x \leqslant 2 \\ 1 & \text{if} \quad x > 2 \end{cases}$$

and let $Y = X^2$. Find

(a) $\mathbf{P}(\frac{1}{2} \leqslant X \leqslant \frac{3}{2})$, (b) $\mathbf{P}(1 \leqslant X < 2)$,
(c) $\mathbf{P}(Y \leqslant X)$, (d) $\mathbf{P}(X \leqslant 2Y)$,
(e) $\mathbf{P}(X + Y \leqslant \frac{3}{4})$, (f) the distribution function of $Z = \sqrt{X}$.

5. Let X have distribution function

$$F(x) = \begin{cases} 0 & \text{if } x < -1 \\ 1 - p & \text{if } -1 \leqslant x < 0 \\ 1 - p + \tfrac{1}{2}xp & \text{if } 0 \leqslant x \leqslant 2 \\ 1 & \text{if } x > 2. \end{cases}$$

Sketch this function, and find

(a) $\mathbf{P}(X = -1)$, (b) $\mathbf{P}(X = 0)$, (c) $\mathbf{P}(X \geqslant 1)$.

6. Buses arrive at ten minute intervals starting at noon. A man arrives at the bus stop a random number X minutes after noon, where X has distribution function

$$\mathbf{P}(X \leqslant x) = \begin{cases} 0 & \text{if } x < 0 \\ x/60 & \text{if } 0 \leqslant x \leqslant 60 \\ 1 & \text{if } x > 60. \end{cases}$$

What is the probability that he waits less than five minutes for a bus?

7. Airlines find that each passenger who reserves a seat fails to turn up with probability $\frac{1}{10}$ independently of the other passengers. So Teeny Weeny Airlines always sell 10 tickets for their 9 seat aeroplane while Blockbuster Airways always sell 20 tickets for their 18 seat aeroplane. Which is more often overbooked?

8. A fairground performer claims the power of telekinesis. The crowd throws coins and he wills them to fall heads up. He succeeds five times out of six. What chance would he have of doing at least as well if he had no supernatural powers?

9. Express the distribution functions of

$$X^+ = \max\{0, X\}, \quad X^- = -\min\{0, X\}, \quad |X| = X^+ + X^-, \quad -X$$

in terms of the distribution function F of the random variable X.

10. Show that $F_X(x)$ is continuous at $x = x_0$ if and only if $\mathbf{P}(X = x_0) = 0$.

11. The real number m is called a *median* of the distribution function F whenever $\lim_{y \uparrow m} F(y) \leqslant \frac{1}{2} \leqslant F(m)$. Show that every distribution function F has at least one median, and that the set of medians of F is a closed interval of $\mathbb{R}$.

12. Show that it is not possible to weight two dice in such a way that the sum of the two numbers shown by these loaded dice is equally likely to take any value between 2 and 12 (inclusive).

13. A function $d: S \times S \to \mathbb{R}$ is called a *metric* on S if

(i) $d(s, t) = d(t, s) \geqslant 0$ for all $s, t \in S$,
(ii) $d(s, t) = 0$ if and only if $s = t$, and
(iii) $d(s, t) \leqslant d(s, u) + d(u, t)$ for all $s, t, u \in S$.

Let F and G be distribution functions and define the 'Lévy' metric

$$d(F, G) = \inf\{\varepsilon > 0: G(x - \varepsilon) - \varepsilon \leqslant F(x) \leqslant G(x + \varepsilon) + \varepsilon \quad \text{for all } x\}.$$

Show that d is indeed a metric on the space of distribution functions.

14. Ascertain in the following cases whether or not F is the joint distribution function of some pair (X, Y) of random variables. If your conclusion is affirmative, find the distribution functions of X and Y separately.

(a) $$F(x, y) = \begin{cases} 1 - e^{-x-y} & \text{if } x, y \geqslant 0 \\ 0 & \text{otherwise.} \end{cases}$$

(b) $$F(x, y) = \begin{cases} 1 - e^{-x} - x e^{-y} & \text{if } 0 \leqslant x \leqslant y \\ 1 - e^{-y} - y e^{-y} & \text{if } 0 \leqslant y \leqslant x \\ 0 & \text{otherwise.} \end{cases}$$

15. It is required to place in order n books $B_1, B_2, \ldots, B_n$ on a library shelf in such a way that readers searching from left to right waste as little time as possible on average. Assuming that each reader requires book B_i with probability p_i, find the ordering of the books which minimizes $\mathbf{P}(T \geqslant n)$ for all n, where T is the (random) number of titles examined by a reader before discovery of the required book.

16. **Transitive coins.** Three coins each show heads with probability $\frac{3}{5}$ and tails otherwise. The first counts 10 points for a head and 2 for a tail, the second counts 4 points for both head and tail, and the third counts 3 points for a head and 20 for a tail.

You and your opponent each choose a coin; you cannot choose the same coin. Each of you tosses your coin and the person with the larger score wins £10^{10}. Would you prefer to be the first to pick a coin or the second?

17. Before the development of radar and inertial navigation, flying to isolated islands (for example, from Los Angeles to Hawaii) was somewhat 'hit or miss'. In heavy cloud or at night it was necessary to fly by dead reckoning, and then to search the surface. With the aid of a radio, the pilot had a good idea of the correct great circle along which to search, but could not be sure which of the two directions along this great circle was correct (since a strong tailwind could have carried the plane over its target). When you are the pilot, you calculate that you can make n searches before your plane will run out of fuel. On each search you will discover the island with probability p (if it is indeed in the direction of the search) independently of the results of other searches; you estimate initially that there is probability α that the island is ahead of you. What policy should you adopt in deciding the directions of your various searches in order to maximize the probability of locating the island?

18. Eight rooks are placed randomly on a chessboard, no more than one to a square. What is the probability that

(a) they are in a straight line (do not forget the diagonals)?
(b) no two are in the same row or column?

3 Discrete random variables

3.1 Probability mass functions

Recall that a random variable X is *discrete* if it takes values in some countable set $\{x_1, x_2, \ldots\}$. Its distribution function $F(x) = \mathbf{P}(X \leqslant x)$ is a jump function; just as important as its distribution function is its mass function.

(1) **Definition.** The **(probability) mass function** of a discrete random variable X is the function $f\colon \mathbb{R} \to [0, 1]$ given by $f(x) = \mathbf{P}(X = x)$.

The distribution and mass functions are related by

$$F(x) = \sum_{i:\, x_i \leqslant x} f(x_i), \qquad f(x) = F(x) - \lim_{y \uparrow x} F(y).$$

(2) **Lemma.** *The probability mass function* $f\colon \mathbb{R} \to [0, 1]$ *satisfies*

(a) $f(x) \neq 0$ *if and only if* x *belongs to some countable set* $\{x_1, x_2, \ldots\}$

(b) $\sum_i f(x_i) = 1$.

Proof. The proof is obvious. ∎

This lemma characterizes probability mass functions.

(3) **Example. Binomial distribution.** A coin is tossed n times, and a head turns up each time with probability $p\ (= 1 - q)$. Then $\Omega = \{H, T\}^n$. Let X be the total number of heads. X takes values in the set $\{0, 1, 2, \ldots, n\}$ and is discrete. Its probability mass function $f(x) = \mathbf{P}(X = x)$ satisfies

$$f(x) = 0 \quad \text{if} \quad x \notin \{0, 1, 2, \ldots, n\}.$$

Let $0 \leqslant k \leqslant n$, and consider $f(k)$. Exactly $\binom{n}{k}$ points in Ω give a total of k heads; each of these points occurs with probability $p^k q^{n-k}$, and so

$$f(k) = \binom{n}{k} p^k q^{n-k} \quad \text{if} \quad 0 \leqslant k \leqslant n.$$

X is said to have the *binomial distribution* with parameters n and p, written

$B(n, p)$. It is the sum $X = Y_1 + Y_2 + \cdots + Y_n$ of n Bernoulli variables (see (2.1.8)). ●

(4) **Example. Poisson distribution.** If a random variable X takes values in the set $\{0, 1, 2, \ldots\}$ with mass function

$$f(k) = \frac{\lambda^k}{k!} e^{-\lambda}, \qquad k = 0, 1, 2, \ldots,$$

where $\lambda > 0$, then X is said to have the *Poisson distribution* with parameter λ. ●

Exercises

(5) For what values of the constant C do the following define mass functions on the positive integers $1, 2, \ldots$?

(a) Geometric: $f(x) = C2^{-x}$.
(b) Logarithmic: $f(x) = C2^{-x}/x$.
(c) Inverse square: $f(x) = Cx^{-2}$.
(d) 'Modifed' Poisson: $f(x) = C2^x/x!$.

(6) For a random variable X having (in turn) each of the four mass functions of Exercise (5), find

(i) $\mathbf{P}(X > 1)$,
(ii) the most probable value of X,
(iii) the probability that X is even.

(7) We toss n coins, and each one shows heads with probability p, independently of each of the others. Each coin which shows heads is tossed again. What is the mass function of the number of heads resulting from the second round of tosses?

(8) Let S_k be the set of positive integers whose base-10 expansion contains exactly k elements (so that, for example, $1024 \in S_4$). A fair coin is tossed until the first head appears, and we write T for the number of tosses required. We pick a random element, N say, from S_T, each such element having equal probability. What is the mass function of N?

3.2 Independence

Remember that events A and B are called 'independent' if the occurrence of A does not change the subsequent probability of B occurring. More rigorously, A and B are independent if and only if $\mathbf{P}(A \cap B) = \mathbf{P}(A)\mathbf{P}(B)$. Similarly, we say that discrete variables X and Y are 'independent' if the numerical value of X does not affect the distribution of Y. With this in mind we make the following definition.

(1) **Definition.** Discrete variables X and Y are **independent** if the events $\{X = x\}$ and $\{Y = y\}$ are independent for all x and y.

Suppose X takes values in $\{x_1, x_2, \ldots\}$ and Y takes values in $\{y_1, y_2, \ldots\}$. Let

$$A_i = \{X = x_i\}, \qquad B_j = \{Y = y_j\}.$$

Notice (see (2.7.2)) that X and Y are linear combinations of the indicator variables I_{A_i}, I_{B_j}, in that

$$X = \sum_i x_i I_{A_i} \quad \text{and} \quad Y = \sum_j y_j I_{B_j}.$$

X and Y are independent if and only if A_i and B_j are independent for all pairs i, j. A similar definition holds for collections $\{X_1, X_2, \ldots, X_n\}$ of discrete variables.

(2) **Example.** A coin is tossed once and heads turns up with probability $p = 1 - q$. Let X and Y be the numbers of heads and tails respectively. It is no surprise that X and Y are not independent. After all,

$$\mathbf{P}(X = Y = 1) = 0, \qquad \mathbf{P}(X = 1)\mathbf{P}(Y = 1) = p(1 - p).$$

Suppose now that the coin is tossed a random number N of times, where N has the Poisson distribution with parameter λ. It is a remarkable fact that the resulting numbers X and Y of heads and tails *are* independent, for

$$\begin{aligned}\mathbf{P}(X = x, Y = y) &= \mathbf{P}(X = x, Y = y \mid N = x + y)\mathbf{P}(N = x + y)\\ &= \binom{x+y}{x} p^x q^y \frac{\lambda^{x+y}}{(x+y)!} e^{-\lambda}\\ &= \frac{(\lambda p)^x (\lambda q)^y}{x!\, y!} e^{-\lambda}.\end{aligned}$$

But, by (1.4.4),

$$\begin{aligned}\mathbf{P}(X = x) &= \sum_{n \geqslant x} \mathbf{P}(X = x \mid N = n)\mathbf{P}(N = n)\\ &= \sum_{n \geqslant x} \binom{n}{x} p^x q^{n-x} \frac{\lambda^n}{n!} e^{-\lambda} = \frac{(\lambda p)^x}{x!} e^{-\lambda p};\end{aligned}$$

a similar result holds for Y, and so

$$\mathbf{P}(X = x, Y = y) = \mathbf{P}(X = x)\mathbf{P}(Y = y). \qquad \bullet$$

If X is a random variable and $g: \mathbb{R} \to \mathbb{R}$, then $Z = g(X)$, defined by $Z(\omega) = g(X(\omega))$, is a random variable also.

(3) **Theorem.** *If X and Y are independent and $g, h: \mathbb{R} \to \mathbb{R}$, then $g(X)$ and $h(Y)$ are independent also.*

Proof. *Exercise.* See Problem (3.11.1). ■

More generally, we say that a family $\{X_i: i \in I\}$ of (discrete) random variables is **independent** if the events $\{X_i = x_i\}$, $i \in I$, are independent for all possible choices of the set $\{x_i: i \in I\}$ of values of the X_i. That is to say, $\{X_i: i \in I\}$ is an independent family if and only if

$$\mathbf{P}(X_i = x_i \text{ for all } i \in J) = \prod_{i \in J} \mathbf{P}(X_i = x_i)$$

for all sets $\{x_i: i \in I\}$ and for all finite subsets J of I. The conditional independence of a family of random variables, given an event C, is defined similarly to the conditional independence of events; see (1.6.5).

Independent families of random variables are very much easier to study than dependent families, as we shall see soon. Note that pairwise-independent families are not necessarily independent.

Exercises

(4) Let X and Y be independent random variables, each taking the values -1 or 1 with probability $\frac{1}{2}$, and let $Z = XY$. Show that X, Y, and Z are pairwise independent. Are they independent?

(5) Let X and Y be independent random variables taking values in the positive integers and having the same mass function $f(x) = 2^{-x}$ for $x = 1, 2, \ldots$. Find

(a) $\mathbf{P}(\min\{X, Y\} \leqslant x)$, (b) $\mathbf{P}(Y > X)$,
(c) $\mathbf{P}(X = Y)$, (d) $\mathbf{P}(X \geqslant kY)$, for a given positive integer k,
(e) $\mathbf{P}(X \text{ divides } Y)$, (f) $\mathbf{P}(X = rY)$, for a given positive rational r.

(6) Let X_1, X_2, X_3 be independent random variables taking values in the positive integers and having mass functions given by $\mathbf{P}(X_i = x) = (1 - p_i)p_i^{x-1}$ for $x = 1, 2, \ldots$, and $i = 1, 2, 3$.

(a) Show that

$$\mathbf{P}(X_1 < X_2 < X_3) = \frac{(1 - p_1)(1 - p_2)p_2 p_3^2}{(1 - p_2 p_3)(1 - p_1 p_2 p_3)}.$$

(b) Find $\mathbf{P}(X_1 \leqslant X_2 \leqslant X_3)$.

(7) Three players, A, B, and C, take turns to roll a die; they do this in the order ABCABCA....

(a) Show that the probability that, of the three players, A is the first to throw a 6, B the second, and C the third, is 216/1001.
(b) Show that the probability that the first 6 to appear is thrown by A, the second 6 to appear is thrown by B, and the third 6 to appear is thrown by C, is 46 656/753 571.

3.3 Expectation

Let $x_1, x_2, \ldots, x_N$ be the numerical outcomes of N repetitions of some experiment. The average of these outcomes is

$$m = \frac{1}{N} \sum_i x_i.$$

In advance of performing these experiments we can represent their outcomes by a sequence $X_1, X_2, \ldots, X_N$ of random variables, and we shall suppose that these variables are discrete with a common mass function f. Then, roughly speaking (see the beginning of Section 1.3), for each possible value x, about $Nf(x)$ of the X_i will take that value x. So the average m is about

$$m \simeq \frac{1}{N} \sum_x xNf(x) = \sum_x xf(x)$$

where the summation here is over all possible values of the X_i. This average is called the 'expectation' or 'mean value' of the underlying distribution with mass function f.

(1) **Definition.** The **mean value**, or **expectation**, or **expected value** of X with mass function f is defined to be

$$\mathbf{E}(X) = \sum_{x:\, f(x) > 0} xf(x)$$

whenever this sum is absolutely convergent.

We require *absolute* convergence in order that $\mathbf{E}(X)$ be unchanged by reordering the x_i. We can, for notational convenience, write $\mathbf{E}(X) = \sum_x xf(x)$. This appears to be an uncountable sum; however, all but countably many of its contributions are zero. If the numbers $f(x)$ are regarded as masses $f(x)$ at points x then $\mathbf{E}(X)$ is just the position of the centre of gravity; we can speak of X as having an 'atom' or 'point mass' of size $f(x)$ at x. We sometimes omit the brackets and simply write $\mathbf{E}X$.

(2) **Example (2.1.5) revisited.** The random variables X and W of this example have mean values

$$\mathbf{E}(X) = \sum_x x\mathbf{P}(X = x) = 0 \cdot \tfrac{1}{4} + 1 \cdot \tfrac{1}{2} + 2 \cdot \tfrac{1}{4} = 1$$

$$\mathbf{E}(W) = \sum_x x\mathbf{P}(W = x) = 0 \cdot \tfrac{3}{4} + 4 \cdot \tfrac{1}{4} = 1.$$ ●

If X is a random variable and $g: \mathbb{R} \to \mathbb{R}$, then $Y = g(X)$, given formally by $Y(\omega) = g(X(\omega))$, is a random variable also. To calculate its expectation we need first to find its probability mass function f_Y. This process can be complicated, and it is avoided by the following lemma (called by some the 'law of the unconscious statistician'!).

(3) **Lemma.** *If X has mass function f and $g: \mathbb{R} \to \mathbb{R}$, then*

$$\mathbf{E}(g(X)) = \sum_x g(x)f(x)$$

whenever this sum is absolutely convergent.

Proof. This is Problem (3.11.3). ■

(4) **Example.** Suppose that X takes values $-2, -1, 1, 3$ with probabilities $\frac{1}{4}, \frac{1}{8}, \frac{1}{4}, \frac{3}{8}$ respectively. Consider the random variable $Y = X^2$; Y takes values 1, 4, 9 with probabilities $\frac{3}{8}, \frac{1}{4}, \frac{3}{8}$ respectively, and so

$$\mathbf{E}(Y) = \sum_x x\mathbf{P}(Y = x) = 1\cdot\tfrac{3}{8} + 4\cdot\tfrac{1}{4} + 9\cdot\tfrac{3}{8} = 19/4.$$

Alternatively, use the law of the unconscious statistician to find that

$$\mathbf{E}(Y) = \mathbf{E}(X^2) = \sum_x x^2\mathbf{P}(X = x) = 4\cdot\tfrac{1}{4} + 1\cdot\tfrac{1}{8} + 1\cdot\tfrac{1}{4} + 9\cdot\tfrac{3}{8} = 19/4. \quad \bullet$$

Lemma (3) provides a method for calculating the 'moments' of a distribution; these are defined as follows.

(5) **Definition.** If k is a positive integer, then the kth **moment** m_k of X is

$$m_k = \mathbf{E}(X^k).$$

The kth **central moment** σ_k is

$$\sigma_k = \mathbf{E}((X - m_1)^k).$$

The two moments of most use are $m_1 = \mathbf{E}(X)$ and $\sigma_2 = \mathbf{E}((X - \mathbf{E}(X))^2)$, called the *mean* (or *expectation*) and *variance* of X. These two quantities are measures of the mean and dispersion of X: m_1 is the average value of X, and σ_2 measures the amount by which X tends to deviate from this average. The mean m_1 is often written as μ, and σ_2 is often written as σ^2 or $\text{var}(X)$; its positive square root σ is called the *standard deviation*. The central moments $\{\sigma_i\}$ can be expressed in terms of the ordinary moments $\{m_i\}$. For example, $\sigma_1 = 0$ and

$$\begin{aligned}\sigma_2 &= \sum_x (x - m_1)^2 f(x)\\ &= \sum_x x^2 f(x) - 2m_1 \sum_x xf(x) + m_1^2 \sum_x f(x)\\ &= m_2 - m_1^2\end{aligned}$$

or

$$\begin{aligned}\operatorname{var}(X) &= \mathbf{E}((X - \mathbf{E}(X))^2)\\ &= \mathbf{E}(X^2) - (\mathbf{E}(X))^2.\end{aligned}$$

Experience with student calculations of variances causes us to stress the following elementary fact: *variances cannot be negative.*

(6) **Example. Bernoulli variables.** Let X be a Bernoulli variable, taking the value 1 with probability p $(=1-q)$. Then

$$\mathbf{E}(X) = \sum_x xf(x) = 0\cdot q + 1\cdot p = p$$

$$\mathbf{E}(X^2) = \sum_x x^2 f(x) = 0\cdot q + 1\cdot p = p$$

$$\operatorname{var}(X) = \mathbf{E}(X)^2 - \mathbf{E}(X)^2 = pq.$$

Thus the indicator variable I_A has expectation $\mathbf{P}(A)$ and variance $\mathbf{P}(A)\mathbf{P}(A^c)$. $\mathbf{E}(X)^2$ means $(\mathbf{E}(X))^2$ and must not be confused with $\mathbf{E}(X^2)$. ●

(7) **Example. Binomial variables.** Let X be $B(n, p)$. Then

$$\mathbf{E}(X) = \sum_{k=0}^{n} kf(k) = \sum_{k=0}^{n} k\binom{n}{k}p^k q^{n-k}.$$

To calculate this, differentiate the identity

$$\sum_{k=0}^{n}\binom{n}{k}x^k = (1+x)^n,$$

multiply by x to obtain

$$\sum_{k=0}^{n} k\binom{n}{k}x^k = nx(1+x)^{n-1},$$

and substitute $x = p/q$ to obtain $\mathbf{E}(X) = np$. A similar argument shows that $\operatorname{var}(X) = npq$. ●

We can think of the process of calculating expectations as a linear operator on the space of random variables.

(8) **Theorem.** *The expectation operator* $\mathbf{E}$ *has the following properties*:

(a) *if* $X \geqslant 0$ *then* $\mathbf{E}(X) \geqslant 0$
(b) *if* $a, b \in \mathbb{R}$ *then* $\mathbf{E}(aX + bY) = a\mathbf{E}(X) + b\mathbf{E}(Y)$
(c) *the random variable* 1, *taking the value* 1 *always, has expectation* $\mathbf{E}(1) = 1$.

Proof. (a) and (c) are obvious.

(b) Let $A_x = \{X = x\}$, $B_y = \{Y = y\}$. Then

$$aX + bY = \sum_{x,y} (ax + by) I_{A_x \cap B_y}$$

and the solution of the first part of Problem (3.11.3) shows that

$$\mathbf{E}(aX + bY) = \sum_{x,y} (ax + by)\mathbf{P}(A_x \cap B_y).$$

However,

$$\sum_y \mathbf{P}(A_x \cap B_y) = \mathbf{P}\left(A_x \cap \left(\bigcup_y B_y\right)\right)$$
$$= \mathbf{P}(A_x \cap \Omega) = \mathbf{P}(A_x)$$

and similarly $\sum_x \mathbf{P}(A_x \cap B_y) = \mathbf{P}(B_y)$, which gives

$$\mathbf{E}(aX + bY) = \sum_x ax \sum_y \mathbf{P}(A_x \cap B_y) + \sum_y by \sum_x \mathbf{P}(A_x \cap B_y)$$
$$= a \sum_x x\mathbf{P}(A_x) + b \sum_y y\mathbf{P}(B_y)$$
$$= a\mathbf{E}(X) + b\mathbf{E}(Y).$$ ∎

It is not in general true that $\mathbf{E}(XY)$ is the same as $\mathbf{E}(X)\mathbf{E}(Y)$.

(9) **Lemma.** *If X and Y are independent then* $\mathbf{E}(XY) = \mathbf{E}(X)\mathbf{E}(Y)$.

Proof. Let A_x and B_y be as in the proof of (8). Then

$$XY = \sum_{x,y} xy I_{A_x \cap B_y}$$

and so

$$\mathbf{E}(XY) = \sum_{x,y} xy\mathbf{P}(A_x)\mathbf{P}(B_y) \quad \text{by independence}$$
$$= \sum_x x\mathbf{P}(A_x) \sum_y y\mathbf{P}(B_y) = \mathbf{E}(X)\mathbf{E}(Y).$$ ∎

(10) **Definition.** X and Y are called **uncorrelated** if $\mathbf{E}(XY) = \mathbf{E}(X)\mathbf{E}(Y)$.

Lemma (9) asserts that independent variables are uncorrelated. The converse is not true, as Problem (3.11.16) indicates.

(11) **Theorem.** *If X and Y are independent then*

(a) *if $a \in \mathbb{R}$ then* $\text{var}(aX) = a^2 \text{var}(X)$
(b) $\text{var}(X + Y) = \text{var}(X) + \text{var}(Y)$.

Proof.

(a) Using the linearity of $\mathbf{E}$,

$$\begin{aligned}\operatorname{var}(aX) &= \mathbf{E}((aX - \mathbf{E}(aX))^2)\\ &= \mathbf{E}(a^2(X - \mathbf{E}(X))^2)\\ &= a^2\mathbf{E}((X - \mathbf{E}(X))^2)\\ &= a^2 \operatorname{var}(X).\end{aligned}$$

(b)
$$\begin{aligned}\operatorname{var}(X + Y) &= \mathbf{E}((X + Y - \mathbf{E}(X + Y))^2)\\ &= \mathbf{E}((X - \mathbf{E}(X))^2 + 2(XY - \mathbf{E}(X)\mathbf{E}(Y)) + (Y - \mathbf{E}(Y))^2)\\ &= \operatorname{var}(X) + 2[\mathbf{E}(XY) - \mathbf{E}(X)\mathbf{E}(Y)] + \operatorname{var}(Y)\\ &= \operatorname{var}(X) + \operatorname{var}(Y) \quad \text{by independence.}\end{aligned}$$
■

Equation (11a) shows that the variance operator 'var' is *not* a linear operator, even when it is applied only to independent variables.

Sometimes the sum $S = \sum xf(x)$ does not converge absolutely, and the mean of the distribution does not exist. If $S = -\infty$ or $S = +\infty$, then we can sometimes speak of the mean as taking these values also. Of course, there exist distributions which do not have a mean value.

(12) **Example. A distribution without a mean.** Let X have mass function

$$f(k) = Ak^{-2} \quad \text{for} \quad k = \pm 1, \pm 2, \ldots$$

where A is chosen so that $\sum f(k) = 1$. The sum

$$\sum kf(k) = A \sum_{k \neq 0} \frac{1}{k}$$

does not converge absolutely, because both the positive and the negative parts diverge. ●

This is a suitable opportunity to point out that we can base probability theory upon the expectation operator $\mathbf{E}$ rather than upon the probability measure $\mathbf{P}$. After all, our intuitions about the notion of 'average' are probably just as well developed as those about quantitative chance. Roughly speaking, the way we proceed is to postulate axioms, such as (a), (b), and (c) of Theorem (8), for a so-called 'expectation operator' $\mathbf{E}$ acting on a space of 'random variables'. The probability of an event can then be recaptured by defining

$$\mathbf{P}(A) = \mathbf{E}(I_A).$$

Whittle (1970) is an able advocate of this approach.

This method can be easily and naturally adapted to deal with probabilistic questions in quantum theory. In this major branch of theoretical physics, questions arise which cannot be formulated entirely within the usual framework of probability theory. However, there still exists an expectation operator $\mathbf{E}$, which is applied to linear operators known as observables (such as square matrices) rather than to random variables. There does not exist a sample space Ω, and nor therefore are there any indicator functions, but nevertheless there exist analogues of other concepts in probability theory. For example, the *variance* of an operator X is defined by

$$\text{var}(X) = \mathbf{E}(X^2) - \mathbf{E}(X)^2.$$

Furthermore, it can be shown that

$$\mathbf{E}(X) = \text{tr}(UX)$$

where tr denotes *trace* and U is a non-negative definite operator with unit trace.

(13) **Example. Wagers.** Historically, there has been confusion amongst probabilists between the price that an individual may be willing to pay in order to play a game, and his expected return from this game. For example, I conceal £2 in one hand and nothing in the other, and then invite a friend to pay a fee which entitles her to choose a hand at random and keep the contents. Other things being equal (my friend is neither a compulsive gambler, nor particularly busy), it would seem that £1 would be a 'fair' fee to ask, since £1 is the expected return to the player. That is to say, faced with a modest (but random) gain, then a fair 'entrance fee' would seem to be the expected value of the gain. However, suppose that I conceal £2^{10} in one hand and nothing in the other; what now is a 'fair' fee? Few persons of modest means can be expected to offer £2^9 for the privilege of playing. There is confusion here between fairness and reasonableness: we do not generally treat large payoffs or penalties in the same way as small ones, even though the relative odds may be unquestionable. The customary resolution of this paradox is to introduce the notion of 'utility'. Writing $u(x)$ for the 'utility' to an individual of £x, it would be fairer to charge a fee of $\frac{1}{2}(u(0) + u(2^{10}))$ for the above prospect. Of course, different individuals have different utility functions, although such functions have presumably various features in common: $u(0) = 0$, u is non-decreasing, $u(x)$ is near to x for small positive x, and u is concave, so that in particular $u(x) \leqslant xu(1)$. ●

Exercises

(14) Is it generally true that $\mathbf{E}(1/X) = 1/\mathbf{E}(X)$? Is it ever true that $\mathbf{E}(1/X) = 1/\mathbf{E}(X)$?

(15) **Coupons.** Every package of some intrinsically dull commodity includes a small and exciting plastic object. There are c different types of object, and each package is equally likely to contain any given type. You buy one package each day.

(a) Find the mean number of days which elapse between the acquisitions of the jth new type of object and the $(j + 1)$th new type.
(b) Find the mean number of days which elapse before you have a full set of objects.

(16) Each member of a group of n players rolls a die.

(a) For any pair of players who throw the same number, the group scores 1 point. Find the mean and variance of the total score of the group.
(b) Find the mean and variance of the total score if any pair of players who throw the same number scores that number.

(17) **St Petersburg paradox.** A fair coin is tossed repeatedly. Let T be the number of tosses until the first head. You are offered the following prospect, which you may accept on payment of a fee. If $T = k$, say, then you will receive £2^k. What would be a 'fair' fee to ask of you? This problem was mentioned by Nicholas Bernoulli in 1713, and Daniel Bernoulli wrote about the question for the St Petersburg Academy.

(18) Let X have mass function

$$f(x) = \begin{cases} \{x(x+1)\}^{-1} & \text{if } \quad x = 1, 2, \ldots \\ 0 & \text{otherwise,} \end{cases}$$

and let $\alpha \in \mathbb{R}$. For what values of α is it the case that $\mathbf{E}(X^{\alpha}) < \infty$? (If α is not integral, than $\mathbf{E}(X^{\alpha})$ is called the *fractional moment of order* α of X. A point concerning notation: for real α and complex $x = r\,\mathrm{e}^{i\theta}$, x^{α} should be interpreted as $r^{\alpha}\,\mathrm{e}^{i\theta\alpha}$, so that $|x^{\alpha}| = r^{\alpha}$. In particular $\mathbf{E}(|X^{\alpha}|) = \mathbf{E}(|X|^{\alpha})$.)

(19) Show that $\mathrm{var}(a + X) = \mathrm{var}(X)$ for any random variable X and constant a.

3.4 Indicators and matching

This section contains light entertainment, in the guise of some illustrations of the uses of indicator functions. These were defined in (2.1.9) and have appeared occasionally since. Recall that

$$I_A(\omega) = \begin{cases} 1 & \text{if} \quad \omega \in A \\ 0 & \text{if} \quad \omega \in A^c \end{cases}$$

and

$$\mathbf{E}I_A = \mathbf{P}(A).$$

(1) **Example. Proofs of (1.3.4c, d).** Note that

$$I_A + I_{A^c} = I_{A \cup A^c} = I_{\Omega} = 1$$

and that $I_{A \cap B} = I_A I_B$. Thus

$$\begin{aligned} I_{A \cup B} &= 1 - I_{(A \cup B)^c} = 1 - I_{A^c \cap B^c} \\ &= 1 - I_{A^c} I_{B^c} = 1 - (1 - I_A)(1 - I_B) \\ &= I_A + I_B - I_A I_B. \end{aligned}$$

Take expectations to obtain

$$\mathbf{P}(A \cup B) = \mathbf{P}(A) + \mathbf{P}(B) - \mathbf{P}(A \cap B).$$

More generally, if $B = \bigcup_{i=1}^{n} A_i$ then

$$I_B = 1 - \prod_{i=1}^{n} (1 - I_{A_i});$$

multiply this out and take expectations to obtain

(2) $$\mathbf{P}(B) = \sum_i \mathbf{P}(A_i) - \sum_{i<j} \mathbf{P}(A_i \cap A_j) + \cdots + (-1)^{n+1}\mathbf{P}(A_1 \cap \cdots \cap A_n). \quad \bullet$$

(3) **Example. Matching problem.** A number of melodramatic applications of (2) are available, of which the following is typical. A secretary types n different letters together with matching envelopes, drops the pile down the stairs, and then places the letters randomly into the available envelopes. Each arrangement is equally likely, and we ask for the probability that exactly r letters are in their correct envelopes. Rather than using (2), we shall proceed directly by way of indicator functions.

Solution. Let $L_1, L_2, \ldots, L_n$ denote the letters. Call a letter *good* if it is correctly addressed, and *bad* otherwise; write X for the number of good letters. Let A_i be the event that L_i is good, and I_i be the indicator function of A_i. Let $j_1, \ldots, j_r, k_{r+1}, \ldots, k_n$ be a permutation of the numbers $1, 2, \ldots, n$ and define

(4) $$S = \sum_{\pi} I_{j_1} \cdots I_{j_r}(1 - I_{k_{r+1}}) \cdots (1 - I_{k_n})$$

where the sum is taken over all such permutations π. Then

$$S = \begin{cases} 0 & \text{if } X \neq r \\ r!(n-r)! & \text{if } X = r. \end{cases}$$

To see this, let $L_{i_1}, \ldots, L_{i_m}$ be the good letters. If $m \neq r$ then each summand in (4) equals 0. If $m = r$ then the summand in (4) equals 1 if and only if $j_1, \ldots, j_r$ is a permutation of $i_1, \ldots, i_r$ and $k_{r+1}, \ldots, k_n$ is a permutation of the remaining numbers; there are $r!(n-r)!$ such permutations. It follows that I, given by

(5) $$I = \frac{1}{r!(n-r)!} S,$$

is the indicator function of the event $\{X = r\}$ that exactly r letters are good. We take expectations of (4) and multiply out to obtain

$$\mathbf{E}(S) = \sum_{\pi} \sum_{s=0}^{n-r} (-1)^s \binom{n-r}{s} \mathbf{E}(I_{j_1} \cdots I_{j_r} I_{k_{r+1}} \cdots I_{k_{r+s}})$$

by a symmetry argument. However,

(6) $$\mathbf{E}(I_{j_1}\cdots I_{j_r}I_{k_{r+1}}\cdots I_{k_{r+s}}) = \frac{(n-r-s)!}{n!}$$

since there are $n!$ possible permutations only $(n-r-s)!$ of which allocate $L_{i_1}, \ldots, L_{j_r}, L_{k_{r+1}}, \ldots, L_{k_{r+s}}$ to their correct envelopes. We combine (4), (5), and (6) to obtain

$$\begin{aligned}\mathbf{P}(X = r) = \mathbf{E}(I) &= \frac{1}{r!(n-r)!}\mathbf{E}(S)\\ &= \frac{1}{r!(n-r)!}\sum_{s=0}^{n-r}(-1)^s\binom{n-r}{s}n!\,\frac{(n-r-s)!}{n!}\\ &= \frac{1}{r!}\sum_{s=0}^{n-r}(-1)^s\frac{1}{s!}\\ &= \frac{1}{r!}\left(\frac{1}{2!}-\frac{1}{3!}+\cdots+\frac{(-1)^{n-r}}{(n-r)!}\right) \quad \text{for } r \leqslant n-2 \text{ and } n \geqslant 2.\end{aligned}$$

In particular, as the number n of letters tends to infinity, we obtain the possibly surprising result that the probability that no letter is put into its correct envelope approaches e^{-1}. It is left as an *exercise* to prove this without using indicators. ●

(7) **Example. Reliability.** When you telephone your friend in Cambridge, your call is routed through the telephone network in a way which depends on the current state of the traffic. For example, if all lines into the Ascot switchboard are in use, then your call may go through the switchboard at Newmarket. Sometimes you may fail to get through at all, owing to a combination of faulty and occupied equipment in the system. We may think of the network as comprising nodes (representing switchboards) joined by edges (representing channels), drawn as 'graphs' in the manner of the examples of Figure 3.1. In each of these examples there is a designated 'source' s and 'sink' t, and we wish to find a path through the network from s to t which uses available channels. As a simple model for such a system in the presence of uncertainty, we suppose that each edge e is 'working' with probability p_e, independently of all other edges. We write $\boldsymbol{p}$ for the vector of edge probabilities p_e, and define the *reliability* $R(\boldsymbol{p})$ of the network to be the probability that there is a path from s to t using only edges which are working. Denoting the network by G, we write $R_G(\boldsymbol{p})$ for $R(\boldsymbol{p})$ when we wish to emphasize the role of G.

We have encountered questions of reliability already. In Example (1.7.2) we were asked for the reliability of the first network in Figure 3.1, and in Problem (1.8.19) of the second, assuming on each occasion that the value of p_e does not depend on the choice of e.

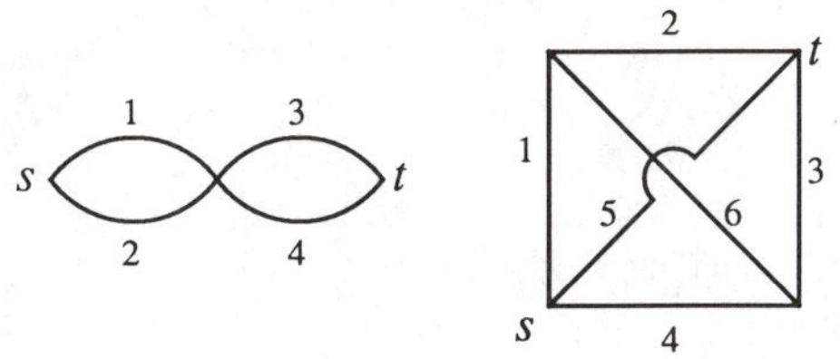

Fig. 3.1 Two networks with source s and sink t.

Let us write

$$X_e = \begin{cases} 1 & \text{if edge } e \text{ is working} \\ 0 & \text{otherwise,} \end{cases}$$

the indicator function of the event that e is working, so that X_e takes the values 0 and 1 with probabilities $1 - p_e$ and p_e respectively. Each realization X of the X_e either includes a working connection from s to t or does not. Thus, there exists a *structure function* ζ taking values 0 and 1 such that

(8) $$\zeta(X) = \begin{cases} 1 & \text{if such a working connection exists} \\ 0 & \text{otherwise;} \end{cases}$$

thus $\zeta(X)$ is the indicator function of the event that a working connection exists. It is immediately seen that $R(\mathbf{p}) = \mathbf{E}(\zeta(X))$. The function ζ may be expressed as

(9) $$\zeta(X) = 1 - \prod_{\pi} I_{\{\pi \text{ not working}\}} = 1 - \prod_{\pi}\left(1 - \prod_{e \in \pi} X_e\right)$$

where π is a typical path in G from s to t, and we say that π is working if and only if every edge in π is working.

For instance, in the case of the first example of Figure 3.1, there are four different paths from s to t. Numbering the edges as indicated, we have that the structure function is given by

(10) $$\zeta(X) = 1 - (1 - X_1X_3)(1 - X_1X_4)(1 - X_2X_3)(1 - X_2X_4).$$

As an *exercise*, expand this and take expectations to calculate the reliability of the network when $p_e = p$ for all edges e. ●

Exercises

(11) A biased coin is tossed n times, and heads shows with probability p on each toss. A *run* is a sequence of throws which result in the same outcome, so that, for example, the sequence $HHTHTTH$ contains five runs. Show that the expected number of runs is $1 + 2(n - 1)p(1 - p)$. Find the variance of the number of runs.

(12) An urn contains n balls numbered $1, 2, \ldots, n$. We remove k balls at random (without replacement) and add up their numbers. Find the mean and variance of the total.

(13) Of the $2n$ people in a given collection of n couples, exactly m die. Assuming that the m have been picked at random, find the mean number of surviving couples. This problem was formulated by Daniel Bernoulli in 1768.

(14) Urn R contains n red balls and urn B contains n blue balls. At each stage, a ball is selected at random from each urn, and they are swapped. Show that the mean number of red balls in urn R after stage k is $\frac{1}{2}n\{1 + (1 - 2/n)^k\}$. This 'diffusion model' was described by Daniel Bernoulli in 1769.

(15) Consider a square with diagonals, with distinct source and sink. Each edge represents a component which is working correctly with probability p, independently of all other components. Write down an expression for the Boolean function which equals 1 if and only if there is a working path from source to sink, in terms of the indicator functions X_i of the events {edge i is working} as i runs over the set of edges. Hence calculate the reliability of the network.

(16) A system is called a 'k out of n' system if it contains n components and it works whenever k or more of these components are working. Suppose that each component is working with probability p, independently of the other components, and let X_c be the indicator function of the event that component c is working. Find, in terms of the X_c, the indicator function of the event that the system works, and deduce the reliability of the system.

3.5 Examples of discrete variables

Exercise. Sketch the mass functions and the distribution functions of the variables discussed here.

(1) **Bernoulli trials.** X takes values 1 and 0 with probabilities p and q $(=1-p)$, respectively. Sometimes we think of these values as representing the 'success' or the 'failure' of the trial. The mass function is

$$f(0) = 1 - p, \qquad f(1) = p,$$

and it follows that $\mathbf{E}X = p$, $\text{var}(X) = p(1-p)$. ●

(2) **Binomial distribution.** We perform n independent Bernoulli trials $X_1, X_2, \ldots, X_n$ and count the total number of successes $Y = X_1 + X_2 + \cdots + X_n$. As in (3.1.3), the mass function of Y is

$$f(k) = \binom{n}{k} p^k (1-p)^{n-k}, \qquad k = 0, 1, \ldots, n.$$

Application of (3.3.8) and (3.3.11) yields immediately

$$\mathbf{E}Y = np, \qquad \text{var}(Y) = np(1-p);$$

the method of (3.3.7) provides a more lengthy derivation of this. ●

(3) **Trinomial distribution.** More generally, suppose we conduct n trials, each of which results in one of three outcomes (red, white, or blue, say), where red occurs with probability p, white with probability q, and blue with probability $1 - p - q$. Then the probability of r reds, w whites, and $n - r - w$ blues is

$$\frac{n!}{r!\,w!\,(n-r-w)!} p^r q^w (1-p-q)^{n-r-w}.$$

This is the *trinomial distribution*, with parameters n, p, and q. The 'multinomial distribution' is the obvious generalization of this distribution to the case of some number, say t, of possible outcomes. ●

(4) **Poisson distribution.** A *Poisson* variable is a random variable with the Poisson mass function

$$f(k) = \frac{\lambda^k}{k!} e^{-\lambda}, \qquad k = 0, 1, 2, \ldots$$

for some $\lambda > 0$. It can be obtained in practice in the following way. Let Y be a $B(n, p)$ variable, and suppose that n is very large and p is very small (an example might be the number Y of misprints on the front page of the *Grauniad*, where n is the total number of characters and p is the probability for each character that the typesetter has made an error). Now, let $n \to \infty$ and $p \to 0$ in such a way that $\mathbf{E}(Y) = np$ approaches a non-zero constant λ. Then for $k = 0, 1, 2, \ldots,$

$$\mathbf{P}(Y = k) = \binom{n}{k} p^k (1-p)^{n-k} \simeq \frac{1}{k!}\left(\frac{np}{1-p}\right)^k (1-p)^n \to \frac{\lambda^k}{k!} e^{-\lambda}.$$

Check that both the mean and the variance of this distribution are equal to λ. Now do Problem (2.7.7) again (*exercise*). ●

(5) **Geometric distribution.** A *geometric* variable is a random variable with the geometric mass function

$$f(k) = p(1-p)^{k-1}, \qquad k = 1, 2, \ldots$$

for some number p in $(0, 1)$. This distribution arises in the following way. Suppose that independent Bernoulli trials (parameter p) are performed at times $1, 2, \ldots$. Let W be the time which elapses before the first success; W is called a *waiting time*. Then

$$\mathbf{P}(W > k) = (1-p)^k$$

and thus

$$\begin{aligned}\mathbf{P}(W = k) &= \mathbf{P}(W > k-1) - \mathbf{P}(W > k) \\ &= p(1-p)^{k-1}.\end{aligned}$$

The reader should calculate the mean and the variance. ●

(6) **Negative binomial distribution.** More generally, in the previous example, let W_r be the waiting time for the rth success. Check that W_r has mass function

$$\mathbf{P}(W_r = k) = \binom{k-1}{r-1} p^r (1-p)^{k-r}, \qquad k = r, r+1, \ldots;$$

it is said to have the *negative binomial distribution* with parameters r and p. W_r is the sum of r independent geometric variables. For, let X_1 be the waiting time for the first success, X_2 the *further* waiting time for the second success, X_3 the *further* waiting time for the third success, and so on. Then $X_1, X_2, \ldots$ are independent and geometric, and

$$W_r = X_1 + X_2 + \cdots + X_r.$$

Apply (3.3.8) and (3.3.11) to find the mean and the variance of W_r. ●

Exercises

(7) Each trial may result in any of t given outcomes, the ith outcome having probability p_i. Let N_i be the number of occurrences of the ith outcome in n independent trials. Show that

$$\mathbf{P}(N_i = n_i \text{ for } 1 \leqslant i \leqslant t) = \frac{n!}{n_1!n_2!\cdots n_t!} p_1^{n_1} p_2^{n_2} \cdots p_t^{n_t}$$

for any collection $n_1, n_2, \ldots, n_t$ of non-negative integers with sum n. The vector N is said to have the *multinomial* distribution.

(8) In your pocket is a random number N of coins, where N has the Poisson distribution with parameter λ. You toss each coin once, with heads showing with probability p each time. Show that the total number of heads has the Poisson distribution with parameter λp.

3.6 Dependence

Probability theory is largely concerned with families of random variables; these families will not in general consist entirely of independent variables.

(1) **Example.** Suppose that we back three horses to win as an accumulator. If our stake is £1 and the starting prices are α, β, and γ, then our total profit is

$$W = (\alpha + 1)(\beta + 1)(\gamma + 1)I_1 I_2 I_3 - 1$$

where I_i denotes the indicator of a win in the ith race by our horse. (In checking this expression remember that a bet of £B on a horse with starting price α brings a return of £$B(\alpha + 1)$, should this horse win.) We lose £1 if some backed horse fails to win. It seems clear that the random variables W and I_1 are *not* independent. If the races are run independently, then

$$\mathbf{P}(W = -1) = \mathbf{P}(I_1 I_2 I_3 = 0),$$

but

$$\mathbf{P}(W = -1 \mid I_1 = 1) = \mathbf{P}(I_2 I_3 = 0)$$

which are different from each other unless the first backed horse is guaranteed victory. ●

We require a tool for studying collections of dependent variables. Knowledge of their individual mass functions is little help by itself. Just as the main tool for studying a random variable is its distribution function, so the study of, say, a pair of random variables is based on its 'joint' distribution function and mass function.

(2) **Definition.** The **joint distribution function** $F: \mathbb{R}^2 \to [0, 1]$ of X and Y, where X and Y are discrete variables, is given by

$$F(x, y) = \mathbf{P}(X \leqslant x \text{ and } Y \leqslant y).$$

Their **joint mass function** $f: \mathbb{R}^2 \to [0, 1]$ is given by

$$f(x, y) = \mathbf{P}(X = x \text{ and } Y = y).$$

Joint distribution functions and joint mass functions of larger collections of variables are defined similarly. The functions F and f can be characterized in much the same way ((2.1.6) and (3.1.2)) as the corresponding functions of a single variable. We omit the details. We write $F_{X,Y}$ and $f_{X,Y}$ when we need to stress the role of X and Y. We think of the joint mass function in the following way. If $A_x = \{X = x\}$ and $B_y = \{Y = y\}$, then

$$f(x, y) = \mathbf{P}(A_x \cap B_y).$$

The definition of independence can now be reformulated in a lemma.

(3) **Lemma.** *X and Y are independent if and only if*

(4) $$f_{X,Y}(x, y) = f_X(x)f_Y(y) \quad \textit{for all } x, y \in \mathbb{R}.$$

More generally, X and Y are independent if and only if $f_{X,Y}(x, y)$ can be factorized as the product $g(x)h(y)$ of a function of x alone and a function of y alone.

Proof. This is Problem (3.11.1). ■

Suppose that X and Y have joint mass function $f_{X,Y}$ and we wish to check whether or not (4) holds. First we need to calculate the *marginal mass functions* f_X and f_Y from our knowledge of $f_{X,Y}$. These are found in the following way:

$$\begin{aligned} f_X(x) = \mathbf{P}(X = x) &= \mathbf{P}\left(\bigcup_y (\{X = x\} \cap \{Y = y\})\right) \\ &= \sum_y \mathbf{P}(X = x, Y = y) = \sum_y f_{X,Y}(x, y), \end{aligned}$$

and similarly

$$f_Y(y) = \sum_x f_{X,Y}(x, y).$$

Having found the marginals, it is a trivial matter to see whether (4) holds or not.

(5) **Example. Calculation of marginals.** In Example (3.2.2) we encountered a pair X, Y of variables with a joint mass function

$$f(x, y) = \frac{\alpha^x \beta^y}{x!\,y!} \exp[-(\alpha + \beta)] \quad \text{for} \quad x, y = 0, 1, 2, \dots$$

where $\alpha, \beta > 0$. The marginal mass function of X is

$$f_X(x) = \sum_y f(x, y) = \frac{\alpha^x}{x!} e^{-\alpha} \sum_{y=0}^{\infty} \frac{\beta^y}{y!} e^{-\beta} = \frac{\alpha^x}{x!} e^{-\alpha}$$

and so X has the Poisson distribution with parameter α. Similarly Y has the Poisson distribution with parameter β, and it is easy to check that (4) holds, giving that X and Y are independent. ●

For any discrete pair X, Y, a real function $g(X, Y)$ is a random variable. We shall often need to find its expectation. To avoid explicit calculation of its mass function, we shall use the following more general form of the law of the unconscious statistician.

(6) **Lemma.** $\mathbf{E}(g(X, Y)) = \sum_{x,y} g(x, y) f_{X,Y}(x, y)$.

Proof. As for (3.3.3). ■

For example, $\mathbf{E}(XY) = \sum_{x,y} xy f_{X,Y}(x, y)$. This formula is particularly useful to statisticians who may need to find simple ways of explaining dependence to laymen. For instance, suppose that the government wishes to announce that the dependence between defence spending and the cost of living is very small. It should *not* publish an estimate of the joint mass function unless its object is obfuscation alone. Most members of the public would prefer to find that this dependence can be represented in terms of a single number on a prescribed scale. Towards this end we make the following definition.

(7) **Definition.** The **covariance** of X and Y is

$$\text{cov}(X, Y) = \mathbf{E}[(X - \mathbf{E}X)(Y - \mathbf{E}Y)].$$

The **correlation (coefficient)** of X and Y is

$$\rho(X, Y) = \frac{\operatorname{cov}(X, Y)}{(\operatorname{var}(X) \cdot \operatorname{var}(Y))^{\frac{1}{2}}}$$

as long as the variances are non-zero.

Expanding the covariance gives

$$\operatorname{cov}(X, Y) = \mathbf{E}(XY) - \mathbf{E}(X)\mathbf{E}(Y).$$

Remember (3.3.10) that X and Y are called *uncorrelated* if $\operatorname{cov}(X, Y) = 0$. Also, independent variables are always uncorrelated, although the converse is not true. Covariance itself is not a satisfactory measure of dependence because the scale of values which $\operatorname{cov}(X, Y)$ may take contains no points which are clearly interpretable in terms of the relationship between X and Y. The following lemma shows that this is not the case for correlations.

(8) **Lemma.** *ρ satisfies*

$$|\rho(X, Y)| \leqslant 1$$

with equality if and only if $\mathbf{P}(Y = aX + b) = 1$ *for some* $a, b \in \mathbb{R}$.

The proof is an application of the following important inequality.

(9) **Theorem. Cauchy–Schwarz inequality.** *For any X and Y*

$$\mathbf{E}(XY)^2 \leqslant \mathbf{E}(X^2)\mathbf{E}(Y^2)$$

with equality if and only if $\mathbf{P}(aX = bY) = 1$ *for some real a and b, at least one of which is non-zero.*

Proof. We can assume that $\mathbf{E}(X^2)$ and $\mathbf{E}(Y^2)$ are strictly positive, since otherwise the result follows immediately from Problem (3.11.2). For $a, b \in \mathbb{R}$, let $Z = aX - bY$. Then

$$0 \leqslant \mathbf{E}(Z^2) = a^2\mathbf{E}(X^2) - 2ab\mathbf{E}(XY) + b^2\mathbf{E}(Y^2).$$

Thus the right-hand side is a quadratic in the variable a with at most one real root. Its discriminant must be non-positive. That is to say, if $b \neq 0$,

$$\mathbf{E}(XY)^2 - \mathbf{E}(X^2)\mathbf{E}(Y^2) \leqslant 0.$$

The discriminant is zero if and only if the quadratic has a real root. This occurs if and only if

$$\mathbf{E}((aX - bY)^2) = 0 \text{ for some } a \text{ and } b,$$

which, by Problem (3.11.2), completes the proof. ■

Proof of (8). Apply (9) to the variables $X - \mathbf{E}X$ and $Y - \mathbf{E}Y$. ■

A more careful treatment than this proof shows that $\rho = +1$ if and only if Y *increases* linearly with X and $\rho = -1$ if and only if Y *decreases* linearly as X increases.

(10) **Example.** Here is a tedious numerical example of the use of joint mass functions. Let X and Y take values in $\{1, 2, 3\}$ and $\{-1, 0, 2\}$ respectively, with joint mass function f where $f(x, y)$ is the appropriate entry in Table 3.1.

Table 3.1 The joint mass function of X and Y.

	$y = -1$	$y = 0$	$y = 2$	f_X
$x = 1$	$\frac{1}{18}$	$\frac{3}{18}$	$\frac{2}{18}$	$\frac{6}{18}$
$x = 2$	$\frac{2}{18}$	0	$\frac{3}{18}$	$\frac{5}{18}$
$x = 3$	0	$\frac{4}{18}$	$\frac{3}{18}$	$\frac{7}{18}$
f_Y	$\frac{3}{18}$	$\frac{7}{18}$	$\frac{8}{18}$	

The indicated row and column sums are the marginal mass functions f_X and f_Y. A quick calculation gives

$$\mathbf{E}(XY) = \sum_{x,y} xyf(x, y) = 29/18$$

$$\mathbf{E}(X) = \sum_{x} xf_X(x) = 37/18, \qquad \mathbf{E}(Y) = 13/18$$

$$\text{var}(X) = \mathbf{E}(X^2) - \mathbf{E}(X)^2 = 233/324, \qquad \text{var}(Y) = 461/324$$

$$\text{cov}(X, Y) = 41/324, \qquad \rho(X, Y) = 41/(107\,413)^{\frac{1}{2}}.$$ ●

Exercises

(11) Show that the collection of random variables having finite variance forms a vector space over the reals.

(12) Find the marginal mass functions of the multinomial distribution of Exercise (3.5.7).

(13) Let X and Y be discrete random variables with joint mass function

$$f(x, y) = \frac{C}{(x + y - 1)(x + y)(x + y + 1)}, \qquad x, y = 1, 2, 3, \ldots.$$

Find the marginal mass functions of X and Y, calculate C, and also the covariance of X and Y.

(14) Let X and Y be discrete random variables with mean 0, variance 1, and covariance ρ. Show that $\mathbf{E}(\max\{X^2, Y^2\}) \leqslant 1 + (1 - \rho^2)^{\frac{1}{2}}$.

3.7 Conditional distributions and conditional expectation

In Section 1.4 we discussed the conditional probability $\mathbf{P}(B \mid A)$. This may be set in the more general context of the conditional distribution of one variable Y given the value of another variable X; this reduces to the definition of the conditional probabilities of events A and B if $X = I_A$ and $Y = I_B$.

Let X and Y be two discrete variables on $(\Omega, \mathcal{F}, \mathbf{P})$.

(1) **Definition.** The **conditional distribution function** of Y given $X = x$, written $F_{Y|X}(\cdot \mid x)$, is defined by

$$F_{Y|X}(y \mid x) = \mathbf{P}(Y \leqslant y \mid X = x)$$

for any x such that $\mathbf{P}(X = x) > 0$. The **conditional (probability) mass function** of Y given $X = x$, written $f_{Y|X}(\cdot \mid x)$, is defined by

(2) $$f_{Y|X}(y \mid x) = \mathbf{P}(Y = y \mid X = x)$$

for any x such that $\mathbf{P}(X = x) > 0$..

Formula (2) is easy to remember as $f_{Y|X} = f_{X,Y}/f_X$. Conditional distribution and mass functions are undefined at values of x for which $\mathbf{P}(X = x) = 0$. Clearly X and Y are independent if and only if $f_{Y|X} = f_Y$.

Suppose we are told that $X = x$. Conditional upon this, the new distribution of Y has mass function $f_{Y|X}(y \mid x)$, which we think of as a function of y. The expected value $\sum_y y f_{Y|X}(y \mid x)$ of this distribution is called the *conditional expectation* of Y given $X = x$ and is written $\psi(x) = \mathbf{E}(Y \mid X = x)$. Now, we observe that the conditional expectation depends on the value x taken by X, and can be thought of as a function $\psi(X)$ of X itself.

(3) **Definition.** Let $\psi(x) = \mathbf{E}(Y \mid X = x)$. Then $\psi(X)$ is called the **conditional expectation** of Y given X, written as $\mathbf{E}(Y \mid X)$.

Although 'conditional expectation' sounds like a number, it is actually a random variable! It has the following important property.

(4) **Theorem.** *The conditional expectation* $\psi(X) = \mathbf{E}(Y \mid X)$ *satisfies*

$$\mathbf{E}(\psi(X)) = \mathbf{E}(Y).$$

Proof. By (3.3.3)

$$\mathbf{E}(\psi(X)) = \sum_x \psi(x) f_X(x) = \sum_{x,y} y f_{Y|X}(y \mid x) f_X(x)$$
$$= \sum_{x,y} y f_{X,Y}(x, y) = \sum_y y f_Y(y) = \mathbf{E}(Y).$$

■

This is an extremely useful theorem, to which we shall make repeated references. It often provides a useful method for calculating $\mathbf{E}(Y)$, since it asserts that

$$\mathbf{E}(Y) = \sum_x \mathbf{E}(Y \mid X = x)\mathbf{P}(X = x).$$

(5) **Example.** A hen lays N eggs, where N has the Poisson distribution with parameter λ. Each egg hatches with probability $p\ (=1 - q)$ independently of the other eggs. Let K be the number of chicks. Find $\mathbf{E}(K \mid N)$, $\mathbf{E}(K)$, and $\mathbf{E}(N \mid K)$.

Solution. We are given that

$$f_N(n) = \frac{\lambda^n}{n!}\,\mathrm{e}^{-\lambda}, \qquad f_{K|N}(k \mid n) = \binom{n}{k} p^k(1 - p)^{n-k}.$$

Therefore

$$\psi(n) = \mathbf{E}(K \mid N = n) = \sum_k k f_{K|N}(k \mid n) = pn.$$

Thus

$$\mathbf{E}(K \mid N) = \psi(N) = pN$$

and

$$\mathbf{E}(K) = \mathbf{E}(\psi(N)) = p\mathbf{E}(N) = p\lambda.$$

To find $\mathbf{E}(N \mid K)$ we need to know the conditional mass function $f_{N|K}$ of N given K. However,

$$\begin{aligned} f_{N|K}(n \mid k) &= \mathbf{P}(N = n \mid K = k) \\ &= \mathbf{P}(K = k \mid N = n)\mathbf{P}(N = n)/\mathbf{P}(K = k) \\ &= \frac{\binom{n}{k} p^k(1 - p)^{n-k}(\lambda^n/n!)\,\mathrm{e}^{-\lambda}}{\sum_{m \geqslant k} \binom{m}{k} p^k(1 - p)^{m-k}(\lambda^m/m!)\,\mathrm{e}^{-\lambda}} \quad \text{if} \quad n \geqslant k \\ &= \frac{(q\lambda)^{n-k}}{(n - k)!}\,\mathrm{e}^{-q\lambda}. \end{aligned}$$

Hence

$$\mathbf{E}(N \mid K = k) = \sum_{n \geqslant k} n \frac{(q\lambda)^{n-k}}{(n - k)!}\,\mathrm{e}^{-q\lambda} = k + q\lambda,$$

giving

$$\mathbf{E}(N \mid K) = K + q\lambda. \qquad \bullet$$

There is a more general version of Theorem (4), and this will be of interest later.

(6) **Theorem.** *The conditional expectation* $\psi(X) = \mathbf{E}(Y \mid X)$ *satisfies*

(7)
$$\mathbf{E}(\psi(X)g(X)) = \mathbf{E}(Yg(X))$$

for any function g *for which both expectations exist.*

Setting $g(x) = 1$ for all x, we obtain the result of (4). Whilst Theorem (6) is useful in its own right, we shall see later that its principal interest lies elsewhere. The conclusion of the theorem may be taken as a *definition* of conditional expectation—as a function $\psi(X)$ of X such that (7) holds for all suitable functions g. Such a definition is convenient when working with a notion of conditional expectation more general than that dealt with here.

Proof. As in the proof of (4),

$$\mathbf{E}(\psi(X)g(X)) = \sum_{x} \psi(x)g(x)f_X(x) = \sum_{x,y} yg(x)f_{Y|X}(y \mid x)f_X(x)$$
$$= \sum_{x,y} yg(x)f_{X,Y}(x, y) = \mathbf{E}(Yg(X)).$$ ■

Exercises

(8) Show the following:

(a) $\mathbf{E}(aY + bZ \mid X) = a\mathbf{E}(Y \mid X) + b\mathbf{E}(Z \mid X)$ for $a, b \in \mathbb{R}$,
(b) $\mathbf{E}(Y \mid X) \geqslant 0$ if $Y \geqslant 0$,
(c) $\mathbf{E}(1 \mid X) = 1$,
(d) if X and Y are independent then $\mathbf{E}(Y \mid X) = \mathbf{E}(Y)$,
(e) $\mathbf{E}(Yg(X) \mid X) = g(X)\mathbf{E}(Y \mid X)$ for any suitable function g,
(f) $\mathbf{E}\{\mathbf{E}(Y \mid X, Z) \mid X\} = \mathbf{E}(Y \mid X) = \mathbf{E}\{\mathbf{E}(Y \mid X) \mid X, Z\}$.

(9) **Uniqueness of conditional expectation.** Suppose that X and Y are random variables, and that $\phi(X)$ and $\psi(X)$ are two functions of X satisfying

$$\mathbf{E}(\phi(X)g(X)) = \mathbf{E}(\psi(X)g(X)) = \mathbf{E}(Yg(X))$$

for any function g for which all the expectations exist. Show that $\phi(X)$ and $\psi(X)$ are almost surely equal, in that $\mathbf{P}(\phi(X) = \psi(X)) = 1$.

(10) Suppose that the conditional expectation of Y given X is defined as the (almost surely) unique function $\psi(X)$ such that $\mathbf{E}(\psi(X)g(X)) = \mathbf{E}(Yg(X))$ for all functions g for which the expectations exist. Show (a)–(f) of Exercise (8) above (with the occasional addition of the expression 'with probability 1').

(11) How should we define $\text{var}(Y \mid X)$, the conditional variance of Y given X? Show that $\text{var}(Y) = \mathbf{E}(\text{var}(Y \mid X)) + \text{var}(\mathbf{E}(Y \mid X))$.

(12) The lifetime of a machine (in days) is a random variable T with mass function f. Given that the machine is working after t days, what is the mean subsequent lifetime

of the machine when

(a) $f(x) = (N+1)^{-1}$ for $x \in \{0, 1, \dots, N\}$, and
(b) $f(x) = 2^{-x}$ for $x = 1, 2, \dots$.

(13) Let $X_1, X_2, \dots$ be identically distributed random variables with mean μ, and let N be a random variable taking values in the non-negative integers and independent of the X_i. Let $S = X_1 + X_2 + \cdots + X_N$. Show that $\mathbf{E}(S \mid N) = \mu N$, and deduce that $\mathbf{E}(S) = \mu \mathbf{E}(N)$.

3.8 Sums of random variables

Much of the classical theory of probability concerns sums of random variables. We have seen already many such sums; the number of heads in n tosses of a coin is one of the simplest such examples, but we shall encounter many situations which are more complicated than this. One particular complication is when the summands are dependent. The first stage in developing a systematic technique is to find a formula for describing the mass function of the sum $Z = X + Y$ of two variables with joint mass function $f(x, y)$.

(1) **Theorem.** $$\mathbf{P}(Z = z) = \sum_x f(x, z - x).$$

Proof. $$\{Z = z\} = \bigcup_x (\{X = x\} \cap \{Y = z - x\}).$$

This is a disjoint union, and at most countably many of its contributors have non-zero probability. Therefore

$$\mathbf{P}(Z = z) = \sum_x \mathbf{P}(X = x, Y = z - x) = \sum_x f(x, z - x). \qquad \blacksquare$$

If X and Y are independent, then

$$\mathbf{P}(Z = z) = f_Z(z) = \sum_x f_X(x) f_Y(z - x) = \sum_y f_X(z - y) f_Y(y).$$

The mass function of Z is called the *convolution* of the mass functions of X and Y, and is written

(2) $$f_Z = f_X * f_Y.$$

(3) **Example (3.5.6) revisited.** Let X_1 and X_2 be independent geometric variables with common mass function

$$f(k) = p(1-p)^{k-1}, \qquad k = 1, 2, \dots.$$

By (2), $Z = X_1 + X_2$ has mass function

$$\begin{aligned}\mathbf{P}(Z = z) &= \sum_k \mathbf{P}(X_1 = k)\mathbf{P}(X_2 = z - k)\\ &= \sum_{k=1}^{z-1} p(1-p)^{k-1}p(1-p)^{z-k-1}\\ &= (z-1)p^2(1-p)^{z-2}, \quad z = 2, 3, \ldots\end{aligned}$$

in agreement with (3.5.6). The general formula for the sum of a number, r say, of geometric variables can easily be verified by induction. ●

Exercises

(4) Let X and Y be independent variables, X being equally likely to take any value in $\{0, 1, \ldots, m\}$, and Y similarly in $\{0, 1, \ldots, n\}$. Find the mass function of $Z = X + Y$. Z is said to have the *trapezoidal* distribution.

(5) Let X and Y have the joint mass function

$$f(x, y) = \frac{C}{(x+y-1)(x+y)(x+y+1)}, \qquad x, y = 1, 2, 3, \ldots.$$

Find the mass functions of $U = X + Y$ and $V = X - Y$.

3.9 Simple random walk

Until now we have dealt largely with general theory; the final two sections of this chapter may provide some lighter relief. One of the simplest random processes is 'simple random walk'; this process arises in many ways, of which the following is traditional. A gambler G plays the following game at the casino. The croupier tosses a (possibly biased) coin repeatedly; each time heads appears, he gives G one franc, and each time tails appears he takes one franc from G. Writing S_n for G's fortune after n tosses of the coin, we have that $S_{n+1} = S_n + X_{n+1}$ where X_{n+1} is a random variable taking the value 1 with some fixed probability p and -1 otherwise; furthermore, X_{n+1} is independent of the results of all previous tosses. Thus

(1) $$S_n = S_0 + \sum_{i=1}^{n} X_i,$$

so that S_n is obtained from the initial fortune S_0 by the addition of n independent random variables. We are assuming here that there are no constraints on G's fortune imposed externally, such as that the game is terminated if his fortune is reduced to zero.

An alternative picture of 'simple random walk' involves the motion of a particle—a particle which inhabits the set of integers and which moves at each step either one step to the right, with probability p, or one step to the left, the directions of different steps being independent of each other. More complicated random walks arise when the steps of the particle are allowed to have some general distribution on the integers, or the reals, so that the position S_n at time n is given by (1) where the X_i are independent and identically distributed random variables having some specified mass function or density function. Even greater generality is obtained by assuming that the X_i take values in $\mathbb{R}^d$ for some $d \geqslant 1$, or even some vector space over the real numbers. Random walks may be used with some success in modelling various practical situations, such as the number of cars in a toll queue at 5 minute intervals, the position of a pollen grain suspended in fluid at 1 second intervals, or the value of the Dow–Jones index each Monday morning. In each case, it may not be too bad a guess that the $(n+1)$th reading differs from the nth by a random quantity which is independent of previous jumps but has the same probability distribution. The theory of random walks is a basic tool in the probabilist's kit, and we shall concern ourselves here with 'simple random walk' only.

At any instant of time a particle inhabits one of the integer points of the real line. At time 0 it starts from some specified point, and at each subsequent epoch of time 1, 2, ... it moves from its current position to a new position according to the following law. With probability p it moves one step to the right and with probability $q = 1 - p$ it moves one step to the left; moves are independent of each other. The walk is called *symmetric* if $p = q = \frac{1}{2}$. Example (1.7.4) concerned a symmetric random walk with 'absorbing' barriers at the points 0 and N. In general, let S_n denote the position of the particle after n moves, and set $S_0 = a$. Then

(2) $$S_n = a + \sum_{i=1}^{n} X_i$$

where $X_1, X_2, \ldots$ is a sequence of independent Bernoulli variables taking values $+1$ and -1 (rather than $+1$ and 0 as before) with probabilities p and q.

We record the motion of the particle as the sequence $\{(n, S_n): n \geqslant 0\}$ of cartesian coordinates of points in the plane. This collection of points, joined by solid lines between neighbours, is called the *path* of the particle. In the example shown in Figure 3.2, the particle has visited the points 0, 1, 0, -1, 0, 1, 2 in succession. This representation has a confusing aspect in that the direction of the particle's steps is parallel to the y-axis, whereas we have previously been specifying the movement in the traditional way as to the right or to the left. In future, any reference to the x-axis or the y-axis will pertain to a diagram of its path as exemplified by Figure 3.2.

The sequence (2) of partial sums has three important properties.

(3) **Lemma.** *The simple random walk is **spatially homogeneous**; that is*

$$\mathbf{P}(S_n = j \mid S_0 = a) = \mathbf{P}(S_n = j + b \mid S_0 = a + b).$$

Proof. Both sides equal $\mathbf{P}(\sum_1^n X_i = j - a)$. ■

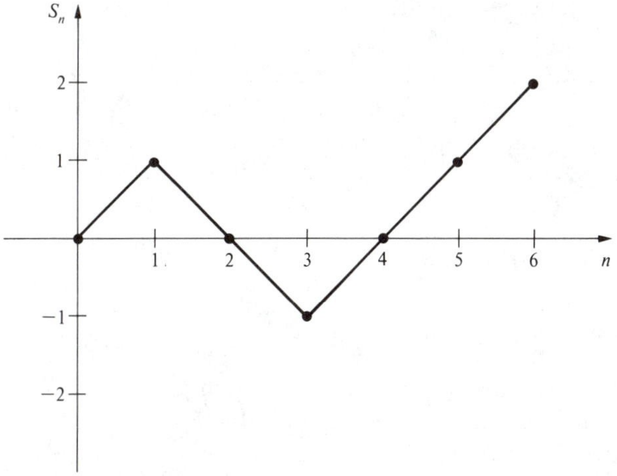

Fig. 3.2 A random walk S_n.

(4) **Lemma.** *The simple random walk is **temporally homogeneous**; that is*

$$\mathbf{P}(S_n = j \mid S_0 = a) = \mathbf{P}(S_{n+m} = j \mid S_m = a).$$

Proof. The left- and right-hand sides satisfy

$$\text{LHS} = \mathbf{P}\left(\sum_1^n X_i = j - a\right) = \mathbf{P}\left(\sum_{m+1}^{m+n} X_i = j - a\right) = \text{RHS}. \qquad \blacksquare$$

(5) **Lemma.** *The simple random walk has the **Markov property**; that is*

$$\mathbf{P}(S_{n+m} = j \mid S_0, S_1, \dots, S_n) = \mathbf{P}(S_{n+m} = j \mid S_n), \qquad m \geqslant 0.$$

Statements such as $\mathbf{P}(S = j \mid X, Y) = \mathbf{P}(S = j \mid X)$ are to be interpreted in the obvious way as meaning that $\mathbf{P}(S = j \mid X = x, Y = y) = \mathbf{P}(S = j \mid X = x)$ for all x and y; this is a slight abuse of notation.

Proof. If one knows the value of S_n, then the distribution of S_{n+m} depends only on the jumps $X_{n+1}, \dots, X_{n+m}$, and cannot depend on further information concerning $S_0, S_1, \dots, S_{n-1}$. $\blacksquare$

This 'Markov property' is often expressed approximately by saying that, conditional upon knowing the value of the process at the nth step, its values after the nth step do not depend on its values before the nth step. More colloquially: conditional upon the present, the future does not depend on the past. We shall meet this property again later.

(6) **Example. Absorbing barriers.** Let us revisit (1.7.4) for general values of p. Equation (1.7.5) gives us the following difference equation for the probabilities $\{p_k\}$ where p_k is the probability of ultimate ruin starting from k:

(7) $$p_k = pp_{k+1} + qp_{k-1} \quad \text{if} \quad 1 \leqslant k \leqslant N - 1$$

with boundary conditions $p_0 = 1$, $p_N = 0$. The solution of such a difference equation proceeds as follows. Look for a solution of the form $p_k = \theta^k$. Substitute this into (7) and cancel out the power θ^{k-1} to obtain $p\theta^2 - \theta + q = 0$, which has roots $\theta_1 = 1$, $\theta_2 = q/p$. If $p \neq \frac{1}{2}$ then these roots are distinct and the general solution of (7) is $p_k = A_1\theta_1^k + A_2\theta_2^k$ for arbitrary constants A_1 and A_2. Use the boundary conditions to obtain

$$p_k = \frac{(q/p)^k - (q/p)^N}{1 - (q/p)^N}.$$

If $p = \frac{1}{2}$ then $\theta_1 = \theta_2 = 1$ and the general solution to (7) is $p_k = A_1 + A_2 k$. Use the boundary conditions to obtain $p_k = 1 - (k/N)$.

A more complicated equation is obtained for the mean number D_k of steps before the particle hits one of the absorbing barriers, starting from k. In this case we use conditional expectations and (3.7.4) to find that

(8) $$D_k = p(1 + D_{k+1}) + q(1 + D_{k-1}) \quad \text{if} \quad 1 \leqslant k \leqslant N - 1$$

with the boundary conditions $D_0 = D_N = 0$. Try solving this; you need to find a general solution and a particular solution, as in the solution of second-order linear differential equations. The answer is

(9) $$D_k = \begin{cases} \dfrac{1}{q-p}\left[k - N\left(\dfrac{1-(q/p)^k}{1-(q/p)^N}\right)\right] & \text{if} \quad p \neq \frac{1}{2} \\ k(N-k) & \text{if} \quad p = \frac{1}{2}. \end{cases}$$ ●

(10) **Example. Retaining barriers.** In (1.7.4), suppose that the Jaguar buyer has a rich uncle who will guarantee all his losses. Then the random walk does not end when the particle hits zero, although it cannot visit a negative integer. Instead $\mathbf{P}(S_{n+1} = 0 \mid S_n = 0) = q$ and $\mathbf{P}(S_{n+1} = 1 \mid S_n = 0) = p$. The origin is said to have a 'retaining' barrier (sometimes called 'reflecting').

What now is the expected duration of the game? The mean duration F_k, starting from k, satisfies the same difference equation (8) as before but subject to different boundary conditions. We leave it as an *exercise* to show that the boundary conditions are $F_N = 0$, $pF_0 = 1 + pF_1$, and hence to find F_k. ●

In such examples the techniques of 'conditioning' are supremely useful. The idea is that in order to calculate a probability $\mathbf{P}(A)$ or expectation $\mathbf{E}(Y)$ we condition either on some partition of Ω (and use (1.4.4)) or on the outcome of some random variable (and use (3.7.4) or the forthcoming (4.6.5)). In this section this technique yielded the difference equations (7) and (8). In later sections the same idea will yield differential equations, integral equations, and functional equations, some of which can be solved.

Exercises

(11) Let T be the time which elapses before a simple random walk is absorbed at either of the absorbing barriers at 0 and N, having started at k where $0 \leqslant k \leqslant N$. Show that

(a) $\mathbf{P}(T < \infty) = 1$, (b) $\mathbf{E}(T^k) < \infty$ for all $k \geqslant 1$.

(12) For simple random walk S with absorbing barriers at 0 and N, let W be the event that the particle is absorbed at 0 rather than at N, and let $p_k = \mathbf{P}(W \mid S_0 = k)$. Show that, if the particle starts at k where $0 < k < N$, the conditional probability that the first step is rightwards, given W, equals pp_{k+1}/p_k. Deduce that the mean duration J_k of the walk, conditional on W, satisfies the equation

$$pp_{k+1}J_{k+1} - p_kJ_k + (p_k - pp_{k+1})J_{k-1} = -p_k, \quad \text{for} \quad 0 < k < N.$$

Show that we may take as boundary condition $J_0 = 0$. Find J_k in the symmetric case, when $p = \frac{1}{2}$.

(13) A coin is tossed repeatedly, heads turning up with probability p on each toss. Player A wins the game if m heads appear before n tails have appeared, and player B wins otherwise. Let p_{mn} be the probability that A wins the game. Set up a difference equation for the p_{mn}; what are the boundary conditions? This has been called the 'problem of the points'.

(14) (a) 'Millionaires should always gamble, poor men never' [J. M. Keynes].
(b) 'If I wanted to gamble, I would buy a casino' [P. Getty].

Discuss.

3.10 Random walk: counting sample paths

In the previous section, our principal technique was to condition on the first step of the walk and then solve the ensuing difference equation. Another primitive but useful technique is to count. Let $X_1, X_2, \ldots$ be independent variables, each taking the values -1 and 1 with probabilities $q = 1 - p$ and p, as before, and let

(1) $$S_n = a + \sum_{i=1}^{n} X_i$$

be the position of the corresponding random walker after n steps, having started at $S_0 = a$. The set of realizations of the walk is the set of vectors $\mathbf{s} = (s_0, s_1, \ldots)$ with $s_0 = a$ and $s_{i+1} - s_i = \pm 1$, and any such vector may be thought of as a 'sample path' of the walk, drawn in the manner of Figure 3.2. The probability that the first n steps of the walk follow a given path $\mathbf{s} = (s_0, s_1, \ldots, s_n)$ is $p^r q^l$ where r is the number of steps of $\mathbf{s}$ to the right and l is the number to the left; that is to say, $r = |\{i: s_{i+1} - s_i = 1\}|$ and $l = |\{i: s_{i+1} - s_i = -1\}|$. Any event may be expressed in terms of an appropriate set of paths, and the probability of the event is the sum of the component probabilities. For example, $\mathbf{P}(S_n = b) = \sum_r M_n^r(a, b)p^r q^{n-r}$ where $M_n^r(a, b)$ is the number of paths $(s_0, s_1, \ldots, s_n)$ with $s_0 = a$, $s_n = b$, and having exactly r rightward steps. It is easy to see that $r + l = n$, the total number of steps, and $r - l = b - a$, the aggregate rightward displacement, so that $r = \frac{1}{2}(n + b - a)$ and $l = \frac{1}{2}(n - b + a)$. Thus

(2) $$\mathbf{P}(S_n = b) = \binom{n}{\frac{1}{2}(n + b - a)} p^{\frac{1}{2}(n+b-a)} q^{\frac{1}{2}(n-b+a)},$$

since there are exactly $\binom{n}{r}$ paths with length n having r rightward steps and $n - r$ leftward steps. Formula (2) is useful only if $\frac{1}{2}(n + b - a)$ is an integer lying in the range $0, 1, \ldots, n$; otherwise, the probability in question equals 0.

Natural questions of interest for the walk include:

(a) when does the first visit of the random walk to a given point occur; and
(b) what is the furthest rightward point visited by the random walk by time n?

Such questions may be answered with the aid of certain elegant results and techniques for counting paths. The first of these is the 'reflection principle'. Here is some basic notation. As in Figure 3.2, we keep a record of the random walk through its path $\{(n, S_n): n \geqslant 0\}$.

Suppose we know that $S_0 = a$ and $S_n = b$. The random walk may or may not have visited the origin in between times 0 and n. Let $N_n(a, b)$ be the number of possible paths from $(0, a)$ to (n, b), and let $N_n^0(a, b)$ be the number of such paths which contain some point $(k, 0)$ on the x-axis.

(3) **Theorem. The reflection principle.** *If $a, b > 0$ then $N_n^0(a, b) = N_n(-a, b)$.*

Proof. Each path from $(0, -a)$ to (n, b) intersects the x-axis at some earliest point $(k, 0)$. Reflect the segment of the path with $0 \leqslant x \leqslant k$ in the x-axis to obtain a path joining $(0, a)$ to (n, b) which intersects the x-axis (see Figure 3.3). This operation gives a one–one correspondence between the collections of such paths, and the theorem is proved. ■

We have, as before, a formula for $N_n(a, b)$.

(4) **Lemma.** $$N_n(a, b) = \binom{n}{\frac{1}{2}(n + b - a)}.$$

Proof. Choose a path from $(0, a)$ to (n, b) and let α and β be the numbers of positive and negative steps, respectively, in this path. Then $\alpha + \beta = n$ and $\alpha - \beta = b - a$, so that $\alpha = \frac{1}{2}(n + b - a)$. The number of such paths is the number of ways of picking α positive steps from the n available. That is

(5) $$N_n(a, b) = \binom{n}{\alpha} = \binom{n}{\frac{1}{2}(n + b - a)}.$$ ■

The famous 'ballot theorem' is a consequence of these elementary results; it was proved first by W. A. Whitworth in 1878.

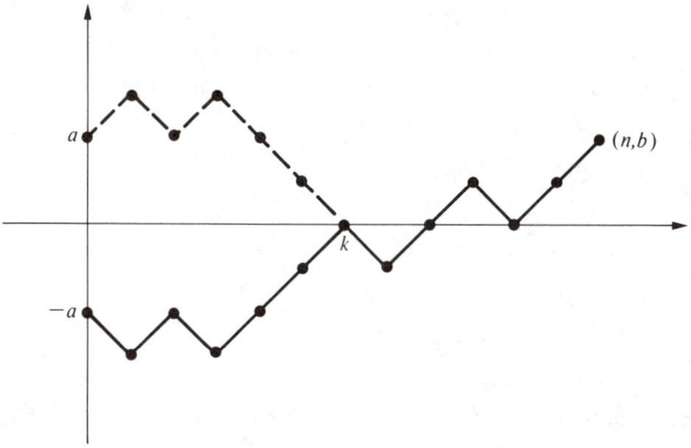

Fig. 3.3 A random walk; the dashed line is the reflection of the first segment of the walk.

(6) **Corollary. Ballot theorem.** *If* $b > 0$ *then the number of paths from* $(0, 0)$ *to* (n, b) *which do not revisit the x-axis equals* $(b/n)N_n(0, b)$.

Proof. The first step of all such paths is to $(1, 1)$, and so the number of such paths is

$$N_{n-1}(1, b) - N^0_{n-1}(1, b) = N_{n-1}(1, b) - N_{n-1}(-1, b)$$

by the reflection principle. We now use (4) and an elementary calculation to obtain the result. ■

As an application, and an explanation of the title of the theorem, we may easily answer the following amusing question. Suppose that, in a ballot, candidate A scores α votes and candidate B scores β votes where $\alpha > \beta$. What is the probability that, during the ballot, A was always ahead of B? Let X_i equal 1 if the ith vote was cast for A, and -1 otherwise. Assuming that each possible combination of α votes for A and β votes for B is equally likely, we have that the probability in question is the proportion of paths from $(0, 0)$ to $(\alpha + \beta, \alpha - \beta)$ which do not revisit the x-axis. Using the ballot theorem, we obtain the answer $(\alpha - \beta)/(\alpha + \beta)$.

Here are some applications of the reflection principle to random walks. First, what is the probability that the walk does not revisit its starting point in the first n steps? We may as well assume that $S_0 = 0$, so that $S_1 \neq 0, \ldots, S_n \neq 0$ if and only if $S_1 S_2 \cdots S_n \neq 0$.

(7) **Theorem.** *If* $S_0 = 0$ *then, for* $n \geqslant 1$,

(8) $$\mathbf{P}(S_1 S_2 \cdots S_n \neq 0,\ S_n = b) = \frac{|b|}{n}\mathbf{P}(S_n = b),$$

and therefore

(9) $$\mathbf{P}(S_1 S_2 \cdots S_n \neq 0) = \frac{1}{n}\mathbf{E}|S_n|.$$

Proof. Suppose that $S_0 = 0$ and $S_n = b$ (>0). The event in question occurs if and only if the path of the random walk does not visit the x-axis in the time interval $[1, n]$. The number of such paths is, by the ballot theorem, $(b/n)N_n(0, b)$, and each such path has $\frac{1}{2}(n+b)$ rightward steps and $\frac{1}{2}(n-b)$ leftward steps. Therefore

$$\mathbf{P}(S_1 S_2 \cdots S_n \neq 0, S_n = b) = \frac{b}{n} N_n(0, b) p^{\frac{1}{2}(n+b)} q^{\frac{1}{2}(n-b)} = \frac{b}{n} \mathbf{P}(S_n = b)$$

as required. A similar calculation is valid if $b < 0$. ■

Another feature of interest is the maximum value attained by the random walk. We write $M_n = \max\{S_i : 0 \leqslant i \leqslant n\}$ for the maximum value up to time n, and shall suppose that $S_0 = 0$, so that $M_n \geqslant 0$. Clearly $M_n \geqslant S_n$, and the first part of the next theorem is therefore trivial.

(10) **Theorem.** *Suppose that* $S_0 = 0$. *Then, for* $r \geqslant 1$,

(11) $$\mathbf{P}(M_n \geqslant r, S_n = b) = \begin{cases} \mathbf{P}(S_n = b) & \text{if } b \geqslant r \\ (q/p)^{r-b} \mathbf{P}(S_n = 2r - b) & \text{if } b < r. \end{cases}$$

It follows that, for $r \geqslant 1$,

(12) $$\mathbf{P}(M_n \geqslant r) = \mathbf{P}(S_n \geqslant r) + \sum_{b=-\infty}^{r-1} (q/p)^{r-b} \mathbf{P}(S_n = 2r - b)$$

$$= \mathbf{P}(S_n = r) + \sum_{c=r+1}^{\infty} [1 + (q/p)^{c-r}] \mathbf{P}(S_n = c),$$

yielding in the symmetric case when $p = q = \frac{1}{2}$ that

(13) $$\mathbf{P}(M_n \geqslant r) = 2\mathbf{P}(S_n \geqslant r + 1) + \mathbf{P}(S_n = r),$$

which is easily expressed in terms of the binomial distribution.

Proof of (10). We may assume that $r \geqslant 1$ and $b < r$. Let $N_n^r(0, b)$ be the number of paths from $(0, 0)$ to (n, b) which include some point having height r, which is to say some point (i, r) with $0 < i < n$; for such a path π, let (i_π, r) be the earliest such point. We may reflect the segment of the path with $i_\pi \leqslant x \leqslant n$ in the line $y = r$ to obtain a path π' joining $(0, 0)$ to $(n, 2r - b)$. Any such path π' is obtained thus from a unique path π, and therefore $N_n^r(0, b) = N_n(0, 2r - b)$. It follows that

$$\begin{aligned} \mathbf{P}(M_n \geqslant r, S_n = b) &= N_n^r(0, b) p^{\frac{1}{2}(n+b)} q^{\frac{1}{2}(n-b)} \\ &= (q/p)^{r-b} N_n(0, 2r - b) p^{\frac{1}{2}(n+2r-b)} q^{\frac{1}{2}(n-2r+b)} \\ &= (q/p)^{r-b} \mathbf{P}(S_n = 2r - b) \end{aligned}$$

as required. ■

What is the chance that the walk reaches a new maximum at a particular time? More precisely, what is the probability that the walk, starting from 0, reaches the point b (>0) for the first time at the nth step? Writing $f_b(n)$ for this probability, we have that

$$\begin{aligned} f_b(n) &= \mathbf{P}(M_{n-1} = S_{n-1} = b-1,\ S_n = b) \\ &= p[\mathbf{P}(M_{n-1} \geqslant b-1,\ S_{n-1} = b-1) - \mathbf{P}(M_{n-1} \geqslant b,\ S_{n-1} = b-1)] \\ &= p[\mathbf{P}(S_{n-1} = b-1) - (q/p)\mathbf{P}(S_{n-1} = b+1)] \quad \text{by (11)} \\ &= \frac{b}{n}\mathbf{P}(S_n = b) \end{aligned}$$

by a simple calculation using (2). A similar conclusion may be reached if $b < 0$, and we arrive at the following.

(14) **Hitting time theorem.** *The probability $f_b(n)$ that a random walk S hits the point b for the first time at the nth step, having started from* 0, *satisfies*

(15) $$f_b(n) = \frac{|b|}{n}\mathbf{P}(S_n = b) \quad \textit{if} \quad n \geqslant 1.$$

The conclusion here has a close resemblance to that of the ballot theorem, and particularly Theorem (7). This is no coincidence: a closer examination of the two results leads to another technique for random walks, the technique of 'reversal'. If the first n steps of the original random walk are

$$\{0, S_1, S_2, \dots, S_n\} = \left\{0, X_1, X_1 + X_2, \dots, \sum_1^n X_i\right\}$$

then the steps of the *reversed* walk, denoted by $0, T_1, \dots, T_n$, are given by

$$\{0, T_1, T_2, \dots, T_n\} = \left\{0, X_n, X_n + X_{n-1}, \dots, \sum_1^n X_i\right\}.$$

Draw a diagram to see how the two walks correspond to each other. The X_i are independent and identically distributed, and it follows that the two walks have identical distributions even if $p \neq \frac{1}{2}$. Notice that the addition of an extra step to the original walk changes *every* step of the reversed walk.

Now, the original walk satisfies $S_n = b$ (>0) and $S_1 S_2 \cdots S_n \neq 0$ if and only if the reversed walk satisfies $T_n = b$ and $T_n - T_{n-i} = X_1 + \cdots + X_i > 0$ for all $i \geqslant 1$, which is to say that the first visit of the reversed walk to the point b takes place at time n. Therefore

(16) $$\mathbf{P}(S_1 S_2 \cdots S_n \neq 0,\ S_n = b) = f_b(n) \quad \text{if} \quad b > 0.$$

This is the 'coincidence' remarked above; a similar argument is valid if $b < 0$. The technique of reversal has other applications. For example, let μ_b be the mean number of visits of the walk to the point b before it returns to its starting point. If $S_0 = 0$ then, by (16),

(17) $$\mu_b = \sum_{n=1}^{\infty} \mathbf{P}(S_1 S_2 \cdots S_n \neq 0,\ S_n = b) = \sum_{n=1}^{\infty} f_b(n) = \mathbf{P}(S_n = b \text{ for some } n),$$

the probability of ultimately visiting b. This leads to the following result.

(18) **Theorem.** *If $p = \frac{1}{2}$ and $S_0 = 0$, then for any b ($\neq 0$) the mean number μ_b of visits of the walk to the point b before returning to the origin equals* 1.

Proof. Let $f_b = \mathbf{P}(S_n = b \text{ for some } n \geq 0)$. We have, by conditioning on the value of S_1, that $f_b = \frac{1}{2}(f_{b+1} + f_{b-1})$ for $b > 0$, with boundary condition $f_0 = 1$. The solution of this difference equation is $f_b = Ab + B$ for constants A and B. The unique such solution lying in $[0, 1]$ with $f_0 = 1$ is given by $f_b = 1$ for all $b \geq 0$. By symmetry, $f_b = 1$ for $b \leq 0$. However, $f_b = \mu_b$ for $b \neq 0$, and the claim follows. ■

'The truly amazing implications of this result appear best in the language of fair games. A perfect coin is tossed until the first equalization of the accumulated numbers of heads and tails. The gambler receives one penny for every time that the accumulated number of heads exceeds the accumulated number of tails by *m. The "fair entrance fee" equals* 1 *independently of m.*' (Feller 1968, p. 367)

We conclude with two celebrated properties of the symmetric random walk.

(19) **Theorem. Arc sine law for last visit to the origin.** *Suppose that $p = \frac{1}{2}$ and $S_0 = 0$. The probability that the last visit to* 0 *up to time $2n$ occurred at time $2k$ is* $\mathbf{P}(S_{2k} = 0)\mathbf{P}(S_{2n-2k} = 0)$.

In advance of proving this, we note some consequences. Writing $\alpha_{2n}(2k)$ for the probability referred to in the theorem, we have that $\alpha_{2n}(2k) = u_{2k}u_{2n-2k}$, where

$$u_{2k} = \mathbf{P}(S_{2k} = 0) = \binom{2k}{k} 2^{-2k}.$$

In order to understand the behaviour of u_{2k} for large values of k, we use Stirling's formula:

(20) $$n! \simeq n^n e^{-n} (2\pi n)^{\frac{1}{2}} \quad \text{as} \quad n \to \infty,$$

which is to say that the ratio of the left-hand side to the right-hand side tends to 1 as $n \to \infty$. Applying this formula, we obtain that $u_{2k} \simeq (\pi k)^{-\frac{1}{2}}$ as $k \to \infty$. This gives rise to the approximation $\alpha_{2n}(2k) \sim \{\pi[k(n-k)]^{\frac{1}{2}}\}^{-1}$, valid for values of k which are close to neither 0 nor n. If T_{2n} is the time of the last visit to 0 up to time $2n$, then it follows that

$$\mathbf{P}(T_{2n} \leqslant 2xn) \sim \sum_{k \leqslant xn} \frac{1}{\pi[k(n-k)]^{\frac{1}{2}}} \sim \int_{u=0}^{xn} \frac{1}{\pi[u(n-u)]^{\frac{1}{2}}}\, du$$
$$= \frac{2}{\pi} \sin^{-1} x^{\frac{1}{2}},$$

which is to say that $T_{2n}/(2n)$ has a distribution function which is approximately $(2/\pi)\sin^{-1}x^{\frac{1}{2}}$ when n is sufficiently large. We have proved a limit theorem.

The arc sine law is rather surprising. One may think that, in a long run of $2n$ tosses of a fair coin, the epochs of time at which there have appeared equal numbers of heads and tails should appear rather frequently. On the contrary, there is for example probability $\frac{1}{2}$ that no such epoch arrived in the final n tosses, and indeed probability approximately $\frac{1}{5}$ that no such epoch occurred after the first $\frac{1}{5}n$ tosses. One may think that, in a long run of $2n$ tosses of a fair coin, the last time at which the numbers of heads and tails were equal tends to be close to the end. On the contrary, the distribution of this time is symmetric around the midpoint.

How much time does a symmetric random walk spend to the right of the origin? More precisely, for how many values of k satisfying $0 \leqslant k \leqslant 2n$ is it the case that $S_k > 0$? Intuitively, one might expect the answer to be around n with large probability, but the truth is quite different. With large probability, the proportion of time spent to the right (or to the left) of the origin is near to 0 or to 1, but not near to $\frac{1}{2}$. That is to say, in a long sequence of tosses of a fair coin, there is large probability that one face (either heads or tails) will lead the other for a disproportionate amount of time.

(21) **Theorem. Arc sine law for sojourn times.** *Suppose that* $p = \frac{1}{2}$ *and* $S_0 = 0$. *The probability that the walk spends exactly* $2k$ *intervals of time, up to time* $2n$, *to the right of the origin equals* $\mathbf{P}(S_{2k} = 0)\mathbf{P}(S_{2n-2k} = 0)$.

We say that the interval $(k, k+1)$ is spent to the right of the origin if either $S_k > 0$ or $S_{k+1} > 0$. It is clear that the number of such intervals is even if the total number of steps is even. The conclusion of this theorem is most striking. First, the answer is the same as that of Theorem (19). Secondly, by the calculations following (19) we have that the probability that the walk spends $2xn$ units of time or less to the right of the origin is approximately $(2/\pi)\sin^{-1} x^{\frac{1}{2}}$.

Proof of (19). The probability in question is

$$\alpha_{2n}(2k) = \mathbf{P}(S_{2k} = 0)\mathbf{P}(S_{2k+1}S_{2k+2}\cdots S_{2n} \neq 0 \mid S_{2k} = 0)$$
$$= \mathbf{P}(S_{2k} = 0)\mathbf{P}(S_1 S_2 \cdots S_{2n-2k} \neq 0).$$

Now, setting $m = n - k$, we have by (8) that

(22)
$$\begin{aligned}\mathbf{P}(S_1S_2\cdots S_{2m} \neq 0) &= 2\sum_{k=1}^{m}\frac{2k}{2m}\mathbf{P}(S_{2m} = 2k) = 2\sum_{k=1}^{m}\frac{2k}{2m}\binom{2m}{m+k}\left(\frac{1}{2}\right)^{2m}\\ &= 2\left(\frac{1}{2}\right)^{2m}\sum_{k=1}^{m}\left[\binom{2m-1}{m+k-1} - \binom{2m-1}{m+k}\right]\\ &= 2\left(\frac{1}{2}\right)^{2m}\binom{2m-1}{m}\\ &= \binom{2m}{m}\left(\frac{1}{2}\right)^{2m} = \mathbf{P}(S_{2m} = 0).\end{aligned}$$ ∎

In passing, note the proof in (22) that

(23)
$$\mathbf{P}(S_1S_2\cdots S_{2m} \neq 0) = \mathbf{P}(S_{2m} = 0)$$

for the simple symmetric random walk.

Proof of (21). Let $\beta_{2n}(2k)$ be the probability in question, and write $u_{2m} = \mathbf{P}(S_{2m} = 0)$ as before. We are claiming that, for all $m \geqslant 1$,

(24)
$$\beta_{2m}(2k) = u_{2k}u_{2m-2k} \quad \text{if} \quad 0 \leqslant k \leqslant m.$$

First,

$$\begin{aligned}\mathbf{P}(S_1S_2\cdots S_{2m} > 0) &= \mathbf{P}(S_1 = 1, S_2 \geqslant 1, \dots, S_{2m} \geqslant 1)\\ &= \tfrac{1}{2}\mathbf{P}(S_1 \geqslant 0, S_2 \geqslant 0, \dots, S_{2m-1} \geqslant 0),\end{aligned}$$

where the second line follows by considering the walk $S_1 - 1, S_2 - 1, \dots, S_{2m} - 1$. Now S_{2m-1} is an odd number, so that $S_{2m-1} \geqslant 0$ implies that $S_{2m} \geqslant 0$ also. Thus

$$\mathbf{P}(S_1S_2\cdots S_{2m} > 0) = \tfrac{1}{2}\mathbf{P}(S_1 \geqslant 0, S_2 \geqslant 0, \dots, S_{2m} \geqslant 0),$$

yielding by (23) that

$$\tfrac{1}{2}u_{2m} = \mathbf{P}(S_1S_2\cdots S_{2m} > 0) = \tfrac{1}{2}\beta_{2m}(2m),$$

and (24) follows for $k = m$, and therefore for $k = 0$ also by symmetry.

Let n be a positive integer, and let T be the time of the first return of the walk to the origin. If $S_{2n} = 0$ then $T \leqslant 2n$; the probability mass function $f_{2r} = \mathbf{P}(T = 2r)$ satisfies

$$\mathbf{P}(S_{2n} = 0) = \sum_{r=1}^{n}\mathbf{P}(S_{2n} = 0 \mid T = 2r)\mathbf{P}(T = 2r) = \sum_{r=1}^{n}\mathbf{P}(S_{2n-2r} = 0)\mathbf{P}(T = 2r),$$

which is to say that

(25)
$$u_{2n} = \sum_{r=1}^{n}u_{2n-2r}f_{2r}.$$

Let $1 \leqslant k \leqslant n-1$, and consider $\beta_{2n}(2k)$. The corresponding event entails that $T = 2r$ for some r satisfying $1 \leqslant r < n$. The time interval $(0, T)$ is spent entirely either to the right or to the left of the origin, and each possibility has probability $\frac{1}{2}$. Therefore

$$\beta_{2n}(2k) = \sum_{r=1}^{k} \tfrac{1}{2}\mathbf{P}(T = 2r)\beta_{2n-2r}(2k-2r) + \sum_{r=1}^{n-k} \tfrac{1}{2}\mathbf{P}(T = 2r)\beta_{2n}(2k-2r). \tag{26}$$

We conclude the proof by using induction. Certainly (24) is valid for all k if $m = 1$. Assume (24) is valid for all k and all $m < n$.

From (26),

$$\begin{aligned}\beta_{2n}(2k) &= \tfrac{1}{2}\sum_{r=1}^{k} f_{2r}u_{2k-2r}u_{2n-2k} + \tfrac{1}{2}\sum_{r=1}^{n-k} f_{2r}u_{2k}u_{2n-2k-2r}\\ &= \tfrac{1}{2}u_{2n-2k}u_{2k} + \tfrac{1}{2}u_{2k}u_{2n-2k} = u_{2k}u_{2n-2k}\end{aligned}$$

by (25), as required. ■

Exercises

(27) Consider a symmetric random walk S with $S_0 = 0$. Let $T = \min\{n \geqslant 1: S_n = 0\}$ be the time of the first return of the walk to its starting point. Show that

$$\mathbf{P}(T = 2n) = \frac{1}{2n-1}\binom{2n}{n}2^{-2n},$$

and deduce that $\mathbf{E}(T^{\alpha}) < \infty$ if and only if $\alpha < \frac{1}{2}$. You may need Stirling's formula: $n! \simeq n^{n+\frac{1}{2}}e^{-n}(2\pi)^{\frac{1}{2}}$.

(28) For a symmetric random walk starting at 0, show that the mass function of the maximum satisfies $\mathbf{P}(M_n = r) = \mathbf{P}(S_n = r) + \mathbf{P}(S_n = r+1)$ for $r \geqslant 0$.

(29) For a symmetric random walk starting at 0, show that the probability that the first visit to S_{2n} takes place at time $2k$ equals the product $\mathbf{P}(S_{2k} = 0)\mathbf{P}(S_{2n-2k} = 0)$, for $0 \leqslant k \leqslant n$.

3.11 Problems

1. (a) Let X and Y be independent discrete random variables, and let $g, h: \mathbb{R} \to \mathbb{R}$. Show that $g(X)$ and $h(Y)$ are independent.
   (b) Show that two discrete random variables X and Y are independent if and only if $f_{X,Y}(x, y) = f_X(x)f_Y(y)$ for all $x, y \in \mathbb{R}$.
   (c) More generally, show that X and Y are independent if and only if $f_{X,Y}(x, y)$ can be factorized as the product $g(x)h(y)$ of a function of x alone and a function of y alone.

2. Show that if $\operatorname{var}(X) = 0$ then X is constant; that is, there exists $a \in \mathbb{R}$ such that $\mathbf{P}(X = a) = 1$. (First show that if $\mathbf{E}(X^2) = 0$ then $\mathbf{P}(X = 0) = 1$.)

3. (a) Let X be a discrete random variable and let $g: \mathbb{R} \to \mathbb{R}$. Show that

$$\mathbf{E}(g(X)) = \sum_{x} g(x)\mathbf{P}(X = x)$$

   if this sum exists.

(b) If X and Y are independent and $g, h: \mathbb{R} \to \mathbb{R}$, show that $\mathbf{E}(g(X)h(Y)) = \mathbf{E}(g(X))\mathbf{E}(h(Y))$ whenever these expectations exist.

4. Let $\Omega = \{\omega_1, \omega_2, \omega_3\}$, with $\mathbf{P}(\omega_1) = \mathbf{P}(\omega_2) = \mathbf{P}(\omega_3) = \frac{1}{3}$. Define $X, Y, Z: \Omega \to \mathbb{R}$ by

$$X(\omega_1) = 1, \quad X(\omega_2) = 2, \quad X(\omega_3) = 3$$

$$Y(\omega_1) = 2, \quad Y(\omega_2) = 3, \quad Y(\omega_3) = 1$$

$$Z(\omega_1) = 2, \quad Z(\omega_2) = 2, \quad Z(\omega_3) = 1.$$

Show that X and Y have the same mass functions. Find the mass functions of $X + Y$, XY, and X/Y. Find the conditional mass functions $f_{Y|Z}$ and $f_{Z|Y}$.

5. For what values of k and α is f a mass function, where

(a) $f(n) = k/\{n(n+1)\}$, $n = 1, 2, \ldots$,
(b) $f(n) = kn^{\alpha}$, $n = 1, 2, \ldots$ (*zeta* or *Zipf* distribution)?

6. Let X and Y be independent Poisson variables with parameters λ and μ. Show that

(a) $X + Y$ is Poisson, parameter $\lambda + \mu$,
(b) the conditional distribution of X, given $X + Y = n$, is binomial, and find its parameters.

7. If X is geometric, show that $\mathbf{P}(X = n + k \mid X > n) = \mathbf{P}(X = k)$ for $k, n \geqslant 1$. Why do you think that this is called the 'lack of memory' property? Does any other distribution on the positive integers have this property?

8. Show that the sum of two independent binomial variables, $B(m, p)$ and $B(n, p)$ respectively, is $B(m + n, p)$.

9. Let N be the number of heads occurring in n tosses of a biased coin. Write down the mass function of N in terms of the probability p of heads turning up on each toss. Prove and utilize the identity

$$\sum_i \binom{n}{2i} x^{2i} y^{n-2i} = \tfrac{1}{2}\{(x+y)^n + (y-x)^n\}$$

in order to calculate the probability p_n that N is even. Compare with Problem (1.8.20).

10. An urn contains N balls, b of which are blue and r $(= N - b)$ of which are red. A random sample of n balls is withdrawn without replacement from the urn. Show that the number B of blue balls in this sample has the mass function

$$\mathbf{P}(B = k) = \binom{b}{k}\binom{N-b}{n-k} \bigg/ \binom{N}{n}.$$

This is called the *hypergeometric distribution* with parameters N, b, and n. Show further that if N, b, and r approach ∞ in such a way that $b/N \to p$ and $r/N \to 1 - p$, then

$$\mathbf{P}(B = k) \to \binom{n}{k} p^k (1-p)^{n-k}.$$

You have shown that, for small n and large N, the distribution of B barely depends on whether or not the balls are replaced in the urn immediately after their withdrawal.

11. Let X and Y be independent $B(n, p)$ variables, and let $Z = X + Y$. Show that the conditional distribution of X given $Z = N$ is the hypergeometric distribution of Problem (10).

12. Suppose X and Y take values in $\{0, 1\}$, with joint mass function $f(x, y)$. Write $f(0, 0) = a$, $f(0, 1) = b$, $f(1, 0) = c$, $f(1, 1) = d$, and find necessary and sufficient conditions for X and Y to be

(a) uncorrelated, (b) independent.

13. (a) If X takes non-negative integer values show that

$$\mathbf{E}(X) = \sum_{n=0}^{\infty} \mathbf{P}(X > n).$$

(b) An urn contains b blue and r red balls. Balls are removed at random until the first blue ball is drawn. Show that the expected number drawn is $(b + r + 1)/(b + 1)$.
(c) The balls are replaced and then removed at random until all the remaining balls are of the same colour. Find the expected number remaining in the urn.

14. Let $X_1, X_2, \ldots, X_n$ be independent random variables, and suppose that X_k is Bernoulli with parameter p_k. Show that $Y = X_1 + X_2 + \cdots + X_n$ has mean and variance given by

$$\mathbf{E}(Y) = \sum_{1}^{n} p_k, \qquad \operatorname{var}(Y) = \sum_{1}^{n} p_k(1 - p_k).$$

Show that, for $\mathbf{E}(Y)$ fixed, $\operatorname{var}(Y)$ is a maximum when $p_1 = p_2 = \cdots = p_n$. That is to say, the variation in the sum is greatest when individuals are most alike. Is this contrary to intuition?

15. Let $\boldsymbol{X} = (X_1, X_2, \ldots, X_n)$ be a vector of random variables. The *covariance matrix* $\boldsymbol{V}(\boldsymbol{X})$ of $\boldsymbol{X}$ is defined to be the symmetric $n \times n$ matrix with entries $(v_{ij}: 1 \leqslant i, j \leqslant n)$ given by $v_{ij} = \operatorname{cov}(X_i, X_j)$. Show that $|\boldsymbol{V}(\boldsymbol{X})| = 0$ if and only if the X_i are linearly dependent with probability one, in that

$$\mathbf{P}(a_1X_1 + a_2X_2 + \cdots + a_nX_n = b) = 1$$

for some $\boldsymbol{a}$ and b. ($|\boldsymbol{V}|$ denotes the determinant of $\boldsymbol{V}$.)

16. Let X and Y be independent Bernoulli random variables with parameter $\frac{1}{2}$. Show that $X + Y$ and $|X - Y|$ are dependent though uncorrelated.

17. A secretary drops n matching pairs of letters and envelopes down the stairs, and then places the letters into the envelopes in a random order. Use indicators to show that the number X of correctly matched pairs has mean and variance 1 for all $n \geqslant 2$. Show that the mass function of X converges to a Poisson mass function as $n \to \infty$.

18. Let $X = (X_1, X_2, \dots, X_n)$ be a vector of independent random variables each having the Bernoulli distribution with parameter p. Let $f: \{0, 1\}^n \to \mathbb{R}$ be *increasing*, which is to say that $f(x) \leqslant f(y)$ whenever $x_i \leqslant y_i$ for each i.

(a) Let $e(p) = \mathbf{E}(f(X))$. Show that $e(p_1) \leqslant e(p_2)$ if $p_1 \leqslant p_2$.
(b) **FKG inequality.** Let f and g be increasing functions from $\{0, 1\}^n$ into $\mathbb{R}$. Show by induction on n that $\operatorname{cov}(f(X), g(X)) \geqslant 0$.

19. Let $R(p)$ be the reliability function of a network G with a given source and sink, each edge of which is working with probability p, and let A be the event that there is a working connection from source to sink. Show that

$$R(p) = \sum_{\omega} I_A(\omega) p^{N(\omega)}(1 - p)^{m - N(\omega)}$$

where ω is a typical realization of the network, $N(\omega)$ is the number of working edges of ω, and m is the total number of edges of G.

Deduce that $R'(p) = \operatorname{cov}(I_A, N)/\{p(1 - p)\}$, and hence that

$$\frac{R(p)(1 - R(p))}{p(1 - p)} \leqslant R'(p) \leqslant \sqrt{\frac{mR(p)(1 - R(p))}{p(1 - p)}}.$$

20. Let $R(p)$ be the reliability function of a network G, each edge of which is working with probability p.

(a) Show that $R(p_1 p_2) \leqslant R(p_1)R(p_2)$ if $0 \leqslant p_1, p_2 \leqslant 1$.
(b) Show that $R(p^\gamma) \leqslant R(p)^\gamma$ for all $0 \leqslant p \leqslant 1$ and $\gamma \geqslant 1$.

21. In a certain style of detective fiction, the sleuth is required to declare 'the criminal has the unusual characteristics . . .; find this person and you have your man'. Assume that any given individual has these unusual characteristics with probability 10^{-7} independently of all other individuals, and that the city in question contains 10^7 inhabitants. Calculate the expected number of such people in the city.

(a) Given that the police inspector finds such a person, what is the probability that there is at least one other?
(b) If the inspector finds two such people, what is the probability that there is at least one more?
(c) How many such people need be found before the inspector can be reasonably confident that he has found them all?
(d) For the given population, how improbable should the characteristics of the criminal be, in order that he (or she) be specified uniquely?

22. In 1710, J. Arbuthnot observed that male births had exceeded female births in London for 82 successive years. Arguing that the two sexes are equally likely, and 2^{-82} is very small, he attributed this run of masculinity to Divine Providence. Let us assume that each birth results in a girl with probability $p = 0.485$, and that the outcomes of different confinements are independent of each other. Ignoring the possibility of twins (and so on), show that the probability that girls outnumber boys in $2n$ live births is no greater than $\binom{2n}{n} p^n q^n \{q/(q - p)\}$, where $q = 1 - p$. Suppose that 20 000 children are born in each of 82 successive years. Show that the probability that boys outnumber girls every year is at least 0.99. You may need Stirling's formula.

23. Consider a symmetric random walk with an absorbing barrier at N and a reflecting barrier at 0 (so that, when the particle is at 0, it moves to 1 at the next step). Let $\alpha_k(j)$ be the probability that the particle, having started at k, visits 0 exactly j times before being absorbed at N; we make the convention that, if $k = 0$, then the starting point counts as one visit. Show that

$$\alpha_k(j) = \frac{N-k}{N^2}\left(1 - \frac{1}{N}\right)^{j-1}, \qquad j \geqslant 1, \quad 0 \leqslant k \leqslant N.$$

24. A coin is tossed repeatedly, heads turning up with probability p on each toss. Player A wins the game if heads appears at least m times before tails has appeared n times; otherwise player B wins the game. Find the probability that A wins the game. This is the 'problem of the points' of Exercise (3.9.13).

25. A coin is tossed repeatedly, heads appearing on each toss with probability p. A gambler starts with initial fortune k (where $0 < k < N$); he wins one point for each head and loses one point for each tail. If his fortune is ever 0 he is bankrupted, whilst if it ever reaches N he stops gambling to buy a Jaguar. Suppose that $p < \frac{1}{2}$. Show that the gambler can increase his chance of winning by doubling the stakes. You may assume that k and N are even.

What is the corresponding strategy if $p \geqslant \frac{1}{2}$?

26. A compulsive gambler is never satisfied. At each stage he wins £1 with probability p and loses £1 otherwise. Find the probability that he is ultimately bankrupted, having started with an initial fortune of £k.

27. Let $(X_n: n \geqslant 1)$ be independent, identically distributed random variables taking integer values. Let $S_0 = 0$, $S_n = \sum_{i=1}^{n} X_i$. The **range** R_n of $S_0, S_1, \ldots, S_n$ is the number of distinct values taken by the sequence. Show that $\mathbf{P}(R_n = R_{n-1} + 1) = \mathbf{P}(S_1 S_2 \cdots S_n \neq 0)$, and deduce that, as $n \to \infty$,

$$\frac{1}{n}\mathbf{E}(R_n) \to \mathbf{P}(S_k \neq 0 \text{ for all } k \geqslant 1).$$

Hence show that, for the simple random walk, $n^{-1}\mathbf{E}(R_n) \to |p - q|$ as $n \to \infty$.

28. **Arc sine law for maxima.** Consider a symmetric random walk S starting from the origin, and let $M_n = \max\{S_i: 0 \leqslant i \leqslant n\}$. Show that, for $i = 2k, 2k+1$, the probability that the walk reaches M_{2n} for the first time at time i equals $\frac{1}{2}\mathbf{P}(S_{2k} = 0)\mathbf{P}(S_{2n-2k} = 0)$.

29. Let S be a symmetric random walk with $S_0 = 0$, and let N_n be the number of points that have been visited by S exactly once up to time n. Show that $\mathbf{E}(N_n) = 2$.

30. **Family planning.** Consider the following fragment of verse entitled 'Note for the scientist'.

People who have three daughters try for more,
And then its fifty–fifty they'll have four,
Those with a son or sons will let things be,
Hence all these surplus women, QED.

(a) What do you think of the argument?
(b) Show that the mean number of children of either sex in a family whose fertile parents have followed this policy equals 1. (You should assume that each delivery yields exactly one child whose sex is equally likely to be male or female.) Discuss.

31. Let $p_1, p_2, \ldots$ denote the prime numbers, and let $N(1), N(2), \ldots$ be independent random variables, $N(i)$ having mass function $\mathbf{P}(N(i) = k) = (1 - \gamma_i)\gamma_i^k$ for $k \geqslant 0$, where $\gamma_i = p_i^{-\beta}$ for all i and some $\beta > 1$. Show that

$$M = \prod_{i=1}^{\infty} p_i^{N(i)}$$

is a random integer with mass function $\mathbf{P}(M = m) = Cm^{-\beta}$ for $m \geqslant 1$ (this is the 'Dirichlet distribution'), where C is a constant satisfying

$$C = \prod_{i=1}^{\infty} \left(1 - \frac{1}{p_i^{\beta}}\right) = \left(\sum_{m=1}^{\infty} \frac{1}{m^{\beta}}\right)^{-1}.$$

32. $N + 1$ plates are laid out around a circular dining table, and a hot cake is passed between them in the manner of a symmetric random walk: each time it arrives on a plate, it is tossed to one of the two neighbouring plates, each possibility having probability $\frac{1}{2}$. The game stops at the moment when the cake has visited every plate at least once. Show that, with the exception of the plate where the cake began, each plate has probability $1/N$ of being the last plate visited by the cake.

4 Continuous random variables

4.1 Probability density functions

Recall that a random variable X is *continuous* if its distribution function $F(x) = \mathbf{P}(X \leqslant x)$ can be written as†

(1) $$F(x) = \int_{-\infty}^{x} f(u)\,du$$

for some integrable $f: \mathbb{R} \to [0, \infty)$.

(2) **Definition.** f is called the **(probability) density function** of X.

The density function of F is not prescribed uniquely by (1) since two integrable functions which take identical values except at some specific point have the same integrals. However, if F is differentiable at u then we shall normally set $f(u) = F'(u)$. We write $f_X(u)$ to stress the role of X.

(3) **Example (2.3.4) revisited.** X and Y have density functions

$$f_X(x) = \begin{cases} (2\pi)^{-1} & \text{if} \quad 0 \leqslant x \leqslant 2\pi \\ 0 & \text{otherwise} \end{cases}$$

$$f_Y(y) = \begin{cases} y^{-\frac{1}{2}}/(4\pi) & \text{if} \quad 0 \leqslant y \leqslant 4\pi^2 \\ 0 & \text{otherwise.} \end{cases}$$ ●

These density functions are non-zero if and only if $x \in [0, 2\pi]$ and $y \in [0, 4\pi^2]$. In such cases in the future, we shall write simply $f_X(x) = (2\pi)^{-1}$ for $0 \leqslant x \leqslant 2\pi$, and similarly for f_Y, with the implicit implication that the functions in question equal zero elsewhere.

Continuous variables contrast starkly with discrete variables in that they satisfy $\mathbf{P}(X = x) = 0$ for all $x \in \mathbb{R}$; this may seem paradoxical since X needs to take *some* value. Very roughly speaking, the resolution of this paradox lies in the observation that there are uncountably many values which X can take; this number is so large that the probability of X taking any particular value cannot exceed zero.

† Never mind what type of integral this is, at this stage.

The numerical value $f(x)$ is *not* a probability. However, we can think of $f(x)\,dx$ as the element of probability $\mathbf{P}(x < X \leqslant x + dx)$, since

$$\mathbf{P}(x < X \leqslant x + dx) = F(x + dx) - F(x) \simeq f(x)\,dx.$$

From equation (1), the probability that X takes a value in the interval $[a, b]$ is

$$\mathbf{P}(a \leqslant X \leqslant b) = \int_a^b f(x)\,dx.$$

(Intuitively, to calculate this probability, we simply add up all the small elements of probability which contribute.) More generally, if B is a sufficiently nice subset of $\mathbb{R}$ (such as an interval, or the countable union of intervals, and so on), then it is reasonable to expect that

(4) $$\mathbf{P}(X \in B) = \int_B f(x)\,dx,$$

and indeed this turns out to be the case.

We have deliberately used the same letter f for mass functions and density functions since these functions perform exactly analogous tasks for the appropriate classes of random variables. In many cases proofs of results for discrete variables can be rewritten for continuous variables by replacing any summation sign by an integral sign, and any probability mass $f(x)$ by the corresponding element of probability $f(x)\,dx$.

(5) **Lemma.** *If X has density function f then*

(a) $\int_{-\infty}^{\infty} f(x)\,dx = 1$
(b) $\mathbf{P}(X = x) = 0$ *for all* $x \in \mathbb{R}$
(c) $\mathbf{P}(a \leqslant X \leqslant b) = \int_a^b f(x)\,dx$.

Proof. *Exercise.* ∎

Part (a) of the lemma characterizes those non-negative integrable functions which are density functions of some random variable.

We conclude this section with a technical note for the more critical reader. For what sets B is (4) meaningful, and why does (5a) characterize density functions? Let $\mathcal{J}$ be the collection of all open intervals in $\mathbb{R}$. By the discussion in Section 1.6, $\mathcal{J}$ can be extended to a unique smallest σ-field $\mathcal{B} = \sigma(\mathcal{J})$ which contains $\mathcal{J}$; $\mathcal{B}$ is called the *Borel σ-field* and contains *Borel sets.* Equation (4) holds for all $B \in \mathcal{B}$. Setting $\mathbf{P}_X(B) = \mathbf{P}(X \in B)$, we can check that $(\mathbb{R}, \mathcal{B}, \mathbf{P}_X)$ is a probability space. Secondly, suppose that $f\colon \mathbb{R} \to [0, \infty)$ is integrable and $\int_{-\infty}^{\infty} f(x)\,dx = 1$. For any $B \in \mathcal{B}$, we define

$$\mathbf{P}(B) = \int_B f(x)\,dx.$$

Then $(\mathbb{R}, \mathscr{B}, \mathbf{P})$ is a probability space and f is the density function of the identity random variable $X: \mathbb{R} \to \mathbb{R}$ given by $X(x) = x$ for any $x \in \mathbb{R}$. The assiduous reader will verify the steps of this argument for his or her own satisfaction (or see Clarke 1975, p. 53).

Exercises

(6) For what values of the parameters are the following functions probability density functions?

(a) $f(x) = C\{x(1-x)\}^{-\frac{1}{2}}, 0 < x < 1.$
(b) $f(x) = C\exp(-x - \mathrm{e}^{-x}), x \in \mathbb{R}.$
(c) $f(x) = C(1 + x^2)^{-m}, x \in \mathbb{R}.$

(7) Find the density function of $Y = aX$, where $a > 0$, in terms of the density function of X. Show that the continuous random variables X and $-X$ have the same distribution function if and only if $f_X(x) = f_X(-x)$ for all $x \in \mathbb{R}$.

(8) If f and g are density functions of random variables X and Y, show that $\alpha f + (1-\alpha)g$ is a density function for $0 \leqslant \alpha \leqslant 1$, and describe a random variable of which it is the density function.

4.2 Independence

This section contains the counterpart of Section 3.2 for continuous variables, though it contains a definition and theorem which hold for any pair of variables, regardless of their types (continuous, discrete, and so on). We cannot continue to define the independence of X and Y in terms of events like $\{X = x\}$ and $\{Y = y\}$, since these events have zero probability and are trivially independent.

(1) **Definition.** X and Y are called **independent** if

(2) $$\{X \leqslant x\} \text{ and } \{Y \leqslant y\} \text{ are independent events for all } x, y \in \mathbb{R}.$$

The reader should verify that discrete variables satisfy (2) if and only if they are independent in the sense of Section 3.2. (1) is the general definition of the independence of any two variables X and Y, regardless of their types. The following general result holds for the independence of functions of random variables. Let X and Y be random variables, and let $g, h: \mathbb{R} \to \mathbb{R}$. Then $g(X)$ and $h(Y)$ are functions which map Ω into $\mathbb{R}$ by

$$g(X)(\omega) = g(X(\omega)), \qquad h(Y)(\omega) = h(Y(\omega))$$

as in (3.2.3). Let us suppose that $g(X)$ and $h(Y)$ are random variables. (This holds if they are $\mathscr{F}$-measurable; it is true for instance if g and h are sufficiently smooth or regular by being, say, continuous or monotonic. The correct condition on g and h is actually that, for all Borel subsets B of $\mathbb{R}$, $g^{-1}(B)$ and $h^{-1}(B)$ are Borel sets also.) In the rest of this book, *we assume that any term of the form '$g(X)$', where g is a function and X is a random variable, is itself a random variable.*

(3) **Theorem.** *If X and Y are independent, then so are $g(X)$ and $h(Y)$.*

Move immediately to the next section unless you want to prove this.

Proof. Some readers may like to try to prove this on their second reading. The proof does not rely on any property like continuity. The key lies in the requirement (2.1.3) that random variables be $\mathscr{F}$-measurable, and in the observation that $g(X)$ is $\mathscr{F}$-measurable if $g: \mathbb{R} \to \mathbb{R}$ is *Borel measurable*, which is to say that $g^{-1}(B) \in \mathscr{B}$, the Borel σ-field, for all $B \in \mathscr{B}$. Complete the proof yourself (*exercise*). ■

Exercises

(4) I am selling my house, and have decided to accept the first offer exceeding £K. Assuming that offers are independent random variables with common distribution function F, find the expected number of offers received before I sell the house.

(5) Let X and Y be independent random variables with common distribution function F and density function f. Show that $V = \max\{X, Y\}$ has distribution function $\mathbf{P}(V \leqslant x) = F(x)^2$ and density function $f_V(x) = 2f(x)F(x)$, $x \in \mathbb{R}$. Find the density function of $U = \min\{X, Y\}$.

4.3 Expectation

The expectation of a discrete variable X is $\mathbf{E}X = \sum_x x\mathbf{P}(X = x)$. This is an average of the possible values of X, each value being weighted by its probability. For continuous variables, expectations are defined as integrals.

(1) **Definition.** The **expectation** of a continuous variable X with density function f is

$$\mathbf{E}X = \int_{-\infty}^{\infty} xf(x)\,\mathrm{d}x$$

whenever this integral exists.

There are various ways of defining the integral of a function $g: \mathbb{R} \to \mathbb{R}$, but it is not appropriate to explore this here. Note that usually we shall allow the existence of $\int g(x)\,\mathrm{d}x$ only if $\int |g(x)|\,\mathrm{d}x < \infty$.

(2) **Examples (2.3.4) and (4.1.3) revisited.** The random variables X and Y of these examples have mean values

$$\mathbf{E}(X) = \int_0^{2\pi} \frac{x}{2\pi}\,dx = \pi,$$

$$\mathbf{E}(Y) = \int_0^{4\pi^2} \frac{y^{\frac{1}{2}}}{4\pi}\,dy = \tfrac{4}{3}\pi^2. \qquad \bullet$$

Roughly speaking, the expectation operator $\mathbf{E}$ has the same properties for continuous variables as it has for discrete variables.

(3) **Theorem.** *If X and $g(X)$ are continuous random variables then*

$$\mathbf{E}(g(X)) = \int_{-\infty}^{\infty} g(x) f_X(x)\,dx.$$

We give a simple proof for the case when g takes only non-negative values, and we leave it to the reader to extend this to the general case. Our proof is a corollary of the next lemma.

(4) **Lemma.** *If X has density function f with $f(x) = 0$ when $x < 0$, and distribution function F, then*

$$\mathbf{E}X = \int_0^{\infty} [1 - F(x)]\,dx.$$

Proof.

$$\int_0^{\infty} [1 - F(x)]\,dx = \int_0^{\infty} \mathbf{P}(X > x)\,dx = \int_0^{\infty} \int_{y=x}^{\infty} f(y)\,dy\,dx.$$

Now change the order of integration in the last term. ∎

Proof of (3) when $g \geqslant 0$. By (4)

$$\mathbf{E}(g(X)) = \int_0^{\infty} \mathbf{P}(g(X) > x)\,dx = \int_0^{\infty} \left(\int_B f_X(y)\,dy \right) dx$$

where $B = \{y: g(y) > x\}$. We interchange the order of integration here to obtain

$$\mathbf{E}(g(X)) = \int_0^{\infty} \int_0^{g(y)} dx\, f_X(y)\,dy = \int_0^{\infty} g(y) f_X(y)\,dy. \qquad ∎$$

(5) **Example (2) continued.** Lemma (4) allows us to find $\mathbf{E}(Y)$ without calculating

f_Y, for

$$\mathbf{E}(Y) = \mathbf{E}(X^2) = \int_0^{2\pi} x^2 f_X(x)\,dx$$
$$= \int_0^{2\pi} \frac{x^2}{2\pi}\,dx = \tfrac{4}{3}\pi^2. \qquad \bullet$$

We were careful to describe many characteristics of discrete variables—such as moments, covariance, correlation, and linearity of $\mathbf{E}$ (see Sections 3.3 and 3.6)—in terms of the operator $\mathbf{E}$ itself. Exactly analogous discussion holds for continuous variables. We do not spell out the details here but only indicate some of the less obvious emendations required to establish these results. For example, (3.3.5) defines the kth moment of the discrete variable X to be

(6) $$m_k = \mathbf{E}(X^k);$$

we define the kth moment of a continuous variable X by the same equation. Of course, the moments of X may not exist since the integral

$$\mathbf{E}(X^k) = \int x^k f(x)\,dx$$

may not converge (see (4.4.7) for an instance of this).

Exercises

(7) For what values of α is $\mathbf{E}(|X|^\alpha)$ finite, if the density function of X is

(a) $f(x) = e^{-x}$ for $x \geqslant 0$,
(b) $f(x) = C(1 + x^2)^{-m}$ for $x \in \mathbb{R}$?

If α is not integral, then $\mathbf{E}(|X|^\alpha)$ is called the *fractional moment of order* α of X, whenever the expectation is well defined; see Exercise (3.3.18).

(8) Let $X_1, X_2, \ldots, X_n$ be independent identically distributed random variables for which $\mathbf{E}(X_1^{-1})$ exists. Show that, if $m \leqslant n$, then $\mathbf{E}(S_m/S_n) = m/n$, where $S_m = X_1 + X_2 + \cdots + X_m$.

(9) Let X be a non-negative random variable with density function f. Show that

$$\mathbf{E}(X^r) = \int_0^\infty r x^{r-1} \mathbf{P}(X > x)\,dx$$

for any $r \geqslant 1$ for which the expectation is finite.

4.4 Examples of continuous variables

Exercise. Sketch all the density and distribution functions which are discussed here, and complete all unfinished calculations of means and variances.

(1) **Uniform distribution.** X is *uniform* on $[a, b]$ if

$$F(x) = \begin{cases} 0 & \text{if } x \leqslant a \\ (x-a)/(b-a) & \text{if } a < x \leqslant b \\ 1 & \text{if } x > b. \end{cases}$$

Roughly speaking, X takes any value between a and b with equal probability. Example (2.3.4) describes a uniform variable X. ●

(2) **Exponential distribution.** X is *exponential* with parameter λ (>0), if

(3) $$F(x) = 1 - e^{-\lambda x}, \qquad x \geqslant 0.$$

This arises as the 'continuous limit' of the waiting time distribution of (3.5.5) and very often occurs in practice as a description of the time elapsing between unpredictable events (such as telephone calls, earthquakes, emissions of radioactive particles, and arrivals of buses, girls, and so on). Suppose, as in (3.5.5), that a sequence of Bernoulli trials is performed at time epochs $\delta, 2\delta, 3\delta, \ldots$ and let W be the waiting time for the first success. Then

$$\mathbf{P}(W > k\delta) = (1-p)^k \quad \text{and} \quad \mathbf{E}W = \delta/p.$$

Now fix a time t. By this time, roughly $k = t/\delta$ trials have been made. We shall let $\delta \downarrow 0$. In order that the limiting distribution $\lim_{\delta \downarrow 0} \mathbf{P}(W > t)$ be non-trivial, we shall need to assume that $p \downarrow 0$ also and p/δ approaches some positive constant λ. Then

$$\mathbf{P}(W > t) = \mathbf{P}\left(W > \left(\frac{t}{\delta}\right)\delta\right) \simeq (1 - \lambda\delta)^{t/\delta} \to e^{-\lambda t}$$

which yields (3).

The exponential distribution (3) has mean

$$\mathbf{E}X = \int_0^\infty [1 - F(x)]\, dx = \lambda^{-1}.$$

Further properties of the exponential distribution will be discussed in Section 4.7 and Problem (4.11.5); this distribution proves to be the cornerstone of the theory of Markov processes in continuous time which will be discussed later. ●

(4) **Normal distribution.** Probably the most important continuous distribution is the *normal* (or *Gaussian*) distribution, which has two parameters μ and σ^2 and density function

$$f(x) = \frac{1}{\sqrt{(2\pi\sigma^2)}} \exp\left(-\frac{(x-\mu)^2}{2\sigma^2}\right), \qquad -\infty < x < \infty.$$

It is denoted by $N(\mu, \sigma^2)$. If $\mu = 0$ and $\sigma^2 = 1$ then

$$f(x) = \frac{1}{\sqrt{(2\pi)}} \exp(-\tfrac{1}{2}x^2)$$

is the density of the *standard* normal distribution. It is an *exercise* in analysis (Problem (4.11.1)) to show that f satisfies (4.1.5a), and is indeed therefore a density function.

The normal distribution arises in many ways. In particular it can be obtained as a continuous limit of the binomial distribution $B(n, p)$ as $n \to \infty$ (this is the de Moivre–Laplace limit theorem). This result is a special case of the central limit theorem to be discussed in Chapter 5; it transpires that in many cases the sum of a large number of independent (or at least not too dependent) random variables is approximately normally distributed. The binomial random variable has this property because it is the sum of Bernoulli variables (see (3.5.2)).

Let X be $N(\mu, \sigma^2)$, where $\sigma > 0$, and let

(5) $$Y = (X - \mu)/\sigma.$$

For the distribution of Y,

$$\begin{aligned}\mathbf{P}(Y \leqslant y) &= \mathbf{P}((X - \mu)/\sigma \leqslant y) = \mathbf{P}(X \leqslant y\sigma + \mu)\\ &= \frac{1}{\sigma\sqrt{(2\pi)}} \int_{-\infty}^{y\sigma+\mu} \exp\left(-\frac{(x-\mu)^2}{2\sigma^2}\right) dx\\ &= \frac{1}{\sqrt{(2\pi)}} \int_{-\infty}^{y} \exp(-\tfrac{1}{2}v^2)\, dv \quad \text{by substituting } x = v\sigma + \mu.\end{aligned}$$

Thus Y is $N(0, 1)$. Routine integrations (see Problem (4.11.1)) show that

$$\mathbf{E}Y = 0, \qquad \operatorname{var}(Y) = 1,$$

and it follows immediately from (5), (3.3.8), and (3.3.11) that the mean and variance of the $N(\mu, \sigma^2)$ distribution are μ and σ^2 respectively, thus explaining the notation.

Traditionally we denote the distribution function of Y by Φ:

$$\Phi(y) = \mathbf{P}(Y \leqslant y) = \int_{-\infty}^{y} \frac{1}{\sqrt{(2\pi)}} \exp(-\tfrac{1}{2}v^2)\, dv.$$ ●

(6) **Gamma distribution.** X has the *gamma* distribution with parameters $\lambda, t > 0$, denoted† by $\Gamma(\lambda, t)$, if it has density

$$f(x) = \frac{1}{\Gamma(t)} \lambda^t x^{t-1} e^{-\lambda x}, \qquad x \geqslant 0.$$

† Do not confuse the order of the parameters. Some authors denote this distribution by $\Gamma(t, \lambda)$.

Here, $\Gamma(t)$ is the *gamma function*

$$\Gamma(t) = \int_0^\infty x^{t-1} e^{-x} \, dx.$$

If $t = 1$ then X is exponentially distributed with parameter λ. We remark that if $\lambda = \frac{1}{2}$, $t = \frac{1}{2}d$, for some integer d, then X is said to have the *chi-squared distribution* $\chi^2(d)$ with d degrees of freedom (see Problem (4.11.12)). ●

(7) **Cauchy distribution.** X is *Cauchy* if

$$f(x) = \frac{1}{\pi(1 + x^2)}, \qquad -\infty < x < \infty.$$

This distribution is notable for having no moments and for its frequent appearances in counter-examples (but see Problem (4.11.4)). ●

(8) **Beta distribution.** X is *beta*, parameters $a, b > 0$, if

$$f(x) = \frac{1}{B(a, b)} x^{a-1}(1 - x)^{b-1}, \qquad 0 \leqslant x \leqslant 1.$$

The constant

$$B(a, b) = \int_0^1 x^{a-1}(1 - x)^{b-1} \, dx$$

ensures that f has total integral equal to one. You may care to prove that

$$B(a, b) = \frac{\Gamma(a)\Gamma(b)}{\Gamma(a + b)}.$$

If $a = b = 1$ then X is uniform on $[0, 1]$. ●

(9) **Weibull distribution.** X is *Weibull*, parameters $\alpha, \beta > 0$, if

$$F(x) = 1 - \exp(-\alpha x^\beta), \qquad x \geqslant 0.$$

Differentiate to find that

$$f(x) = \alpha\beta x^{\beta-1} \exp(-\alpha x^\beta), \qquad x \geqslant 0.$$

Set $\beta = 1$ to obtain the exponential distribution. ●

Exercises

(10) Prove that $\Gamma(t) = (t - 1)\Gamma(t - 1)$ for $t > 1$, and deduce that $\Gamma(n) = (n - 1)!$ for $n = 1, 2, \ldots$. Show that $\Gamma(\frac{1}{2}) = \sqrt{\pi}$ and deduce a closed form for $\Gamma(n + \frac{1}{2})$ for $n = 0, 1, 2, \ldots$.

(11) Show, as claimed in (8) above, that the beta function satisfies $B(a, b) = \Gamma(a)\Gamma(b)/\Gamma(a+b)$.

(12) Let X have the uniform distribution on $[0, 1]$. For what function g does $Y = g(X)$ have the exponential distribution with parameter 1?

(13) Find the distribution function of a random variable X with the Cauchy distribution. For what values of α does $|X|$ have a finite (possibly fractional) moment of order α?

(14) **Log-normal distribution.** Let $Y = e^X$ where X is $N(0, 1)$. Find the density function of Y.

4.5 Dependence

Many interesting probabilistic statements about a pair X, Y of variables concern the way X and Y vary together as functions on the same domain Ω.

(1) **Definition.** The **joint distribution function** of X and Y is the function $F: \mathbb{R}^2 \to [0, 1]$ given by

$$F(x, y) = \mathbf{P}(X \leqslant x, Y \leqslant y).$$

If X and Y are continuous then we cannot talk of their joint mass function (see (3.6.2)) since this is identically zero. Instead we need another density function.

(2) **Definition.** X and Y are **(jointly) continuous** with **joint (probability) density function** $f: \mathbb{R}^2 \to [0, \infty)$ if

$$F(x, y) = \int_{-\infty}^{y} \int_{-\infty}^{x} f(u, v)\, du\, dv \quad \text{for each } x, y \in \mathbb{R}.$$

If F is sufficiently differentiable at the point (x, y), then we normally specify

$$f(x, y) = \frac{\partial^2}{\partial x\, \partial y} F(x, y).$$

The properties of joint distribution and density functions are very much the same as those of the corresponding functions of a single variable, and the reader is left to find them. We note the following facts. Let X and Y have joint distribution function F and joint density function f. (Sometimes we write $F_{X,Y}$ and $f_{X,Y}$ to stress the roles of X and Y.)

(3) **Probabilities.**

$$\mathbf{P}(a \leqslant X \leqslant b, c \leqslant Y \leqslant d) = F(b, d) - F(a, d) - F(b, c) + F(a, c)$$
$$= \int_{c}^{d} \int_{a}^{b} f(x, y)\, dx\, dy.$$

Think of $f(x, y)\, dx\, dy$ as the element of probability $\mathbf{P}(x < X \leqslant x + dx, y < Y \leqslant y + dy)$, so that if B is a sufficiently nice subset of $\mathbb{R}^2$ (such as a rectangle or a union of rectangles and so on) then

(4) $$\mathbf{P}((X, Y) \in B) = \iint_B f(x, y)\, dx\, dy.$$

We can think of (X, Y) as a point chosen randomly from the plane; then $\mathbf{P}((X, Y) \in B)$ is the probability that the outcome of this random choice lies in B.

(5) **Marginal distributions.** The *marginal distribution functions* of X and Y are

$$F_X(x) = \mathbf{P}(X \leqslant x) = F(x, \infty), \qquad F_Y(y) = \mathbf{P}(Y \leqslant y) = F(\infty, y)$$

($F(x, \infty)$ is shorthand for $\lim_{y \to \infty} F(x, y)$); now

$$F_X(x) = \int_{-\infty}^{x} \left(\int_{-\infty}^{\infty} f(u, y)\, dy \right) du$$

and it follows that the *marginal density function* of X is

$$f_X(x) = \int_{-\infty}^{\infty} f(x, y)\, dy.$$

Similarly, the *marginal density function* of Y is

$$f_Y(y) = \int_{-\infty}^{\infty} f(x, y)\, dx.$$

(6) **Expectation.** If $g: \mathbb{R}^2 \to \mathbb{R}$ is a sufficiently nice function (see the proof of (4.2.3) for an idea of what this means) then

$$\mathbf{E}(g(X, Y)) = \int_{-\infty}^{\infty} \int_{-\infty}^{\infty} g(x, y) f(x, y)\, dx\, dy;$$

in particular, setting $g(x, y) = ax + by$,

$$\mathbf{E}(aX + bY) = a\mathbf{E}X + b\mathbf{E}Y.$$

(7) **Independence.** X and Y are *independent* if and only if

$$F(x, y) = F_X(x) F_Y(y) \quad \text{for all } x, y \in \mathbb{R},$$

or if and only if

$$f(x, y) = f_X(x) f_Y(y)$$

whenever F is differentiable at (x, y) (see Problem (4.11.6) also) where f, f_X, f_Y are taken to be the appropriate derivatives of F, F_X, and F_Y.

(8) **Example. Buffon's needle.** A plane is ruled by the lines $y = n$ ($n = 0, \pm 1, \pm 2, \ldots$) and a needle of unit length is cast randomly on to the plane. What is the probability that it intersects some line? We suppose that the needle shows no preference for position or direction.

Solution. Let (X, Y) be the coordinates of the centre of the needle and let Θ be the angle, modulo π, made by the needle and the x-axis. Denote the distance from the needle's centre and the nearest line beneath it by Z $(= Y - \lfloor Y \rfloor$, where $\lfloor Y \rfloor$ is the greatest integer not greater than Y). We need to interpret the statement 'a needle is cast randomly', and do this by assuming that

(a) Z is uniformly distributed on $[0, 1]$, so that $f_Z(z) = 1$ if $0 \leqslant z \leqslant 1$,
(b) Θ is uniformly distributed on $[0, \pi]$, so that $f_\Theta(\theta) = 1/\pi$ if $0 \leqslant \theta \leqslant \pi$,
(c) Z and Θ are independent, so that $f_{Z,\Theta}(z, \theta) = f_Z(z) f_\Theta(\theta)$.

Thus the pair Z, Θ has joint density function

$$f(z, \theta) = \frac{1}{\pi} \quad \text{if} \quad 0 \leqslant z \leqslant 1, 0 \leqslant \theta \leqslant \pi.$$

Draw a diagram to see that an intersection occurs if and only if $(Z, \Theta) \in B$ where $B \subseteq [0, 1] \times [0, \pi]$ is given by

$$B = \{(z, \theta): z \leqslant \tfrac{1}{2} \sin \theta \text{ or } 1 - z \leqslant \tfrac{1}{2} \sin \theta\}.$$

Hence

$$\begin{aligned} \mathbf{P}(\text{intersection}) &= \iint_B f(z, \theta)\, dz\, d\theta \\ &= \frac{1}{\pi} \int_0^\pi \left(\int_0^{\frac{1}{2}\sin\theta} dz + \int_{1-\frac{1}{2}\sin\theta}^1 dz \right) d\theta \\ &= \frac{2}{\pi}. \end{aligned}$$

Buffon designed the experiment in order to estimate the numerical value of π. Try it if you have time. ●

(9) **Example. Bivariate normal distribution.** Let $f: \mathbb{R}^2 \to \mathbb{R}$ be given by

(10)
$$f(x, y) = \frac{1}{2\pi\sqrt{(1-\rho^2)}} \exp\left(-\frac{1}{2(1-\rho^2)} (x^2 - 2\rho xy + y^2) \right)$$

where ρ is a constant satisfying $-1 < \rho < 1$. Check that f is a joint density function by verifying that

$$f(x, y) \geqslant 0, \qquad \int_{-\infty}^{\infty} \int_{-\infty}^{\infty} f(x, y)\, dx\, dy = 1;$$

f is called the *standard bivariate normal* density function of some pair X and Y. Calculation of its marginals shows that X and Y are $N(0, 1)$ variables (*exercise*). Furthermore, the covariance

$$\text{cov}(X, Y) = \mathbf{E}(XY) - \mathbf{E}(X)\mathbf{E}(Y)$$

is given by

$$\operatorname{cov}(X, Y) = \int_{-\infty}^{\infty}\int_{-\infty}^{\infty} xyf(x, y)\,\mathrm{d}x\,\mathrm{d}y = \rho;$$

check this. Remember that independent variables are uncorrelated, but the converse is not true in general. In this case, however, if $\rho = 0$ then

$$f(x, y) = \left(\frac{1}{\sqrt{(2\pi)}}\exp(-\tfrac{1}{2}x^2)\right)\left(\frac{1}{\sqrt{(2\pi)}}\exp(-\tfrac{1}{2}y^2)\right) = f_X(x)f_Y(y)$$

and so X and Y are independent. We reach the following important conclusion. *Standard bivariate normal variables are independent if and only if they are uncorrelated.*

The general bivariate normal distribution is more complicated. We say that the pair X, Y has the bivariate normal distribution with means μ_1 and μ_2, variances σ_1^2 and σ_2^2, and correlation ρ if their joint density function is

$$f(x, y) = \frac{1}{2\pi\sigma_1\sigma_2\sqrt{(1-\rho^2)}}\exp[-\tfrac{1}{2}Q(x, y)]$$

where $\sigma_1, \sigma_2 > 0$ and Q is the following quadratic form:

$$Q(x, y) = \frac{1}{(1-\rho^2)}\left[\left(\frac{x-\mu_1}{\sigma_1}\right)^2 - 2\rho\left(\frac{x-\mu_1}{\sigma_1}\right)\left(\frac{y-\mu_2}{\sigma_2}\right) + \left(\frac{y-\mu_2}{\sigma_2}\right)^2\right].$$

Routine integrations (*exercise*) show that

(a) X is $N(\mu_1, \sigma_1^2)$ and Y is $N(\mu_2, \sigma_2^2)$,
(b) the correlation between X and Y is ρ,
(c) *X and Y are independent if and only if $\rho = 0$.*

Finally, here is a hint about calculating integrals associated with normal density functions. It is an analytical exercise (Problem (4.11.1)) to show that

$$\int_{-\infty}^{\infty}\exp(-\tfrac{1}{2}x^2)\,\mathrm{d}x = \sqrt{(2\pi)}$$

and hence that

$$f(x) = \frac{1}{\sqrt{(2\pi)}}\exp(-\tfrac{1}{2}x^2)$$

is indeed a density function. Similarly, a change of variables in the integral shows that the more general function

$$f(x) = \frac{1}{\sigma\sqrt{(2\pi)}}\exp\left[-\frac{1}{2}\left(\frac{x-\mu}{\sigma}\right)^2\right]$$

is itself a density function. This knowledge can often be used to shorten calculations. For example, let X and Y have joint density function given by (10). Then, by completing the square in the exponent of the integrand, we see that

$$\operatorname{cov}(X, Y) = \iint xyf(x, y)\,\mathrm{d}x\,\mathrm{d}y$$

$$= \int y \frac{1}{\sqrt{(2\pi)}} \exp(-\tfrac{1}{2}y^2)\left(\int xg(x, y)\,\mathrm{d}x\right)\mathrm{d}y$$

where

$$g(x, y) = \frac{1}{\sqrt{[2\pi(1-\rho^2)]}} \exp\left(-\frac{1}{2}\frac{(x-\rho y)^2}{(1-\rho^2)}\right)$$

is the density function of the $N(\rho y, 1-\rho^2)$ distribution. Therefore $\int xg(x, y)\,\mathrm{d}x$ is the mean, ρy, of this distribution, giving

$$\operatorname{cov}(X, Y) = \rho \int y^2 \frac{1}{\sqrt{(2\pi)}} \exp(-\tfrac{1}{2}y^2)\,\mathrm{d}y.$$

However, the integral here is, in turn, the variance of the $N(0, 1)$ distribution, and so

$$\operatorname{cov}(X, Y) = \rho$$

as was asserted previously. ●

(11) **Example.** Here is another example of how to manipulate density functions. Let X and Y have joint density function

$$f(x, y) = \frac{1}{y}\exp\left(-y - \frac{x}{y}\right), \qquad 0 < x, y < \infty.$$

Find the marginal density function of Y.

Solution.

$$f_Y(y) = \int_{-\infty}^{\infty} f(x, y)\,\mathrm{d}x$$

$$= \int_0^{\infty} \frac{1}{y}\exp\left(-y - \frac{x}{y}\right)\mathrm{d}x = \mathrm{e}^{-y}, \qquad y > 0,$$

and hence Y is exponentially distributed. ●

Following the final paragraph of Section 4.3, we should note that the expectation operator $\mathbf{E}$ has similar properties when applied to a family of continuous variables as when applied to discrete variables. Consider just one example of this.

(12) **Theorem. Cauchy–Schwarz inequality.** *For any pair X, Y of jointly continuous variables, we have that*

$$\{\mathbf{E}(XY)\}^2 \leqslant \mathbf{E}(X^2)\mathbf{E}(Y^2),$$

with equality if and only if $\mathbf{P}(aX = bY) = 1$ *for some real* a *and* b, *at least one of which is non-zero.*

Proof. Exactly as for (3.6.9). ■

Exercises

(13) Let

$$f(x, y) = \frac{|x|}{\sqrt{(8\pi)}} \exp\{-|x| - \tfrac{1}{2}x^2y^2\}, \qquad x, y \in \mathbb{R}.$$

Show that f is a continuous joint density function, but that the (first) marginal density function $g(x) = \int_{-\infty}^{\infty} f(x, y)\,dy$ is not continuous. Let $Q = \{q_n : n \geqslant 1\}$ be a set of real numbers, and define

$$f_Q(x, y) = \sum_{n=1}^{\infty} (\tfrac{1}{2})^n f(x - q_n, y).$$

Show that f_Q is a continuous joint density function whose first marginal density function is discontinuous at the points in Q. Can you construct a continuous joint density function whose first marginal density function is continuous nowhere?

(14) **Buffon's needle revisited.** Two grids of parallel lines are superimposed: the first grid contains lines distance a apart, and the second contains lines distance b apart which are perpendicular to those of the first set. A needle of length r ($<\min\{a, b\}$) is dropped at random. Show that the probability it intersects a line equals $r(2a + 2b - r)/(\pi ab)$.

(15) **Buffon's cross.** The plane is ruled by the lines $y = n$, for $n = 0, \pm 1, \dots$, and on to this plane we drop a cross formed by welding together two unit needles perpendicularly at their midpoints. Let Z be the number of intersections of the cross with the grid of parallel lines. Show that $\mathbf{E}(\tfrac{1}{2}Z) = 2/\pi$ and that

$$\text{var}(\tfrac{1}{2}Z) = \frac{3 - \sqrt{2}}{\pi} - \frac{4}{\pi^2}.$$

If you had the choice of using either a needle of unit length, or the cross, in estimating $2/\pi$, which would you use?

(16) Let X and Y be independent random variables each having the uniform distribution on $[0, 1]$. Let $U = \min\{X, Y\}$ and $V = \max\{X, Y\}$. Find $\mathbf{E}(U)$, and hence calculate $\text{cov}(U, V)$.

(17) Let X and Y be independent continuous random variables. Show that

$$\mathbf{E}(g(X)h(Y)) = \mathbf{E}(g(X))\mathbf{E}(h(Y)),$$

whenever these expectations exist. If X and Y have the exponential distribution with parameter 1, find $\mathbf{E}\{\exp[\tfrac{1}{2}(X + Y)]\}$.

4.6 Conditional distributions and conditional expectation

Suppose that X and Y have joint density function f. We wish to discuss the conditional distribution of Y given that X takes the value x. However, the probability $\mathbf{P}(Y \leqslant y \mid X = x)$ is undefined since (see (1.4.1)) we may only condition on events which have strictly positive probability. We proceed as follows. If $f_X(x) > 0$, then, by (4.5.4),

$$\mathbf{P}(Y \leqslant y \mid x \leqslant X \leqslant x + \mathrm{d}x) = \frac{\mathbf{P}(Y \leqslant y, x \leqslant X \leqslant x + \mathrm{d}x)}{\mathbf{P}(x \leqslant X \leqslant x + \mathrm{d}x)}$$

$$\simeq \frac{\displaystyle\int_{v=-\infty}^{y} f(x, v)\,\mathrm{d}x\,\mathrm{d}v}{f_X(x)\,\mathrm{d}x}$$

$$= \int_{v=-\infty}^{y} \frac{f(x, v)}{f_X(x)}\,\mathrm{d}v.$$

As $\mathrm{d}x \downarrow 0$, the left-hand side of this equation approaches our intuitive notion of the probability that $Y \leqslant y$ given $X = x$, and it is appropriate to make the following definition.

(1) **Definition.** The **conditional distribution function** of Y given $X = x$, written $F_{Y|X}(y \mid x)$ or $\mathbf{P}(Y \leqslant y \mid X = x)$, is defined to be

$$F_{Y|X}(y \mid x) = \int_{v=-\infty}^{y} \frac{f(x, v)}{f_X(x)}\,\mathrm{d}v$$

for any x such that $f_X(x) > 0$.

Remembering that distribution functions are integrals of density functions, we are led to the following definition.

(2) **Definition.** The **conditional density function** of $F_{Y|X}$, written $f_{Y|X}$, is given by

$$f_{Y|X}(y \mid x) = \frac{f(x, y)}{f_X(x)}$$

for any x such that $f_X(x) > 0$.

Of course,

$$f_X(x) = \int_{-\infty}^{\infty} f(x, y)\,\mathrm{d}y,$$

and so

$$f_{Y|X}(y \mid x) = \frac{f(x, y)}{\displaystyle\int_{-\infty}^{\infty} f(x, y)\,\mathrm{d}y}.$$

Definition (2) is easily remembered as $f_{Y|X} = f_{X,Y}/f_X$. Here is an example of a conditional density function in action.

(3) **Example.** Let X and Y have joint density function

$$f_{X,Y}(x, y) = \frac{1}{x}, \qquad 0 \leqslant y \leqslant x \leqslant 1.$$

Show for yourself (*exercise*) that

$$f_X(x) = 1 \quad \text{if} \quad 0 \leqslant x \leqslant 1, \quad f_{Y|X}(y \mid x) = \frac{1}{x} \quad \text{if} \quad 0 \leqslant y \leqslant x \leqslant 1,$$

which is to say that X is uniformly distributed on $[0, 1]$ and, conditional on the event $\{X = x\}$, Y is uniform on $[0, x]$. To calculate probabilities such as $\mathbf{P}(X^2 + Y^2 \leqslant 1 \mid X = x)$, say, we proceed as follows. If $x > 0$, define

$$A(x) = \{y \in \mathbb{R}: 0 \leqslant y \leqslant x, x^2 + y^2 \leqslant 1\};$$

clearly $A(x) = [0, \min\{x, (1 - x^2)^{\frac{1}{2}}\}]$. Also,

$$\begin{aligned} \mathbf{P}(X^2 + Y^2 \leqslant 1 \mid X = x) &= \int_{A(x)} f_{Y|X}(y \mid x)\, dy \\ &= \frac{1}{x} \min\{x, (1 - x^2)^{\frac{1}{2}}\} \\ &= \min\{1, (x^{-2} - 1)^{\frac{1}{2}}\}. \end{aligned}$$

Next, let us calculate $\mathbf{P}(X^2 + Y^2 \leqslant 1)$. Let

$$A = \{(x, y): 0 \leqslant y \leqslant x \leqslant 1, x^2 + y^2 \leqslant 1\}.$$

Then

(4)
$$\begin{aligned} \mathbf{P}(X^2 + Y^2 \leqslant 1) &= \iint_A f_{X,Y}(x, y)\, dx\, dy \\ &= \int_{x=0}^{1} f_X(x) \int_{y \in A(x)} f_{Y|X}(y \mid x)\, dy\, dx \\ &= \int_0^1 \min\{1, (x^{-2} - 1)^{\frac{1}{2}}\}\, dx = \log(1 + \sqrt{2}). \end{aligned}$$
●

From Definitions (1) and (2) it is easy to see that the *conditional expectation* of Y given X can be defined as in Section 3.7 by

$$\mathbf{E}(Y \mid X) = \psi(X)$$

where

$$\psi(x) = \mathbf{E}(Y \mid X = x) = \int_{-\infty}^{\infty} y f_{Y|X}(y \mid x)\, dy;$$

once again, $\mathbf{E}(Y \mid X)$ has the following important property.

(5) **Theorem.** *The conditional expectation* $\psi(X) = \mathbf{E}(Y \mid X)$ *satisfies*

$$\mathbf{E}(\psi(X)) = \mathbf{E}(Y).$$

We shall use this result repeatedly; it is normally written as

$$\mathbf{E}(\mathbf{E}(Y \mid X)) = \mathbf{E}(Y),$$

and it provides a useful method for calculating $\mathbf{E}(Y)$ since it asserts that

$$\mathbf{E}(Y) = \int_{-\infty}^{\infty} \mathbf{E}(Y \mid X = x) f_X(x) \, dx.$$

The proof of (5) proceeds exactly as for discrete variables (see (3.7.4)); indeed the theorem holds for all pairs of random variables, regardless of their types. For example, in the special case when X is continuous and Y is the discrete random variable I_B, the indicator function of an event B, the theorem asserts that

(6) $$\mathbf{P}(B) = \mathbf{E}(\psi(X)) = \int_{-\infty}^{\infty} \mathbf{P}(B \mid X = x) f_X(x) \, dx$$

(see equation (4) above for an application of (6)).

(7) **Example.** Let X and Y have the standard bivariate normal distribution of (4.5.9). Then

$$\begin{aligned} f_{Y|X}(y \mid x) &= f_{X,Y}(x, y)/f_X(x) \\ &= \frac{1}{\sqrt{[2\pi(1-\rho^2)]}} \exp\left(-\frac{(y-\rho x)^2}{2(1-\rho^2)}\right) \end{aligned}$$

is the density function of a $N(\rho x, 1-\rho^2)$ distribution. Thus $\mathbf{E}(Y \mid X = x) = \rho x$, giving that $\mathbf{E}(Y \mid X) = \rho X$. ●

(8) **Example.** Continuous and discrete variables have mean values, but what can we say about variables which are neither continuous nor discrete, such as X in Example (2.3.5)? In that example, let A be the event that a tail turns up. Then

$$\begin{aligned} \mathbf{E}(X) &= \mathbf{E}(\mathbf{E}(X \mid I_A)) \\ &= \mathbf{E}(X \mid I_A = 1)\mathbf{P}(I_A = 1) + \mathbf{E}(X \mid I_A = 0)\mathbf{P}(I_A = 0) \\ &= \mathbf{E}(X \mid \text{tail})\mathbf{P}(\text{tail}) + \mathbf{E}(X \mid \text{head})\mathbf{P}(\text{head}) \\ &= -1 \cdot q + \pi \cdot p = \pi p - q \end{aligned}$$

since X is uniformly distributed on $[0, 2\pi]$ if a head turns up. ●

(9) **Example (3) revisited.** Suppose, in the notation of (3), that we wish to calculate $\mathbf{E}(Y)$. Use (5) to obtain

$$\mathbf{E}(Y) = \int_0^1 \mathbf{E}(Y \mid X = x) f_X(x)\, dx$$

$$= \int_0^1 \tfrac{1}{2}x\, dx = \tfrac{1}{4}$$

since, conditional on $\{X = x\}$, Y is uniformly distributed on $[0, x]$. ●

There is a more general version of Theorem (5) which will be of interest later.

(10) **Theorem.** *The conditional expectation* $\psi(X) = \mathbf{E}(Y \mid X)$ *satisfies*

(11) $$\mathbf{E}(\psi(X)g(X)) = \mathbf{E}(Yg(X))$$

for any function g for which both expectations exist.

As in Section 3.7, we recapture Theorem (5) by setting $g(x) = 1$ for all x. We omit the proof, which is an elementary *exercise*. Conclusion (11) may be taken as a definition of the conditional expectation of Y given X, that is as a function $\psi(X)$ such that (11) holds for all appropriate functions g. We shall return to this discussion in later chapters.

Exercises

(12) A point is picked uniformly at random on the surface of a unit sphere. Writing Θ and Φ for its longitude and latitude, find the conditional density functions of Θ given Φ, and of Φ given Θ.

(13) Show that the conditional expectation $\psi(X) = \mathbf{E}(Y \mid X)$ of Y given X satisfies $\mathbf{E}(\psi(X)g(X)) = \mathbf{E}(Yg(X))$, for any function g for which both expectations exist.

(14) Construct an example of two random variables X and Y for which $\mathbf{E}(Y) = \infty$ but $\mathbf{E}(Y \mid X) < \infty$ (almost surely).

(15) Find the conditional density function and expectation of Y given X when they have joint density function

(a) $f(x, y) = \lambda^2 e^{-\lambda y}$ for $0 \leqslant x \leqslant y$,
(b) $f(x, y) = x e^{-x(y+1)}$ for $x, y \geqslant 0$.

4.7 Functions of random variables

Let X be a random variable with density function f, and let $g: \mathbb{R} \to \mathbb{R}$ be a sufficiently nice function (in the sense of the discussion after (4.2.3)). Then $Y = g(X)$ is a random variable also. To calculate the distribution of Y, proceed thus:†

$$\mathbf{P}(Y \leqslant y) = \mathbf{P}(g(X) \leqslant y) = \mathbf{P}(g(X) \in (-\infty, y])$$
$$= \mathbf{P}(X \in g^{-1}(-\infty, y]) = \int_{g^{-1}(-\infty, y]} f(x)\,\mathrm{d}x.$$

Example (2.3.4) contains an instance of this calculation, when $g(x) = x^2$.

(1) **Example.** Let X be $N(0, 1)$ and let $g(x) = x^2$. Then $Y = g(X) = X^2$ has distribution function

$$\mathbf{P}(Y \leqslant y) = \mathbf{P}(X^2 \leqslant y) = \mathbf{P}(-\sqrt{y} \leqslant X \leqslant +\sqrt{y})$$
$$= \Phi(\sqrt{y}) - \Phi(-\sqrt{y}) = 2\Phi(\sqrt{y}) - 1 \quad \text{if} \quad y \geqslant 0,$$

by the fact that $\Phi(x) = 1 - \Phi(-x)$. Differentiate to obtain

$$f_Y(y) = 2\frac{\mathrm{d}}{\mathrm{d}y}\Phi(\sqrt{y}) = y^{-\frac{1}{2}}\Phi'(\sqrt{y}) = \frac{1}{\sqrt{(2\pi y)}}\exp(-\tfrac{1}{2}y)$$

for $y \geqslant 0$. Compare with (4.4.6) to see that X^2 is $\Gamma(\frac{1}{2}, \frac{1}{2})$, or chi-squared with one degree of freedom. See Problem (4.11.12) also. ●

(2) **Example.** Let $g(x) = ax + b$ for fixed $a, b \in \mathbb{R}$. Then $Y = g(X) = aX + b$ has distribution function

$$\mathbf{P}(Y \leqslant y) = \mathbf{P}(aX + b \leqslant y) = \begin{cases} \mathbf{P}(X \leqslant (y - b)/a) & \text{if} \quad a > 0 \\ \mathbf{P}(X \geqslant (y - b)/a) & \text{if} \quad a < 0. \end{cases}$$

Differentiate to obtain $f_Y(y) = |a|^{-1} f_X((y - b)/a)$. ●

More generally, if X_1 and X_2 have joint density function f, and $g, h\colon \mathbb{R}^2 \to \mathbb{R}$, then what is the joint density function of the pair $Y_1 = g(X_1, X_2)$, $Y_2 = h(X_1, X_2)$? Recall how to change variables within an integral. Let $y_1 = y_1(x_1, x_2)$, $y_2 = y_2(x_1, x_2)$ be a one–one mapping $T\colon (x_1, x_2) \mapsto (y_1, y_2)$ taking some domain $D \subseteq \mathbb{R}^2$ onto some range $R \subseteq \mathbb{R}^2$. The transformation can be inverted as $x_1 = x_1(y_1, y_2)$, $x_2 = x_2(y_1, y_2)$; the *Jacobian* of this inverse is defined to be the determinant

$$J = \begin{vmatrix} \dfrac{\partial x_1}{\partial y_1} & \dfrac{\partial x_2}{\partial y_1} \\[2ex] \dfrac{\partial x_1}{\partial y_2} & \dfrac{\partial x_2}{\partial y_2} \end{vmatrix} = \frac{\partial x_1}{\partial y_1}\frac{\partial x_2}{\partial y_2} - \frac{\partial x_1}{\partial y_2}\frac{\partial x_2}{\partial y_1}$$

which we express as a function $J = J(y_1, y_2)$. We assume that these partial derivatives are continuous.

† If $A \subseteq \mathbb{R}$ then $g^{-1}(A) = \{x \in \mathbb{R}\colon g(x) \in A\}$.

(3) **Theorem.** *If $g: \mathbb{R}^2 \to \mathbb{R}$, and T maps the set $A \subseteq D$ onto the set $B \subseteq R$ then*

$$\iint_A g(x_1, x_2)\,dx_1\,dx_2 = \iint_B g(x_1(y_1, y_2), x_2(y_1, y_2))|J(y_1, y_2)|\,dy_1\,dy_2.$$

(4) **Corollary.** *If X_1, X_2 have joint density function f, then the pair Y_1, Y_2 given by $(Y_1, Y_2) = T(X_1, X_2)$ has joint density function*

$$f_{Y_1,Y_2}(y_1, y_2) = \begin{cases} f(x_1(y_1, y_2), x_2(y_1, y_2))|J(y_1, y_2)| & \\ & \text{if } (y_1, y_2) \text{ is in the range of } T \\ 0 & \text{otherwise.} \end{cases}$$

A similar result holds for mappings of $\mathbb{R}^n$ into $\mathbb{R}^n$. This technique is sometimes referred to as the method of *change of variables.*

Proof of Corollary. Let $A \subseteq D$, $B \subseteq R$ be typical sets such that $T(A) = B$. Then $(X_1, X_2) \in A$ if and only if $(Y_1, Y_2) \in B$. Thus

$$\begin{aligned} \mathbf{P}((Y_1, Y_2) \in B) = \mathbf{P}((X_1, X_2) \in A) &= \iint_A f(x_1, x_2)\,dx_1\,dx_2 \\ &= \iint_B f(x_1(y_1, y_2), x_2(y_1, y_2))|J(y_1, y_2)|\,dy_1\,dy_2 \end{aligned}$$

by (4.5.4) and (3). Compare this with the definition of the joint density function of Y_1 and Y_2,

$$\mathbf{P}((Y_1, Y_2) \in B) = \iint_B f_{Y_1,Y_2}(y_1, y_2)\,dy_1\,dy_2 \quad \text{for suitable sets } B \subseteq \mathbb{R}^2,$$

to obtain the result. ∎

(5) **Example.** Suppose that

$$X_1 = aY_1 + bY_2$$
$$X_2 = cY_1 + dY_2$$

and $ad - bc \neq 0$. Check that

$$f_{Y_1,Y_2}(y_1, y_2) = |ad - bc|\,f_{X_1,X_2}(ay_1 + by_2, cy_1 + dy_2). \quad \bullet$$

(6) **Example.** If X and Y have joint density function f, show that the density function of $U = XY$ is

$$f_U(u) = \int_{-\infty}^{\infty} f(x, u/x)\,|x|^{-1}\,dx.$$

Solution. Let T map (x, y) onto (u, v) by

$$u = xy, \qquad v = x.$$

The inverse T^{-1} maps (u, v) onto (x, y) by

$$x = v, \qquad y = u/v$$

and the Jacobian is

$$J(u, v) = \begin{vmatrix} \dfrac{\partial x}{\partial u} & \dfrac{\partial y}{\partial u} \\ \dfrac{\partial x}{\partial v} & \dfrac{\partial y}{\partial v} \end{vmatrix} = -v^{-1}.$$

Thus $f_{U,V}(u, v) = f(v, u/v)\,|v|^{-1}$. Integrate over v to obtain the result. ●

(7) **Example.** Let X_1 and X_2 be independent exponential variables, parameter λ. Find the joint density function of

$$Y_1 = X_1 + X_2, \qquad Y_2 = X_1/X_2$$

and show that they are independent.

Solution. Let T map (x_1, x_2) onto (y_1, y_2) by

$$y_1 = x_1 + x_2, \qquad y_2 = x_1/x_2, \qquad x_1, x_2, y_1, y_2 \geqslant 0.$$

The inverse T^{-1} maps (y_1, y_2) onto (x_1, x_2) by

$$x_1 = y_1 y_2/(1 + y_2), \qquad x_2 = y_1/(1 + y_2)$$

and the Jacobian is

$$J(y_1, y_2) = -y_1/(1 + y_2)^2,$$

giving

$$f_{Y_1, Y_2}(y_1, y_2) = f_{X_1, X_2}(y_1 y_2/(1 + y_2), y_1/(1 + y_2))\,|y_1|/(1 + y_2)^2.$$

However, X_1 and X_2 are independent and exponential, so that

$$\begin{aligned} f_{X_1, X_2}(x_1, x_2) &= f_{X_1}(x_1) f_{X_2}(x_2) \\ &= \lambda^2 \exp[-\lambda(x_1 + x_2)] \quad \text{if} \quad x_1, x_2 \geqslant 0. \end{aligned}$$

Thus

$$f_{Y_1, Y_2}(y_1, y_2) = \lambda^2 \exp(-\lambda y_1) y_1/(1 + y_2)^2 \quad \text{if} \quad y_1, y_2 \geqslant 0$$

is the joint density function of Y_1 and Y_2. However,

$$f_{Y_1, Y_2}(y_1, y_2) = [\lambda^2 y_1 \exp(-\lambda y_1)](1 + y_2)^{-2}$$

factorizes as the product of a function of y_1 and a function of y_2; therefore, by Problem (4.11.6) they are independent. Suitable normalization of the functions in this product gives

$$f_{Y_1}(y_1) = \lambda^2 y_1 \exp(-\lambda y_1),$$

$$f_{Y_2}(y_2) = (1 + y_2)^{-2}. \qquad \bullet$$

(8) **Example.** Let X_1 and X_2 be given by the previous example and let

$$X = X_1, \qquad S = X_1 + X_2.$$

By (4), X and S have joint density function

$$f(x, s) = \lambda^2 \mathrm{e}^{-\lambda s} \quad \text{if} \quad 0 \leqslant x \leqslant s.$$

This may look like the product of a function of x with a function of s, implying that X and S are independent; a glance at the domain of f shows this to be false. Suppose we know that $S = s$. What now is the conditional distribution of X, given $S = s$?

Solution.

$$\begin{aligned}\mathbf{P}(X \leqslant x \mid S = s) &= \int_{-\infty}^{x} f(u, s)\, \mathrm{d}u \Big/ \int_{-\infty}^{\infty} f(u, s)\, \mathrm{d}u \\ &= x\lambda^2 \mathrm{e}^{-\lambda s}/(s\lambda^2 \mathrm{e}^{-\lambda s}) \\ &= x/s \quad \text{if} \quad 0 \leqslant x \leqslant s.\end{aligned}$$

So, conditional on $S = s$, X is uniformly distributed on $[0, s]$. This result, and its later generalization, is of great interest to statisticians. ●

(9) **Example. A warning.** Let X_1 and X_2 be independent exponential variables (as in (7) and (8)). What is the conditional density function of $X_1 + X_2$ given $X_1 = X_2$?

'*Solution*' 1. Let $Y_1 = X_1 + X_2$ and $Y_2 = X_1/X_2$. Now $X_1 = X_2$ if and only if $Y_2 = 1$. We have from (7) that Y_1 and Y_2 are independent, and it follows that the conditional density function of Y_1 is its marginal density function

(10) $$f_{Y_1}(y_1) = \lambda^2 y_1 \exp(-\lambda y_1) \quad \text{for} \quad y_1 \geqslant 0.$$

'*Solution*' 2. Let $Y_1 = X_1 + X_2$ and $Y_3 = X_1 - X_2$. It is an *exercise* to show that $f_{Y_1, Y_3}(y_1, y_3) = \frac{1}{2}\lambda^2 \exp(-\lambda y_1)$ for $|y_3| \leqslant y_1$, and therefore the conditional density function of Y_1 given Y_3 is

$$f_{Y_1 \mid Y_3}(y_1 \mid y_3) = \lambda \exp[-\lambda(y_1 - |y_3|)] \quad \text{for} \quad |y_3| \leqslant y_1.$$

Now $X_1 = X_2$ if and only if $Y_3 = 0$, and the required conditional density function is therefore

(11) $$f_{Y_1 \mid Y_3}(y_1 \mid 0) = \lambda \exp(-\lambda y_1) \quad \text{for} \quad y_1 \geqslant 0.$$

Something is wrong: (10) and (11) are different. The error derives from the original question: what does it mean to condition on the event $\{X_1 = X_2\}$, an event having probability 0? As we have seen, the answer depends upon how we do the conditioning—one cannot condition on such events quite so blithely as one may on events having strictly positive probability. In solution 1, we are essentially conditioning on the event $\{X_1 \leqslant X_2 \leqslant (1+h)X_1\}$ for small h, whereas in solution 2 we are conditioning on $\{X_1 \leqslant X_2 \leqslant X_1 + h\}$; these two events contain different sets of information. ●

The technology above is satisfactory when the change of variables is one–one, but a problem can arise if the transformation is many–one. The simplest examples arise of course for one-dimensional transformations. For example, if $Y = X^2$ then the associated transformation $T: x \mapsto x^2$ is not one–one, since it loses the sign of x. This complication is easily dealt with for transformations which are piecewise one–one (and sufficiently smooth). For example, the above transformation T maps $(-\infty, 0)$ smoothly onto $(0, \infty)$ and similarly for $[0, \infty)$: there are two contributions to the density function of $Y = X^2$, one from each of the intervals $(-\infty, 0)$ and $[0, \infty)$. Arguing similarly but more generally, one arrives at the following conclusion, the proof of which is left as an *exercise*.

Let $I_1, I_2, \dots, I_n$ be intervals which partition $\mathbb{R}$ (it is not important whether or not these intervals contain their endpoints), and suppose that $Y = g(X)$ where, for each $1 \leqslant i \leqslant n$, g is strictly monotone and continuously differentiable on I_i. For each i, the function $g: I_i \to \mathbb{R}$ is invertible on $g(I_i)$, and we write h_i for the inverse function. Then

(12)
$$f_Y(y) = \sum_{i=1}^{n} f_X(h_i(y))|h_i'(y)|$$

with the convention that the ith summand is 0 if h_i is not defined at y. There is a natural extension of this formula to transformations in two and more dimensions.

Exercises

(13) Let X, Y, and Z be independent and uniformly distributed on $[0, 1]$. Find the joint density function of XY and Z^2, and show that $\mathbf{P}(XY < Z^2) = \frac{5}{9}$.

(14) Let X and Y be independent exponential random variables with parameter 1. Find the joint density function of $U = X + Y$ and $V = X/(X+Y)$, and deduce that V is uniformly distributed on $[0, 1]$.

(15) Let X be uniformly distributed on $[0, \frac{1}{2}\pi]$. Find the density function of $Y = \sin X$.

(16) Find the density function of $Y = \sin^{-1} X$ when

(a) X is uniformly distributed on $[0, 1]$,
(b) X is uniformly distributed on $[-1, 1]$.

(17) Let X and Y have the bivariate normal density function

$$f(x, y) = \frac{1}{2\pi\sqrt{(1-\rho^2)}} \exp\left\{-\frac{1}{2(1-\rho^2)}(x^2 - 2\rho xy + y^2)\right\}.$$

Show that X and $Z = (Y - \rho X)/(1-\rho^2)^{\frac{1}{2}}$ are independent $N(0, 1)$ variables, and deduce that

$$\mathbf{P}(X > 0, Y > 0) = \frac{1}{4} + \frac{1}{2\pi}\sin^{-1}\rho.$$

4.8 Sums of random variables

This section contains an important result which is a very simple application of the change of variable technique.

(1) **Theorem.** *If X and Y have joint density function f then $Z = X + Y$ has density function*

$$f_Z(z) = \int_{-\infty}^{\infty} f(x, z - x)\,dx.$$

Proof. Let $A = \{(x, y): x + y \leqslant z\}$. Then

$$\mathbf{P}(Z \leqslant z) = \iint_A f(u, v)\,du\,dv = \int_{u=-\infty}^{\infty}\int_{v=-\infty}^{z-u} f(u, v)\,dv\,du$$

$$= \int_{x=-\infty}^{\infty}\int_{y=-\infty}^{z} f(x, y - x)\,dy\,dx$$

by the substitution $x = u$, $y = v + u$. Reverse the order of integration to obtain the result. ■

If X and Y are independent, the result becomes

$$f_{X+Y}(z) = \int_{-\infty}^{\infty} f_X(x) f_Y(z - x)\,dx = \int_{-\infty}^{\infty} f_X(z - y) f_Y(y)\,dy.$$

f_{X+Y} is called the *convolution* of f_X and f_Y, and is written

(2) $$f_{X+Y} = f_X * f_Y.$$

(3) **Example.** Let X and Y be independent $N(0, 1)$ variables. Then $Z = X + Y$ has density function

$$f_Z(z) = \frac{1}{2\pi}\int_{-\infty}^{\infty} \exp[-\tfrac{1}{2}x^2 - \tfrac{1}{2}(z - x)^2]\,dx$$

$$= \frac{1}{2\sqrt{\pi}}\exp(-\tfrac{1}{4}z^2)\int_{-\infty}^{\infty}\frac{1}{\sqrt{(2\pi)}}\exp(-\tfrac{1}{2}v^2)\,dv$$

by the substitution $v = (x - \frac{1}{2}z)\sqrt{2}$. Therefore

$$f_Z(z) = \frac{1}{2\sqrt{\pi}}\exp(-\tfrac{1}{4}z^2)$$

showing that Z is $N(0, 2)$. More generally, if X is $N(\mu_1, \sigma_1^2)$ and Y is $N(\mu_2, \sigma_2^2)$, and X and Y are independent, then $Z = X + Y$ is $N(\mu_1 + \mu_2, \sigma_1^2 + \sigma_2^2)$. You should check this. ●

(4) **Example (4.6.3) revisited.** You must take great care in applying (1) when the domain of f depends on x and y. For example, in the notation of (4.6.3)

$$f_{X+Y}(z) = \int_A \frac{1}{x}\,dx, \qquad 0 \leqslant z \leqslant 2,$$

where $A = \{x\colon 0 \leqslant z - x \leqslant x \leqslant 1\} = [\frac{1}{2}z, \min\{z, 1\}]$. Thus

$$f_{X+Y}(z) = \begin{cases} \log 2 & 0 \leqslant z \leqslant 1 \\ \log(2/z) & 1 \leqslant z \leqslant 2. \end{cases}$$ ●

Exercises

(5) Let X and Y be independent variables having the exponential distribution with parameters λ and μ respectively. Find the density function of $X + Y$.

(6) Let X and Y be independent variables with the Cauchy distribution. Find the density function of $\alpha X + \beta Y$ where $\alpha\beta \neq 0$. (Do you know about contour integration?)

(7) Find the density function of $Z = X + Y$ when X and Y have joint density function $f(x, y) = \frac{1}{2}(x + y)\,e^{-(x+y)}$, $x, y \geqslant 0$.

4.9 Multivariate normal distribution

The centrepiece of the normal density function is the function $\exp(-x^2)$, and of the bivariate normal density function the function $\exp(-x^2 - bxy - y^2)$ for suitable b. Both cases feature a quadratic in the exponent, and there is a natural generalization to functions of n variables which is of great value in statistics. Roughly speaking, we say that $X_1, X_2, \ldots, X_n$ have the multivariate normal distribution if their joint density function is obtained by 'rescaling' the function $\exp(-\sum_i x_i^2 - 2\sum_{i<j} b_{ij}x_ix_j)$ of the n real variables $x_1, x_2, \ldots, x_n$. The exponent here is a 'quadratic form', but not all quadratic forms give rise to density functions. A *quadratic form* is a function $Q\colon \mathbb{R}^n \to \mathbb{R}$ of the form

(1) $$Q(\mathbf{x}) = \sum_{1 \leqslant i,j \leqslant n} a_{ij}x_ix_j = \mathbf{x}\mathbf{A}\mathbf{x}'$$

where $\mathbf{x} = (x_1, x_2, \ldots, x_n)$, $\mathbf{x}'$ is the transpose of $\mathbf{x}$, and $\mathbf{A} = (a_{ij})$ is a real symmetric matrix with non-zero determinant. A well-known theorem about diagonalizing matrices states that there exists an orthogonal matrix $\mathbf{B}$ such that

(2) $$\mathbf{A} = \mathbf{B}\mathbf{\Lambda}\mathbf{B}'$$

where $\mathbf{\Lambda}$ is the diagonal matrix with the eigenvalues $\lambda_1, \lambda_2, \ldots, \lambda_n$ of $\mathbf{A}$ on its diagonal. Substitute (2) into (1) to obtain

(3) $$Q(\mathbf{x}) = \mathbf{y}\mathbf{\Lambda}\mathbf{y}' = \sum_i \lambda_i y_i^2$$

where $\boldsymbol{y} = \boldsymbol{x}\boldsymbol{B}$. Q (respectively $\boldsymbol{A}$) is called a *positive definite quadratic form* (respectively *matrix*) if $Q(\boldsymbol{x}) > 0$ for all vectors $\boldsymbol{x}$ with some non-zero co-ordinate, and we write $Q > 0$ (respectively $\boldsymbol{A} > 0$) if this holds. From (3), $Q > 0$ if and only if $\lambda_i > 0$ for all i. This is all elementary matrix theory. We are concerned with the following question: when is the function $f: \mathbb{R}^n \to \mathbb{R}$ given by

$$f(\boldsymbol{x}) = K \exp(-\tfrac{1}{2}Q(\boldsymbol{x})), \qquad \boldsymbol{x} \in \mathbb{R}^n$$

the joint density function of some collection of n random variables? It is necessary and sufficient that

(a) $f(\boldsymbol{x}) \geqslant 0$ for all $\boldsymbol{x} \in \mathbb{R}^n$
(b) $\int_{\mathbb{R}^n} f(\boldsymbol{x})\, d\boldsymbol{x} = 1$

(this integral is shorthand for $\int \cdots \int f(x_1, \ldots, x_n)\, dx_1 \cdots dx_n$).

It is clear that (a) holds whenever $K > 0$. Next we investigate (b). First note that Q must be positive definite, since otherwise f has an infinite integral. If $Q > 0$,

$$\begin{aligned}\int_{\mathbb{R}^n} f(\boldsymbol{x})\, d\boldsymbol{x} &= \int_{\mathbb{R}^n} K \exp(-\tfrac{1}{2}Q(\boldsymbol{x}))\, d\boldsymbol{x}\\ &= \int_{\mathbb{R}^n} K \exp\left(-\tfrac{1}{2}\sum_i \lambda_i y_i^2\right) d\boldsymbol{y}\\ &\quad \text{by (4.7.3) and (3), since } |J| = 1 \text{ for orthogonal transformations}\\ &= K \prod_i \int_{-\infty}^{\infty} \exp(-\tfrac{1}{2}\lambda_i y_i^2)\, dy_i\\ &= K[(2\pi)^n/(\lambda_1\lambda_2 \cdots \lambda_n)]^{\frac{1}{2}} = K[(2\pi)^n/|\boldsymbol{A}|]^{\frac{1}{2}}\end{aligned}$$

where $|\boldsymbol{A}|$ denotes the determinant of $\boldsymbol{A}$. Hence (b) holds whenever $K = [(2\pi)^{-n}|\boldsymbol{A}|]^{\frac{1}{2}}$.

We have seen that

$$f(\boldsymbol{x}) = [(2\pi)^{-n}|\boldsymbol{A}|]^{\frac{1}{2}} \exp(-\tfrac{1}{2}\boldsymbol{x}\boldsymbol{A}\boldsymbol{x}'), \qquad \boldsymbol{x} \in \mathbb{R}^n,$$

is a joint density function if and only if $\boldsymbol{A}$ is positive definite. Suppose that $\boldsymbol{A} > 0$ and that $\boldsymbol{X} = (X_1, \ldots, X_n)$ is a sequence of variables with density function f. It is easy to see that each X_i has zero mean; just note that

$$f(\boldsymbol{x}) = f(-\boldsymbol{x})$$

and so $(X_1, \ldots, X_n)$ and $(-X_1, \ldots, -X_n)$ are identically distributed random vectors; however, $\mathbf{E}|X_i| < \infty$ and so $\mathbf{E}(X_i) = \mathbf{E}(-X_i)$, giving $\mathbf{E}(X_i) = 0$. $\boldsymbol{X}$ is said to have the *multivariate normal distribution* with zero means. More generally, if $\boldsymbol{Y} = (Y_1, \ldots, Y_n)$ is given by

$$\boldsymbol{Y} = \boldsymbol{X} + \boldsymbol{\mu}$$

for some vector $\boldsymbol{\mu} = (\mu_1, \dots, \mu_n)$ of constants, then $\boldsymbol{Y}$ is said to have the *multivariate normal distribution*.

(4) **Definition.** $\boldsymbol{X} = (X_1, \dots, X_n)$ has the **multivariate normal distribution** (or **multinormal distribution**), written $N(\boldsymbol{\mu}, \boldsymbol{V})$, if its joint density function is

$$f(\boldsymbol{x}) = [(2\pi)^n |\boldsymbol{V}|]^{-\frac{1}{2}} \exp[-\tfrac{1}{2}(\boldsymbol{x} - \boldsymbol{\mu})\boldsymbol{V}^{-1}(\boldsymbol{x} - \boldsymbol{\mu})']$$

where $\boldsymbol{V}$ is a positive definite symmetric matrix.

We have replaced $\boldsymbol{A}$ by $\boldsymbol{V}^{-1}$ in this definition. The reason for this is part (b) of the following theorem.

(5) **Theorem.** *If $\boldsymbol{X}$ is $N(\boldsymbol{\mu}, \boldsymbol{V})$ then*

(a) $\mathbf{E}(\boldsymbol{X}) = \boldsymbol{\mu}$, *which is to say that* $\mathbf{E}(X_i) = \mu_i$ *for all* i,
(b) $\boldsymbol{V} = (v_{ij})$ *is called the* ***covariance matrix***, *because* $v_{ij} = \text{cov}(X_i, X_j)$.

Proof. Part (a) follows by the argument before (4). Part (b) may be proved by performing an elementary integration, and more elegantly by the forthcoming method of characteristic functions; see (5.8.6). ■

We often write

$$\boldsymbol{V} = \mathbf{E}((\boldsymbol{X} - \boldsymbol{\mu})'(\boldsymbol{X} - \boldsymbol{\mu}))$$

since $(\boldsymbol{X} - \boldsymbol{\mu})'(\boldsymbol{X} - \boldsymbol{\mu})$ is a matrix with (i, j)th entry $(X_i - \mu_i)(X_j - \mu_j)$.

A very important property of this distribution is its invariance of type under linear changes of variables.

(6) **Theorem.** *If* $\boldsymbol{X} = (X_1, X_2, \dots, X_n)$ *is* $N(\boldsymbol{0}, \boldsymbol{V})$ *and* $\boldsymbol{Y} = (Y_1, Y_2, \dots, Y_m)$ *is given by* $\boldsymbol{Y} = \boldsymbol{XD}$ *for some matrix* $\boldsymbol{D}$ *of rank* $m \leqslant n$, *then* $\boldsymbol{Y}$ *is* $N(\boldsymbol{0}, \boldsymbol{D}'\boldsymbol{V}\boldsymbol{D})$.

Proof when $m = n$. The mapping $T\colon \boldsymbol{x} \mapsto \boldsymbol{y} = \boldsymbol{xD}$ is non-singular and can be inverted as $T^{-1}\colon \boldsymbol{y} \mapsto \boldsymbol{x} = \boldsymbol{yD}^{-1}$. Use this change of variables in (4.7.3) to show that, if $A, B \subseteq \mathbb{R}^n$ and $B = T(A)$, then

$$\begin{aligned}\mathbf{P}(\boldsymbol{Y} \in B) &= \int_A f(\boldsymbol{x})\,\mathrm{d}\boldsymbol{x} = \int_A [(2\pi)^n |\boldsymbol{V}|]^{-\frac{1}{2}} \exp(-\tfrac{1}{2}\boldsymbol{x}\boldsymbol{V}^{-1}\boldsymbol{x}')\,\mathrm{d}\boldsymbol{x} \\ &= \int_B [(2\pi)^n |\boldsymbol{W}|]^{-\frac{1}{2}} \exp(-\tfrac{1}{2}\boldsymbol{y}\boldsymbol{W}^{-1}\boldsymbol{y}')\,\mathrm{d}\boldsymbol{y}\end{aligned}$$

where $\boldsymbol{W} = \boldsymbol{D}'\boldsymbol{V}\boldsymbol{D}$ as required. The proof for values of m strictly smaller than n is more difficult and is omitted (but see Kingman and Taylor 1966, p. 372). ■

A similar result holds for linear transformations of $N(\boldsymbol{\mu}, \boldsymbol{V})$ variables.

There are various (essentially equivalent) ways of defining the multivariate normal distribution, of which the above way is perhaps neither the neatest nor the most useful. Here is another.

(7) **Definition.** The vector $\boldsymbol{X} = (X_1, X_2, \ldots, X_n)$ of random variables is said to have the **multivariate normal distribution** whenever, for all $\boldsymbol{a} \in \mathbb{R}^n$, $\boldsymbol{Xa}' = a_1X_1 + a_2X_2 + \cdots + a_nX_n$ has a normal distribution.

That is to say, $\boldsymbol{X}$ is multivariate normal if and only if every linear combination of the X_i is univariate normal. It is often easier to work with this definition, which differs in one important respect from the earlier one. Using (6), it is easy to see that vectors $\boldsymbol{X}$ satisfying (4) also satisfy (7). Definition (7) is, however, slightly more general than (4) as the following indicates. Suppose that $\boldsymbol{X}$ satisfies (7), and in addition there exists $\boldsymbol{a} \in \mathbb{R}^n$, $b \in \mathbb{R}$ such that $\boldsymbol{a} \neq \boldsymbol{0}$ and

(8) $$\mathbf{P}(\boldsymbol{Xa}' = b) = 1,$$

so that the X_i are linearly related; in this case there are strictly fewer than n 'degrees of freedom' in the vector $\boldsymbol{X}$, and we say that $\boldsymbol{X}$ has a *singular* multivariate normal distribution. It may be shown (see Exercise (5.8.14)) that, if $\boldsymbol{X}$ satisfies (7) and in addition its distribution is non-singular, then $\boldsymbol{X}$ satisfies (4) for appropriate $\boldsymbol{\mu}$ and $\boldsymbol{V}$. The singular case is, however, not covered by (4). If (8) holds, then $0 = \text{var}(\boldsymbol{Xa}') = \boldsymbol{aVa}'$, where $\boldsymbol{V}$ is the covariance matrix of $\boldsymbol{X}$. Hence $\boldsymbol{V}$ is a singular matrix, and therefore possesses no inverse. In particular, Definition (4) cannot apply.

Exercises

(9) A symmetric matrix is called *non-negative* (respectively *positive*) *definite* if its eigenvalues are non-negative (respectively strictly positive). Show that a non-negative definite symmetric matrix $\boldsymbol{V}$ has a square root, in that there exists a symmetric matrix $\boldsymbol{W}$ satisfying $\boldsymbol{W}^2 = \boldsymbol{V}$. Show further that $\boldsymbol{W}$ is non-singular if and only if $\boldsymbol{V}$ is positive definite.

(10) If $\boldsymbol{X}$ has the $N(\boldsymbol{\mu}, \boldsymbol{V})$ distribution where $\boldsymbol{V}$ is non-singular, show that $\boldsymbol{Y} = (\boldsymbol{X} - \boldsymbol{\mu})\boldsymbol{W}^{-1}$ has the $N(\boldsymbol{0}, \boldsymbol{I})$ distribution, where $\boldsymbol{I}$ is the identity matrix and $\boldsymbol{W}$ is a symmetric matrix satisfying $\boldsymbol{W}^2 = \boldsymbol{V}$. The random vector $\boldsymbol{Y}$ is said to have the *standard* multivariate normal distribution.

(11) Let $\boldsymbol{X} = (X_1, X_2, \ldots, X_n)$ have the $N(\boldsymbol{\mu}, \boldsymbol{V})$ distribution, and show that $Y = a_1X_1 + a_2X_2 + \cdots + a_nX_n$ has the (univariate) $N(\mu, \sigma^2)$ distribution where

$$\mu = \sum_{i=1}^{n} a_i \mathbf{E}(X_i), \qquad \sigma^2 = \sum_{i=1}^{n} a_i^2 \,\text{var}(X_i) + 2 \sum_{i<j} a_i a_j \,\text{cov}(X_i, X_j).$$

(12) Let X and Y have the bivariate normal distribution with zero means, unit variances, and correlation ρ. Find the joint density function of $X + Y$ and $X - Y$, and their marginal density functions.

(13) Let X have the $N(0, 1)$ distribution and let $a > 0$. Show that the random variable Y given by

$$Y = \begin{cases} X & \text{if } |X| < a \\ -X & \text{if } |X| \geqslant a \end{cases}$$

has the $N(0, 1)$ distribution, and find an expression for $\rho(a) = \text{cov}(X, Y)$ in terms of the density function ϕ of X. Does the pair (X, Y) have a bivariate normal distribution?

4.10 Distributions arising from the normal distribution

This section contains some distributional results which have applications in statistics. The reader may omit it without prejudicing his or her understanding of the rest of the book.

Statisticians are frequently faced with a collection $X_1, X_2, \ldots, X_n$ of random variables arising from a sequence of experiments. They might be prepared to make a general assumption about the unknown distributions of these variables without specifying the numerical values of certain parameters. Commonly they might suppose that $X_1, \ldots, X_n$ is a collection of independent $N(\mu, \sigma^2)$ variables for some fixed but unknown values for μ and σ^2; this assumption is often a very close approximation to reality. They might then proceed to estimate the values of μ and σ^2 by using functions of $X_1, \ldots, X_n$. For reasons which are explained in statistics textbooks, they will commonly use the *sample mean*

$$\bar{X} = \frac{1}{n} \sum_1^n X_i$$

as a guess at the value of μ, and the *sample variance*†

$$S^2 = \frac{1}{n-1} \sum_1^n (X_i - \bar{X})^2$$

as a guess at the value of σ^2; these at least have the property that $\mathbf{E}(\bar{X}) = \mu$ and $\mathbf{E}(S^2) = \sigma^2$. The pair $\bar{X}$, S^2 are related in a striking and important way.

(1) **Theorem.** *If $X_1, X_2, \ldots$ are independent $N(\mu, \sigma^2)$ variables then $\bar{X}$ and S^2 are independent. $\bar{X}$ is $N(\mu, \sigma^2/n)$ and $(n-1)S^2/\sigma^2$ is $\chi^2(n-1)$.*

Remember from (4.4.6) that $\chi^2(d)$ denotes the chi-squared distribution with d degrees of freedom.

Proof. Define $Y_i = (X_i - \mu)/\sigma$, and

$$\bar{Y} = \frac{1}{n} \sum_1^n Y_i = \frac{\bar{X} - \mu}{\sigma}.$$

From (4.4.5) Y_i is $N(0, 1)$, and clearly

$$\sum_1^n (Y_i - \bar{Y})^2 = (n-1)S^2/\sigma^2.$$

† In some texts the sample variance is defined with n in place of $(n-1)$.

The joint density function of $Y_1, \ldots, Y_n$ is

$$f(\mathbf{y}) = (2\pi)^{-\frac{1}{2}n} \exp\left(-\tfrac{1}{2}\sum_1^n y_i^2\right).$$

f has spherical symmetry in the sense that, if $A = (a_{ij})$ is an orthogonal rotation of $\mathbb{R}^n$ and

(2) $$Y_i = \sum_{j=1}^n Z_j a_{ji} \quad \text{and} \quad \sum_1^n Y_i^2 = \sum_1^n Z_i^2,$$

then $Z_1, Z_2, \ldots, Z_n$ are independent $N(0, 1)$ variables also. Now choose

(3) $$Z_1 = n^{-\frac{1}{2}} \sum_1^n Y_i = n^{\frac{1}{2}} \bar{Y}.$$

It is left to the reader to check that Z_1 is $N(0, 1)$. Then let $Z_2, \ldots, Z_n$ be any collection of variables such that (2) holds, where A is orthogonal. From (2) and (3)

(4) $$\begin{aligned}\sum_2^n Z_i^2 &= \sum_1^n Y_i^2 - \frac{1}{n}\left(\sum_1^n Y_i\right)^2 \\ &= \sum_1^n Y_i^2 - \frac{2}{n}\sum_{i=1}^n \sum_{j=1}^n Y_i Y_j + \frac{1}{n^2}\sum_{i=1}^n \left(\sum_{j=1}^n Y_j\right)^2 \\ &= \sum_{i=1}^n \left(Y_i - \frac{1}{n}\sum_1^n Y_j\right)^2 = \frac{(n-1)S^2}{\sigma^2}.\end{aligned}$$

Now, Z_1 is independent of $Z_2, \ldots, Z_n$, and so by (3) and (4), $\bar{Y}$ is independent of $(n-1)S^2/\sigma^2$. By (3) and (4.4.4), $\bar{Y}$ is $N(0, 1/n)$ and so $\bar{X}$ is $N(\mu, \sigma^2/n)$. Finally, $(n-1)S^2/\sigma^2$ is the sum of the squares of $n-1$ independent $N(0, 1)$ variables, and the result of Problem (4.11.12) completes the proof. ■

We may observe that σ is only a scaling factor for $\bar{X}$ and S $(=\sqrt{S^2})$. That is to say

$$U = \frac{n-1}{\sigma^2} S^2 \quad \text{is} \quad \chi^2(n-1)$$

which does not depend on σ, and

$$V = \frac{n^{\frac{1}{2}}}{\sigma}(\bar{X} - \mu) \quad \text{is} \quad N(0, 1)$$

which does not depend on σ. Hence the random variable

$$T = \frac{V}{[U/(n-1)]^{\frac{1}{2}}}$$

has a distribution which does not depend on σ. T is the ratio of two independent random variables, the numerator being $N(0, 1)$ and the denominator the square root of $(n-1)^{-1}$ times a $\chi^2(n-1)$ variable; T is said to have the *t distribution* with $n-1$ degrees of freedom, written $t(n-1)$. It is sometimes called 'Student's t distribution' in honour of a famous experimenter at the Guinness factory in Dublin. Let us calculate its density function. The joint density of U and V is

$$f(u, v) = \frac{(\frac{1}{2})^r e^{-\frac{1}{2}u} u^{\frac{1}{2}r-1}}{\Gamma(\frac{1}{2}r)} \frac{1}{\sqrt{(2\pi)}} \exp(-\tfrac{1}{2}v^2)$$

where $r = n-1$. Then map (u, v) onto (s, t) by

$$s = u, \qquad t = v(u/r)^{-\frac{1}{2}}.$$

Use (4.7.4) to obtain

$$f_{U,T}(s, t) = \left(\frac{s}{r}\right)^{\frac{1}{2}} f\left(s, t\left(\frac{s}{r}\right)^{\frac{1}{2}}\right)$$

and integrate over s to obtain

$$f_T(t) = \frac{\Gamma(\frac{1}{2}(r+1))}{(\pi r)^{\frac{1}{2}}\Gamma(\frac{1}{2}r)} \left(1 + \frac{t^2}{r}\right)^{-\frac{1}{2}(r+1)}, \qquad -\infty < t < \infty$$

as the density function of the $t(r)$ distribution.

Another important distribution in statistics is the F distribution which arises as follows. Let U and V be independent variables with the $\chi^2(r)$ and $\chi^2(s)$ distributions respectively. Then

$$F = \frac{U/r}{V/s}$$

is said to have the *F distribution* with r and s degrees of freedom, written $F(r, s)$. The following properties are obvious:

(a) F^{-1} is $F(s, r)$,
(b) T^2 is $F(1, r)$ if T is $t(r)$.

As an *exercise* in the techniques of Section 4.7, show that the density function of the $F(r, s)$ distribution is

$$f(x) = \frac{r\Gamma(\frac{1}{2}(r+s))}{s\Gamma(\frac{1}{2}r)\Gamma(\frac{1}{2}s)} \frac{(rx/s)^{\frac{1}{2}r-1}}{[1 + (rx/s)]^{\frac{1}{2}(r+s)}}, \qquad x > 0.$$

In Exercises (5.7.17,18) we shall encounter more general forms of the χ^2, t, and F distributions; these are the (so-called) 'non-central' versions of these distributions.

Exercises

(5) Let X_1 and X_2 be independent variables with the $\chi^2(m)$ and $\chi^2(n)$ distributions respectively. Show that $X_1 + X_2$ has the $\chi^2(m+n)$ distribution.

(6) Show that the mean of the $t(r)$ distribution is 0, and that the mean of the $F(r, s)$ distribution is $s/(s-2)$ if $s > 2$. What happens if $s \leqslant 2$?

(7) Show that the $t(1)$ distribution and the Cauchy distribution are the same.

(8) Let X and Y be independent variables having the exponential distribution with parameter 1. Show that X/Y has an F distribution, Which F distribution?

4.11 Problems

1. (a) Show that $\int_{-\infty}^{\infty} \exp(-x^2)\, dx = \sqrt{\pi}$, and deduce that

$$f(x) = \frac{1}{\sigma\sqrt{(2\pi)}} \exp\left\{-\frac{(x-\mu)^2}{2\sigma^2}\right\}, \qquad -\infty < x < \infty$$

is a density function if $\sigma > 0$.

(b) Calculate the mean and variance of a standard normal variable.

(c) Show that the $N(0, 1)$ distribution function Φ satisfies

$$(x^{-1} - x^{-3}) \exp(-\tfrac{1}{2}x^2) < (2\pi)^{\frac{1}{2}}\{1 - \Phi(x)\} < x^{-1} \exp(-\tfrac{1}{2}x^2),$$

for $x > 0$. These bounds are of interest because Φ has no closed form.

2. Let X be continuous with density function $f(x) = C(x - x^2)$, where $\alpha < x < \beta$ and $C > 0$.

(a) What are the possible values of α and β?
(b) What is C?

3. Let X be a random variable which takes non-negative values only. Let $A_i = \{i - 1 \leqslant X < i\}$ and show that

$$\sum_{i=1}^{\infty} (i-1)I_{A_i} \leqslant X < \sum_{i=1}^{\infty} iI_{A_i}.$$

Deduce directly that

$$\sum_{i=1}^{\infty} \mathbf{P}(X \geqslant i) \leqslant \mathbf{E}(X) < 1 + \sum_{i=1}^{\infty} \mathbf{P}(X \geqslant i).$$

4. (a) Let X have a continuous distribution function F. Show that
   (i) $F(X)$ is uniformly distributed on $[0, 1]$,
   (ii) $-\log F(X)$ is exponentially distributed.

(b) A straight line ℓ touches a circle with unit diameter at the point P which is diametrically opposed on the circle to another point Q. A straight line QR joins Q to some point R on ℓ. If the angle $\widehat{PQR}$ between the lines PQ and QR is a random variable with the uniform distribution on $[-\frac{1}{2}\pi, \frac{1}{2}\pi]$, show that the length of PR has the Cauchy distribution (this length is measured positive or negative depending upon which side of P the point R lies).

5. Let X have an exponential distribution. Show that $\mathbf{P}(X > s + x \mid X > s) = \mathbf{P}(X > x)$, for $x, s \geqslant 0$. This is the 'lack of memory' property again. Show that the exponential distribution is the only continuous distribution with this property. You may need to use the fact that the only non-negative monotonic solutions of the functional equation $g(s + t) = g(s)g(t)$ for $s, t \geqslant 0$, with $g(0) = 1$, are of the form $g(s) = e^{\mu s}$. Can you prove this?

6. Show that X and Y are independent continuous variables if and only if their joint density function f factorizes as the product $f(x, y) = g(x)h(y)$ of functions of the single variables x and y alone.

7. Let X and Y have joint density function $f(x, y) = 2\,\mathrm{e}^{-x-y}$, $0 < x < y < \infty$. Are they independent? Find their marginal density functions and their covariance.

8. **Bertrand's paradox.** A chord of the unit circle is picked at random. What is the probability that an equilateral triangle with the chord as base can fit inside the circle if

   (a) the perpendicular distance from the chord to the centre of the circle is uniform on $[0, 1]$?

   (b) the acute angle between the chord and a tangent at one of its endpoints is uniform on $[0, \frac{1}{2}\pi]$?

9. **Monte Carlo.** It is required to estimate $J = \int_0^1 g(x)\,\mathrm{d}x$ where $0 \leqslant g(x) \leqslant 1$ for all x, as in (2.6.3). Let X and Y be independent random variables with common density function $f(x) = 1$ if $0 < x < 1$, $f(x) = 0$ otherwise. Let $U = I_{\{Y \leqslant g(X)\}}$, the indicator function of the event that $Y \leqslant g(X)$, and let $V = g(X)$, $W = \frac{1}{2}\{g(X) + g(1 - X)\}$. Show that $\mathbf{E}(U) = \mathbf{E}(V) = \mathbf{E}(W) = J$, and that $\mathrm{var}(W) \leqslant \mathrm{var}(V) \leqslant \mathrm{var}(U)$, so that, of the three, W is the most 'efficient' estimator of J.

10. Let $X_1, X_2, \ldots, X_n$ be independent exponential variables, parameter λ. Show by induction that $S = X_1 + X_2 + \cdots + X_n$ has the $\Gamma(\lambda, n)$ distribution.

11. Let X and Y be independent variables, $\Gamma(\lambda, m)$ and $\Gamma(\lambda, n)$ respectively.

    (a) Use the result of (10) to show that $X + Y$ is $\Gamma(\lambda, m + n)$ when m and n are integral (the same conclusion is actually valid for non-integral m and n).

    (b) Find the joint density function of $X + Y$ and $X/(X + Y)$, and deduce that they are independent.

    (c) If Z is Poisson with parameter λt, and m is integral, show that $\mathbf{P}(Z < m) = \mathbf{P}(X > t)$.

    (d) If $0 < m < n$ and B is independent of Y with the beta distribution with parameters m and $n - m$, show that YB has the same distribution as X.

12. Let $X_1, X_2, \ldots, X_n$ be independent $N(0, 1)$ variables.

    (a) Show that X_1^2 is $\chi^2(1)$.

    (b) Show that $X_1^2 + X_2^2$ is $\chi^2(2)$ by expressing its distribution function as an integral and changing to polar co-ordinates.

    (c) More generally, show that $X_1^2 + X_2^2 + \cdots + X_n^2$ is $\chi^2(n)$.

13. Let X and Y have the bivariate normal distributions with means μ_1, μ_2, variances σ_1^2, σ_2^2, and correlation ρ. Show that

    (a) $\mathbf{E}(X \mid Y) = \mu_1 + \rho\sigma_1(Y - \mu_2)/\sigma_2$,

    (b) the variance of the conditional density function $f_{X|Y}$ is $\mathrm{var}(X \mid Y) = \sigma_1^2(1 - \rho^2)$.

14. Let X and Y have joint density function f. Find the density function of Y/X.

15. Let X and Y be independent variables with common density function f. Show that $\tan^{-1}(Y/X)$ has the uniform distribution on $(-\frac{1}{2}\pi, \frac{1}{2}\pi)$ if and only if

$$\int_{-\infty}^{\infty} f(x)f(xy)|x|\,\mathrm{d}x = \frac{1}{\pi(1 + y^2)}, \qquad y \in \mathbb{R}.$$

Verify that this is valid if either f is the $N(0, 1)$ density function or $f(x) = a(1 + x^4)^{-1}$ for some constant a.

16. Let X and Y be independent $N(0, 1)$ variables, and think of (X, Y) as a random point in the plane. Change to polar co-ordinates (R, Θ) given by $R^2 = X^2 + Y^2$, $\tan \Theta = Y/X$; show that R^2 is $\chi^2(2)$, $\tan \Theta$ has the Cauchy distribution, and R and Θ are independent. Find the density of R.
Find $\mathbf{E}(X^2/R^2)$ and

$$\mathbf{E}\left\{\frac{\min\{|X|, |Y|\}}{\max\{|X|, |Y|\}}\right\}.$$

17. If X and Y are independent random variables, show that $U = \min\{X, Y\}$ and $V = \max\{X, Y\}$ have distribution functions

$$F_U(u) = 1 - \{1 - F_X(u)\}\{1 - F_Y(u)\}, \qquad F_V(v) = F_X(v)F_Y(v).$$

Let X and Y be independent exponential variables, parameter 1. Show that

(a) U is exponential, parameter 2,
(b) V has the same distribution as $X + \frac{1}{2}Y$. Hence find the mean and variance of V.

18. Let X and Y be independent variables having the exponential distribution with parameters λ and μ respectively. Let $U = \min\{X, Y\}$, $V = \max\{X, Y\}$, and $W = V - U$.

(a) Find $\mathbf{P}(U = X) = \mathbf{P}(X \leqslant Y)$.
(b) Show that U and W are independent.

19. Let X and Y be independent non-negative random variables with continuous density functions on $(0, \infty)$.

(a) If, given $X + Y = u$, X is uniformly distributed on $[0, u]$ whatever the value of u, show that X and Y have the exponential distribution.
(b) If, given that $X + Y = u$, X/u has a given beta distribution (parameters α and β, say) whatever the value of u, show that X and Y have gamma distributions.

You may need the fact that the only non-negative continuous solutions of the functional equation $g(s + t) = g(s)g(t)$ for $s, t \geqslant 0$, with $g(0) = 1$, are of the form $g(s) = e^{\mu s}$. Remember Problem (5).

20. Show that it cannot be the case that $U = X + Y$ where U is uniformly distributed on $[0, 1]$ and X and Y are independent and identically distributed. You should not assume that X and Y are continuous variables.

21. **Order statistics.** Let $X_1, X_2, \ldots, X_n$ be independent identically distributed variables with a common density function f. Such a collection is called a *random sample.* For each $\omega \in \Omega$, arrange the sample values $X_1(\omega), \ldots, X_n(\omega)$ in non-decreasing order $X_{(1)}(\omega) \leqslant X_{(2)}(\omega) \leqslant \cdots \leqslant X_{(n)}(\omega)$, where $(1), (2), \ldots, (n)$ is a (random) permutation of $1, 2, \ldots, n$. The new variables $X_{(1)}, X_{(2)}, \ldots, X_{(n)}$ are called the *order statistics.* Show, by a symmetry argument, that the joint distribution function of the order

statistics satisfies

$$\begin{aligned}\mathbf{P}(X_{(1)} \leqslant y_1, \ldots, X_{(n)} \leqslant y_n) &= n!\,\mathbf{P}(X_1 \leqslant y_1, \ldots, X_n \leqslant y_n, X_1 < X_2 < \cdots < X_n) \\ &= \int \cdots \int_{x_1 \leqslant y_1, x_2 \leqslant y_2, \ldots, x_n \leqslant y_n} L(x_1, \ldots, x_n) n! f(x_1) \cdots f(x_n)\, dx_1 \cdots dx_n\end{aligned}$$

where L is given by

$$L(\mathbf{x}) = \begin{cases} 1 & \text{if } x_1 < x_2 < \cdots < x_n \\ 0 & \text{otherwise,} \end{cases}$$

and $\mathbf{x} = (x_1, x_2, \ldots, x_n)$. Deduce that the joint density function of $X_{(1)}, \ldots, X_{(n)}$ is $g(\mathbf{y}) = n!\,L(\mathbf{y}) f(y_1) \cdots f(y_n)$.

22. Find the marginal density function of the kth order statistic $X_{(k)}$ of a sample with size n

(a) by integrating the result of Problem (21),
(b) directly.

23. Find the joint density function of the order statistics of n independent uniform variables on $[0, T]$.

24. Let $X_1, X_2, \ldots, X_n$ be independent and uniformly distributed on $[0, 1]$, with order statistics $X_{(1)}, X_{(2)}, \ldots, X_{(n)}$.

(a) Show that, for fixed k, the density function of $nX_{(k)}$ converges as $n \to \infty$, and find and identify the limit function.
(b) Show that $\log X_{(k)}$ has the same distribution as $-\sum_{i=k}^{n} i^{-1} Y_i$, where the Y_i are independent random variables having the exponential distribution with parameter 1.
(c) Show that $Z_1, Z_2, \ldots, Z_n$, defined by $Z_k = (X_{(k)}/X_{(k+1)})^k$ for $k < n$ and $Z_n = (X_{(n)})^n$, are independent random variables with the uniform distribution on $[0, 1]$.

25. Let X_1, X_2, X_3 be independent variables with the uniform distribution on $[0, 1]$. What is the probability that rods of lengths X_1, X_2, and X_3 may be used to make a triangle? Generalize your answer to n rods used to form a polygon.

26. Let X_1 and X_2 be independent variables with the uniform distribution on $[0, 1]$. A stick of unit length is broken at points distance X_1 and X_2 from one of the ends. What is the probability that the three pieces may be used to make a triangle? Generalize your answer to a stick broken in n places.

27. Let X, Y be a pair of jointly continuous variables.

(a) **Hölder's inequality.** Show that if $p, q > 1$ and $p^{-1} + q^{-1} = 1$ then

$$\mathbf{E}|XY| \leqslant \{\mathbf{E}|X^p|\}^{1/p}\{\mathbf{E}|Y^q|\}^{1/q}.$$

Set $p = q = 2$ to deduce the Cauchy–Schwarz inequality (4.5.12).

(b) **Minkowski's inequality.** Show that, if $p \geqslant 1$, then

$$\{\mathbf{E}(|X + Y|^p)\}^{1/p} \leqslant \{\mathbf{E}|X^p|\}^{1/p} + \{\mathbf{E}|Y^p|\}^{1/p}.$$

Note that in both cases your proof need not depend on the continuity of X and Y; deduce that the same inequalities hold for all random variables.

28. Let Z be a random variable. Choose X and Y appropriately in the Cauchy–Schwarz (or Hölder) inequality to show that $g(p) = \log \mathbf{E}|Z^p|$ is a convex function of p on the interval of values of p such that $\mathbf{E}|Z^p| < \infty$. Deduce **Lyapunov's inequality**

$$\{\mathbf{E}|Z^r|\}^{1/r} \geqslant \{\mathbf{E}|Z^s|\}^{1/s} \quad \text{whenever} \quad r \geqslant s > 0.$$

You have shown in particular that, if Z has finite rth moment, then Z has finite sth moment for all positive $s \leqslant r$.

29. Show that, using the obvious notation, $\mathbf{E}\{\mathbf{E}(X \mid Y, Z) \mid Y\} = \mathbf{E}(X \mid Y)$.

30. Motor cars of unit length park randomly in a street in such a way that the centre of each car, in turn, is positioned uniformly at random in the space available to it. Let $m(x)$ be the expected number of cars which are able to park in a street of length x. Show that

$$m(x+1) = \frac{1}{x}\int_0^x \{m(y) + m(x-y) + 1\}\, dy.$$

It is possible to deduce that $m(x)$ is about as big as $\frac{3}{4}x$ when x is large.

31. **Buffon's needle revisited: Buffon's noodle**

(a) A plane is ruled by the lines $y = nd$ $(n = 0, \pm 1, \ldots)$. A needle with length L $(<d)$ is cast randomly on to the plane. Show that the probability that the needle intersects a line is $2L/(\pi d)$.
(b) Now fix the needle and let C be a circle with diameter d centred at the midpoint of the needle. Let λ be a line whose direction and distance from the centre of C are independent and uniformly distributed on $[0, 2\pi]$ and $[0, \frac{1}{2}d]$ respectively. This is equivalent to 'casting the ruled plane at random'. Show that the probability of an intersection between the needle and λ is $2L/(\pi d)$.
(c) Let S be a curve within C having finite length $L(S)$. Use indicators to show that the expected number of intersections between S and λ is $2L(S)/(\pi d)$.

This type of result is used in stereology, which seeks knowledge of the contents of a cell by studying its cross-sections.

32. **Buffon's needle ingested.** In the excitement of calculating π, Mr Buffon (no relation) inadvertently swallows the needle and is X-rayed. If the needle exhibits no preference for direction in the gut, what is the distribution of the length of its image on the X-ray plate? If he swallowed Buffon's cross (see Exercise (4.5.15)) also, what would be the joint distribution of the lengths of the images of the two arms of the cross?

33. Let $X_1, X_2, \ldots, X_n$ be independent exponential variables with parameter λ, and let $X_{(1)} \leqslant \cdots \leqslant X_{(n)}$ be their order statistics. Show that

$$Y_1 = nX_{(1)}, \qquad Y_r = (n+1-r)(X_{(r)} - X_{(r-1)}), \qquad 1 < r \leqslant n$$

are also independent and have the same joint distribution as the X_r.

34. Let $X_{(1)}, \ldots, X_{(n)}$ be the order statistics of a family of independent variables with common continuous distribution function F. Show that

$$Y_n = \{F(X_{(n)})\}^n, \qquad Y_r = \left\{\frac{F(X_{(r)})}{F(X_{(r+1)})}\right\}^r, \qquad 1 \leqslant r < n$$

are independent and uniformly distributed on [0, 1]. This is equivalent to Problem (33). Why?

5 Generating functions and their applications

5.1 Generating functions

A sequence $a = \{a_i : i = 0, 1, 2, \ldots\}$ of real numbers contains a lot of information. One concise way of storing this information is to wrap up the numbers together in a 'generating function'. For example, the (ordinary) *generating function* of the sequence a is the function G_a defined by

(1) $$G_a(s) = \sum_{i=0}^{\infty} a_i s^i \quad \text{for} \quad s \in \mathbb{R} \quad \text{for which the sum converges.}$$

The sequence a may in principle be reconstructed from the function G_a by setting $a_i = G_a^{(i)}(0)/i!$, where $f^{(i)}$ denotes the ith derivative of the function f. In many circumstances it is easier to work with the generating function G_a than with the original sequence a.

(2) **Example. De Moivre's theorem.** The sequence $a_n = (\cos\theta + \mathrm{i}\sin\theta)^n$ has generating function

$$G_a(s) = \sum_{n=0}^{\infty} [s(\cos\theta + \mathrm{i}\sin\theta)]^n = \frac{1}{1 - s(\cos\theta + \mathrm{i}\sin\theta)}$$

if $|s| < 1$; here $\mathrm{i} = \sqrt{(-1)}$. It is easily checked by examining the coefficient of s^n that

$$[1 - s(\cos\theta + \mathrm{i}\sin\theta)] \sum_{n=0}^{\infty} s^n[\cos(n\theta) + \mathrm{i}\sin(n\theta)] = 1$$

when $|s| < 1$. Thus

$$\sum_{n=0}^{\infty} s^n[\cos(n\theta) + \mathrm{i}\sin(n\theta)] = \frac{1}{1 - s(\cos\theta + \mathrm{i}\sin\theta)}$$

if $|s| < 1$. Equating the coefficients of s^n we obtain $\cos(n\theta) + \mathrm{i}\sin(n\theta) = (\cos\theta + \mathrm{i}\sin\theta)^n$. ●

There are several different types of generating function, of which G_a is perhaps the simplest. Another is the *exponential generating function* E_a

given by

(3) $$E_a(s) = \sum_{i=0}^{\infty} \frac{a_i s^i}{i!} \quad \text{for} \quad s \in \mathbb{R} \quad \text{for which the sum converges.}$$

Whilst such generating functions have many uses in mathematics, the ordinary generating function (1) is of most value when the a_i are probabilities. This is because 'convolutions' are common in probability theory, and (ordinary) generating functions provide an invaluable tool for studying them.

(4) **Convolution.** The *convolution* of the real sequences $a = \{a_i: i \geqslant 0\}$ and $b = \{b_i: i \geqslant 0\}$ is the sequence $c = \{c_i: i \geqslant 0\}$ defined by

(5) $$c_n = a_0 b_n + a_1 b_{n-1} + \cdots + a_n b_0;$$

we write $c = a * b$. If a and b have generating functions G_a and G_b, then the generating function of c is

(6) $$G_c(s) = \sum_{n=0}^{\infty} c_n s^n = \sum_{n=0}^{\infty} \left(\sum_{i=0}^{n} a_i b_{n-i} \right) s^n$$

$$= \sum_{i=0}^{\infty} a_i s^i \sum_{n=i}^{\infty} b_{n-i} s^{n-i} = G_a(s) G_b(s).$$

Thus we learn that, if $c = a * b$, then $G_c(s) = G_a(s)G_b(s)$; convolutions are numerically complicated operations, and it is often easier to work with generating functions.

(7) **Example.** The combinatorial identity $\sum_i \binom{n}{i}^2 = \binom{2n}{n}$ may be obtained as follows. The left-hand side is the convolution of the sequence $a_i = \binom{n}{i}$, $i = 0, 1, 2, \ldots$, with itself. However, $G_a(s) = \sum_i \binom{n}{i} s^i = (1 + s)^n$, so that

$$G_{a*a}(s) = G_a(s)^2 = (1 + s)^{2n} = \sum_i \binom{2n}{i} s^i.$$

Equating the coefficients of s^n yields the required identity. ●

(8) **Example.** Let X and Y be independent random variables having the Poisson distribution with parameters λ and μ respectively. What is the distribution of $Z = X + Y$?

We have from (3.8.2) that the mass function of Z is the convolution of the mass functions of X and Y: $f_Z = f_X * f_Y$. The generating function of the sequence $\{f_X(i): i \geq 0\}$ is

(9) $$G_X(s) = \sum_{i=0}^{\infty} \frac{\lambda^i e^{-\lambda}}{i!} s^i = \exp[\lambda(s-1)],$$

and similarly $G_Y(s) = \exp[\mu(s-1)]$. Hence the generating function G_Z of $\{f_Z(i): i \geq 0\}$ satisfies $G_Z(s) = G_X(s)G_Y(s) = \exp[(\lambda + \mu)(s-1)]$, which we recognize from (9) as the generating function of the Poisson mass function with parameter $\lambda + \mu$. ●

The last example is canonical: generating functions provide a basic technique for dealing with sums of independent random variables. With this example in mind, we make an important definition. Suppose that X is a discrete random variable taking values in the non-negative integers $\{0, 1, 2, \ldots\}$; its distribution is specified by the sequence of probabilities $f(i) = \mathbf{P}(X = i)$.

(10) **Definition.** The (**probability**) **generating function** of the random variable X is defined to be the generating function $G(s) = \mathbf{E}(s^X)$ of its probability mass function.

Note that G does indeed generate the sequence $\{f(i): i \geq 0\}$ since $\mathbf{E}(s^X) = \sum_i s^i \mathbf{P}(X = i) = \sum_i s^i f(i)$ by (3.3.3). We write G_X when we wish to stress the role of X. If X takes values in the non-negative integers, its generating function G_X converges at least when $|s| \leq 1$ and sometimes in a larger interval. Generating functions can be defined for random variables taking negative as well as positive integer values. Such generating functions generally converge for values of s satisfying $\alpha < |s| < \beta$ for some α, β such that $\alpha \leq 1 \leq \beta$. We shall make occasional use of such generating functions, but do not develop their theory systematically.

In advance of giving examples and applications of the method of generating functions, we recall some basic properties of power series. Let $G_a(s) = \sum_0^\infty a_i s^i$ where $a = \{a_i: i \geq 0\}$ is a real sequence.

(11) **Convergence.** There exists a *radius of convergence* R (≥ 0) such that the sum converges absolutely if $|s| < R$ and diverges if $|s| > R$. The sum is uniformly convergent on sets of the form $\{s: |s| \leq R'\}$ for any $R' < R$.

(12) **Differentiation.** $G_a(s)$ may be differentiated or integrated term by term any number of times when $|s| < R$.

(13) **Uniqueness.** If $G_a(s) = G_b(s)$ for $|s| < R'$ where $0 < R' \leq R$ then $a_n = b_n$ for all n. Furthermore

(14) $$a_n = \frac{1}{n!} G_a^{(n)}(0).$$

(15) **Abel's theorem.** If $a_i \geqslant 0$ for all i and $G_a(s)$ is finite for $|s| < 1$, then $\lim_{s\uparrow 1} G_a(s) = \sum_{i=0}^{\infty} a_i$, whether this sum is finite or equals $+\infty$. This standard result is useful when the radius of convergence R satisfies $R = 1$, since then one has no *a priori* right to take the limit as $s\uparrow 1$.

Returning to the discrete random variable X taking values in $\{0, 1, 2, \ldots\}$ we have that $G(s) = \sum_0^\infty s^i \mathbf{P}(X = i)$, so that

(16) $$G(0) = \mathbf{P}(X = 0), \qquad G(1) = 1.$$

In particular, the radius of convergence of a probability generating function is at least 1. Here are some examples of probability generating functions.

(17) **Examples.**

(a) **Constant variables.** If $\mathbf{P}(X = c) = 1$ then $G(s) = \mathbf{E}(s^X) = s^c$.

(b) **Bernoulli variables.** If $\mathbf{P}(X = 1) = p$ and $\mathbf{P}(X = 0) = 1 - p$ then

$$G(s) = \mathbf{E}(s^X) = (1 - p) + ps.$$

(c) **Geometric distribution.** If X is geometrically distributed with parameter $\frac{1}{2}$ then

$$G(s) = \mathbf{E}(s^X) = \sum_{k=1}^{\infty} s^k (\tfrac{1}{2})^k = \frac{s}{2 - s}.$$

(d) **Poisson distribution.** If X is Poisson distributed with parameter λ then

$$G(s) = \mathbf{E}(s^X) = \sum_{k=0}^{\infty} s^k \frac{\lambda^k}{k!} e^{-\lambda} = \exp[\lambda(s - 1)].$$ ●

Generating functions are useful when working with integer-valued random variables. Problems arise when random variables take negative or non-integer values. Later in this chapter we shall see how to construct another function, called a 'characteristic function', which is very closely related to G_X but which exists for all random variables regardless of their types.

There are two major applications of probability generating functions—in calculating moments, and in calculating the distributions of *sums* of independent random variables. We begin with moments.

(18) **Theorem.** *If X has generating function $G(s)$ then*

(a) $\mathbf{E}(X) = G'(1)$,
(b) *more generally*, $\mathbf{E}[X(X - 1)\cdots(X - k + 1)] = G^{(k)}(1)$.

Of course $G^{(k)}(1)$ is shorthand for $\lim_{s\uparrow 1} G^{(k)}(s)$ whenever the radius of convergence of G is 1. The quantity $\mathbf{E}[X(X - 1)\cdots(X - k + 1)]$ is known as the *kth factorial moment* of X.

Proof of (b). Take $s < 1$ and calculate the kth derivative of G to obtain

$$G^{(k)}(s) = \sum_i s^{i-k} i(i-1)\cdots(i-k+1)f(i)$$
$$= \mathbf{E}[s^{X-k}X(X-1)\cdots(X-k+1)].$$

Let $s \uparrow 1$ and use Abel's theorem (15) to obtain

$$G^{(k)}(s) \to \sum_i i(i-1)\cdots(i-k+1)f(i) = \mathbf{E}[X(X-1)\cdots(X-k+1)]. \quad \blacksquare$$

In order to calculate the variance of X in terms of G, we proceed as follows:

(19) $$\text{var}(X) = \mathbf{E}(X^2) - \mathbf{E}(X)^2 = \mathbf{E}(X(X-1) + X) - \mathbf{E}(X)^2$$
$$= \mathbf{E}(X(X-1)) + \mathbf{E}(X) - \mathbf{E}(X)^2 = G''(1) + G'(1) - G'(1)^2.$$

Exercise. Find the means and variances of the distributions in (17) by this method.

(20) **Example.** Recall the hypergeometric distribution (3.11.10) with mass function $f(k) = \binom{b}{k}\binom{N-b}{n-k}\Big/\binom{N}{n}$. Then $G(s) = \sum_k s^k f(k)$, which can be recognized as the coefficient of x^n in

$$Q(s, x) = (1 + sx)^b(1 + x)^{N-b}\Big/\binom{N}{n}.$$

Hence the mean $G'(1)$ is the coefficient of x^n in

$$\frac{\partial Q}{\partial s}(1, x) = xb(1 + x)^{N-1}\Big/\binom{N}{n}$$

and so $G'(1) = bn/N$. Now calculate the variance yourself. ●

If you are more interested in the moments of X than in its mass function, then you may prefer to work not with G_X but with the function M_X defined by $M_X(t) = G_X(e^t)$. This change of variable is convenient for the following reason. Expanding $M_X(t)$ as a power series in t, we obtain

(21) $$M_X(t) = \sum_{k=0}^{\infty} e^{tk}\mathbf{P}(X = k) = \sum_{k=0}^{\infty}\sum_{n=0}^{\infty} \frac{(tk)^n}{n!}\mathbf{P}(X = k)$$
$$= \sum_{n=0}^{\infty} \frac{t^n}{n!}\left(\sum_{k=0}^{\infty} k^n\mathbf{P}(X = k)\right) = \sum_{n=0}^{\infty} \frac{t^n}{n!}\mathbf{E}(X^n),$$

the exponential generating function of the moments $\mathbf{E}(X^0), \mathbf{E}(X^1), \ldots$ of X. The function M_X is called the *moment generating function* of the random variable X. We have assumed in (21) that the series in question converge. Some complications can arise in using moment generating functions unless the series $\sum_n t^n\mathbf{E}(X^n)/n!$ has a strictly positive radius of convergence.

(22) **Example.** We have from (9) that the moment generating function of the Poisson distribution with parameter λ is $M(t) = \exp[\lambda(e^t - 1)]$. ●

We turn next to sums and convolutions. Much of probability theory is concerned with sums of random variables. To study such a sum we need a useful way of describing its distribution in terms of the distributions of its summands, and generating functions prove to be an invaluable asset in this respect. The formula (3.8.1) for the mass function of the sum of two independent discrete variables, $\mathbf{P}(X + Y = z) = \sum_x \mathbf{P}(X = x)\mathbf{P}(Y = z - x)$, involves a complicated calculation; the corresponding generating functions provide a more economical way of specifying the distribution of this sum.

(23) **Theorem.** *If X and Y are independent then*

$$G_{X+Y}(s) = G_X(s)G_Y(s).$$

Proof. The direct way of doing this is to use (3.8.2) to find that $f_Z = f_X * f_Y$, so that the generating function of $\{f_Z(i): i \geqslant 0\}$ is the product of the generating functions of $\{f_X(i): i \geqslant 0\}$ and $\{f_Y(i): i \geqslant 0\}$, by (4). Alternatively, $g(X) = s^X$ and $h(Y) = s^Y$ are independent, by (3.2.3), and so $\mathbf{E}(g(X)h(Y)) = \mathbf{E}(g(X))\mathbf{E}(h(Y))$, as required. ■

(24) **Example. Binomial distribution.** Let $X_1, X_2, \ldots, X_n$ be independent Bernoulli variables, parameter p, with sum $S = X_1 + \cdots + X_n$. Each X_i has generating function $G(s) = qs^0 + ps^1 = q + ps$. Apply (23) repeatedly to find that the $B(n, p)$ variable S has generating function

$$G_S(s) = (G(s))^n = (q + ps)^n.$$

The sum $S_1 + S_2$ of two independent variables, $B(n, p)$ and $B(m, p)$ respectively, has generating function

$$G_{S_1+S_2}(s) = G_{S_1}(s)G_{S_2}(s) = (q + ps)^{m+n}$$

and is thus $B(m + n, p)$. This was Problem (3.11.8). ●

Theorem (23) tells us that the sum

$$S = X_1 + X_2 + \cdots + X_n$$

of independent variables taking values in the non-negative integers has generating function given by

$$G_S = G_{X_1}G_{X_2}\cdots G_{X_n}.$$

If n is itself the outcome of a random experiment then the answer is not quite so simple.

(25) **Theorem.** *If $X_1, X_2, \ldots$ is a sequence of independent identically distributed variables with common generating function G_X, and N ($\geqslant 0$) is a random variable which is independent of the X_i and has generating function G_N, then*

$$S = X_1 + X_2 + \cdots + X_N$$

has generating function given by

(26)
$$G_S(s) = G_N(G_X(s)).$$

This has many important applications, one of which we shall meet in Section 5.4. It is an example of a process known as 'compounding' with respect to a parameter. Formula (26) is easily remembered; possible confusion about the order in which the functions G_N and G_X are compounded is avoided by remembering that if $\mathbf{P}(N = n) = 1$ then $G_N(s) = s^n$ and $G_S(s) = (G_X(s))^n$. Incidentally, we adopt the usual convention that, in the case when $N = 0$, the sum $X_1 + X_2 + \cdots + X_N$ is the 'empty' sum, and equals 0 also.

Proof. Use conditional expectation and (3.7.4) to find that

$$\begin{aligned} G_S(s) &= \mathbf{E}(s^S) = \mathbf{E}(\mathbf{E}(s^S \mid N)) \\ &= \sum_n \mathbf{E}(s^S \mid N = n)\mathbf{P}(N = n) \\ &= \sum_n \mathbf{E}(s^{X_1 + \cdots + X_n})\mathbf{P}(N = n) \\ &= \sum_n \mathbf{E}(s^{X_1}) \cdots \mathbf{E}(s^{X_n})\mathbf{P}(N = n) \quad \text{by independence} \\ &= \sum_n (G_X(s))^n \mathbf{P}(N = n) = G_N(G_X(s)). \end{aligned}$$ ■

(27) **Example (3.7.5) revisited.** A hen lays N eggs, where N is Poisson distributed with parameter λ. Each egg hatches with probability p, independently of all other eggs. Let K be the number of chicks. Then

$$K = X_1 + X_2 + \cdots + X_N$$

where $X_1, X_2, \ldots$ are independent Bernoulli variables with parameter p. How is K distributed? Clearly

$$G_N(s) = \sum_{n=0}^{\infty} s^n \frac{\lambda^n}{n!} e^{-\lambda} = \exp[\lambda(s - 1)]$$

$$G_X(s) = q + ps$$

and so

$$G_K(s) = G_N(G_X(s)) = \exp[\lambda p(s - 1)]$$

which, by comparison with G_N, we see to be the generating function of a Poisson variable with parameter λp. ●

Just as information about a mass function can be encapsulated in a generating function, so may joint mass functions be similarly described.

(28) **Definition.** The **joint (probability) generating function** of variables X_1 and X_2 taking values in the non-negative integers is defined by

$$G_{X_1,X_2}(s_1, s_2) = \mathbf{E}(s_1^{X_1} s_2^{X_2}).$$

There is a similar definition for the joint generating function of an arbitrary family of random variables. Joint generating functions also have important uses, one of which is the following characterization of independence.

(29) **Theorem.** *X_1 and X_2 are independent if and only if*

$$G_{X_1,X_2}(s_1, s_2) = G_{X_1}(s_1)G_{X_2}(s_2) \quad \textit{for all } s_1 \textit{ and } s_2.$$

Proof. If X_1 and X_2 are independent then so are $g(X_1) = s_1^{X_1}$ and $h(X_2) = s_2^{X_2}$; then proceed as in the proof of (23). To prove the converse, equate the coefficients of terms like $s_1^i s_2^j$ to deduce after some manipulation that

$$\mathbf{P}(X_1 = i, X_2 = j) = \mathbf{P}(X_1 = i)\mathbf{P}(X_2 = j). \qquad \blacksquare$$

So far we have only considered random variables X which always take finite values, and consequently their generating functions G_X satisfy $G_X(1) = 1$. In the near future we shall encounter variables which can take the value $+\infty$ (see the first passage time T_0 of Section 5.3 for example). For such variables X we note that $G_X(s) = \mathbf{E}(s^X)$ converges so long as $|s| < 1$, and

(30) $$\lim_{s \uparrow 1} G_X(s) = \sum_k \mathbf{P}(X = k) = 1 - \mathbf{P}(X = \infty).$$

We can no longer find the moments of X in terms of G_X; of course, they all equal $+\infty$. If $\mathbf{P}(X = \infty) > 0$ then we say that X is 'defective' with defective distribution function F_X.

Exercises

(31) Find the generating functions of the following mass functions, and state where they converge. Hence calculate their means and variances.

(a) $f(m) = \binom{n+m-1}{m} p^n (1-p)^m$, for $m \geqslant 0$.
(b) $f(m) = \{m(m+1)\}^{-1}$, for $m \geqslant 1$.
(c) $f(m) = (1-p)p^{|m|}/(1+p)$, for $m = \ldots, -1, 0, 1, \ldots$.

The constant p satisfies $0 < p < 1$.

(32) Let X have probability generating function G and write $t(n) = \mathbf{P}(X > n)$ for the 'tail' probabilities of X. Show that the generating function of the sequence $\{t(n): n \geqslant 0\}$ is $T(s) = (1 - G(s))/(1 - s)$. Show that $\mathbf{E}(X) = T(1)$ and $\text{var}(X) = 2T'(1) + T(1) - T(1)^2$.

(33) Let $G_{X,Y}(s, t)$ be the joint probability generating function of X and Y. Show that $G_X(s) = G_{X,Y}(s, 1)$ and $G_Y(t) = G_{X,Y}(1, t)$. Show that

$$\mathbf{E}(XY) = \frac{\partial^2}{\partial s\,\partial t} G_{X,Y}(s, t)\bigg|_{s=t=1}$$

(34) Find the joint generating functions of the following joint mass functions, and state for what values of the variables the series converge.

(a) $f(j, k) = (1 - \alpha)(\beta - \alpha)\alpha^j\beta^{k-j-1}$, for $0 \leqslant k \leqslant j$, where $0 < \alpha < 1$, $\alpha < \beta$.
(b) $f(j, k) = (e - 1)\,e^{-(2k+1)}k^j/j!$, for $j, k \geqslant 0$.
(c) $f(j, k) = \binom{k}{j} p^{j+k}(1 - p)^{k-j}/[k \log\{1/(1 - p)\}]$, for $0 \leqslant j \leqslant k$, $k \geqslant 1$, where $0 < p < 1$.

Deduce the marginal probability generating functions and the covariances.

(35) A coin is tossed n times, and heads turns up with probability p on each toss. Assuming the usual independence, show that the joint probability generating function of the numbers H and T of heads and tails is $G_{H,T}(x, y) = \{px + (1 - p)y\}^n$. Generalize this conclusion to find the joint probability generating function of the multinomial distribution of Exercise (3.5.7).

5.2 Some applications

Generating functions provide a powerful tool, particularly in the presence of difference equations and convolutions. This section contains a variety of examples of this tool in action.

(1) **Example. Problem of the points.** A coin is tossed repeatedly and heads turns up with probability p on each toss. Player A wins if m heads appear before n tails, and player B wins otherwise. We have seen, in (3.9.13) and (3.11.24), two approaches to the problem of determining the probability that A wins. It is elementary, by conditioning on the outcome of the first toss, that the probability p_{mn}, that A wins, satisfies

(2)
$$p_{mn} = pp_{m-1,n} + qp_{m,n-1}, \quad \text{for} \quad m, n \geqslant 1,$$

where $p + q = 1$. The boundary conditions are $p_{m0} = 0$, $p_{0n} = 1$ for $m, n > 0$. We may solve (2) by introducing the generating function

$$G(x, y) = \sum_{m=0}^{\infty} \sum_{n=0}^{\infty} p_{mn} x^m y^n$$

subject to the convention that $p_{00} = 0$. Multiplying throughout (2) by $x^m y^n$ and summing over $m, n \geqslant 1$, we obtain

(3)
$$G(x, y) - \sum_{m=1}^{\infty} p_{m0} x^m - \sum_{n=1}^{\infty} p_{0n} y^n = px \sum_{m,n=1}^{\infty} p_{m-1,n} x^{m-1} y^n + qy \sum_{m,n=1}^{\infty} p_{m,n-1} x^m y^{n-1},$$

and hence, by the boundary conditions,

$$G(x, y) - \frac{y}{1-y} = pxG(x, y) + qy\left(G(x, y) - \frac{y}{1-y}\right), \quad |y| < 1.$$

Therefore,

(4)
$$G(x, y) = \frac{y(1-qy)}{(1-y)(1-px-qy)},$$

from which one may derive the required information by expanding in powers of x and y and finding the coefficient of $x^m y^n$. A cautionary note: in passing from (2) to (3), one should be very careful with the limits of the summations. ●

(5) **Example. Matching revisited.** The famous (mis)matching problem of (3.4.3) involves the random placing of n different letters into n differently addressed envelopes. What is the probability p_n that no letter is placed in the correct envelope? Let M be the event that the first letter is put into its correct envelope, and let N be the event that no match occurs. Then

(6)
$$p_n = \mathbf{P}(N) = \mathbf{P}(N \mid M^c)\mathbf{P}(M^c),$$

where $\mathbf{P}(M^c) = 1 - n^{-1}$. It is convenient to think of $\alpha_n = \mathbf{P}(N \mid M^c)$ in the following way. It is the probability that, given $n-2$ pairs of matching white letters and envelopes together with a non-matching red letter and blue envelope, there are no colour matches when the letters are inserted randomly into the envelopes. Either the red letter is placed into the blue envelope or it is not, and a consideration of these two cases gives that

(7)
$$\alpha_n = \frac{1}{n-1}p_{n-2} + \left(1 - \frac{1}{n-1}\right)\alpha_{n-1}.$$

Combining (6) and (7) we obtain, for $n \geqslant 3$,

(8)
$$p_n = \left(1 - \frac{1}{n}\right)\alpha_n = \left(1 - \frac{1}{n}\right)\left[\frac{1}{n-1}p_{n-2} + \left(1 - \frac{1}{n-1}\right)\alpha_{n-1}\right]$$
$$= \left(1 - \frac{1}{n}\right)\left(\frac{1}{n-1}p_{n-2} + p_{n-1}\right) = \frac{1}{n}p_{n-2} + \left(1 - \frac{1}{n}\right)p_{n-1},$$

a difference relation subject to the boundary condition $p_1 = 0$, $p_2 = \frac{1}{2}$. We may solve this difference relation by using the generating function

(9)
$$G(s) = \sum_{n=1}^{\infty} p_n s^n.$$

We multiply throughout (8) by ns^{n-1} and sum over all suitable values of n to obtain

$$\sum_{n=3}^{\infty} ns^{n-1}p_n = s\sum_{n=3}^{\infty} s^{n-2}p_{n-2} + s\sum_{n=3}^{\infty} (n-1)s^{n-2}p_{n-1}$$

which we recognize as yielding

$$G'(s) - p_1 - 2p_2 s = sG(s) + s[G'(s) - p_1]$$

or $(1-s)G'(s) = sG(s) + s$, since $p_1 = 0$ and $p_2 = \frac{1}{2}$. This differential equation is easily solved subject to the boundary condition $G(0) = 0$ to obtain $G(s) = (1-s)^{-1}e^{-s} - 1$. Expanding as a power series in s and comparing with (9), we arrive at the conclusion

(10) $$p_n = \frac{(-1)^n}{n!} + \frac{(-1)^{n-1}}{(n-1)!} + \cdots + \frac{(-1)}{1!} + 1, \quad \text{for} \quad n \geq 1,$$

as in the conclusion of (3.4.3) with $r = 0$. ●

(11) **Example. Matching and occupancy.** The matching problem above is one of the simplest of a class of problems involving putting objects randomly into containers. In a general approach to such questions, we suppose that we are given a collection $\mathcal{A} = \{A_i: 1 \leq i \leq n\}$ of events, and we ask for properties of the random number X of these events which occur (in the previous example, A_i is the event that the ith letter is in the correct envelope, and X is the number of correctly placed letters). The problem is to express the mass function of X in terms of probabilities of the form $\mathbf{P}(A_{i_1} \cap A_{i_2} \cap \cdots \cap A_{i_m})$. We introduce the notation

$$S_m = \sum_{i_1 < i_2 < \cdots < i_m} \mathbf{P}(A_{i_1} \cap A_{i_2} \cap \cdots \cap A_{i_m}),$$

the sum of the probabilities of the intersections of exactly m of the events in question. We make the convention that $S_0 = 1$. It is easily seen as follows that

(12) $$S_m = \mathbf{E}\binom{X}{m},$$

the mean value of the (random) binomial coefficient $\binom{X}{m}$: writing N_m for the number of sub-families of $\mathcal{A}$ having size m, all of whose component events occur, we have that

$$S_m = \sum_{i_1 < \cdots < i_m} \mathbf{E}(I_{A_{i_1}} I_{A_{i_2}} \cdots I_{A_{i_m}}) = \mathbf{E}(N_m),$$

whereas $N_m = \binom{X}{m}$. It follows from (12) that

(13) $$S_m = \sum_{i=0}^{n} \binom{i}{m} \mathbf{P}(X = i).$$

We introduce the generating functions

$$G_S(x) = \sum_{m=0}^{n} x^m S_m, \qquad G_X(x) = \sum_{i=0}^{n} x^i \mathbf{P}(X = i),$$

and we then multiply throughout (13) by x^m and sum over m, obtaining

$$G_S(x) = \sum_i \mathbf{P}(X = i) \sum_m x^m \binom{i}{m} = \sum_i (1 + x)^i \mathbf{P}(X = i) = G_X(1 + x).$$

Hence $G_X(x) = G_S(x - 1)$, and equating coefficients of x^i yields

(14) $$\mathbf{P}(X = i) = \sum_{j=i}^{n} (-1)^{j-i} \binom{j}{i} S_j, \quad \text{for} \quad 0 \leqslant i \leqslant n.$$

This formula, sometimes known as 'Waring's theorem', is a complete generalization of certain earlier results, including (10). It may be derived without using generating functions, but at considerable personal cost. ●

(15) **Example. Recurrent events.** Meteorites fall from the sky, your car runs out of fuel, there is a power failure, you fall ill. Each such event recurs at regular or irregular intervals; one cannot generally predict just when such an event will happen next, but one may be prepared to hazard guesses. A simplistic mathematical model is the following. We call the happening in question H, and suppose that at each time point $1, 2, \ldots$, either H occurs or H does not occur. We write X_1 for the first time at which H occurs, $X_1 = \min\{n: H$ occurs at time $n\}$, and X_m for the time which elapses between the $(m - 1)$th and mth occurrence of H. Thus the mth occurrence of H takes place at time

(16) $$T_m = X_1 + X_2 + \cdots + X_m.$$

Here are our main assumptions. We assume that the 'inter-occurrence' times $X_1, X_2, \ldots$ are independent random variables taking values in $\{1, 2, \ldots\}$, and furthermore that $X_2, X_3, \ldots$ are identically distributed. That is to say, whilst we assume that *inter*-occurrence times are independent and identically distributed, we allow the time to the *first* occurrence to have a special distribution.

Given the distributions of the X_i, how may we calculate the probability that H occurs at some given time? Define $u_n = \mathbf{P}(H$ occurs at time $n)$. We have by conditioning on X_1 that

(17) $$u_n = \sum_{i=1}^{n} \mathbf{P}(H_n \mid X_1 = i)\mathbf{P}(X_1 = i),$$

where H_n is the event that H occurs at time n. Now $\mathbf{P}(H_n \mid X_1 = i) = \mathbf{P}(H_{n-i+1} \mid X_1 = 1) = \mathbf{P}(H_{n-i+1} \mid H_1)$, using the 'translation invariance' entailed by the assumption that the X_i, $i \geqslant 2$, are independent and identically distributed. A similar conditioning on X_2 yields

(18) $$\begin{aligned}\mathbf{P}(H_m \mid H_1) &= \sum_{j=1}^{m-1} \mathbf{P}(H_m \mid H_1, X_2 = j)\mathbf{P}(X_2 = j)\\ &= \sum_{j=1}^{m-1} \mathbf{P}(H_{m-j} \mid H_1)\mathbf{P}(X_2 = j)\end{aligned}$$

for $m \geq 2$, by translation invariance once again. Multiplying through (18) by x^{m-1} and summing over m, we obtain

(19) $$\sum_{m=2}^{\infty} x^{m-1}\mathbf{P}(H_m \mid H_1) = \mathbf{E}(x^{X_2}) \sum_{n=1}^{\infty} x^{n-1}\mathbf{P}(H_n \mid H_1),$$

so that $G_H(x) = \sum_{m=1}^{\infty} x^{m-1}\mathbf{P}(H_m \mid H_1)$ satisfies $G_H(x) - 1 = F(x)G_H(x)$, where $F(x)$ is the common probability generating function of the inter-occurrence times, and hence

(20) $$G_H(x) = \frac{1}{1 - F(x)}.$$

Returning to (17), we obtain similarly that $U(x) = \sum_{n=1}^{\infty} x^n u_n$ satisfies

(21) $$U(x) = D(x)G_H(x) = \frac{D(x)}{1 - F(x)}$$

where $D(x)$ is the probability generating function of X_1. Equation (21) contains much of the information relevant to the process, since it relates the occurrences of H to the generating functions of the elements of the sequence $X_1, X_2, \ldots$. We should like to extract information out of (21) about $u_n = \mathbf{P}(H_n)$, the coefficient of x^n in $U(x)$, particularly for large values of n.

In principle, one may expand $D(x)/[1 - F(x)]$ as a polynomial in x in order to find u_n, but this is difficult in practice. There is one special situation in which this may be done with ease, and this is the situation when $D(x)$ is the function $D = D^*$ given by

(22) $$D^*(x) = \frac{1 - F(x)}{\mu(1 - x)} \quad \text{for} \quad |x| < 1,$$

and $\mu = \mathbf{E}(X_2)$ is the mean inter-occurrence time. Let us first check that D^* is indeed a suitable probability generating function. The coefficient of x^n in D^* is easily seen to be $(1 - f_1 - f_2 - \cdots - f_n)/\mu$, where $f_i = \mathbf{P}(X_2 = i)$. This coefficient is non-negative since the f_i form a mass function; furthermore, by L'Hôpital's rule,

$$D^*(1) = \lim_{x \uparrow 1} \frac{1 - F(x)}{\mu(1 - x)} = \lim_{x \uparrow 1} \frac{-F'(x)}{-\mu} = 1$$

since $F'(1) = \mu$, the mean inter-occurrence time. Hence $D^*(x)$ is indeed a probability generating function, and with this choice for D we obtain that $U = U^*$ where

(23) $$U^*(x) = \frac{1}{\mu(1 - x)}$$

from (21). Writing $U^*(x) = \sum_n u_n^* x^n$ we find that $u_n^* = \mu^{-1}$ for all n. That is to say, for the special choice of D^*, the corresponding sequence of the u_n^* is *constant*, so that the density of occurrences of H is constant as time passes. This special process is called a *stationary* recurrent-event process.

How relevant is the choice of D to the behaviour of u_n for large n? Intuitively speaking, the choice of distribution of X_1 should not affect greatly the behaviour of the process over long time periods, and so one might expect that $u_n \to \mu^{-1}$ as $n \to \infty$, irrespective of the choice of D. This is indeed the case, so long as we rule out the possibility that there is 'periodicity' in the process. We call the process *non-arithmetic* if $\gcd\{n: \mathbf{P}(X_2 = n) > 0\} = 1$; certainly the process is non-arithmetic if, for example, $\mathbf{P}(X_2 = 1) > 0$. Note that gcd stands for greatest common divisor.

(24) **Renewal theorem.** *If the mean inter-occurrence time μ is finite and the process is non-arithmetic, then $u_n = \mathbf{P}(H_n)$ satisfies $u_n \to \mu^{-1}$ as $n \to \infty$.*

Sketch proof. The classical proof of this theorem is a purely analytical approach to the equation (21) (see Feller 1971, pp. 335–8). There is a much neater probabilistic proof using the technique of 'coupling'. We do not give a complete proof at this stage, but merely a sketch. The main idea is to introduce a second recurrent-event process, which is stationary and independent of the first. Let $X = (X_i: i \geqslant 1)$ be the first and inter- occurrence times of the original process, and let $X^* = (X_i^*: i \geqslant 1)$ be another sequence of independent random variables, independent of X, such that $X_2^*, X_3^*, \ldots$ have the common distribution of $X_2, X_3, \ldots$, and X_1^* has probability generating function D^*. Let H_n and H_n^* be the events that H occurs at time n in the first and second process (respectively), and let $T = \min\{n: H_n \cap H_n^* \text{ occurs}\}$ be the earliest time at which H occurs simultaneously in both processes. It may be shown that $T < \infty$ with probability 1, using the assumptions that $\mu < \infty$ and that the processes are non-arithmetic; it is intuitively natural that a coincidence occurs sooner or later, but this is not quite so easy to prove, and we omit a rigorous proof at this point, returning to complete the job in (5.10.21). The point is that, once the time T has passed, the non-stationary and stationary recurrent-event processes are indistinguishable from each other, since they have had simultaneous occurrences of H. That is to say, we have that

$$\begin{aligned} u_n &= \mathbf{P}(H_n \mid T \leqslant n)\mathbf{P}(T \leqslant n) + \mathbf{P}(H_n \mid T > n)\mathbf{P}(T > n) \\ &= \mathbf{P}(H_n^* \mid T \leqslant n)\mathbf{P}(T \leqslant n) + \mathbf{P}(H_n \mid T > n)\mathbf{P}(T > n) \end{aligned}$$

since, if $T \leqslant n$, then the two processes have coincided and the (conditional) probability of H_n equals that of H_n^*. Similarly

$$u_n^* = \mathbf{P}(H_n^* \mid T \leqslant n)\mathbf{P}(T \leqslant n) + \mathbf{P}(H_n^* \mid T > n)\mathbf{P}(T > n),$$

so that $|u_n - u_n^*| \leqslant \mathbf{P}(T > n) \to 0$ as $n \to \infty$. However, $u_n^* = \mu^{-1}$ for all n, so that $u_n \to \mu^{-1}$ as $n \to \infty$. ■

Exercises

(25) Let X be the number of events in the sequence $A_1, A_2, \ldots, A_n$ which occur. Let $S_m = \mathbf{E}\binom{X}{m}$, the mean value of the random binomial coefficient $\binom{X}{m}$, and show that

$$\mathbf{P}(X \geqslant i) = \sum_{j=i}^{n} (-1)^{j-i}\binom{j-1}{i-1}S_j, \quad \text{for} \quad 1 \leqslant i \leqslant n,$$

and

$$S_m = \sum_{j=m}^{n} \binom{j-1}{m-1}\mathbf{P}(X \geqslant j), \quad \text{for} \quad 1 \leqslant m \leqslant n.$$

(26) Each person in a group of n people chooses another at random. Find the probability

(a) that exactly k people are chosen by nobody,
(b) that at least k people are chosen by nobody.

(27) **Compounding.**
(a) Let X have the Poisson distribution with parameter Y, where Y has the Poisson distribution with parameter μ. Show that $G_{X+Y}(x) = \exp\{\mu(x\,e^{x-1} - 1)\}$.
(b) Let $X_1, X_2, \ldots$ be independent identically distributed random variables with mass function

$$f(k) = \frac{(1-p)^k}{k\log(1/p)}, \qquad k \geqslant 1,$$

where $0 < p < 1$. If N is independent of the X_i and has the Poisson distribution with parameter μ, show that $Y = \sum_{i=1}^{N} X_i$ has a negative binomial distribution.

(28) Let X have the binomial distribution with parameters n and p, and show that

$$\mathbf{E}\left(\frac{1}{1+X}\right) = \frac{1-(1-p)^{n+1}}{(n+1)p}.$$

Find the limit of this expression as $n \to \infty$ and $p \to 0$, the limit being taken in such a way that $np \to \lambda$ where $0 < \lambda < \infty$. Comment.

(29) A coin is tossed repeatedly, and heads turns up with probability p on each toss. Let h_n be the probability of an even number of heads in the first n tosses, with the convention that 0 is an even number. Find a difference equation for the h_n and deduce that they have generating function $\frac{1}{2}\{(1+2ps-s)^{-1} + (1-s)^{-1}\}$.

5.3 Random walk

Generating functions are particularly valuable when studying random walks. As before, we suppose that $X_1, X_2, \ldots$ are independent random variables, each taking the value 1 with probability p, and -1 otherwise, and we write $S_n = \sum_{i=1}^{n} X_i$; the sequence $S = \{S_i : i \geqslant 0\}$ is a simple random walk starting at the origin. Natural questions of interest concern the sequence of random times at which the particle subsequently returns to the origin. To describe this sequence we need only find the distribution of the time until the particle returns for the first time, since subsequent times between consecutive visits to the origin are independent copies of this.

Let $p_0(n) = \mathbf{P}(S_n = 0)$ be the probability of being at the origin after n steps, and let $f_0(n) = \mathbf{P}(S_1 \neq 0, \ldots, S_{n-1} \neq 0, S_n = 0)$ be the probability that the first return occurs after n steps. Denote the generating functions of these sequences by

$$P_0(s) = \sum_0^\infty p_0(n)s^n \qquad F_0(s) = \sum_1^\infty f_0(n)s^n.$$

F_0 is the probability generating function of the random time T_0 until the particle makes its first return to the origin. That is

$$F_0(s) = \mathbf{E}(s^{T_0}).$$

Take care here; T_0 may be defective, and so it may be the case that $F_0(1) = \mathbf{P}(T_0 < \infty)$ satisfies $F_0(1) < 1$.

(1) **Theorem.**

(a) $P_0(s) = 1 + P_0(s)F_0(s)$
(b) $P_0(s) = (1 - 4pqs^2)^{-\frac{1}{2}}$
(c) $F_0(s) = 1 - (1 - 4pqs^2)^{\frac{1}{2}}$.

Proof. (a) Let A be the event that $S_n = 0$, and let B_k be the event that the first return to the origin happens at the kth step. Clearly the B_k are disjoint and so, by (1.4.4),

$$\mathbf{P}(A) = \sum_1^n \mathbf{P}(A \mid B_k)\mathbf{P}(B_k).$$

However, $\mathbf{P}(B_k) = f_0(k)$ and

$$\mathbf{P}(A \mid B_k) = p_0(n-k) \quad \text{by temporal homogeneity,}$$

giving

(2) $$p_0(n) = \sum_{k=1}^n p_0(n-k)f_0(k) \quad \text{if} \quad n \geqslant 1.$$

Multiply (2) by s^n, sum over n remembering that $p_0(0) = 1$, and use the convolution property of generating functions to obtain

$$P_0(s) = 1 + P_0(s)F_0(s).$$

(b) $S_n = 0$ if and only if the particle takes equal numbers of steps to the left and to the right during its first n steps. The number of ways in which it can do this is $\binom{n}{\frac{1}{2}n}$ and each such way occurs with probability $(pq)^{n/2}$, giving

(3) $$p_0(n) = \binom{n}{\frac{1}{2}n}(pq)^{n/2}.$$

Of course $p_0(n) = 0$ if n is odd. This sequence (3) has generating function

$$P_0(s) = (1 - 4pqs^2)^{-\frac{1}{2}}.$$

(c) This follows immediately from (a) and (b). ∎

(4) **Corollary.** *The probability that the particle ever returns to the origin is*

$$\sum_1^{\infty} f_0(n) = F_0(1) = 1 - |p - q|.$$

(b) *If eventual return is certain, that is* $F_0(1) = 1$ *and* $p = \frac{1}{2}$, *then the expected time to the first return is*

$$\sum_1^{\infty} nf_0(n) = F'_0(1) = \infty.$$

We call the process *persistent* (or *recurrent*) if eventual return to the origin is (almost) certain; otherwise it is called *transient*. It is immediately obvious from (4a) that the process is persistent if and only if $p = \frac{1}{2}$. This is consistent with our intuition, which suggests that if $p > \frac{1}{2}$ or $p < \frac{1}{2}$, then the particle tends to stray a long way to the right or to the left of the origin respectively. Even when $p = \frac{1}{2}$ the time until first return has infinite mean.

Proof. (a) Let $s \uparrow 1$ in (1c), and remember (5.1.30).

(b) Eventual return is certain if and only if $p = \frac{1}{2}$. But then the generating function of the time T_0 to the first return is $F_0(s) = 1 - (1 - s^2)^{\frac{1}{2}}$ and

$$\mathbf{E}(T_0) = \lim_{s \uparrow 1} F'_0(s) = \infty.$$ ∎

Now let us consider the times of visits to the point r. Define $f_r(n) = \mathbf{P}(S_1 \neq r, \ldots, S_{n-1} \neq r, S_n = r)$ to be the probability that the first such visit occurs at the nth step, with generating function

$$F_r(s) = \sum_1^{\infty} f_r(n)s^n.$$

(5) **Theorem.** (a) $F_r(s) = [F_1(s)]^r$ *if* $r \geq 1$.
(b) $F_1(s) = [1 - (1 - 4pqs^2)^{\frac{1}{2}}]/(2qs)$.

Proof. (a) The same argument which yields (2) also shows that

$$f_r(n) = \sum_{k=1}^{n-1} f_{r-1}(n-k)f_1(k) \quad \text{if} \quad r > 1.$$

Multiply by s^n and sum over n to obtain

$$F_r(s) = F_{r-1}(s)F_1(s) = (F_1(s))^r.$$

We could have written this out in terms of random variables instead of probabilities, and then used (5.1.23). For, let $T_r = \min\{n: S_n = r\}$ be the number of steps taken before the particle reaches r for the first time (T_r may equal $+\infty$ if $r > 0$ and $p < \frac{1}{2}$ or if $r < 0$ and $p > \frac{1}{2}$). In order to visit r, the particle must first visit the point 1; this requires T_1 steps. After visiting 1 the particle requires a further number, T_{1r} say, of steps to reach r; T_{1r} is distributed like T_{r-1} by 'spatial homogeneity'. Thus

$$T_r = \begin{cases} \infty & \text{if} \quad T_1 = \infty \\ T_1 + T_{1r} & \text{if} \quad T_1 < \infty, \end{cases}$$

and the result follows from (5.1.23). Some difficulties arise from the possibility that $T_1 = \infty$, but these are resolved fairly easily (*exercise*).

(b) Condition on X_1 to obtain, for $n > 1$,

$$\begin{aligned} \mathbf{P}(T_1 = n) &= \mathbf{P}(T_1 = n \mid X_1 = 1)p + \mathbf{P}(T_1 = n \mid X_1 = -1)q \\ &= 0 \cdot p + \mathbf{P}(\text{first visit to 1 takes } n-1 \text{ steps} \mid S_0 = -1) \cdot q \\ &\qquad\qquad \text{by temporal homogeneity} \\ &= \mathbf{P}(T_2 = n-1)q \quad \text{by spatial homogeneity} \\ &= qf_2(n-1). \end{aligned}$$

Therefore $f_1(n) = qf_2(n-1)$ if $n > 1$, and $f_1(1) = p$. Multiply by s^n and sum to obtain

$$F_1(s) = ps + sqF_2(s) = qs + qs(F_1(s))^2$$

by (a). Solve this quadratic to find its two roots. Only one can be a probability generating function; why? (Hint: $F_1(0) = 0$.) ■

(6) **Corollary.** *The probability that the walk ever visits the positive part of the real axis is*

$$F_1(1) = (1 - |p-q|)/(2q) = \min\{1, p/q\}.$$

Knowledge of Theorem (5) enables us to calculate $F_0(s)$ directly without recourse to (1). The method of doing this relies upon a symmetry within the collection of paths which may be followed by a random walk. Condition on the value of X_1 as usual to obtain

$$f_0(n) = qf_1(n-1) + pf_{-1}(n-1)$$

and thus

$$F_0(s) = qsF_1(s) + psF_{-1}(s).$$

We need to find $F_{-1}(s)$. Consider any possible path π that the particle may have taken to arrive at the point -1 and replace each step in the path by its mirror image, positive steps becoming negative and negative becoming positive, to obtain a path π^* which ends at $+1$. This operation of reflection provides a one–one correspondence between the collection of paths ending at -1 and the collection of paths ending at $+1$. If $\mathbf{P}(\pi; p, q)$ is the probability that the particle follows π when each step is to the right with probability p, then $\mathbf{P}(\pi; p, q) = \mathbf{P}(\pi^*; q, p)$; thus

$$F_{-1}(s) = [1 - (1 - 4pqs^2)^{\frac{1}{2}}]/(2ps),$$

giving that $F_0(s) = 1 - (1 - 4pqs^2)^{\frac{1}{2}}$ as before.

We made use in the last paragraph of a version of the reflection principle discussed in Section 3.10. Generally speaking, results obtained using the reflection principle may also be obtained using generating functions, sometimes in greater generality than before. Consider for example the hitting time theorem (3.10.14): the mass function of the time T_b of the first visit of S to the point b is given by

$$\mathbf{P}(T_b = n) = \frac{|b|}{n}\mathbf{P}(S_n = b) \quad \text{if} \quad n \geqslant 1.$$

We shall state and prove a version of this for random walks of a more general nature. Consider a sequence $X_1, X_2, \dots$ of independent identically distributed random variables taking values in the integers (positive and negative). We may think of $S_n = X_1 + X_2 + \cdots + X_n$ as being the nth position of a random walk which takes steps X_i; for simple random walk, each X_i is required to take the values ± 1 only. We call a random walk *right-continuous* (respectively *left-continuous*) if $\mathbf{P}(X_i \leqslant 1) = 1$ (respectively $\mathbf{P}(X_i \geqslant -1) = 1$), which is to say that the maximum rightward (respectively leftward) step is no greater than 1. In order to avoid certain situations of no interest, we shall consider only right-continuous walks (respectively left-continuous walks) for which $\mathbf{P}(X_i = 1) > 0$ (respectively $\mathbf{P}(X_i = -1) > 0$).

(7) **Hitting time theorem.** *Assume that S is a right-continuous random walk, and let T_b be the first hitting time of the point b. Then*

$$\mathbf{P}(T_b = n) = \frac{b}{n}\mathbf{P}(S_n = b) \quad \textit{for} \quad b, n \geqslant 1.$$

For left-continuous walks of course, the conclusion becomes

(8) $$\mathbf{P}(T_{-b} = n) = \frac{b}{n}\mathbf{P}(S_n = -b) \quad \text{for} \quad b, n \geqslant 1.$$

Proof. We introduce the functions

$$G(z) = \mathbf{E}(z^{-X_1}) = \sum_{n=-\infty}^{1} z^{-n}\mathbf{P}(X_1 = n), \qquad F_b(z) = \mathbf{E}(z^{T_b}) = \sum_{n=0}^{\infty} z^n \mathbf{P}(T_b = n).$$

These are functions of the complex variable z. $G(z)$ has a simple pole at the origin, and $F_b(z)$ converges for $|z| < 1$.

Since the walk is assumed to be right-continuous, in order to reach b (where $b > 0$) it must pass through the points $1, 2, \ldots, b-1$. The argument leading to (5a) may therefore be applied, and we find that

(9) $$F_b(z) = (F_1(z))^b \quad \text{for} \quad b \geqslant 1.$$

The argument leading to (5b) may be expressed as

$$F_1(z) = \mathbf{E}(z^{T_1}) = \mathbf{E}(\mathbf{E}(z^{T_1} \mid X_1)) = \mathbf{E}(z^{1+T_J}) \quad \text{where } J = 1 - X_1$$

since, conditional on X_1, the further time required to reach 1 has the same distribution as T_{1-X_1}. Now $1 - X_1 \geqslant 0$, and therefore

$$F_1(z) = z\mathbf{E}(F_{1-X_1}(z)) = z\mathbf{E}(F_1(z)^{1-X_1}) = zF_1(z)G(F_1(z)),$$

yielding

(10) $$z = \frac{1}{G(w)}$$

where

(11) $$w = w(z) = F_1(z).$$

Inverting (10) to find $F_1(z)$, and hence $F_b(z) = F_1(z)^b$, is a standard exercise in complex analysis using what is called Lagrange's inversion formula.

(12) **Lagrange's inversion formula.** *Let $z = w/f(w)$ where $w/f(w)$ is an analytic function of w on a neighbourhood of the origin. If g is infinitely differentiable, then*

(13) $$g(w(z)) = g(0) + \sum_{n=1}^{\infty} \frac{1}{n!} z^n \left[\frac{d^{n-1}}{du^{n-1}} [g'(u)f(u)^n] \right]_{u=0}.$$

We apply this as follows. Define $w = F_1(z)$ and $f(w) = wG(w)$, so that (10) becomes $z = w/f(w)$. Note that $f(w) = \mathbf{E}(w^{1-X_1})$ which, by the right-continuity of the walk, is a power series in w which converges for $|w| < 1$. Also $f(0) = \mathbf{P}(X_1 = 1) > 0$, and hence $w/f(w)$ is analytic on a neighbourhood of the origin. We set $g(w) = w^b$ $(= F_1(z)^b = F_b(z)$, by (9)). The inversion formula now yields

(14) $$F_b(z) = g(w(z)) = g(0) + \sum_{n=1}^{\infty} \frac{1}{n!} z^n D_n$$

where

$$D_n = \frac{d^{n-1}}{du^{n-1}} [bu^{b-1}u^n G(u)^n] \bigg|_{u=0}.$$

We pick out the coefficient of z^n in (14) to obtain

(15) $$\mathbf{P}(T_b = n) = \frac{1}{n!} D_n \quad \text{for} \quad n \geqslant 1.$$

Now $G(u)^n = \sum_{i=-\infty}^{n} u^{-i}\mathbf{P}(S_n = i)$, so that

$$D_n = \frac{d^{n-1}}{du^{n-1}} \left(b \sum_{i=-\infty}^{n} u^{b+n-1-i}\mathbf{P}(S_n = i) \right) \bigg|_{u=0} = b(n-1)!\mathbf{P}(S_n = b),$$

which may be combined with (15) as required. ∎

Once we have the hitting time theorem, we are in a position to derive a magical result called Spitzer's identity, relating the distributions of the maxima of a random walk to those of the walk itself. This identity is valid in considerable generality; the proof given here uses the hitting time theorem, and is therefore valid only for right-continuous walks (and *mutatis mutandis* for left-continuous walks and their minima).

(16) **Theorem. Spitzer's identity.** *Assume that S is a right-continuous random walk, and let $M_n = \max\{S_i : 0 \leqslant i \leqslant n\}$ be the maximum of the walk up to time n. Then, for $|s|, |t| < 1$,*

(17) $$\log\left(\sum_{n=0}^{\infty} t^n \mathbf{E}(s^{M_n}) \right) = \sum_{n=1}^{\infty} \frac{1}{n} t^n \mathbf{E}(s^{S_n^+})$$

where $S_n^+ = \max\{0, S_n\}$ as usual.

This curious and remarkable identity relates the generating function of the probability generating functions of the maxima M_n to the corresponding object for S_n^+. It contains full information about the distributions of the maxima.

Proof. Writing $f_j(n) = \mathbf{P}(T_j = n)$ as in Section 3.10, we have that

(18) $$\mathbf{P}(M_n = k) = \sum_{j=0}^{n} f_k(j)\mathbf{P}(T_1 > n - j) \quad \text{for} \quad k \geqslant 0,$$

since $M_n = k$ if the passage to k occurs at some time j ($\leqslant n$), and in addition the walk does not rise above k during the next $n - j$ steps; remember that $T_1 = \infty$ if no visit to 1 takes place. Multiply throughout (18) by $s^k t^n$ (where $|s|, |t| \leqslant 1$) and sum over $k, n \geqslant 0$ to obtain

$$\sum_{n=0}^{\infty} t^n \mathbf{E}(s^{M_n}) = \sum_{k=0}^{\infty} s^k \left(\sum_{n=0}^{\infty} t^n \mathbf{P}(M_n = k) \right) = \sum_{k=0}^{\infty} s^k F_k(t) \left(\frac{1 - F_1(t)}{1 - t} \right),$$

by the convolution formula for generating functions. We have used the result of Exercise (5.1.32) here; as usual, $F_k(t) = \mathbf{E}(t^{T_k})$. Now $F_k(t) = F_1(t)^k$, by (9), and therefore

(19) $$\sum_{n=0}^{\infty} t^n \mathbf{E}(s^{M_n}) = D(s, t)$$

where

(20) $$D(s, t) = \frac{1 - F_1(t)}{(1-t)(1 - sF_1(t))}.$$

We shall find $D(s, t)$ by finding an expression for $\partial D/\partial t$ and integrating with respect to t.

By the hitting time theorem, for $n \geqslant 0$,

(21) $$n\mathbf{P}(T_1 = n) = \mathbf{P}(S_n = 1) = \sum_{j=0}^{n} \mathbf{P}(T_1 = j)\mathbf{P}(S_{n-j} = 0),$$

as usual; multiply throughout by t^n and sum over n: $tF_1'(t) = F_1(t)P_0(t)$. Hence

(22) $$\frac{\partial}{\partial t}\log[1 - sF_1(t)] = \frac{-sF_1'(t)}{1 - sF_1(t)} = -\frac{s}{t}F_1(t)P_0(t)\sum_{k=0}^{\infty} s^k F_1(t)^k$$
$$= -\sum_{k=1}^{\infty} \frac{s^k}{t} F_k(t)P_0(t)$$

by (9). Now $F_k(t)P_0(t)$ is the generating function of the sequence

$$\sum_{j=0}^{n} \mathbf{P}(T_k = j)\mathbf{P}(S_{n-j} = 0) = \mathbf{P}(S_n = k)$$

as in (21), which implies that

$$\frac{\partial}{\partial t}\log[1 - sF_1(t)] = -\sum_{n=1}^{\infty} t^{n-1}\sum_{k=1}^{\infty} s^k \mathbf{P}(S_n = k).$$

Hence

$$\frac{\partial}{\partial t}\log D(s, t) = -\frac{\partial}{\partial t}\log(1-t) + \frac{\partial}{\partial t}\log[1 - F_1(t)] - \frac{\partial}{\partial t}\log[1 - sF_1(t)]$$
$$= \sum_{n=1}^{\infty} t^{n-1}\left(1 - \sum_{k=1}^{\infty}\mathbf{P}(S_n = k) + \sum_{k=1}^{\infty} s^k\mathbf{P}(S_n = k)\right)$$
$$= \sum_{n=1}^{\infty} t^{n-1}\left(\mathbf{P}(S_n \leqslant 0) + \sum_{k=1}^{\infty} s^k \mathbf{P}(S_n = k)\right) = \sum_{n=1}^{\infty} t^{n-1}\mathbf{E}(s^{S_n^+}).$$

Integrate over t, noting that both sides of (19) equal 1 when $t = 0$, to obtain (17). ∎

For our final example of the use of generating functions, we return to simple random walk, for which each jump equals 1 or -1 with probabilities p and $q = 1 - p$. Suppose that we are told that $S_{2n} = 0$, so that the walk is 'tied down', and we ask about the number L_{2n} of steps of the walk which were not within the negative half-line. In the language of gambling, L_{2n} is the amount of time that the gambler was ahead. In the arc sine law for sojourn times (3.10.21), we explored the distribution of L_{2n} without imposing the condition that $S_{2n} = 0$. Given that $S_{2n} = 0$, we might think that L_{2n} would be about n, but, as often happens, the contrary turns out to be the case.

(23) **Theorem. Leads for tied-down random walk.** *For the simple random walk S,*

$$\mathbf{P}(L_{2n} = 2k \mid S_{2n} = 0) = \frac{1}{n+1}, \quad k = 0, 1, 2, \ldots, n.$$

Thus each possible value of L_{2n} is equally likely. Unlike the related results of Section 3.10, we prove this using generating functions. Note that the distribution of L_{2n} does not depend on the value of p. This is not surprising since, conditional on $\{S_{2n} = 0\}$, the joint distribution of $S_0, S_1, \ldots, S_{2n}$ does not depend on p (*exercise*).

Proof. Assume $|s|, |t| < 1$, and define $G_{2n}(s) = \mathbf{E}(s^{L_{2n}} \mid S_{2n} = 0)$, $F_0(s) = \mathbf{E}(s^{T_0})$, and the bivariate generating function

$$H(s, t) = \sum_{n=0}^{\infty} t^{2n}\mathbf{P}(S_{2n} = 0)G_{2n}(s).$$

By conditioning on the time of the first return to the origin,

(24) $$G_{2n}(s) = \sum_{r=1}^{n} \mathbf{E}(s^{L_{2n}} \mid S_{2n} = 0, T_0 = 2r)\mathbf{P}(T_0 = 2r \mid S_{2n} = 0).$$

We may assume without loss of generality that $p = q = \frac{1}{2}$, so that

$$\mathbf{E}(s^{L_{2n}} \mid S_{2n} = 0, T_0 = 2r) = G_{2n-2r}(s)(\tfrac{1}{2} + \tfrac{1}{2}s^{2r}),$$

since, under these conditions, L_{2r} has (conditional) probability $\frac{1}{2}$ of being equal to either 0 or $2r$. Also

$$\mathbf{P}(T_0 = 2r \mid S_{2n} = 0) = \frac{\mathbf{P}(T_0 = 2r)\mathbf{P}(S_{2n-2r} = 0)}{\mathbf{P}(S_{2n} = 0)},$$

so that (24) becomes

$$G_{2n}(s) = \sum_{r=1}^{n} [G_{2n-2r}(s)\mathbf{P}(S_{2n-2r} = 0)][\tfrac{1}{2}(1 + s^{2r})\mathbf{P}(T_0 = 2r)]/\mathbf{P}(S_{2n} = 0).$$

Multiply throughout by $t^{2n}\mathbf{P}(S_{2n} = 0)$ and sum over $n \geqslant 1$, to find that

$$H(s, t) - 1 = \tfrac{1}{2}H(s, t)[F_0(t) + F_0(st)].$$

Hence

$$H(s,t) = \frac{2}{(1-t^2)^{\frac{1}{2}} + (1-s^2t^2)^{\frac{1}{2}}} = \frac{2[(1-s^2t^2)^{\frac{1}{2}} - (1-t^2)^{\frac{1}{2}}]}{t^2(1-s^2)}$$

$$= \sum_{n=0}^{\infty} t^{2n}\mathbf{P}(S_{2n}=0)\left(\frac{1-s^{2n+2}}{(n+1)(1-s^2)}\right)$$

after a little work using (1b). Hence $G_{2n}(s) = \sum_{k=0}^{n} (n+1)^{-1}s^{2k}$ and the proof is finished. ■

Exercises

(25) For a simple random walk S with $S_0 = 0$ and $p = 1 - q < \frac{1}{2}$, show that the maximum $M = \max\{S_n : n \geqslant 0\}$ satisfies $\mathbf{P}(M \geqslant r) = (p/q)^r$ for $r \geqslant 0$.

(26) Use generating functions to show that, for a symmetric random walk,

(a) $2kf_0(2k) = \mathbf{P}(S_{2k-2} = 0)$ for $k \geqslant 1$, and
(b) $\mathbf{P}(S_1S_2 \cdots S_{2n} \neq 0) = \mathbf{P}(S_{2n} = 0)$ for $n \geqslant 1$.

(27) A particle performs a random walk on the corners of the square ABCD. At each step, the probability of moving from corner i to corner j equals ρ_{ij}, where

$$\rho_{AB} = \rho_{BA} = \rho_{CD} = \rho_{DC} = \alpha, \qquad \rho_{AD} = \rho_{DA} = \rho_{BC} = \rho_{CB} = \beta,$$

and $\alpha,\ \beta > 0$, $\alpha + \beta = 1$. Let $G_A(s)$ be the generating function of the sequence $\{p_{AA}(n) : n \geqslant 0\}$, where $p_{AA}(n)$ is the probability that the particle is at A after n steps, having started at A. Show that

$$G_A(s) = \frac{1}{2}\left\{\frac{1}{1-s^2} + \frac{1}{1-\{|\beta - \alpha|s\}^2}\right\}.$$

Hence find the probability generating function of the time of the first return to A.

(28) A particle performs a symmetric random walk in two dimensions starting at the origin: each step is of unit length and has equal probability $\frac{1}{4}$ of being northwards, southwards, eastwards, and westwards. The particle first reaches the line $x + y = m$ at the point (X, Y) and at the time T. Find the probability generating functions of T and $X - Y$, and state where they converge.

(29) Derive the arc sine law for sojourn times (3.10.21) using generating functions. That is to say, let L_{2n} be the length of time spent (up to time $2n$) by a simple symmetric random walk to the right of its starting point. Show that

$$\mathbf{P}(L_{2n} = 2k) = \mathbf{P}(S_{2k} = 0)\mathbf{P}(S_{2n-2k} = 0), \quad \text{for} \quad 0 \leqslant k \leqslant n.$$

5.4 Branching processes

Besides gambling, many probabilists have been interested in reproduction. Accurate models for the evolution of a population are notoriously difficult to handle, but there are simpler non-trivial models which are both tractable and mathematically interesting. The branching process is such a model. Suppose that a population evolves in generations, and let Z_n be the number of members of the nth generation. Each member of the nth generation gives

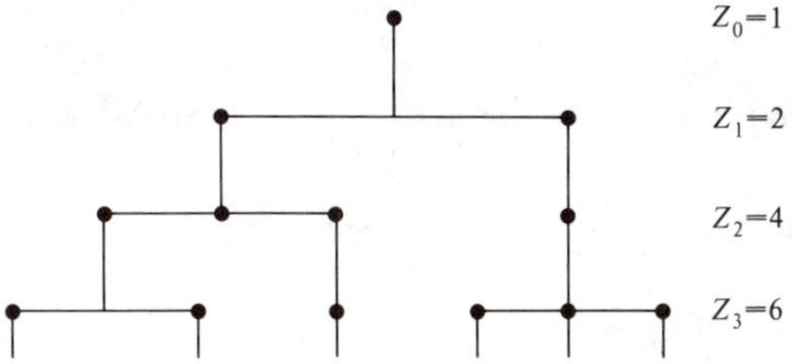

Fig. 5.1 The family tree of a branching process.

birth to a family, possibly empty, of members of the $(n+1)$th generation; the size of this family is a random variable. We shall make the following assumptions about these family sizes:

(a) the family sizes of the individuals of the branching process form a collection of independent random variables;
(b) all family sizes have the same probability mass function f and generating function G.

These assumptions, together with information about the distribution of the number Z_0 of founding members, specify the random evolution of the process. We assume here that $Z_0 = 1$. There is nothing notably human about this model, which may be just as suitable a description for the growth of a population of cells, or for the increase of neutrons in a reactor, or for the spread of an epidemic in some population. See Figure 5.1 for a picture of a branching process.

We are interested in the random sequence $Z_0, Z_1, \ldots$ of generation sizes. Let $G_n(s) = \mathbf{E}(s^{Z_n})$ be the generating function of Z_n.

(1) **Theorem.** *$G_{m+n}(s) = G_m(G_n(s))$, and thus $G_n(s) = G(G(\ldots(G(s))\ldots))$ is the n-fold iterate of G.*

Proof. Each member of the $(m+n)$th generation has a unique ancestor in the mth generation. Thus

$$Z_{m+n} = X_1 + X_2 + \cdots + X_{Z_m}$$

where X_i is the number of members of the $(m+n)$th generation which stem from the ith member of the mth generation. This is the sum of a random number Z_m of variables. These variables are independent by assumption (a); furthermore, by assumption (b) they are identically distributed with the same distribution as the number Z_n of nth-generation offspring of the first individual in the process. Now use (5.1.25) to obtain $G_{m+n}(s) = G_m(G_{X_1}(s))$ where $G_{X_1}(s) = G_n(s)$. Iterate this relation to obtain

$$G_n(s) = G_1(G_{n-1}(s)) = G_1(G_1(G_{n-2}(s))) = G_1(G_1(\ldots(G_1(s))\ldots))$$

and notice that $G_1(s)$ is what we called $G(s)$. ■

In principle, (1) tells us all about Z_n and its distribution, but in practice $G_n(s)$ may be hard to evaluate. The moments of Z_n, at least, may be routinely computed in terms of the moments of a typical family size Z_1. For example:

(2) **Lemma.** *Let* $\mu = \mathbf{E}(Z_1)$, $\sigma^2 = \text{var}(Z_1)$. *Then*

$$\mathbf{E}(Z_n) = \mu^n$$

$$\text{var}(Z_n) = \begin{cases} n\sigma^2 & \text{if } \mu = 1 \\ \sigma^2(\mu^n - 1)\mu^{n-1}(\mu-1)^{-1} & \text{if } \mu \neq 1. \end{cases}$$

Proof. Differentiate $G_n(s) = G(G_{n-1}(s))$ once at $s = 1$ to obtain

$$\mathbf{E}(Z_n) = \mu\mathbf{E}(Z_{n-1})$$

and iterate to obtain $\mathbf{E}(Z_n) = \mu^n$. Differentiate twice to obtain

$$G_n''(1) = G''(1)(G_{n-1}'(1))^2 + G'(1)G_{n-1}''(1)$$

and use (5.1.19) to obtain the second result. ∎

(3) **Example.** Suppose that each family size has the mass function $f(k) = qp^k$ $(k \geq 0)$ where $q = 1 - p$. Then $G(s) = q(1 - ps)^{-1}$, and each family size is one member less than a geometric variable. We can show by induction that

$$G_n(s) = \begin{cases} \dfrac{n - (n-1)s}{n + 1 - ns} & \text{if } p = q = \frac{1}{2} \\[2ex] \dfrac{q[p^n - q^n - ps(p^{n-1} - q^{n-1})]}{p^{n+1} - q^{n+1} - ps(p^n - q^n)} & \text{if } p \neq q. \end{cases}$$

This result can be useful in providing inequalities for more general distributions. What can we say about the behaviour of this process after many generations? In particular, does it eventually become extinct, or, conversely, do all generations have non-zero size? For this example, we can answer this question from a position of strength since we know all about $G_n(s)$. In fact

$$\mathbf{P}(Z_n = 0) = G_n(0) = \begin{cases} \dfrac{n}{n+1} & \text{if } p = q \\[2ex] \dfrac{q(p^n - q^n)}{p^{n+1} - q^{n+1}} & \text{if } p \neq q. \end{cases}$$

Let $n \to \infty$ to obtain

$$\mathbf{P}(Z_n = 0) \to \mathbf{P}(\text{ultimate extinction}) = \begin{cases} 1 & \text{if } p \leq q \\ q/p & \text{if } p > q. \end{cases}$$

We have used (1.3.5) here surreptitiously, since

(4) $$\{\text{ultimate extinction}\} = \bigcup_n \{Z_n = 0\}$$

and $A_n = \{Z_n = 0\}$ satisfies $A_n \subseteq A_{n+1}$. ●

We saw in this example that extinction is certain if and only if $\mu = \mathbf{E}(Z_1) = p/q$ satisfies $\mathbf{E}(Z_1) \leqslant 1$. This is a very natural condition; it seems reasonable that if $\mathbf{E}(Z_n) = (\mathbf{E}(Z_1))^n \leqslant 1$ then $Z_n = 0$ sooner or later. Actually this result holds in general.

(5) **Theorem.** $\mathbf{P}(Z_n = 0) \to \mathbf{P}$(*ultimate extinction*) $= \eta$, *say, where* η *is the smallest non-negative root of the equation* $s = G(s)$. *Also,* $\eta = 1$ *if* $\mu < 1$, *and* $\eta < 1$ *if* $\mu > 1$. *If* $\mu = 1$ *then* $\eta = 1$ *so long as the family-size distribution has strictly positive variance.*

Proof. Let $\eta_n = \mathbf{P}(Z_n = 0)$. Then, by (1),

$$\eta_n = G_n(0) = G(G_{n-1}(0)) = G(\eta_{n-1}).$$

In the light of the remarks about equation (4) we know that $\eta_n \uparrow \eta$, and the continuity of G guarantees that

$$\eta = G(\eta).$$

We show that if e is any non-negative root of the equation $s = G(s)$ then $\eta \leqslant e$. For, G is non-decreasing on $[0, 1]$ and so

$$\eta_1 = G(0) \leqslant G(e) = e.$$

Similarly

$$\eta_2 = G(\eta_1) \leqslant G(e) = e$$

and hence, by induction, $\eta_n \leqslant e$ for all n, giving $\eta \leqslant e$. Thus η is the smallest non-negative root of the equaton $s = G(s)$.

To verify the second assertion of the theorem, we need the fact that G is convex on $[0, 1]$. This holds because

$$G''(s) = \mathbf{E}[Z_1(Z_1 - 1)s^{Z_1 - 2}] \geqslant 0 \quad \text{if} \quad s \geqslant 0.$$

So G is convex and non-decreasing on $[0, 1]$ with $G(1) = 1$. We can verify that the two curves $y = G(s)$ and $y = s$ generally have two intersections in $[0, 1]$, and these occur at $s = \eta$ and $s = 1$. A glance at Figure 5.2 (and a more analytical verification) tells us that these intersections are coincident if $\mu = G'(1) < 1$. On the other hand, if $\mu > 1$ then these two intersections are not coincident. In the special case when $\mu = 1$ we need to distinguish between the non-random case in which $\sigma^2 = 0$, $G(s) = s$, and $\eta = 0$, and the random case in which $\sigma^2 > 0$, $G(s) > s$ for $0 \leqslant s < 1$, and $\eta = 1$. ■

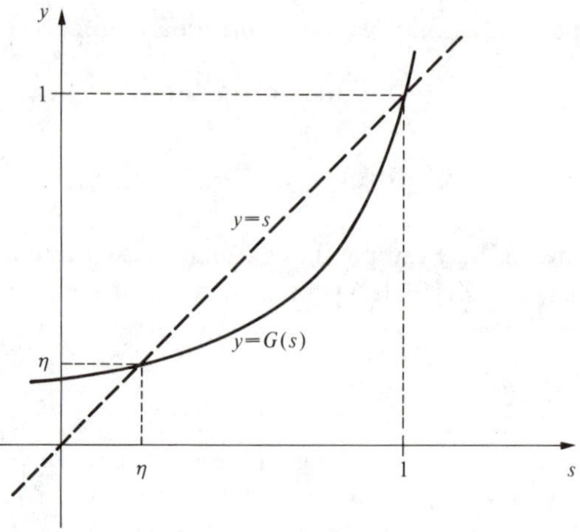

Fig. 5.2 A sketch of $G(s)$ showing the roots of the equation $G(s) = s$.

We have seen that, for large n, the nth generation is empty with probability approaching η. However, what if the process does *not* die out? If $\mathbf{E}(Z_1) > 1$ then $\eta < 1$ and extinction is not certain. Indeed $\mathbf{E}(Z_n)$ grows geometrically as $n \to \infty$, and it can be shown that

$$\mathbf{P}(Z_n \to \infty \mid \text{non-extinction}) = 1$$

when this conditional probability is suitably interpreted. To see just how fast Z_n grows, we define

$$W_n = Z_n / \mathbf{E}(Z_n)$$

where $\mathbf{E}(Z_n) = \mu^n$ and suppose that $\mu > 1$. Easy calculations show that $\mathbf{E}(W_n) = 1$, $\text{var}(W_n) = \sigma^2(1 - \mu^{-n})(\mu^2 - \mu)^{-1} \to \sigma^2(\mu^2 - \mu)^{-1}$, and it seems that W_n may have some non-trivial limit, called W say. (Actually we are asserting that the sequence $\{W_n\}$ of variables converges to a limit variable W. The convergence of random variables is a complicated topic and is described later. Neglect the details for the moment.) To study W, define

$$g_n(s) = \mathbf{E}(s^{W_n}).$$

Then

$$g_n(s) = \mathbf{E}(s^{Z_n \mu^{-n}}) = G_n(s^{\mu^{-n}})$$

and (1) shows that g_n satisfies the functional recurrence relation

$$g_n(s) = G(g_{n-1}(s^{1/\mu})).$$

Now, as $n \to \infty$, $W_n \to W$, $g_n(s) \to g(s) = \mathbf{E}(s^W)$, and we obtain

(6) $$g(s) = G(g(s^{1/\mu}))$$

by abandoning many of our current notions of mathematical rigour. This functional equation can be established rigorously (see (7.8.5)) and has various uses. For example, although we cannot solve it for g, we can reach such conclusions as 'if $\mathbf{E}(Z_1^2) < \infty$ then W is continuous, apart from a point mass of size η at zero'.

We have made considerable progress with the theory of branching processes. They are reasonably tractable because they satisfy the Markov condition (see (3.9.5)). Can you formulate and prove this property?

Exercises

(7) Show that, for a branching process Z with mean family-size μ, $\mathbf{E}(Z_m Z_n) = \mu^{n-m}\mathbf{E}(Z_m^2)$ for $m \leqslant n$.

(8) Consider a branching process Z in which $Z_0 = 1$ and $\mathbf{P}(Z_1 = 0) = 0$. Pick two individuals at random (with replacement) from the nth generation and let L be the index of the generation which contains their most recent common ancestor. Show that $\mathbf{P}(L = r) = \mathbf{E}(Z_r^{-1}) - \mathbf{E}(Z_{r+1}^{-1})$ for $0 \leqslant r < n$. What can be said if $\mathbf{P}(Z_1 = 0) > 0$?

(9) Consider a branching process Z whose family sizes have the geometric mass function $f(k) = qp^k$, $k \geqslant 0$, where $p + q = 1$. Let $T = \min\{n\colon Z_n = 0\}$ be the extinction time of the branching process, and suppose that $Z_0 = 1$. Find $\mathbf{P}(T = n)$. For what values of p is it the case that $\mathbf{E}(T) < \infty$?

(10) For a branching process Z with $Z_0 = 1$, find an expression for the generating function G_n of Z_n, in the cases when Z_1 has generating function given by

(a) $G(s) = 1 - \alpha(1 - s)^\beta$, $0 < \alpha, \beta < 1$,
(b) $G(s) = f^{-1}(P(f(s)))$, where P is a probability generating function, and f is a suitable function satisfying $f(1) = 1$.
(c) In the latter case, calculate your answer explicitly when $f(x) = x^m$ and $P(s) = s\{\gamma - (\gamma - 1)s\}^{-1}$ where $\gamma > 1$.

(11) **Branching with immigration.** Each generation of a branching process (with a single progenitor) is augmented by a random number of immigrants who are indistinguishable from the other members of the population. Suppose that the numbers of immigrants in different generations are independent of each other and of the past history of the branching process, each such number having probability generating function $H(s)$. Show that the probability generating function G_n of the size of the nth generation satisfies $G_{n+1}(s) = G_n(G(s))H(s)$, where G is the probability generating function of a typical family of offspring.

5.5 Age-dependent branching processes

Here is a more general model for the growth of a population. It incorporates the observation that generations are not contemporaneous in most populations, with individuals in the same generation giving birth to families at different times. To model this we attach another random variable, called 'age', to each individual; we shall suppose that the collection of all ages is a set of variables which are independent of each other and of all family sizes, and which are continuous, positive, and have the common density function f_T. Each individual lives for a period of time, equal to its 'age', before it

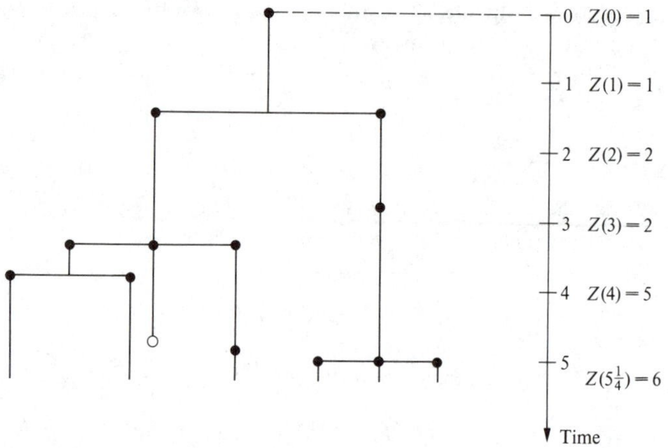

Fig. 5.3 The family tree of an age-dependent branching process; ● indicates the birth of an individual, and ○ indicates the death of an individual which has no descendants.

gives birth to its family of next generation descendants as before. See Figure 5.3 for a picture of an age-dependent branching process.

Let $Z(t)$ denote the size of the population at time t; we shall assume that $Z(0) = 1$. The population-size generating function $G_t(s) = \mathbf{E}(s^{Z(t)})$ is now a function of t as well. As usual, we hope to find an expression involving G_t by conditioning on some suitable event. In this case we condition on the age of the initial individual in the population.

(1) **Theorem.** $$G_t(s) = \int_0^t G(G_{t-u}(s))f_T(u)\,du + \int_t^\infty sf_T(u)\,du.$$

Proof. Let T be the age of the initial individual. By the use of conditional expectation

(2) $$\begin{aligned} G_t(s) &= \mathbf{E}(s^{Z(t)}) = \mathbf{E}(\mathbf{E}(s^{Z(t)} \mid T)) \\ &= \int_0^\infty \mathbf{E}(s^{Z(t)} \mid T = u)f_T(u)\,du. \end{aligned}$$

If $T = u$, then at time u the initial individual dies and is replaced by a random number N of offspring, where N has generating function G. Each of these offspring behaves in the future as their ancestor did in the past, and the effect of their ancestor's death is to replace the process by the sum of N independent copies of the process displaced in time by an amount u. Now if $u > t$ then $Z(t) = 1$ and $\mathbf{E}(s^{Z(t)} \mid T = u) = s$, whilst if $u < t$ then $Z(t) = Y_1 + Y_2 + \cdots + Y_N$ is the sum of N independent copies of $Z(t-u)$ and so $\mathbf{E}(s^{Z(t)} \mid T = u) = G(G_{t-u}(s))$ by (5.1.25). Substitute into (2) to obtain the result. ■

Unfortunately we cannot solve equation (1) except in certain special cases.

Possibly the most significant case with which we can make some progress arises when the ages are exponentially distributed. Then

$$f_T(t) = \lambda e^{-\lambda t} \quad \text{for} \quad t \geq 0$$

and the reader may show (*exercise*) that

(3)
$$\frac{\partial}{\partial t} G_t(s) = \lambda[G(G_t(s)) - G_t(s)].$$

It is no mere coincidence that this case is more tractable. In this very special instance, and in no other, $Z(t)$ satisfies a Markov condition; it is called a Markov process, and we shall return to the general theory of such processes later.

Some information about the moments of $Z(t)$ is fairly readily available from (1). For example,

$$m(t) = \mathbf{E}(Z(t)) = \lim_{s \uparrow 1} \frac{\partial}{\partial s} G_t(s)$$

satisfies the integral equation

(4)
$$m(t) = \mu \int_0^t m(t-u) f_T(u)\, du + \int_t^\infty f_T(u)\, du \quad \text{where} \quad \mu = G'(1).$$

We can find the general solution to this equation only by numerical or series methods. It is reasonably amenable to Laplace transform methods and produces a closed expression for the Laplace transform of m. Later we shall use renewal theory arguments (see (10.4.22)) to show that there exist $\delta > 0$ and $\beta > 0$ such that $m(t) \sim \delta e^{\beta t}$ as $t \to \infty$ whenever $\mu > 1$.

Finally observe that, in some sense, the age-dependent process $Z(t)$ contains the old process Z_n. We say that Z_n is 'imbedded' in $Z(t)$ in that we can recapture Z_n by aggregating the generation sizes of $Z(t)$. This imbedding enables us to use properties of Z_n to derive corresponding properties of the less tractable $Z(t)$. For instance, $Z(t)$ dies out if and only if Z_n dies out, and so (5.4.5) provides us immediately with the extinction probability of the age-dependent process. This technique has uses elsewhere as well. With any non-Markov process we can try to find an imbedded Markov process which provides information about the original process. We consider examples of this later.

Exercises

(5) Let Z be an age-dependent branching process, the lifetime distribution of which is exponential with parameter λ. If $Z(0) = 1$, show that the probability generating function $G_t(s)$ of $Z(t)$ satisfies

$$\frac{\partial}{\partial t} G_t(s) = \lambda\{G(G_t(s)) - G_t(s)\}.$$

Show in the case of 'exponential binary fission', when $G(s) = s^2$, that

$$G_t(s) = \frac{s\,\mathrm{e}^{-\lambda t}}{1 - s(1 - \mathrm{e}^{-\lambda t})}$$

and hence derive the mass function of the population size $Z(t)$ at time t.

(6) Solve the differential equation of Exercise (5) when $\lambda = 1$, $G(s) = \frac{1}{2}(1 + s^2)$, to obtain

$$G_t(s) = \frac{2s + t(1 - s)}{2 + t(1 - s)}.$$

Hence find $\mathbf{P}(Z(t) \geq k)$, and deduce that

$$\mathbf{P}\left(\frac{1}{t} Z(t) \geq x \,\middle|\, Z(t) > 0\right) \to \mathrm{e}^{-2x} \quad \text{as} \quad t \to \infty.$$

5.6 Expectation revisited

This section is divided into parts A and B. All readers must read A before they proceed to the next section; B is for people with a keener appreciation of detailed technique. We are about to extend the definition of probability generating functions to more general types of variables than those concentrated on the non-negative integers, and it is a suitable moment to insert some discussion of the expectation of an arbitrary random variable regardless of its type (discrete, continuous, and so on). Up to now we have made only guarded remarks about such variables.

(A) Notation

Remember that the expectations of discrete and continuous variables are given respectively by

(1) $$\mathbf{E}X = \sum x f(x) \qquad \text{if } X \text{ has mass function } f$$

(2) $$\mathbf{E}X = \int x f(x)\,\mathrm{d}x \quad \text{if } X \text{ has density function } f.$$

We require a single piece of notation which incorporates both these cases. Suppose X has distribution function F. Subject to a trivial and unimportant condition, (1) and (2) can be rewritten as

(3) $$\mathbf{E}X = \sum x\,\mathrm{d}F(x) \text{ where } \mathrm{d}F(x) = F(x) - \lim_{y \uparrow x} F(y) = f(x)$$

(4) $$\mathbf{E}X = \int x\,\mathrm{d}F(x) \text{ where } \mathrm{d}F(x) = \frac{\mathrm{d}F}{\mathrm{d}x}\,\mathrm{d}x = f(x)\,\mathrm{d}x.$$

This suggests that we denote $\mathbf{E}X$ by

(5) $$\mathbf{E}X = \int x\,\mathrm{d}F \quad \text{or} \quad \int x\,\mathrm{d}F(x)$$

whatever the type of X, where (5) is interpreted as (3) for discrete variables and as (4), of course, for continuous variables. We adopt this notation forthwith. Those readers who fail to conquer an aversion to this notation should read $\mathrm{d}F$ as $f(x)\,\mathrm{d}x$. Previous properties of expectation received two statements and proofs which can now be unified. For instance, (3.3.3) and (4.3.3) become

(6) $$\text{if}\quad g\colon \mathbb{R} \to \mathbb{R} \quad\text{then}\quad \mathbf{E}(g(X)) = \int g(x)\,\mathrm{d}F.$$

(B) Abstract integration

The expectation of a random variable X is specified by its distribution function F. But F itself is describable in terms of X and the underlying probability space, and it follows that $\mathbf{E}X$ can be thus described also. This part contains a brief sketch of how to integrate on a probability space $(\Omega, \mathscr{F}, \mathbf{P})$. It contains no details, and the reader is left to check up on his intuition elsewhere (see Clarke 1975 or Williams 1991 for example). Let $(\Omega, \mathscr{F}, \mathbf{P})$ be some probability space.

(7) The random variable $X\colon \Omega \to \mathbb{R}$ is called *simple* if it takes only finitely many distinct values. Simple variables can be written

$$X = \sum_{i=1}^{n} x_i I_{A_i}$$

for some partition $A_1, A_2, \dots, A_n$ of Ω; we define the *integral* of X, written $\mathbf{E}(X)$, to be

$$\mathbf{E}(X) = \sum_{i=1}^{n} x_i \mathbf{P}(A_i).$$

(8) Any non-negative random variable $X\colon \Omega \to [0, \infty)$ is the limit of some increasing sequence $\{X_n\}$ of simple variables. That is, $X_n(\omega) \uparrow X(\omega)$ for all $\omega \in \Omega$. We define the *integral* of X, written $\mathbf{E}(X)$, to be

$$\mathbf{E}(X) = \lim_{n\to\infty} \mathbf{E}(X_n).$$

This is well defined in the sense that two increasing sequences of simple functions, both converging to X, have the same limit for their sequences of integrals. $\mathbf{E}(X)$ can be $+\infty$.

(9) Any random variable $X\colon \Omega \to \mathbb{R}$ can be written as the difference $X = X^+ - X^-$ of non-negative random variables

$$X^+(\omega) = \max\{X(\omega), 0\}, \qquad X^-(\omega) = -\min\{X(\omega), 0\}.$$

If at least one of $\mathbf{E}(X^+)$ and $\mathbf{E}(X^-)$ is finite, then we define the *integral* of X, written $\mathbf{E}(X)$, to be

$$\mathbf{E}(X) = \mathbf{E}(X^+) - \mathbf{E}(X^-).$$

(10) Thus, $\mathbf{E}(X)$ is well defined, at least for any variable X such that

$$\mathbf{E}|X| = \mathbf{E}(X^+ + X^-) < \infty.$$

(11) In the language of measure theory $\mathbf{E}(X)$ is denoted by

$$\mathbf{E}(X) = \int_\Omega X(\omega)\,\mathrm{d}\mathbf{P} \quad \text{or} \quad \int_\Omega X(\omega)\mathbf{P}(\mathrm{d}\omega).$$

The *expectation operator* $\mathbf{E}$ defined in this way has all the properties which were described in detail for discrete and continuous variables.

(12) **Continuity of E.** Important further properties are the following. If $\{X_n\}$ is a sequence of variables with $X_n(\omega) \to X(\omega)$ for all $\omega \in \Omega$ then

(a) (*monotone convergence*) *if* $X_n(\omega) \geqslant 0$ *and* $X_n(\omega) \leqslant X_{n+1}(\omega)$ *for all n and* ω, *then* $\mathbf{E}(X_n) \to \mathbf{E}(X)$,
(b) (*dominated convergence*) *if* $|X_n(\omega)| \leqslant Y(\omega)$ *for all n and* ω, *and* $\mathbf{E}|Y| < \infty$ *then* $\mathbf{E}(X_n) \to \mathbf{E}(X)$,
(c) (*bounded convergence*) this is a special case of dominated convergence; *if* $|X_n(\omega)| \leqslant c$ *for some constant c and all n and* ω *then* $\mathbf{E}(X_n) \to \mathbf{E}(X)$.

Rather more is true. Events having zero probability (that is, null events) make no contributions to expectations, and may therefore be ignored. Consequently, it suffices to assume above that $X_n(\omega) \to X(\omega)$ for all ω *except possibly on some null event*, with a similar weakening of the hypotheses of (a), (b), and (c). For example, the bounded convergence theorem is normally stated as follows: if $\{X_n\}$ is a sequence of random variables satisfying $X_n \to X$ a.s. and $|X_n| \leqslant c$ a.s. for some constant c, then $\mathbf{E}(X_n) \to \mathbf{E}(X)$. The expression 'a.s.' is an abbreviation for 'almost surely', and means 'except possibly on an event of zero probability'.

One further property of expectation is called *Fatou's lemma*: if $\{X_n\}$ is a sequence of random variables such that $X_n \geqslant Y$ a.s. for all n and some Y with $\mathbf{E}|Y| < \infty$, then

(13) $$\mathbf{E}\left(\liminf_{n\to\infty} X_n\right) \leqslant \liminf_{n\to\infty} \mathbf{E}(X_n).$$

This inequality is often applied in practice with $Y = 0$.

(14) **Lebesgue–Stieltjes integral.** Let X have distribution function F. F gives rise to a probability measure μ_F on the Borel sets of $\mathbb{R}$ as follows:

(a) define $\mu_F((a, b]) = F(b) - F(a)$,
(b) as in the discussion after (4.1.5), the domain of μ_F can be extended to include the Borel σ-field $\mathscr{B}$, which is the smallest σ-field containing all half-open intervals $(a, b]$.

So $(\mathbb{R}, \mathscr{B}, \mu_F)$ is a probability space; its completion (see Section 1.6) is denoted by $(\mathbb{R}, \mathscr{L}_F, \mu_F)$, where $\mathscr{L}_F$ is the smallest σ-field containing $\mathscr{B}$ and all subsets of μ_F-null sets. If $g: \mathbb{R} \to \mathbb{R}$ (is $\mathscr{L}_F$-measurable) then the abstract integral

$$\int g \, \mathrm{d}\mu_F$$

is called the *Lebesgue–Stieltjes integral* of g with respect to μ_F, and we normally denote it by

$$\int g(x) \, \mathrm{d}F \quad \text{or} \quad \int g(x) \, \mathrm{d}F(x).$$

Think of it as a special case of the abstract integral (11). The purpose of this discussion is the assertion that if $g: \mathbb{R} \to \mathbb{R}$ (and g is suitably measurable) then $g(X)$ is random variable and

$$\mathbf{E}(g(X)) = \int g(x) \, \mathrm{d}F,$$

and we adopt this forthwith as the official notation for expectation. Here is a final word of caution. If $g(x) = I_B(x)h(x)$ where I_B is the indicator function of some $B \subseteq \mathbb{R}$ then

$$\int g(x) \, \mathrm{d}F = \int_B h(x) \, \mathrm{d}F.$$

We do not in general obtain the same result when we integrate over $B_1 = [a, b]$ and $B_2 = (a, b)$ unless F is continuous at a and b, and so we do not use the notation

$$\int_a^b h(x) \, \mathrm{d}F$$

unless there is no danger of ambiguity.

Exercises

(15) **Jensen's inequality.** A function $u: \mathbb{R} \to \mathbb{R}$ is called *convex* if for all real a there exists λ, depending on a, such that $u(x) \geqslant u(a) + \lambda(x - a)$ for all x. (Draw a diagram to illustrate this definition.) Show that, if u is convex and X is a random variable with finite mean, then $\mathbf{E}(u(X)) \geqslant u(\mathbf{E}(X))$.

(16) Let $X_1, X_2, \dots$ be random variables satisfying

$$\mathbf{E}\left(\sum_{i=1}^{\infty} |X_i|\right) < \infty.$$

Show that

$$\mathbf{E}\left(\sum_{i=1}^{\infty} X_i\right) = \sum_{i=1}^{\infty} \mathbf{E}(X_i).$$

(17) Let $\{X_n\}$ be a sequence of random variables satisfying $X_n \leqslant Y$ a.s. for some Y with $\mathbf{E}|Y| < \infty$. Show that

$$\mathbf{E}\left(\limsup_{n\to\infty} X_n\right) \geqslant \limsup_{n\to\infty} \mathbf{E}(X_n).$$

(18) Suppose that $\mathbf{E}|X^r| < \infty$ where $r > 0$. Deduce that $x^r\mathbf{P}(|X| \geqslant x) \to 0$ as $x \to \infty$. Conversely, suppose that $x^r\mathbf{P}(|X| \geqslant x) \to 0$ as $x \to \infty$ where $r \geqslant 0$, and show that $\mathbf{E}|X^s| < \infty$ for $0 \leqslant s < r$.

(19) Show that $\mathbf{E}(X) < \infty$ if and only if the following holds: for all $\varepsilon > 0$, there exists $\delta > 0$, such that $\mathbf{E}(|X|I_A) < \varepsilon$ for all A such that $\mathbf{P}(A) < \delta$.

5.7 Characteristic functions

Probability generating functions proved to be very useful in handling non-negative integral random variables. For more general variables X it is natural to make the substitution $s = e^t$ in the quantity $G_X(s) = \mathbf{E}(s^X)$.

(1) **Definition.** The **moment generating function** of a variable X is a function $M: \mathbb{R} \to [0, \infty)$ given by

$$M(t) = \mathbf{E}(e^{tX}).$$

Moment generating functions are related to Laplace transforms, since

$$M(t) = \int e^{tx}\, dF(x) = \int e^{tx} f(x)\, dx$$

if X is continuous with denstiy function f. They have properties similar to those of probability generating functions. For example, if $M(t) < \infty$ on some open interval containing the origin then

(a) $\mathbf{E}X = M'(0)$, $\mathbf{E}(X^k) = M^{(k)}(0)$,

(b) (*Taylor's theorem*)

$$M(t) = \sum_k \frac{\mathbf{E}(X^k)}{k!} t^k;$$

that is, M is the 'exponential generating function' of the sequence of moments of X.

(c) (*Laplace convolution theorem*) If X and Y are independent then

$$M_{X+Y}(t) = M_X(t)M_Y(t).$$

This is essentially the assertion that the Laplace transform of a convolution (see (4.8.2)) is the product of the Laplace transforms.

Moment generating functions provide a very useful technique but suffer the disadvantage that the integrals which define them may not always be finite. Rather than explore their properties in detail we move on immediately to another class of functions which are equally useful and whose finiteness is guaranteed.

(2) **Definition.** The **characteristic function** of X is the function $\phi: \mathbb{R} \to \mathbb{C}$ defined by

$$\phi(t) = \mathbf{E}(e^{itX}) \quad \text{where} \quad i = \sqrt{(-1)}.$$

Characteristic functions are related to Fourier transforms, since

$$\phi(t) = \int e^{itx}\, dF(x).$$

In the notation of Section 5.6, ϕ is the abstract integral of a complex-valued random variable. It is well defined in the terms of Section 5.6 by

$$\phi(t) = \mathbf{E}(\cos tX) + i\mathbf{E}(\sin tX).$$

Furthermore, $\phi(t)$ is better behaved than $M(t)$.

(3) **Theorem.** *The characteristic function ϕ satisfies*

(a) $\phi(0) = 1$, $|\phi(t)| \leqslant 1$ *for all* t
(b) ϕ *is uniformly continuous on* $\mathbb{R}$
(c) ϕ *is non-negative definite, which is to say that*

$$\sum_{j,k} \phi(t_j - t_k) z_j \bar{z}_k \geqslant 0$$

for all real $t_1, \ldots, t_n$ *and complex* $z_1, \ldots, z_n$.

Proof.

(a) $\phi(0) = \mathbf{E}(1) = 1$. Furthermore,

$$|\phi(t)| \leqslant \int |e^{itx}|\, dF = \int dF = 1.$$

(b)
$$\begin{aligned}|\phi(t+h) - \phi(t)| &= |\mathbf{E}(e^{i(t+h)X} - e^{itX})| \\ &\leqslant \mathbf{E}|e^{itX}(e^{ihX} - 1)| = \mathbf{E}(Y(h))\end{aligned}$$

where $Y(h) = |e^{itX} - 1|$. However, $|Y(h)| \leqslant 2$ and $Y(h) \to 0$ as $h \to 0$ and so $\mathbf{E}(Y(h)) \to 0$ (by bounded convergence (5.6.12)).

(c)
$$\begin{aligned}\sum_{j,k} \phi(t_j - t_k) z_j \bar{z}_k &= \sum_{j,k} \int [z_j \exp(it_j x)][\bar{z}_k \exp(-it_k x)]\, dF \\ &= \mathbf{E}\left(\left|\sum_j z_j \exp(it_j X)\right|^2\right) \geqslant 0.\end{aligned}$$

■

Actually, (3) characterizes characteristic functions in the sense that ϕ is a characteristic function if and only if it satisfies (a), (b), and (c). This is Bochner's theorem, for which we offer no proof. Many of the properties of characteristic functions rely for their proofs on a knowledge of complex analysis. This is a textbook on probability theory, and will not include such proofs unless they indicate some essential technique. We have asserted that the method of characteristic functions is very useful; however, we warn the reader that we shall not make use of them until Section 5.10. In the meantime we shall establish some of their properties.

First and foremost, from a knowledge of ϕ_X we can recapture the distribution of X. The full power of this statement is deferred until the next section; here we concern ourselves only with the moments of X. Many of the interesting characteristic functions are not very well behaved, and we must move carefully.

(4) **Theorem.**

(a) *If* $\phi^{(k)}(0)$ *exists then* $\begin{cases} \mathbf{E}|X^k| < \infty & \text{if } k \text{ is even} \\ \mathbf{E}|X^{k-1}| < \infty & \text{if } k \text{ is odd}. \end{cases}$

(b) *If* $\mathbf{E}|X^k| < \infty$ *then*†

$$\phi(t) = \sum_{j=0}^{k} \frac{\mathbf{E}(X^j)}{j!}(\mathrm{i}t)^j + o(t^k),$$

and so $\phi^{(k)}(0) = \mathrm{i}^k \mathbf{E}(X^k)$.

Proof. This is essentially Taylor's theorem for a function of a complex variable. For the proof, see Moran (1968) and Kingman and Taylor (1966). ■

One of the useful properties of characteristic functions is that they enable us to handle sums of independent variables with the minimum of fuss.

(5) **Theorem.** *If X and Y are independent then*

$$\phi_{X+Y}(t) = \phi_X(t)\phi_Y(t).$$

Proof.

$$\phi_{X+Y}(t) = \mathbf{E}(e^{\mathrm{i}t(X+Y)}) = \mathbf{E}(e^{\mathrm{i}tX} e^{\mathrm{i}tY}).$$

† See Subsection (10) of Appendix I for a reminder about Landau's O–o notation.

Expand each exponential term into cosines and sines, multiply out, use independence, and put back together to obtain the result. ■

(6) **Theorem.** *If $a, b \in \mathbb{R}$ and $Y = aX + b$ then*

$$\phi_Y(t) = e^{itb}\phi_X(at).$$

Proof.

$$\begin{aligned}\phi_Y(t) &= \mathbf{E}(e^{it(aX+b)}) = \mathbf{E}(e^{itb}\, e^{i(at)X}) \\ &= e^{itb}\mathbf{E}(e^{i(at)X}) = e^{itb}\phi_X(at).\end{aligned}$$ ■

We shall make repeated use of these last two theorems. We sometimes need to study collections of variables which may be dependent.

(7) **Definition.** The **joint characteristic function** of X and Y is the function $\phi: \mathbb{R}^2 \to \mathbb{R}$ given by

$$\phi(s, t) = \mathbf{E}(e^{isX}\, e^{itY}).$$

Notice that $\phi(s, t) = \phi_{sX+tY}(1)$. As usual we shall be interested mostly in independent variables.

(8) **Theorem.** *X and Y are independent if and only if*

$$\phi_{X,Y}(s, t) = \phi_X(s)\phi_Y(t) \quad \textit{for all } s \textit{ and } t.$$

Proof. If X and Y are independent then the result follows by the argument of (5). The converse is proved by extending the inversion theorem of the next section to deal with joint distributions and showing that the joint distribution function factorizes. ■

Note particularly that for X and Y to be independent it is not sufficient that

(9) $$\phi_{X,Y}(t, t) = \phi_X(t)\phi_Y(t), \quad \text{for all } t.$$

Exercise. Can you find an example of dependent variables which satisfy (9)?

We have seen (4) that it is an easy calculation to find the moments of X by differentiating its characteristic function $\phi_X(t)$ at $t = 0$. A similar calculation gives the 'joint moments' $\mathbf{E}(X^j Y^k)$ of two variables from a knowledge of their joint characteristic function $\phi_{X,Y}(s, t)$ (see Problem (5.12.30) for details).

The properties of moment generating functions are closely related to those of characteristic functions. In the rest of the text we shall use the latter whenever possible, but it will be appropriate to use the former for any topic whose analysis employs Laplace transforms; for example, this is the case for the queueing theory of Chapter 11.

(10) **Remark. Moment problem.** If I am given a distribution function F, then I can calculate the corresponding moments $m_k(F) = \int_{-\infty}^{\infty} x^k \, dF(x)$, $k = 1, 2, \ldots$, whenever these integrals exist. Is the converse true: does the collection of moments $(m_k(F): k = 1, 2, \ldots)$ specify F uniquely? In other words, do there exist distribution functions F and G, all of whose moments exist, such that $F \neq G$ but $m_k(F) = m_k(G)$ for all k? The answer is *yes*, and the usual example is obtained by using the log-normal distribution (see Problem (5.12.43)).

Under what conditions on F is it the case that no such G exists? Various sets of conditions are known which guarantee that F is specified by its moments, but there is no attractive known condition which is both necessary and sufficient. Perhaps the simplest sufficient condition is that the moment generating function of F, $M(t) = \int_{-\infty}^{\infty} e^{tx} \, dF(x)$, be finite in some neighbourhood of the point $t = 0$. Those familiar with the theory of Laplace transforms will understand why this is sufficient. ●

Exercises

(11) Find two dependent random variables X and Y such that $\phi_{X+Y}(t) = \phi_X(t)\phi_Y(t)$ for all t.

(12) If ϕ is a characteristic function, show that $\text{Re}\{1 - \phi(t)\} \geq \frac{1}{4}\text{Re}\{1 - \phi(2t)\}$, and deduce that $1 - |\phi(2t)| \leq 8\{1 - |\phi(t)|\}$.

(13) The **cumulant generating function** $K_X(\theta)$ of the random variable X is defined by $K_X(\theta) = \log \mathbf{E}(e^{\theta X})$, the logarithm of the moment generating function of X. If the latter is finite in a neighbourhood of the origin, then K_X has a convergent Taylor expansion:

$$K_X(\theta) = \sum_{n=1}^{\infty} \frac{1}{n!} k_n(X)\theta^n$$

and $k_n(X)$ is called the nth *cumulant* (or *semi-invariant*) of X.

(a) Express $k_1(X)$, $k_2(X)$, and $k_3(X)$ in terms of the moments of X.
(b) If X and Y are independent random variables, show that $k_n(X + Y) = k_n(X) + k_n(Y)$.

(14) Let X be $N(0, 1)$, and show that the cumulants of X are $k_2(X) = 1$, $k_m(X) = 0$ for $m \neq 2$.

(15) The random variable X is said to have a *lattice distribution* if there exist a and b such that X takes values in the set $L(a, b) = \{a + bm: m = 0, \pm 1, \ldots\}$. The *span* of such a variable X is the maximal value of b for which there exists a such that X takes values in $L(a, b)$.

(a) Suppose that X has a lattice distribution with span b. Show that $|\phi_X(2\pi/b)| = 1$, and that $|\phi_X(t)| < 1$ for $0 < t < 2\pi/b$.
(b) Suppose that $|\phi_X(T)| = 1$ for some $T \neq 0$. Show that X has a lattice distribution with span $2\pi k/T$ for some integer k.

(16) Let X be a random variable with density function f. Show that $|\phi_X(t)| \to 0$ as $t \to \pm\infty$.

(17) Let $X_1, X_2, \ldots, X_n$ be independent variables, X_i being $N(\mu_i, 1)$, and let $Y = X_1^2 + X_2^2 + \cdots + X_n^2$. Show that the characteristic function of Y is

$$\phi_Y(t) = \frac{1}{(1-2it)^{n/2}} \exp\left(\frac{it\theta}{1-2it}\right)$$

where $\theta = \mu_1^2 + \mu_2^2 + \cdots + \mu_n^2$. Y is said to have the *non-central chi-squared distribution* with n degrees of freedom and non-centrality parameter θ, written $\chi^2(n; \theta)$.

(18) Let X be $N(\mu, 1)$ and let Y be $\chi^2(n)$, and suppose that X and Y are independent. The random variable $T = X/\sqrt{(Y/n)}$ is said to have the *non-central t distribution* with n degrees of freedom and non-centrality parameter μ. If U and V are independent, U being $\chi^2(m; \theta)$ and V being $\chi^2(n)$, then $F = (U/m)/(V/n)$ is said to have the *non-central F distribution* with m and n degrees of freedom and non-centrality parameter θ, written $F(m, n; \theta)$.

(a) Show that T^2 is $F(1, n; \mu^2)$.
(b) Show that

$$\mathbf{E}(F) = \frac{n(m+\theta)}{m(n-2)} \quad \text{if} \quad n > 2.$$

5.8 Examples of characteristic functions

Those who feel daunted by $\sqrt{(-1)}$ should find it a useful exercise to work through this section using $M(t) = \mathbf{E}(e^{tX})$ in place of $\phi(t) = \mathbf{E}(e^{itX})$. Many calculations here are left as *exercises*.

(1) **Example. Bernoulli distribution.** If X is Bernoulli parameter p then

$$\phi(t) = \mathbf{E}(e^{itX}) = e^{it0} \cdot q + e^{it1} \cdot p = q + p\,e^{it}. \qquad \bullet$$

(2) **Example. Binomial distribution.** If X is $B(n, p)$ then X has the same distribution as the sum of n independent Bernoulli variables $Y_1, Y_2, \ldots, Y_n$. Thus

$$\phi_X(t) = \phi_{Y_1}(t) \cdots \phi_{Y_n}(t) = (q + p\,e^{it})^n. \qquad \bullet$$

(3) **Example. Exponential distribution.** If $f(x) = \lambda e^{-\lambda x}$ for $x \geqslant 0$ then

$$\phi(t) = \int_0^\infty e^{itx} \lambda e^{-\lambda x}\,dx.$$

This is a complex integral and its solution relies on a knowledge of how to integrate around contours in $\mathbb{R}^2$ (the appropriate contour is a sector). If you know about this then do it. Do not fall into the trap of treating i as if it were a real number, even though this malpractice yields the correct answer in this case:

$$\phi(t) = \frac{\lambda}{\lambda - it}. \qquad \bullet$$

(4) **Example. Cauchy distribution.** If $f(x) = \{\pi(1 + x^2)\}^{-1}$ then

$$\phi(t) = \frac{1}{\pi}\int_{-\infty}^{\infty} \frac{e^{itx}}{1 + x^2}\,dx.$$

Treating i as a real number will not help you to avoid the contour integral this time. The answer is

$$\phi(t) = e^{-|t|}.$$

Those who are interested should try integrating around a semicircle with diameter $[-R, R]$ on the real axis. ●

(5) **Example. Normal distribution.** If X is $N(0, 1)$ then

$$\phi(t) = \mathbf{E}(e^{itX}) = \int_{-\infty}^{\infty} \frac{1}{\sqrt{(2\pi)}} \exp(itx - \tfrac{1}{2}x^2)\,dx.$$

Again, do not treat i as a real number. Consider instead the moment generating function of X

$$M(s) = \mathbf{E}(e^{sX}) = \int_{-\infty}^{\infty} \frac{1}{\sqrt{(2\pi)}} \exp(sx - \tfrac{1}{2}x^2)\,dx.$$

Complete the square in the integrand and use the hint at the end of Example (4.5.9) to obtain

$$M(s) = e^{s^2/2}.$$

We may not substitute $s = it$ without justification. In this particular instance the theory of analytic continuation of functions of a complex variable provides this justification, and we deduce that

$$\phi(t) = e^{-t^2/2}.$$

By (5.7.6), the characteristic function of the $N(\mu, \sigma^2)$ variable $Y = \sigma X + \mu$ is

$$\phi_Y(t) = e^{it\mu}\phi_X(\sigma t) = \exp(i\mu t - \tfrac{1}{2}\sigma^2 t^2).$$ ●

(6) **Example. Multivariate normal distribution.** If $X_1, \ldots, X_n$ has the multivariate normal distribution $N(\mathbf{0}, V)$ then its joint density function is

$$f(\mathbf{x}) = \{(2\pi)^n|V|\}^{-\frac{1}{2}} \exp(-\tfrac{1}{2}\mathbf{x}V^{-1}\mathbf{x}').$$

The joint characteristic function of $X_1, \ldots, X_n$ is

$$\phi(\mathbf{t}) = \mathbf{E}(e^{i\mathbf{t}\mathbf{X}'})$$

where $\mathbf{t} = (t_1, \ldots, t_n)$ and $\mathbf{X} = (X_1, \ldots, X_n)$. One way to proceed is to use the fact that $\mathbf{t}\mathbf{X}'$ is univariate normal. Alternatively,

(7) $$\phi(\mathbf{t}) = \int_{\mathbb{R}^n} [(2\pi)^n|V|]^{-\frac{1}{2}} \exp(i\mathbf{t}\mathbf{x}' - \tfrac{1}{2}\mathbf{x}V^{-1}\mathbf{x}')\,d\mathbf{x}.$$

As in the discussion of Section 4.9, there is a linear transformation $\boldsymbol{y} = \boldsymbol{x}\boldsymbol{B}$ such that

$$\boldsymbol{x}V^{-1}\boldsymbol{x}' = \sum_j \lambda_j y_j^2$$

just as in (4.9.3). Make this transformation in (7) to see that the integrand factorizes into the product of functions of the single variables $y_1, y_2, \ldots, y_n$. Then use (5) to obtain

$$\phi(\boldsymbol{t}) = \exp(-\tfrac{1}{2}\boldsymbol{t}V\boldsymbol{t}').$$

It is now an easy *exercise* to prove Theorem (4.9.5), that V is the covariance matrix of $\boldsymbol{X}$, by using the result of Problem (5.12.30). ●

(8) **Example. Gamma distribution.** If X is $\Gamma(\lambda, s)$ then

$$\phi(t) = \int_0^\infty \frac{1}{\Gamma(s)} \lambda^s x^{s-1} \exp(itx - \lambda x)\, dx.$$

As for the exponential distribution (3), routine methods of complex analysis give

$$\phi(t) = \left(\frac{\lambda}{\lambda - it}\right)^s.$$

Why is this similar to the result of (3)? This example includes the chi-squared distribution because a $\chi^2(d)$ variable is $\Gamma(\frac{1}{2}, \frac{1}{2}d)$ and thus has characteristic function

$$\phi(t) = (1 - 2it)^{-d/2}.$$

You may try to prove this from the result of Problem (4.11.12). ●

Exercises

(9) If ϕ is a characteristic function, show that $\bar{\phi}$, ϕ^2, $|\phi|^2$, $\mathrm{Re}(\phi)$ are characteristic functions. Show that $|\phi|$ is not necessarily a characteristic function.

(10) Show that

$$\mathbf{P}(X \geqslant x) \leqslant \inf_{t \geqslant 0} \{e^{-tx} M_X(t)\},$$

where M_X is the moment generating function of X.

(11) Let X have the $\Gamma(\lambda, m)$ distribution and let Y be independent of X with the beta distribution with parameters n and $m - n$, where m and n are non-negative integers satisfying $n \leqslant m$. Show that $Z = XY$ has the $\Gamma(\lambda, n)$ distribution.

(12) Find the characteristic function of X^2 when X has the $N(\mu, \sigma^2)$ distribution.

(13) Let $X_1, X_2, \ldots$ be independent $N(0, 1)$ variables. Use characteristic functions to find

the distributions of

(a) X_1^2, (b) $\sum_{i=1}^{n} X_i^2$,

(c) X_1/X_2, (d) X_1X_2,

(e) $X_1X_2 + X_3X_4$.

(14) Let $X_1, X_2, \dots, X_n$ be such that, for all $a_1, a_2, \dots, a_n \in \mathbb{R}$, the linear combination $a_1X_1 + a_2X_2 + \cdots + a_nX_n$ has a normal distribution. Show that the joint characteristic function of the X_i is $\exp(i\boldsymbol{t\mu}' - \frac{1}{2}\boldsymbol{t}\mathbf{V}\boldsymbol{t}')$, for some vector $\boldsymbol{\mu}$ and matrix $\mathbf{V}$. Deduce that the vector $(X_1, X_2, \dots, X_n)$ has a multivariate normal *density function* so long as $\mathbf{V}$ is invertible.

5.9 Inversion and continuity theorems

This section contains two major reasons why characteristic functions are useful. The first of these states that the distribution of a random variable is specified by its characteristic function. That is to say, if X and Y have the same characteristic function then they have the same distribution. Furthermore, there is a formula which tells us how to recapture the distribution function F corresponding to the characteristic function ϕ. Here is a special case first.

(1) **Theorem.** *If X is continuous with density function f and characteristic function ϕ then*

$$f(x) = \frac{1}{2\pi}\int_{-\infty}^{\infty} e^{-itx}\phi(t)\,dt$$

at every point x at which f is differentiable.

Proof. This is the Fourier inversion theorem and can be found in any introduction to Fourier transforms. If the integral fails to converge absolutely then we interpret it as its principal value (see Apostol 1957, p. 487). ■

A sufficient, but not necessary, condition that a characteristic function ϕ be the characteristic function of a continuous variable is that

$$\int_{-\infty}^{\infty} |\phi(t)|\,dt < \infty.$$

The general case is more complicated, and is contained in the next theorem.

(2) **Inversion theorem.** *Let X have distribution function F and characteristic function ϕ. Define $\bar{F}: \mathbb{R} \to [0, 1]$ by*

$$\bar{F}(x) = \frac{1}{2}\left\{F(x) + \lim_{y \uparrow x} F(y)\right\}.$$

Then

$$\bar{F}(b) - \bar{F}(a) = \lim_{N \to \infty} \int_{-N}^{N} \frac{e^{-iat} - e^{-ibt}}{2\pi it} \phi(t) \, dt.$$

Proof. See Kingman and Taylor (1966). ■

(3) **Corollary.** *X and Y have the same characteristic function if and only if they have the same distribution function.*

Proof. If $\phi_X = \phi_Y$ then, by (2),

$$\bar{F}_X(b) - \bar{F}_X(a) = \bar{F}_Y(b) - \bar{F}_Y(a).$$

Let $a \to -\infty$ to obtain

$$\bar{F}_X(b) = \bar{F}_Y(b);$$

now, for any fixed $x \in \mathbb{R}$, let $b \downarrow x$ and use right-continuity (2.1.6c) to obtain

$$F_X(x) = F_Y(x).$$ ■

Exactly similar results hold for jointly distributed random variables. For example, if X and Y have joint density function f and joint characteristic function ϕ then whenever f is differentiable at (x, y)

$$f(x, y) = \frac{1}{4\pi^2} \iint_{\mathbb{R}^2} e^{-isx} e^{-ity} \phi(s, t) \, ds \, dt$$

and (5.7.8) follows straight away for this special case.

The second result of this section deals with a sequence $X_1, X_2, \ldots$ of random variables. Roughly speaking it asserts that if the distribution functions $F_1, F_2, \ldots$ of the sequence approach some limit F then the characteristic functions $\phi_1, \phi_2, \ldots$ of the sequence approach the characteristic function of the distribution function F.

(4) **Definition.** We say that the sequence $F_1, F_2, \ldots$ of distribution functions **converges** to the distribution function F, written $F_n \to F$, if $F(x) = \lim_{n\to\infty} F_n(x)$ at each point x where F is continuous.

The reason for the condition of continuity of F at x is indicated by the following. Define the distribution functions F_n and G_n by

$$F_n(x) = \begin{cases} 0 & \text{if } x < \frac{1}{n} \\ 1 & \text{if } x \geq \frac{1}{n} \end{cases} \qquad G_n(x) = \begin{cases} 0 & \text{if } x < -\frac{1}{n} \\ 1 & \text{if } x \geq -\frac{1}{n}. \end{cases}$$

As $n \to \infty$

$$F_n(x) \to F(x) \quad \text{if } x \neq 0,\ F_n(0) \to 0$$
$$G_n(x) \to F(x) \quad \text{for all } x$$

where $F(x)$ is the distribution function of a random variable which is constantly zero. Indeed $\lim_{n\to\infty} F_n(x)$ is not even a distribution function since it is not right-continuous at zero. It is intuitively reasonable to demand that the sequences $\{F_n\}$ and $\{G_n\}$ have the same limit, and so we drop the requirement that $F_n(x) \to F(x)$ at the point of discontinuity of F.

(5) **Continuity theorem.** *Suppose that $F_1, F_2, \ldots$ is a sequence of distribution functions with corresponding characteristic functions $\phi_1, \phi_2, \ldots$.*

(a) *If $F_n \to F$ for some distribution function F with characteristic function ϕ, then $\phi_n(t) \to \phi(t)$ for all t.*
(b) *Conversely, if $\phi(t) = \lim_{n\to\infty} \phi_n(t)$ exists and is continuous at $t = 0$, then ϕ is the characteristic function of some distribution function F, and $F_n \to F$.*

Proof. As for (2). See also (5.12.35). ∎

(6) **Example. Stirling's formula.** This well-known formula, due to de Moivre, states that $n! \simeq n^n e^{-n}\sqrt{(2\pi n)}$ as $n \to \infty$, which is to say that

$$\frac{n!}{n^{n+\frac{1}{2}} e^{-n}\sqrt{(2\pi)}} \to 1 \quad \text{as} \quad n \to \infty.$$

A more general form of this relation states that

(7)
$$\frac{\Gamma(t)}{t^{t-\frac{1}{2}} e^{-t}\sqrt{(2\pi)}} \to 1 \quad \text{as} \quad t \to \infty$$

where Γ is the gamma function, $\Gamma(t) = \int_0^\infty x^{t-1} e^{-x}\,dx$. Remember that $\Gamma(t) = (t-1)!$ if t is a positive integer; see (4.4.6) and (4.4.10). To prove (7) is an 'elementary' exercise in analysis, but it is perhaps amusing to see how simply it follows from the Fourier inversion theorem (1).

Let Y be a random variable with the $\Gamma(1, t)$ distribution. Then $X = t^{-\frac{1}{2}}(Y - t)$ has density function

(8)
$$f_t(x) = \frac{1}{\Gamma(t)} t^{\frac{1}{2}}(xt^{\frac{1}{2}} + t)^{t-1} \exp[-(xt^{\frac{1}{2}} + t)], \quad -t^{\frac{1}{2}} \leqslant x < \infty,$$

and characteristic function

$$\phi_t(u) = \mathbf{E}(e^{iuX}) = \exp(-iut^{\frac{1}{2}})\left(1 - \frac{iu}{t^{\frac{1}{2}}}\right)^{-t}.$$

Now f_t is differentiable on $(-t^{\frac{1}{2}}, \infty)$. We apply Theorem (1) at $x = 0$ to obtain

(9) $$f_t(0) = \frac{1}{2\pi}\int_{-\infty}^{\infty} \phi_t(u)\, du.$$

However, $f_t(0) = t^{t-\frac{1}{2}} e^{-t}/\Gamma(t)$ from (8); also

$$\begin{aligned}\phi_t(u) &= \exp\left[-iut^{\frac{1}{2}} - t\log\left(1 - \frac{iu}{t^{\frac{1}{2}}}\right)\right]\\ &= \exp\left[-iut^{\frac{1}{2}} - t\left(-\frac{iu}{t^{\frac{1}{2}}} + \frac{u^2}{2t} + O(u^3 t^{-\frac{3}{2}})\right)\right]\\ &= \exp[-\tfrac{1}{2}u^2 + O(u^3 t^{-\frac{1}{2}})] \to \exp(-u^2/2) \quad \text{as} \quad t \to \infty.\end{aligned}$$

Taking the limit in (9) as $t \to \infty$, we find that

$$\begin{aligned}\lim_{t\to\infty}\left(\frac{1}{\Gamma(t)} t^{t-\frac{1}{2}} e^{-t}\right) &= \lim_{t\to\infty} \frac{1}{2\pi}\int_{-\infty}^{\infty} \phi_t(u)\, du\\ &= \frac{1}{2\pi}\int_{-\infty}^{\infty}\left(\lim_{t\to\infty}\phi_t(u)\right) du\\ &= \frac{1}{2\pi}\int_{-\infty}^{\infty} e^{-u^2/2}\, du = \frac{1}{\sqrt{(2\pi)}}\end{aligned}$$

as required for (7). A spot of rigour is needed to justify the interchange of the limit and the integral sign above, and this may be provided by the dominated convergence theorem. ●

Exercises

(10) Let X_n be a discrete random variable taking values in $\{1, 2, \ldots, n\}$, each possible value having probability n^{-1}. Show that, as $n \to \infty$, $\mathbf{P}(n^{-1}X_n \leqslant y) \to y$, for $0 \leqslant y \leqslant 1$.

(11) Let X_n have distribution function

$$F_n(x) = x - \frac{\sin(2n\pi x)}{2n\pi}, \qquad 0 \leqslant x \leqslant 1.$$

(a) Show that F_n is indeed a distribution function, and that X_n has a density function.

(b) Show that, as $n \to \infty$, F_n converges to the uniform distribution function, but that the density function of F_n does not converge to the uniform density function.

(12) A coin is tossed repeatedly, with heads turning up with probability p on each toss. Let N be the minimum number of tosses required to obtain k heads. Show that, as $p \downarrow 0$, the distribution function of $2Np$ converges to that of a gamma distribution.

(13) If X is an integer-valued random variable with characteristic function ϕ, show that

$$\mathbf{P}(X = k) = \frac{1}{2\pi}\int_{-\pi}^{\pi} e^{-itk}\phi(t)\,dt.$$

What is the corresponding result for a random variable whose distribution is arithmetic with span λ (that is, there is probability one that X is a multiple of λ, and λ is the largest positive number with this property)?

5.10 Two limit theorems

We are now in a position to prove two very celebrated theorems in probability theory, the 'law of large numbers' and the 'central limit theorem'. The first of these explains the remarks of Sections 1.1 and 1.3, where we discussed a heuristic foundation of probability theory. Part of our intuition about chance is that if we perform many repetitions of an experiment which has numerical outcomes then the average of all the outcomes settles down to some fixed number. This observation deals once again in the convergence of sequences of random variables, the general theory of which is dealt with later. Here it suffices to introduce only one new definition.

(1) **Definition.** If $X_1, X_2, \ldots, X$ is a collection of random variables with distributions $F_1, F_2, \ldots, F$, then we say that X_n **converges in distribution** to X, written $X_n \xrightarrow{D} X$, if $F_n \to F$.

This is just (5.9.4) rewritten in terms of random variables.

(2) **Theorem. Law of large numbers.** *Let $X_1, X_2, \ldots$ be a sequence of independent identically distributed random variables with finite means μ. Their partial sums*

$$S_n = X_1 + X_2 + \cdots + X_n$$

satisfy

$$\frac{1}{n}S_n \xrightarrow{D} \mu \quad \textit{as} \quad n \to \infty.$$

Proof. The theorem asserts that

$$\mathbf{P}(S_n \leqslant x) \to \begin{cases} 0 & \text{if} \quad x < \mu \\ 1 & \text{if} \quad x > \mu \end{cases} \quad \text{as} \quad n \to \infty.$$

The method of proof is clear. By the continuity theorem (5.9.5) we need to show that the characteristic function of S_n approaches the characteristic function of the constant random variable μ. This is easy. Let ϕ_X be the common characteristic function of the X_i, and let ϕ_n be the characteristic function of $(1/n)S_n$. By (5.7.5) and (5.7.6),

(3) $$\phi_n(t) = (\phi_X(t/n))^n.$$

The behaviour of $\phi_X(t/n)$ for large n is given by (5.7.4):

$$\phi_X(t) = 1 + \mathrm{i}t\mu + \mathrm{o}(t).$$

Substitute into (3) to obtain

$$\phi_n(t) = \left(1 + \frac{\mathrm{i}\mu t}{n} + \mathrm{o}\left(\frac{t}{n}\right)\right)^n \to \mathrm{e}^{\mathrm{i}t\mu} \quad \text{as} \quad n \to \infty.$$

However, this limit is the characteristic function of the constant μ and the result is proved. ■

So, for large n, S_n is about as big as $n\mu$. What can we say about the difference $S_n - n\mu$? There is an extraordinary answer to this question so long as the X_i have finite variance:

(a) $S_n - n\mu$ is about as big as $n^{\frac{1}{2}}$,
(b) the distribution of $n^{-\frac{1}{2}}(S_n - n\mu)$ approaches the normal distribution as $n \to \infty$ *irrespective* of the distribution of the X_i.

(4) **Central limit theorem.** *Let $X_1, X_2, \ldots$ be a sequence of independent identically distributed random variables with finite means μ and finite non-zero variances σ^2, and let*

$$S_n = X_1 + X_2 + \cdots + X_n.$$

Then

$$\frac{S_n - n\mu}{\sqrt{(n\sigma^2)}} \xrightarrow{\mathrm{D}} N(0, 1) \quad \textit{as} \quad n \to \infty.$$

Note that the assertion of the theorem is an abuse of notation, since $N(0, 1)$ is a distribution and not a random variable; it is admissible because convergence in distribution involves only the corresponding distribution functions. The method of proof is the same as for the law of large numbers.

Proof. First, write $Y_i = (X_i - \mu)/\sigma$, and let ϕ_Y be the characteristic function of the Y_i. By (5.7.4)

$$\phi_Y(t) = 1 - \tfrac{1}{2}t^2 + \mathrm{o}(t^2).$$

Also, the characteristic function ψ_n of

$$U_n = \frac{S_n - n\mu}{\sqrt{(n\sigma^2)}} = \frac{1}{\sqrt{n}} \sum_1^n Y_i$$

satisfies, by (5.7.5) and (5.7.6),

$$\begin{aligned}\psi_n(t) &= (\phi_Y(tn^{-\frac{1}{2}}))^n\\ &= \left(1 - \frac{t^2}{2n} + \mathrm{o}\left(\frac{t^2}{n}\right)\right)^n\\ &\to \mathrm{e}^{-t^2/2} \quad \text{as} \quad n \to \infty.\end{aligned}$$

However, this is the characteristic function of the $N(0, 1)$ distribution and the continuity theorem (5.9.5) completes the proof. ■

Numerous generalizations of the law of large numbers and the central limit theorem are available. For example, in Chapter 7 we shall meet two stronger versions of (2), involving weaker assumptions on the X_i and more powerful conclusions. The central limit theorem can be generalized in several directions, two of which deal with dependent variables and differently distributed variables respectively. Some of these are within the reader's grasp. Here is an example.

(5) **Theorem.** *Let $X_1, X_2, \ldots$ be independent variables satisfying*

$$\mathbf{E}X_j = 0, \qquad \operatorname{var}(X_j) = \sigma_j^2, \qquad \mathbf{E}|X_j^3| < \infty$$

and such that

$$\frac{1}{\sigma(n)^3}\sum_{j=1}^{n} \mathbf{E}|X_j^3| \to 0 \quad \text{as} \quad n \to \infty$$

where

$$\sigma(n)^2 = \operatorname{var}\left(\sum_1^n X_j\right) = \sum_1^n \sigma_j^2.$$

Then

$$\frac{1}{\sigma(n)}\sum_1^n X_j \xrightarrow{\text{D}} N(0, 1).$$

Proof. See Loève (1977, p. 287), and also Problem (5.12.40). ■

The roots of central limit theory are at least 250 years old. The first proof of (4) was found by de Moivre around 1733 for the special case of Bernoulli variables with $p = \frac{1}{2}$. General values of p were treated later by Laplace. Their methods involved the direct estimation of sums like

$$\sum_{k \leqslant np + x\sqrt{(npq)}} \binom{n}{k} p^k q^{n-k} \quad \text{where} \quad p + q = 1.$$

The first rigorous proof of (4) was discovered by Lyapunov around 1901, thereby confirming a less rigorous proof of Laplace. A glance at these old proofs confirms that the method of characteristic functions is outstanding in its elegance and brevity.

The central limit theorem (4) asserts that the *distribution function* of S_n, suitably normalized to have mean 0 and variance 1, converges to the distribution function of the $N(0, 1)$ distribution. Is the corresponding result valid at the level of density functions and mass functions? Broadly speaking the answer is yes, but some condition of smoothness is necessary; after all, if $F_n(x) \to F(x)$ as $n \to \infty$ for all x, it is not necessarily the case that the derivatives satisfy $F_n'(x) \to F'(x)$. The result which follows is called a 'local limit theorem' since it deals in the local rather than the cumulative behaviour of the random variables in question. In order to simplify the statement of the theorem, we shall assume that the X_i have zero mean and unit variance.

(6) **Local limit theorem.** *Let $X_1, X_2, \dots$ be independent identically distributed random variables with zero mean and unit variance, and suppose further that their common characteristic function ϕ satisfies*

$$\int_{-\infty}^{\infty} |\phi(t)|^r \, dt < \infty \tag{7}$$

for some integer $r \geq 1$. The density function g_n of $U_n = n^{-\frac{1}{2}}(X_1 + X_2 + \cdots + X_n)$ exists for $n \geq r$, and furthermore

$$g_n(x) \to \frac{1}{\sqrt{(2\pi)}} e^{-x^2/2} \quad as \quad n \to \infty \tag{8}$$

uniformly in $x \in \mathbb{R}$.

A similar result is valid for sums of lattice-valued random variables, suitably adjusted to have zero mean and unit variance. We state this here, leaving its proof as an *exercise*. In place of (7) we assume that the X_i are restricted to take the values a, $a \pm h$, $a \pm 2h, \dots$, where h is the largest positive number for which such a restriction holds. Then U_n is restricted to values of the form $x = (na + kh)n^{-\frac{1}{2}}$ for $k = 0, \pm 1, \dots$. For such a number x, we write $g_n(x) = \mathbf{P}(U_n = x)$ and leave $g_n(y)$ undefined for other values of y. It is the case that

$$\frac{\sqrt{n}}{h} g_n(x) \to \frac{1}{\sqrt{(2\pi)}} e^{-x^2/2} \quad \text{as} \quad n \to \infty \tag{9}$$

uniformly in all appropriate x.

Proof of (6). A certain amount of analysis is inevitable here. First, the assumption that $|\phi|^r$ is integrable for some $r \geq 1$ implies that $|\phi|^n$ is integrable for $n \geqslant r$, since $|\phi(t)| \leqslant 1$; hence g_n exists and is given by the Fourier inversion formula

(10) $$g_n(x) = \frac{1}{2\pi}\int_{-\infty}^{\infty} e^{-itx}\psi_n(t)\,dt,$$

where $\psi_n(t) = \phi(t/\sqrt{n})^n$ is the characteristic function of U_n. The Fourier inversion theorem is valid for the normal distribution, and therefore

(11) $$\left|g_n(x) - \frac{1}{\sqrt{(2\pi)}}e^{-x^2/2}\right| \leqslant \frac{1}{2\pi}\left|\int_{-\infty}^{\infty} e^{-itx}[\phi(t/\sqrt{n})^n - e^{-t^2/2}]\,dt\right| \leqslant I_n$$

where

$$I_n = \frac{1}{2\pi}\int_{-\infty}^{\infty} |\phi(t\sqrt{n})^n - e^{-t^2/2}|\,dt.$$

It suffices to show that $I_n \to 0$ as $n \to \infty$. We have from (5.7.4) that $\phi(t) = 1 - \frac{1}{2}t^2 + o(t^2)$ as $t \to 0$, and therefore there exists δ (>0) such that

(12) $$|\phi(t)| \leqslant e^{-t^2/4} \quad \text{if} \quad |t| \leqslant \delta.$$

Now, for any $a > 0$, $\phi(t/\sqrt{n})^n \to e^{-t^2/2}$ as $n \to \infty$ *uniformly* in $t \in [-a, a]$ (to see this, investigate the proof of (4) slightly more carefully), so that

(13) $$\int_{-a}^{a} |\phi(t/\sqrt{n})^n - e^{-t^2/2}|\,dt \to 0 \quad \text{as} \quad n \to \infty,$$

for any a. Also, by (12),

(14) $$\int_{a<|t|<\delta\sqrt{n}} |\phi(t/\sqrt{n})^n - e^{-t^2/2}|\,dt \leqslant 2\int_a^{\infty} 2\,e^{-t^2/4}\,dt$$

which tends to zero as $a \to \infty$.

It remains to deal with the contribution to I_n arising from $|t| > \delta\sqrt{n}$. From the fact that g_n exists for $n \geqslant r$, we have from Exercises (5.7.15) and (5.7.16) that $|\phi(t)^r| < 1$ for $t \neq 0$, and $|\phi(t)^r| \to 0$ as $t \to \pm\infty$. Hence $|\phi(t)| < 1$ for $t \neq 0$, and $|\phi(t)| \to 0$ as $t \to \pm\infty$, and therefore $\eta = \sup\{|\phi(t)| : |t| \geqslant \delta\}$ satisfies $\eta < 1$. Now, for $n \geqslant r$,

(15) $$\begin{aligned}\int_{|t|>\delta\sqrt{n}} |\phi(t/\sqrt{n})^n - e^{-t^2/2}|\,dt &\leqslant \eta^{n-r}\int_{-\infty}^{\infty} |\phi(t/\sqrt{n})|^r\,dt + 2\int_{\delta\sqrt{n}}^{\infty} e^{-t^2/2}\,dt \\ &= \eta^{n-r}\sqrt{n}\int_{-\infty}^{\infty} |\phi(u)|^r\,du + 2\int_{\delta\sqrt{n}}^{\infty} e^{-t^2/2}\,dt \\ &\to 0 \quad \text{as} \quad n \to \infty.\end{aligned}$$

Combining (13)–(15), we deduce that

$$\lim_{n\to\infty} I_n \leqslant 4\int_a^\infty e^{-t^2/4}\,dt \to 0 \quad \text{as} \quad a \to \infty,$$

so that $I_n \to 0$ as $n \to \infty$ as required. ■

(16) **Example. Random walks.** Here is an application of the law of large numbers to the persistence of random walks. A simple random walk performs steps of size 1, to the right or left with probabilities p and $1 - p$. We saw in Section 5.3 that a simple random walk is persistent (that is, returns to its starting point with probability 1) if and only if it is symmetric (which is to say that $p = 1 - p = \frac{1}{2}$). Think of this as saying that the walk is persistent if and only if the mean value of a typical step X satisfies $\mathbf{E}(X) = 0$, that is, each step is 'unbiased'. This conclusion is valid in much greater generality.

Let $X_1, X_2, \ldots$ be independent identically distributed integer-valued random variables, and let $S_n = X_1 + X_2 + \cdots + X_n$. We think of X_i as being the ith jump of a random walk, so that S_n is the position of the random walker after n jumps, having started at $S_0 = 0$. We call the walk *persistent* (or *recurrent*) if $\mathbf{P}(S_n = 0 \text{ for some } n \geqslant 1) = 1$ and *transient* otherwise.

(17) **Theorem.** *The random walk is persistent if the mean size of jumps is* 0.

The converse is valid also: the walk is transient if the mean size of jumps is non-zero (Problem (5.12.44)).

Proof. Suppose that $\mathbf{E}(X_i) = 0$ and let V_i denote the mean number of visits of the walk to the point i,

$$V_i = \mathbf{E}|\{n \geqslant 0 : S_n = i\}| = \mathbf{E}\left(\sum_{n=0}^{\infty} I_{\{S_n = i\}}\right) = \sum_{n=0}^{\infty} \mathbf{P}(S_n = i),$$

where I_A is the indicator function of the event A, as usual. We shall prove first that $V_0 = \infty$, and from this we shall deduce the persistence of the walk. Let T be the time of the first visit of the walk to i, with the convention that $T = \infty$ if i is never visited. Then

$$V_i = \sum_{n=0}^{\infty} \mathbf{P}(S_n = i) = \sum_{n=0}^{\infty}\sum_{t=0}^{\infty} \mathbf{P}(S_n = i \mid T = t)\mathbf{P}(T = t)$$
$$= \sum_{t=0}^{\infty}\left(\sum_{n=t}^{\infty} \mathbf{P}(S_n = i \mid T = t)\right)\mathbf{P}(T = t)$$

since $S_n \neq i$ for $n < T$. Now we use the spatial homogeneity of the walk to deduce that

(18) $$V_i = \sum_{t=0}^{\infty} V_0 \mathbf{P}(T = t) = V_0 \mathbf{P}(T < \infty) \leqslant V_0.$$

The mean number of time points n for which $|S_n| \leqslant K$ satisfies

$$\sum_{n=0}^{\infty} \mathbf{P}(|S_n| \leqslant K) = \sum_{i=-K}^{K} V_i \leqslant (2K+1)V_0$$

by (18), and hence

(19) $$V_0 \geqslant \frac{1}{2K+1} \sum_{n=0}^{\infty} \mathbf{P}(|S_n| \leqslant K).$$

Now we use the law of large numbers. For $\varepsilon > 0$, it is the case that $\mathbf{P}(|S_n| \leqslant n\varepsilon) \to 1$ as $n \to \infty$, so that there exists m such that $\mathbf{P}(|S_n| \leqslant n\varepsilon) > \frac{1}{2}$ for $n \geqslant m$. If $n\varepsilon \leqslant K$ then $\mathbf{P}(|S_n| \leqslant n\varepsilon) \leqslant \mathbf{P}(|S_n| \leqslant K)$, so that

(20) $$\mathbf{P}(|S_n| \leqslant K) > \tfrac{1}{2} \quad \text{for} \quad m \leqslant n \leqslant K/\varepsilon.$$

Substituting (20) into (19), we obtain

$$V_0 \geqslant \frac{1}{2K+1} \sum_{m \leqslant n \leqslant K/\varepsilon} \mathbf{P}(|S_n| \leqslant K) > \frac{1}{2(2K+1)}\left(\frac{K}{\varepsilon} - m - 1\right).$$

This is valid for all large K, and we may therefore let $K \to \infty$ and $\varepsilon \downarrow 0$ in that order, finding that $V_0 = \infty$ as claimed.

It is now fairly straightforward to deduce that the walk is persistent. Let $T(1)$ be the time of the first return to 0, with the convention that $T(1) = \infty$ if this never occurs. If $T(1) < \infty$, we write $T(2)$ for the subsequent time which elapses until the next visit to 0. It is clear from the homogeneity of the process that, conditional on $\{T(1) < \infty\}$, $T(2)$ has the same distribution as $T(1)$. Continuing likewise, we see that the times of returns to 0 are distributed in the same way as the sequence $T_1, T_1 + T_2, \ldots$, where $T_1, T_2, \ldots$ are independent identically distributed random variables having the same distribution as $T(1)$. We wish to exclude the possibility that $\mathbf{P}(T(1) = \infty) > 0$. There are several ways of doing this, one of which is to make use of the recurrent-event analysis of Example (5.2.15). We shall take a slightly more direct route here. Suppose that $\beta = \mathbf{P}(T(1) = \infty)$ satisfies $\beta > 0$, and let $I = \min\{i\colon T_i = \infty\}$ be the earliest i for which T_i is infinite. The event $\{I = i\}$ corresponds to exactly $i - 1$ returns to the origin. Thus, the mean number of returns is $\sum_{i=1}^{\infty} (i-1)\mathbf{P}(I = i)$. However, $I = i$ if and only if $T_j < \infty$ for $1 \leqslant j < i$ and $T_i = \infty$, an event with probability $(1-\beta)^{i-1}\beta$. Hence the mean number of returns to 0 is $\sum_{i=1}^{\infty} (i-1)(1-\beta)^{i-1}\beta = (1-\beta)/\beta$, which is finite. This contradicts the infiniteness of V_0, and hence $\beta = 0$. The proof is complete. ■

We have proved that a walk whose jumps have zero mean must (with probability 1) return to its starting point. It follows that it must return *infinitely often*, since otherwise there exists some T_i which equals infinity, an event having zero probability. ●

(21) **Example. Recurrent events.** The renewal theorem of Example (5.2.15) is one of the basic results of applied probability, and it will recur in various forms through this book. Our 'elementary' proof in (5.2.15) was incomplete, but we may now complete it with the aid of the last theorem (17) concerning the persistence of random walks.

Suppose that we are provided with two sequences $X_1, X_2, \ldots$ and $X_1^*, X_2^*, \ldots$ of independent identically distributed random variables taking values in the positive integers $\{1, 2, \ldots\}$. Let $Y_n = X_n - X_n^*$ and $S_n = \sum_{i=1}^n Y_i = \sum_{i=1}^n X_i - \sum_{i=1}^n X_i^*$. Then $S = (S_n: n \geqslant 0)$ may be thought of as a random walk on the integers with steps $Y_1, Y_2, \ldots$; the mean step-size satisfies $\mathbf{E}(Y_1) = \mathbf{E}(X_1) - \mathbf{E}(X_1^*) = 0$, and therefore this walk is persistent, by Theorem (17). Furthermore, the walk must revisit its starting point *infinitely often* (with probability 1), which is to say that $\sum_{i=1}^n X_i = \sum_{i=1}^n X_i^*$ for infinitely many values of n.

What have we proved about recurrent-event processes? Consider two independent recurrent-event processes for which the first occurrence times, X_1 and X_1^*, have the same distribution as the inter-occurrence times. Not only does there exist some finite time T at which the event H occurs simultaneously in both processes, but also: (i) there exist infinitely many such times T, and (ii) there exist infinitely many such times T even if one insists that, by time T, the event H has occurred the *same number of times* in the two processes.

We need to relax the assumption that X_1 and X_1^* have the same distribution as the inter-occurrence times, and it is here that we require that the process be non-arithmetic. Suppose that $X_1 = u$ and $X_1^* = v$. Now $S_n = S_1 + \sum_{i=2}^n Y_i$ is a random walk with zero mean jump size and starting point $S_1 = u - v$. By the foregoing argument, there exist (with probability 1) infinitely many values of n such that $S_n = u - v$, which is to say that

(22)
$$\sum_{i=2}^{n} X_i = \sum_{i=2}^{n} X_i^*;$$

we denote these (random) times by the increasing sequence $N_1, N_2, \ldots$.

The process is non-arithmetic, and it follows that, for any integer x, there exist integers r and s such that

(23)
$$\gamma(r, s; x) = \mathbf{P}\big((X_2 + \cdots + X_r) - (X_2^* + \cdots + X_s^*) = x\big) > 0.$$

To check this is an elementary *exercise* (27) in number theory. The reader may be satisfied with the following proof for the special case when $\beta = \mathbf{P}(X_2 = 1)$ satisfies $\beta > 0$. Then

$$\mathbf{P}(X_2 + \cdots + X_{x+1} = x) \geqslant \mathbf{P}(X_i = 1 \text{ for } 2 \leqslant i \leqslant x + 1) = \beta^x > 0$$

if $x \geqslant 0$, and

$$\mathbf{P}(-X_2^* - \cdots - X_{|x|+1}^* = x) \geqslant \mathbf{P}(X_i^* = 1 \text{ for } 2 \leqslant i \leqslant |x| + 1) = \beta^{|x|} > 0$$

if $x < 0$, so that (23) is valid with $r = x + 1$, $s = 1$ and $r = 1$, $s = |x| + 1$ in these two respective cases. Without more ado we shall accept that such r, s exist under the assumption that the process is non-arithmetic. We set $x = -(u - v)$, choose r and s accordingly, and write $\gamma = \gamma(r, s; x)$.

Suppose now that (22) occurs for some value of n. Then

$$\sum_{i=1}^{n+r-1} X_i - \sum_{i=1}^{n+s-1} X_i^* = (X_1 - X_1^*) + \left(\sum_{i=n+1}^{n+r-1} X_i - \sum_{i=n+1}^{n+s-1} X_i^*\right)$$

which equals $(u - v) - (u - v) = 0$ with strictly positive probability (since the contents of the final parentheses have, by (23), strictly positive probability of equalling $-(u - v)$). Therefore, for each n satisfying (22), there is a strictly positive probability γ that the $(n + r - 1)$th recurrence of the first process coincides with the $(n + s - 1)$th recurrence of the second. There are infinitely many such values N_i for n, and one of infinitely many shots at a target must succeed! More rigorously, define $M_1 = N_1$, and $M_{i+1} = \min\{N_j: N_j > M_i + \max\{r, s\}\}$; the sequence of the M_i is an infinite subsequence of the N_j satisfying $M_{i+1} - M_i > \max\{r, s\}$. Call M_i a *failure* if the $(M_i + r - 1)$th recurrence of the first process does not coincide with the $(M_i + s - 1)$th of the second. Then the events $F_I = \{M_i \text{ is a failure for } 1 \leqslant i \leqslant I\}$ satisfy

$$\mathbf{P}(F_{I+1}) = \mathbf{P}(M_{I+1} \text{ is a failure} \mid F_I)\mathbf{P}(F_I) = (1 - \gamma)\mathbf{P}(F_i),$$

so that $\mathbf{P}(F_I) = (1 - \gamma)^I \to 0$ as $I \to \infty$. However, $\{F_I: I \geqslant 1\}$ is a decreasing sequence of events with limit $\{M_i \text{ is a failure for all } i\}$, which event therefore has zero probability. Thus one of the M_i is *not* a failure, with probability 1, implying that some recurrence of the first process coincides with some recurrence of the second, as required.

The above argument is valid for all 'initial values' u and v for X_1 and X_1^*, and therefore for all choices of the distribution of X_1 and X_1^*:

$$\begin{aligned}\mathbf{P}(\text{coincident recurrences}) &= \sum_{u,v} \mathbf{P}(\text{coincident recurrences} \mid X_1 = u, X_1^* = v) \\ &\qquad \times \mathbf{P}(X_1 = u)\mathbf{P}(X_1^* = v) \\ &= \sum_{u,v} 1 \cdot \mathbf{P}(X_1 = u)\mathbf{P}(X_1^* = v) = 1.\end{aligned}$$

In particular, the conclusion is valid when X_1^* has probability generating function D^* given by (5.2.22); the proof of the renewal theorem is thereby completed. ●

Exercises

(24) Prove that, for $x \geqslant 0$, as $n \to \infty$,

(a) $$\sum_{\substack{k:\\ |k - \frac{1}{2}n| \leqslant \frac{1}{2}x\sqrt{n}}} \binom{n}{k} \sim 2^n \int_{-x}^{x} \frac{1}{\sqrt{(2\pi)}} e^{-\frac{1}{2}u^2}\, du,$$

(b) $$\sum_{\substack{k:\\ |k - n| \leqslant x\sqrt{n}}} \frac{n^k}{k!} \sim e^n \int_{-x}^{x} \frac{1}{\sqrt{(2\pi)}} e^{-\frac{1}{2}u^2}\, du.$$

(25) It is well known that infants born to mothers who smoke tend to be small and prone to a range of ailments. It is conjectured that also they look abnormal. Nurses were shown selections of photographs of babies, half of whom had smokers as mothers; the nurses were asked to judge from a baby's appearance whether or not the mother smoked. In 1500 trials the correct answer was given 910 times. Is the conjecture plausible? If so, why?

(26) Let X have the $\Gamma(1, s)$ distribution; given that $X = x$, let Y have the Poisson distribution with parameter x. Find the characteristic function of Y, and show that

$$\frac{Y - \mathbf{E}(Y)}{\sqrt{\operatorname{var}(Y)}} \xrightarrow{D} N(0, 1) \quad \text{as} \quad s \to \infty.$$

Explain the connection with the central limit theorem.

(27) Let $X_1, X_2, \ldots$ be independent random variables taking values in the positive integers, whose common distribution is non-arithmetic, in that $\gcd\{n: \mathbf{P}(X_1 = n) > 0\} = 1$. Prove that, for all integers x, there exist non-negative integers $r = r(x)$, $s = s(x)$, such that

$$\mathbf{P}(X_1 + \cdots + X_r - X_{r+1} - \cdots - X_{r+s} = x) > 0.$$

(28) Prove the local limit theorem for sums of random variables taking integer values. You may assume for simplicity's sake that the summands have span 1, in that they satisfy $\gcd\{|x|: \mathbf{P}(X = x) > 0\} = 1$.

(29) Let $X_1, X_2, \ldots$ be independent random variables having common density function $f(x) = \{2|x|(\log|x|)^2\}^{-1}$ for $|x| < e^{-1}$. Show that the X_i have zero mean and finite variance, and that the density function f_n of $X_1 + X_2 + \cdots + X_n$ satisfies $f_n(x) \to \infty$ as $x \to 0$. Deduce that the X_i do not satisfy the local limit theorem.

5.11 Large deviations

The law of large numbers asserts that the sum S_n of n independent identically distributed variables is approximately $n\mu$, where μ is a typical mean. The central limit theorem asserts that the deviations of S_n from $n\mu$ are typically of the order of $n^{\frac{1}{2}}$, that is, small compared with the mean. Of course, S_n may deviate from $n\mu$ by quantities of greater order than $n^{\frac{1}{2}}$, say n^α where $\alpha > \frac{1}{2}$, but such 'large deviations' have probabilities which tend to zero as $n \to \infty$. It is often necessary to estimate such probabilities, in order to control the chances of errors. The theory of large deviations studies the asymptotic behaviour of $\mathbf{P}(|S_n - n\mu| > n^\alpha)$ as $n \to \infty$, for values of α satisfying $\alpha > \frac{1}{2}$; of particular interest is the case when $\alpha = 1$, corresponding to deviations of S_n from its mean $n\mu$ having the same order as the mean. The behaviour of such quantities is somewhat delicate, depending on rather more than the mean and variance of a typical summand.

Let $X_1, X_2, \ldots$ be a sequence of independent identically distributed random variables having means μ and partial sums

(1) $$S_n = X_1 + X_2 + \cdots + X_n.$$

It is our target to estimate $\mathbf{P}(S_n - n\mu > na)$ where $a > 0$. This question is only interesting if $\mathbf{P}(X_1 - \mu > a) > 0$, since otherwise $\mathbf{P}(S_n - n\mu \leqslant na) = 1$. In answering this question, we may assume without loss of generality that $\mu = 0$; if $\mu \neq 0$, replace X_i by $X_i - \mu$.

(2) **Theorem. Large deviations.** *Let $X_1, X_2, \ldots$ be a sequence of independent identically distributed random variables with mean 0, and suppose that their moment generating function $M(t) = \mathbf{E}(\exp(tX_1))$ is finite in some neighbourhood of the point $t = 0$. If $a > 0$ and $\mathbf{P}(X_1 > a) > 0$ then*

(3) $$\mathbf{P}(S_n > na)^{1/n} \to e^{-\psi(a)} \quad \text{as} \quad n \to \infty,$$

where

(4) $$\psi(a) = -\log\left(\inf_{t>0} \{e^{-at}M(t)\}\right)$$

satisfies $\psi(a) > 0$.

This remarkably useful result asserts that $\mathbf{P}(S_n > na)$ decays to 0 exponentially fast as $n \to \infty$, and moreover it identifies the rate of convergence explicitly as being that of $e^{-n\psi(a)}$, where $\psi(a)$ is given by (4).

The theorem appears to deal only with deviations of S_n in *excess* of its mean; the corresponding result for deviations of S_n below its mean is obtained by replacing X_i by $-X_i$.

Proof. Often the most useful part of the theorem is the upper bound for $\mathbf{P}(S_n > na)$ therein. We consider this first, using the same method as for Bernstein's inequality (2.2.4). For $t > 0$, $e^{tS_n} > e^{tx}I_{\{S_n > x\}}$, so that, setting $x = na$,

$$\mathbf{P}(S_n > na) \leqslant e^{-tan}\mathbf{E}(e^{tS_n}) = \{e^{-at}M(t)\}^n.$$

This is valid for all $t > 0$, and therefore

(5) $$\mathbf{P}(S_n > na)^{1/n} \leqslant \inf_{t>0} \{e^{-at}M(t)\} = e^{-\psi(a)},$$

providing half of (3).

Before continuing with the other half of (3), we note that the bound in (5) is interesting only if $\psi(a) > 0$. To see that $\psi(a) > 0$, it suffices to show that $e^{-at}M(t) < 1$ for some $t > 0$. For small positive values of t, we have from the remarks after (5.7.1) that

$$e^{-at}M(t) = \frac{1 + \frac{1}{2}\sigma^2 t^2 + o(t^2)}{1 + at + o(t)}$$

where $\sigma^2 = \text{var}(X_1)$; it is here that we use the assumption that $M(t) < \infty$ near the origin. For sufficiently small t, the denominator exceeds the numerator, so that the ratio is less than 1 as required.

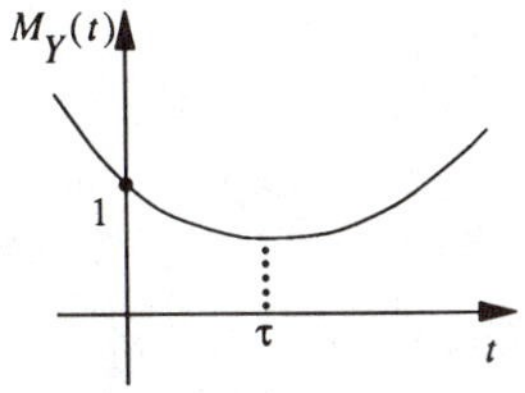

Fig. 5.4 A sketch of the moment generating function $M_Y(t)$.

We turn now to the more difficult matter of establishing a respectable lower bound for $\mathbf{P}(S_n > na)$. It is convenient to consider the 'shifted' random variables $Y_i = X_i - a$, with means $\mathbf{E}(Y_1) = -a < 0$. The moment generating function $M_Y(t) = \mathbf{E}(e^{tY_1})$ is given by

(6) $$M_Y(t) = e^{-at}M(t).$$

Now M_Y is a convex function of t (since it is an average over y of e^{ty}, which is itself convex in t), and in particular M_Y is continuous on any interval where it is finite. Also $M'_Y(0) = \mathbf{E}(Y_1) < 0$, so that the function M_Y looks something like the sketch in Figure 5.4. Within the circle of convergence of $M_Y(t)$, thought of as an infinite series of powers of t, M_Y is infinitely differentiable term by term, so that $M''_Y(t) = \mathbf{E}(Y_1^2 e^{tY_1}) > 0$. Hence $M_Y(t)$ has a unique minimum for $t > 0$, at the point $t = \tau$ say. Furthermore

(7) $$M_Y(\tau) = \inf_{t>0} \{M_Y(t)\} = e^{-\psi(a)}$$

by (4) and (6).

It is a device of great convenience to introduce an ancillary distribution function. Let F_n be the distribution function of $T_n = Y_1 + Y_2 + \cdots + Y_n$, and let G_n be the distribution function satisfying

(8) $$dG_n(x) = \frac{e^{\tau x}}{M_Y(\tau)^n} dF_n(x).$$

You may prefer to define G_n by

(9) $$G_n(y) = \frac{\mathbf{E}(\exp(\tau T_n); T_n \leqslant y)}{M_Y(\tau)^n}$$

where $\mathbf{E}(R; A)$, the mean value of the random variable R on the event A, is defined by $\mathbf{E}(R; A) = \mathbf{E}(RI_A)$; you should check that (9) is equivalent to (8). It is immediate from (9) that G_n is indeed a distribution function.

Finally we introduce an ancillary sequence $Z_1, Z_2, \ldots$ of independent random variables having common distribution function G_1 and partial sum $U_n = Z_1 + Z_2 + \cdots + Z_n$. The moment generating function M_Z of the Z_i is

(10) $$M_Z(t) = \int_{-\infty}^{\infty} e^{tz}\, dG_1(z) = \int_{-\infty}^{\infty} \frac{\exp[(t+\tau)z]}{M_Y(\tau)} dF_1(z) = \frac{M_Y(t+\tau)}{M_Y(\tau)}$$

so that

(11) $$\mathbf{E}(Z_1) = M'_Z(0) = \frac{M'_Y(\tau)}{M_Y(\tau)} = 0$$

since τ marks the minimum of M_Y, and

(12) $$\mathbf{E}(Z_1^2) = M''_Z(0) = \frac{M''_Y(\tau)}{M_Y(\tau)} > 0.$$

By a calculation similar to (10), we find that G_n has the same moment generating function as the variable U_n, whence U_n has distribution function G_n. Hence

(13) $$\mathbf{E}(U_n) = 0, \qquad \mathbf{E}(U_n^2) = ns^2$$

where $s^2 = \mathbf{E}(Z_1^2)$. Furthermore

(14) $$\mathbf{P}(U_n > 0) = \mathbf{P}(U_n/(s\sqrt{n}) > 0) \to \tfrac{1}{2}$$

as $n \to \infty$, by the central limit theorem.

We require a lower bound for $\mathbf{P}(T_n > 0)$. Now

(15) $$\mathbf{P}(T_n > 0) = \int_{(0,\infty)} \mathrm{d}F_n(u) = \mathrm{e}^{-n\psi(a)} \int_{(0,\infty)} \mathrm{e}^{-\tau x}\, \mathrm{d}G_n(x)$$

by (7) and (8). However,

$$\int_{(0,\infty)} \mathrm{e}^{-\tau x}\, \mathrm{d}G_n(x) = \mathbf{P}(U_n > 0)\mathbf{E}[\exp(-\tau U_n) \mid U_n > 0]$$
$$\geq \mathbf{P}(U_n > 0) \exp[-\tau \mathbf{E}(U_n \mid U_n > 0)]$$

by Jensen's inequality (5.6.15). Also,

$$\mathbf{E}(U_n \mid U_n > 0) = \frac{\mathbf{E}(U_n I_{\{U_n > 0\}})}{\mathbf{P}(U_n > 0)} \leq \frac{\mathbf{E}|U_n|}{\mathbf{P}(U_n > 0)}$$
$$\leq \frac{\mathbf{E}(U_n^2)^{\frac{1}{2}}}{\mathbf{P}(U_n > 0)} = \frac{s\sqrt{n}}{\mathbf{P}(U_n > 0)}$$

by the Cauchy–Schwarz inequality and (13). Substituting into (15), we obtain

(16) $$\mathbf{P}(T_n > 0)^{1/n} \geq \mathrm{e}^{-\psi(a)} A_n$$

where

$$A_n = \mathbf{P}(U_n > 0)^{1/n} \exp\left(-\frac{\tau s}{n^{\frac{1}{2}}\mathbf{P}(U_n > 0)}\right).$$

We use the fact (14) that $\mathbf{P}(U_n > 0) \to \frac{1}{2}$ to find that $A_n \to 1$ as $n \to \infty$.
In the notation of the theorem, (16) amounts to

$$\mathbf{P}(S_n > na)^{1/n} \geqslant e^{-\psi(a)} A_n \to e^{-\psi(a)}$$

as $n \to \infty$, which completes the proof. ■

Exercises

(17) A fair coin is tossed n times, showing heads H_n times and tails T_n times. Let $S_n = H_n - T_n$. Show that

$$\mathbf{P}(S_n > an)^{1/n} \to \frac{1}{\sqrt{\{(1+a)^{1+a}(1-a)^{1-a}\}}} \quad \text{if} \quad 0 < a < 1.$$

What happens if $a \geqslant 1$?

(18) Show that

$$T_n^{1/n} \to \frac{4}{\sqrt{\{(1+a)^{1+a}(1-a)^{1-a}\}}}$$

as $n \to \infty$, where

$$T_n = \sum_{\substack{k: \\ |k - \frac{1}{2}n| > \frac{1}{2}an}} \binom{n}{k}$$

and $0 < a < 1$.
Find the asymptotic behaviour of $T_n^{1/n}$ when

$$T_n = \sum_{\substack{k: \\ k > n(1+a)}} \frac{n^k}{k!}, \quad \text{where} \quad a > 0.$$

(19) Show that the moment generating function of X is finite in a neighbourhood of the origin if and only if X has exponentially decaying tails, in the sense that there exist positive constants λ and μ such that $\mathbf{P}(|X| \geqslant a) \leqslant \mu e^{-\lambda a}$ for $a > 0$. Seen in the light of this observation, the condition of the large deviation theorem (2) is very natural.

(20) Let $X_1, X_2, \ldots$ be independent random variables having the Cauchy distribution, and let $S_n = X_1 + X_2 + \cdots + X_n$. Find $\mathbf{P}(S_n > an)$.

5.12 Problems

1. A die is thrown ten times. What is the probability that the sum of the scores is 27?

2. A coin is tossed repeatedly, heads appearing with probability p on each toss.

   (a) Let X be the number of tosses until the first occasion by which three heads have appeared successively. Write down a difference equation for $f(k) = \mathbf{P}(X = k)$ and solve it. Now write down an equation for $\mathbf{E}(X)$ using conditional expectation. (Try the same thing for the first occurrence of HTH.)

   (b) Let N be the number of heads in n tosses of the coin. Write down $G_N(s)$. Hence find the probability that:
   (i) N is divisible by 2, (ii) N is divisible by 3.

3. A coin is tossed repeatedly, heads occurring on each toss with probability p. Find the probability generating function of the number T of tosses before a run of n heads has appeared for the first time.

4. Find the generating function of the negative binomial mass function

$$f(k) = \binom{k-1}{r-1} p^r (1-p)^{k-r}, \quad k = r, r+1, \ldots,$$

where $0 < p < 1$ and r is a positive integer. Deduce the mean and variance.

5. For the simple random walk, show that the probability $p_0(2n)$ that the particle returns to the origin at the $(2n)$th step satisfies $p_0(2n) \simeq (4pq)^n/\sqrt{(\pi n)}$, and use this to prove that the walk is persistent if and only if $p = \frac{1}{2}$. You will need Stirling's formula: $n! \simeq n^{n+\frac{1}{2}} e^{-n} \sqrt{(2\pi)}$.

6. A symmetric random walk in two dimensions is defined to be a sequence of points $\{(X_n, Y_n): n \geqslant 0\}$ which evolves in the following way: if $(X_n, Y_n) = (x, y)$ then (X_{n+1}, Y_{n+1}) is one of the four points $(x \pm 1, y)$, $(x, y \pm 1)$, each being picked with equal probability $\frac{1}{4}$. If $(X_0, Y_0) = (0, 0)$,

   (a) show that $\mathbf{E}(X_n^2 + Y_n^2) = n$,
   (b) find the probability $p_0(2n)$ that the particle is at the origin after the $(2n)$th step, and deduce that the probability of ever returning to the origin is 1.

7. Consider the one-dimensional random walk $\{S_n\}$ given by

$$S_{n+1} = \begin{cases} S_n + 2 & \text{with probability } p \\ S_n - 1 & \text{with probability } q = 1 - p \end{cases}$$

where $0 < p < 1$. What is the probability of ever reaching the origin starting from $S_0 = a$ where $a > 0$?

8. Let X and Y be independent variables taking values in the positive integers such that

$$\mathbf{P}(X = k \mid X + Y = n) = \binom{n}{k} p^k (1-p)^{n-k}$$

for some p and all $0 \leqslant k \leqslant n$. Show that X and Y have Poisson distributions.

9. In a branching process whose family sizes have mean μ and variance σ^2, find the variance of Z_n, the size of the nth generation, given that $Z_0 = 1$.

10. **Waldegrave's problem.** A group $\{A_1, A_2, \ldots, A_r\}$ of $r\,(>2)$ people play the following game. A_1 and A_2 wager on the toss of a fair coin. The loser puts £1 in the pool, the winner goes on to play A_3. In the next wager, the loser puts £1 in the pool, the winner goes on to play A_4, and so on. The winner of the $(r-1)$th wager goes on to play A_1, and the cycle recommences. The first person to beat all the others in sequence takes the pool.

    (a) Find the probability generating function of the duration of the game.
    (b) Find an expression for the probability that A_k wins.
    (c) Find an expression for the expected size of the pool at the end of the game, given that A_k wins.
    (d) Find an expression for the probability that the pool is intact after the nth spin of the coin.

This problem was discussed by Montmort, Bernoulli, de Moivre, Laplace, and others.

11. Show that the generating function H_n of the *total* number of individuals in the first n generations of a branching process satisfies $H_n(s) = sG(H_{n-1}(s))$.

12. Show that the number Z_n of individuals in the nth generation of a branching process satisfies $\mathbf{P}(Z_n > N \mid Z_m = 0) \leqslant (G_m(0))^N$ for $n < m$.

13. (a) A hen lays N eggs where N is Poisson with parameter λ. The weight of the nth egg is W_n, where $W_1, W_2, \ldots$ are independent identically distributed variables with common probability generating function $G(s)$. Show that the generating function G_W of the total weight $W = \sum_{i=1}^N W_i$ is given by $G_W(s) = \exp\{-\lambda + \lambda G(s)\}$. W is said to have a *compound Poisson distribution*. Show further that, for any positive integral value of n, $\{G_W(s)\}^{1/n}$ is the probability generating function of some random variable; W (or its distribution) is said to be *infinitely divisible* in this regard.

(b) Show that if $H(s)$ is the probability generating function of some infinitely divisible distribution on the non-negative integers then $H(s) = \exp\{-\lambda + \lambda G(s)\}$ for some λ (>0) and some probability generating function $G(s)$.

14. The distribution of a random variable X is called *infinitely divisible* if, for all positive integers n, there exists a sequence $Y_1^{(n)}, Y_2^{(n)}, \ldots, Y_n^{(n)}$ of independent identically distributed random variables such that X and $Y_1^{(n)} + Y_2^{(n)} + \cdots + Y_n^{(n)}$ have the same distribution.

(a) Show that the normal, Poisson, and gamma distributions are infinitely divisible.
(b) Show that the characteristic function ϕ of an infinitely divisible distribution has no real zeros, in that $\phi(t) \neq 0$ for all real t.

15. Let $X_1, X_2, \ldots$ be independent variables each taking the values 0 or 1 with probabilities $1 - p$ and p, where $0 < p < 1$. Let N be a random variable taking values in the positive integers, independent of the X_i, and write $S = X_1 + X_2 + \cdots + X_N$. Write down the conditional generating function of N given that $S = N$, in terms of the probability generating function G of N. Show that N has a Poisson distribution if and only if $\{\mathbf{E}(x^N)\}^p = \mathbf{E}(x^N \mid S = N)$ for all p and x.

16. If X and Y have joint probability generating function

$$G_{X,Y}(s,t) = \mathbf{E}(s^X t^Y) = \frac{\{1-(p_1+p_2)\}^n}{\{1-(p_1 s + p_2 t)\}^n} \quad \text{where} \quad p_1 + p_2 \leqslant 1,$$

find the marginal mass functions of X and Y, and the mass function of $X + Y$. Find also the conditional probability generating function $G_{X|Y}(s \mid y) = \mathbf{E}(s^X \mid Y = y)$ of X given that $Y = y$. The pair X, Y is said to have the *bivariate negative binomial distribution*.

17. If X and Y have joint probability generating function

$$G_{X,Y}(s,t) = \exp\{\alpha(s-1) + \beta(t-1) + \gamma(st-1)\}$$

find the marginal distributions of X, Y, and the distribution of $X + Y$, showing that X and Y have the Poisson distribution, but that $X + Y$ does not unless $\gamma = 0$.

18. Define

$$I(a, b) = \int_0^\infty \exp(-a^2u^2 - b^2u^{-2})\,du$$

for $a, b > 0$. Show that

(a) $I(a, b) = a^{-1}I(1, ab)$,
(b) $\partial I/\partial b = -2I(1, ab)$,
(c) $I(a, b) = \pi^{\frac{1}{2}}\,e^{-2ab}/(2a)$.
(d) If X has density function $(d/\sqrt{x})\,e^{-c/x-gx}$ for $x > 0$, then X has moment generating function given by $\mathbf{E}(e^{-tX}) = d\sqrt{\pi/(g+t)}\exp(-2\sqrt{c(g+t)})$, $t > -g$.
(e) If X has density function $(2\pi x^3)^{-\frac{1}{2}}\,e^{-1/(2x)}$ for $x > 0$, then X has moment generating function given by $\mathbf{E}(e^{-tX}) = \exp\{-\sqrt{2t}\}$, $t \geq 0$.

19. Let X, Y, Z be independent $N(0, 1)$ variables. Use characteristic functions and moment generating functions to find the distributions of

(a) $U = X/Y$,
(b) $V = X^{-2}$,
(c) $W = XYZ/\sqrt{(X^2Y^2 + Y^2Z^2 + Z^2X^2)}$.

20. Let X have density function f and characteristic function ϕ, and suppose that $\int_{-\infty}^{\infty} |\phi(t)|\,dt < \infty$. Deduce that

$$f(x) = \frac{1}{2\pi}\int_{-\infty}^{\infty} e^{-itx}\phi(t)\,dt.$$

21. **Conditioned branching process.** Consider a branching process Z in which $Z_0 = 1$, and whose family-sizes have the geometric mass function $f(k) = qp^k$, $k \geq 0$, where $\mu = p/q > 1$. Show that the conditional distribution of Z_n/μ^n, given that $Z_n > 0$, converges as $n \to \infty$ to the exponential distribution with parameter $1 - \mu^{-1}$.

22. A random variable X is called *symmetric* if X and $-X$ are identically distributed. Show that X is symmetric if and only if the imaginary part of its characteristic function is identically zero.

23. Let X and Y be independent identically distributed variables with means 0 and variances 1. Let $\phi(t)$ be their common characteristic function, and suppose that $X + Y$ and $X - Y$ are independent. Show that $\phi(2t) = \{\phi(t)\}^3\phi(-t)$, and deduce that X and Y are $N(0, 1)$ variables.

More generally, suppose that X and Y are independent and identically distributed with means 0 and variances 1, and furthermore that $\mathbf{E}(X - Y \mid X + Y) = 0$ and $\text{var}(X - Y \mid X + Y) = 2$. Deduce that $\phi(s)^2 = \phi'(s)^2 - \phi(s)\phi''(s)$, and hence that X and Y are independent $N(0, 1)$ variables.

24. Show that the average $Z = n^{-1}\sum_{i=1}^{n} X_i$ of n independent Cauchy variables has the Cauchy distribution too. Why does this not violate the law of large numbers?

25. Let X and Y be independent random variables each having the Cauchy density function $f(x) = \{\pi(1 + x^2)\}^{-1}$, and let $Z = \frac{1}{2}(X + Y)$.

(a) Show by using characteristic functions that Z has the Cauchy distribution also.

(b) Show by the convolution formula that Z has the Cauchy density function. You may find it helpful to check first that

$$f(x)f(y-x) = \frac{f(x)+f(y-x)}{\pi(4+y^2)} + g(y)\{xf(x)+(y-x)f(y-x)\}$$

where $g(y) = 2/\{\pi y(4+y^2)\}$.

26. Let $X_1, X_2, \ldots, X_n$ be independent variables with characteristic functions $\phi_1, \phi_2, \ldots, \phi_n$. Describe random variables which have the following characteristic functions:

(a) $\phi_1(t)\phi_2(t)\cdots\phi_n(t)$, (b) $|\phi_1(t)|^2$,
(c) $\sum_1^n p_j\phi_j(t)$ where $p_j \geqslant 0$ and $\sum_1^n p_j = 1$, (d) $(2-\phi_1(t))^{-1}$,
(e) $\int_0^\infty \phi_1(ut)\,\mathrm{e}^{-u}\,du$.

27. Find the characteristic functions corresponding to the following density functions on $(-\infty, \infty)$.

(a) $1/\cosh(\pi x)$, (b) $(1-\cos x)/(\pi x^2)$,
(c) $\exp(-x-\mathrm{e}^{-x})$, (d) $\frac{1}{2}\mathrm{e}^{-|x|}$.

28. Which of the following are characteristic functions:

(a) $\phi(t) = 1-|t|$ if $|t| \leqslant 1$, $\phi(t) = 0$ otherwise,
(b) $\phi(t) = (1+t^4)^{-1}$, (c) $\phi(t) = \exp(-t^4)$,
(d) $\phi(t) = \cos t$, (e) $\phi(t) = 2(1-\cos t)/t^2$?

29. Show that the characteristic function ϕ of a random variable X satisfies $|1-\phi(t)| \leqslant \mathbf{E}|tX|$.

30. Suppose X and Y have joint characteristic function $\phi(s,t)$. Show that, subject to the appropriate conditions of differentiability,

$$\mathrm{i}^{m+n}\mathbf{E}(X^m Y^n) = \left.\frac{\partial^{m+n}\phi}{\partial s^m\,\partial t^n}\right|_{s=t=0}$$

for any positive integers m and n.

31. If X has distribution function F and characteristic function ϕ, show that for $t > 0$

(a)
$$\int_{[-t^{-1},t^{-1}]} x^2\,\mathrm{d}F \leqslant \frac{3}{t^2}\{1-\mathrm{Re}\,\phi(t)\}$$

(b)
$$\mathbf{P}\left(|X| \geqslant \frac{1}{t}\right) \leqslant \frac{7}{t}\int_0^t \{1-\mathrm{Re}\,\phi(v)\}\,\mathrm{d}v.$$

32. Let $X_1, X_2, \ldots$ be independent variables which are uniformly distributed on $[0, 1]$. Let $M_n = \max\{X_1, X_2, \ldots, X_n\}$ and show that $n(1-M_n) \xrightarrow{\mathrm{D}} X$ where X is exponentially distributed with parameter 1. You need not use characteristic functions.

33. If X is either (a) Poisson with parameter λ, or (b) $\Gamma(1, \lambda)$, show that the distribution of $Y_\lambda = (X - \mathbf{E}X)(\operatorname{var} X)^{-\frac{1}{2}}$ approaches the $N(0, 1)$ distribution as $\lambda \to \infty$.

(c) Show that

$$e^{-n}\left(1 + n + \frac{n^2}{2!} + \cdots + \frac{n^n}{n!}\right) \to \frac{1}{2} \quad \text{as} \quad n \to \infty.$$

34. **Coupon collecting.** Recall that you regularly buy quantities of some ineffably dull commodity. To attract your attention, the manufacturers add to each packet a small object which is also dull, and in addition useless, but there are n different types. Assume that each packet is equally likely to contain any one of the different types, as usual. Let T_n be the number of packets bought before you acquire a complete set of n objects. Show that $n^{-1}(T_n - n \log n) \xrightarrow{D} T$, where T is a random variable with distribution function $\mathbf{P}(T \leqslant x) = \exp(-e^{-x})$, $-\infty < x < \infty$.

35. Find a sequence (ϕ_n) of characteristic functions with the property that the limit $\phi(t) = \lim_{n\to\infty} \phi_n(t)$ exists for all t, but such that ϕ is not itself a characteristic function.

36. Use generating functions to show that it is not possible to load two dice in such a way that the sum of the values which they show is equally likely to take any value between 2 and 12. Compare with your method for Problem (2.7.12).

37. A biased coin is tossed N times, where N is a random variable which is Poisson distributed with parameter λ. Prove that the total number of heads shown is independent of the total number of tails. Show conversely that if the numbers of heads and tails are independent, then N has the Poisson distribution.

38. A *binary tree* is a tree (as in the section on branching processes) in which each node has exactly two descendants. Suppose that each node of the tree is coloured black with probability p, and white otherwise, independently of all other nodes. For any path π containing n nodes beginning at the root of the tree, let $B(\pi)$ be the number of black nodes in π, and let $X_n(k)$ be the number of such paths π for which $B(\pi) \geqslant k$. Show that there exists β_c such that

$$\mathbf{E}\{X_n(\beta n)\} \to \begin{cases} 0 & \text{if } \beta > \beta_c \\ \infty & \text{if } \beta < \beta_c \end{cases}$$

and show how to determine the value β_c.

Prove that

$$\mathbf{P}(X_n(\beta n) \geqslant 1) \to \begin{cases} 0 & \text{if } \beta > \beta_c \\ 1 & \text{if } \beta < \beta_c. \end{cases}$$

39. Use the continuity theorem (5.9.5) to show that, as $n \to \infty$,

(a) if X_n is $B(n, \lambda/n)$ then the distribution of X_n converges to a Poisson distribution.
(b) if Y_n is geometric with parameter $p = \lambda/n$ then the distribution of Y_n/n converges to an exponential distribution.

40. Let $X_1, X_2, \ldots$ be independent random variables with zero means and such that $\mathbf{E}|X_j^3| < \infty$ for all j. Show that $S_n = X_1 + X_2 + \cdots + X_n$ satisfies $S_n/\sqrt{\mathrm{var}(S_n)} \xrightarrow{\mathrm{D}} N(0, 1)$ as $n \to \infty$ if

$$\sum_{j=1}^{n} \mathbf{E}|X_j^3| = \mathrm{o}(\{\mathrm{var}(S_n)\}^{-3/2}).$$

The following steps may be useful. Let $\sigma_j^2 = \mathrm{var}(X_j)$, $\sigma(n)^2 = \mathrm{var}(S_n)$, $\rho_j = \mathbf{E}|X_j^3|$, and ϕ_j and ψ_n be the characteristic functions of X_j and $S_n/\sigma(n)$ respectively.

(i) Use Taylor's theorem to show that $|\phi_j(t) - 1| \leqslant 2t^2\sigma_j^2$ and $|\phi_j(t) - 1 + \frac{1}{2}\sigma_j^2 t^2| \leqslant |t|^3 \rho_j$ for $j \geqslant 1$.

(ii) Show that $|\log(1 + z) - z| \leqslant |z|^2$ if $|z| \leqslant \frac{1}{2}$, where the logarithm has its principal value.

(iii) Show that $\sigma_j^3 \leqslant \rho_j$, and deduce from the condition given in the question that $\max_{1 \leqslant j \leqslant n} \sigma_j/\sigma(n) \to 0$ as $n \to \infty$, implying that $\max_{1 \leqslant j \leqslant n} |\phi_j(t/\sigma(n)) - 1| \to 0$.

(iv) Deduce an upper bound for $|\log \phi_j(t/\sigma(n)) - \frac{1}{2}t^2\sigma_j^2/\sigma(n)^2|$, and sum to obtain that $\log \psi_n(t) \to -\frac{1}{2}t^2$.

41. Let $X_1, X_2, \ldots$ be independent variables each taking values $+1$ or -1 with probabilities $\frac{1}{2}$ and $\frac{1}{2}$. Show that

$$\sqrt{\frac{3}{n^3}} \sum_{k=1}^{n} kX_k \xrightarrow{\mathrm{D}} N(0, 1) \qquad \text{as } n \to \infty.$$

42. **Normal sample.** Let $X_1, X_2, \ldots, X_n$ be independent $N(\mu, \sigma^2)$ variables. Define $\bar{X} = \sum_1^n X_i/n$ and $Z_i = X_i - \bar{X}$. Find the joint characteristic function of $\bar{X}, Z_1, Z_2, \ldots, Z_n$, and hence prove that $\bar{X}$ and $S^2 = (n-1)^{-1}\sum_1^n (X_i - \bar{X})^2$ are independent.

43. Let X be $N(0, 1)$, and let $Y = \mathrm{e}^X$; Y is said to have the *log-normal* distribution. Show that the density function of Y is $f(x) = (x\sqrt{2\pi})^{-1}\exp\{-\frac{1}{2}(\log x)^2\}$ for $x > 0$. For $|a| \leqslant 1$, define $f_a(x) = \{1 + a\sin(2\pi \log x)\}f(x)$. Show that f_a is a density function with finite moments of all (positive) orders, none of which depends on the value of a. The family $\{f_a \colon |a| \leqslant 1\}$ contains density functions which are not specified by their moments.

44. Consider a random walk whose steps are independent and identically distributed integer-valued random variables with non-zero mean. Prove that the walk is transient.

6 Markov chains

6.1 Markov processes

The simple random walk (5.3) and the branching process (5.4) are two examples of collections of random variables, $\{S_0, S_1, \dots\}$ and $\{Z_0, Z_1, \dots\}$ respectively, which evolve in some random but prescribed manner. Such collections are called† 'random processes'. A typical random process X is a family $\{X_t: t \in T\}$ of random variables indexed by some set T. In the above examples $T = \{0, 1, 2, \dots\}$ and we call the process a 'discrete-time' process; in other important examples $T = \mathbb{R}$ or $T = [0, \infty)$ and we call these 'continuous-time' processes. In either case we think of a random process as a family of variables which evolve as time passes. These variables may even be independent of each other, but then the evolution is not very surprising and this very special case is of little interest to us in this chapter. Rather, we are concerned with more general, and we hope realistic, models for random evolution. Simple random walks and branching processes shared the following property: conditional on their values at the nth step, their future values did not depend on their previous values. This property proved to be very useful in their analysis, and it is to the general theory of processes with this property that we turn our attention now.

Until further notice we shall be interested in discrete-time processes. Let $\{X_0, X_1, \dots\}$ be a sequence of random variables‡ which take values in some countable set S, called the *state space*. Each X_n is a discrete random variable which takes one of N possible values, where $N = |S|$ (N may equal $+\infty$).

(1) **Definition.** The process X is a **Markov chain** if it satisfies the **Markov condition**:

$$\mathbf{P}(X_n = s \mid X_0 = x_0, X_1 = x_1, \dots, X_{n-1} = x_{n-1}) = \mathbf{P}(X_n = s \mid X_{n-1} = x_{n-1})$$

for all $n \geq 1$ and all $s, x_0, x_1, \dots, x_{n-1} \in S$.

A proof that the random walk is a Markov chain was given in (3.9.5). The reader can check that the Markov property is equivalent to each of the stipulations (2) and (3) below: for each $s \in S$ and for every sequence $\{x_i: i \geq 0\}$

† Such collections are often called 'stochastic' processes; the verbal form of the Greek stem of the word 'stochastic' means 'to get at' or 'to aim at with an arrow'.

‡ There is, of course, an underlying probability space $(\Omega, \mathcal{F}, \mathbf{P})$, and each X_n is an $\mathcal{F}$-measurable function which maps Ω into S.

in S,

(2) $$\mathbf{P}(X_{n+1} = s \mid X_{n_1} = x_{n_1}, X_{n_2} = x_{n_2}, \dots, X_{n_k} = x_{n_k}) = \mathbf{P}(X_{n+1} = s \mid X_{n_k} = x_{n_k})$$
for all $n_1 < n_2 < \cdots < n_k \leqslant n$,

(3) $$\mathbf{P}(X_{n+m} = s \mid X_0 = x_0, X_1 = x_1, \dots, X_n = x_n) = \mathbf{P}(X_{n+m} = s \mid X_n = x_n)$$
for any $m, n \geqslant 0$.

We have assumed that X takes values in some *countable* set S. The reason for this is essentially the same as the reason for treating discrete and continuous variables separately. Since S is assumed countable, it can be put in one–one correspondence with some subset S' of the integers, and without loss of generality we can assume that S *is* this set S' of integers. If $X_n = i$, then we say that the chain is in the 'ith state at the nth step'; we can also talk of the chain as 'having the value i', 'visiting i', or 'being in state i', depending upon the context of the remark.

The evolution of a chain is described by its 'transition probabilities' $\mathbf{P}(X_{n+1} = j \mid X_n = i)$; it can be quite complicated in general since these probabilities depend upon the three quantities n, i, and j. We shall restrict our attention to the case when they do not depend on n but only upon i and j.

(4) **Definition.** The chain X is called **homogeneous** if

$$\mathbf{P}(X_{n+1} = j \mid X_n = i) = \mathbf{P}(X_1 = j \mid X_0 = i)$$

for all n, i, j. The **transition matrix** $\boldsymbol{P} = (p_{ij})$ is the $|S| \times |S|$ matrix of **transition probabilities**

$$p_{ij} = \mathbf{P}(X_{n+1} = j \mid X_n = i).$$

Some authors write p_{ji} in place of p_{ij} here, so beware; sometimes we write $p_{i,j}$ for p_{ij}. Henceforth, *all Markov chains are assumed homogeneous* unless otherwise specified; we assume that the process X is a Markov chain, and we denote the transition matrix of such a chain by $\boldsymbol{P}$.

(5) **Theorem.** $\boldsymbol{P}$ *is a* ***stochastic matrix****, which is to say that*

(a) $\boldsymbol{P}$ *has non-negative entries, or* $p_{ij} \geqslant 0$
(b) $\boldsymbol{P}$ *has row sums equal to one, or* $\sum_j p_{ij} = 1$.

Proof. An easy *exercise.* ■

We can easily see that (5) characterizes transition matrices.

Broadly speaking, we are interested in the evolution of X over two different time scales, the 'short term' and the 'long term'. In the short term the random evolution of X is described by $\boldsymbol{P}$, whilst long-term changes are described in the following way.

(6) **Definition.** The ***n*-step transition matrix** $\boldsymbol{P}_n = (p_{ij}(n))$ is the matrix of ***n*-step transition probabilities**

$$p_{ij}(n) = \mathbf{P}(X_{m+n} = j \mid X_m = i).$$

Of course, $\boldsymbol{P}_1 = \boldsymbol{P}$.

(7) **Theorem. Chapman–Kolmogorov equations.**

$$p_{ij}(m+n) = \sum_k p_{ik}(m)p_{kj}(n).$$

Hence $\boldsymbol{P}_{m+n} = \boldsymbol{P}_m \boldsymbol{P}_n$, *and so* $\boldsymbol{P}_n = \boldsymbol{P}^n$, *the nth power of* $\boldsymbol{P}$.

Proof.

$$\begin{aligned} p_{ij}(m+n) &= \mathbf{P}(X_{m+n} = j \mid X_0 = i) \\ &= \sum_k \mathbf{P}(X_{m+n} = j, X_m = k \mid X_0 = i) \\ &= \sum_k \mathbf{P}(X_{m+n} = j \mid X_m = k, X_0 = i)\mathbf{P}(X_m = k \mid X_0 = i) \\ &= \sum_k \mathbf{P}(X_{m+n} = j \mid X_m = k)\mathbf{P}(X_m = k \mid X_0 = i) \end{aligned}$$

as required, where we have used the result of Exercise (1.4.9),

$$\mathbf{P}(A \cap B \mid C) = \mathbf{P}(A \mid B \cap C)\mathbf{P}(B \mid C)$$

and the Markov property (2). The rest of the theorem follows immediately. ●

This theorem relates long-term development to short-term development, and tells us how X_n depends on the initial variable X_0. Let $\mu_i^{(n)} = \mathbf{P}(X_n = i)$ be the mass function of X_n, and write $\boldsymbol{\mu}^{(n)}$ for the row vector with entries $(\mu_i^{(n)} : i \in S)$.

(8) **Lemma.** $\boldsymbol{\mu}^{(m+n)} = \boldsymbol{\mu}^{(m)}\boldsymbol{P}_n$, *and hence* $\boldsymbol{\mu}^{(n)} = \boldsymbol{\mu}^{(0)}\boldsymbol{P}^n$.

Proof.

$$\begin{aligned} \mu_j^{(m+n)} &= \mathbf{P}(X_{m+n} = j) = \sum_i \mathbf{P}(X_{m+n} = j \mid X_m = i)\mathbf{P}(X_m = i) \\ &= \sum_i \mu_i^{(m)} p_{ij}(n) = (\boldsymbol{\mu}^{(m)}\boldsymbol{P}_n)_j \end{aligned}$$

and the result follows from (7). ■

Thus we reach the important conclusion that the random evolution of the chain is determined by the transition matrix $\boldsymbol{P}$ and the initial mass function

$\boldsymbol{\mu}^{(0)}$. Many questions about the chain can be expressed in terms of these quantities, and the study of the chain is thus largely reducible to the study of algebraic properties of matrices.

(9) **Example. Simple random walk.** $S = \{0, \pm 1, \pm 2, \ldots\}$ and

$$p_{ij} = \begin{cases} p & \text{if } \; j = i + 1 \\ q = 1 - p & \text{if } \; j = i - 1 \\ 0 & \text{otherwise.} \end{cases}$$

The argument of (3.10.2) shows that

$$p_{ij}(n) = \begin{cases} \dbinom{n}{\frac{1}{2}(n + j - i)} p^{\frac{1}{2}(n+j-i)} q^{\frac{1}{2}(n-j+i)} & \text{if } \; n + j - i \text{ is even} \\ 0 & \text{otherwise.} \end{cases}$$ ●

(10) **Example. Branching process (Section 5.4).** $S = \{0, 1, 2, \ldots\}$ and p_{ij} is the coefficient of s^j in $(G(s))^i$. Also $p_{ij}(n)$ is the coefficient of s^j in $(G_n(s))^i$. ●

(11) **Example. Gene frequencies.** One of the most interesting and extensive applications of probability theory is to genetics, and particularly to the study of gene frequencies. The problem may be inadequately and superficially described as follows. For definiteness suppose the population is human. Genetic information is contained in chromosomes, which are strands of chemicals grouped in cell nuclei. In humans ordinary cells carry 46 chromosomes, 44 of which are homologous pairs. For our purposes a chromosome can be regarded as an ordered set of n sites, the states of which can be thought of as a sequence of random variables $C_1, C_2, \ldots, C_n$. The possible values of each C_i are certain combinations of chemicals, and these values influence (or determine) some characteristic of the owner such as hair colour or leg length.

Now, suppose that A is a possible value of C_1, say, and let X_n be the number of individuals in the nth generation for which C_1 has the value A. What is the behaviour of the sequence $X_1, X_2, \ldots, X_n, \ldots$? The first important (and obvious) point is that the sequence is random, because of the following factors.

(a) The value A for C_1 may affect the owner's chances of contributing to the next generation. If A gives you short legs, you stand a better chance of being caught by a sabre-toothed tiger. The breeding population is randomly selected from those born, but there may be bias for or against the gene A.

(b) The breeding population is randomly combined into pairs to produce offspring. Each parent contributes 23 chromosomes to its offspring, but here again, if A gives you short legs you may have a smaller (or larger) chance of catching a mate.

(c) Sex cells having half the normal complement of chromosomes are produced by a special and complicated process called 'meiosis'. We shall not go into details, but essentially the homologous pairs of the parent are shuffled to produce new and different chromosomes for offspring. The sex cells from each parent (with 23 chromosomes) are then combined to give a new cell (with 46 chromosomes).

(d) Since meiosis involves a large number of complex chemical operations it is hardly surprising that things go wrong occasionally, producing a new value for C_1, $\hat{A}$ say. This is a 'mutation'.

The reader can now see that if generations are segregated (in a laboratory, say), then we can suppose that $X_1, X_2, \ldots$ is a Markov chain with a finite state space. If generations are not segregated and $X(t)$ is the frequency of A in the population at time t, then $X(t)$ may be a continuous-time Markov chain.

For a simple example, suppose that the population size is N, a constant. If $X_n = i$, then it may seem reasonable that any member of the $(n+1)$th generation carries A with probability i/N, independently of the others. Then

$$p_{ij} = \mathbf{P}(X_{n+1} = j \mid X_n = i) = \binom{N}{j}\left(\frac{i}{N}\right)^j\left(1 - \frac{i}{N}\right)^{N-j}.$$

Even more simply, suppose that at each stage exactly one individual dies and is replaced by a new individual; each individual is picked for death with probability $1/N$. If $X_n = i$, we assume that the probability that the replacement carries A is i/N. Then

$$p_{ij} = \begin{cases} \dfrac{i(N-i)}{N^2} & \text{if } j = i \pm 1 \\[2ex] 1 - 2\dfrac{i(N-i)}{N^2} & \text{if } j = i \\[2ex] 0 & \text{otherwise.} \end{cases}$$ ●

(12) **Example. Recurrent events.** Suppose that X is a Markov chain on S, with $X_0 = i$. Let $T(1)$ be the time of the first return of the chain to i: that is, $T(1) = \min\{n \geq 1: X_n = i\}$, with the convention that $T(1) = \infty$ if $X_n \neq i$ for all $n \geq 1$. Suppose that you tell me that $T(1) = 3$, say, which is to say that $X_n \neq i$ for $n = 1, 2$, and $X_3 = i$. The future evolution of the chain $\{X_3, X_4, \ldots\}$ depends, by the Markov property, only on the fact that the new starting point X_3 equals i, and does not depend further on the values of X_0, X_1, X_2. Thus the future process $\{X_3, X_4, \ldots\}$ has the same distribution as had the original process $\{X_0, X_1, \ldots\}$. The same argument is valid for any given value of $T(1)$, and we are therefore led to the following observation. Having returned to its starting point for the first time, the future of the chain has the same distribution as had the original chain. Let $T(2)$ be the time which elapses between the first and second return of the chain to its starting point.

Then $T(1)$ and $T(2)$ must be independent and identically distributed random variables. Arguing similarly for future returns, we deduce that the time of the nth return of the chain to its starting point may be represented as $T(1) + T(2) + \cdots + T(n)$, where $T(1), T(2), \ldots$ are independent identically distributed random variables. That is to say, the return times of the chain form a 'recurrent-event process'; see Example (5.2.15).

A problem arises with the above argument if $T(1)$ takes the value ∞ with strictly positive probability, which is to say that the chain is not (almost) certain to return to its starting point. For the moment we overlook this difficulty, and suppose not only that $\mathbf{P}(T(1) < \infty) = 1$, but also that $\mu = \mathbf{E}(T(1))$ satisfies $\mu < \infty$. It is now an immediate consequence of the renewal theorem (5.2.24) that

$$p_{ii}(n) = \mathbf{P}(X_n = i \mid X_0 = i) \to \frac{1}{\mu} \quad \text{as} \quad n \to \infty$$

so long as the distribution of $T(1)$ is non-arithmetic; the latter condition is certainly satisfied if, say, $p_{ii} > 0$. ●

(13) **Example. Bernoulli process.** Let $S = \{0, 1, 2, \ldots\}$ and define the Markov chain Y by $Y_0 = 0$ and

$$\mathbf{P}(Y_{n+1} = s + 1 \mid Y_n = s) = p, \qquad \mathbf{P}(Y_{n+1} = s \mid Y_n = s) = 1 - p,$$

for all $n \geq 0$, where $0 < p < 1$. You may think of Y_n as the number of heads thrown in n tosses of a coin. It is easy to see that

$$\mathbf{P}(Y_{m+n} = j \mid Y_m = i) = \binom{n}{j-i} p^{j-i}(1-p)^{n-j+i}, \quad 0 \leq j - i \leq n.$$

Viewed as a Markov chain, Y is not a very interesting process. Suppose, however, that the value of Y_n is counted using a conventional digital decimal meter, and let X_n be the final digit of the reading, $X_n = Y_n$ modulo 10. It may be checked that $X = \{X_n : n \geq 0\}$ is a Markov chain on the state space $S' = \{0, 1, 2, \ldots, 9\}$ with transition matrix

$$\boldsymbol{P} = \begin{pmatrix} 1-p & p & 0 & \ldots & 0 \\ 0 & 1-p & p & \ldots & 0 \\ \vdots & \vdots & \vdots & & \vdots \\ p & 0 & 0 & \ldots & 1-p \end{pmatrix}.$$

There are various ways of studying the behaviour of X. If we are prepared to use the renewal theorem (5.2.24), then we might argue as follows. X passes through the values $0, 1, 2, \ldots, 9, 0, 1, \ldots$ sequentially. Consider the times at which X takes the value i, say. These times form a recurrent-event process for which a typical inter-occurrence time T satisfies

$$T = \begin{cases} 1 & \text{with probability } 1 - p \\ 1 + Z & \text{with probability } p \end{cases}$$

where Z has the negative binomial distribution with parameters 9 and p. Therefore $\mathbf{E}(T) = 1 + p\mathbf{E}(Z) = 1 + p(9/p) = 10$. It is now an immediate consequence of the renewal theorem that $\mathbf{P}(X_n = i) \to 10^{-1}$ for $i = 0, 1, \ldots, 9$, as $n \to \infty$. ●

(14) **Example. Markov's other chain (1910).** Let $Y_1, Y_3, Y_5, \ldots$ be a sequence of independent identically distributed random variables such that

(15) $$\mathbf{P}(Y_{2k+1} = -1) = \mathbf{P}(Y_{2k+1} = 1) = \tfrac{1}{2}, \quad k = 0, 1, 2, \ldots,$$

and define $Y_{2k} = Y_{2k-1}Y_{2k+1}$, for $k = 1, 2, \ldots$. You may check that $Y_2, Y_4, \ldots$ is a sequence of independent identically distributed variables with the same distribution (15). Now $\mathbf{E}(Y_{2k}Y_{2k+1}) = \mathbf{E}(Y_{2k-1}Y_{2k+1}^2) = \mathbf{E}(Y_{2k-1}) = 0$, and so (by the result of (3.11.12)) the sequence $Y_1, Y_2, \ldots$ is pairwise independent. Hence $p_{ij}(n) = \mathbf{P}(Y_{m+n} = j \mid Y_m = i)$ satisfies $p_{ij}(n) = \frac{1}{2}$ for all n and $i, j = \pm 1$, and it follows easily that the Chapman–Kolmogorov equations are satisfied.

Is Y a Markov chain? *No*, because $\mathbf{P}(Y_{2k+1} = 1 \mid Y_{2k} = -1) = \frac{1}{2}$, but

$$\mathbf{P}(Y_{2k+1} = 1 \mid Y_{2k} = -1, Y_{2k-1} = 1) = 0.$$

Thus, whilst the Chapman–Kolmogorov equations are *necessary* for the Markov property, they are not *sufficient*; this is for much the same reason that pairwise independence is weaker than independence.

Although Y is not a Markov chain, we can find a Markov chain by enlarging the state space. Let $Z_n = (Y_n, Y_{n+1})$, taking values in $S = \{-1, +1\}^2$. It is an *exercise* to check that Z is a (non-homogeneous) Markov chain with, for example,

$$\mathbf{P}(Z_{n+1} = (1, 1) \mid Z_n = (1, 1)) = \begin{cases} \frac{1}{2} & \text{if} \quad n \text{ even} \\ 1 & \text{if} \quad n \text{ odd}. \end{cases}$$

This technique of 'imbedding' Y in a Markov chain on a larger state space turns out to be useful in many contexts of interest. ●

Exercises

(16) Show that any sequence of independent random variables taking values in the countable set S is a Markov chain. Under what condition is this chain homogeneous?

(17) Let X_n be the maximum reading obtained in the first n throws of fair die. Show that X is a Markov chain, and find the transition probabilities $p_{ij}(n)$.

(18) Let $\{S_n : n \geq 0\}$ be a simple random walk with $S_0 = 0$, and show that $X_n = |S_n|$ defines a Markov chain; find the transition probabilities of this chain. Let $M_n = \max\{S_k : 0 \leq k \leq n\}$, and show that $Y_n = M_n - S_n$ defines a Markov chain.

(19) Let X be a Markov chain and let $\{n_r : r \geq 0\}$ be an unbounded increasing sequence of positive integers. Show that $Y_r = X_{n_r}$ constitutes a (possibly inhomogeneous) Markov chain. Find the transition matrix of Y when $n_r = 2r$ and X is: (a) simple random walk, and (b) a branching process.

(20) Let X be a Markov chain on S, and let $I: S^n \to \{0, 1\}$. Show that the distribution of $X_n, X_{n+1}, \ldots$ conditional on $\{I(X_1, \ldots, X_n) = 1\} \cap \{X_n = i\}$, is identical to the distribution of $X_n, X_{n+1}, \ldots$ conditional on $\{X_n = i\}$.

(21) **Strong Markov property.** Let X be a Markov chain on S, and let T be a random variable taking values in $\{0, 1, 2, \dots\}$ with the property that the indicator function $I_{\{T=n\}}$, of the event that $T = n$, is a function of the variables $X_1, X_2, \dots, X_n$. Such a random variable T is called a *stopping time*, and the above definition requires that it is decidable whether or not $T = n$ with a knowledge only of the past and present, $X_0, X_1, \dots, X_n$, and with no further information about the future.

Show that

$$\mathbf{P}(X_{T+m} = j \mid X_k = x_k \text{ for } 0 \leqslant k < T, X_T = i) = \mathbf{P}(X_{T+m} = j \mid X_T = i)$$

for $m \geqslant 0$, $i, j \in S$, and all sequences (x_k) of states.

(22) Let X be a Markov chain with state space S, and suppose that $h: S \to T$ is one–one. Show that $Y_n = h(X_n)$ defines a Markov chain on T. Must this be so if h is not one–one?

(23) Let X and Y be Markov chains on the set $\mathbb{Z}$ of integers. Is the sequence $Z_n = X_n + Y_n$ necessarily a Markov chain?

6.2 Classification of states

We can think of the development of the chain as the motion of a notional particle which jumps between the states of S at each epoch of time. As in Section 5.3, we may be interested in the (possibly infinite) time which elapses before the particle returns to its starting point. We saw there that it sufficed to find the distribution of the length of time until the particle returns for the first time since other interarrival times are merely independent copies of this. However, need the particle ever return to its starting point? With this question in mind we make the following definition.

(1) **Definition.** State i is called **persistent** (or **recurrent**) if

$$\mathbf{P}(X_n = i \text{ for some } n \geqslant 1 \mid X_0 = i) = 1,$$

which is to say that the probability of eventual return to i, having started from i, is 1. If this probability is strictly less than 1, i is called **transient**.

As in Section 5.3, we are interested in the *first passage times* of the chain. Let

$$f_{ij}(n) = \mathbf{P}(X_1 \neq j, X_2 \neq j, \dots, X_{n-1} \neq j, X_n = j \mid X_0 = i)$$

be the probability that the first visit to state j, starting from i, takes place at the nth step. Define

(2) $$f_{ij} = \sum_{n=1}^{\infty} f_{ij}(n)$$

to be the probability that the chain ever visits j, starting from i. Of course, j is persistent if and only if $f_{jj} = 1$. We seek a criterion for a state to be persistent in terms of the n-step transition probabilities. Following our random walk experience, define the generating functions

$$P_{ij}(s) = \sum_n s^n p_{ij}(n), \qquad F_{ij}(s) = \sum_n s^n f_{ij}(n),$$

with the conventions that $p_{ij}(0) = \delta_{ij}$, the Kronecker delta, and $f_{ij}(0) = 0$ for all i and j. Clearly $f_{ij} = F_{ij}(1)$.

(3) **Theorem.**

(a) $P_{ii}(s) = 1 + F_{ii}(s)P_{ii}(s)$
(b) $P_{ij}(s) = F_{ij}(s)P_{jj}(s) \quad \textit{if} \quad i \neq j.$

Proof. The proof is exactly as that of Theorem (5.3.1). Fix $i, j \in S$ and let $A_m = \{X_m = j\}$ and B_m be the event that the first visit to j (after time 0) takes place at time m, $B_m = \{X_r \neq j \text{ for } 1 \leqslant r < m,\ X_m = j\}$. The B_m are disjoint, so that

$$\mathbf{P}(A_m \mid X_0 = i) = \sum_{r=1}^{m} \mathbf{P}(A_m \cap B_r \mid X_0 = i).$$

Now, using the Markov condition (in the form of (6.2.20) or (6.2.21)),

$$\begin{aligned}\mathbf{P}(A_m \cap B_r \mid X_0 = i) &= \mathbf{P}(A_m \mid B_r, X_0 = i)\mathbf{P}(B_r \mid X_0 = i)\\ &= \mathbf{P}(A_m \mid X_r = j)\mathbf{P}(B_r \mid X_0 = i).\end{aligned}$$

Hence

$$p_{ij}(m) = \sum_{r=1}^{m} f_{ij}(r)p_{jj}(m-r), \qquad m = 1, 2, \ldots .$$

Multiply throughout by s^m, where $|s| < 1$, and sum over m ($\geqslant 1$) to find that $P_{ij}(s) - \delta_{ij} = F_{ij}(s)P_{jj}(s)$ as required. ■

(4) **Corollary.**

(a) *j is persistent if $\sum_n p_{jj}(n) = \infty$, and if this holds then $\sum_n p_{ij}(n) = \infty$ for all i such that $f_{ij} > 0$.*
(b) *j is transient if $\sum_n p_{jj}(n) < \infty$, and if this holds then $\sum_n p_{ij}(n) < \infty$ for all i.*

Proof. First we show that j is persistent if and only if $\sum_n p_{jj}(n) = \infty$. From (3a), $P_{jj}(s) = [1 - F_{jj}(s)]^{-1}$ if $|s| < 1$. Hence, as $s \uparrow 1$, $P_{jj}(s) \to \infty$ if and only if $f_{jj} = F_{jj}(1) = 1$. Now use Abel's theorem to obtain $\lim_{s\uparrow 1} P_{jj}(s) = \sum_n p_{jj}(n)$ and our claim is shown. Use (3b) to complete the proof. ■

(5) **Corollary.** *If j is transient then $p_{ij}(n) \to 0$ as $n \to \infty$ for all i.*

Proof. This is immediate from (4). ■

An application of (4) to the random walk is given in Problem (4.12.5).

Thus each state is either persistent or transient. It is intuitively clear that the number $N(i)$ of times which the chain visits its starting point i satisfies

(6) $$\mathbf{P}(N(i) = \infty) = \begin{cases} 1 & \text{if } \quad i \text{ is persistent} \\ 0 & \text{if } \quad i \text{ is transient,} \end{cases}$$

since after each such visit, subsequent return is assured if and only if $f_{ii} = 1$ (see Problem (6.13.5) for a more detailed argment).

Here is another important classification of states. Let

$$T_j = \min\{n \geqslant 1 \colon X_n = j\}$$

be the time of the first visit to j, with the convention that $T_j = \infty$ if this visit never occurs; $\mathbf{P}(T_i = \infty \mid X_0 = i) > 0$ if and only if i is transient, and in this case $\mathbf{E}(T_i \mid X_0 = i) = \infty$.

(7) **Definition.** The **mean recurrence time** μ_i of a state i is defined as

$$\mu_i = \mathbf{E}(T_i \mid X_0 = i) = \begin{cases} \sum_n n f_{ii}(n) & \text{if } i \text{ is persistent} \\ \infty & \text{if } i \text{ is transient.} \end{cases}$$

μ_i may be infinite even if i is persistent.

(8) **Definition.**

The persistent state i is called $\begin{cases} \textbf{null} & \text{if} \quad \mu_i = \infty \\ \textbf{non-null (or positive)} & \textit{if} \quad \mu_i < \infty. \end{cases}$

There is a simple criterion for nullity in terms of the transition probabilities.

(9) **Theorem.** *A persistent state is null if and only if $p_{ii}(n) \to 0$ as $n \to \infty$; if this holds then $p_{ji}(n) \to 0$ for all j.*

Proof. We defer this until note (a) after (6.4.17). ∎

Finally, for technical reasons we shall sometimes be interested in the epochs of time at which return to the starting point is possible.

(10) **Definition.** The **period** $d(i)$ of a state i is defined by

$$d(i) = \gcd\{n \colon p_{ii}(n) > 0\},$$

the greatest common divisor of the epochs at which return is possible. We call i **periodic** if $d(i) > 1$ and **aperiodic** if $d(i) = 1$.

That is to say, $p_{ii}(n) = 0$ unless n is a multiple of $d(i)$.

(11) **Definition.** A state is called **ergodic** if it is persistent, non-null, and aperiodic.

(12) **Example. Random walk.** (5.3.4) and (5.12.5) show that the states of the simple random walk are all

(a) transient, if $p \neq \frac{1}{2}$
(b) null persistent, if $p = \frac{1}{2}$.

All states have period $d = 2$. ●

(13) **Example. Branching process.** Consider the branching process of Section 5.4 and suppose that $\mathbf{P}(Z_1 = 0) > 0$. Then 0 is called an *absorbing* state, because the chain never leaves it once it has visited it; all other states are transient. ●

Exercises

(14) **Last exits.** Let $l_{ij}(n) = \mathbf{P}(X_n = j, X_k \neq i \text{ for } 1 \leqslant k < n \mid X_0 = i)$, the probability that the chain passes from i to j in n steps without revisiting i. Writing

$$L_{ij}(s) = \sum_{n=1}^{\infty} s^n l_{ij}(n),$$

show that $P_{ij}(s) = P_{ii}(s)L_{ij}(s)$ if $i \neq j$. Deduce that the first passage times and last exit times have the same distribution for any Markov chain for which $P_{ii}(s) = P_{jj}(s)$ for all i and j. Give an example of such a chain.

(15) Let X be a Markov chain containing an absorbing state s with which all other states communicate (see (6.3.1)). Show that all states other than s are transient.

(16) Show that a state i is persistent if and only if the mean number of visits of the chain to i, having started at i, is infinite.

6.3 Classification of chains

Next, we consider the way in which the states of a Markov chain are related to each other. This investigation will help us to achieve a full classification of the states in the language of the previous section.

(1) **Definition.** We say i **communicates with** j, written $i \to j$, if the chain may ever visit state j with positive probability, starting from i. That is, $i \to j$ if $p_{ij}(m) > 0$ for some $m \geqslant 0$. We say i and j **intercommunicate** if $i \to j$ and $j \to i$, in which case we write $i \leftrightarrow j$.

If $i \neq j$, then $i \to j$ if and only if $f_{ij} > 0$. Clearly $i \to i$ since $p_{ii}(0) = 1$, and it follows that $\leftrightarrow$ is an equivalence relation (*exercise*: if $i \leftrightarrow j$ and $j \leftrightarrow k$, show that $i \leftrightarrow k$). The state space S can be partitioned into the equivalence classes of $\leftrightarrow$. Within each equivalence class all states are of the same type.

(2) **Theorem.** *If $i \leftrightarrow j$ then*

(a) *i and j have the same period*
(b) *i is transient if and only if j is transient*
(c) *i is null persistent if and only if j is null persistent.*

Proof. (b) If $i \leftrightarrow j$ then there exist $m, n \geqslant 0$ such that

$$\alpha = p_{ij}(m)p_{ji}(n) > 0.$$

By the Chapman–Kolmogorov equations (6.1.7),

$$p_{ii}(m+r+n) \geqslant p_{ij}(m)p_{jj}(r)p_{ji}(n) = \alpha p_{jj}(r),$$

for any non-negative integer r. Now sum over r to obtain

$$\sum_r p_{jj}(r) < \infty \quad \text{if} \quad \sum_r p_{ii}(r) < \infty.$$

Thus, by (6.2.4), j is transient if i is transient. The converse holds similarly and (b) is shown.

(a) This proof is similar and proceeds by way of (6.2.10).

(c) We defer this until the next section. A possible route is by way of (6.2.9), but we prefer to proceed differently in order to avoid the danger of using a circular argument. ∎

(3) **Definition.** A set C of states is called

(a) **closed** if $p_{ij} = 0$ for all $i \in C, j \notin C$

(b) **irreducible** if $i \leftrightarrow j$ for all $i, j \in C$.

Once the chain takes a value in a closed set C of states then it never leaves C subsequently. A closed set containing exactly one state is called *absorbing*; for example, the state 0 is absorbing for the branching process. It is clear that the equivalence classes of $\leftrightarrow$ are irreducible. We call an irreducible set C *aperiodic* (or *persistent*, *null*, and so on) if all the states in C have this property; (2) ensures that this is meaningful. If the whole state space S is irreducible, then we speak of the chain itself as having the property in question.

(4) **Decomposition theorem.** *The state space S can be partitioned uniquely as*

$$S = T \cup C_1 \cup C_2 \cup \cdots$$

where T is the set of transient states, and the C_i are irreducible closed sets of persistent states.

Proof. Let $C_1, C_2, \ldots$ be the persistent equivalence classes of $\leftrightarrow$. We need only show that each C_r is closed. Suppose on the contrary that there exist $i \in C_r, j \notin C_r$ such that $p_{ij} > 0$. Now $j \nrightarrow i$, and therefore

$$\mathbf{P}(X_n \neq i \text{ for all } n \geqslant 1 \mid X_0 = i) \geqslant \mathbf{P}(X_1 = j \mid X_0 = i) > 0,$$

in contradiction of the assumption that i is persistent. ∎

The decomposition theorem clears the air a little. For, if $X_0 \in C_r$, say, then the chain never leaves C_r and we might as well take C_r to be the whole state space. On the other hand, if $X_0 \in T$ then the chain either stays in T for ever or moves eventually to one of the C_k where it subsequently remains. Thus, either the chain always takes values in the set of transient states or it lies eventually in some irreducible closed set of persistent states. For the special case when S is finite the first of these possibilities cannot occur.

(5) **Lemma.** *If S is finite, then at least one state is persistent and all persistent states are non-null.*

Proof. If all states are transient, then take the limit through the summation sign to obtain the contradiction

$$1 = \lim_{n \to \infty} \sum_j p_{ij}(n) = 0$$

by (6.2.5). The same contradiction arises by (6.2.9) for the closed set of all null persistent states, should this set be non-empty. ■

(6) **Example.** Let $S = \{1, 2, 3, 4, 5, 6\}$ and

$$\mathbf{P} = \begin{pmatrix} \frac{1}{2} & \frac{1}{2} & 0 & 0 & 0 & 0 \\ \frac{1}{4} & \frac{3}{4} & 0 & 0 & 0 & 0 \\ \frac{1}{4} & \frac{1}{4} & \frac{1}{4} & \frac{1}{4} & 0 & 0 \\ \frac{1}{4} & 0 & \frac{1}{4} & \frac{1}{4} & 0 & \frac{1}{4} \\ 0 & 0 & 0 & 0 & \frac{1}{2} & \frac{1}{2} \\ 0 & 0 & 0 & 0 & \frac{1}{2} & \frac{1}{2} \end{pmatrix}.$$

$\{1, 2\}$ and $\{5, 6\}$ are irreducible closed sets and therefore contain persistent non-null states. States 3 and 4 are transient because $3 \to 4 \to 6$ but return from 6 is impossible. All states have period 1 because $p_{ii}(1) > 0$ for all i. Hence, 3 and 4 are transient, and 1, 2, 5, and 6 are ergodic. Easy calculations give

$$f_{11}(n) = \begin{cases} p_{11} = \frac{1}{2} & \text{if } n = 1 \\ p_{12}(p_{22})^{n-2}p_{21} = \frac{1}{2}(\frac{3}{4})^{n-2}\frac{1}{4} & \text{if } n \geq 2 \end{cases}$$

and hence $\mu_1 = \sum_n nf_{11}(n) = 3$. Other mean recurrence times can be found similarly. The next section gives another way of finding the μ_i which usually requires less computation. ●

Exercises

(7) Let X be a Markov chain on $\{0, 1, 2, \dots\}$ with transition matrix given by $p_{0j} = a_j$ for $j \geq 0$, $p_{ii} = r$ and $p_{i,i-1} = 1 - r$ for $i \geq 1$. Classify the states of the chain, and find their mean recurrence times.

(8) Determine whether or not the random walk on the integers having transition probabilities $p_{i,i+2} = p$, $p_{i,i-1} = 1 - p$, for all i, is persistent.

(9) Classify the states of the Markov chains with transition matrices

(a)
$$\begin{pmatrix} 1-2p & 2p & 0 \\ p & 1-2p & p \\ 0 & 2p & 1-2p \end{pmatrix}$$

(b)
$$\begin{pmatrix} 0 & p & 0 & 1-p \\ 1-p & 0 & p & 0 \\ 0 & 1-p & 0 & p \\ p & 0 & 1-p & 0 \end{pmatrix}$$

In each case, calculate $p_{ij}(n)$ and the mean recurrence times of each state.

(10) A particle performs a random walk on the vertices of a cube. At each step it remains where it is with probability $\frac{1}{4}$, and moves to each of its neighbouring vertices with probability $\frac{1}{4}$. Let v and w be two diametrically opposite vertices. If the walk starts at v, find

(a) the mean number of steps until its first return to v,
(b) the mean number of steps until its first visit to w,
(c) the mean number of visits to w before its first return to v.

6.4 Stationary distributions and the limit theorem

How does X_n behave after a long time n has elapsed? The sequence $\{X_n\}$ cannot generally, of course, converge to some particular state s since it enjoys the inherent random fluctuation which is specified by the transition matrix. However, we might hold out some hope that the *distribution* of X_n settles down. Indeed, subject to certain conditions this turns out to be the case. The classical study of limiting distributions proceeds by algebraic manipulation of the generating functions of (6.2.3); we shall avoid this here, contenting ourselves for the moment with results which are not quite the best possible but which have attractive probabilistic proofs. This section is in two parts, which deal with stationary distributions and limit theorems respectively.

(A) **Stationary distributions.** We shall see that the existence of a limiting distribution for X_n, as $n \to \infty$, is closely bound up with the existence of so-called 'stationary distributions'.

(1) **Definition.** The vector $\boldsymbol{\pi}$ is called a **stationary distribution** of the chain if $\boldsymbol{\pi}$ has entries $(\pi_j: j \in S)$ such that

(a) $\pi_j \geqslant 0$ for all j, and $\sum_j \pi_j = 1$
(b) $\boldsymbol{\pi} = \boldsymbol{\pi P}$, which is to say that $\pi_j = \sum_i \pi_i p_{ij}$ for all j.

Such a distribution is called stationary for the following reason. Iterate (1b) to obtain

$$\boldsymbol{\pi}\mathbf{P}^2 = (\boldsymbol{\pi}\mathbf{P})\mathbf{P} = \boldsymbol{\pi}\mathbf{P} = \boldsymbol{\pi}$$

and so

(2) $$\boldsymbol{\pi}\mathbf{P}^n = \boldsymbol{\pi} \quad \text{for all } n \geqslant 0.$$

Now use (6.1.8) to see that if X_0 has distribution $\boldsymbol{\pi}$ then X_n has distribution $\boldsymbol{\pi}$ for all n, showing that the distribution of X_n is 'stationary' as time passes; in such a case, of course, $\boldsymbol{\pi}$ is also the limiting distribution of X_n as $n \to \infty$.

Following the discussion after the decomposition theorem (6.3.4), we shall assume henceforth that the chain is irreducible and shall investigate the existence of stationary distributions.

(3) **Theorem.** *An irreducible chain has a stationary distribution $\boldsymbol{\pi}$ if and only if all the states are non-null persistent; in this case, $\boldsymbol{\pi}$ is the unique stationary distribution and is given by $\pi_i = \mu_i^{-1}$ for each $i \in S$, where μ_i is the mean recurrence time of i.*

Stationary distributions $\boldsymbol{\pi}$ satisfy $\boldsymbol{\pi} = \boldsymbol{\pi}\mathbf{P}$. We may display a root $\mathbf{x}$ of the matrix equation $\mathbf{x} = \mathbf{x}\mathbf{P}$ explicitly as follows, whenever the chain is irreducible and persistent. Fix a state k and let $\rho_i(k)$ be the mean number of visits of the chain to the state i between two successive visits to state k; that is, $\rho_i(k) = \mathbf{E}(N_i \mid X_0 = k)$ where

$$N_i = \sum_{n=1}^{\infty} I_{\{X_n = i\} \cap \{T_k \geqslant n\}}$$

and T_k is the time of the first return to state k, as before. Note that $N_k = 1$, so that $\rho_k(k) = 1$, and that

$$\rho_i(k) = \sum_{n=1}^{\infty} \mathbf{P}(X_n = i, T_k \geqslant n \mid X_0 = k).$$

We write $\boldsymbol{\rho}(k)$ for the vector $(\rho_i(k): i \in S)$. Clearly $T_k = \sum_{i \in S} N_i$, since the time between visits to k must be spent somewhere; taking expectations, we find that

(4) $$\mu_k = \sum_{i \in S} \rho_i(k),$$

so that the vector $\boldsymbol{\rho}(k)$ contains terms with sum equal to the mean recurrence time μ_k.

(5) **Lemma.** *For any state k of an irreducible persistent chain, the vector $\boldsymbol{\rho}(k)$ satisfies $\rho_i(k) < \infty$ for all i, and furthermore $\boldsymbol{\rho}(k) = \boldsymbol{\rho}(k)\mathbf{P}$.*

Proof. We show first that $\rho_i(k) < \infty$ when $i \neq k$. Write $\ell_{ki}(n) = \mathbf{P}(X_n = i, T_k \geqslant n \mid X_0 = k)$, the probability that the chain reaches i in n steps but with no intermediate return to its starting point k. Clearly $f_{kk}(m+n) \geqslant \ell_{ki}(m) f_{ik}(n)$, since the first return time to k equals $m+n$ if: (a) $X_m = i$, (b) there is no return to k prior to time m, and (c) the next subsequent visit to k takes place after another n steps. By the irreducibility of the chain, there exists n such that $f_{ik}(n) > 0$. With this choice of n, we have that $\ell_{ki}(m) \leqslant f_{kk}(m+n)/f_{ik}(n)$, and so

$$\rho_i(k) = \sum_{m=1}^{\infty} \ell_{ki}(m) \leqslant \frac{1}{f_{ik}(n)} \sum_{m=1}^{\infty} f_{kk}(m+n) \leqslant \frac{1}{f_{ik}(n)} < \infty$$

as required.

For the second statement of the lemma, we argue as follows. We have that $\rho_i(k) = \sum_{n=1}^{\infty} \ell_{ki}(n)$. Now $\ell_{ki}(1) = p_{ki}$, and

$$\ell_{ki}(n) = \sum_{j \neq k} \ell_{kj}(n-1) p_{ji} \quad \text{for} \quad n \geqslant 2,$$

by conditioning on the value of X_{n-1}. Summing over $n \geqslant 2$, we obtain

$$\rho_i(k) = p_{ki} + \sum_{j \neq k} \left(\sum_{n \geqslant 2} \ell_{kj}(n-1) \right) p_{ji} = \rho_k(k) p_{ki} + \sum_{j \neq k} \rho_j(k) p_{ji}$$

since $\rho_k(k) = 1$. The lemma is proved. ■

We have from (4) and Lemma (5) that, for any irreducible persistent chain, the vector $\boldsymbol{\rho}(k)$ satisfies $\boldsymbol{\rho}(k) = \boldsymbol{\rho}(k)\mathbf{P}$, and furthermore that the components of $\boldsymbol{\rho}(k)$ are non-negative with sum μ_k. Hence, if $\mu_k < \infty$, then the vector $\boldsymbol{\pi}$ with entries $\pi_i = \rho_i(k)/\mu_k$ satisfies $\boldsymbol{\pi} = \boldsymbol{\pi}\mathbf{P}$ and furthermore has non-negative entries which sum to 1; that is to say, $\boldsymbol{\pi}$ is a stationary distribution. We have proved that every non-null persistent irreducible chain has a stationary distribution, an important step towards the proof of the main theorem (3).

Before continuing with the rest of the proof of (3), we note a consequence of the results so far. Lemma (5) implies the existence of a root of the equation $\mathbf{x} = \mathbf{x}\mathbf{P}$, whenever the chain is irreducible and persistent. Furthermore, there exists a root whose components are strictly positive (certainly there exists a non-negative root, and it is not difficult—see the argument after (8)—to see that this root may be taken strictly positive). It may be shown that this root is unique up to a multiplicative constant (Problem (6.13.7)), and we arrive therefore at the following useful conclusion.

(6) **Theorem.** *If the chain is irreducible and persistent, there exists a positive root* $\mathbf{x}$ *of the equation* $\mathbf{x} = \mathbf{x}\mathbf{P}$, *which is unique up to a multiplicative constant. The chain is non-null if* $\sum_i x_i < \infty$ *and null if* $\sum_i x_i = \infty$.

Proof of (3). Suppose that $\boldsymbol{\pi}$ is a stationary distribution of the chain. If all states are transient then $p_{ij}(n) \to 0$, as $n \to \infty$, for all i and j by (6.2.5). From (2),

(7) $$\pi_j = \sum_i \pi_i p_{ij}(n) \to 0 \quad \text{as} \quad n \to \infty, \quad \text{for all } i \text{ and } j,$$

which contradicts (1a). Thus all states are persistent. To see the limit in (7)†, let F be a finite subset of S and write

$$\begin{aligned}\sum_i \pi_i p_{ij}(n) &\leqslant \sum_{i \in F} \pi_i p_{ij}(n) + \sum_{i \notin F} \pi_i \\ &\to \sum_{i \notin F} \pi_i \quad \text{as} \quad n \to \infty, \quad \text{since } F \text{ is finite} \\ &\to 0 \quad \text{as} \quad F \uparrow S.\end{aligned}$$

We show next that the existence of $\boldsymbol{\pi}$ implies that all states are non-null and that $\pi_i = \mu_i^{-1}$ for each i. Suppose that X_0 has distribution $\boldsymbol{\pi}$, so that $\mathbf{P}(X_0 = i) = \pi_i$ for each i. Then, by (3.11.13a),

$$\begin{aligned}\pi_j \mu_j &= \sum_{n=1}^{\infty} \mathbf{P}(T_j \geqslant n \mid X_0 = j)\mathbf{P}(X_0 = j) \\ &= \sum_{n=1}^{\infty} \mathbf{P}(T_j \geqslant n, X_0 = j).\end{aligned}$$

However, $\mathbf{P}(T_j \geqslant 1, X_0 = j) = \mathbf{P}(X_0 = j)$, and for $n \geqslant 2$

$$\begin{aligned}\mathbf{P}(T_j \geqslant n, X_0 = j) &= \mathbf{P}(X_0 = j, X_m \neq j \text{ for } 1 \leqslant m \leqslant n-1) \\ &= \mathbf{P}(X_m \neq j \text{ for } 1 \leqslant m \leqslant n-1) \\ &\quad - \mathbf{P}(X_m \neq j \text{ for } 0 \leqslant m \leqslant n-1) \\ &= \mathbf{P}(X_m \neq j \text{ for } 0 \leqslant m \leqslant n-2) \\ &\quad - \mathbf{P}(X_m \neq j \text{ for } 0 \leqslant m \leqslant n-1) \quad \text{by homogeneity} \\ &= a_{n-2} - a_{n-1}\end{aligned}$$

where $a_n = \mathbf{P}(X_m \neq j \text{ for } 0 \leqslant m \leqslant n)$. Sum over n to obtain

$$\pi_j \mu_j = \mathbf{P}(X_0 = j) + \mathbf{P}(X_0 \neq j) - \lim_{n \to \infty} a_n = 1 - \lim_{n \to \infty} a_n.$$

However,

$$\lim_{n \to \infty} a_n = \mathbf{P}(X_m \neq j \text{ for all } m) = 0$$

by the persistence of j. We have shown that

(8) $$\pi_j \mu_j = 1,$$

† Actually this argument is a form of the bounded convergence theorem (5.6.12) applied to sums instead of to integrals. We shall make repeated use of this technique.

so that $\mu_j = \pi_j^{-1} < \infty$ if $\pi_j > 0$. To see that $\pi_j > 0$ for all j, suppose on the contrary that $\pi_j = 0$ for some j. Then

$$0 = \pi_j = \sum_i \pi_i p_{ij}(n) \geqslant \pi_i p_{ij}(n) \quad \text{for all } i \text{ and } n,$$

yielding that $\pi_i = 0$ whenever $i \to j$. The chain is assumed irreducible, so that $\pi_i = 0$ for all i in contradiction of the fact that the π_i have sum 1. Hence $\mu_j < \infty$ and all states of the chain are non-null. Furthermore, (8) specifies π_j uniquely as μ_j^{-1}.

Thus, if $\boldsymbol{\pi}$ exists then it is unique and all the states of the chain are non-null persistent. Conversely, if the states of the chain are non-null persistent then the chain has a stationary distribution given by (5). ■

We may now complete the proof of (6.3.2c).

Proof of (6.3.2c). Let $C(i)$ be the irreducible closed equivalence class of states which contains the non-null persistent state i. Suppose that $X_0 \in C(i)$. Then $X_n \in C(i)$ for all n, and (5) and (3) combine to tell us that all states in $C(i)$ are non-null. ■

(9) **Example (6.3.6) revisited.** To find μ_1 and μ_2 consider the irreducible closed set $C = \{1, 2\}$. If $X_0 \in C$, then solve the equation $\boldsymbol{\pi} = \boldsymbol{\pi} \mathbf{P}_C$ for $\boldsymbol{\pi} = (\pi_1, \pi_2)$ in terms of

$$\mathbf{P}_C = \begin{pmatrix} \frac{1}{2} & \frac{1}{2} \\ \frac{1}{4} & \frac{3}{4} \end{pmatrix}$$

to find the unique stationary distribution $\boldsymbol{\pi} = (\frac{1}{3}, \frac{2}{3})$, giving that

$$\mu_1 = \pi_1^{-1} = 3, \qquad \mu_2 = \pi_2^{-1} = \tfrac{3}{2}.$$

Now find the other mean recurrence times yourself (*exercise*). ●

Theorem (3) provides a useful criterion for deciding whether or not an irreducible chain is non-null persistent: just look for a stationary distribution. There is a similar criterion for the transience of irreducible chains.

(10) **Theorem.** *Let $s \in S$ be any state of an irreducible chain. The chain is transient if and only if there exists a non-zero solution $\{y_j : j \neq s\}$ to the equations*

(11) $$y_i = \sum_{j \neq s} p_{ij} y_j, \qquad i \neq s,$$

such that $|y_j| \leqslant 1$ for all j.

Proof. The chain is transient if and only if s is transient. First suppose s is transient and define

(12)
$$\tau_i(n) = \mathbf{P}(\text{no visit to } s \text{ in first } n \text{ steps} \mid X_0 = i)$$
$$= \mathbf{P}(X_m \neq s, 1 \leqslant m \leqslant n \mid X_0 = i).$$

Then

$$\tau_i(1) = \sum_{j \neq s} p_{ij}, \qquad \tau_i(n+1) = \sum_{j \neq s} p_{ij}\tau_j(n).$$

Furthermore, $\tau_i(n) \geqslant \tau_i(n+1)$, and so

$$\tau_i = \lim_{n \to \infty} \tau_i(n) = \mathbf{P}(\text{no visit to } s \text{ ever} \mid X_0 = i) = 1 - f_{is}$$

satisfies (11). (Can *you* prove this? Use the method of proof of (7).) Also $\tau_i > 0$ for some i, since otherwise $f_{is} = 1$ for all $i \neq s$, and so

$$f_{ss} = p_{ss} + \sum_{i \neq s} p_{si} f_{is} = \sum_i p_{si} = 1$$

by conditioning on X_1; this contradicts the transience of s.

Conversely, let $\mathbf{y}$ satisfy (11) with $|y_i| \leqslant 1$. Then

$$|y_i| \leqslant \sum_{j \neq s} p_{ij}|y_j| \leqslant \sum_{j \neq s} p_{ij} = \tau_i(1)$$

$$|y_i| \leqslant \sum_{j \neq s} p_{ij}\tau_j(1) = \tau_j(2),$$

and so on, where the $\tau_i(n)$ are given by (12). Thus

$$|y_i| \leqslant \tau_i(n) \quad \text{for all } n.$$

Let $n \to \infty$ to show that

$$\tau_i = \lim_{n \to \infty} \tau_i(n) > 0$$

for some i, which shows that s is transient by the result of (6.13.6). ■

This theorem provides a necessary and sufficient condition for persistence: an irreducible chain is persistent if and only if the only bounded solution to (11) is the zero solution. This combines with (3) to give a condition for null persistence. Another condition is the following (see Cox and Miller 1965, p. 113 for a proof); a corresponding result holds for any countably infinite state space S.

(13) **Theorem.** *Let $s \in S$ be any state of an irreducible chain on $S = \{0, 1, 2, \ldots\}$. The chain is persistent if there exists a solution $\{y_j : j \neq s\}$ to the inequalities*

(14)
$$y_i \geqslant \sum_{j \neq s} p_{ij} y_j, \qquad i \neq s,$$

such that $y_i \to \infty$ as $i \to \infty$.

(15) **Example. Random walk with retaining barrier.** A particle performs a random walk on the non-negative integers with a retaining barrier at 0. The transition probabilities are

$$p_{i,i+1} = p \quad \text{if} \quad i \geqslant 0$$

$$p_{0,0} = q, \qquad p_{i,i-1} = q \quad \text{if} \quad i \geqslant 1$$

where $p + q = 1$. Let $\rho = p/q$.

(a) If $q < p$, take $s = 0$ to see that $y_j = 1 - \rho^{-j}$ satisfies (11), and so the chain is transient.
(b) Solve the equation $\boldsymbol{\pi} = \boldsymbol{\pi} \mathbf{P}$ to find that there exists a stationary distribution, with $\pi_j = \rho^j(1 - \rho)$, if and only if $q > p$. Thus the chain is non-null persistent if and only if $q > p$.
(c) If $q = p = \frac{1}{2}$, take $s = 0$ in (13) and check that $y_j = j$ $(j \geqslant 1)$ solves (14). Thus the chain is null persistent. Alternatively, argue as follows. The chain is persistent since symmetric random walk is persistent (just reflect negative excursions of a symmetric random walk into the positive half-line). Solve the equation $\mathbf{x} = \mathbf{x}\mathbf{P}$ to find that $x_i = 1$ for all i provides a root, unique up to a multiplicative constant. However, $\sum_i x_i = \infty$ so that the chain is null, by Theorem (6).

These conclusions match our intuitions well. ●

(B) Limit theorems. Next we explore the link between the existence of a stationary distribution and the limiting behaviour of the probabilities $p_{ij}(n)$ as $n \to \infty$. The following example indicates a difficulty which arises from periodicity.

(16) **Example.** If $S = \{1, 2\}$ and $p_{12} = p_{21} = 1$, then

$$p_{11}(n) = p_{22}(n) = \begin{cases} 0 & \text{if} \quad n \text{ is odd} \\ 1 & \text{if} \quad n \text{ is even.} \end{cases}$$

Clearly $p_{ii}(n)$ does not converge as $n \to \infty$; the reason is that both states are periodic with period 2. ●

Until further notice we shall deal only with irreducible *aperiodic* chains. The principal result is the following theorem.

(17) **Theorem.** *For an irreducible aperiodic chain, we have that*

$$p_{ij}(n) \to \frac{1}{\mu_j} \quad as \quad n \to \infty, \quad for\ all\ i\ and\ j.$$

We make the following remarks.

(a) If the chain is *transient* or *null persistent* then $p_{ij}(n) \to 0$ for all i and j, since $\mu_j = \infty$. Now we can prove (6.2.9). Let $C(i)$ be the irreducible closed set of states which contains the persistent state i. If $C(i)$ is aperiodic then the result is an immediate consequence of (17); the periodic case can be treated similarly, but with slightly more difficulty (see note (d) following).

(b) If the chain is *non-null persistent* then $p_{ij}(n) \to \pi_j = \mu_j^{-1}$, where $\boldsymbol{\pi}$ is the unique stationary distribution by (3).

(c) It follows from (17) that the limit probability, $\lim_{n\to\infty} p_{ij}(n)$, does not depend on the starting point $X_0 = i$; that is, the chain forgets its origin. It is now easy to check that

$$\mathbf{P}(X_n = j) = \sum_i \mathbf{P}(X_0 = i)p_{ij}(n) \to \frac{1}{\mu_j} \quad \text{as} \quad n \to \infty$$

by (6.1.8), irrespective of the distribution of X_0.

(d) If $X = \{X_n\}$ is an irreducible chain with period d, then $Y = \{Y_n = X_{nd}: n \geqslant 0\}$ is an aperiodic chain, and it follows that

$$p_{jj}(nd) = \mathbf{P}(Y_n = j \mid Y_0 = j) \to \frac{d}{\mu_j} \quad \text{as} \quad n \to \infty.$$

Proof. If the chain is transient then the result holds from (6.2.5). The persistent case is treated by an important technique known as 'coupling'. Construct a 'coupled chain' $Z = (X, Y)$, being an ordered pair $X = \{X_n: n \geqslant 0\}$, $Y = \{Y_n: n \geqslant 0\}$ of *independent* Markov chains, each with transition matrix $\boldsymbol{P}$. Then $Z = \{Z_n = (X_n, Y_n): n \geqslant 0\}$ takes values in $S \times S$, and it is easy to check that Z is a Markov chain with transition probabilities

$$\begin{aligned} p_{ij,kl} &= \mathbf{P}(Z_{n+1} = (k, l) \mid Z_n = (i, j)) \\ &= \mathbf{P}(X_{n+1} = k \mid X_n = i)\mathbf{P}(Y_{n+1} = l \mid Y_n = j) \quad \text{by independence} \\ &= p_{ik}\, p_{jl}. \end{aligned}$$

Since X is irreducible and aperiodic, for any states i, j, k, l there exists $N = N(i, j, k, l)$ such that

$$p_{ik}(n)p_{jl}(n) > 0 \quad \text{for all } n \geqslant N;$$

thus Z also is irreducible (see Problem (6.13.4); *only here* do we require that X be aperiodic).

Suppose that X is non-null persistent. Then X has a unique stationary distribution $\boldsymbol{\pi}$, by (3), and it is easy to see that Z has a stationary distribution $\boldsymbol{\nu} = (\nu_{ij}: i, j \in S)$ given by $\nu_{ij} = \pi_i \pi_j$; thus Z is also non-null persistent, by (3). Now, suppose that $X_0 = i$ and $Y_0 = j$, so that $Z_0 = (i, j)$. Choose any state $s \in S$ and let

$$T = \min\{n \geqslant 1: Z_n = (s, s)\}$$

denote the time of the first passage of Z to (s, s); from Problem (6.13.6) and the persistence of Z, $\mathbf{P}(T < \infty) = 1$. The central idea of the proof is the following observation. If $m \leqslant n$ and $X_m = Y_m$, then X_n and Y_n are identically distributed since the distributions of X_n and Y_n depend only upon the shared transition matrix $\boldsymbol{P}$ and upon the shared value of the chains at the mth stage. Thus, conditional on $\{T \leqslant n\}$, X_n and Y_n have the same distribution. We shall use this fact, together with the finiteness of T, to show that the ultimate distributions of X and Y are independent of their starting points. More precisely, starting from $Z_0 = (X_0, Y_0) = (i, j)$,

$$\begin{aligned} p_{ik}(n) &= \mathbf{P}(X_n = k) \\ &= \mathbf{P}(X_n = k,\ T \leqslant n) + \mathbf{P}(X_n = k,\ T > n) \\ &= \mathbf{P}(Y_n = k,\ T \leqslant n) + \mathbf{P}(X_n = k,\ T > n) \\ &\qquad \text{because, given that } T \leqslant n,\ X_n \text{ and } Y_n \text{ are identically distributed} \\ &\leqslant \mathbf{P}(Y_n = k) + \mathbf{P}(T > n) \\ &= p_{jk}(n) + \mathbf{P}(T > n). \end{aligned}$$

This, and the related inequality with i and j interchanged, yields

$$|p_{ik}(n) - p_{jk}(n)| \leqslant \mathbf{P}(T > n) \to 0 \quad \text{as} \quad n \to \infty$$

because $\mathbf{P}(T < \infty) = 1$; therefore

(18) $$p_{ik}(n) - p_{jk}(n) \to 0 \quad \text{as} \quad n \to \infty \quad \text{for all } i, j, \text{ and } k.$$

Thus, if $\lim_{n\to\infty} p_{ik}(n)$ exists, then it does not depend on i. To show that it exists, write

(19) $$\pi_k - p_{jk}(n) = \sum_i \pi_i (p_{ik}(n) - p_{jk}(n)) \to 0 \quad \text{as} \quad n \to \infty,$$

giving the result. To see that the limit in (19) follows from (18), use the bounded convergence argument in the proof of (7): for any finite subset F of S,

$$\begin{aligned} \sum_i \pi_i |p_{ik}(n) - p_{jk}(n)| &\leqslant \sum_{i \in F} |p_{ik}(n) - p_{jk}(n)| + 2 \sum_{i \notin F} \pi_i \\ &\to 2 \sum_{i \notin F} \pi_i \quad \text{as} \quad n \to \infty \end{aligned}$$

which in turn tends to zero as $F \uparrow S$.

Finally, suppose that X is null persistent; the argument is a little trickier in this case. If Z is transient, then from (6.2.5) applied to Z

$$\mathbf{P}(Z_n = (j, j) \mid Z_0 = (i, i)) = p_{ij}(n)^2 \to 0 \quad \text{as} \quad n \to \infty$$

and the result holds. If Z is non-null persistent then, starting from $Z_0 = (i, i)$, the epoch T_{ii}^Z of the first return of Z to (i, i) is no smaller than the epoch T_i of the first return of X to i; however, $\mathbf{E}(T_i) = \infty$ and $\mathbf{E}(T_{ii}^Z) < \infty$ which is a contradiction. Lastly, suppose that Z is null persistent. The argument which leads to (18) still holds, and we wish to deduce that

$$p_{ij}(n) \to 0 \quad \text{as} \quad n \to \infty \quad \text{for all } i \text{ and } j.$$

If this does not hold then there exists a subsequence $n_1, n_2, \ldots$ along which

(20) $$p_{ij}(n_r) \to \alpha_j \quad \text{as} \quad r \to \infty \quad \text{for all } i \text{ and } j,$$

for some $\boldsymbol{\alpha}$, where the α_j are not all zero and are independent of i by (18); this is an application of the principle of 'diagonal selection' (see Billingsley 1986, p. 292, Feller 1968, p. 336, or Exercise (28)). For any finite set F of states,

$$\sum_{j \in F} \alpha_j = \lim_{r \to \infty} \sum_{j \in F} p_{ij}(n_r) \leqslant 1$$

and so $\alpha = \sum_j \alpha_j$ satisfies $0 < \alpha \leqslant 1$. Furthermore

$$\sum_{k \in F} p_{ik}(n_r) p_{kj} \leqslant p_{ij}(n_r + 1) = \sum_k p_{ik} p_{kj}(n_r);$$

let $r \to \infty$ here to deduce from (20) and bounded convergence (as used in the proof of (19)) that

$$\sum_{k \in F} \alpha_k p_{kj} \leqslant \sum_k p_{ik} \alpha_j = \alpha_j,$$

and so, letting $F \uparrow S$, we obtain $\sum_k \alpha_k p_{kj} \leqslant \alpha_j$ for each $j \in S$. But equality must hold here, since if strict inequality holds for some j then

$$\sum_k \alpha_k = \sum_{k,j} \alpha_k p_{kj} < \sum_j \alpha_j,$$

which is a contradiction. Therefore

$$\sum_k \alpha_k p_{kj} = \alpha_j \quad \text{for each } j \in S,$$

giving that $\boldsymbol{\pi} = \{\alpha_j/\alpha : j \in S\}$ is a stationary distribution for X; this contradicts the nullity of X by (3). ■

The original and more general version of the ergodic theorem (17) for Markov chains does *not* assume that the chain is irreducible. We state it here; for a proof see (5.2.24) or (10.4.20).

(21) **Theorem.** *For any aperiodic state j of a Markov chain,*

$$p_{jj}(n) \to \frac{1}{\mu_j} \quad \textit{as } n \to \infty.$$

Furthermore, if i is any other state then

$$p_{ij}(n) \to \frac{1}{\mu_j} f_{ij} \quad \textit{as} \quad n \to \infty.$$

(22) **Corollary.** *Let*

$$\tau_{ij}(n) = \frac{1}{n} \sum_{m=1}^{n} p_{ij}(m)$$

be the mean proportion of elapsed time up to the nth step during which the chain was in state j, starting from i. Then, if j is aperiodic,

$$\tau_{ij}(n) \to \frac{1}{\mu_j} f_{ij} \quad \textit{as} \quad n \to \infty.$$

Proof. *Exercise*: prove and use the fact that, as $n \to \infty$,

$$\frac{1}{n} \sum_{1}^{n} x_i \to x \quad \text{if} \quad x_n \to x. \qquad \blacksquare$$

(23) **Example. The coupling game.** You may be able to amaze your friends and break the ice at parties with the following card 'trick'. A pack of cards is shuffled, and you deal the cards (face up) one by one. You instruct the audience as follows. Each person is to select some card, secretly, chosen from the first six or seven, say. If the face value of this card is m (aces count 1 and court cards count 10), let the next $m-1$ cards pass and note the face value of the mth. If it is n, let the next $n-1$ cards pass by and note the face value of the nth. Continuing according to this rule, there will arrive a last card in this sequence, face value X say, with fewer than X cards remaining. Call X the 'score'. Each person's score is known to that person but not to you, and can generally be any number between 1 and 10. At the end of the game, using an apparently fiendishly clever method you announce to the audience a number between 1 and 10. If few errors have been made, the majority of the audience will find that your number agrees with their score. Your popularity will then be assured, for a short while at least.

This is the 'trick'. You follow the same rules as the audience, beginning for the sake of simplicity with the first card. You will obtain a 'score' of Y, say, and it happens that there is a large probability that any given person obtains the score Y also; therefore you announce the score Y.

Why does the game often work? Suppose that someone picks the m_1th card, m_2th card, and so on, and you pick the $n_1(=1)$th, n_2th, etc. If $n_i = m_j$ for some i and j, then the two of you are 'stuck together' forever after, since the rules of the game require you to follow the same pattern henceforth; when this happens first, we say that 'coupling' has occurred. Prior to coupling, each time you read the value of a card, there is a positive probability that you will arrive at the next stage on exactly the same card as the other person. If the pack of cards were infinitely large, then coupling would certainly take place sooner or later, and it turns out that there is a good chance that coupling takes place before the last card of a regular pack has been dealt.

You may recognize the argument above as being closely related to that used near the beginning of the proof of Theorem (3). ●

Exercises

(24) The proof copy of a book is read by an infinite sequence of editors checking for mistakes. Each mistake is detected with probability p at each reading; between readings the printer corrects the detected mistakes but introduces a random number of new errors (errors may be introduced even if no mistakes were detected). Assuming as much independence as usual, and that the numbers of new errors after different readings are identically distributed, find an expression for the probability generating function of the stationary distribution of the number X_n of errors after the nth editor–printer cycle, whenever this exists. Find it explicitly when the printer introduces a Poisson-distributed number of errors at each stage.

(25) Do the appropriate parts of Exercises (6.3.7)–(6.3.10) again, making use of the new techniques at your disposal.

(26) **Dams.** Let X_n be the amount of water in a reservoir at noon on day n. During the 24 hour period beginning at this time, a quantity Y_n of water flows into the reservoir, and just before noon on each day exactly one unit of water is removed (if this amount can be found). The maximum capacity of the reservoir is K, and excessive inflows are spilled and lost. Assume that the Y_n are independent and identically distributed random variables and that, by some miracle, all numbers in this exercise are non-negative integers. Show that (X_n) is a Markov chain, and find its transition matrix and an expression for its stationary distribution in terms of the probability generating function G of the Y_n.

Find the stationary distribution explicitly when Y has probability generating function $G(s) = p(1 - qs)^{-1}$.

(27) Show by example that chains which are not irreducible may have many different stationary distributions.

(28) Let $(x_i(n)\colon i, n \geqslant 1)$ be a bounded collection of real numbers. Show that there exists an increasing sequence $n_1, n_2, \ldots$ of positive integers such that $\lim_{r\to\infty} x_i(n_r)$ exists for all i. Use this result to prove that, for an irreducible Markov chain, if it is not the case that $p_{ij}(n) \to 0$ as $n \to \infty$ for all i and j, then there exists a sequence $(n_r\colon r \geqslant 1)$ and a vector $\boldsymbol{\alpha}$ ($\neq \mathbf{0}$) such that $p_{ij}(n_r) \to \alpha_j$ as $r \to \infty$ for all i and j.

6.5 Time-reversibility

Most laws of physics have the property that they would make the same assertions if the universal clock were reversed and time were made to run backwards. It may be thought to be implausible that nature works in such ways (have *you* ever seen the fragments of a shattered teacup re-assemble themselves on the table from which they fell?), and so one might postulate a non-decreasing quantity called 'entropy'. However, never mind such objections; let us think about the reversal of the time scale of a Markov chain.

Suppose that $\{X_n: -\infty < n < \infty\}$ is an irreducible non-null persistent Markov chain, with transition matrix $\boldsymbol{P}$ and unique stationary distribution $\boldsymbol{\pi}$. Suppose further that X_n has distribution $\boldsymbol{\pi}$ for every $n \in (-\infty, \infty)$. (In order that this hold, it is *not* sufficient to assume that X_0 has distribution $\boldsymbol{\pi}$; certainly in this case X_n has distribution $\boldsymbol{\pi}$ for all $n \geqslant 0$, by (6.1.8), but it does not follow that X_{-1} has distribution $\boldsymbol{\pi}$ also.) Define the 'reversed chain' Y by

$$Y_n = X_{-n}, \qquad -\infty < n < \infty.$$

It is not difficult to show that Y is a Markov chain also, and of course Y_n has distribution $\boldsymbol{\pi}$ for each n.

(1) **Definition.** X is called **time-reversible** if the transition matrices of X and Y are the same.

(2) **Theorem.** *X is time-reversible if and only if*

$$\pi_i p_{ij} = \pi_j p_{ji} \quad \textit{for all } i, j \in S.$$

Proof. The transition probabilities of Y are

$$\begin{aligned} q_{ij} &= \mathbf{P}(Y_{n+1} = j \mid Y_n = i) \\ &= \mathbf{P}(X_{-n-1} = j \mid X_{-n} = i) \\ &= \mathbf{P}(X_m = i \mid X_{m-1} = j)\mathbf{P}(X_{m-1} = j)/\mathbf{P}(X_m = i) \quad \text{where } m = -n \\ &= p_{ji}\frac{\pi_j}{\pi_i} \end{aligned}$$

by the identity

$$\mathbf{P}(A \mid B) = \mathbf{P}(B \mid A)\mathbf{P}(A)/\mathbf{P}(B).$$

Thus $p_{ij} = q_{ij}$ if and only if $\pi_i p_{ij} = \pi_j p_{ji}$. ■

The equations (2) of time-reversibility provide a useful way of finding the stationary distributions of some chains.

(3) **Theorem.** *For an irreducible chain, if there exists $\boldsymbol{\pi}$ such that*

$$0 \leqslant \pi_i \leqslant 1, \qquad \sum_i \pi_i = 1, \qquad \pi_i p_{ij} = \pi_j p_{ji} \quad \textit{for all } i, j,$$

then the chain is time-reversible (in equilibrium) and non-null persistent, with stationary distribution $\boldsymbol{\pi}$.

Proof. Suppose that $\boldsymbol{\pi}$ satisfies the conditions of the theorem. Then

$$\sum_i \pi_i p_{ij} = \sum_i \pi_j p_{ji} = \pi_j \sum_i p_{ji} = \pi_j$$

and so $\boldsymbol{\pi} = \boldsymbol{\pi P}$ and the result follows from (6.4.3). ■

(4) **Example. Ehrenfest model of diffusion.** Two containers A and B are placed adjacently to each other and gas is allowed to pass through a small aperture joining them. A total of m gas molecules is distributed between the containers. We assume that at each epoch of time one molecule, picked uniformly at random from the m available, passes through this aperture. Let X_n be the number of molecules in container A after n units of time have passed. Clearly $\{X_n\}$ is a Markov chain with transition matrix

$$p_{i,i+1} = 1 - \frac{i}{m}, \qquad p_{i,i-1} = \frac{i}{m} \quad \text{if} \quad 0 \leqslant i \leqslant m.$$

Rather than solve the equation $\boldsymbol{\pi} = \boldsymbol{\pi P}$ to find the stationary distribution, we note that such a reasonable diffusion model should be time-reversible. Look for solutions of

$$\pi_i p_{ij} = \pi_j p_{ji}$$

to obtain $\pi_i = \binom{m}{i}(\frac{1}{2})^m$. ●

Here is a way of thinking about time-reversibility and the equations $\pi_i p_{ij} = \pi_j p_{ji}$. Suppose we are provided with a Markov chain with state space S and stationary distribution $\boldsymbol{\pi}$. To this chain there corresponds a 'network' as follows. The nodes of the network are the states in S, and arrows are added between certain pairs of nodes; an arrow is added pointing from state i to state j whenever $p_{ij} > 0$. We are provided with one unit of material (disease, water, or sewage, perhaps) which is distributed about the nodes of the network and allowed to flow along the arrows. The transportation rule is as follows: at each epoch of time a proportion p_{ij} of the amount of material at node i is transported to node j. Initially the material is distributed in such a way that exactly π_i of it is at node i, for each i. It is a simple calculation that the amount at node i after one epoch of time is $\sum_j \pi_j p_{ji}$, which equals π_i, since $\boldsymbol{\pi} = \boldsymbol{\pi P}$. Therefore the system is in equilibrium: there is a 'global balance' in the sense that the total quantity leaving each node equals the total quantity arriving there. There may or may not be a 'local balance', in the sense that, for all i, j, the amount flowing from i to j equals the amount flowing from j to i. Local balance occurs if and only if $\pi_i p_{ij} = \pi_j p_{ji}$ for all i, j, which is to say that the chain is time-reversible. With this motivation the equations $\pi_i p_{ij} = \pi_j p_{ji}$ are usually referred to as the 'balance equations'.

Exercises

(5) A random walk on the set $\{0, 1, 2, \ldots, b\}$ has transition matrix given by $p_{00} = 1 - \lambda_0$, $p_{bb} = 1 - \mu_b$, $p_{i,i+1} = \lambda_i$ and $p_{i+1,i} = \mu_{i+1}$ for $0 \leqslant i < b$, where $0 < \lambda_i, \mu_i < 1$ for all i, and $\lambda_i + \mu_i = 1$ for $1 \leqslant i < b$. Show that this process is time-reversible in equilibrium.

(6) Let X be an irreducible non-null persistent aperiodic Markov chain. Show that X is time-reversible in equilibrium if and only if

$$p_{j_1 j_2} p_{j_2 j_3} \cdots p_{j_{n-1} j_n} p_{j_n j_1} = p_{j_1 j_n} p_{j_n j_{n-1}} \cdots p_{j_2 j_1}$$

for all n and all finite sequences $j_1, j_2, \ldots, j_n$ of states.

(7) Let X be a time-reversible Markov chain, and let C be a non-empty subset of the state space S. Define the Markov chain Y on S by the transition matrix $\boldsymbol{Q} = (q_{ij})$ where

$$q_{ij} = \begin{cases} \beta p_{ij} & \text{if } i \in C \text{ and } j \notin C \\ p_{ij} & \text{otherwise} \end{cases}$$

for $i \neq j$, and where β is a constant satisfying $0 < \beta < 1$. The diagonal terms q_{ii} are arranged so that $\boldsymbol{Q}$ is a stochastic matrix. Show that Y is time-reversible in equilibrium, and find its stationary distribution. Describe the situation in the limit as $\beta \downarrow 0$.

6.6 Chains with finitely many states

The theory of Markov chains is much simplified by the condition that S be finite. By (6.3.5), if S is irreducible then it is necessarily non-null persistent. It may even be possible to calculate the n-step transition probabilities explicitly. Of central importance here is the following algebraic theorem, in which $\mathrm{i} = \sqrt{(-1)}$. Let N denote the cardinality of S.

(1) **Theorem. (Perron–Frobenius).** *If $\boldsymbol{P}$ is the transition matrix of a finite irreducible chain with period d then*

(a) $\lambda_1 = 1$ *is an eigenvalue of* $\boldsymbol{P}$

(b) *the d complex roots of unity*

$$\lambda_1 = \omega^0, \lambda_2 = \omega^1, \ldots, \lambda_d = \omega^{d-1} \quad \textit{where } \omega = \exp(2\pi \mathrm{i}/d)$$

are eigenvalues of $\boldsymbol{P}$

(c) *the remaining eigenvalues $\lambda_{d+1}, \ldots, \lambda_N$ satisfy* $|\lambda_j| < 1$.

If the eigenvalues $\lambda_1, \ldots, \lambda_N$ are distinct then it is well known that there exists a matrix $\boldsymbol{B}$ such that

$$\boldsymbol{P} = \boldsymbol{B}^{-1} \boldsymbol{\Lambda} \boldsymbol{B}$$

where $\boldsymbol{\Lambda}$ is the diagonal matrix with entries $\lambda_1, \ldots, \lambda_N$. Thus

$$\boldsymbol{P}^n = \boldsymbol{B}^{-1} \boldsymbol{\Lambda}^n \boldsymbol{B} = \boldsymbol{B}^{-1} \begin{pmatrix} \lambda_1^n & & 0 \\ & \ddots & \\ 0 & & \lambda_N^n \end{pmatrix} \boldsymbol{B}.$$

The rows of $\boldsymbol{B}$ are left eigenvectors of $\boldsymbol{P}$. We can use the Perron–Frobenius theorem to explore the properties of $\boldsymbol{P}^n$ for large n. For example, if the chain is aperiodic then $d = 1$ and

$$\boldsymbol{P}^n \to \boldsymbol{B}^{-1}\begin{pmatrix} 1 & & 0 \\ & 0 & \\ & & \ddots \\ 0 & & 0 \end{pmatrix}\boldsymbol{B} \quad \text{as} \quad n \to \infty.$$

When the eigenvalues of $\boldsymbol{P}$ are not distinct, then $\boldsymbol{P}$ cannot always be reduced to the diagonal canonical form in this way. The best that we may be able to do is to rewrite $\boldsymbol{P}$ in its 'Jordan canonical form'

$$\boldsymbol{P} = \boldsymbol{B}^{-1}\boldsymbol{M}\boldsymbol{B}$$

where

$$\boldsymbol{M} = \begin{pmatrix} \boldsymbol{J}_1 & & & 0 \\ & \boldsymbol{J}_2 & & \\ & & \boldsymbol{J}_3 & \\ 0 & & & \ddots \end{pmatrix}$$

and $\boldsymbol{J}_1, \boldsymbol{J}_2, \ldots$ are square matrices given as follows. Let $\lambda_1, \lambda_2, \ldots, \lambda_m$ be the distinct eigenvalues of $\boldsymbol{P}$ and let k_i be the multiplicity of λ_i. Then

$$\boldsymbol{J}_i = \begin{pmatrix} \lambda_i & 1 & 0 & 0 & \cdots \\ 0 & \lambda_i & 1 & 0 & \cdots \\ 0 & 0 & \lambda_i & 1 & \cdots \\ \vdots & \vdots & \vdots & \vdots & \end{pmatrix}$$

is a $k_i \times k_i$ matrix with each diagonal term λ_i, each superdiagonal term 1, and all other terms 0. Once again we have that $\boldsymbol{P}^n = \boldsymbol{B}^{-1}\boldsymbol{M}^n\boldsymbol{B}$, where $\boldsymbol{M}^n$ has quite a simple form (see Cox and Miller (1965, p. 118 *et seq.*) for more details).

(2) **Example. Inbreeding.** Consider the genetic model described in (6.1.11c) and suppose that C_1 can take the values A or a on each of two homologous chromosomes. Then the possible types of individuals can be denoted by

$$AA,\ Aa\ (\equiv aA),\ aa,$$

and mating between types is denoted by

$$AA \times AA,\ AA \times Aa, \text{ and so on.}$$

As described in (6.1.11c), meiosis selects the offspring's chromosomes randomly from each parent; in the simplest case (since there are two choices for each of two places) each outcome has probability $\frac{1}{4}$. Thus for the offspring of $AA \times Aa$ the four possible outcomes are

$$AA,\ Aa,\ AA,\ Aa$$

and

$$\mathbf{P}(AA) = \mathbf{P}(Aa) = \tfrac{1}{2}.$$

For the cross $Aa \times Aa$,

$$\mathbf{P}(AA) = \mathbf{P}(aa) = \tfrac{1}{2}\mathbf{P}(Aa) = \tfrac{1}{4}.$$

Clearly the offspring of $AA \times AA$ can only be AA, and those of $aa \times aa$ can only be aa.

We now construct a Markov chain by mating an individual with itself, then crossing a single resulting offspring with itself, and so on. (This scheme is possible with plants.) Then the genetic types of this sequence of individuals constitute a Markov chain with three states, AA, Aa, aa. In view of the above discussion, the transition matrix is

$$\boldsymbol{P} = \begin{pmatrix} 1 & 0 & 0 \\ \frac{1}{4} & \frac{1}{2} & \frac{1}{4} \\ 0 & 0 & 1 \end{pmatrix}$$

and the reader can verify that

$$\boldsymbol{P}^n = \begin{pmatrix} 1 & 0 & 0 \\ \frac{1}{2} - (\frac{1}{2})^{n+1} & (\frac{1}{2})^n & \frac{1}{2} - (\frac{1}{2})^{n+1} \\ 0 & 0 & 1 \end{pmatrix} \rightarrow \begin{pmatrix} 1 & 0 & 0 \\ \frac{1}{2} & 0 & \frac{1}{2} \\ 0 & 0 & 1 \end{pmatrix} \quad \text{as} \quad n \rightarrow \infty.$$

Thus, ultimately, inbreeding produces a pure (AA or aa) line for which all subsequent offspring have the same type. In like manner one can consider the progress of many different breeding schemes which include breeding with rejection of unfavourable genes, back-crossing to encourage desirable genes, and so on. ●

Exercises

The first two exercises provide proofs that a Markov chain with finitely many states has a stationary distribution.

(3) The Markov–Kakutani theorem asserts that, for any convex compact subset C of $\mathbb{R}^n$ and any linear continuous mapping T of C into C, T has a fixed point (in the sense that $T(x) = x$ for some $x \in C$). Use this to prove that a finite stochastic matrix has a non-negative non-zero left eigenvector corresponding to the eigenvalue 1.

(4) Let $\boldsymbol{T}$ be an $m \times n$ matrix and let $\boldsymbol{v} \in \mathbb{R}^n$. Farkas's theorem asserts that exactly one of the following holds:

(i) there exists $\boldsymbol{x} \in \mathbb{R}^m$ such that $\boldsymbol{x} \geqslant \mathbf{0}$ and $\boldsymbol{x}\boldsymbol{T} = \boldsymbol{v}$,
(ii) there exists $\boldsymbol{y} \in \mathbb{R}^n$ such that $\boldsymbol{y}\boldsymbol{v}' < 0$ and $\boldsymbol{T}\boldsymbol{y}' \geqslant \mathbf{0}$.

Use this to prove that a finite stochastic matrix has a non-negative non-zero left eigenvector corresponding to the eigenvalue 1.

(5) Let X be a Markov chain with state space $S = \{1, 2, 3\}$ and transition matrix

$$\mathbf{P} = \begin{pmatrix} 1-p & p & 0 \\ 0 & 1-p & p \\ p & 0 & 1-p \end{pmatrix}$$

where $0 < p < 1$. Prove that

$$\mathbf{P}^n = \begin{pmatrix} a_{1n} & a_{2n} & a_{3n} \\ a_{3n} & a_{1n} & a_{2n} \\ a_{2n} & a_{3n} & a_{1n} \end{pmatrix}$$

where $a_{1n} + \omega a_{2n} + \omega^2 a_{3n} = (1 - p + p\omega)^n$, ω being a complex cube root of 1.

6.7 Branching processes revisited

The foregoing general theory is an attractive and concise account of the evolution through time of a Markov chain. Unfortunately, it is an inadequate description of many specific Markov chains. Consider for example a branching process $\{Z_0, Z_1, \ldots\}$ where $Z_0 = 1$. If there is strictly positive probability $\mathbf{P}(Z_1 = 0)$ that each family is empty then 0 is an absorbing state. Hence 0 is persistent non-null, and all other states are transient. The chain is not irreducible but there exists a unique stationary distribution $\boldsymbol{\pi}$ given by $\pi_0 = 1$, $\pi_i = 0$ if $i > 0$. These facts tell us next to nothing about the behaviour of the process, and we must look elsewhere for detailed information. The difficulty is that the process may behave in one of various qualitatively different ways depending, for instance, on whether or not it ultimately becomes extinct. One way of approaching the problem is to study the behaviour of the process *conditional* upon the occurrence of some event, like extinction, or the value of some random variable, such as the total number $\sum_i Z_i$ of progeny. This section contains an outline of such a method.

Let f and G be the mass function and generating function of a typical family size Z_1:

$$f(k) = \mathbf{P}(Z_1 = k), \qquad G(s) = \mathbf{E}(s^{Z_1}).$$

Let

$$T = \begin{cases} \inf\{n: Z_n = 0\} & \text{if extinction occurs} \\ \infty & \text{otherwise} \end{cases}$$

be the time until extinction. Roughly speaking, if $T = \infty$ then the process will grow beyond all possible bounds, whilst if $T < \infty$ then the size of the process never becomes very large and subsequently reduces to zero. Think of $\{Z_n\}$ as a fluctuating sequence which either becomes so large that it escapes to ∞ or is absorbed at 0 during one of its fluctuations. From Section 5.4, the probability $\mathbf{P}(T < \infty)$ of ultimate extinction is the smallest non-negative root of the equation $s = G(s)$. Now let

$$E_n = \{n < T < \infty\}$$

be the event that extinction occurs at some time after n. We shall study the distribution of Z_n conditional upon the occurrence of E_n. Let

$$_0p_j^{(n)} = \mathbf{P}(Z_n = j \mid E_n)$$

be the conditional probability that $Z_n = j$ given the future extinction of Z. We are interested in the limiting value

$$_0\pi_j = \lim_{n\to\infty} {_0p_j(n)},$$

if this limit exists. To avoid certain trivial cases we assume henceforth that

$$0 < f(0) + f(1) < 1, \qquad f(0) > 0;$$

these conditions imply for example that $0 < \mathbf{P}(E_n) < 1$ and that the probability η of ultimate extinction satisfies $0 < \eta \leqslant 1$.

(1) **Lemma.** *If* $\mathbf{E}(Z_1) < \infty$ *then* $\lim_{n\to\infty} {_0p_j(n)} = {_0\pi_j}$ *exists. The generating function*

$$G^\pi(s) = \sum_j {_0\pi_j} s^j$$

satisfies the functional equation

(2) $$G^\pi(\eta^{-1}G(s\eta)) = mG^\pi(s) + 1 - m$$

where η *is the probability of ultimate extinction and* $m = G'(\eta)$.

Note that if $\mu = \mathbf{E}Z_1 \leqslant 1$ then $\eta = 1$ and $m = \mu$. Thus (2) reduces to

$$G^\pi(G(s)) = G^\pi(s) + 1 - \mu.$$

Whatever the value of μ, we have that $G'(\eta) \leqslant 1$, with equality if and only if $\mu = 1$.

Proof. For $s \in [0, 1)$, let

$$\begin{aligned} G_n^\pi(s) = \mathbf{E}(s^{Z_n} \mid E_n) &= \sum_j {_0p_j(n)} s^j \\ &= \sum_j s^j \frac{\mathbf{P}(Z_n = j, E_n)}{\mathbf{P}(E_n)} \\ &= \frac{G_n(s\eta) - G_n(0)}{\eta - G_n(0)} \end{aligned}$$

where $G_n(s) = \mathbf{E}(s^{Z_n})$ as before, since

$$\begin{aligned} \mathbf{P}(Z_n = j, E_n) &= \mathbf{P}(Z_n = j \text{ and all subsequent lines die out}) \\ &= \mathbf{P}(Z_n = j)\eta^j \quad \text{if} \quad j \geqslant 1, \end{aligned}$$

and

$$\mathbf{P}(E_n) = \mathbf{P}(T < \infty) - \mathbf{P}(T \leqslant n) = \eta - G_n(0).$$

Let

$$H_n(s) = \frac{\eta - G_n(s)}{\eta - G_n(0)}, \qquad h(s) = \frac{\eta - G(s)}{\eta - s}, \qquad 0 \leqslant s < \eta,$$

so that

(3) $$G_n^\pi(s) = 1 - H_n(s\eta).$$

Note that H_n has domain $[0, \eta)$ and G_n^π has domain $[0, 1)$. By (5.4.1),

$$\frac{H_n(s)}{H_{n-1}(s)} = \frac{h(G_{n-1}(s))}{h(G_{n-1}(0))}.$$

However, G_{n-1} is non-decreasing, and h is non-decreasing because G is convex on $[0, \eta)$, giving that

$$H_n(s) \geqslant H_{n-1}(s) \quad \text{for} \quad s < \eta.$$

Hence, by (3),

$$\lim_{n \to \infty} G_n^\pi(s) = G^\pi(s) \quad \text{and} \quad \lim_{n \to \infty} H_n(s\eta) = H(s\eta)$$

exist for $s \in [0, 1)$ and satisfy

(4) $$G^\pi(s) = 1 - H(s\eta) \quad \text{if} \quad 0 \leqslant s < 1.$$

Thus the coefficient ${}_0\pi_j$ of s^j in $G^\pi(s)$ exists for all j as required. Furthermore, if $0 \leqslant s < \eta$,

(5) $$H_n(G(s)) = \frac{\eta - G_n(G(s))}{\eta - G_n(0)} = \frac{\eta - G(G_n(0))}{\eta - G_n(0)} \frac{\eta - G_{n+1}(s)}{\eta - G_{n+1}(0)}$$
$$= h(G_n(0))H_{n+1}(s).$$

As $n \to \infty$, $G_n(0) \uparrow \eta$ and so

$$h(G_n(0)) \to \lim_{s \uparrow \eta} \frac{\eta - G(s)}{\eta - s} = G'(\eta).$$

Let $n \to \infty$ in (5) to obtain

(6) $$H(G(s)) = G'(\eta)H(s) \quad \text{if} \quad 0 \leqslant s < \eta$$

and (2) follows from (4). ■

(7) **Corollary.** *If* $\mu \neq 1$ *then* $\sum_j {}_0\pi_j = 1$.
If $\mu = 1$ *then* ${}_0\pi_j = 0$ *for all* j.

Proof. $\mu = 1$ if and only if $G'(\eta) = 1$. If $\mu \neq 1$ then $G'(\eta) \neq 1$ and letting s increase to η in (6) gives

$$\lim_{s \uparrow \eta} H(s) = 0;$$

so, from (4), $\lim_{s\uparrow 1} G^{\pi}(s) = 1$, or

$$\sum_j {}_0\pi_j = 1.$$

If $\mu = 1$ then $G'(\eta) = 1$ and (2) becomes

$$G^{\pi}(G(s)) = G^{\pi}(s).$$

However, $G(s) > s$ for all $s < 1$ and so

$$G^{\pi}(s) = G^{\pi}(0) = 0 \quad \text{for all } s < 1.$$

Thus ${}_0\pi_j = 0$ for all j. ∎

So long as $\mu \neq 1$, the distribution of Z_n, conditional on future extinction, converges as $n \to \infty$ to some limit $\{{}_0\pi_j\}$ which is a proper distribution. The so-called 'critical' branching process with $\mu = 1$ is more difficult in that, for $j \geqslant 1$,

$$\mathbf{P}(Z_n = j) \to 0 \qquad \text{because extinction is certain}$$

$$\mathbf{P}(Z_n = j \mid E_n) \to 0 \quad \text{because } Z_n \to \infty, \text{ conditional on } E_n.$$

However, it is possible to show, in the spirit of the discussion at the end of Section 5.4, that the distribution of

$$Y_n = \frac{Z_n}{n\sigma^2} \quad \text{where} \quad \sigma^2 = \operatorname{var} Z_1,$$

conditional on E_n, converges as $n \to \infty$.

(8) **Theorem.** *If $\mu = 1$ and $G''(1) < \infty$ then*

$$\mathbf{P}(Y_n \leqslant y \mid E_n) \to 1 - \exp(-2y), \quad \text{as} \quad n \to \infty.$$

Proof. See Athreya and Ney (1972, p. 20). ∎

So, if $\mu = 1$, then the distribution of Y_n, given E_n, is asymptotically exponential with parameter 2. In this case, the branching process is called *critical*; the cases $\mu < 1$ and $\mu > 1$ are called *subcritical* and *supercritical* respectively. See Athreya and Ney (1972) for further details.

Exercises

(9) Let Z be a branching process with $Z_0 = 1$ and $\mathbf{P}(Z_1 = k) = 2^{-k}$ for $k \geqslant 0$. Show directly that, as $n \to \infty$, $\mathbf{P}(Z_n \leqslant 2yn \mid Z_n > 0) \to 1 - e^{-2y}$, $y > 0$, in agreement with Theorem (8).

(10) Let Z be a supercritical branching process with $Z_0 = 1$ and family-size generating function G. Assume that the probability η of extinction satisfies $0 < \eta < 1$. Find a way of describing the process Z, *conditioned on its ultimate extinction.*

(11) Let Z be a branching process with $Z_0 = 1$ and $\mathbf{P}(Z_1 = k) = qp^k$ for $k \geqslant 0$, where $p + q = 1$ and $p > \frac{1}{2}$. Use your answer to Exercise (10) to show that, if we condition on the ultimate extinction of Z, then the process grows in the manner of a branching process $\tilde{Z}$ with $\tilde{Z}_0 = 1$ and $\mathbf{P}(\tilde{Z}_1 = k) = pq^k$ for $k \geqslant 0$.

(12) (a) Show that $\mathbf{E}(X \mid X > 0) \leqslant \mathbf{E}(X^2)/\mathbf{E}(X)$ for any random variable X taking non-negative values.

(b) Let Z be a branching process with $Z_0 = 1$ and $\mathbf{P}(Z_1 = k) = qp^k$ for $k \geqslant 0$, where $p > \frac{1}{2}$. Use part (a) to show that $\mathbf{E}(Z_n/\mu^n \mid Z_n > 0) \leqslant 2p/(p - q)$, where $\mu = p/q$.

(c) Show that, in the notation of part (b), $\mathbf{E}(Z_n/\mu^n \mid Z_n > 0) \to p/(p - q)$ as $n \to \infty$.

6.8 Birth processes and the Poisson process

Many processes in nature may change their values at any instant of time rather than at certain specified epochs only. Such a process is a family $\{X(t): t \geqslant 0\}$ of random variables taking values in a state space S. Depending on the underlying random mechanism, X may or may not be a Markov process. Before attempting to study any general theory of continuous-time processes we explore one simple but non-trivial example in detail.

Given the right equipment, we should have no difficulty in observing that the process of emission of particles from a radioactive source seems to behave in a manner which is not totally predictable. If we switch on our Geiger counter at time zero, then the reading $N(t)$ which it shows at a later time t is the outcome of some random process. This process $\{N(t): t \geqslant 0\}$ has certain obvious properties, such as

(a) $N(0) = 0$, $N(t) \in \{0, 1, 2, \dots\}$
(b) if $s < t$ then $N(s) \leqslant N(t)$,

but it is not so easy to specify more detailed properties. We might use the following description. In the time interval $(t, t + h)$ there may or may not be some emissions. If h is small then the likelihood of an emission is roughly proportional to h; it is not very likely that two or more emissions will occur in a small interval. More formally, we make the following definition.

(1) **Definition. A Poisson process with intensity** λ is a process $N = \{N(t): t \geqslant 0\}$ taking values in $S = \{0, 1, 2, \dots\}$ such that

(a) $N(0) = 0$; if $s < t$ then $N(s) \leqslant N(t)$

(b) $$\mathbf{P}(N(t+h) = n + m \mid N(t) = n) = \begin{cases} \lambda h + o(h) & \text{if } m = 1 \\ o(h) & \text{if } m > 1 \\ 1 - \lambda h + o(h) & \text{if } m = 0 \end{cases}$$

(c) if $s < t$ then the number $N(t) - N(s)$ of emissions in the interval $(s, t]$ is independent of the times of emissions during $[0, s]$.

We speak of $N(t)$ as the number of 'arrivals' or 'occurrences', or in this example 'emissions', of the process by time t. N is called a 'counting process' and is one of the simplest examples of continuous-time Markov chains. We shall consider the general theory of such processes in the next section; here we study special properties of Poisson processes and their generalizations.

We are interested first in the distribution of $N(t)$.

(2) **Theorem.** *$N(t)$ has the Poisson distribution with parameter λt; that is to say*

$$\mathbf{P}(N(t) = j) = \frac{(\lambda t)^j}{j!} e^{-\lambda t}, \qquad j = 0, 1, 2, \ldots .$$

Proof. Condition $N(t + h)$ on $N(t)$ to obtain

$$\begin{aligned}\mathbf{P}(N(t+h) = j) &= \sum_i \mathbf{P}(N(t) = i)\mathbf{P}(N(t+h) = j \mid N(t) = i) \\ &= \sum_i \mathbf{P}(N(t) = i)\mathbf{P}((j-i) \text{ arrivals in } (t, t+h]) \\ &= \mathbf{P}(N(t) = j-1)\mathbf{P}(\text{one arrival}) \\ &\qquad + \mathbf{P}(N(t) = j)\mathbf{P}(\text{no arrivals}) + o(h).\end{aligned}$$

Thus $p_j(t) = \mathbf{P}(N(t) = j)$ satisfies

$$p_j(t+h) = \lambda h p_{j-1}(t) + (1 - \lambda h)p_j(t) + o(h) \quad \text{if} \quad j \neq 0$$

$$p_0(t+h) = (1 - \lambda h)p_0(t) + o(h).$$

Subtract $p_j(t)$ from each side of the first of these equations, divide by h, and let $h \downarrow 0$ to obtain

(3) $$p_j'(t) = \lambda p_{j-1}(t) - \lambda p_j(t) \quad \text{if} \quad j \neq 0;$$

likewise

(4) $$p_0'(t) = -\lambda p_0(t).$$

The boundary condition is

(5) $$p_j(0) = \delta_{j0}.$$

These form a collection of differential–difference equations for the $p_j(t)$. Here are two methods of solution, both of which have applications elsewhere.

Method A. Induction. Solve (4) subject to the condition $p_0(0) = 1$ to obtain

$$p_0(t) = e^{-\lambda t}.$$

Substitute into (3) with $j = 1$ to obtain

$$p_1(t) = \lambda t \, e^{-\lambda t}$$

and continue, to prove

$$p_j(t) = \frac{(\lambda t)^j}{j!} e^{-\lambda t}$$

by induction.

Method B. Generating functions. Define the generating function

$$G(s, t) = \sum_{j=0}^{\infty} p_j(t) s^j = \mathbf{E}(s^{N(t)}).$$

Multiply (3) by s^j and sum over j to obtain

$$\frac{\partial G}{\partial t} = \lambda(s-1)G$$

with the boundary condition $G(s, 0) = 1$. The solution is

(6) $$G(s, t) = \exp[\lambda(s-1)t] = e^{-\lambda t} \sum_{j=0}^{\infty} \frac{(\lambda t)^j}{j!} s^j$$

as required. ■

This result seems very like the account (3.5.4) that the binomial $B(n, p)$ distribution approaches the Poisson distribution if $n \to \infty$ and $np \to \lambda$. Why is this no coincidence?

There is an important alternative and equivalent formulation of a Poisson process which provides much insight into its behaviour. Let $T_0, T_1, \ldots$ be given by

(7) $$T_0 = 0, \qquad T_n = \inf\{t: N(t) = n\}.$$

Then T_n is the time of the nth arrival. The *interarrival times* are the random variables $X_1, X_2, \ldots$ given by

(8) $$X_n = T_n - T_{n-1}.$$

From knowledge of N, we can find the values of $X_1, X_2, \ldots$ by (7) and (8). Conversely, we can reconstruct N from a knowledge of the X_i by

(9) $$T_n = \sum_{1}^{n} X_i, \qquad N(t) = \max\{n: T_n \leqslant t\}.$$

Figure 6.1 is an illustration of this.

(10) **Theorem.** *$X_1, X_2, \ldots$ are independent exponential random variables with parameter λ.*

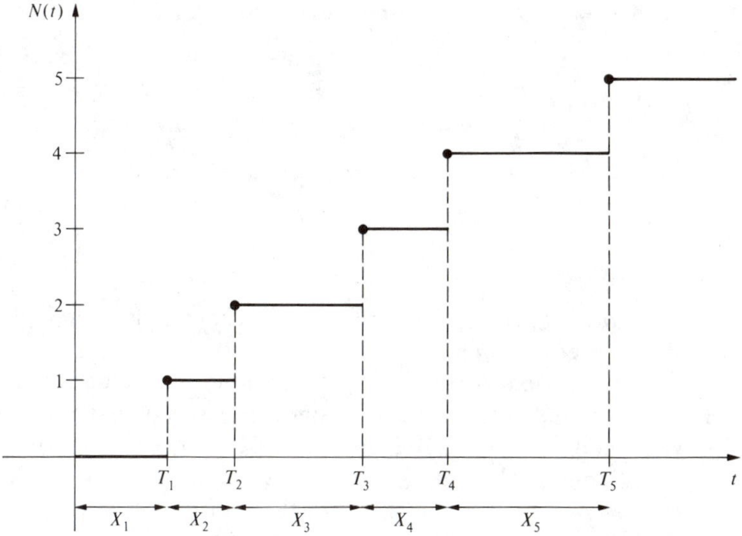

Fig. 6.1 A typical realization of a Poisson process $N(t)$.

There is an important generalization of this result to arbitrary continuous-time Markov chains with countable state space. We shall investigate this in the next section.

Proof. First consider X_1:

$$\mathbf{P}(X_1 > t) = \mathbf{P}(N(t) = 0) = e^{-\lambda t}$$

and so X_1 is exponential. Now, conditional on X_1,

$$\mathbf{P}(X_2 > t \mid X_1 = t_1) = \mathbf{P}(\text{no arrival in } (t_1, t_1 + t] \mid X_1 = t_1).$$

The event $\{X_1 = t_1\}$ relates to arrivals during the time interval $[0, t_1]$, whereas the event $\{\text{no arrival in } (t_1, t_1 + t]\}$ relates to arrivals after t_1. These events are independent, by (1c), and therefore

$$\mathbf{P}(X_2 > t \mid X_1 = t_1) = \mathbf{P}(\text{no arrival in } (t_1, t_1 + t]) = e^{-\lambda t}.$$

Thus X_2 is independent of X_1, and has the same distribution. Similarly,

$$\mathbf{P}(X_{n+1} > t \mid X_1 = t_1, \dots, X_n = t_n) = \mathbf{P}(\text{no arrival in } (T, T + t])$$

where $T = t_1 + t_2 + \cdots + t_n$, and the claim of the theorem follows by induction on n. ■

It is not difficult to see that the process N, constructed by (9) from a sequence $X_1, X_2, \dots,$ is a Poisson process if and only if the X_i are independent identically distributed exponential variables (*exercise*: use the lack of memory property (4.11.5)). If the X_i form such a sequence, then it is a simple matter to deduce the distribution of $N(t)$ directly, as follows. In this case, $T_n = \sum_1^n X_i$ is $\Gamma(\lambda, n)$ and $N(t)$ is specified by the useful remark that

$$N(t) \geqslant j \quad \text{if and only if} \quad T_j \leqslant t.$$

Therefore

$$\begin{aligned} \mathbf{P}(N(t) = j) &= \mathbf{P}(T_j \leqslant t < T_{j+1}) \\ &= \mathbf{P}(T_j \leqslant t) - \mathbf{P}(T_{j+1} \leqslant t) \\ &= \frac{(\lambda t)^j}{j!} e^{-\lambda t} \end{aligned}$$

using the properties of gamma variables and integration by parts. (This was Problem (4.11.11c).)

The Poisson process is a very satisfactory model for radioactive emissions from a sample of uranium-235 since this isotope has a half-life of 7×10^8 years and decays fairly slowly. However, for a newly produced sample of strontium-92, which has a half-life of 2.7 hours, we need a more sophisticated process which takes into account the retardation in decay rate over short time intervals. We might suppose that the rate λ at which emissions are detected depends on the number detected already.

(11) **Definition.** A **birth process with intensities** $\lambda_0, \lambda_1, \ldots$ is a process $\{N(t): t \geqslant 0\}$ taking values in $S = \{0, 1, 2, \ldots\}$ such that

(i) $N(0) = 0$; if $s < t$ then $N(s) \leqslant N(t)$

(ii) $\mathbf{P}(N(t+h) = n + m \mid N(t) = n) = \begin{cases} \lambda_n h + o(h) & \text{if} \quad m = 1 \\ o(h) & \text{if} \quad m > 1 \\ 1 - \lambda_n h + o(h) & \text{if} \quad m = 0 \end{cases}$

(iii) if $s < t$ then $N(t) - N(s)$ is independent of all arrivals prior to s.

Here are some interesting special cases.

(a) **Poisson process.** $\lambda_n = \lambda$ for all n. ●

(b) **Simple birth.** $\lambda_n = n\lambda$. This models the growth of a population in which each living individual may give birth to a new individual with probability $\lambda h + o(h)$ in the interval $(t, t+h)$. No individuals may die. The number M of births in the interval $(t, t+h)$ satisfies.

$$\begin{aligned} \mathbf{P}(M = m \mid N(t) = n) &= \binom{n}{m} (\lambda h)^m (1 - \lambda h)^{n-m} + o(h) \\ &= \begin{cases} 1 - n\lambda h + o(h) & \text{if} \quad m = 0 \\ n\lambda h + o(h) & \text{if} \quad m = 1 \\ o(h) & \text{if} \quad m > 1. \end{cases} \end{aligned}$$

●

(c) **Simple birth with immigration.** $\lambda_n = n\lambda + \nu$. This models a simple birth process which experiences immigration at constant rate ν from elsewhere. ●

Suppose that $\{N(t)\}$ is a birth process with positive intensities $\lambda_0, \lambda_1, \ldots$. Let us proceed as for the Poisson process. Define the transition probabilities

$$\begin{aligned} p_{ij}(t) &= \mathbf{P}(N(s+t) = j \mid N(s) = i) \\ &= \mathbf{P}(N(t) = j \mid N(0) = i); \end{aligned}$$

now condition $N(t+h)$ on $N(t)$ and let $h \downarrow 0$ as we did for (3) and (4), to obtain the so-called

(12) **Forward system of equations**

$$p'_{ij}(t) = \lambda_{j-1} p_{i,j-1}(t) - \lambda_j p_{ij}(t), \qquad j \geqslant i$$

with the convention that $\lambda_{-1} = 0$, and the boundary condition $p_{ij}(0) = \delta_{ij}$. Alternatively we might condition $N(t+h)$ on $N(h)$ and let $h \downarrow 0$ to obtain the so-called

(13) **Backward system of equations**

$$p'_{ij}(t) = \lambda_i p_{i+1,j}(t) - \lambda_i p_{ij}(t), \qquad j \geqslant i$$

with the boundary condition $p_{ij}(0) = \delta_{ij}$.

Can we solve these equations as we did for the Poisson process?

(14) **Theorem.** *The forward system has a unique solution, which satisfies the backward system.*

Proof. Note first that $p_{ij}(t) = 0$ if $j < i$. Solve the forward equation with $j = i$ to obtain

$$p_{ii}(t) = \exp(-\lambda_i t).$$

Substitute into the forward equation with $j = i + 1$ to find $p_{i,i+1}(t)$. Continue this operation to deduce that the forward system has a unique solution. To obtain more information about this solution, define the Laplace transforms†

$$\hat{p}_{ij}(\theta) = \int_0^\infty \exp(-\theta t) p_{ij}(t)\, dt.$$

Transform the forward system to obtain

$$(\theta + \lambda_j)\hat{p}_{ij}(\theta) = \delta_{ij} + \lambda_{j-1}\hat{p}_{i,j-1}(\theta);$$

this is a difference equation which is readily solved to give

(15) $$\hat{p}_{ij}(\theta) = \frac{1}{\lambda_j} \frac{\lambda_i}{\theta + \lambda_i} \frac{\lambda_{i+1}}{\theta + \lambda_{i+1}} \cdots \frac{\lambda_j}{\theta + \lambda_j} \quad \text{for} \quad j \geqslant i.$$

This determines $p_{ij}(t)$ uniquely by the inversion theorem for Laplace transforms.

† See Appendix I for some properties of Laplace transforms.

To see that this solution satisfies the backward system, transform this system similarly to obtain that any solution $\pi_{ij}(t)$ to the backward equation, with Laplace transform

$$\hat{\pi}_{ij}(\theta) = \int_0^\infty \exp(-\theta t)\pi_{ij}(t)\,dt,$$

satisfies

$$(\theta + \lambda_i)\hat{\pi}_{ij}(\theta) = \delta_{ij} + \lambda_i\hat{\pi}_{i+1,j}(\theta).$$

The $\hat{p}_{ij}$, given by (15), satisfy this equation, and so the p_{ij} satisfy the backward system. ■

We have not been able to show that the backward system has a unique solution, for the very good reason that this may not be true. All we can show is that it has a minimal solution.

(16) **Theorem.** *If* $\{p_{ij}(t)\}$ *is the unique solution of the forward system, then any solution* $\{\pi_{ij}(t)\}$ *of the backward system satisfies*

$$p_{ij}(t) \leqslant \pi_{ij}(t) \quad \textit{for all } i, j, t.$$

Proof. See Feller (1968, pp. 475–477). ■

There is something wrong here, because the condition

$$\sum_j p_{ij}(t) = 1 \tag{17}$$

in conjunction with the result of (16) would constrain $\{p_{ij}(t)\}$ to be the *unique* solution of the backward system which is a proper distribution. The point is that (17) may fail to hold.

(18) **Definition.** We call N **honest** if $\mathbf{P}(N(t) < \infty) = 1$ for all t.

N is honest if and only if (17) holds for all t.

(19) **Theorem.** *N is honest if and only if* $\sum_n \frac{1}{\lambda_n} = \infty$.

This attractive theorem asserts that if the birth rates are small enough then the process is always finite, but if they are so large that $\sum \lambda_n^{-1}$ converges then births occur so frequently that there is positive probability of infinitely many births occurring in a finite interval of time. Thus $N(t)$ may take the value $+\infty$ instead of a non-negative integer. Think of the deficit $1 - \sum_j p_{ij}(t)$ as the probability of escaping to infinity in time t, starting from i. The condition of (19) is a very natural one for the following reason. Arguing as we did for the interarrival times of the Poisson process, it is easy to see that if

$$T_0 = 0, \qquad T_n = \inf\{t\colon N(t) = n\}, \qquad X_n = T_n - T_{n-1}$$

then $X_1, X_2, \ldots$ are independent exponential variables and X_i has parameter λ_{i-1}. Thus the mean time until the nth arrival is

$$\mathbf{E}(T_n) = \sum_1^n \mathbf{E}(X_i) = \sum_0^{n-1} \frac{1}{\lambda_i}$$

which tends to infinity as $n \to \infty$ if and only if this sum diverges.

Proof of (19). Sum the forward system over j to obtain

$$\sum_{j=i}^n p'_{ij}(t) = -\lambda_n p_{in}(t);$$

take Laplace transforms and use (15) to obtain

$$\theta \sum_{j=i}^n \hat{p}_{ij}(\theta) = 1 - \lambda_n \hat{p}_{in}(\theta)$$

$$= 1 - \prod_{j=i}^n \left(1 + \frac{\theta}{\lambda_j}\right)^{-1}.$$

Now let $n \to \infty$ for suitable values of θ to find that†

$$\sum_{j=i}^n \hat{p}_{ij}(\theta) \to \frac{1}{\theta} \quad \text{if and only if} \quad \sum_0^\infty \frac{1}{\lambda_j} = \infty.$$

Invert the first relation to obtain

$$\sum_j p_{ij}(t) = 1 \quad \text{if and only if} \quad \sum_0^\infty \frac{1}{\lambda_j} = \infty. \qquad \blacksquare$$

In summary, we have considered several random processes, indexed by continuous time, which model phenomena occurring in nature. However, certain dangers arise unless we take care in the construction of such processes. They may even find a way to the so-called 'boundary' of the state space by exploding in finite time.

We terminate this section with a brief discussion of the Markov property for birth processes. Recall that a sequence $X = \{X_n: n \geq 0\}$ is said to satisfy the Markov property if, conditional on the event $\{X_n = i\}$, events relating to the collection $\{X_m: m > n\}$ are independent of events relating to $\{X_m: m < n\}$. Birth processes have a similar property. Let N be a birth process and let T be a fixed time. Conditional on the event $\{N(T) = i\}$, the evolution of the process subsequent to time T is independent of that prior to T; this is an immediate consequence of (11), and is called the 'weak Markov property'. It is often desirable to make use of a stronger property, in which T is allowed to be a *random variable* rather than merely a constant. On the other hand, such a conclusion cannot be valid for all random T, since if T 'looks into the future' as well as the past, then information about the past may generally

† See Subsection (8) of Appendix I for some notes about infinite products.

be relevant to the future (*exercise*: find a random variable T for which the desired conclusion is false). A useful class of random times are those whose values depend only on the past, and here is a formal definition. We call the random time T a *stopping time* for the process N if, for all $t \geqslant 0$, the indicator function of the event $\{T \leqslant t\}$ is a function of the values $\{N(s): s \leqslant t\}$ of the process up to time t; that is to say, we require that it be decidable whether or not T has occurred by time t knowing only the values of the process up to time t. Examples of stopping times are the times $T_1, T_2, \ldots$ of arrivals; examples of times which are not stopping times are $T_4 - 2$, $\frac{1}{2}(T_1 + T_2)$, and other random variables which 'look into the future'.

(20) **Theorem. Strong Markov property.** *Let N be a birth process and let T be a stopping time for N. Let A be an event which depends on $\{N(s): s > T\}$ and B be an event which depends on $\{N(s): s \leqslant T\}$. Then*

(21) $$\mathbf{P}(A \mid N(T) = i, B) = \mathbf{P}(A \mid N(T) = i) \quad \textit{for all } i.$$

Proof. The following argument may be made rigorous. The event B contains information about the process N prior to T; the 'worst' such event is one which tells everything. Assume then that B is a complete description of $\{N(s): s \leqslant T\}$ (problems of measurability may arise here, but these are not serious in this case since birth processes have only countably many arrivals). Since B is a complete description, knowledge of B carries with it knowledge of the value of the stopping time T, which we write as $T = T(B)$. Therefore

$$\mathbf{P}(A \mid N(T) = i, B) = \mathbf{P}(A \mid N(T) = i, B, T = T(B)).$$

The event $\{N(T) = i\} \cap B \cap \{T = T(B)\}$ specifies: (i) the value of T, (ii) the value of $N(T)$, and (iii) the history of the process up to time T; it is by virtue of the fact that T is a stopping time that this event is defined in terms of $\{N(s): s \leqslant T(B)\}$. By the *weak* Markov property, since T is constant on this event, we may discount information in (iii), so that

$$\mathbf{P}(A \mid N(T) = i, B) = \mathbf{P}(A \mid N(T) = i, T = T(B)).$$

Now, the process is temporally homogeneous, and A is defined in terms of $\{N(s): s > T\}$; it follows that the (conditional) probability of A depends only on the value of $N(T)$, which is to say that

$$\mathbf{P}(A \mid N(T) = i, T = T(B)) = \mathbf{P}(A \mid N(T) = i)$$

and (21) follows.

To obtain (21) for more general events B than that given above requires a small spot of measure theory. For those readers who want to see this, we note that, for general B,

$$\begin{aligned}\mathbf{P}(A \mid N(T) = i, B) &= \mathbf{E}(I_A \mid N(T) = i, B) \\ &= \mathbf{E}\big(\mathbf{E}(I_A \mid N(T) = i, B, H) \mid N(T) = i, B\big)\end{aligned}$$

where $H = \{N(s): s \leqslant T\}$. The inner expectation equals $\mathbf{P}(A \mid N(T) = i)$, by the argument above, and the claim follows. ∎

We used two properties of birth processes in our proof of the strong Markov property: temporal homogeneity and the weak Markov property. The strong Markov property plays an important role in the study of continuous-time Markov chains and processes, and we shall encounter it in a more general form later. When applied to a birth process N, it implies that the new process N', defined by $N'(t) = N(t+T) - N(T)$, $t \geqslant 0$, conditional on $\{N(T) = i\}$, is also a birth process, whenever T is a stopping time for N; it is easily seen that this new birth process has intensities $\lambda_i, \lambda_{i+1}, \dots$. In the case of the Poisson process, we have that $N'(t) = N(t+T) - N(T)$ is also a Poisson process.

(22) **Example.** A Poisson process N is said to have 'stationary independent increments', since: (a) the distribution of $N(t) - N(s)$ depends only on $t - s$, and (b) the increments $\{N(t_i) - N(s_i): i = 1, \dots, n\}$ are independent if $s_1 \leqslant t_1 \leqslant s_2 \leqslant t_2 \leqslant \cdots \leqslant t_n$. This property is nearly a characterization of the Poisson process. Suppose that $M = \{M(t): t \geqslant 0\}$ is a non-decreasing right-continuous integer-valued process with $M(0) = 0$, having stationary independent increments, and with the extra property that M has only jump discontinuities of size 1. Note first that, for $u, v \geqslant 0$,

$$\mathbf{E}M(u+v) = \mathbf{E}M(u) + \mathbf{E}[M(u+v) - M(u)] = \mathbf{E}M(u) + \mathbf{E}M(v)$$

by the assumption of stationary increments. Now $\mathbf{E}M(u)$ is non-decreasing in u, so that there exists λ such that

(23) $$\mathbf{E}M(u) = \lambda u, \qquad u \geqslant 0.$$

Let $T = \sup\{t: M(t) = 0\}$ be the time of the first jump of M. We have from the right-continuity of M that $M(T) = 1$ (almost surely), so that T is a stopping time for M. Now

(24) $$\mathbf{E}M(s) = \mathbf{E}(\mathbf{E}(M(s) \mid T)).$$

Certainly $\mathbf{E}(M(s) \mid T) = 0$ if $s < T$, and for $s \geqslant t$

$$\begin{aligned}\mathbf{E}(M(s) \mid T = t) &= \mathbf{E}(M(t) \mid T = t) + \mathbf{E}(M(s) - M(t) \mid T = t)\\ &= 1 + \mathbf{E}\big(M(s) - M(t) \mid M(t) = 1, M(u) = 0 \text{ for } u < t\big)\\ &= 1 + \mathbf{E}M(s-t)\end{aligned}$$

by the assumption of stationary independent increments. We substitute this into (24) to obtain

$$\mathbf{E}M(s) = \int_0^s [1 + \mathbf{E}M(s-t)]\, \mathrm{d}F(t)$$

where F is the distribution function of T. Now $\mathbf{E}M(s) = \lambda s$ for all s, so that

(25) $$\lambda s = F(s) + \lambda \int_0^s (s-t)\, \mathrm{d}F(t),$$

an integral equation for the unknown function F. One of the standard ways of solving such an equation is to use Laplace transforms. We leave it as an *exercise* to deduce from (25) that $F(t) = 1 - e^{-\lambda t}$, $t \geqslant 0$, so that T has the exponential distribution. An argument similar to that used for Theorem (10) now shows that the 'inter-jump' times of M are independent and have the exponential distribution. Hence M is a Poisson process with intensity λ. ●

Exercises

(26) **Superposition.** Flies and wasps land on your dinner plate in the manner of independent Poisson processes with respective intensities λ and μ. Show that the arrivals of flying objects form a Poisson process with intensity $\lambda + \mu$.

(27) **Thinning.** Insects land in the soup in the manner of a Poisson process with intensity λ, and each such insect is green with probability p, independently of the colours of all other insects. Show that the arrivals of green insects form a Poisson process with intensity λp.

(28) Let T_n be the time of the nth arrival in a Poisson process N with intensity λ, and define the excess lifetime process $E(t) = T_{N(t)+1} - t$, being the time one must wait subsequent to t before the next arrival. Show by conditioning on T_1 that

$$\mathbf{P}(E(t) > x) = e^{-\lambda(t+x)} + \int_0^t \mathbf{P}(E(t-u) > x)\lambda\, e^{-\lambda u}\, du.$$

Solve this integral equation in order to find the distribution function of $E(t)$. Explain your conclusion.

(29) Let B be a simple birth process (11b) with $B(0) = I$; the birth rates are $\lambda_n = n\lambda$. Write down the forward system of equations for the process and deduce that

$$\mathbf{P}(B(t) = k) = \binom{k-1}{I-1} e^{-I\lambda t}(1 - e^{-\lambda t})^{k-I}, \qquad k \geqslant I.$$

Show also that $\mathbf{E}B(t) = I\, e^{\lambda t}$ and $\operatorname{var}(B(t)) = I\, e^{2\lambda t}(1 - e^{-\lambda t})$.

(30) Let B be a process of simple birth with immigration (11c) with parameters λ and ν, and with $B(0) = 0$; the birth rates are $\lambda_n = n\lambda + \nu$. Write down the sequence of differential–difference equations for $p_n(t) = \mathbf{P}(B(t) = n)$. Without solving these equations, use them to show that $m(t) = \mathbf{E}B(t)$ satisfies $m'(t) = \lambda m(t) + \nu$, and solve for $m(t)$.

(31) Let N be a birth process with intensities $\lambda_0, \lambda_1, \ldots$, and let $N(0) = 0$. Show that $p_n(t) = \mathbf{P}(N(t) = n)$ is given by

$$p_n(t) = \frac{1}{\lambda_n} \sum_{i=0}^{n} \lambda_i\, e^{-\lambda_i t} \prod_{\substack{j=0 \\ j \neq i}}^{n} \frac{\lambda_j}{\lambda_j - \lambda_i}$$

provided that $\lambda_i \neq \lambda_j$ whenever $i \neq j$.

(32) Suppose that the general birth process of the previous exercise is such that $\sum_n \lambda_n^{-1} < \infty$. Show that $\lambda_n p_n(t) \to f(t)$ as $n \to \infty$ where f is the density function of the random variable $T = \sup\{t\colon N(t) < \infty\}$. Deduce that $\mathbf{E}(N(t) \mid N(t) < \infty)$ is finite or infinite depending on the convergence or divergence of $\sum_n n\lambda_n^{-1}$.

Find the Laplace transform of f in closed form for the case when $\lambda_n = (n + \frac{1}{2})^2$, and deduce an expression for f.

6.9 Continuous-time Markov chains

Let $X = \{X(t): t \geqslant 0\}$ be a family of random variables taking values in some countable state space S and indexed by the half-line $[0, \infty)$. As before, we shall assume that S is a subset of the integers. X is called a (continuous-time) *Markov chain* if it satisfies the following condition.

(1) **Definition.** The process X satisfies the **Markov property** if

$$\mathbf{P}(X(t_n) = j \mid X(t_1) = i_1, \ldots, X(t_{n-1}) = i_{n-1}) = \mathbf{P}(X(t_n) = j \mid X(t_{n-1}) = i_{n-1})$$

for all $j, i_1, \ldots, i_{n-1} \in S$ and any sequence $t_1 < t_2 < \cdots < t_n$ of times.

The evolution of continuous-time Markov chains can be described in very much the same terms as those used for discrete-time processes. Various difficulties may arise in the analysis, especially when S is infinite. The way out of these difficulties is too difficult to describe in detail here, and the reader should look elsewhere (see Chung 1960, or Freedman 1971 for example). The general scheme is as follows. For discrete-time processes we wrote the n-step transition probabilities in matrix form and expressed them in terms of the one-step matrix $\boldsymbol{P}$. In continuous time there is no exact analogue for $\boldsymbol{P}$ since there is no implicit unit length of time. The infinitesimal calculus enables us to plug this gap; we shall see that there exists a matrix $\boldsymbol{G}$, called the *generator* of the chain, which takes over the role of $\boldsymbol{P}$.

(2) **Definition.** The **transition probability** $p_{ij}(s, t)$ is defined to be

$$p_{ij}(s, t) = \mathbf{P}(X(t) = j \mid X(s) = i) \quad \text{for} \quad s \leqslant t.$$

The chain is called **homogeneous** if

$$p_{ij}(s, t) = p_{ij}(0, t - s) \quad \text{for all } i, j, s, t,$$

and we write $p_{ij}(t - s)$ for $p_{ij}(s, t)$.

Henceforth we suppose that X is a homogeneous chain, and we write $\boldsymbol{P}_t$ for the $|S| \times |S|$ matrix with entries $p_{ij}(t)$.

(3) **Theorem.** *The family $\{\boldsymbol{P}_t: t \geqslant 0\}$ is a* **stochastic semigroup**; *that is, it satisfies the following*:

(a) $\boldsymbol{P}_0 = \mathbf{I}$, *the identity matrix*;
(b) $\boldsymbol{P}_t$ *is stochastic, that is* $\boldsymbol{P}_t$ *has non-negative entries and row sums* 1;
(c) *the* **Chapman–Kolmogorov equations**, $\boldsymbol{P}_{s+t} = \boldsymbol{P}_s\boldsymbol{P}_t$ *if* $s, t \geqslant 0$.

Proof.

(a) Obvious.

(b) Let **1** denote a row vector of ones.

$$(\boldsymbol{P}_t\mathbf{1}')_i = \sum_j p_{ij}(t) = \mathbf{P}\left(\bigcup_j \{X(t)=j\} \,\middle|\, X(0)=i\right) = 1.$$

(c)
$$\begin{aligned} p_{ij}(t+s) &= \mathbf{P}(X(t+s)=j \mid X(0)=i) \\ &= \sum_k \mathbf{P}(X(t+s)=j \mid X(s)=k)\mathbf{P}(X(s)=k \mid X(0)=i) \\ &= \sum_k p_{ik}(s)p_{kj}(t) \quad \text{as for (6.1.7).} \end{aligned}$$
■

As before, the evolution of $X(t)$ is specified by the stochastic semigroup $\{\boldsymbol{P}_t\}$ and the distribution of $X(0)$. Most questions about X can be rephrased in terms of these matrices and their properties.

Many readers will not be very concerned with the general theory of these processes, but will be much more interested in specific examples and their stationary distributions. Therefore, we present only a broad outline of the theory in the remaining part of this section and hope that it is sufficient for most applications. Technical conditions are usually omitted, with the consequence that *some of the statements which follow are false in general*; such statements are marked with an asterisk. Indications of how to fill in the details are given in the next section. We shall always suppose that the transition probabilities are continuous.

(4) **Definition.** The semigroup $\{\boldsymbol{P}_t\}$ is called **standard** if

$$\boldsymbol{P}_t \to \mathbf{I} \quad \text{as} \quad t \downarrow 0,$$

which is to say that $p_{ii}(t) \to 1$ and $p_{ij}(t) \to 0$ for $i \neq j$ as $t \downarrow 0$.

Note that the semigroup is standard if and only if its elements $p_{ij}(t)$ are continuous functions of t. For, it is not difficult to see that $p_{ij}(t)$ is continuous for all t whenever the semigroup is standard; we just use the Chapman–Kolmogorov equations (3c) (see Problem (6.13.14)). Henceforth we consider only Markov chains with standard semigroups of transition probabilities.

Suppose that the chain is in state $X(t) = i$ at time t. Various things may happen in the small time interval $(t, t+h)$:

(a) nothing may happen, with probability $p_{ii}(h) + o(h)$, the error term taking into account the possibility that the chain moves out of i and back to i in the interval;

(b) the chain may move to a new state j with probability

$$p_{ij}(h) + o(h).$$

We are assuming here that the probability of two or more transitions in the interval $(t, t+h)$ is $o(h)$; this can be proved. Following (a) and (b), we are interested in the behaviour of $p_{ij}(h)$ for small h; it turns out that $p_{ij}(h)$

is approximately linear in h when h is small. That is, there exist constants $\{g_{ij}: i, j \in S\}$ such that

(5) $$p_{ij}(h) \simeq g_{ij}h \quad \text{if} \quad i \neq j, \qquad p_{ii}(h) \simeq 1 + g_{ii}h.$$

Clearly $g_{ij} \geqslant 0$ for $i \neq j$ and $g_{ii} \leqslant 0$ for all i; the matrix $\mathbf{G} = (g_{ij})$ is called the *generator* of the chain and takes over the role of the transition matrix $\mathbf{P}$ for discrete-time chains. Combine (5) with (a) and (b) above to find that, starting from $X(t) = i$,

(a) nothing happens in $(t, t + h)$ with probability $1 + g_{ii}h + o(h)$,
(b) the chain jumps to state j ($\neq i$) with probability $g_{ij}h + o(h)$.

One may expect that $\sum_j p_{ij}(t) = 1$, and so

$$1 = \sum_j p_{ij}(h) \simeq 1 + h \sum_j g_{ij}$$

giving that

(6)* $$\sum_j g_{ij} = 0 \quad \text{for all } i, \quad \text{or} \quad \mathbf{G1}' = \mathbf{0}',$$

where **1** and **0** are row vectors of ones and zeros. Treat (6) with care; there are some chains for which it fails to hold.

(7) **Example. Birth process (6.8.11).** From the definition of this process, it is clear that

$$g_{ii} = -\lambda_i, \quad g_{i,i+1} = \lambda_i, \quad g_{ij} = 0 \quad \text{if} \quad j < i \quad \text{or} \quad j > i + 1.$$

Thus

$$\mathbf{G} = \begin{pmatrix} -\lambda_0 & \lambda_0 & 0 & 0 & 0 & \cdots \\ 0 & -\lambda_1 & \lambda_1 & 0 & 0 & \cdots \\ 0 & 0 & -\lambda_2 & \lambda_2 & 0 & \cdots \\ \vdots & \vdots & \vdots & \vdots & \vdots & \end{pmatrix}.$$ ●

Relation (5) is usually written as

(8) $$\lim_{h \downarrow 0} \frac{1}{h}(\mathbf{P}_h - \mathbf{I}) = \mathbf{G},$$

and amounts to saying that $\mathbf{P}_t$ is differentiable at $t = 0$. It is clear that $\mathbf{G}$ can be found from knowledge of $\{\mathbf{P}_t\}$. The converse also is usually true. We argue roughly as follows. Suppose that $X(0) = i$, and condition $X(t + h)$ on $X(t)$

to find that

$$p_{ij}(t+h) = \sum_k p_{ik}(t)p_{kj}(h)$$

$$\simeq p_{ij}(t)(1+g_{jj}h) + \sum_{k \neq j} p_{ik}(t)g_{kj}h \quad \text{by (5)}$$

$$= p_{ij}(t) + h\sum_k p_{ik}(t)g_{kj},$$

giving that

$$\frac{1}{h}[p_{ij}(t+h) - p_{ij}(t)] \simeq \sum_k p_{ik}(t)g_{kj} = (\boldsymbol{P}_t\boldsymbol{G})_{ij}.$$

Let $h \downarrow 0$ to obtain the

(9)* **forward equations:** $p'_{ij}(t) = \sum_k p_{ik}(t)g_{kj}$, or $\boldsymbol{P}'_t = \boldsymbol{P}_t\boldsymbol{G}$,

where $\boldsymbol{P}'_t$ denotes the matrix with entries $p'_{ij}(t)$.

A similar argument, by conditioning $X(t+h)$ on $X(h)$, yields the

(10)* **backward equations:** $p'_{ij}(t) = \sum_k g_{ik}p_{kj}(t)$, or $\boldsymbol{P}'_t = \boldsymbol{G}\boldsymbol{P}_t$.

These equations are general forms of (6.8.12) and (6.8.13) and relate $\{\boldsymbol{P}_t\}$ to $\boldsymbol{G}$. Subject to the boundary condition $\boldsymbol{P}_0 = \mathbf{I}$, they often have a unique solution given by the infinite sum

(11)* $$\boldsymbol{P}_t = \sum_{n=0}^{\infty} \frac{t^n}{n!}\boldsymbol{G}^n$$

of powers of matrices (remember that $\boldsymbol{G}^0 = \mathbf{I}$). (11) is deducible from (9) or (10) in very much the same way as we might show that the function of the single variable $p(t) = e^{gt}$ solves the differential equation $p'(t) = gp(t)$. The representation (11) for $\{\boldsymbol{P}_t\}$ is very useful and is usually written as

(12)* $$\boldsymbol{P}_t = e^{t\boldsymbol{G}} \quad \text{or} \quad \boldsymbol{P}_t = \exp(t\boldsymbol{G}),$$

where $e^{\boldsymbol{A}}$ is the natural abbreviation for $\sum_{n=0}^{\infty}(1/n!)\boldsymbol{A}^n$ whenever $\boldsymbol{A}$ is a square matrix.

So, subject to certain technical conditions, a continuous-time chain has a generator $\boldsymbol{G}$ which specifies the transition probabilities. Several examples of such generators are given in Section 6.11. In the last section we saw that a Poisson process (this is Example (7) with $\lambda_i = \lambda$ for all $i \geqslant 0$) can be described in terms of its interarrival times; an equivalent remark holds here. Suppose that the chain is in state $X(t) = i$ at time t. The future development of

$X(t+s)$, for $s \geqslant 0$, goes roughly as follows. Let

$$T = \inf\{s \geqslant 0: X(t+s) \neq i\}$$

be the further time until the chain changes its state; T is called a 'holding time'.

(13)* **Claim.** *T is exponentially distributed with parameter $-g_{ii}$.*

This is a remarkable result which explains all earlier remarks about the importance of the exponential distribution.

Sketch proof. The distribution of T has the 'lack of memory' property (see Problem (4.11.5)) because

$$\begin{aligned}\mathbf{P}(T > x+y \mid T > x) &= \mathbf{P}(T > x+y \mid X(t+x) = i)\\ &= \mathbf{P}(T > y) \quad \text{if} \quad x, y \geqslant 0\end{aligned}$$

by the Markov property and the homogeneity of the chain. It follows that the distribution function F_T of T satisfies

$$1 - F_T(x+y) = [1 - F_T(x)][1 - F_T(y)]$$

and so

$$1 - F_T(x) = \exp(-\lambda x)$$

where $\lambda = F'_T(0) = -g_{ii}$. ■

Therefore, if $X(t) = i$, then the chain remains in state i for an exponentially distributed time T, after which it jumps to some other state j.

(14)* **Claim.** *The probability that the chain jumps to j $(\neq i)$ is $-g_{ij}/g_{ii}$.*

Sketch proof. Roughly speaking, suppose that $x < T \leqslant x + h$ and suppose that the chain jumps only once in $(x, x+h]$. Then

$$\mathbf{P}(\text{jumps to } j \mid \text{it jumps}) \simeq \frac{p_{ij}(h)}{1 - p_{ii}(h)} \to -\frac{g_{ij}}{g_{ii}} \quad \text{as} \quad h \downarrow 0. \qquad ■$$

(15) **Example.** Consider a two-state chain X with $S = \{1, 2\}$; $X(t)$ jumps between 1 and 2 as time passes. There are two equivalent ways of describing the chain, depending on whether we specify $\boldsymbol{G}$ or we specify the holding times:

(a) X has generator $\boldsymbol{G} = \begin{pmatrix} -\alpha & \alpha \\ \beta & -\beta \end{pmatrix}$.

(b) If the chain is in state 1 (or 2), then it stays in this state for a length of time which is exponentially distributed with parameter α (or β) before jumping to 2 (or 1).

The forward equations (9), $\boldsymbol{P}_t' = \boldsymbol{P}_t\boldsymbol{G}$, take the form

$$p'_{11}(t) = -\alpha p_{11}(t) + \beta p_{12}(t)$$

and are easily solved to find the transition probabilities of the chain (*exercise*). ●

We move on to the classification of states; this is not such a chore as it was for discrete-time chains. It turns out that for any pair i, j of states

(16) $$\text{either } p_{ij}(t) = 0 \text{ for all } t > 0, \text{ or } p_{ij}(t) > 0 \text{ for all } t > 0,$$

and this leads to a definition of irreducibility.

(17) **Definition.** The chain is called **irreducible** if for any pair i, j of states we have that $p_{ij}(t) > 0$ for some t.

Any time $t > 0$ will suffice in (17), because of (16). The birth process is *not* irreducible, since it is non-decreasing. See Problem (6.13.15) for a condition for irreducibility in terms of the generator $\boldsymbol{G}$ of the chain.

As before, the asymptotic behaviour of $X(t)$ for large t is closely bound up with the existence of stationary distributions. Compare their definition with (6.4.1).

(18) **Definition.** The vector $\boldsymbol{\pi}$ is a **stationary distribution** of the chain if $\pi_j \geqslant 0$, $\sum_j \pi_j = 1$, and $\boldsymbol{\pi} = \boldsymbol{\pi}\boldsymbol{P}_t$ for all $t \geqslant 0$.

If $X(0)$ has distribution $\boldsymbol{\mu}^{(0)}$ then the distribution $\boldsymbol{\mu}^{(t)}$ of $X(t)$ is given by

(19) $$\boldsymbol{\mu}^{(t)} = \boldsymbol{\mu}^{(0)}\boldsymbol{P}_t.$$

If $\boldsymbol{\mu}^{(0)} = \boldsymbol{\pi}$, a stationary distribution, then $X(t)$ has distribution $\boldsymbol{\pi}$ for all t. For discrete-time chains we found stationary distributions by solving the equations $\boldsymbol{\pi} = \boldsymbol{\pi}\boldsymbol{P}$; the corresponding equations $\boldsymbol{\pi} = \boldsymbol{\pi}\boldsymbol{P}_t$ for continuous-time chains may seem complicated but they amount to a simple condition relating $\boldsymbol{\pi}$ and $\boldsymbol{G}$.

(20)* **Claim.** *$\boldsymbol{\pi} = \boldsymbol{\pi}\boldsymbol{P}_t$ for all t if and only if $\boldsymbol{\pi}\boldsymbol{G} = \boldsymbol{0}$.*

Sketch proof. From (11),

$$\begin{aligned}
\boldsymbol{\pi}\boldsymbol{G} = \boldsymbol{0} &\Leftrightarrow \boldsymbol{\pi}\boldsymbol{G}^n = \boldsymbol{0} && \text{for all } n \geqslant 1\\
&\Leftrightarrow \sum_1^\infty \frac{t^n}{n!}\boldsymbol{\pi}\boldsymbol{G}^n = \boldsymbol{0} && \text{for all } t\\
&\Leftrightarrow \boldsymbol{\pi}\sum_0^\infty \frac{t^n}{n!}\boldsymbol{G}^n = \boldsymbol{\pi} && \text{for all } t\\
&\Leftrightarrow \boldsymbol{\pi}\boldsymbol{P}_t = \boldsymbol{\pi} && \text{for all } t.
\end{aligned}$$

■

This provides a useful collection of equations which specify stationary distributions, whenever they exist. The ergodic theorem for continuous-time chains is as follows; it holds exactly as stated, and requires no extra conditions.

(21) **Theorem.** *Let X be irreducible with a standard semigroup $\{\mathbf{P}_t\}$ of transition probabilities.*

(a) *If there exists a stationary distribution $\boldsymbol{\pi}$ then it is unique and*

$$p_{ij}(t) \to \pi_j \quad \text{as} \quad t \to \infty, \quad \text{for all } i \text{ and } j.$$

(b) *If there is no stationary distribution then $p_{ij}(t) \to 0$ as $t \to \infty$, for all i and j.*

Sketch proof. Fix $h > 0$ and let $Y_n = X(nh)$. Then $Y = \{Y_n\}$ is an irreducible aperiodic discrete-time Markov chain; Y is called a *skeleton* of X. If Y is non-null persistent, then it has a unique stationary distribution $\boldsymbol{\pi}^h$ and $p_{ij}(nh) = \mathbf{P}(Y_n = j \mid Y_0 = i) \to \pi_j^h$ as $n \to \infty$; otherwise $p_{ij}(nh) \to 0$ as $n \to \infty$. Use this argument for two rational values h_1 and h_2 of h and observe that the sequences $\{nh_1: n \geqslant 0\}$, $\{nh_2: n \geqslant 0\}$ have infinitely many points in common to deduce that $\boldsymbol{\pi}^{h_1} = \boldsymbol{\pi}^{h_2}$ in the non-null persistent case. Thus the limit of $p_{ij}(t)$ exists along all sequences $\{nh: n \geqslant 0\}$ of times, for rational h; now use the continuity of $p_{ij}(t)$ to fill in the gaps. The proof is essentially complete. ■

Exercises

(22) Let X be a Markov chain on $\{1, 2\}$ with generator

$$\mathbf{G} = \begin{pmatrix} -\mu & \mu \\ \lambda & -\lambda \end{pmatrix}$$

where $\lambda\mu > 0$.

(a) Write down the forward equations and solve them for the transition probabilities $p_{ij}(t)$, $i, j = 1, 2$.
(b) Calculate $\mathbf{G}^n$ and hence find $\sum_{n=0}^{\infty} (t^n/n!)\mathbf{G}^n$. Compare your answer with that to part (a).
(c) Solve the equation $\boldsymbol{\pi}\mathbf{G} = \mathbf{0}$ in order to find the stationary distribution. Verify that $p_{ij}(t) \to \pi_j$ as $t \to \infty$.

(23) As a continuation of the previous exercise, find

(a) $\mathbf{P}(X(t) = 2 \mid X(0) = 1, X(3t) = 1)$,
(b) $\mathbf{P}(X(t) = 2 \mid X(0) = 1, X(3t) = 1, X(4t) = 1)$.

(24) Jobs arrive in a computer queue in the manner of a Poisson process with intensity λ. The central processor handles them one by one in the order of their arrival, and each has an exponentially distributed runtime with parameter μ, the runtimes of different jobs being independent of each other and of the arrival process. Let $X(t)$ be the number of jobs in the system (either running or waiting) at time t, where $X(0) = 0$. Explain why X is a Markov chain, and write down its generator. Show that a stationary distribution exists if and only if $\lambda < \mu$, and find it in this case.

(25) Let $X = \{X(t): t \geqslant 0\}$ be a Markov chain having stationary distribution $\boldsymbol{\pi}$. We may sample X at the times of a Poisson process: let N be a Poisson process with intensity λ, independent of X, and define $Y_n = X(T_n)$, the value taken by X immediately after the epoch T_n of the nth arrival of N. Show that $Y = \{Y_n: n \geqslant 0\}$ is a discrete-time Markov chain with the same stationary distribution as X. (This is called the 'Pasta' property: Poisson arrivals see time averages.)

6.10 Uniform semigroups

This section is not for lay readers and may be omitted; it indicates where some of the difficulties lie in the heuristic discussion of the last section (see Chung (1960) or Freedman (1971) for the proofs of the following results).

Perhaps the most important claim is (6.9.5), that $p_{ij}(h)$ is approximately linear in h when h is small.

(1) **Theorem.** *If $\{\boldsymbol{P}_t\}$ is a standard stochastic semigroup then there exists an $|S| \times |S|$ matrix $\boldsymbol{G} = (g_{ij})$ such that, as $t \downarrow 0$*

(a) $p_{ij}(t) = g_{ij}t + o(t)$ *if* $i \neq j$
(b) $p_{ii}(t) = 1 + g_{ii}t + o(t)$.

*Also, $0 \leqslant g_{ij} < \infty$ if $i \neq j$, and $0 \geqslant g_{ii} \geqslant -\infty$. $\boldsymbol{G}$ is called the **generator** of $\{\boldsymbol{P}_t\}$.*

(1b) is fairly easy to demonstrate (see Problem (6.13.14)); the proof of (1a) is considerably more difficult. $\boldsymbol{G}$ is a matrix with non-negative entries off the diagonal and non-positive entries (*which may be* $-\infty$) on the diagonal. We normally write

(2) $$\boldsymbol{G} = \lim_{t \downarrow 0} \frac{1}{t}(\boldsymbol{P}_t - \mathbf{I}).$$

If S is finite then

$$\boldsymbol{G}\mathbf{1}' = \lim_{t \downarrow 0} \frac{1}{t}(\boldsymbol{P}_t - \mathbf{I})\mathbf{1}' = \lim_{t \downarrow 0} \frac{1}{t}(\boldsymbol{P}_t\mathbf{1}' - \mathbf{1}') = \mathbf{0}'$$

from (6.9.3b), and so the row sums of $\boldsymbol{G}$ equal 0. If S is infinite, all we can assert is that

$$\sum_j g_{ij} \leqslant 0.$$

In the light of (6.9.13), states i with $g_{ii} = -\infty$ are called *instantaneous*, since the chain leaves them at the same instant that it arrives in them. Otherwise, state i is called *stable* if $0 > g_{ii} > -\infty$ and *absorbing* if $g_{ii} = 0$.

We cannot proceed much further unless we impose a stronger condition on $\{\mathbf{P}_t\}$ than that it be standard.

(3) **Definition.** We call $\{\mathbf{P}_t\}$ **uniform** if

$$\mathbf{P}_t \to \mathbf{I} \quad \text{uniformly as} \quad t \downarrow 0,$$

which is to say that

(4) $$p_{ii}(t) \to 1 \quad \text{as} \quad t \downarrow 0, \text{ uniformly in } i \in S.$$

Clearly (4) implies that $p_{ij}(t) \to 0$ for $i \neq j$, since $p_{ij}(t) \leqslant 1 - p_{ii}(t)$. A uniform semigroup is standard; the converse is not generally true, but holds if S is finite. The uniformity of the semigroup depends upon the sizes of the diagonal elements of its generator $\mathbf{G}$.

(5) **Theorem.** *$\{\mathbf{P}_t\}$ is uniform if and only if* $\sup_i\{-g_{ii}\} < \infty$.

We consider uniform semigroups only for the rest of this section; they are so well behaved as to be rather dull. Here is the main result, which vindicates (6.9.9), (6.9.10), and (6.9.11).

(6) **Theorem. Kolmogorov's equations.** *If $\{\mathbf{P}_t\}$ is a uniform semigroup with generator $\mathbf{G}$, then it is the unique solution to the*

(7) ***forward equation**: $\mathbf{P}'_t = \mathbf{P}_t\mathbf{G}$*

(8) ***backward equation**: $\mathbf{P}'_t = \mathbf{G}\mathbf{P}_t$*

subject to the boundary condition $\mathbf{P}_0 = \mathbf{I}$. Furthermore,

(9) $$\mathbf{P}_t = \exp(t\mathbf{G}) \quad \textit{and} \quad \mathbf{G}\mathbf{1}' = \mathbf{0}'.$$

The backward equation is more fundamental than the forward equation since it can be derived subject to the condition that $\mathbf{G}\mathbf{1}' = \mathbf{0}'$, which is a weaker condition than that the semigroup be uniform. This remark has some bearing on the discussion of dishonesty in Section 6.8. (Of course, a dishonest birth process is not even a Markov chain in our sense, unless we augment the state space $\{0, 1, 2, \ldots\}$ by adding on the point $\{\infty\}$.) You can prove (6) yourself. Just use the argument which established (6.9.9) and (6.9.10) with an eye to rigour; then show that (9) gives a solution to (7) and (8), and finally prove uniqueness.

Thus uniform semigroups are characterized by their generators; but which matrices are generators of uniform semigroups? Let $\mathcal{M}$ be the collection of $|S| \times |S|$ matrices $\mathbf{A} = (a_{ij})$ for which

$$\|\mathbf{A}\| = \sup_i \sum_{j \in S} |a_{ij}| < \infty.$$

(10) **Theorem.** *$A \in \mathcal{M}$ is the generator of a uniform semigroup $\boldsymbol{P}_t = \exp(t\boldsymbol{A})$ if and only if*

$$a_{ij} \geqslant 0 \quad \text{for} \quad i \neq j, \qquad \text{and} \qquad \sum_j a_{ij} = 0 \quad \text{for all } i.$$

Next we discuss irreducibility. Observation (6.9.16) amounts to the following.

(11) **Theorem.** *If $\{\boldsymbol{P}_t\}$ is standard (but not necessarily uniform) then*

(a) *$p_{ii}(t) > 0$ for all $t \geqslant 0$.*
(b) ***Lévy dichotomy***: *If $i \neq j$ then*

$$\begin{aligned} \textit{either} \quad &\text{(i)}\ p_{ij}(t) = 0 \quad \text{for all } t > 0 \\ \textit{or} \quad &\text{(ii)}\ p_{ij}(t) > 0 \quad \text{for all } t > 0. \end{aligned}$$

Partial proof. (a) $\{\boldsymbol{P}_t\}$ is assumed standard, so $p_{ii}(t) \to 1$ as $t \downarrow 0$. Pick $h > 0$ such that $p_{ii}(s) > 0$ for all $s \leqslant h$. For any real t pick n large enough so that $t \leqslant hn$. By the Chapman–Kolmogorov equations

$$p_{ii}(t) \geqslant (p_{ii}(t/n))^n > 0 \quad \text{because} \quad t/n \leqslant h.$$

(b) The proof of this is quite difficult, though the method of (a) can easily be adapted to show that if

$$\alpha = \inf\{t \colon p_{ij}(t) > 0\}$$

then $p_{ij}(t) > 0$ for all $t > \alpha$. The full result asserts that either $\alpha = 0$ or $\alpha = \infty$. ■

(12) **Example (6.9.15) revisited.** If $\alpha > 0$, $\beta > 0$, and $S = \{1, 2\}$, then

$$\boldsymbol{G} = \begin{pmatrix} -\alpha & \alpha \\ \beta & -\beta \end{pmatrix}$$

is the generator of a uniform stochastic semigroup $\{\boldsymbol{P}_t\}$ given by the following calculation. Diagonalize $\boldsymbol{G}$ to obtain $\boldsymbol{G} = \boldsymbol{B}\boldsymbol{\Lambda}\boldsymbol{B}^{-1}$ where

$$\boldsymbol{B} = \begin{pmatrix} \alpha & 1 \\ -\beta & 1 \end{pmatrix}, \qquad \boldsymbol{\Lambda} = \begin{pmatrix} -(\alpha+\beta) & 0 \\ 0 & 0 \end{pmatrix}.$$

Therefore

$$\begin{aligned} \boldsymbol{P}_t &= \sum_0^\infty \frac{t^n}{n!} \boldsymbol{G}^n = \boldsymbol{B}\left(\sum_0^\infty \frac{t^n}{n!} \boldsymbol{\Lambda}^n\right)\boldsymbol{B}^{-1} \\ &= \boldsymbol{B}\begin{pmatrix} h(t) & 0 \\ 0 & 1 \end{pmatrix}\boldsymbol{B}^{-1} \quad \text{since} \quad \boldsymbol{\Lambda}^0 = \mathbf{I} \\ &= \frac{1}{\alpha+\beta}\begin{pmatrix} \alpha h(t) + \beta & \alpha[1 - h(t)] \\ \beta[1 - h(t)] & \alpha + \beta h(t) \end{pmatrix} \end{aligned}$$

where $h(t) = \exp[-t(\alpha + \beta)]$. Let $t \to \infty$ to obtain

$$\boldsymbol{P}_t \to \begin{pmatrix} 1-\rho & \rho \\ 1-\rho & \rho \end{pmatrix} \quad \text{where} \quad \rho = \alpha/(\alpha+\beta)$$

and so

$$\mathbf{P}(X(t) = i) \to \begin{cases} 1-\rho & \text{if} \quad i = 1 \\ \rho & \text{if} \quad i = 2 \end{cases}$$

irrespective of the initial distribution of $X(0)$. This shows that $\boldsymbol{\pi} = (1-\rho, \rho)$ is the limiting distribution. Check that $\boldsymbol{\pi G} = \mathbf{0}$. The method of (6.9.15) provides an alternative and easier route to these results. ●

(13) **Example. Birth process.** Recall the birth process of (6.8.11), and suppose that $\lambda_i > 0$ for all i. The process is uniform if and only if

$$\sup_i \{-g_{ii}\} = \sup_i \{\lambda_i\} < \infty$$

and this is a sufficient condition for both the forward and backward equations to have unique solutions. We saw in Section 6.8 that the weaker condition

$$\sum_i \frac{1}{\lambda_i} = \infty$$

is a necessary and sufficient condition for this to hold. ●

6.11 Birth–death processes and imbedding

A birth process is a non-decreasing Markov chain for which the probability of moving from state n to state $n+1$ in the time interval $(t, t+h)$ is $\lambda_n h + o(h)$. More realistic continuous-time models for population growth incorporate death also. Suppose then that the number $X(t)$ of individuals alive in some population at time t evolves in the following way:

(a) X is a Markov chain taking values in $\{0, 1, 2, \dots\}$
(b) the infinitesimal transition probabilities are given by

(1)
$$\mathbf{P}\big(X(t+h) = n+m \mid X(t) = n\big) = \begin{cases} \lambda_n h + o(h) & \text{if} \quad m = 1 \\ \mu_n h + o(h) & \text{if} \quad m = -1 \\ o(h) & \text{if} \quad |m| > 1 \end{cases}$$

(c) the 'birth rates' $\lambda_0, \lambda_1, \dots$ and the 'death rates' $\mu_0, \mu_1, \dots$ satisfy

$$\lambda_i \geqslant 0, \quad \mu_i \geqslant 0, \quad \mu_0 = 0.$$

Then X is called a *birth–death process*. It has generator $\boldsymbol{G} = (g_{ij}: i, j \geqslant 0)$

given by

$$G = \begin{pmatrix} -\lambda_0 & \lambda_0 & 0 & 0 & 0 & \cdots \\ \mu_1 & -(\lambda_1+\mu_1) & \lambda_1 & 0 & 0 & \cdots \\ 0 & \mu_2 & -(\lambda_2+\mu_2) & \lambda_2 & 0 & \cdots \\ 0 & 0 & \mu_3 & -(\lambda_3+\mu_3) & \lambda_3 & \cdots \\ \vdots & \vdots & \vdots & \vdots & \vdots & \end{pmatrix}.$$

It is uniform if and only if $\sup_i\{\lambda_i + \mu_i\} < \infty$. In many particular cases we have that $\lambda_0 = 0$, and then 0 is an absorbing state and the chain is not irreducible.

The transition probabilities $p_{ij}(t) = \mathbf{P}(X(t) = j \mid X(0) = i)$ may in principle be calculated from a knowledge of the birth and death rates, although in practice these functions rarely have nice forms. It is an easier matter to determine the asymptotic behaviour of the process as $t \to \infty$. Suppose that $\lambda_i > 0$ and $\mu_i > 0$ for all relevant i. A stationary distribution $\boldsymbol{\pi}$ would satisfy $\boldsymbol{\pi}\boldsymbol{G} = \mathbf{0}$, which is to say that

$$-\lambda_0\pi_0 + \mu_1\pi_1 = 0,$$

$$\lambda_{n-1}\pi_{n-1} - (\lambda_n + \mu_n)\pi_n + \mu_{n+1}\pi_{n+1} = 0 \quad \text{if} \quad n \geqslant 1.$$

A simple induction yields that

(2) $$\pi_n = \frac{\lambda_0\lambda_1\cdots\lambda_{n-1}}{\mu_1\mu_2\cdots\mu_n}\pi_0, \qquad n \geqslant 1.$$

Such a vector $\boldsymbol{\pi}$ is a stationary distribution if and only if $\sum_n \pi_n = 1$; this may happen if and only if

(3) $$\sum_{n=0}^{\infty} \frac{\lambda_0\lambda_1\cdots\lambda_{n-1}}{\mu_1\mu_2\cdots\mu_n} < \infty,$$

where the term in $n = 0$ is interpreted as 1; if this holds, then

(4) $$\pi_0 = \left(\sum_{n=0}^{\infty} \frac{\lambda_0\lambda_1\cdots\lambda_{n-1}}{\mu_1\mu_2\cdots\mu_n}\right)^{-1}.$$

We have from Theorem (6.9.21) that the process settles into equilibrium (with stationary distribution given by (2) and (4)) if and only if the summation in (3) is finite, a condition requiring that the birth rates are not too large relative to the death rates.

Here are some examples of birth–death processes.

(5) **Example. Pure birth.** $\mu_n = 0$ for all n. ●

(6) **Example. Simple death with immigration.** Let us model a population which evolves in the following way. At time zero the size $X(0)$ of the population equals I. Individuals do not reproduce, but new individuals immigrate into the population at the arrival times of a Poisson process with intensity $\lambda > 0$. Each individual may die in the time interval $(t, t+h)$ with probability $\mu h + o(h)$, where $\mu > 0$. The transition probabilities of $X(t)$ satisfy

$$p_{ij}(h) = \mathbf{P}(X(t+h) = j \mid X(t) = i)$$
$$= \begin{cases} \mathbf{P}(j-i \text{ arrivals, no deaths}) + o(h) & \text{if } j \geqslant i \\ \mathbf{P}(i-j \text{ deaths, no arrivals}) + o(h) & \text{if } j < i \end{cases}$$

since the probability of two or more changes occurring in $(t, t+h)$ is $o(h)$. Therefore

$$p_{i,i+1}(h) = \lambda h(1-\mu h)^i + o(h) = \lambda h + o(h)$$
$$p_{i,i-1}(h) = i(\mu h)(1-\mu h)^{i-1}(1-\lambda h) + o(h) = (i\mu)h + o(h)$$
$$p_{ij}(h) = o(h) \quad \text{if} \quad |j-i| > 1$$

and we recognize $X(t)$ as a birth–death process with parameters

(7) $$\lambda_n = \lambda, \qquad \mu_n = n\mu.$$

It is an irreducible continuous-time Markov chain; (6.10.5) shows that it is not uniform. We may ask for the distribution of $X(t)$ and for the limiting distribution of the chain as $t \to \infty$. The former question is answered by solving the forward equations; this is Problem (6.13.18). The latter question is answered by the following.

(8) **Theorem.** *$X(t)$ is asymptotically Poisson distributed with parameter $\rho = \lambda/\mu$. That is*

$$\mathbf{P}(X(t) = n) \to \frac{\rho^n}{n!}e^{-\rho}, \qquad n = 0, 1, 2, \ldots .$$

Proof. Either substitute (7) into (2) and (4), or solve the equation $\boldsymbol{\pi G} = \mathbf{0}$ directly. ■●

(9) **Example. Simple birth–death.** Assume that each individual who is alive in the population at time t either dies in the interval $(t, t+h)$ with probability $\mu h + o(h)$ or splits into two in the interval with probability $\lambda h + o(h)$. The transition probabilities satisfy equations like

$$p_{i,i+1}(h) = \mathbf{P}(\text{one birth, no deaths}) + o(h)$$
$$= i(\lambda h)(1-\lambda h)^{i-1}(1-\mu h)^i + o(h)$$
$$= (i\lambda)h + o(h)$$

and it is easy to check that the number $X(t)$ of living individuals at time t satisfies (1) with

$$\lambda_n = n\lambda, \quad \mu_n = n\mu.$$

We shall explore this model in detail. The chain $X = \{X(t)\}$ is standard but not uniform. We shall assume that $X(0) = I > 0$; 0 is an absorbing state. We find the distribution of $X(t)$ through its generating function.

(10) **Theorem.** *The generating function of $X(t)$ is*

$$G(s, t) = \mathbf{E}(s^{X(t)}) = \begin{cases} \left(\dfrac{\lambda t(1-s)+s}{\lambda t(1-s)+1}\right)^I & \text{if } \mu = \lambda \\ \left(\dfrac{\mu(1-s)-(\mu-\lambda s)\exp[-t(\lambda-\mu)]}{\lambda(1-s)-(\mu-\lambda s)\exp[-t(\lambda-\mu)]}\right)^I & \text{if } \mu \neq \lambda. \end{cases}$$

Proof. This is like Proof B of (6.8.2). Write $p_j(t) = \mathbf{P}(X(t) = j)$ and condition $X(t+h)$ on $X(t)$ to obtain the forward equations

$$p_j'(t) = \lambda(j-1)p_{j-1}(t) - (\lambda+\mu)jp_j(t) + \mu(j+1)p_{j+1}(t) \quad \text{if} \quad j \geqslant 1$$

$$p_0'(t) = \mu p_1(t).$$

Multiply the jth equation by s^j and sum to obtain

$$\sum_0^\infty s^j p_j'(t) = \lambda s^2 \sum_1^\infty (j-1)s^{j-2}p_{j-1}(t) - (\lambda+\mu)s\sum_0^\infty js^{j-1}p_j(t) + \mu\sum_0^\infty (j+1)s^j p_{j+1}(t).$$

Put $G(s, t) = \sum_0^\infty s^j p_j(t) = \mathbf{E}(s^{X(t)})$ to obtain

(11)
$$\frac{\partial G}{\partial t} = \lambda s^2 \frac{\partial G}{\partial s} - (\lambda+\mu)s\frac{\partial G}{\partial s} + \mu\frac{\partial G}{\partial s} = (\lambda s - \mu)(s-1)\frac{\partial G}{\partial s}$$

with boundary condition $G(s, 0) = s^I$. The solution to this partial differential equation is given by (10); to see this either solve (11) by standard methods, see Hildebrand (1962, Chapter 8), or substitute the conclusion of (10) into (11). ■

Note that X is honest for all λ and μ since $G(1, t) = 1$ for all t. To find the mean and variance of $X(t)$, differentiate G:

$$\mathbf{E}(X(t)) = I\,e^{(\lambda-\mu)t}$$

$$\operatorname{var}(X(t)) = \begin{cases} 2I\lambda t & \text{if } \lambda = \mu \\ I\dfrac{\lambda+\mu}{\lambda-\mu}e^{(\lambda-\mu)t}(e^{(\lambda-\mu)t}-1) & \text{if } \lambda \neq \mu. \end{cases}$$

Write $\rho = \lambda/\mu$ and notice that

$$\mathbf{E}(X(t)) \to \begin{cases} 0 & \text{if } \rho < 1 \\ \infty & \text{if } \rho > 1. \end{cases}$$

(12) **Corollary.** *The extinction probabilities* $\eta(t) = \mathbf{P}(X(t) = 0)$ *satisfy*

$$\eta(t) \to \begin{cases} 1 & \textit{if } \rho \leqslant 1 \\ \rho^{-I} & \textit{if } \rho > 1 \end{cases} \quad \textit{as} \quad t \to \infty.$$

Proof. $\eta(t) = G(0, t)$. Substitute $s = 0$ in $G(s, t)$ to find $\eta(t)$ explicitly. ■

The observant reader will have noticed that these results are almost identical to those obtained for the branching process, except in that they pertain to a process in continuous time. There are (at least) two discrete Markov chains imbedded in X.

(*A*) *Imbedded random walk.* We saw in (6.9.13) and (6.9.14) that if $X(t) = n$, say, then the length of time

$$T = \inf\{s > 0: X(t+s) \neq n\}$$

until the next birth or death is exponentially distributed with parameter $-g_{nn} = n(\lambda + \mu)$. When this time is up, X moves from state n to state $n + M$ where

$$\mathbf{P}(M = 1) = -g_{n,n+1}/g_{nn} = \frac{\lambda}{\lambda+\mu}$$

$$\mathbf{P}(M = -1) = \frac{\mu}{\lambda+\mu}.$$

Think of this transition as the movement of a particle from the integer n to the new integer $n + M$, where $M = \pm 1$. Such a particle performs a simple random walk with parameter $p = \lambda/(\lambda + \mu)$ and initial position I. We know already (see (3.9.6)) that the probability of ultimate absorption at 0 is given by (12). Other properties of random walks (see Sections 3.9 and 5.3) are applicable also.

(*B*) *Imbedded branching process.* We can think of the birth–death process in the following way. After birth an individual lives for a certain length of time which is exponentially distributed with parameter $\lambda + \mu$. When this period is over it dies, leaving behind it either no individuals, with probability $\mu/(\lambda + \mu)$, or two individuals, with probability $\lambda/(\lambda + \mu)$. This is just an age-dependent branching process with age density function

(13) $$f_T(u) = (\lambda + \mu)\,\mathrm{e}^{-(\lambda+\mu)u}, \qquad u \geqslant 0$$

and family-size generating function

(14) $$G(s) = \frac{\mu + \lambda s^2}{\mu + \lambda}$$

in the notation of Section 5.5 (do not confuse G in (14) with $G(s, t) = \mathbf{E}(s^{X(t)})$). Thus if $I = 1$, the generating function $G(s, t) = \mathbf{E}(s^{X(t)})$ satisfies the differential equation

(15) $$\frac{\partial G}{\partial t} = \lambda G^2 - (\lambda + \mu)G + \mu.$$

After (11), this is the *second* differential equation for $G(s, t)$. Needless to say, (15) is really just the backward equation of the process; the reader should check this and verify that it has the same solution as the forward equation (11). Suppose we lump together the members of each generation of this age-dependent branching process. Then we obtain an ordinary branching process with family-size generating function $G(s)$ given by (14). From the general theory, the extinction probability of the process is the smallest non-negative root of the equation $s = G(s)$, and we can verify easily that this is given by (12) with $I = 1$. ●

(16) **Example. A more general branching process.** Finally, we consider a more general type of age-dependent branching process than that above, and investigate its honesty. Suppose that each individual in a population lives for an exponentially distributed time with parameter λ say. After death it leaves behind it a (possibly empty) family of offspring: the size N of this family has mass function $f(k) = \mathbf{P}(N = k)$ and generating function G_N. Let $X(t)$ be the size of the population at time t; we assume that $X(0) = 1$. From Section 5.5 the backward equation for $G(s, t) = \mathbf{E}(s^{X(t)})$ is just

$$\frac{\partial G}{\partial t} = \lambda(G_N(G) - G)$$

with boundary condition $G(s, 0) = s$; the solution is given by

(17) $$\int_s^{G(s,t)} \frac{\mathrm{d}u}{G_N(u) - u} = \lambda t$$

provided that $G_N(u) - u$ has no zeros within the domain of the integral.

There are many interesting questions about this process; for example, is it honest in the sense that

$$\sum_{j=0}^{\infty} \mathbf{P}(X(t) = j) = 1?$$

(18) **Theorem.** *X is honest if and only if*

(19) $$\int_{1-\varepsilon}^{1} \frac{du}{G_N(u) - u} \quad \textit{diverges for all } \varepsilon > 0.$$

Proof. See Harris (1963, p. 107). ■

If condition (19) fails then the population size may explode to $+\infty$ in finite time.

(20) **Corollary.** *X is honest if* $\mathbf{E}(N) < \infty$.

Proof. Expand $G_N(u) - u$ about $u = 1$ to find that

$$G_N(u) - u = [\mathbf{E}(N) - 1](u - 1) + o(u - 1) \quad \text{as} \quad u \uparrow 1. \qquad ■ ●$$

Exercises

(21) Let $X = \{X(t): t \geq 0\}$ be an irreducible Markov chain with generator $\mathbf{G}$, and let $Y = \{Y_n: n \geq 0\}$ be the imbedded chain given by $Y_0 = X(0)$ and Y_n is the value of X just after its nth jump.

(a) If X has stationary distribution $\boldsymbol{\pi}$, show that Y has stationary distribution $\hat{\boldsymbol{\pi}}$ where

$$\hat{\pi}_k = \frac{\pi_k g_{kk}}{\sum_i \pi_i g_{ii}},$$

provided $\sum_i \pi_i g_{ii} < \infty$. When is it the case that $\hat{\boldsymbol{\pi}} = \boldsymbol{\pi}$?

(b) Show that a state is persistent for X if and only if it is persistent for Y.

(c) Let Z be an irreducible discrete-time Markov chain on a countably infinite state space S, having transition matrix $\mathbf{Q} = (q_{ij})$ satisfying $q_{ii} = 0$ for all states i, and with stationary distribution $\boldsymbol{\nu}$. Construct a continuous-time process X on S for which Z is the imbedded chain, such that X has no stationary distribution.

(22) Describe the imbedded chain (as in the previous exercise) for a birth–death process with rates λ_n and μ_n.

(23) Consider an immigration–death process X, being a birth–death process with birth rates $\lambda_n = \lambda$ and death rates $\mu_n = n\mu$; let Y be the imbedded chain (see Exercise (21)). Find the transition matrix of Y, and show that it has as stationary distribution

$$\pi_n = \frac{1}{2(n!)}\left(1 + \frac{n}{\rho}\right)\rho^n e^{-\rho}$$

where $\rho = \lambda/\mu$. Explain why this differs from the stationary distribution of X.

(24) Consider the birth–death process X with $\lambda_n = n\lambda$ and $\mu_n = n\mu$ for all $n \geq 0$. Suppose $X(0) = 1$ and let $\eta(t) = \mathbf{P}(X(t) = 0)$. Show that η satisfies the differential equation $\eta'(t) + (\lambda + \mu)\eta(t) = \mu + \lambda\eta(t)^2$. Hence find $\eta(t)$. Hence calculate $\mathbf{P}(X(t) = 0 \mid X(u) = 0)$ for $0 < t < u$.

(25) For the birth–death process of the previous exercise with $\lambda < \mu$, show that the distribution of $X(t)$, conditional on the event $\{X(t) > 0\}$, converges as $t \to \infty$ to a geometric distribution.

6.12 Special processes

There are many more general formulations of the processes which we modelled in Sections 6.8 and 6.11. Here is a very small selection of some of them, with some details of the areas in which they have been found useful.

(1) **Non-homogeneous chains.** Relax the assumption that the transition probabilities $p_{ij}(s, t) = \mathbf{P}(X(t) = j \mid X(s) = i)$ satisfy

$$p_{ij}(s, t) = p_{ij}(0, t - s).$$

This leads to some very difficult problems. We may make some progress in the special case when $X(t)$ is the simple birth–death process of the previous section, for which

$$\lambda_n = n\lambda, \qquad \mu_n = n\mu.$$

The parameters λ and μ are now assumed to be non-constant functions of t. (After all, most populations have birth and death rates which vary from season to season.) It is easy to check that the forward equation (6.11.11) remains unchanged:

$$\frac{\partial G}{\partial t} = [\lambda(t)s - \mu(t)](s - 1)\frac{\partial G}{\partial s}.$$

The solution is

$$G(s, t) = \left[1 + \left(\frac{\exp[r(t)]}{s - 1} - \int_0^t \lambda(u)\exp[r(u)]\,du\right)^{-1}\right]^I$$

where $I = X(0)$ and

$$r(t) = \int_0^t [\mu(u) - \lambda(u)]\,du.$$

The extinction probability of $X(t)$ is the coefficient of s^0 in $G(s, t)$, and it is left as an *exercise* for the reader to prove the next result.

(2) **Theorem.** $\mathbf{P}(X(t) = 0) \to 1$ *if and only if*

$$\int_0^T \mu(u)\exp[r(u)]\,du \to \infty \quad \textit{as} \quad T \to \infty. \qquad \bullet$$

(3) **A bivariate branching process.** We advertised the branching process as a feasible model for the growth of cell populations; we should also note one of its inadequacies in this role. Even the age-dependent process cannot meet the main objection, which is that the time of division of a cell may depend rather more on the *size* of the cell than on its *age*. So here is a model for the growth and degradation of long-chain polymers†.

A population comprises *particles*. Let $N(t)$ be the number of particles present at time t, and suppose that $N(0) = 1$. We suppose that the $N(t)$ particles are divided into $W(t)$ groups of size $N_1, N_2, \dots, N_W$, $(\sum_1^{W(t)} N_i = N(t))$, such that the particles in each group are aggregated into a *cell*. Think of the cells as a collection of $W(t)$ polymers, containing $N_1, N_2, \dots, N_W$ particles respectively. As time progresses each cell grows and divides. We suppose that each cell can accumulate one particle from outside the system with probability $\lambda h + o(h)$ in the time interval $(t, t + h)$. As cells become larger they are more likely to divide. We assume that the probability that a cell of size N divides into two cells of sizes M and $N - M$, for some $0 < M < N$, in the interval $(t, t + h)$ is $\mu(N - 1)h + o(h)$. The assumption that the probability of division is a *linear* function of the cell size N is reasonable for polymer degradation since the particles are strung together in a line and any of the $N - 1$ 'links' between pairs of particles may sever. At time t there are $N(t)$ particles and $W(t)$ cells, and the process is said to be in state $X(t) = (N(t), W(t))$. In the interval $(t, t + h)$ various transitions for $X(t)$ are possible. Either some cell grows or some cell divides, or more than one such event occurs. The probability that some cell grows is $\lambda Wh + o(h)$ since there are W chances of this happening; the probability of a division is $\mu(N_1 + \cdots + N_W - W)h + o(h) = \mu(N - W)h + o(h)$ since there are $N - W$ links in all; the probability of more than one such occurrence is $o(h)$. Putting this information together gives a Markov chain $X(t) = (N(t), W(t))$ with state space $\{1, 2, \dots\}^2$ and transition probabilities

$$\mathbf{P}\big(X(t + h) = (n, w) + \varepsilon \mid X(t) = (n, w)\big)$$
$$= \begin{cases} \lambda wh + o(h) & \text{if } \varepsilon = (1, 0) \\ \mu(n - w)h + o(h) & \text{if } \varepsilon = (0, 1) \\ 1 - [w(\lambda - \mu) + \mu n]h + o(h) & \text{if } \varepsilon = (0, 0) \\ o(h) & \text{otherwise.} \end{cases}$$

Write down the forward equations as usual to obtain that the joint generating function

$$G(x, y; t) = \mathbf{E}(x^{N(t)} y^{W(t)})$$

satisfies the partial differential equation

$$\frac{\partial G}{\partial t} = \mu x(y - 1)\frac{\partial G}{\partial x} + y[\lambda(x - 1) - \mu(y - 1)]\frac{\partial G}{\partial y}$$

† In physical chemistry, a *polymer* is a chain of molecules, neighbouring pairs of which are joined by bonds.

with $G(x, y; 0) = xy$. The joint moments of N and W are easily derived from this equation. More sophisticated techniques show that $N(t) \to \infty$, $W(t) \to \infty$, and $N(t)/W(t)$ approaches some constant as $t \to \infty$.

Unfortunately, most cells in nature are irritatingly non-Markovian! ●

(4) **A non-linear epidemic.** Consider a population of constant size $N + 1$, and watch the spread of a disease about its members. Let $X(t)$ be the number of healthy individuals at time t and suppose that $X(0) = N$. We assume that if $X(t) = n$ then the probability of a new infection in $(t, t + h)$ is proportional to the number of possible encounters between ill folk and healthy folk. That is,

$$\mathbf{P}\big(X(t + h) = n - 1 \mid X(t) = n\big) = \lambda n(N + 1 - n)h + o(h).$$

Nobody ever gets better. In the usual way, the reader can show that

$$G(s, t) = \mathbf{E}(s^{X(t)}) = \sum_{n=0}^{N} s^n \mathbf{P}(X(t) = n)$$

satisfies

$$\frac{\partial G}{\partial t} = \lambda(1 - s)\left(N \frac{\partial G}{\partial s} - s \frac{\partial^2 G}{\partial s^2}\right)$$

with $G(s, 0) = s^N$. There is no simple way of solving this equation, though a lot of information is available about approximate solutions. ●

(5) **Birth–death with immigration.** We saw in Example (6.11.6) that populations are not always closed and that there is sometimes a chance that a new process will be started by an arrival from outside. This may be due to mutation (if we are counting genes), or leakage (if we are counting neutrons), or irresponsibility (if we are counting cases of rabies).

Suppose that there is one individual in the population at time zero; this individual is the founding member of some birth–death process $N(t)$ with fixed but unspecified parameters. Suppose further that other individuals immigrate into the population like a Poisson process $I(t)$ with intensity ν. Each immigrant starts a new birth–death process which is an independent identically distributed copy of the original process N but displaced in time according to its time of arrival. Let $T_0\ (= 0), T_1, T_2, \ldots$ be the times at which immigrants arrive, and let $X_1, X_2, \ldots$ be the interarrival times

$$X_n = T_n - T_{n-1}.$$

The total population at time t is the aggregate of the processes generated by the $I(t) + 1$ immigrants up to time t. Call this total $Y(t)$ to obtain

(6) $$Y(t) = \sum_{i=0}^{I(t)} N_i(t - T_i)$$

where $N_1, N_2, \ldots$ are independent copies of $N = N_0$. The problem is to find how the distribution of Y depends on the typical process N and the immigration rate ν; this is an example of the problem of compounding discussed in (5.1.25).

First we prove an interesting result about order statistics. Remember that $I(t)$ is a Poisson process and $T_n = \inf\{t: I(t) = n\}$ is the time of the nth immigration.

(7) **Theorem.** *The conditional joint distribution of $T_1, T_2, \ldots, T_n$, conditional on the event $\{I(t) = n\}$, is the same as the joint distribution of the order statistics of a family of n independent variables which are uniformly distributed on* $[0, t]$.

This is something of a mouthful, and asserts that if we know that n immigrants have arrived by time t then their actual arrival times are indistinguishable from a collection of n points chosen uniformly at random in the interval $[0, t]$.

Proof. We want the conditional density function of $\boldsymbol{T} = (T_1, \ldots, T_n)$ given $I = I(t) = n$. First note that $X_1, \ldots, X_n$ are independent exponential variables with parameters ν so that

$$f_{\boldsymbol{X}}(\boldsymbol{x}) = \nu^n \exp\left(-\nu \sum_1^n x_i\right).$$

Make the transformation $\boldsymbol{X} \mapsto \boldsymbol{T}$ and use the change of variable formula (4.7.4) to find that

$$f_{\boldsymbol{T}}(\boldsymbol{t}) = \nu^n \exp(-\nu t_n) \quad \text{if} \quad t_1 < t_2 < \cdots < t_n.$$

Let $C \subset \mathbb{R}^n$. Then

(8) $$\mathbf{P}(\boldsymbol{T} \in C \mid I = n) = \frac{\mathbf{P}(I = n \text{ and } \boldsymbol{T} \in C)}{\mathbf{P}(I = n)},$$

but

(9) $$\begin{aligned}\mathbf{P}(I = n \text{ and } \boldsymbol{T} \in C) &= \int_C \mathbf{P}(I = n \mid \boldsymbol{T} = \boldsymbol{t}) f_{\boldsymbol{T}}(\boldsymbol{t})\, d\boldsymbol{t} \\ &= \int_C \mathbf{P}(I = n \mid T_n = t_n) f_{\boldsymbol{T}}(\boldsymbol{t})\, d\boldsymbol{t}\end{aligned}$$

and

(10) $$\mathbf{P}(I = n \mid T_n = t_n) = \mathbf{P}(X_{n+1} > t - t_n) = \exp[-\nu(t - t_n)]$$

so long as $t_n \leqslant t$. Substitute (10) into (9) and (9) into (8) to obtain

$$\mathbf{P}(\boldsymbol{T} \in C \mid I = n) = \int_C L(\boldsymbol{t}) n!\, t^{-n}\, d\boldsymbol{t}$$

where

$$L(\boldsymbol{t}) = \begin{cases} 1 & \text{if} \quad t_1 < t_2 < \cdots < t_n \\ 0 & \text{otherwise.} \end{cases}$$

We recognize $g(t) = L(t)n!\,t^{-n}$ from the result of Problem (4.11.23) as the joint density function of the order statistics of n independent uniform variables on $[0, t]$. ■

We are now ready to describe $Y(t)$ in terms of the constituent processes N_i.

(11) **Theorem.** *If $N(t)$ has generating function $G_N(s, t) = \mathbf{E}(s^{N(t)})$ then the generating function $G(s, t) = \mathbf{E}(s^{Y(t)})$ of $Y(t)$ satisfies*

$$G(s, t) = G_N(s, t)\exp\left(\nu \int_0^t [G_N(s, u) - 1]\,\mathrm{d}u\right).$$

Proof. Let $U_1, U_2, \dots$ be a sequence of independent uniform variables on $[0, t]$. By (6),

$$\mathbf{E}(s^{Y(t)}) = \mathbf{E}(s^{N_0(t) + N_1(t - T_1) + \cdots + N_I(t - T_I)})$$

where $I = I(t)$. By independence, conditional expectation, and (7),

(12)
$$\begin{aligned}\mathbf{E}(s^{Y(t)}) &= \mathbf{E}(s^{N_0(t)})\mathbf{E}(\mathbf{E}(s^{N_1(t-T_1)+\cdots+N_I(t-T_I)} \mid I))\\ &= G_N(s, t)\mathbf{E}(\mathbf{E}(s^{N_1(t-U_1)+\cdots+N_I(t-U_I)} \mid I))\\ &= G_N(s, t)\mathbf{E}([\mathbf{E}(s^{N_1(t-U_1)})]^I).\end{aligned}$$

However,

(13)
$$\begin{aligned}\mathbf{E}(s^{N_1(t-U_1)}) &= \mathbf{E}(\mathbf{E}(s^{N_1(t-U_1)} \mid U_1))\\ &= \int_0^t \frac{1}{t} G_N(s, t-u)\,\mathrm{d}u = H(s, t), \quad \text{say,}\end{aligned}$$

and

(14)
$$\mathbf{E}(H^I) = \sum_{k=0}^{\infty} H^k \frac{(\nu t)^k}{k!} \mathrm{e}^{-\nu t} = \exp[\nu t(H - 1)].$$

Substitute (13) and (14) into (12) to obtain the result. ■ ●

(15) **Branching random walk.** Another characteristic of many interesting populations is their distribution about the space which they inhabit. We introduce this spatial aspect gently, by assuming that each individual lives at some point on the real line. (With the help of our imagination, this may seem to be a fair description of a sewer, river, or hedge.) Let us suppose that the evolution proceeds as follows. After its birth, a typical individual inhabits a randomly determined spot X in $\mathbb{R}$ for a random time T. After this time has elapsed it dies, leaving behind a family containing N offspring which it distributes at points $X + Y_1, X + Y_2, \dots, X + Y_N$ where $Y_1, Y_2, \dots$ are independent and identically distributed. These individuals then behave as their ancestor did, producing the next generation offspring after random times at points $X + Y_i + Y_{ij}$, where Y_{ij} is the displacement of the jth offspring of the ith individual, and the Y_{ij} are independent and identically distributed. We shall be interested in the way that living individuals are distributed about $\mathbb{R}$ at some time t.

Suppose that the process begins with a single newborn individual at the point 0. We require some notation. Write $G_N(s)$ for the generating function of a typical family size N and let F be the distribution function of a typical Y. Let $Z(x, t)$ be the number of living individuals at points in the interval $(-\infty, x]$ at time t. We shall study the generating function

$$G(s; x, t) = \mathbf{E}(s^{Z(x,t)}).$$

Let T be the lifetime of the initial individual, N its family size, and $Y_1, Y_2, \dots, Y_N$ the positions of its offspring. We shall condition Z on all these variables to obtain a type of backward equation. We must be careful about the order in which we do this conditioning, for the length of the sequence $Y_1, Y_2, \dots$ depends on N. Hold your breath, and note from (4.11.29) that $G(s; x, t) = \mathbf{E}(\mathbf{E}(\mathbf{E}(\mathbf{E}(s^Z \mid T, N, \mathbf{Y}) \mid T, N) \mid T))$ or

$$G(s; x, t) = \mathbf{E}_T\left(\mathbf{E}_N\left(\mathbf{E}_{\mathbf{Y}}(\mathbf{E}(s^Z \mid T, N, \mathbf{Y}) \mid N)\right)\right)$$

where $\mathbf{E}_T$, for example, denotes the operation of averaging over the variable T. Clearly

$$Z(x, t) = \begin{cases} Z(x, 0) & \text{if} \quad T > t \\ \sum_{i=1}^{N} Z_i(x - Y_i, t - T) & \text{if} \quad T \leqslant t \end{cases}$$

where the processes $Z_1, Z_2, \dots$ are independent copies of Z. Hence

$$\mathbf{E}(s^Z \mid T, N, \mathbf{Y}) = \begin{cases} G(s; x, 0) & \text{if} \quad T > t \\ \sum_{i=1}^{N} G(s; x - Y_i, t - T) & \text{if} \quad T \leqslant t. \end{cases}$$

Thus, if $T \leqslant t$ then

$$\begin{aligned} \mathbf{E}_N\left(\mathbf{E}_{\mathbf{Y}}(\mathbf{E}(s^Z \mid T, N, \mathbf{Y}) \mid N)\right) &= \mathbf{E}_N\left(\left(\int_{-\infty}^{\infty} G(s; x - y, t - T)\, \mathrm{d}F(y)\right)^N\right) \\ &= G_N\left(\int_{-\infty}^{\infty} G(s; x - y, t - T)\, \mathrm{d}F(y)\right). \end{aligned}$$

Now breathe again. We consider here only the Markovian case when T is exponentially distributed with some parameter μ. Then

$$G(s; x, t) = \int_0^t \mu\, \mathrm{e}^{-\mu u} G_N\left(\int_{-\infty}^{\infty} G(s; x - y, t - u)\, \mathrm{d}F(y)\right) \mathrm{d}u + \mathrm{e}^{-\mu t} G(s; x, 0).$$

Substitute $v = t - u$ inside the integral and differentiate with respect to t to obtain

$$\frac{\partial G}{\partial t} + \mu G = \mu G_N\left(\int_{-\infty}^{\infty} G(s; x - y, t)\, \mathrm{d}F(y)\right).$$

It is not immediately clear that this is useful. However, differentiate with respect to s at $s = 1$ to find that

$$m(x, t) = \mathbf{E}(Z(x, t))$$

satisfies

$$\frac{\partial m}{\partial t} + \mu m = \mu \mathbf{E}(N) \int_{-\infty}^{\infty} m(x - y, t)\, \mathrm{d}F(y)$$

which is approachable by Laplace transform techniques. Such results can easily be generalized to higher dimensions. ●

(16) **Spatial growth.** Here is a simple model for skin cancer. Suppose that each point (x, y) of the two-dimensional square lattice $\{(x, y): x, y = 0, \pm 1, \pm 2, \dots\}$ is a skin cell. There are two types of cell, called b-cells (*benign* cells) and m-cells (*malignant* cells). Each cell lives for an exponentially distributed period of time, parameter β for b-cells and parameter μ for m-cells, after which it splits into two similar cells, one of which remains at the point of division and the other displaces one of the four nearest neighbours, each chosen at random with probability $\frac{1}{4}$. The displaced cell moves out of the system. Thus there are two competing types of cell. We assume that m-cells divide at least as fast as b-cells; the ratio

$$\kappa = \mu/\beta \geqslant 1$$

is the 'carcinogenic advantage'.

Suppose that there is only one m-cell initially and that all other cells are benign. What happens to the resulting tumour of malignant cells?

(17) **Theorem.** *If $\kappa = 1$ then the m-cells die out with probability 1, but the mean time until extinction is infinite. If $\kappa > 1$ then there is probability κ^{-1} that the m-cells die out, and probability $1 - \kappa^{-1}$ that their number grows beyond all bounds.*

Thus there is strictly positive probability of the malignant cells becoming significant if and only if the carcinogenic advantage exceeds one.

Proof. Let $X(t)$ be the number of m-cells at time t, and let T_0 $(=0)$, $T_1, T_2, \dots$ be the sequence of times at which X changes its value. Consider the imbedded discrete-time process $X = \{X_n\}$, where

$$X_n = X(T_n +)$$

is the number of m-cells just after the nth transition; X is a Markov chain taking values in $\{0, 1, 2, \dots\}$. Remember the imbedded random walk of the birth–death process (6.11.9); in the case under consideration a little thought shows that X has transition probabilities

$$p_{i,i+1} = \frac{\mu}{\mu + \beta} = \frac{\kappa}{\kappa + 1}, \qquad p_{i,i-1} = \frac{1}{\kappa + 1} \quad \text{if} \quad i \neq 0, \qquad p_{0,0} = 1.$$

Therefore X_n is just a random walk with parameter $p = \kappa/(\kappa + 1)$ and with an absorbing barrier at 0. The probability of ultimate extinction from the starting point $X(0) = 1$ is κ^{-1}. The walk is symmetric null persistent if $\kappa = 1$ and all non-zero states are transient if $\kappa > 1$. ■

If $\kappa = 1$ then the same argument shows that the m-cells certainly die out whenever there is a finite number of them to start with. However, suppose that they are distributed initially at the points of some (possibly infinite) set. It is possible to decide what happens after a long length of time; roughly speaking this depends on the relative densities of benign and malignant cells over large distances. One striking result is the following.

(18) **Theorem.** *If* $\kappa = 1$, *the probability that a specified finite collection of points contains only one type of cell approaches one as* $t \to \infty$.

Sketch proof. If two cells have a common ancestor then they are of the same type. Since offspring displace any neighbour with equal probability, the line of ancestors of any cell performs a symmetric random walk in two dimensions stretching backwards in time. Therefore, given any two cells at time t, the probability that they have a common ancestor is the probability that two symmetric and independent random walks S_1 and S_2 which originate at these points have met by time t. The difference $S_1 - S_2$ is also a type of symmetric random walk, and, as in Theorem (5.10.17), $S_1 - S_2$ almost certainly visits the origin sooner or later, implying that $\mathbf{P}(S_1(t) = S_2(t) \text{ for some } t) = 1$. ■

These large connected groups of cells of the same type may be called 'empires'. ●

(19) **Simple queue.** Here is a simple model for a queueing system. Customers enter a shop like a Poisson process, parameter λ. They are served in the order of their arrival by a single assistant; each service period is a random variable which we assume to be exponential with parameter μ and which is independent of all other considerations. Let $X(t)$ be the length of the waiting line at time t. It is easy to see that X is a birth–death process with parameters

$$\lambda_n = \lambda \ (n \geqslant 0), \qquad \mu_n = \mu \ (n \geqslant 1).$$

The server would be very unhappy indeed if the queue length $X(t)$ were to tend to infinity as $t \to \infty$, since then he or she would have very few tea breaks. It is not difficult to see that the distribution of $X(t)$ settles down to a limit distribution, as $t \to \infty$, if and only if $\lambda < \mu$, which is to say that arrivals occur more slowly than departures on average (see condition (6.11.3)). We consider this process in detail in Chapter 11, together with other more complicated queueing models. The techniques of this chapter find many applications there. ●

6.13 Problems

1. Classify the states of the discrete-time Markov chains with state space $S = \{1, 2, 3, 4\}$ and transition matrices

$$\text{(a)} \begin{pmatrix} \frac{1}{3} & \frac{2}{3} & 0 & 0 \\ \frac{1}{2} & \frac{1}{2} & 0 & 0 \\ \frac{1}{4} & 0 & \frac{1}{4} & \frac{1}{2} \\ 0 & 0 & 0 & 1 \end{pmatrix} \qquad \text{(b)} \begin{pmatrix} 0 & \frac{1}{2} & \frac{1}{2} & 0 \\ \frac{1}{3} & 0 & 0 & \frac{2}{3} \\ 1 & 0 & 0 & 0 \\ 0 & 0 & 1 & 0 \end{pmatrix}.$$

In case (a), calculate $f_{34}(n)$, and deduce that the probability of ultimate absorption in state 4, starting from 3, equals $\frac{2}{3}$. Find the mean recurrence times of the states in case (b).

2. A transition matrix is called *doubly stochastic* if all its column sums equal 1: that is, if $\sum_i p_{ij} = 1$ for all $j \in S$.

   (a) Show that if a finite chain has a doubly stochastic transition matrix, then all its states are non-null persistent, and that if it is, in addition, irreducible and aperiodic then $p_{ij}(n) \to N^{-1}$ as $n \to \infty$, where N is the number of states.
   (b) Show that, if an infinite irreducible chain has a doubly stochastic transition matrix, then its states are either all null persistent or all transient.

3. Prove that intercommunicating states of a Markov chain have the same period.

4. (a) Show that for each pair i, j of states of an irreducible aperiodic chain, there exists $N = N(i, j)$ such that $p_{ij}(n) > 0$ for all $n \geqslant N$.
   (b) Let X and Y be independent irreducible aperiodic chains with the same state space S and transition matrix $\boldsymbol{P}$. Show that the bivariate chain $Z_n = (X_n, Y_n)$, $n \geqslant 0$, is irreducible and aperiodic.
   (c) Show that the bivariate chain Z may be reducible if X and Y are periodic.

5. Suppose $\{X_n: n \geqslant 0\}$ is a discrete-time Markov chain with $X_0 = i$. Let N be the total number of visits made subsequently by the chain to the state j. Show that

$$\mathbf{P}(N = n) = \begin{cases} 1 - f_{ij} & \text{if } n = 0 \\ f_{ij}(f_{jj})^{n-1}(1 - f_{jj}) & \text{if } n \geqslant 1, \end{cases}$$

and deduce that $\mathbf{P}(N = \infty) = 1$ if and only if $f_{ij} = f_{jj} = 1$.

6. Let i and j be two states of a discrete-time Markov chain. Show that if i communicates with j, then there is positive probability of reaching j from i without revisiting i in the meantime. Deduce that, if the chain is irreducible and persistent, then the probability f_{ij} of ever reaching j from i equals 1 for all i and j.

7. Let $\{X_n: n \geqslant 0\}$ be a persistent irreducible discrete-time Markov chain on the state space S with transition matrix $\boldsymbol{P}$, and let $\boldsymbol{x}$ be a positive solution of the equation $\boldsymbol{x} = \boldsymbol{x}\boldsymbol{P}$.

   (a) Show that

$$q_{ij}(n) = \frac{x_j}{x_i} p_{ji}(n), \quad i, j \in S, \; n \geqslant 1,$$

defines the n-step transition probabilities of a persistent irreducible Markov chain on S whose first-passage probabilities are given by

$$g_{ij}(n) = \frac{x_j}{x_i} l_{ji}(n), \quad i \neq j,\ n \geqslant 1,$$

where $l_{ji}(n) = \mathbf{P}(X_n = i, T > n \mid X_0 = j)$ and $T = \min\{m > 0\colon X_m = j\}$.

(b) Show that $\boldsymbol{x}$ is unique up to a multiplicative constant.

8. The sequence $u = \{u_n\colon n \geqslant 0\}$ is called a 'renewal sequence' if

$$u_0 = 1, \qquad u_n = \sum_{i=1}^{n} f_i u_{n-i} \quad \text{for} \quad n \geqslant 1,$$

for some collection $f = \{f_n\colon n \geqslant 1\}$ of non-negative numbers summing to 1.

(a) Show that u is a renewal sequence if and only if there exists a Markov chain X on a countable state space S such that $u_n = \mathbf{P}(X_n = s \mid X_0 = s)$, for some persistent $s \in S$ and all $n \geqslant 1$.

(b) Show that if u and v are renewal sequences then so is $\{u_n v_n\colon n \geqslant 0\}$.

9. Consider the symmetric random walk in three dimensions on the set of points $\{(x, y, z)\colon x, y, z = 0, \pm 1, \pm 2, \ldots\}$; this process is a sequence $\{X_n\colon n \geqslant 0\}$ of points such that $\mathbf{P}(X_{n+1} = X_n + \varepsilon) = \frac{1}{6}$ for $\varepsilon = (\pm 1, 0, 0), (0, \pm 1, 0), (0, 0, \pm 1)$. Suppose that $X_0 = (0, 0, 0)$. Show that

$$\mathbf{P}(X_{2n} = (0,0,0)) = \left(\frac{1}{6}\right)^{2n} \sum_{i+j+k=n} \frac{(2n)!}{(i!\,j!\,k!)^2}$$

$$= \left(\frac{1}{2}\right)^{2n} \binom{2n}{n} \sum_{i+j+k=n} \left(\frac{n!}{3^n i!\,j!\,k!}\right)^2$$

and deduce by Stirling's formula that the origin is a transient state.

10. Consider the three-dimensional version of the cancer model (6.12.16). If $\kappa = 1$, are the empires inevitable in this case?

11. Let X_n be a discrete-time Markov chain with state space $S = \{1, 2\}$, and transition matrix

$$\boldsymbol{P} = \begin{pmatrix} 1-\alpha & \alpha \\ \beta & 1-\beta \end{pmatrix}.$$

Classify the states of the chain. Suppose that $\alpha\beta > 0$ and $\alpha\beta \neq 1$. Find the n-step transition probabilities and show directly that they converge to the unique stationary distribution as $n \to \infty$. For what values of α and β is the chain time-reversible in equilibrium?

12. **Another diffusion model.** N black balls and N white balls are placed in two urns so that each contains N balls. After each unit of time one ball is selected at random from each urn, and the two balls thus selected are interchanged. Let the number of black balls in the first urn denote the state of the system. Write down the transition matrix of this Markov chain and find the unique stationary distribution. Is the chain time-reversible in equilibrium?

13. Consider a Markov chain on the set $S = \{0, 1, 2, \ldots\}$ with transition probabilities $p_{i,i+1} = a_i$, $p_{i,0} = 1 - a_i$, $i \geq 0$, where $(a_i: i \geq 0)$ is a sequence of constants which satisfy $0 < a_i < 1$ for all i. Let $b_0 = 1$, $b_i = a_0 a_1 \cdots a_{i-1}$ for $i \geq 1$. Show that the chain is

(a) persistent if and only if $b_i \to 0$ as $i \to \infty$,
(b) non-null persistent if and only if $\sum_i b_i < \infty$,

and write down the stationary distribution if the latter condition holds.

Let A and β be positive constants and suppose that $a_i = 1 - Ai^{-\beta}$ for all large i. Show that the chain is

(c) transient if $\beta > 1$,
(d) non-null persistent if $\beta < 1$.

Finally, if $\beta = 1$ show that the chain is

(e) non-null persistent if $A > 1$,
(f) null persistent if $A \leq 1$.

14. Let X be a continuous-time Markov chain with countable state space S and standard semigroup $\{\boldsymbol{P}_t\}$. Show that $p_{ij}(t)$ is a continuous function of t. Let $g(t) = -\log p_{ii}(t)$; show that g is a continuous function, $g(0) = 0$, and $g(s + t) \leq g(s) + g(t)$. We say that g is 'subadditive', and a well-known theorem gives the result that

$$\lim_{t \downarrow 0} \frac{g(t)}{t} = \lambda \quad \text{exists and} \quad \lambda = \sup_{t > 0} \frac{g(t)}{t} \leq \infty.$$

Deduce that $g_{ii} = \lim_{t \downarrow 0} t^{-1}\{p_{ii}(t) - 1\}$ exists, but may be $-\infty$.

15. Let X be a continuous-time Markov chain with generator $\boldsymbol{G} = (g_{ij})$ and suppose that the transition semigroup $\boldsymbol{P}_t$ satisfies $\boldsymbol{P}_t = \exp(t\boldsymbol{G})$. Show that X is irreducible if and only if for any pair i, j of states there exists a sequence $k_1, k_2, \ldots, k_n$ of states such that $g_{i,k_1} g_{k_1,k_2} \cdots g_{k_n,j} \neq 0$.

16. (a) Let $X = \{X(t): -\infty < t < \infty\}$ be a Markov chain with stationary distribution $\boldsymbol{\pi}$, and suppose that $X(0)$ has distribution $\boldsymbol{\pi}$. We call X *time-reversible* if X and Y have the same joint distributions, where $Y(t) = X(-t)$. If the transition semigroup $\{\boldsymbol{P}_t\}$ of X is standard with generator $\boldsymbol{G}$, show that $\pi_i g_{ij} = \pi_j g_{ji}$ (for all i and j) is a necessary condition for X to be time-reversible. If $\boldsymbol{P}_t = \exp(t\boldsymbol{G})$, show that $X(t)$ has distribution $\boldsymbol{\pi}$ for all t and that the above condition is sufficient for the chain to be time-reversible.
(b) Show that every irreducible chain X with exactly two states is time-reversible in equilibrium.
(c) Show that every birth–death process X having a stationary distribution is time-reversible in equilibrium.

17. Show that not every discrete-time Markov chain can be imbedded in a continuous-time chain. More precisely, let

$$\boldsymbol{P} = \begin{pmatrix} \alpha & 1 - \alpha \\ 1 - \alpha & \alpha \end{pmatrix} \quad \text{for some } 0 < \alpha < 1$$

be a transition matrix. Show that there exists a uniform semigroup $\{\boldsymbol{P}_t\}$ of transition probabilities in continuous time such that $\boldsymbol{P}_1 = \boldsymbol{P}$, if and only if $\frac{1}{2} < \alpha < 1$. In this case show that $\{\boldsymbol{P}_t\}$ is unique and calculate it in terms of α.

18. Consider an immigration–death process $X(t)$, being a birth–death process with rates $\lambda_n = \lambda$, $\mu_n = n\mu$. Show that its generating function $G(s, t) = \mathbf{E}(s^{X(t)})$ is given by

$$G(s, t) = \{1 + (s - 1)\,e^{-\mu t}\}^I \exp\{\rho(s - 1)(1 - e^{-\mu t})\}$$

where $\rho = \lambda/\mu$ and $X(0) = I$. Deduce the limiting distribution of $X(t)$ as $t \to \infty$.

19. A *non-homogeneous* Poisson process is a process $N(t)$ which is defined in the same way as a Poisson process except in that the likelihood of an arrival in $(t, t + h)$ is $\lambda(t)h + o(h)$ where $\lambda(t)$ is a smooth function of t. Write down the forward and backward equations for N, and solve them.

Let $N(0) = 0$, and find the density function of the time T until the first arrival in the process. If $\lambda(t) = c/(1 + t)$, show that $\mathbf{E}T < \infty$ if and only if $c > 1$.

20. Successive offers for my house are independent identically distributed random variables $X_1, X_2, \ldots$, having density function f and distribution function F. Let $Y_1 = X_1$, let Y_2 be the first offer exceeding Y_1, and generally let Y_{n+1} be the first offer exceeding Y_n. Show that $Y_1, Y_2, \ldots$ are the times of arrivals in a non-homogeneous Poisson process with intensity function $\lambda(t) = f(t)/(1 - F(t))$. The Y_i are called 'record values'.

Now let Z_1 be the first offer received which is the second largest to date, and let Z_2 be the second such offer, and so on. Show that the Z_i are the arrival times of a non-homogeneous Poisson process with intensity function $\lambda(t)$.

21. Let N be a Poisson process with intensity λ, and let $Y_1, Y_2, \ldots$ be independent random variables with common characteristic function ϕ and density function f. The process

$$N^*(t) = \sum_{n=1}^{N(t)} Y_n$$

is called a *compound* Poisson process. Y_n is the change in the value of N^* at the nth arrival of the Poisson process N. Think of it like this. A 'random alarm clock' rings at the arrival times of a Poisson process. At the nth ring the process N^* accumulates an extra quantity Y_n. Write down a forward equation for N^* and hence find the characteristic function of $N^*(t)$. Can you see directly why it has the form which you have found?

22. If the intensity function $\lambda(t)$ of a non-homogeneous Poisson process N is itself a random process, then N is called a *doubly stochastic* Poisson process (or *Cox process*). Consider the case when $\lambda(t) = \Lambda$ for all t, and Λ is a random variable taking either of two values λ_1 or λ_2, each being picked with equal probability $\frac{1}{2}$. Find the probability generating function of $N(t)$, and deduce its mean and variance.

23. Show that a simple birth process X with parameter λ is a doubly stochastic Poisson process with intensity function $\lambda(t) = \lambda X(t)$.

24. The Markov chain $X = \{X(t): t \geqslant 0\}$ is a birth process whose intensities $\lambda_k(t)$ depend also on the time t and are given by

$$\mathbf{P}\big(X(t + h) = k + 1 \mid X(t) = k\big) = \frac{1 + \mu k}{1 + \mu t}\, h + o(h)$$

as $h \downarrow 0$. Show that the probability generating function $G(s, t) = \mathbf{E}(s^{X(t)})$ satisfies

$$\frac{\partial G}{\partial t} = \frac{s-1}{1+\mu t}\left\{G + \mu s \frac{\partial G}{\partial s}\right\}, \qquad 0 < s < 1.$$

Hence find the mean and variance of $X(t)$ when $X(0) = I$.

25. Let X be a birth–death process with strictly positive birth rates $\lambda_0, \lambda_1, \ldots$ and death rates $\mu_1, \mu_2, \ldots$. Let η_i be the probability that $X(t)$ ever takes the value 0 starting from $X(0) = i$. Show that

$$\lambda_j \eta_{j+1} - (\lambda_j + \mu_j)\eta_j + \mu_j \eta_{j-1} = 0, \qquad j \geqslant 1,$$

and deduce that $\eta_i = 1$ for all i so long as $\sum_1^\infty e_j = \infty$ where $e_j = \mu_1\mu_2 \cdots \mu_j/(\lambda_1\lambda_2 \cdots \lambda_j)$.

26. Find a good necessary condition and a good sufficient condition for the birth–death process $X(t)$ of Problem (25) to be honest.

27. Let X be a simple symmetric birth–death process with $\lambda_n = \mu_n = n\lambda$, and let T be the time until extinction. Show that

$$\mathbf{P}(T \leqslant x \mid X(0) = I) = \left(\frac{\lambda x}{1 + \lambda x}\right)^I,$$

and deduce that extinction is certain if $\mathbf{P}(X(0) < \infty) = 1$.

Show that $\mathbf{P}(\lambda T/I \leqslant x \mid X(0) = I) \to e^{-1/x}$ as $I \to \infty$.

28. **Immigration–death with disasters.** Let X be an immigration–death–disaster process, that is, a birth–death process with parameters $\lambda_i = \lambda$, $\mu_i = i\mu$, and with the additional possibility of 'disasters' which reduce the population to 0. Disasters occur at the times of a Poisson process with intensity δ, independently of all previous births and deaths.

(a) Show that X has a stationary distribution, and find an expression for the generating function of this distribution.
(b) Show that, in equilibrium, the mean of $X(t)$ is $\lambda/(\delta + \mu)$.

29. With any sufficiently nice (Lebesgue-measurable, say) subset B of the real line $\mathbb{R}$ is associated a random variable $X(B)$ such that

(i) $X(B)$ takes values in $\{0, 1, 2, \ldots\}$,
(ii) if $B_1, B_2, \ldots, B_n$ are disjoint then $X(B_1), X(B_2), \ldots, X(B_n)$ are independent, and $X(B_1 \cup B_2) = X(B_1) + X(B_2)$,
(iii) the distribution of $X(B)$ depends only on B through its Lebesgue measure ('length') $|B|$, and

$$\frac{\mathbf{P}(X(B) \geqslant 1)}{\mathbf{P}(X(B) = 1)} \to 1 \quad \text{as} \quad |B| \to 0.$$

Show that X is a Poisson process.

30. **Spatial Poisson process.** Let $\lambda\colon \mathbb{R}^n \to [0, \infty)$ be an integrable function. An *n-dimensional Poisson process* X with intensity function λ is defined as follows. To each nice subset (say Lebesgue-measurable) B of $\mathbb{R}^n$ is associated a random variable $X(B)$ such

that

(i) $X(B)$ has the Poisson distribution with parameter $\int_B \lambda(\boldsymbol{u})\,d\boldsymbol{u}$,
(ii) if $B_1, B_2, \dots, B_n$ are disjoint then $X(B_1), X(B_2), \dots, X(B_n)$ are independent, and $X(B_1 \cup B_2) = X(B_1) + X(B_2)$.

Let $g: \mathbb{R}^n \to \mathbb{R}^n$ be measurable with the property that $g^{-1}(B)$ has Lebesgue measure 0 whenever B has Lebesgue measure 0. Show that $Y(B) = X(g^{-1}(B))$ defines a spatial Poisson process, and express its intensity function in terms of λ.

31. Let X be an n-dimensional Poisson process with constant intensity λ. Show that the volume of the largest (n-dimensional) sphere centred at the origin which contains no point of X is exponentially distributed. Deduce the density function of the distance R from the origin to the nearest point of X.

32. A village of $N + 1$ people suffers an epidemic. Let $X(t)$ be the number of ill people at time t, and suppose that $X(0) = 1$ and X is a birth process with rates $\lambda_i = \lambda i(N + 1 - i)$. Let T be the length of time required until every member of the population has succumbed to the illness. Show that

$$\mathbf{E}(T) = \frac{1}{\lambda}\sum_{k=1}^{N} \frac{1}{k(N+1-k)}$$

and deduce that

$$\mathbf{E}(T) = \frac{2(\log N + \gamma)}{\lambda(N+1)} + O(N^{-2})$$

where γ is Euler's constant. It is striking that $\mathbf{E}(T)$ decreases with N, for large N.

33. A particle has velocity $V(t)$ at time t, where $V(t)$ is assumed to take values in $\{n + \frac{1}{2}: n \geq 0\}$. Transitions during $(t, t + h)$ are possible as follows:

$$\mathbf{P}(V(t+h) = w \mid V(t) = v) = \begin{cases} (v + \frac{1}{2})h + o(h) & \text{if } w = v + 1, \\ 1 - 2vh + o(h) & \text{if } w = v, \\ (v - \frac{1}{2})h + o(h) & \text{if } w = v - 1. \end{cases}$$

Initially $V(0) = \frac{1}{2}$. Let

$$G(s, t) = \sum_{n=0}^{\infty} s^n \mathbf{P}(V(t) = n + \tfrac{1}{2}).$$

(a) Show that

$$\frac{\partial G}{\partial t} = (1 - s)^2 \frac{\partial G}{\partial s} - (1 - s)G$$

and deduce that $G(s, t) = \{1 + (1 - s)t\}^{-1}$.

(b) Show that the expected length $m_n(T)$ of time for which $V = n + \frac{1}{2}$ during the time interval $[0, T]$ is given by

$$m_n(T) = \int_0^T \mathbf{P}(V(t) = n + \tfrac{1}{2})\,dt$$

and that, for fixed k,

$$m_k(T) - \log T \to -\sum_{i=1}^{k} \frac{1}{i} \quad \text{as} \quad T \to \infty.$$

(c) What is the expected velocity of the particle at time t?

7 Convergence of random variables

7.1 Introduction

Expressions such as 'in the long run' and 'on the average' are common-place in everyday usage, and express our faith that the averages of the results of repeated experimentation show less and less random fluctuation as they settle down to some limit.

(1) **Example. Buffon's needle (4.5.8).** In order to estimate the numerical value of π, Buffon devised the following experiment. Fling a needle a large number n of times on to a ruled plane and count the number S_n of times that the needle meets a line. In accordance with the result of (4.5.8), the proportion S_n/n of intersections is indeed found to be near to the probability $2/\pi$. Thus $X_n = 2n/S_n$ is a plausible estimate for π; this estimate converges as $n \to \infty$, and it seems reasonable to write

$$X_n \to \pi \quad \text{as} \quad n \to \infty. \qquad \bullet$$

(2) **Example.** Any number y satisfying $0 \leqslant y < 1$ has a decimal expansion

$$y = 0 \cdot y_1 y_2 \cdots = \sum_{j=1}^{\infty} y_j 10^{-j},$$

where each y_j takes some value in $\{0, 1, 2, \ldots, 9\}$. Now think of y_j as the outcome of a random variable Y_j where $\{Y_j\}$ is a family of independent variables each of which may take any value in $\{0, 1, 2, \ldots, 9\}$ with equal probability $\frac{1}{10}$. The quantity

$$Y = \sum_{j=1}^{\infty} Y_j 10^{-j}$$

is a random variable taking values in $[0, 1]$. It seems likely that Y is uniformly distributed on $[0, 1]$, and this turns out to be the case (see Problem (7.11.4)). More rigorously, this amounts to asserting that the sequence $\{X_n\}$ given by

$$X_n = \sum_{j=1}^{n} Y_j 10^{-j}$$

converges in some sense as $n \to \infty$ to a limit Y, and that this limit random variable is uniform on $[0, 1]$. $\bullet$

In both these examples we encountered a sequence $\{X_n\}$ of random variables together with the assertion that

(3) $$X_n \to X \quad \text{as} \quad n \to \infty$$

for some other random variable X. However, random variables are real-valued functions on some sample space, and so (3) is a statement about the convergence of a sequence of *functions*. It is not immediately clear how such convergence is related to our experience of the theory of convergence of sequences $\{x_n\}$ of real numbers, and so we digress briefly to discuss sequences of functions.

Suppose for example that $f_1(\cdot)$, $f_2(\cdot), \ldots$ is a sequence of functions mapping $[0, 1]$ into $\mathbb{R}$. In what manner may they converge to some limit function f?

(4) **Convergence pointwise.** If, for all $x \in [0, 1]$, the sequence $\{f_n(x)\}$ of real numbers satisfies

$$f_n(x) \to f(x) \quad \text{as} \quad n \to \infty,$$

then we say that $f_n \to f$ *pointwise.* ●

(5) **Norm convergence.** Let V be a collection of functions mapping $[0, 1]$ into $\mathbb{R}$. Subject to certain conditions on the members of V, we can endow V with a function $\|\cdot\|: V \to \mathbb{R}$ satisfying

(a) $\|f\| \geqslant 0$ for all $f \in V$
(b) $\|f\| = 0$ if and only if f is the zero function (or equivalent to it, in some sense to be specified)
(c) $\|af\| = |a| \, \|f\|$ for all $a \in \mathbb{R}$, $f \in V$
(d) $\|f + g\| \leqslant \|f\| + \|g\|$ (this is called the *triangle inequality*).

The function $\|\cdot\|$ is called a *norm.* If $\{f_n\}$ is a sequence of members of V then we say that $f_n \to f$ *with respect to this norm* if

$$\|f_n - f\| \to 0 \quad \text{as} \quad n \to \infty.$$

Certain special and important norms are given by

$$\|g\|_p = \left(\int_0^1 |g(x)|^p \, dx \right)^{1/p}$$

for $p \geqslant 1$ and any suitable function g. ●

(6) **Convergence in measure.** Let $\varepsilon > 0$ be prescribed, and define the 'distance' between two functions g and h by

$$d_\varepsilon(g, h) = \int_E dx$$

where $E = \{u \in [0, 1]: |g(u) - h(u)| > \varepsilon\}$. We say that $f_n \to f$ *in measure* if

$$d_\varepsilon(f_n, f) \to 0 \quad \text{as} \quad n \to \infty \quad \text{for all } \varepsilon > 0. \qquad \bullet$$

The convergence of $\{f_n\}$ according to one definition does not necessarily imply its convergence according to another. For example, we shall see later that

(a) if $f_n \to f$ pointwise then $f_n \to f$ in measure, but the converse is not generally true

(b) there exist sequences which converge pointwise but not with respect to $\|\cdot\|_1$, and vice versa.

In this chapter we shall see how to adapt these modes of convergence to suit families of *random variables*. Major applications of the ensuing theory include the study of the sequence

(7) $$S_n = X_1 + X_2 + \cdots + X_n$$

of partial sums of an independent identically distributed sequence $\{X_i\}$; the law of large numbers of Section 5.10 will appear as a special case.

It will be clear, from our discussion and the reader's experience, that probability theory is indispensable in descriptions of many processes which occur naturally in the world. Often in such cases we are interested in the future values of the process, and thus in the long-term behaviour of the mathematical model; this is why we need to prove limit theorems for sequences of random variables. Many of these sequences are generated by less tractable operations than, say, the partial sums in (7), and general results such as the law of large numbers may not be useful. It turns out that many other types of sequence are guaranteed to converge; in particular we shall consider later the remarkable theory of 'martingales' which has important applications throughout the field of applied probability. This chapter continues with a simple account of the convergence theorem for martingales, together with some examples of its use; these include the asymptotic behaviour of the branching process and provide rigorous derivations of certain earlier remarks (such as (5.4.6)). Conditional expectation is put on a firm footing in Section 7.9.

All readers should follow the chapter up to and including Section 7.4. The subsequent material may be omitted at the first reading.

Exercises

(8) Let $r \geq 1$, and define $\|X\|_r = \{\mathbf{E}|X^r|\}^{1/r}$. Show that

(a) $\|cX\|_r = |c| \cdot \|X\|_r$ for $c \in \mathbb{R}$,

(b) $\|X + Y\|_r \leq \|X\|_r + \|Y\|_r$,

(c) $\|X\|_r = 0$ if and only if $\mathbf{P}(X = 0) = 1$.

This amounts to saying that $\|\cdot\|_r$ is a norm on the set of equivalence classes of random variables on a given probability space with finite rth mean, the equivalence relation being given by $X \sim Y$ if and only if $\mathbf{P}(X = Y) = 1$.

(9) Define $\langle X, Y\rangle = \mathbf{E}(XY)$ for random variables X and Y having finite variance, and define $\|X\| = \sqrt{\langle X, X\rangle}$. Show that

(a) $\langle aX + bY, Z\rangle = a\langle X, Z\rangle + b\langle Y, Z\rangle$,
(b) $\|X + Y\|^2 + \|X - Y\|^2 = 2(\|X\|^2 + \|Y\|^2)$, the *parallelogram property*,
(c) if $\langle X_i, X_j\rangle = 0$ for all $i \neq j$ then

$$\left\|\sum_{i=1}^{n} X_i\right\|^2 = \sum_{i=1}^{n} \|X_i\|^2.$$

(10) Let $\varepsilon > 0$. Let $g, h\colon [0, 1] \to \mathbb{R}$, and define $d_\varepsilon(g, h) = \int_E dx$ where

$$E = \{u \in [0, 1]\colon |g(u) - h(u)| > \varepsilon\}.$$

Show that d_ε does not satisfy the triangle inequality.

(11) **Lévy metric.** For two distribution functions F and G, let

$$d(F, G) = \inf\{\delta > 0\colon F(x - \delta) - \delta \leqslant G(x) \leqslant F(x + \delta) + \delta \text{ for all } x \in \mathbb{R}\}.$$

Show that d is a metric on the space of distribution functions.

(12) Find random variables $X, X_1, X_2, \ldots$ such that $\mathbf{E}(|X_n - X|^2) \to 0$ as $n \to \infty$, but $\mathbf{E}|X_n| = \infty$ for all n.

7.2 Modes of convergence

There are four principal ways of interpreting the statement '$X_n \to X$ as $n \to \infty$'. Three of these are related to (7.1.4), (7.1.5), and (7.1.6), and the fourth is already familiar to us.

(1) **Definition.** Let $X_1, X_2, \ldots, X$ be random variables on some probability space $(\Omega, \mathscr{F}, \mathbf{P})$. We say that

(a) $X_n \to X$ **almost surely**, written $X_n \xrightarrow{\text{a.s.}} X$, if $\{\omega \in \Omega\colon X_n(\omega) \to X(\omega)$ as $n \to \infty\}$ is an event whose probability is 1

(b) $X_n \to X$ **in rth mean**, where $r \geqslant 1$, written $X_n \xrightarrow{r} X$, if $\mathbf{E}|X_n^r| < \infty$ for all n and

$$\mathbf{E}(|X_n - X|^r) \to 0 \quad \text{as} \quad n \to \infty$$

(c) $X_n \to X$ **in probability**, written $X_n \xrightarrow{\text{P}} X$, if

$$\mathbf{P}(|X_n - X| > \varepsilon) \to 0 \quad \text{as} \quad n \to \infty \quad \text{for all } \varepsilon > 0$$

(d) $X_n \to X$ **in distribution**, written† $X_n \xrightarrow{\text{D}} X$, if

$$\mathbf{P}(X_n \leqslant x) \to \mathbf{P}(X \leqslant x) \quad \text{as} \quad n \to \infty$$

for all points x at which $F_X(x) = \mathbf{P}(X \leqslant x)$ is continuous.

† Many authors avoid this notation since convergence in distribution pertains only to the *distribution function* of X and not to the variable X itself. We use it here for the sake of uniformity of notation, but refer the reader to note (d) below.

It is appropriate to make some remarks about the four sections of this potentially bewildering definition.

(a) The natural adaptation of (7.1.4) is to say that $X_n \to X$ *pointwise* if the set $A = \{\omega \in \Omega: X_n(\omega) \to X(\omega) \text{ as } n \to \infty\}$ satisfies

$$A = \Omega.$$

Such a condition is of little interest to probabilists since it contains no reference to probabilities. In part (a) of (1) we do not require that A is the whole of Ω, but rather that its complement A^c is a null set. There are several notations for this mode of convergence, and we shall use these later. They include

$X_n \to X$ *almost everywhere*, or $X_n \xrightarrow{\text{a.e.}} X$
$X_n \to X$ *with probability* 1, or $X_n \to X$ w.p.l.

(b) It is easy to check by Minkowski's inequality (4.11.27) that

$$\|Y\|_r = (\mathbf{E}|Y^r|)^{1/r} = \left(\int |y|^r \, \mathrm{d}F_Y\right)^{1/r}$$

defines a norm on the collection of random variables with finite rth moment, for any value of $r \geqslant 1$. Rewrite (7.1.5) with this norm to obtain Definition (1b). Here we shall only consider positive integral values of r, though the subsequent theory can be extended without difficulty to deal with any real r not smaller than 1. Of most use are the values $r = 1$ and $r = 2$, in which cases we write respectively

$$X_n \xrightarrow{1} X, \text{ or } X_n \to X \textit{ in mean}, \text{ or l.i.m. } X_n = X$$

and

$$X_n \xrightarrow{2} X, \text{ or } X_n \to X \textit{ in mean square}, \text{ or } X_n \xrightarrow{\text{m.s.}} X.$$

(c) The functions of (7.1.6) had a common domain [0, 1]; the X have a common domain Ω, and the distance function d_ε is naturally adapted to become

$$d_\varepsilon(Y, Z) = \mathbf{P}(|Y - Z| > \varepsilon) = \int_E \mathrm{d}\mathbf{P}$$

where $E = \{\omega \in \Omega: |Y(\omega) - Z(\omega)| > \varepsilon\}$. This notation will be familiar to those readers with knowledge of the abstract integral of Section 5.6.

(d) We have seen this already in Section 5.9 where we discussed the continuity condition. Further examples of convergence in distribution are to be found in Chapter 6, where we saw, for example, that an irreducible ergodic Markov chain converges in distribution to its unique stationary distribution. Note that if $X_n \xrightarrow{\text{D}} X$ then $X_n \xrightarrow{\text{D}} X'$ for any X' which has the same distribution as X.

It is no surprise to learn that the four modes of convergence are not equivalent to each other. You may guess after some reflection that convergence in distribution is the weakest, since it is a condition only on the *distribution functions* of the X_n; it contains no reference to the *sample space* Ω and no information about, say, the dependence or independence of the X_n. The following example is a partial confirmation of this.

(2) **Example.** Let X be a Bernoulli variable taking values 0 and 1 with equal probability $\frac{1}{2}$. Let $X_1, X_2, \dots$ be identical random variables given by

$$X_n = X \quad \text{for all } n.$$

The X_n are certainly not independent, but $X_n \xrightarrow{\text{D}} X$. Let $Y = 1 - X$. Clearly $X_n \xrightarrow{\text{D}} Y$ also, since X and Y have the same distribution. However, X_n cannot converge to Y in any other mode because $|X_n - Y| = 1$ always. ●

Cauchy convergence. As in the case of sequences of real numbers, it is often convenient to work with a definition of convergence which does not make explicit reference to the limit. For example, we say that the sequence $\{X_n : n \geq 1\}$ of random variables on the probability space $(\Omega, \mathcal{F}, \mathbf{P})$ is *almost surely Cauchy convergent* if the set of points ω of the sample space for which the real sequence $\{X_n(\omega) : n \geq 1\}$ is Cauchy convergent is an event having probability 1, which is to say that

$$\mathbf{P}(\{\omega \in \Omega : X_m(\omega) - X_n(\omega) \to 0 \text{ as } m, n \to \infty\}) = 1.$$

(See Appendix I for a brief discussion of the Cauchy convergence of a sequence of real numbers.) Now, a sequence of reals converges if and only if it is Cauchy convergent. Thus $\{X_n(\omega) : n \geq 1\}$ converges if and only if it is Cauchy convergent, implying that $\{X_n : n \geq 1\}$ converges almost surely if and only if it is almost surely Cauchy convergent. Other modes of Cauchy convergence appear in Exercise (7.3.19) and Problem (7.11.11).

Here is the chart of implications between the modes of convergence. Learn it well. Statements such as

$$(X_n \xrightarrow{\text{P}} X) \Rightarrow (X_n \xrightarrow{\text{D}} X)$$

mean that any sequence which converges in probability also converges in distribution to the same limit.

(3) **Theorem.** *The following implications hold*:

$$\begin{array}{l} (X_n \xrightarrow{\text{a.s.}} X) \searrow \\ \qquad\qquad\qquad (X_n \xrightarrow{\text{P}} X) \Rightarrow (X_n \xrightarrow{\text{D}} X) \\ (X_n \xrightarrow{r} X) \nearrow \end{array}$$

for any $r \geq 1$. *Also, if* $r > s \geq 1$ *then*

$$(X_n \xrightarrow{r} X) \Rightarrow (X_n \xrightarrow{s} X).$$

No other implications hold in general.†

† But see (14).

The four basic implications of this theorem are of the general form 'if A holds, then B holds'. The converse implications are false in general, but become true if certain extra conditions are imposed; such partial converses take the form 'if B holds together with C, then A holds'. These two types of statement are sometimes said to be of the 'Abelian' and 'Tauberian' types, respectively; these titles are derived from the celebrated theory of the summability of series. Usually, there are many possible choices for appropriate sets C of extra conditions, and it is often difficult to establish attractive 'corrected converses'.

(4) **Theorem.**

(a) *If* $X_n \xrightarrow{\text{D}} c$, *where* c *is constant, then* $X_n \xrightarrow{\text{P}} c$.
(b) *If* $X_n \xrightarrow{\text{P}} X$ *and* $\mathbf{P}(|X_n| \leqslant k) = 1$ *for all* n *and some* k, *then* $X_n \xrightarrow{r} X$ *for all* $r \geqslant 1$.
(c) *If* $P_n(\varepsilon) = \mathbf{P}(|X_n - X| > \varepsilon)$ *satisfies* $\sum_n P_n(\varepsilon) < \infty$ *for all* $\varepsilon > 0$, *then* $X_n \xrightarrow{\text{a.s.}} X$.

You should become well acquainted with Theorems (3) and (4). The proofs follow as a series of lemmas. These lemmas contain some other relevant and useful results.

Consider briefly the first and principal part of Theorem (3). We may already anticipate some way of showing that convergence in probability implies convergence in distribution, since both modes involve probabilities of the form $\mathbf{P}(Y \leqslant y)$ for some random variable Y and real y. The other two implications require intermediate steps. Specifically, the relation between convergence in rth mean and convergence in probability requires a link between expectations and distributions. We have to move very carefully in this context; even apparently 'natural' statements may be false. For example, if $X_n \xrightarrow{\text{a.s.}} X$ (and therefore $X_n \xrightarrow{\text{P}} X$ also) then it does *not* necessarily follow that $\mathbf{E}X_n \to \mathbf{E}X$ (see (9) for an instance of this); this matter is explored fully in Section 7.10. The proof of the appropriate stage of Theorem (3) requires Markov's inequality (7).

(5) **Lemma.** *If* $X_n \xrightarrow{\text{P}} X$ *then* $X_n \xrightarrow{\text{D}} X$. *The converse assertion fails in general.*†

Proof. Suppose $X_n \xrightarrow{\text{P}} X$ and write

$$F_n(x) = \mathbf{P}(X_n \leqslant x), \qquad F(x) = \mathbf{P}(X \leqslant x)$$

for the distribution functions of X_n and X respectively. Then, if $\varepsilon > 0$,

$$\begin{aligned} F_n(x) = \mathbf{P}(X_n \leqslant x) &= \mathbf{P}(X_n \leqslant x,\ X \leqslant x + \varepsilon) + \mathbf{P}(X_n \leqslant x,\ X > x + \varepsilon) \\ &\leqslant F(x + \varepsilon) + \mathbf{P}(|X_n - X| > \varepsilon). \end{aligned}$$

† But see (14).

Similarly

$$F(x-\varepsilon) = \mathbf{P}(X \leqslant x-\varepsilon) = \mathbf{P}(X \leqslant x-\varepsilon,\ X_n \leqslant x) + \mathbf{P}(X \leqslant x-\varepsilon,\ X_n > x)$$
$$\leqslant F_n(x) + \mathbf{P}(|X_n - X| > \varepsilon).$$

Thus

$$F(x-\varepsilon) - \mathbf{P}(|X_n - X| > \varepsilon) \leqslant F_n(x) \leqslant F(x+\varepsilon) + \mathbf{P}(|X_n - X| > \varepsilon).$$

Let $n \to \infty$ to obtain

$$F(x-\varepsilon) \leqslant \liminf_{n\to\infty} F_n(x) \leqslant \limsup_{n\to\infty} F_n(x) \leqslant F(x+\varepsilon)$$

for all $\varepsilon > 0$. If F is continuous at x then

$$F(x-\varepsilon)\uparrow F(x) \quad \text{and} \quad F(x+\varepsilon)\downarrow F(x) \quad \text{as} \quad \varepsilon\downarrow 0,$$

and the result is proved.

Example (2) shows that the converse is false. ■

(6) **Lemma.**

(a) *If* $r > s \geqslant 1$ *and* $X_n \xrightarrow{r} X$ *then* $X_n \xrightarrow{s} X$.
(b) *If* $X_n \xrightarrow{1} X$ *then* $X_n \xrightarrow{\mathrm{P}} X$.

The converse assertions fail in general.

This includes the fact that convergence in rth mean implies convergence in probability. Here is a useful inequality which we shall use in the proof of this lemma.

(7) **Lemma. Markov's inequality.** *If X is any random variable with finite mean then*

$$\mathbf{P}(|X| \geqslant a) \leqslant \frac{\mathbf{E}|X|}{a} \quad \textit{for any } a > 0.$$

Proof. Let $A = \{|X| \geqslant a\}$. Then

$$|X| \geqslant aI_A$$

where I_A is the indicator function of A. Take expectations to obtain the result. ■

Proof of Lemma (6).

(a) By the result of Problem (4.11.28),

$$[\mathbf{E}(|X_n - X|^s)]^{1/s} \leqslant [\mathbf{E}(|X_n - X|^r)]^{1/r}$$

and the result follows immediately. To see that the converse fails, define an independent sequence $X_1, X_2, \dots$ by

(8) $$X_n = \begin{cases} n & \text{with probability } n^{-(r+s)/2} \\ 0 & \text{with probability } 1 - n^{-(r+s)/2}. \end{cases}$$

It is an easy *exercise* to check that

$$\mathbf{E}|X_n^s| = n^{(s-r)/2} \to 0, \qquad \mathbf{E}|X_n^r| = n^{(r-s)/2} \to \infty.$$

(b) By Markov's inequality (7)

$$\mathbf{P}(|X_n - X| > \varepsilon) \leqslant \frac{\mathbf{E}|X_n - X|}{\varepsilon} \quad \text{for all } \varepsilon > 0$$

and the result follows immediately. To see that the converse fails, define an independent sequence $\{X_n\}$ by

(9) $$X_n = \begin{cases} n^3 & \text{with probability } n^{-2} \\ 0 & \text{with probability } 1 - n^{-2}. \end{cases}$$

Then $\mathbf{P}(|X_n| > \varepsilon) = n^{-2}$ for all large n, and so $X_n \xrightarrow{\text{P}} 0$. However, $\mathbf{E}|X_n| = n \to \infty$. ■

(10) **Lemma.** *Let* $A_n(\varepsilon) = \{|X_n - X| > \varepsilon\}$ *and* $B_m(\varepsilon) = \bigcup_{n \geqslant m} A_n(\varepsilon)$. *Then*

(a) $X_n \xrightarrow{\text{a.s.}} X$ *if and only if* $\mathbf{P}(B_m(\varepsilon)) \to 0$ *as* $m \to \infty$, *for all* $\varepsilon > 0$

(b) $X_n \xrightarrow{\text{a.s.}} X$ *if* $\sum_n \mathbf{P}(A_n(\varepsilon)) < \infty$ *for all* $\varepsilon > 0$

(c) *if* $X_n \xrightarrow{\text{a.s.}} X$ *then* $X_n \xrightarrow{\text{P}} X$, *but the converse fails in general.*

Proof.

(a) Let $C = \{\omega \in \Omega: X_n(\omega) \to X(\omega) \text{ as } n \to \infty\}$ and let

$$A(\varepsilon) = \{\omega \in \Omega: \omega \in A_n(\varepsilon) \text{ for infinitely many values of } n\}.$$

Now $X_n(\omega) \to X(\omega)$ if and only if $\omega \notin A(\varepsilon)$ for all $\varepsilon > 0$. Hence $\mathbf{P}(C) = 1$ implies $\mathbf{P}(A(\varepsilon)) = 0$ for all $\varepsilon > 0$. On the other hand, if $\mathbf{P}(A(\varepsilon)) = 0$ for all $\varepsilon > 0$, then

$$\mathbf{P}(C^c) = \mathbf{P}\left(\bigcup_{\varepsilon > 0} A(\varepsilon)\right) = \mathbf{P}\left(\bigcup_{m \geqslant 1} A(1/m)\right) \quad \text{since } A(\varepsilon) \subseteq A(\varepsilon') \text{ if } \varepsilon \geqslant \varepsilon'$$
$$\leqslant \sum_{m \geqslant 1} \mathbf{P}(A(1/m)) = 0.$$

It follows that $\mathbf{P}(C) = 1$ if and only if $\mathbf{P}(A(\varepsilon)) = 0$ for all $\varepsilon > 0$.

In addition, $\{B_m(\varepsilon): m \geqslant 1\}$ is a decreasing sequence of events with limit $A(\varepsilon)$ (see Problem (1.8.16)), and therefore $\mathbf{P}(A(\varepsilon)) = 0$ if and only if $\mathbf{P}(B_m(\varepsilon)) \to 0$ as $m \to \infty$.

(b) From the definition of $B_m(\varepsilon)$

$$\mathbf{P}(B_m(\varepsilon)) \leqslant \sum_{n=m}^{\infty} \mathbf{P}(A_n(\varepsilon))$$

and so $\mathbf{P}(B_m(\varepsilon)) \to 0$ as $m \to \infty$ whenever

$$\sum_n \mathbf{P}(A_n(\varepsilon)) < \infty.$$

(c) $A_n(\varepsilon) \subseteq B_n(\varepsilon)$ and so $\mathbf{P}(|X_n - X| > \varepsilon) = \mathbf{P}(A_n(\varepsilon)) \to 0$ whenever $\mathbf{P}(B_n(\varepsilon)) \to 0$. To see that the converse fails, define an independent sequence $\{X_n\}$ by

(11) $$X_n = \begin{cases} 1 & \text{with probability } n^{-1} \\ 0 & \text{with probability } 1 - n^{-1}. \end{cases}$$

Clearly $X_n \xrightarrow{\text{P}} 0$. However, if $0 < \varepsilon < 1$,

$$\begin{aligned} \mathbf{P}(B_m(\varepsilon)) &= 1 - \lim_{r\to\infty} \mathbf{P}(X_n = 0 \text{ for all } n \text{ such that } m \leqslant n \leqslant r) \quad \text{by (1.3.5)} \\ &= 1 - \left(1 - \frac{1}{m}\right)\left(1 - \frac{1}{m+1}\right)\cdots \quad \text{by independence} \\ &= 1 - \lim_{M\to\infty}\left(\frac{m-1}{m}\,\frac{m}{m+1}\,\frac{m+1}{m+2}\cdots\frac{M}{M+1}\right) \\ &= 1 - \lim_{M\to\infty}\frac{m-1}{M+1} = 1 \quad \text{for all } m, \end{aligned}$$

and so $\{X_n\}$ does not converge almost surely. ■

(12) **Lemma.** *There exist sequences which*

(a) *converge almost surely but not in mean*
(b) *converge in mean but not almost surely.*

Proof.

(a) Consider Example (9). Use (10b) to show that $X_n \xrightarrow{\text{a.s.}} 0$.
(b) Consider Example (11). ■

This completes the proof of Theorem (3), and we move to Theorem (4).

Proof of Theorem (4).

(a) $$\begin{aligned} \mathbf{P}(|X_n - c| > \varepsilon) &= \mathbf{P}(X_n < c - \varepsilon) + \mathbf{P}(X_n > c + \varepsilon) \\ &\to 0 \quad \text{if} \quad X_n \xrightarrow{\text{D}} c. \end{aligned}$$

(b) If $X_n \xrightarrow{\text{P}} X$ and $\mathbf{P}(|X_n| \leqslant k) = 1$ then $\mathbf{P}(|X| \leqslant k) = 1$ also, since

$$\mathbf{P}(|X| \leqslant k + \varepsilon) = \lim_{n\to\infty} \mathbf{P}(|X_n| \leqslant k + \varepsilon) = 1$$

for all $\varepsilon > 0$. Now, let $A_n(\varepsilon) = \{|X_n - X| > \varepsilon\}$, with complement $A_n^c(\varepsilon)$. Then

$$|X_n - X|^r \leqslant \varepsilon^r I_{A_n^c(\varepsilon)} + (2k)^r I_{A_n(\varepsilon)}$$

with probability 1. Take expectations to obtain

$$\begin{aligned}\mathbf{E}(|X_n - X|^r) &\leqslant \varepsilon^r + [(2k)^r - \varepsilon^r]\mathbf{P}(A_n(\varepsilon))\\ &\to \varepsilon^r \quad \text{as} \quad n \to \infty.\end{aligned}$$

Let $\varepsilon \downarrow 0$ to obtain that $X_n \xrightarrow{r} X$.

(c) This is just (10b). ■

Note that any sequence $\{X_n\}$ which satisfies $X_n \xrightarrow{\mathrm{P}} X$ contains a subsequence $\{X_{n_i}: 1 \leqslant i < \infty\}$ which converges almost surely.

(13) **Theorem.** *If $X_n \xrightarrow{\mathrm{P}} X$ then there exists a non-random increasing sequence of integers $n_1, n_2, \ldots$ such that $X_{n_i} \xrightarrow{\text{a.s.}} X$ as $i \to \infty$.*

Proof. Since $X_n \xrightarrow{\mathrm{P}} X$, we have that

$$\mathbf{P}(|X_n - X| > \varepsilon) \to 0 \quad \text{as} \quad n \to \infty, \quad \text{for all } \varepsilon > 0.$$

Pick an increasing sequence $n_1, n_2, \ldots$ of positive integers such that

$$\mathbf{P}\left(|X_{n_i} - X| > \frac{1}{i}\right) \leqslant \frac{1}{i^2}.$$

Then, for any $\varepsilon > 0$

$$\sum_{i > \varepsilon^{-1}} \mathbf{P}(|X_{n_i} - X| > \varepsilon) \leqslant \sum_{i > \varepsilon^{-1}} \mathbf{P}(|X_{n_i} - X| > i^{-1}) < \infty$$

and the result follows from (10b). ■

We have seen that convergence in distribution is the weakest mode of convergence since it involves distribution functions only and makes no reference to an underlying probability space (see (5.9.4) for an equivalent formulation of convergence in distribution which involves distribution functions alone). However, assertions of the form '$X_n \xrightarrow{\mathrm{D}} X$' (or equivalently '$F_n \to F$', where F_n and F are the distribution functions of X_n and X) have important and useful representations in terms of almost sure convergence.

(14) **Skorokhod's representation theorem.** *If $\{X_n\}$ and X, with distribution functions $\{F_n\}$ and F, are such that*

$$X_n \xrightarrow{\mathrm{D}} X \text{ (or, equivalently, } F_n \to F) \text{ as } n \to \infty,$$

then there exists a probability space $(\Omega', \mathcal{F}', \mathbf{P}')$ and random variables $\{Y_n\}$ and Y, which map Ω' into $\mathbb{R}$, such that

(a) *$\{Y_n\}$ and Y have distribution functions $\{F_n\}$ and F*

(b) *$Y_n \xrightarrow{\text{a.s.}} Y$ as $n \to \infty$.*

Therefore although X_n may fail to converge to X in any mode other than in distribution, there exists a sequence $\{Y_n\}$, distributed identically to $\{X_n\}$, which converges almost surely to a copy of X. The proof is elementary, but may be omitted.

Proof. Let $\Omega' = (0, 1)$, $\mathscr{F}'$ be the Borel σ-field generated by the intervals of Ω' (see the discussion at the end of Section 4.1), and let $\mathbf{P}'$ be the probability measure induced on $\mathscr{F}'$ by the requirement that, for any interval $I = (a, b) \subseteq \Omega'$, $\mathbf{P}'(I) = (b - a)$; $\mathbf{P}'$ is called *Lebesgue measure.* For $\omega \in \Omega'$, define

$$Y_n(\omega) = \inf\{x: \omega \leqslant F_n(x)\}$$
$$Y(\omega) = \inf\{x: \omega \leqslant F(x)\}.$$

Note that Y_n and Y are essentially the inverse functions of F_n and F since

(15)
$$\omega \leqslant F_n(x) \Leftrightarrow Y_n(\omega) \leqslant x$$
$$\omega \leqslant F(x) \Leftrightarrow Y(\omega) \leqslant x.$$

It follows immediately that Y_n and Y satisfy (14a) since, for example, from (15)

$$\mathbf{P}'(Y \leqslant y) = \mathbf{P}'((0, F(y)]) = F(y).$$

To show (14b), proceed as follows. Given $\varepsilon > 0$ and $\omega \in \Omega'$, pick a point x of continuity of F such that

$$Y(\omega) - \varepsilon < x < Y(\omega).$$

By (15), $F(x) < \omega$, but $F_n(x) \to F(x)$ as $n \to \infty$ and so $F_n(x) < \omega$ for all large n, giving that

$$Y(\omega) - \varepsilon < x < Y_n(\omega) \quad \text{for all large } n;$$

now let $n \to \infty$ and $\varepsilon \downarrow 0$ to obtain

(16)
$$\liminf_{n\to\infty} Y_n(\omega) \geqslant Y(\omega) \quad \text{for all } \omega.$$

Finally, if $\omega < \omega' < 1$, pick a point x of continuity of F such that

$$Y(\omega') < x < Y(\omega') + \varepsilon.$$

By (15), $\omega < \omega' \leqslant F(x)$ and so $\omega < F_n(x)$ for all large n, giving that

$$Y_n(\omega) \leqslant x < Y(\omega') + \varepsilon \quad \text{for all large } n;$$

now let $n \to \infty$ and $\varepsilon \downarrow 0$ to obtain

(17)
$$\limsup_{n\to\infty} Y_n(\omega) \leqslant Y(\omega') \quad \text{whenever } \omega < \omega'.$$

Combine this with (16) to see that $Y_n(\omega) \to Y(\omega)$ for all points ω of continuity of Y. However, Y is monotone non-decreasing and so the set D of discontinuities of Y is countable; thus $\mathbf{P}'(D) = 0$ and the proof is complete. ■

We complete this section with two elementary applications of the representation theorem (14). The results in question are standard, but the usual classical proofs are tedious.

(18) **Theorem.** *If $X_n \xrightarrow{\text{D}} X$ and $g\colon \mathbb{R} \to \mathbb{R}$ is continuous then $g(X_n) \xrightarrow{\text{D}} g(X)$.*

Proof. Let $\{Y_n\}$ and Y be given as in (14). By the continuity of g

$$\{\omega\colon g(Y_n(\omega)) \to g(Y(\omega))\} \supseteq \{\omega\colon Y_n(\omega) \to Y(\omega)\},$$

and so $g(Y_n) \xrightarrow{\text{a.s.}} g(Y)$ as $n \to \infty$. Therefore $g(Y_n) \xrightarrow{\text{D}} g(Y)$; however, $\{g(Y_n)\}$ and $g(Y)$ have the same distributions as $\{g(X_n)\}$ and $g(X)$. ■

(19) **Theorem.** *The following three statements are equivalent.*

(a) $X_n \xrightarrow{\text{D}} X$.

(b) $\mathbf{E}(g(X_n)) \to \mathbf{E}(g(X))$ *for all bounded continuous functions g.*

(c) $\mathbf{E}(g(X_n)) \to \mathbf{E}(g(X))$ *for all functions g of the form $g(x) = f(x)I_{[a,b]}(x)$ where f is continuous on $[a, b]$ and a and b are points of continuity of the distribution function of X.*

It is not important in (c) that g be continuous on the *closed* interval $[a, b]$. The same proof is valid if g in part (c) is of the form $g(x) = f(x)I_{(a,b)}(x)$ where f is bounded and continuous on the open interval (a, b).

Proof. First we prove that (a) implies (b). Suppose that $X_n \xrightarrow{\text{D}} X$ and g is bounded and continuous. By the Skorokhod representation theorem (14), there exist random variables $Y, Y_1, Y_2, \ldots$ having the same distributions as $X, X_1, X_2, \ldots$ and such that $Y_n \xrightarrow{\text{a.s.}} Y$. Therefore $g(Y_n) \xrightarrow{\text{a.s.}} g(Y)$ by the continuity of g, and furthermore $\{g(Y_n)\}$ are uniformly bounded random variables. We apply the bounded convergence theorem (5.6.12) to deduce that $\mathbf{E}(g(Y_n)) \to \mathbf{E}(g(Y))$, and (b) follows since $\mathbf{E}(g(Y_n)) = \mathbf{E}(g(X_n))$ and $\mathbf{E}(g(Y)) = \mathbf{E}(g(X))$.

We write C for the set of points of continuity of F_X. Now F_X is monotone and has therefore at most countably many points of discontinuity; hence C^c is countable.

Suppose now that (b) holds. For (c), it suffices to prove that $\mathbf{E}(h(X_n)) \to \mathbf{E}(h(X))$ for all functions h of the form $h(x) = f(x)I_{(-\infty,b]}(x)$, where f is bounded and continuous, and $b \in C$; the general result follows by an exactly analogous argument. Suppose then that $h(x) = f(x)I_{(-\infty,b]}(x)$ as prescribed. The idea is to approximate to h by a continuous function. For $\delta > 0$, define the continuous functions h' and h'' by

$$h'(x) = \begin{cases} h(x) & \text{if} \quad x \notin (b, b+\delta) \\ \left(1 + \dfrac{b-x}{\delta}\right)h(b) & \text{if} \quad x \in (b, b+\delta) \end{cases}$$

$$h''(x) = \begin{cases} \left(1 + \dfrac{x-b}{\delta}\right)h(b) & \text{if} \quad x \in (b-\delta, b) \\ \left(1 + \dfrac{b-x}{\delta}\right)h(b) & \text{if} \quad x \in [b, b+\delta) \\ 0 & \text{otherwise.} \end{cases}$$

It may be helpful to draw a picture. Now

$$|\mathbf{E}(h(X_n) - h'(X_n))| \leqslant |\mathbf{E}(h''(X_n))|, \qquad |\mathbf{E}(h(X) - h'(X))| \leqslant |\mathbf{E}(h''(X))|,$$

so that

$$\begin{aligned} |\mathbf{E}(h(X_n)) - \mathbf{E}(h(X))| &\leqslant |\mathbf{E}(h''(X_n))| + |\mathbf{E}(h''(X))| + |\mathbf{E}(h'(X_n)) - \mathbf{E}(h'(X))| \\ &\to 2|\mathbf{E}(h''(X))| \quad \text{as} \quad n \to \infty \end{aligned}$$

by assumption (b). We now observe that

$$|\mathbf{E}(h''(X))| \leqslant |h(b)|\mathbf{P}(b - \delta < X < b + \delta) \to 0 \quad \text{as} \quad \delta \downarrow 0,$$

by the assumption that $\mathbf{P}(X = b) = 0$. Hence (c) holds.

Suppose finally that (c) holds, and that b is such that $\mathbf{P}(X = b) = 0$. By considering the function $f(x) = 1$ for all x, we have that, if $a \in C$,

(20)
$$\begin{aligned} \mathbf{P}(X_n \leqslant b) \geqslant \mathbf{P}(a \leqslant X_n \leqslant b) &\to \mathbf{P}(a \leqslant X \leqslant b) \quad \text{as} \quad n \to \infty \\ &\to \mathbf{P}(X \leqslant b) \quad \text{as} \quad a \to -\infty \text{ through } C. \end{aligned}$$

A similar argument, but taking the limit in the other direction, yields for $b' \in C$

(21)
$$\begin{aligned} \mathbf{P}(X_n \geqslant b') \geqslant \mathbf{P}(b' \leqslant X_n \leqslant c) &\quad \text{if} \quad c \geqslant b' \\ \to \mathbf{P}(b' \leqslant X \leqslant c) &\quad \text{as} \quad n \to \infty, \text{ if } c \in C \\ \to \mathbf{P}(X \geqslant b') &\quad \text{as} \quad c \to \infty \text{ through } C. \end{aligned}$$

It follows from (20) and (21) that, if $b, b' \in C$ and $b < b'$, then for any $\varepsilon > 0$ there exists N such that

$$\mathbf{P}(X \leqslant b) - \varepsilon \leqslant \mathbf{P}(X_n \leqslant b) \leqslant \mathbf{P}(X_n < b') \leqslant \mathbf{P}(X < b') + \varepsilon$$

for all $n \geqslant N$. Now take the limits as $n \to \infty$, $\varepsilon \downarrow 0$, and $b' \downarrow b$ through C, in that order, to obtain that $\mathbf{P}(X_n \leqslant b) \to \mathbf{P}(X \leqslant b)$ as $n \to \infty$ if $b \in C$, the required result. ■

Exercises

(22) (a) Suppose $X_n \xrightarrow{r} X$ where $r \geqslant 1$. Show that $\mathbf{E}|X_n^r| \to \mathbf{E}|X^r|$.
(b) Suppose $X_n \xrightarrow{1} X$. Show that $\mathbf{E}(X_n) \to \mathbf{E}(X)$. Is the converse true?
(c) Suppose $X_n \xrightarrow{2} X$. Show that $\text{var}(X_n) \to \text{var}(X)$.

(23) **Dominated convergence.** Suppose $|X_n| \leqslant Z$ for all n, where $\mathbf{E}Z < \infty$. Prove that if $X_n \xrightarrow{\text{P}} X$ then $X_n \xrightarrow{1} X$.

(24) Give a rigorous proof that $\mathbf{E}(XY) = \mathbf{E}(X)\mathbf{E}(Y)$ for any pair X, Y of independent non-negative random variables on $(\Omega, \mathscr{F}, \mathbf{P})$ with finite means. [Hint: For $k \geqslant 0$, $n \geqslant 1$, define $X_n = k/n$ if $k/n \leqslant X < (k+1)/n$, and similarly for Y_n. Show that X_n and Y_n are independent, and $X_n \leqslant X$, and $Y_n \leqslant Y$. Deduce that $\mathbf{E}X_n \to \mathbf{E}X$ and $\mathbf{E}Y_n \to \mathbf{E}Y$, and also $\mathbf{E}(X_n Y_n) \to \mathbf{E}(XY)$.]

(25) Show that convergence in distribution is equivalent to convergence with respect to the Lévy metric of Exercise (7.1.11).

(26) (a) Suppose that $X_n \xrightarrow{\text{D}} X$ and $Y_n \xrightarrow{\text{P}} c$, where c is a constant. Show that $X_n Y_n \xrightarrow{\text{D}} cX$, and that $X_n/Y_n \xrightarrow{\text{D}} X/c$ if $c \neq 0$.
(b) Suppose that $X_n \xrightarrow{\text{D}} 0$ and $Y_n \xrightarrow{\text{P}} Y$, and let $g\colon \mathbb{R}^2 \to \mathbb{R}$ be such that $g(x, y)$ is a continuous function of y for all x, and $g(x, y)$ is continuous at $x = 0$ for all y. Show that $g(X_n, Y_n) \xrightarrow{\text{P}} g(0, Y)$.
[These results are sometimes attributed to Slutsky.]

(27) Let $X_1, X_2, \ldots$ be random variables on the probability space $(\Omega, \mathscr{F}, \mathbf{P})$. Show that the set $A = \{\omega \in \Omega\colon \text{the sequence } X_n(\omega) \text{ converges}\}$ is an event (that is, lies in $\mathscr{F}$), and that there exists a random variable X (that is, an $\mathscr{F}$-measurable function $X\colon \Omega \to \mathbb{R}$) such that $X_n(\omega) \to X(\omega)$ for $\omega \in A$.

(28) Let $\{X_n\}$ be a sequence of random variables, and let $\{c_n\}$ be a sequence of reals converging to the limit c. For convergence almost surely, in rth mean, in probability, and in distribution, show that the convergence of X_n to X entails the convergence of $c_n X_n$ to cX.

(29) Let $\{X_n\}$ be a sequence of independent random variables which converges in probability to the limit X. Show that X is almost surely constant.

7.3 Some ancillary results

Next we shall develop some refinements of the methods of the last section; these will prove to be of great value later. There are two areas of interest. The first deals with inequalities and generalizes Markov's inequality (7.2.7). The second deals with infinite families of events and the Borel–Cantelli lemmas; it is related to the result of (7.2.4c).

Markov's inequality is easily generalized.

(1) **Theorem.** *Let $h\colon \mathbb{R} \to [0, \infty)$ be a non-negative function. Then*

$$\mathbf{P}(h(X) \geqslant a) \leqslant \frac{\mathbf{E}(h(X))}{a} \quad \textit{for all } a > 0.$$

Proof. Let $A = \{h(X) \geqslant a\}$. Then

$$h(X) \geqslant a I_A.$$

Take expectations to obtain the result. ■

Note some special cases of this.

(2) **Example. Markov's inequality.** Set $h(x) = |x|$. ●

(3) **Example†. Chebyshov's inequality.** Set $h(x) = x^2$ to obtain

$$\mathbf{P}(|X| \geqslant a) \leqslant \frac{\mathbf{E}(X^2)}{a^2} \quad \text{if} \quad a > 0.$$

This inequality was also discovered by Bienaymé and others. ●

(4) **Example.** More generally, let $g\colon [0, \infty) \to [0, \infty)$ be a strictly increasing non-negative function, and set $h(x) = g(|x|)$ to obtain

$$\mathbf{P}(|X| \geqslant a) \leqslant \frac{\mathbf{E}(g(|X|))}{g(a)} \quad \text{if} \quad a > 0.$$ ●

Theorem (1) provides an upper bound for the probability $\mathbf{P}(h(X) \geqslant a)$. Lower bounds are harder to find in general, but pose no difficulty in the case when h is a uniformly bounded function.

(5) **Theorem.** *If $h\colon \mathbb{R} \to [0, M]$ is a non-negative function taking values bounded by some number M, then*

$$\mathbf{P}(h(X) \geqslant a) \geqslant \frac{\mathbf{E}(h(X)) - a}{M - a} \quad \textit{whenever} \quad 0 \leqslant a < M.$$

Proof. Let $A = \{h(X) \geqslant a\}$ as before and note that

$$h(X) \leqslant MI_A + aI_{A^c}.$$ ■

The reader is left to apply this result to the special cases (2), (3), and (4). This is an appropriate moment to note three other important inequalities. Let X and Y be random variables.

(6) **Theorem. Hölder's inequality.** *If $p, q > 1$ and $p^{-1} + q^{-1} = 1$ then*

$$\mathbf{E}|XY| \leqslant (\mathbf{E}|X^p|)^{1/p}(\mathbf{E}|Y^q|)^{1/q}.$$

(7) **Theorem. Minkowski's inequality.** *If $p \geqslant 1$ then*

$$[\mathbf{E}(|X + Y|^p)]^{1/p} \leqslant (\mathbf{E}|X^p|)^{1/p} + (\mathbf{E}|Y^p|)^{1/p}.$$

Proof of (6) and (7). You did these for Problem (4.11.27). ■

† Our transliteration of Чебышёв (Chebyshov) is at odds with common practice, but dispenses with the need for clairvoyance in pronunciation.

(8) **Theorem.** *$\mathbf{E}(|X+Y|^p) \leqslant C_p(\mathbf{E}|X^p| + \mathbf{E}|Y^p|)$ where $p > 0$ and*

$$C_p = \begin{cases} 1 & \text{if } 0 < p \leqslant 1 \\ 2^{p-1} & \text{if } p > 1. \end{cases}$$

Proof. It is not difficult to show that

$$|x+y|^p \leqslant C_p(|x|^p + |y|^p)$$

for all $x, y \in \mathbb{R}$ and $p > 0$. Now complete the details. ■

Of course, (6) and (7) assert that

$$\|XY\|_1 \leqslant \|X\|_p \|Y\|_q \qquad \text{if} \quad p^{-1} + q^{-1} = 1$$

$$\|X+Y\|_p \leqslant \|X\|_p + \|Y\|_p \qquad \text{if} \quad p \geqslant 1$$

where $\|\cdot\|_p$ is given by

$$\|X\|_p = (\mathbf{E}|X^p|)^{1/p}.$$

Here is an application of these inequalities. It is related to the fact that if $x_n \to x$ and $y_n \to y$ then $x_n + y_n \to x + y$.

(9) **Theorem.**

(a) *If $X_n \xrightarrow{\text{a.s.}} X$ and $Y_n \xrightarrow{\text{a.s.}} Y$ then $X_n + Y_n \xrightarrow{\text{a.s.}} X + Y$.*

(b) *If $X_n \xrightarrow{r} X$ and $Y_n \xrightarrow{r} Y$ then $X_n + Y_n \xrightarrow{r} X + Y$.*

(c) *If $X_n \xrightarrow{\text{P}} X$ and $Y_n \xrightarrow{\text{P}} Y$ then $X_n + Y_n \xrightarrow{\text{P}} X + Y$.*

(d) *It is not in general true that $X_n + Y_n \xrightarrow{\text{D}} X + Y$ if $X_n \xrightarrow{\text{D}} X$ and $Y_n \xrightarrow{\text{D}} Y$.*

Proof. *You* do it. You will need either (7) or (8) to prove part (b). ■

Theorem (7.2.4) contains a criterion for a sequence to converge almost surely. It is a special case of two very useful results called the 'Borel–Cantelli lemmas'. Let $A_1, A_2, \ldots$ be an infinite sequence of events from some probability space $(\Omega, \mathscr{F}, \mathbf{P})$. We shall often be interested in finding out how many of the A_n occur. Recall (Problem (1.8.16)) that the event that infinitely many of the A_n occur, sometimes written $\{A_n \text{ infinitely often}\}$ or $\{A_n \text{ i.o.}\}$, satisfies

$$\{A_n \text{ i.o.}\} = \limsup_{n\to\infty} A_n = \bigcap_n \bigcup_{m=n}^{\infty} A_m.$$

(10) **Theorem. Borel–Cantelli lemmas.** *Let $A = \bigcap_n \bigcup_{m=n}^{\infty} A_m$ be the event that infinitely many of the A_n occur. Then*

(a) $\mathbf{P}(A) = 0$ *if* $\sum_n \mathbf{P}(A_n) < \infty$

(b) $\mathbf{P}(A) = 1$ *if* $\sum_n \mathbf{P}(A_n) = \infty$ *and* $A_1, A_2, \dots$ *are independent events.*

It is easy to see that the following assertion, similar to (b),

$$\mathbf{P}(A) = 1 \quad \text{if} \quad \sum_n \mathbf{P}(A_n) = \infty,$$

is false unless we impose an extra condition, such as independence. Just consider some event E with $0 < \mathbf{P}(E) < 1$ and define

$$A_n = E \quad \text{for all } n.$$

Then $A = E$ and $\mathbf{P}(A) = \mathbf{P}(E)$.

Proof.

(a) For any n

$$A \subseteq \bigcup_{m=n}^{\infty} A_m$$

and so

$$\mathbf{P}(A) \leqslant \sum_{m=n}^{\infty} \mathbf{P}(A_m) \to 0 \quad \text{as} \quad n \to \infty$$

whenever $\sum_n \mathbf{P}(A_n) < \infty$.

(b) It is an easy *exercise* in set theory to check that

$$A^c = \bigcup_n \bigcap_{m=n}^{\infty} A_m^c.$$

However,

$$\begin{aligned}
\mathbf{P}\left(\bigcap_{m=n}^{\infty} A_m^c\right) &= \lim_{r\to\infty} \mathbf{P}\left(\bigcap_{m=n}^{r} A_m^c\right) && \text{by (1.3.5)} \\
&= \prod_{m=n}^{\infty} [1 - \mathbf{P}(A_m)] && \text{by independence} \\
&\leqslant \prod_{m=n}^{\infty} \exp[-\mathbf{P}(A_m)] && \text{since} \quad 1 - x \leqslant e^{-x} \quad \text{if} \quad x \geqslant 0 \\
&= \exp\left(-\sum_{m=n}^{\infty} \mathbf{P}(A_m)\right) \\
&= 0
\end{aligned}$$

whenever $\sum_n \mathbf{P}(A_n) = \infty$. Thus

$$\mathbf{P}(A^c) = \lim_{n\to\infty} \mathbf{P}\left(\bigcap_{m=n}^{\infty} A_m^c\right) = 0,$$

giving $\mathbf{P}(A) = 1$ as required. ■

(11) **Example. Markov chains.** Let $\{X_n\}$ be a Markov chain with $X_0 = i$ for some state i. Let

$$A_n = \{X_n = i\}$$

be the event that the chain returns to i after n steps. State i is persistent if and only if

$$\mathbf{P}(A_n \text{ i.o.}) = 1.$$

By the first Borel–Cantelli lemma

$$\mathbf{P}(A_n \text{ i.o.}) = 0 \quad \text{if} \quad \sum_n \mathbf{P}(A_n) < \infty$$

and it follows that

$$i \text{ is transient if} \sum_n p_{ii}(n) < \infty$$

which is part of an earlier result (6.2.4). We cannot establish the converse by this method since the A_n are not independent. ●

The remaining part of this section may be omitted without sustaining too much damage.

If the events $A_1, A_2, \ldots$ of (10) are independent then $\mathbf{P}(A)$ is either 0 or 1 depending on whether or not $\sum \mathbf{P}(A_n)$ converges. This is an example of a general theorem called a 'zero–one law'. There are many such results, of which the following is a simple example.

(12) **Theorem. Zero–one law.** *Let $A_1, A_2, \ldots$ be a collection of events, and let $\mathscr{A}$ be the smallest σ-field of subsets of Ω which contains all of them. If $A \in \mathscr{A}$ is an event which is independent of the finite collection $A_1, A_2, \ldots, A_n$ for each value of n, then*

$$\textit{either} \quad \mathbf{P}(A) = 0 \quad \textit{or} \quad \mathbf{P}(A) = 1.$$

Proof. Roughly speaking, the assertion that A is in $\mathscr{A}$ means that A is definable in terms of $A_1, A_2, \ldots$. Examples of such events include B_1, B_2, and B_3 defined by

$$B_1 = A_7 \backslash A_9, \qquad B_2 = A_3 \cup A_6 \cup A_9 \cup \cdots, \qquad B_3 = \bigcup_n \bigcap_{m=n}^{\infty} A_m.$$

A standard result of measure theory asserts that if $A \in \mathscr{A}$ then there exists a sequence of events $\{C_n\}$ such that

(13) $$C_n \in \mathscr{A}_n \quad \text{and} \quad \mathbf{P}(A \triangle C_n) \to 0 \quad \text{as} \quad n \to \infty$$

where $\mathscr{A}_n$ is the smallest σ-field which contains the finite collection $A_1, A_2, \ldots, A_n$. But A is independent of this collection, and so is independent of C_n for all n. From (13)

(14) $$\mathbf{P}(A \cap C_n) \to \mathbf{P}(A).$$

However, by independence,

$$\mathbf{P}(A \cap C_n) = \mathbf{P}(A)\mathbf{P}(C_n) \to \mathbf{P}(A)^2$$

which combines with (14) to give

$$\mathbf{P}(A) = \mathbf{P}(A)^2$$

and so $\mathbf{P}(A)$ is 0 or 1. ■

Read on for another zero–one law. Let $X_1, X_2, \ldots$ be a collection of random variables on the probability space $(\Omega, \mathscr{F}, \mathbf{P})$. For any subcollection $\{X_i: i \in I\}$, write $\sigma(X_i: i \in I)$ for the smallest σ-field with respect to which each of the variables X_i $(i \in I)$ is measurable. This σ-field exists by the argument of Section 1.6. It contains events which are 'defined in terms of $\{X_i: i \in I\}$'. Let

$$\mathscr{H}_n = \sigma(X_{n+1}, X_{n+2}, \ldots).$$

Then $\mathscr{H}_n \supseteq \mathscr{H}_{n+1} \supseteq \ldots$; write

$$\mathscr{H}_\infty = \bigcap_n \mathscr{H}_n.$$

$\mathscr{H}_\infty$ is called the *tail σ-field* of the X_n and contains events such as

$$\{X_n > 0 \text{ i.o.}\}, \quad \left\{\limsup_{n\to\infty} X_n = \infty\right\}, \quad \left\{\sum_n X_n \text{ converges}\right\}$$

the definitions of which need never refer to any finite subcollection such as $\{X_1, X_2, \ldots, X_n\}$. Events in $\mathscr{H}_\infty$ are called *tail events.*

(15) **Theorem. Kolmogorov's zero–one law.** *If $X_1, X_2, \ldots$ are independent variables then all events $H \in \mathscr{H}_\infty$ satisfy either $\mathbf{P}(H) = 0$ or $\mathbf{P}(H) = 1$.*

Such a σ-field $\mathscr{H}_\infty$ is called *trivial* since it contains only null events and their complements. You may try to prove this theorem using the techniques in the proof of (12); it is not difficult.

(16) **Example.** Let $X_1, X_2, \ldots$ be independent random variables and let

$$H_1 = \left\{\omega \in \Omega: \sum_n X_n(\omega) \text{ converges}\right\}$$

$$H_2 = \left\{\omega \in \Omega: \limsup_{n\to\infty} X_n(\omega) = \infty\right\}.$$

Each H_i ($i = 1, 2$) has either probability 0 or probability 1. ●

We can associate many other random variables with the sequence $X_1, X_2, \ldots$; these include

$$Y_1 = \tfrac{1}{2}(X_3 + X_6), \qquad Y_2 = \limsup_{n\to\infty} X_n, \qquad Y_3 = Y_1 + Y_2.$$

We call such a variable Y a *tail function* if it is $\mathscr{H}_\infty$-measurable, where $\mathscr{H}_\infty$ is the tail σ-field of the X_n. Roughly speaking, Y is a tail function if its definition includes no essential reference to any finite subsequence $X_1, X_2, \ldots, X_n$. Y_1 and Y_3 are *not* tail functions; can you see why Y_2 *is* a tail function? More rigorously (see the discussion after (2.1.3)) Y is a tail function if and only if

$$\{\omega \in \Omega: Y(\omega) \leqslant y\} \in \mathscr{H}_\infty \quad \text{for all } y \in \mathbb{R}.$$

Thus, if $\mathscr{H}_\infty$ is trivial then the distribution function

$$F_Y(y) = \mathbf{P}(Y \leqslant y)$$

takes the values 0 and 1 only. Such a function is the distribution function of a random variable which is constant (see (2.1.7)), and we have shown the following useful result.

(17) **Theorem.** *Let Y be a tail function of the independent sequence $X_1, X_2, \ldots$. Then there exists a real number k ($-\infty \leqslant k \leqslant \infty$) such that*

$$\mathbf{P}(Y = k) = 1.$$

Proof. Let $k = \inf\{y: \mathbf{P}(Y \leqslant y) = 1\}$, with the convention that the infimum of an empty set is $+\infty$. Then

$$\mathbf{P}(Y \leqslant y) = \begin{cases} 0 & \text{if } y < k \\ 1 & \text{if } y \geqslant k. \end{cases}$$ ■

(18) **Example.** Let $X_1, X_2, \ldots$ be independent variables, with partial sums $S_n = \sum_{i=1}^n X_i$. Then

$$Z_1 = \liminf_{n\to\infty} \frac{1}{n} S_n, \qquad Z_2 = \limsup_{n\to\infty} \frac{1}{n} S_n$$

are almost surely constant (but possibly infinite). To see this, note that if

$m \leqslant n$ then

$$\frac{1}{n} S_n = \frac{1}{n} \sum_{i=1}^{m} X_i + \frac{1}{n} \sum_{i=m+1}^{n} X_i = S(1) + S(2), \text{ say.}$$

However, $S(1) \to 0$ pointwise as $n \to \infty$, and so Z_1 and Z_2 depend in no way upon the values of $X_1, \ldots, X_m$. It follows that the event

$$\left\{\frac{1}{n} S_n \text{ converges}\right\} = \{Z_1 = Z_2\}$$

has either probability 1 or probability 0. That is, $n^{-1}S_n$ converges either almost everywhere or almost nowhere; this was, of course, deducible from (15) since $\{Z_1 = Z_2\} \in \mathscr{H}_\infty$. ●

Exercises

(19) (a) Suppose that $X_n \xrightarrow{P} X$. Show that $\{X_n\}$ is *Cauchy convergent in probability* in that, for all $\varepsilon > 0$, $\mathbf{P}(|X_n - X_m| > \varepsilon) \to 0$ as $n, m \to \infty$. In what sense is the converse true?

(b) Let $\{X_n\}$ and $\{Y_n\}$ be sequences of random variables such that the pairs (X_i, X_j) and (Y_i, Y_j) have the same distributions for all i, j. If $X_n \xrightarrow{P} X$, show that Y_n converges in probability to some limit Y having the same distribution as X.

(20) Show that the probability that infinitely many of the events $\{A_n : n \geqslant 1\}$ occur satisfies $\mathbf{P}(A_n \text{ i.o.}) \geqslant \limsup_{n\to\infty} \mathbf{P}(A_n)$.

(21) Let $\{S_n : n \geqslant 0\}$ be a simple random walk which moves to the right with probability p at each step, and suppose that $S_0 = 0$. Write $X_n = S_n - S_{n-1}$.

(a) Show that $\{S_n = 0 \text{ i.o.}\}$ is not a tail event of the sequence $\{X_n\}$.
(b) Show that $\mathbf{P}(S_n = 0 \text{ i.o.}) = 0$ if $p \neq \frac{1}{2}$.
(c) Let $T_n = S_n/\sqrt{n}$, and show that

$$\left\{\liminf_{n\to\infty} T_n \leqslant -x\right\} \cap \left\{\limsup_{n\to\infty} T_n \geqslant x\right\}$$

is a tail event of the sequence $\{X_n\}$, for all $x > 0$, and deduce directly that $\mathbf{P}(S_n = 0 \text{ i.o.}) = 1$ if $p = \frac{1}{2}$.

(22) **Hewitt–Savage zero–one law.** Let $X_1, X_2, \ldots$ be independent identically distributed random variables. The event A, defined in terms of the X_n, is called *exchangeable* if A is invariant under finite permutations of the coordinates, which is to say that its indicator function I_A satisfies $I_A(X_1, X_2, \ldots, X_n, \ldots) = I_A(X_{i_1}, X_{i_2}, \ldots, X_{i_n}, X_{n+1}, \ldots)$ for all $n \geqslant 1$ and all permutations $(i_1, i_2, \ldots, i_n)$ of $(1, 2, \ldots, n)$. Show that all exchangeable events A are such that either $\mathbf{P}(A) = 0$ or $\mathbf{P}(A) = 1$.

(23) Returning to the simple random walk of Exercise (21), show that $\{S_n = 0 \text{ i.o.}\}$ is an exchangeable event with respect to the steps of the walk, and deduce from the Hewitt–Savage zero–one law that it has either probability 0 or 1.

(24) **Weierstrass's approximation theorem.** Let $f : [0, 1] \to \mathbb{R}$ be a continuous function, and let S_n be a random variable having the binomial distribution with parameters n and x. Using the formula $\mathbf{E}(Z) = \mathbf{E}(ZI_A) + \mathbf{E}(ZI_{A^c})$ with $Z = f(x) - f(n^{-1}S_n)$ and

$A = \{|n^{-1}S_n - x| > \delta\}$, show that

$$\lim_{n\to\infty} \sup_{0\leqslant x\leqslant 1} \left| f(x) - \sum_{k=0}^{n} f(k/n)\binom{n}{k}x^k(1-x)^{n-k} \right| = 0.$$

You have proved Weierstrass's approximation theorem, which states that every continuous function on [0, 1] may be approximated by a polynomial uniformly over the interval.

(25) **Complete convergence.** A sequence $X_1, X_2, \ldots$ of random variables is said to be *completely convergent* to X if

$$\sum_n \mathbf{P}(|X_n - X| > \varepsilon) < \infty \quad \text{for all } \varepsilon > 0.$$

Show that, for sequences of independent variables, complete convergence is equivalent to a.s. convergence. Find a sequence of (dependent) random variables which converges a.s. but not completely.

(26) Let $X_1, X_2, \ldots$ be independent identically distributed random variables with common mean μ and finite variance. Show that

$$\binom{n}{2}^{-1} \sum_{1\leqslant i<j\leqslant n} X_iX_j \xrightarrow{\text{P}} \mu^2 \quad \text{as} \quad n \to \infty.$$

7.4 Laws of large numbers

Let $\{X_n\}$ be a sequence of random variables with partial sums

$$S_n = \sum_1^n X_i.$$

We are interested in the asymptotic behaviour of S_n as $n \to \infty$; this long-term behaviour depends crucially upon the original sequence of the X_i. The general problem may be described as follows. Under what conditions does the following convergence occur?

(1)
$$\frac{S_n}{b_n} - a_n \to S \quad \text{as} \quad n \to \infty$$

where $a = \{a_n\}$ and $b = \{b_n\}$ are sequences of real numbers, S is a random variable, and the convergence takes place in some mode to be specified.

(2) **Example.** Let $X_1, X_2, \ldots$ be independent identically distributed variables with mean μ and variance σ^2. By (5.10.2) and (5.10.4) we have that

$$\frac{S_n}{n} \xrightarrow{\text{D}} \mu \quad \text{and} \quad \frac{S_n}{\sigma n^{\frac{1}{2}}} - \frac{\mu}{\sigma} n^{\frac{1}{2}} \xrightarrow{\text{D}} N(0, 1).$$

So there may not be a *unique* collection a, b, S such that (1) occurs. ●

The convergence problem (1) can often be simplified by setting $a_n = 0$ for all n, whenever the X_i have finite means. Just rewrite the problem in terms of

$$X'_i = X_i - \mathbf{E}X_i, \quad S'_n = S_n - \mathbf{E}S_n.$$

The general theory of relations like (1) is well established and extensive. We shall restrict our attention here to a small but significant part of the theory when the X_i are independent and identically distributed random variables. Suppose for the moment that this is true. We saw in Example (2) that (at least) two types of convergence may be established for such sequences, so long as they have finite second moments. The law of large numbers admits stronger forms than that given in (2). For example, notice that $n^{-1}S_n$ converges in distribution to a constant limit, and use (7.2.4) to see that $n^{-1}S_n$ converges in probability also. Perhaps we can strengthen this further to include convergence in rth mean, for some r, or almost sure convergence. Indeed, this turns out to be possible when suitable conditions are imposed on the common distribution of the X_i. We shall not use the method of characteristic functions of Chapter 5, preferring to approach the problem more directly in the spirit of Section 7.2.

We shall say that the sequence $\{X_n\}$ obeys the 'weak law of large numbers' if there exists a constant μ such that

$$\frac{1}{n} S_n \xrightarrow{\mathrm{P}} \mu.$$

If the stronger result

$$\frac{1}{n} S_n \xrightarrow{\text{a.s.}} \mu$$

holds, then we call it the 'strong law of large numbers'. We seek sufficient, and if possible necessary, conditions on the common distribution of the X_i for the weak and strong laws to hold. As the title suggests, the weak law is implied by the strong law, since convergence in probability is implied by almost sure convergence. A sufficient condition for the strong law is given by the following theorem.

(3) **Theorem.** *Let $X_1, X_2, \ldots$ be independent identically distributed random variables with $\mathbf{E}(X_1^2) < \infty$. Then*

$$\frac{1}{n} \sum_{i=1}^{n} X_i \to \mu \quad \textit{almost surely and in mean square,}$$

where $\mu = \mathbf{E}X_1$.

So the strong law holds whenever the X_i have finite second moment. The proof of mean square convergence is very easy; almost sure convergence is harder to demonstrate (but see Problem (7.11.6) for an easy proof of almost sure convergence subject to the stronger condition that $\mathbf{E}(X_1^4) < \infty$).

Proof. To show mean square convergence, calculate

$$\begin{aligned}\mathbf{E}\left(\left(\frac{1}{n}S_n-\mu\right)^2\right)&=\mathbf{E}\left(\frac{1}{n^2}(S_n-\mathbf{E}S_n)^2\right)\\&=\frac{1}{n^2}\operatorname{var}\left(\sum_1^n X_i\right)\\&=\frac{1}{n^2}\sum_1^n\operatorname{var}(X_i)\quad\text{by independence and (3.3.11)}\\&=\frac{1}{n}\operatorname{var}(X_1)\to 0\quad\text{as}\quad n\to\infty,\end{aligned}$$

since $\operatorname{var}(X_1)<\infty$ by virtue of the assumption that $\mathbf{E}(X_1^2)<\infty$.

Next we show almost sure convergence. We saw in (7.2.13) that there necessarily exists a subsequence $n_1, n_2, \ldots$ along which $n^{-1}S_n$ converges to μ almost surely; we can find such a subsequence explicitly. Write $n_i = i^2$ and use Chebyshov's inequality (7.3.3) to find that

$$\mathbf{P}\left(\frac{1}{i^2}|S_{i^2}-i^2\mu|>\varepsilon\right)\leqslant\frac{\operatorname{var}(S_{i^2})}{i^4\varepsilon^2}=\frac{\operatorname{var}(X_1)}{i^2\varepsilon^2}.$$

Sum over i and use (7.2.4c) to find that

(4) $$\frac{1}{i^2}S_{i^2}\xrightarrow{\text{a.s.}}\mu\quad\text{as}\quad i\to\infty.$$

We need to fill in the gaps in this limit process. Suppose for the moment that the X_i are *non-negative*. Then $\{S_n\}$ is monotonic non-decreasing, and so

$$S_{i^2}\leqslant S_n\leqslant S_{(i+1)^2}\quad\text{if}\quad i^2\leqslant n\leqslant(i+1)^2.$$

Divide by n to find that

$$\frac{1}{(i+1)^2}S_{i^2}\leqslant\frac{1}{n}S_n\leqslant\frac{1}{i^2}S_{(i+1)^2}\quad\text{if}\quad i^2\leqslant n\leqslant(i+1)^2;$$

now let $n\to\infty$ and use (4), remembering that $i^2/(i+1)^2\to 1$ as $i\to\infty$, to deduce that

(5) $$\frac{1}{n}S_n\xrightarrow{\text{a.s.}}\mu\quad\text{as}\quad n\to\infty$$

as required, whenever the X_i are non-negative. Finally, we lift the non-negativity condition. For the general X_i, define random variables X_n^+, X_n^- by

$$X_n^+(\omega)=\max\{X_n(\omega),0\},\qquad X_n^-(\omega)=-\min\{X_n(\omega),0\};$$

then X_n^+ and X_n^- are non-negative and

$$X_n = X_n^+ - X_n^-, \qquad \mathbf{E}(X_n) = \mathbf{E}(X_n^+) - \mathbf{E}(X_n^-).$$

Furthermore, $X_n^+ \leqslant |X_n|$ and $X_n^- \leqslant |X_n|$, so that $\mathbf{E}((X_1^+)^2) < \infty$ and $\mathbf{E}((X_1^-)^2) < \infty$. Now apply (5) to the sequences $\{X_n^+\}$ and $\{X_n^-\}$ to find that

$$\frac{1}{n} S_n = \frac{1}{n}\left(\sum_1^n X_i^+ - \sum_1^n X_i^-\right)$$
$$\xrightarrow{\text{a.s.}} \mathbf{E}(X_1^+) - \mathbf{E}(X_1^-) = \mathbf{E}(X_1) \quad \text{as} \quad n \to \infty,$$

by (7.3.9a). ■

Is the result of Theorem (3) as sharp as possible? It is not difficult to see that the condition that $\mathbf{E}(X_1^2) < \infty$ is both necessary and sufficient for mean square convergence to hold. For almost sure convergence the weaker condition that

(6) $$\mathbf{E}|X_1| < \infty$$

will turn out to be necessary and sufficient, but the proof of this is slightly more difficult and is deferred until the next section. There exist sequences which satisfy the weak law but not the strong law. Indeed, the characteristic function technique (see Section 5.10) can be used to prove the following necessary and sufficient condition for the weak law. We offer no proof, but see Laha and Rohatgi (1979, p. 320), Feller (1971, p. 565), and Problem (7.11.15).

(7) **Theorem.** *The independent identically distributed sequence $\{X_n\}$, with shared distribution function F, satisfies*

$$\frac{1}{n}\sum_{i=1}^n X_i \xrightarrow{\text{P}} \mu$$

for some constant μ, if and only if one of the following conditions (8) *or* (9) *holds:*

(8) $$n\mathbf{P}(|X_1| > n) \to 0 \textit{ and } \int_{[-n,n]} x \, \mathrm{d}F \to \mu \textit{ as } n \to \infty$$

(9) *the characteristic function $\phi(t)$ of the X_j is differentiable at $t = 0$ and $\phi'(0) = \mathrm{i}\mu$.*

Of course, the integral in (8) can be rewritten as

$$\int_{[-n,n]} x \, \mathrm{d}F = \mathbf{E}(X_1 \mid |X_1| \leqslant n)\mathbf{P}(|X_1| \leqslant n) = \mathbf{E}(X_1 I_{\{|X_1| \leqslant n\}}).$$

Thus, a sequence satisfies the weak law but not the strong law whenever (8) holds without (6); as an example of this, suppose the X_j are symmetric (so that X_1 and $-X_1$ have the same distribution) but their distribution function F satisfies

$$F(x) \simeq 1 - (x \log x)^{-1} \quad \text{as} \quad x \to \infty.$$

Some distributions fail even to satisfy (8).

(10) **Example.** Let the X_j have the Cauchy distribution with density function

$$f(x) = \frac{1}{\pi(1 + x^2)}.$$

Then the first part of (8) is violated. Indeed, the characteristic function of $U_n = n^{-1}S_n$ is

$$\begin{aligned}\phi_{U_n}(t) &= \phi_{X_1}\left(\frac{t}{n}\right)\cdots\phi_{X_n}\left(\frac{t}{n}\right)\\ &= \left[\exp\left(-\frac{|t|}{n}\right)\right]^n = \exp(-|t|)\end{aligned}$$

and so U_n itself has the Cauchy distribution for all values of n. In particular, (1) holds with $b_n = n$, $a_n = 0$, where S is Cauchy, and the convergence is in distribution. ●

Exercise

(11) Let $X_2, X_3, \ldots$ be independent random variables such that

$$\mathbf{P}(X_n = n) = \mathbf{P}(X_n = -n) = \frac{1}{2n \log n}, \qquad \mathbf{P}(X_n = 0) = 1 - \frac{1}{n \log n}.$$

Show that this sequence obeys the weak law but not the strong law, in the sense that $n^{-1}\sum_1^n X_i$ converges to 0 in probability but not almost surely.

7.5 The strong law

This section is devoted to the proof of the strong law of large numbers.

(1) **Theorem. Strong law of large numbers.** *Let $X_1, X_2, \ldots$ be independent identically distributed random variables. Then*

$$\frac{1}{n}\sum_{i=1}^{n} X_i \to \mu \text{ almost surely, as } n \to \infty,$$

for some constant μ, if and only if $\mathbf{E}|X_1| < \infty$. In this case $\mu = \mathbf{E}X_1$.

The traditional proof of this theorem is long and difficult, and proceeds by a generalization of Chebyshov's inequality. We avoid that here, and give a relatively elementary proof which is an adaptation of the method used to prove (7.4.3). We need only one new technique, called the method of *truncation*.

Proof.† Suppose first that the X_i are *non-negative* random variables with $\mathbf{E}|X_1| = \mathbf{E}(X_1) < \infty$, and write $\mu = \mathbf{E}(X_1)$. We 'truncate' the X_i to obtain a new sequence $\{Y_n\}$ given by

(2) $$Y_n = X_n I_{\{X_n < n\}} = \begin{cases} X_n & \text{if} \quad X_n < n \\ 0 & \text{if} \quad X_n \geqslant n. \end{cases}$$

Note that

$$\sum_n \mathbf{P}(X_n \neq Y_n) = \sum_n \mathbf{P}(X_n \geqslant n) \leqslant \mathbf{E}(X_1) < \infty$$

by the result of (4.11.3). Of course, $\mathbf{P}(X_n \geqslant n) = \mathbf{P}(X_1 \geqslant n)$ since the X_i are identically distributed. By the first Borel–Cantelli lemma (7.3.10a), $\mathbf{P}(X_n \neq Y_n$ for infinitely many values of $n) = 0$, and so

(3) $$\frac{1}{n}\sum_{i=1}^{n} (X_i - Y_i) \xrightarrow{\text{a.s.}} 0 \quad \text{as} \quad n \to \infty;$$

thus it will suffice to show that

(4) $$\frac{1}{n}\sum_{i=1}^{n} Y_i \xrightarrow{\text{a.s.}} \mu \quad \text{as} \quad n \to \infty.$$

We shall need the following elementary observation. If $\alpha > 1$ and $\beta_k = \lfloor \alpha^k \rfloor$, the integer part of α^k, then there exists $A > 0$ such that

(5) $$\sum_{k=m}^{\infty} \frac{1}{\beta_k^2} \leqslant \frac{A}{\beta_m^2} \quad \text{for} \quad m \geqslant 1.$$

This holds because, for large m, the convergent series on the left-hand side is 'nearly' geometric with first term β_m^{-2}. Note also that

(6) $$\beta_{k+1}/\beta_k \to \alpha \quad \text{as} \quad k \to \infty.$$

Write $S'_n = \sum_{i=1}^{n} Y_i$. For $\alpha > 1$, $\varepsilon > 0$, use Chebyshov's inequality to find that

(7) $$\begin{aligned} \sum_{n=1}^{\infty} \mathbf{P}\left(\frac{1}{\beta_n}|S'_{\beta_n} - \mathbf{E}(S'_{\beta_n})| > \varepsilon\right) &\leqslant \frac{1}{\varepsilon^2}\sum_{n=1}^{\infty} \frac{1}{\beta_n^2}\operatorname{var}(S'_{\beta_n}) \\ &= \frac{1}{\varepsilon^2}\sum_{n=1}^{\infty}\frac{1}{\beta_n^2}\sum_{i=1}^{\beta_n}\operatorname{var}(Y_i) \quad \text{by independence} \\ &\leqslant \frac{A}{\varepsilon^2}\sum_{i=1}^{\infty}\frac{1}{i^2}\mathbf{E}(Y_i^2) \end{aligned}$$

by changing the order of summation and using (5).

† This method was found by N. Etemadi.

Let $B_{ij} = \{j - 1 \leqslant X_i < j\}$, and note that $\mathbf{P}(B_{ij}) = \mathbf{P}(B_{1j})$. Now

(8)
$$\sum_{i=1}^{\infty} \frac{1}{i^2} \mathbf{E}(Y_i^2) = \sum_{i=1}^{\infty} \frac{1}{i^2} \sum_{j=1}^{i} \mathbf{E}(Y_i^2 I_{B_{ij}}) \quad \text{by (2)}$$
$$\leqslant \sum_{i=1}^{\infty} \frac{1}{i^2} \sum_{j=1}^{i} j^2 \mathbf{P}(B_{ij})$$
$$\leqslant \sum_{j=1}^{\infty} j^2 \mathbf{P}(B_{1j}) \frac{2}{j} \leqslant 2[\mathbf{E}(X_1) + 1] < \infty.$$

Combine (7) and (8) and use (7.2.4c) to deduce that

(9)
$$\frac{1}{\beta_n} [S'_{\beta_n} - \mathbf{E}(S'_{\beta_n})] \xrightarrow{\text{a.s.}} 0 \quad \text{as} \quad n \to \infty.$$

Also,

$$\mathbf{E}(Y_n) = \mathbf{E}(X_n I_{\{X_n < n\}}) = \mathbf{E}(X_1 I_{\{X_1 < n\}}) \to \mathbf{E}(X_1) = \mu$$

as $n \to \infty$, by monotone convergence (5.6.12). Thus

$$\frac{1}{\beta_n} \mathbf{E}(S'_{\beta_n}) = \frac{1}{\beta_n} \sum_{i=1}^{\beta_n} \mathbf{E}(Y_i) \to \mu \quad \text{as} \quad n \to \infty$$

(remember the hint in the proof of (6.4.22)), yielding from (9) that

(10)
$$\frac{1}{\beta_n} S'_{\beta_n} \xrightarrow{\text{a.s.}} \mu \quad \text{as} \quad n \to \infty;$$

this is a partial demonstration of (4). To fill in the gaps, use the fact that the Y_i are non-negative, implying that the sequence $\{S'_n\}$ is monotonic non-decreasing, to deduce that

(11)
$$\frac{1}{\beta_{n+1}} S'_{\beta_n} \leqslant \frac{1}{m} S'_m \leqslant \frac{1}{\beta_n} S'_{\beta_{n+1}} \quad \text{if} \quad \beta_n \leqslant m \leqslant \beta_{n+1}.$$

Let $m \to \infty$ in (11) and remember (6) to find that

(12)
$$\alpha^{-1}\mu \leqslant \liminf_{m \to \infty} \frac{1}{m} S'_m \leqslant \limsup_{m \to \infty} \frac{1}{m} S'_m \leqslant \alpha\mu \text{ almost surely.}$$

But this holds for all $\alpha > 1$; let $\alpha \downarrow 1$ to obtain (4), and deduce by (3) that

(13)
$$\frac{1}{n} \sum_{i=1}^{n} X_i \xrightarrow{\text{a.s.}} \mu \quad \text{as} \quad n \to \infty$$

whenever the X_i are non-negative. Now proceed exactly as in the proof of (7.4.3) in order to lift the non-negativity condition. Note that we have proved the main part of the theorem without using the full strength of the independence assumption; we have used only the fact that the X_i are *pairwise* independent.

To prove the converse, suppose that

$$\frac{1}{n}\sum_{i=1}^{n} X_i \xrightarrow{\text{a.s.}} \mu.$$

Then $n^{-1}X_n \xrightarrow{\text{a.s.}} 0$ by the theory of convergent real series, and the second Borel–Cantelli lemma (7.3.10b) gives

$$\sum_n \mathbf{P}(|X_n| \geqslant n) < \infty,$$

since the divergence of this sum would imply that $\mathbf{P}(n^{-1}|X_n| \geqslant 1$ infinitely often$) = 1$ (only here do we use the full assumption of independence). But (4.10.3) shows that

$$\mathbf{E}|X_1| \leqslant 1 + \sum_{n=1}^{\infty} \mathbf{P}(|X_1| \geqslant n) = 1 + \sum_{n=1}^{\infty} \mathbf{P}(|X_n| \geqslant n),$$

and hence $\mathbf{E}|X_1| < \infty$, which completes the proof of the theorem. ■

Exercises

(14) The interval [0, 1] is partitioned into n disjoint sub-intervals with lengths $p_1, p_2, \ldots, p_n$, and the *entropy* of this partition is defined to be

$$h = -\sum_{i=1}^{n} p_i \log p_i.$$

Let $X_1, X_2, \ldots$ be independent random variables having the uniform distribution on [0, 1], and let $Z_m(i)$ be the number of the $X_1, X_2, \ldots, X_m$ which lie in the ith interval of the partition above. Show that

$$R_m = \prod_{i=1}^{n} p_i^{Z_m(i)}$$

satisfies $m^{-1}\log R_m \to -h$ almost surely as $m \to \infty$.

(15) **Recurrent events.** Catastrophes occur at the times $T_1, T_2, \ldots$ where $T_i = X_1 + X_2 + \cdots + X_i$ and the X_j are independent identically distributed positive random variables. Let $N(t) = \max\{n: T_n \leqslant t\}$ be the number of catastrophes which have occurred by time t. Prove that if $\mathbf{E}X_1 < \infty$ then $N(t) \to \infty$ and $N(t)/t \to 1/\mathbf{E}X_1$ as $t \to \infty$, almost surely.

(16) **Random walk.** Let $X_1, X_2, \ldots$ be independent identically distributed random variables taking values in the integers $\mathbb{Z}$ and having a finite mean. Show that the Markov chain $S = \{S_n\}$ given by $S_n = \sum_1^n X_i$ is transient if $\mathbf{E}X_1 \neq 0$.

7.6 The law of the iterated logarithm

Let S_n be the partial sum

$$S_n = \sum_{i=1}^{n} X_i$$

of independent identically distributed variables, as usual, and suppose further that $\mathbf{E}(X_i) = 0$ and $\text{var}(X_i) = 1$ for all i. To date, we have two results about the growth rate of $\{S_n\}$.

Law of large numbers: $\frac{1}{n} S_n \to 0$ a.s. and in mean square.

Central limit theorem: $\frac{1}{\sqrt{n}} S_n \xrightarrow{\text{D}} N(0, 1)$.

Thus the sequence

$$U_n = \frac{1}{\sqrt{n}} S_n$$

enjoys a random fluctuation which is asymptotically regularly distributed. Apart from this long-term trend towards the normal distribution, the sequence $\{U_n\}$ may suffer some large but rare fluctuations. The law of the iterated logarithm is an extraordinary result which tells us exactly how big these fluctuations are. First note that, in the language of Section 7.3 (if you have read this),

$$U = \limsup_{n\to\infty} \frac{U_n}{\sqrt{(2 \log \log n)}}$$

is a tail function of the sequence of the X_i. The zero–one law (7.3.17) tells us that there exists a number k, possibly infinite, such that

$$\mathbf{P}(U = k) = 1.$$

The next theorem asserts that $k = 1$!

(1) **Theorem. Law of the iterated logarithm.** *If $X_1, X_2, \ldots$ are independent identically distributed random variables with mean* 0 *and variance* 1 *then*

$$\mathbf{P}\left(\limsup_{n\to\infty} \frac{S_n}{\sqrt{(2n \log \log n)}} = 1\right) = 1.$$

The proof is long and difficult and is omitted (but see the discussion in Billingsley (1986) or Laha and Rohatgi (1979)). The theorem amounts to the assertion that

$$A_n = \{S_n \geqslant c(2n \log \log n)^{\frac{1}{2}}\}$$

occurs for infinitely many values of n if $c < 1$ and for only finitely many values of n if $c > 1$, with probability 1. It is an immediate corollary of (1) that

$$\mathbf{P}\left(\liminf_{n\to\infty} \frac{S_n}{\sqrt{(2n \log \log n)}} = -1\right) = 1;$$

just apply (1) to the sequence $-X_1, -X_2, \ldots$.

Exercise

(2) A function $\phi(x)$ is said to belong to the 'upper class' if, in the notation of this section, $\mathbf{P}(S_n > \phi(n)\sqrt{n} \text{ i.o.}) = 0$. A consequence of the law of the iterated logarithm is that $(\alpha \log \log x)^{\frac{1}{2}}$ is in the upper class for all $\alpha > 2$. Use the first Borel–Cantelli lemma to prove the much weaker fact that $\phi(x) = (\alpha \log x)^{\frac{1}{2}}$ is in the upper class for all $\alpha > 2$, in the special case when the X_i are independent $N(0, 1)$ variables.

7.7 Martingales

Many probabilists specialize in limit theorems, and much of applied probability is devoted to finding such results. The accumulated literature is vast and the techniques multifarious. One of the most useful skills for establishing such results is that of martingale divination, because the convergence of martingales is guaranteed.

(1) **Example.** It is appropriate to discuss an example of the use of the word 'martingale' which pertains to gambling, a favourite source of probabilistic illustrations. We are all familiar with the following gambling strategy. A gambler has a large fortune. He wages £1 on an evens bet. If he loses then he wagers £2 on the next play. If he loses on the nth play then he wagers £2^n on the next. Each sum is calculated so that his inevitable ultimate win will cover his lost stakes and profit him by £1. This strategy is called a 'martingale'. Nowadays casinos do not allow its use, and croupiers have instructions to refuse the bets of those who are seen to practise it. Thackeray's advice was to avoid its use at all costs, and his reasoning may have had something to do with the following calculation. Suppose the gambler wins for the first time at the Nth play. N is a random variable with mass function

$$\mathbf{P}(N = n) = (\tfrac{1}{2})^n$$

and so $\mathbf{P}(N < \infty) = 1$; the gambler is almost surely guaranteed a win in the long run. However, by this time he will have lost an amount £L with mean value

$$\mathbf{E}(L) = \sum_{n=1}^{\infty} (\tfrac{1}{2})^n(1 + 2 + \cdots + 2^{n-2}) = \infty.$$

He must be prepared to lose a lot of money! And so, of course, must the proprietor of the casino.

The perils of playing the martingale are illustrated by the following two excerpts from the memoirs of G. Casanova recalling his stay in Venice in 1754 (Casanova 1922, Chapter 7).

> Playing the martingale, continually doubling my stake, I won every day during the rest of the carnival. I was fortunate enough never to lose the sixth card, and if I had lost it, I should have been without money to play, for I had 2000 sequins on that card. I congratulated myself on having increased the fortune of my dear mistress.

However, some days later:

> I still played the martingale, but with such bad luck that I was soon left without a sequin. As I shared my property with my mistress, I was obliged to tell her of my losses, and at her request sold all her diamonds, losing what I got for them; she had now only 500 sequins. There was no more talk of her escaping from the convent, for we had nothing to live on.

Shortly after these events, Casanova was imprisoned by the authorities, until he escaped to organize a lottery for the benefit of both himself and the French treasury in Paris. Before it became merely a spangle, the sequin was an Italian gold coin. ●

In the spirit of this diversion, suppose a gambler wagers repeatedly with an initial capital S_0, and let S_n be his capital after n plays. We shall think of $S_0, S_1, \ldots$ as a sequence of dependent random variables. Before his $(n+1)$th wager the gambler knows the numerical values of $S_0, S_1, \ldots, S_n$, but can only guess at the future $S_{n+1}, \ldots$. If the game is fair then, conditional upon the past information, he will expect no change in his present capital on average. That is,

(2)† $$\mathbf{E}(S_{n+1} \mid S_0, S_1, \ldots, S_n) = S_n.$$

Most casinos need to pay at least their overheads, and will find a way of changing this equation to

$$\mathbf{E}(S_{n+1} \mid S_0, S_1, \ldots, S_n) \leqslant S_n.$$

The gambler is fortunate indeed if this inequality is reversed. Sequences satisfying (2) are called 'martingales', and they have very special and well studied properties of convergence. They may be discovered within many

† Such conditional expectations appear often in this section. Make sure you understand their meanings. This one is the mean value of S_{n+1}, calculated as though $S_0, \ldots, S_n$ were already known. Clearly this mean value depends on $S_0, \ldots, S_n$; so it is a *function* of $S_0, \ldots, S_n$. Assertion (2) is that it has the value S_n. Any detailed account of conditional expectations would probe into the guts of measure theory. We shall avoid that here, but describe some important properties at the end of this section and in Section 7.9.

probabilistic models, and their general theory may be used to establish limit theorems. We shall now abandon the gambling example, and refer disappointed readers to *How to gamble if you must* by L. Dubins and L. Savage, where they may find an account of the gamblers' ruin theorem.

(3) **Definition.** A sequence $\{S_n: n \geqslant 1\}$ is a **martingale** with respect to the sequence $\{X_n: n \geqslant 1\}$ if, for all $n \geqslant 1$,

(a) $\mathbf{E}|S_n| < \infty$
(b) $\mathbf{E}(S_{n+1} \mid X_1, X_2, \ldots, X_n) = S_n$.

Equation (2) shows that the sequence of gambler's fortunes is a martingale with respect to itself. The extra generality, introduced by the sequence $\{X_n\}$ in (3), is useful for martingales which arise in the following way. A specified sequence $\{X_n\}$ of random variables, such as a Markov chain, may itself *not* be a martingale. However, it is often possible to find some function ϕ such that $\{S_n = \phi(X_n): n \geqslant 1\}$ *is* a martingale. In this case, the martingale property (2) becomes the assertion that, given the values of $X_1, X_2, \ldots, X_n$, the mean value of $S_{n+1} = \phi(X_{n+1})$ is just $S_n = \phi(X_n)$; that is

(4) $$\mathbf{E}(S_{n+1} \mid X_1, \ldots, X_n) = S_n.$$

Of course, condition (b) of (3) is without meaning unless S_n is some function, say ϕ_n, of $X_1, \ldots, X_n$ (that is, $S_n = \phi_n(X_1, \ldots, X_n)$) since the conditional expectation in (3) is itself a function of $X_1, \ldots, X_n$. We shall often omit reference to the underlying sequence $\{X_n\}$, asserting merely that $\{S_n\}$ is a martingale.

(5) **Example. Branching processes–two martingales.** Let Z_n be the size of the nth generation of a branching process. Recall that the probability η that the process ultimately becomes extinct is the smallest non-negative root of the equation

$$s = G(s)$$

where G is the probability generating function of Z_1. There are two martingales associated with the process. First, conditional on $Z_n = z_n$, Z_{n+1} is the sum of z_n independent family sizes, and so

$$\mathbf{E}(Z_{n+1} \mid Z_n = z_n) = z_n\mu$$

where $\mu = G'(1)$ is the mean family size. Thus, by the Markov property,

$$\mathbf{E}(Z_{n+1} \mid Z_1, Z_2, \ldots, Z_n) = Z_n\mu.$$

Now define

$$W_n = Z_n/\mathbf{E}(Z_n)$$

and remember that $\mathbf{E}(Z_n) = \mu^n$ to obtain

$$\mathbf{E}(W_{n+1} \mid Z_1, \ldots, Z_n) = W_n,$$

and so $\{W_n\}$ is a martingale (with respect to $\{Z_n\}$). It is not the only martingale which arises from the branching process. Let

$$V_n = \eta^{Z_n}$$

where η is the probability of ultimate extinction. Surprisingly perhaps, $\{V_n\}$ is a martingale also. For, as in the proof of (5.4.1), write

$$Z_{n+1} = X_1 + \cdots + X_{Z_n}$$

in terms of the family sizes of the members of the nth generation to obtain

$$\begin{aligned} \mathbf{E}(V_{n+1} \mid Z_1, \ldots, Z_n) &= \mathbf{E}(\eta^{(X_1 + \cdots + X_{Z_n})} \mid Z_1, \ldots, Z_n) \\ &= \prod_{i=1}^{Z_n} \mathbf{E}(\eta^{X_i} \mid Z_1, \ldots, Z_n) \quad \text{by independence} \\ &= \prod_{i=1}^{Z_n} \mathbf{E}(\eta^{X_i}) \\ &= \prod_{i=1}^{Z_n} G(\eta) = \eta^{Z_n} = V_n, \end{aligned}$$

since $\eta = G(\eta)$. These facts are very significant in the study of the long-term behaviour of the branching process. ●

(6) **Example.** Let $X_1, X_2, \ldots$ be independent variables with zero means. We claim that the sequence of partial sums

$$S_n = X_1 + \cdots + X_n$$

is a martingale (with respect to $\{X_n\}$). For

$$\begin{aligned} \mathbf{E}(S_{n+1} \mid X_1, \ldots, X_n) &= \mathbf{E}(S_n + X_{n+1} \mid X_1, \ldots, X_n). \\ &= \mathbf{E}(S_n \mid X_1, \ldots, X_n) + \mathbf{E}(X_{n+1} \mid X_1, \ldots, X_n) \\ &= S_n + 0, \quad \text{by independence.} \end{aligned}$$ ●

(7) **Example. Markov chains.** Let $X_0, X_1, \ldots$ be a discrete-time Markov chain taking values in some countable state space S with transition matrix $\boldsymbol{P}$. Suppose that $\psi: S \to \mathbb{R}$ is a bounded function which satisfies

(8) $$\sum_{j \in S} p_{ij}\psi(j) = \psi(i) \quad \text{for all } i \in S.$$

We claim that $S_n = \psi(X_n)$ constitutes a martingale (with respect to $\{X_n\}$).

For,

$$\begin{aligned}\mathbf{E}(S_{n+1} \mid X_1, \ldots, X_n) &= \mathbf{E}(\psi(X_{n+1}) \mid X_1, \ldots, X_n)\\ &= \mathbf{E}(\psi(X_{n+1}) \mid X_n) \quad \text{by the Markov property}\\ &= \sum_{j \in S} p_{X_n, j}\psi(j)\\ &= \psi(X_n) = S_n \text{ by (8).}\end{aligned}$$

●

(9) **Example.** Let $X_1, X_2, \ldots$ be independent variables with zero means, finite variances, and partial sums $S_n = \sum_{i=1}^{n} X_i$. Define

$$T_n = S_n^2 = \left(\sum_{i=1}^{n} X_i\right)^2.$$

Then

$$\begin{aligned}\mathbf{E}(T_{n+1} \mid X_1, \ldots, X_n) &= \mathbf{E}(S_n^2 + 2S_n X_{n+1} + X_{n+1}^2 \mid X_1, \ldots, X_n)\\ &= T_n + 2\mathbf{E}(X_{n+1})\mathbf{E}(S_n \mid X_1, \ldots, X_n) + \mathbf{E}(X_{n+1}^2)\\ &\qquad\qquad \text{by independence}\\ &= T_n + \mathbf{E}(X_{n+1}^2) \geqslant T_n.\end{aligned}$$

$\{T_n\}$ is not a martingale, since it only satisfies (4) with $\geqslant$ in place of $=$; it is called a 'submartingale', and has properties similar to those of a martingale. ●

These examples show that martingales are all around us. They are extremely useful because, subject to a condition on their moments, they always converge; this is 'Doob's convergence theorem' and is the main result of the next section. Martingales are explored in considerable detail in Chapter 12.

Finally, here are some properties of conditional expectation. Do not read them straightaway, but refer back to them when necessary. Recall that the conditional expectation of X given Y is defined by

$$\mathbf{E}(X \mid Y) = \psi(Y) \quad \text{where} \quad \psi(y) = \mathbf{E}(X \mid Y = y)$$

is the mean of the conditional distribution of X given that $Y = y$. Most of the conditional expectations in this chapter take the form $\mathbf{E}(X \mid \boldsymbol{Y})$, the mean value of X conditional on the values of the variables in the random vector $\boldsymbol{Y} = (Y_1, Y_2, \ldots, Y_n)$. We stress that $\mathbf{E}(X \mid \boldsymbol{Y})$ is a function of $\boldsymbol{Y}$ alone. Expressions like '$\mathbf{E}(X \mid \boldsymbol{Y}) = Z$' should sometimes be qualified by 'almost surely'; we omit this qualification always.

(10) **Lemma.**

(a) $\mathbf{E}(X_1 + X_2 \mid \boldsymbol{Y}) = \mathbf{E}(X_1 \mid \boldsymbol{Y}) + \mathbf{E}(X_2 \mid \boldsymbol{Y})$
(b) $\mathbf{E}(Xg(\boldsymbol{Y}) \mid \boldsymbol{Y}) = g(\boldsymbol{Y})\mathbf{E}(X \mid \boldsymbol{Y})$ *for nice functions* $g\colon \mathbb{R}^n \to \mathbb{R}$
(c) $\mathbf{E}(X \mid h(\boldsymbol{Y})) = \mathbf{E}(X \mid \boldsymbol{Y})$ *if* $h\colon \mathbb{R}^n \to \mathbb{R}^n$ *is one–one.*

Proof.

(a) This depends on the linearity of expectation only.

(b) $\mathbf{E}(Xg(Y) \mid Y = y) = g(y)\mathbf{E}(X \mid Y = y)$.

(c) Roughly speaking, knowledge of Y is interchangeable with knowledge of $h(Y)$, in that

$$Y(\omega) = y \text{ if and only if } h(Y(\omega)) = h(y), \quad \text{for any } \omega \in \Omega. \qquad \blacksquare$$

(11) **Lemma.** $\mathbf{E}[\mathbf{E}(X \mid Y_1, Y_2) \mid Y_1] = \mathbf{E}(X \mid Y_1)$.

Proof. Just write down these expectations as integrals involving conditional distributions to see that the result holds. It is a more general version of Problem (4.11.29). ■

Sometimes we consider the mean value $\mathbf{E}(X \mid A)$ of a random variable X conditional upon the occurrence of some event A. This is just the mean of the corresponding distribution function

$$F_{X|A}(x) = \mathbf{P}(X \leqslant x \mid A).$$

We can think of $\mathbf{E}(X \mid A)$ as a constant random variable with domain $A \subseteq \Omega$; it is undefined at points $\omega \in A^c$. The following result is an application of (1.4.4).

(12) **Lemma.** *If* $\{B_i: 1 \leqslant i \leqslant n\}$ *is a partition of* A *then*

$$\mathbf{E}(X \mid A)\mathbf{P}(A) = \sum_{i=1}^{n} \mathbf{E}(X \mid B_i)\mathbf{P}(B_i).$$

You may like the following proof:

$$\mathbf{E}(XI_A) = \mathbf{E}\left(X \sum_i I_{B_i}\right) = \sum_i \mathbf{E}(XI_{B_i}).$$

Sometimes we consider mixtures of these two types of conditional expectation. These are of the form $\mathbf{E}(X \mid Y, A)$ where $X, Y_1, \dots, Y_n$ are random variables and A is an event. Such quantities are defined in the obvious way and have the usual properties. For example, (11) becomes

(13) $$\mathbf{E}(X \mid A) = \mathbf{E}[\mathbf{E}(X \mid Y, A) \mid A].$$

We shall make some use of the following fact soon. If, in (13), A is an event which is defined in terms of the Y_i (such as $A = \{Y_1 \leqslant 1\}$ or $A = \{|Y_2 Y_3 - Y_4| > 2\}$) then it is not difficult to see that

(14) $$\mathbf{E}[\mathbf{E}(X \mid Y) \mid A] = \mathbf{E}[\mathbf{E}(X \mid Y, A) \mid A];$$

just note that evaluating the random variable $\mathbf{E}(X \mid Y, A)$ at a point $\omega \in \Omega$ yields

$$\mathbf{E}(X \mid Y, A)(\omega)\begin{cases} = \mathbf{E}(X \mid Y)(\omega) & \text{if} \quad \omega \in A \\ \text{is undefined} & \text{if} \quad \omega \notin A. \end{cases}$$

The sequences $\{S_n\}$ of this section satisfy

(15) $$\mathbf{E}|S_n| < \infty, \qquad \mathbf{E}(S_{n+1} \mid X_1, \dots, X_n) = S_n.$$

(16) **Lemma.** *If* $\{S_n\}$ *satisfies* (15) *then*

(a) $\mathbf{E}(S_{n+m} \mid X_1, \dots, X_n) = S_n$ *for all* $n, m \geqslant 1$
(b) $\mathbf{E}(S_n) = \mathbf{E}(S_1)$ *for all* n.

Proof.

(a) Use (11) with $X = S_{n+m}$, $Y_1 = (X_1, \dots, X_n)$, and

$$Y_2 = (X_{n+1}, \dots, X_{n+m-1})$$

to obtain

$$\begin{aligned}\mathbf{E}(S_{n+m} \mid X_1, \dots, X_n) &= \mathbf{E}[\mathbf{E}(S_{n+m} \mid X_1, \dots, X_{n+m-1}) \mid X_1, \dots, X_n] \\ &= \mathbf{E}(S_{n+m-1} \mid X_1, \dots, X_n)\end{aligned}$$

and iterate to obtain the result.

(b) $\mathbf{E}(S_n) = \mathbf{E}(\mathbf{E}(S_n \mid X_1)) = \mathbf{E}(S_1)$ by (a). ■

For a more satisfactory account of conditional expectation, see Section 7.9.

Exercises

(17) Let $X_1, X_2, \dots$ be random variables such that the partial sums $S_n = X_1 + X_2 + \cdots + X_n$ determine a martingale. Show that $\mathbf{E}(X_i X_j) = 0$ if $i \neq j$.

(18) Let Z_n be the size of the nth generation of a branching process with immigration, in which the family-sizes have mean μ ($\neq 1$) and the mean number of immigrants in each generation is m. Suppose that $\mathbf{E}(Z_0) < \infty$, and show that

$$S_n = \mu^{-n}\left\{Z_n - m\left(\frac{1-\mu^n}{1-\mu}\right)\right\}$$

is a martingale with respect to a suitable sequence of random variables.

(19) Let $X_0, X_1, X_2, \dots$ be a sequence of random variables with finite means and satisfying $\mathbf{E}(X_{n+1} \mid X_0, X_1, \dots, X_n) = aX_n + bX_{n-1}$ for $n \geqslant 1$, where $0 < a, b < 1$ and $a + b = 1$. Find a value of α for which $S_n = \alpha X_n + X_{n-1}$, $n \geqslant 1$, defines a martingale with respect to the sequence X.

(20) Let X_n be the net profit to the gambler of betting a unit stake on the nth play in a casino; the X_n may be dependent, but the game is fair in the sense that

$$\mathbf{E}(X_{n+1} \mid X_1, X_2, \dots, X_n) = 0$$

for all n. The gambler stakes Y on the first play, and henceforth stakes $f_n(X_1, X_2, \ldots, X_n)$ on the $(n+1)$th play, where $f_1, f_2, \ldots$ are given functions. Show that his profit after n plays is

$$S_n = \sum_{i=1}^{n} X_i f_{i-1}(X_1, \ldots, X_{i-1}),$$

where $f_0 = Y$, and show further that S satisfies the martingale condition $\mathbf{E}(S_{n+1} \mid X_1, X_2, \ldots, X_n) = S_n$, $n \geqslant 1$, if Y is assumed to be known throughout.

7.8 Martingale convergence theorem

This section is devoted to the proof and subsequent applications of the following theorem. It receives a section to itself by virtue of its wealth of applications.

(1) **Theorem.** *If $\{S_n\}$ is a martingale with $\mathbf{E}(S_n^2) < M < \infty$ for some M and all n, then there exists a random variable S such that S_n converges to S almost surely and in mean square.*

This result has a more general version which, amongst other things,

(i) deals with submartingales
(ii) imposes weaker moment conditions
(iii) explores convergence in mean also

but the proof of this is more difficult. On the other hand, the proof of (1) is within our grasp, and is only slightly more difficult than the proof of the strong law (7.5.1) for independent sequences; it mimics the traditional proof of the strong law and begins with a generalization of Chebyshov's inequality. We return to the theory of martingales in much greater generality in Chapter 12.

(2) **Theorem. Doob–Kolmogorov inequality.** *If $\{S_n\}$ is a martingale with respect to $\{X_n\}$ then*

$$\mathbf{P}\left(\max_{1 \leqslant i \leqslant n} |S_i| \geqslant \varepsilon\right) \leqslant \frac{1}{\varepsilon^2}\mathbf{E}(S_n^2) \quad \textit{whenever} \quad \varepsilon > 0.$$

Proof of (2). Let $A_0 = \Omega$, $A_k = \{|S_i| < \varepsilon \text{ for all } i \leqslant k\}$, and let $B_k = A_{k-1} \cap \{|S_k| \geqslant \varepsilon\}$ be the event that $|S_i| \geqslant \varepsilon$ for the first time when $i = k$. Then

$$A_k \cup \left(\bigcup_{i=1}^{k} B_i\right) = \Omega.$$

Therefore

(3)
$$\mathbf{E}(S_n^2) = \sum_{i=1}^{n} \mathbf{E}(S_n^2 I_{B_i}) + \mathbf{E}(S_n^2 I_{A_n})$$
$$\geqslant \sum_{i=1}^{n} \mathbf{E}(S_n^2 I_{B_i}).$$

However,

$$\begin{aligned}\mathbf{E}(S_n^2 I_{B_i}) &= \mathbf{E}((S_n - S_i + S_i)^2 I_{B_i}) \\ &= \mathbf{E}((S_n - S_i)^2 I_{B_i}) + 2\mathbf{E}((S_n - S_i)S_i I_{B_i}) + \mathbf{E}(S_i^2 I_{B_i}) \\ &= \alpha + \beta + \gamma, \text{ say.}\end{aligned}$$

Note that $\alpha \geqslant 0$ and $\gamma \geqslant \varepsilon^2 \mathbf{P}(B_i)$, because $|S_i| \geqslant \varepsilon$ if B_i occurs. To deal with β, note that

$$\begin{aligned}\mathbf{E}((S_n - S_i)S_i I_{B_i}) &= \mathbf{E}[S_i I_{B_i} \mathbf{E}(S_n - S_i \mid X_1, \dots, X_i)] \text{ by (7.7.10b)} \\ &= 0 \quad \text{by (7.7.16a),}\end{aligned}$$

since B_i concerns $X_1, \dots, X_i$ only, by the discussion after (7.7.4). Thus (3) becomes

$$\mathbf{E}(S_n^2) \geqslant \sum_{i=1}^{n} \varepsilon^2 \mathbf{P}(B_i) = \varepsilon^2 \mathbf{P}\left(\max_{1 \leqslant i \leqslant n} |S_i| \geqslant \varepsilon\right)$$

and the result is shown. ■

Proof of (1). First note that S_m and $(S_{m+n} - S_m)$ are uncorrelated whenever $m, n \geqslant 1$; for,

$$\mathbf{E}(S_m(S_{m+n} - S_m)) = \mathbf{E}[S_m \mathbf{E}(S_{m+n} - S_m \mid X_1, \dots, X_m)] = 0$$

by (7.7.16). Thus

(4)
$$\mathbf{E}(S_{m+n}^2) = \mathbf{E}(S_m^2) + \mathbf{E}((S_{m+n} - S_m)^2).$$

It follows that $\{\mathbf{E}(S_n^2)\}$ is a non-decreasing sequence, which is bounded above, by the assumption in (1); hence we may suppose that the constant M is chosen such that

$$\mathbf{E}(S_n^2) \uparrow M \quad \text{as} \quad n \to \infty.$$

We shall show that the sequence $\{S_n(\omega): n \geqslant 1\}$ is Cauchy convergent for all ω belonging to some event C having probability 1 (see the notes on Cauchy convergence after (7.2.2)). On this event C, $\{S_n(\omega)\}$ converges to a limit $S(\omega)$, and therefore almost-sure convergence will have been proved. Let $C = \{\omega \in \Omega: \{S_n(\omega)\} \text{ is Cauchy convergent}\}$. Then

$$C = \{\forall \varepsilon > 0, \exists m \text{ such that } |S_{m+i} - S_{m+j}| < \varepsilon \text{ for all } i, j \geqslant 1\}.$$

By the triangle inequality

$$|S_{m+i} - S_{m+j}| \leqslant |S_{m+i} - S_m| + |S_{m+j} - S_m|,$$

so that

$$C = \{\forall \varepsilon > 0, \exists m \text{ such that } |S_{m+i} - S_m| < \varepsilon \text{ for all } i \geqslant 1\}$$
$$= \bigcap_{\varepsilon>0} \bigcup_m \{|S_{m+i} - S_m| < \varepsilon \text{ for all } i \geqslant 1\}.$$

Therefore the complement of C may be expressed as

$$C^c = \bigcup_{\varepsilon>0} \bigcap_m \{|S_{m+i} - S_m| \geqslant \varepsilon \text{ for some } i \geqslant 1\} = \bigcup_{\varepsilon>0} \bigcap_m A_m(\varepsilon)$$

where $A_m(\varepsilon) = \{|S_{m+i} - S_m| \geqslant \varepsilon \text{ for some } i \geqslant 1\}$. Now $A_m(\varepsilon) \subseteq A_m(\varepsilon')$ if $\varepsilon \geqslant \varepsilon'$, so that

$$\mathbf{P}(C^c) = \lim_{\varepsilon \downarrow 0} \mathbf{P}\left(\bigcap_m A_m(\varepsilon)\right) \leqslant \lim_{\varepsilon \downarrow 0} \lim_{m\to\infty} \mathbf{P}(A_m(\varepsilon)).$$

In order to prove that $\mathbf{P}(C^c) = 0$ as required, it suffices to show that $\mathbf{P}(A_m(\varepsilon)) \to 0$ as $m \to \infty$ for all $\varepsilon > 0$. To this end we shall use the Doob–Kolmogorov inequality.

For a given choice of m, define the sequence $Y = \{Y_n : n \geqslant 1\}$ by $Y_n = S_{m+n} - S_m$. It may be checked that Y is a martingale with respect to itself:

$$\mathbf{E}(Y_{n+1} \mid Y_1, \dots, Y_n) = \mathbf{E}[\mathbf{E}(Y_{n+1} \mid X_1, \dots, X_{m+n}) \mid Y_1, \dots, Y_n]$$
$$= \mathbf{E}(Y_n \mid Y_1, \dots, Y_n) = Y_n$$

by (7.7.11) and the martingale property. We apply the Doob–Kolmogorov inequality (2) to this martingale to find that

$$\mathbf{P}(|S_{m+i} - S_m| \geqslant \varepsilon \text{ for some } 1 \leqslant i \leqslant n) \leqslant \frac{1}{\varepsilon^2} \mathbf{E}((S_{m+n} - S_m)^2).$$

Letting $n \to \infty$ and using (4) we obtain $\mathbf{P}(A_m(\varepsilon)) \leqslant \varepsilon^{-2}[M - \mathbf{E}(S_m^2)]$, and hence $\mathbf{P}(A_m(\varepsilon)) \to 0$ as $m \to \infty$ as required for almost-sure convergence. We have proved that there exists a random variable S such that $S_n \xrightarrow{\text{a.s.}} S$.

It remains only to prove convergence of S_n to S in mean square. For this we need Fatou's lemma (5.6.13). It is the case that

$$\mathbf{E}((S_n - S)^2) = \mathbf{E}\left(\liminf_{m\to\infty} (S_n - S_m)^2\right) \leqslant \liminf_{m\to\infty} \mathbf{E}((S_n - S_m)^2)$$
$$= M - \mathbf{E}(S_n^2) \to 0 \quad \text{as} \quad n \to \infty,$$

and the proof is finished. ■

Here are some applications of the martingale convergence theorem.

(5) **Example. Branching processes.** Recall (7.7.5). By (5.4.2),

$$W_n = Z_n / \mathbf{E}(Z_n)$$

has second moment

$$\mathbf{E}(W_n^2) = 1 + \frac{\sigma^2(1-\mu^{-n})}{\mu(\mu-1)} \quad \text{if} \quad \mu \neq 1$$

where $\sigma^2 = \text{var}(Z_1)$. Thus, if $\mu \neq 1$, there exists a random variable W such that

$$W_n \xrightarrow{\text{a.s.}} W$$

and so $W_n \xrightarrow{\text{D}} W$ also; their characteristic functions satisfy

$$\phi_{W_n}(t) \to \phi_W(t)$$

by (5.9.5). This makes the discussion at the end of Section 5.4 fully rigorous, and we can rewrite (5.4.6) as

$$\phi_W(\mu t) = G(\phi_W(t)). \qquad \bullet$$

(6) **Example. Markov chains.** Suppose that the chain $X_0, X_1, \ldots$ of (7.7.7) is irreducible and persistent, and let ψ be a bounded function mapping S into $\mathbb{R}$ which satisfies (7.7.8). Then the sequence $\{S_n\}$, given by $S_n = \psi(X_n)$, is a martingale and satisfies the condition

$$\mathbf{E}(S_n^2) \leqslant M$$

for some M, by the boundedness of ψ. For any state i, the event $\{X_n = i\}$ occurs for infinitely many values of n with probability 1. However, $\{S_n = \psi(i)\} \supseteq \{X_n = i\}$ and so

$$S_n \xrightarrow{\text{a.s.}} \psi(i) \quad \text{for all } i,$$

which is clearly impossible unless $\psi(i)$ is the same for all i. We have shown that any bounded solution of (7.7.8) is constant. $\bullet$

(7) **Example. Genetic model.** Recall Example (6.1.11), which dealt with gene frequencies in the evolution of a population. We encountered there a Markov chain $X_0, X_1, \ldots$ taking values in $\{0, 1, \ldots, N\}$ with transition probabilities given by

(8) $$p_{ij} = \mathbf{P}(X_{n+1} = j \mid X_n = i) = \binom{N}{j}\left(\frac{i}{N}\right)^j\left(1 - \frac{i}{N}\right)^{N-j}.$$

Then

$$\begin{aligned}\mathbf{E}(X_{n+1} \mid X_0, \ldots, X_n) &= \mathbf{E}(X_{n+1} \mid X_n) \quad \text{by the Markov property} \\ &= \sum_j j p_{X_n, j} = X_n\end{aligned}$$

by (8). Thus $X_0, X_1, \ldots$ is a martingale. Also, let Y_n be defined by

$$Y_n = X_n(N - X_n),$$

and suppose that $N > 1$. Then

$$\mathbf{E}(Y_{n+1} \mid X_0, \ldots, X_n) = \mathbf{E}(Y_{n+1} \mid X_n) \quad \text{by the Markov property}$$

and

$$\mathbf{E}(Y_{n+1} \mid X_n = i) = \sum_j j(N-j)p_{ij} = i(N-i)(1 - N^{-1})$$

by (8). Thus

(9) $$\mathbf{E}(Y_{n+1} \mid X_0, \ldots, X_n) = Y_n(1 - N^{-1}),$$

and we see that $\{Y_n\}$ is not itself a martingale. However, set $S_n = Y_n/(1 - N^{-1})^n$ to obtain from (9) that

$$\mathbf{E}(S_{n+1} \mid X_0, \ldots, X_n) = S_n;$$

deduce that $\{S_n\}$ is a martingale.

The martingale $\{X_n\}$ has uniformly bounded second moments, and so there exists an X such that

$$X_n \xrightarrow{\text{a.s.}} X.$$

Unlike the previous example, this chain is not irreducible. In fact, 0 and N are absorbing states, and X takes these values only. Can you find the probability $\mathbf{P}(X = 0)$ that the chain is ultimately absorbed at 0? The results of the next section will help you with this.

Finally, what happens when we apply the convergence theorem to $\{S_n\}$? ●

Exercises

(10) **Kolmogorov's inequality.** Let $X_1, X_2, \ldots$ be independent random variables with zero means and finite variances, and let $S_n = X_1 + X_2 + \cdots + X_n$. Use the Doob–Kolmogorov inequality to show that

$$\mathbf{P}\left(\max_{1 \leqslant j \leqslant n} |S_j| > \varepsilon\right) \leqslant \frac{1}{\varepsilon^2} \sum_{j=1}^{n} \text{var}(X_j) \quad \text{for} \quad \varepsilon > 0.$$

(11) Let $X_1, X_2, \ldots$ be independent random variables such that

$$\sum_n \frac{1}{n^2} \text{var}(X_n) < \infty.$$

Use Kolmogorov's inequality to prove that

$$\sum_{i=1}^{n} \frac{X_i - \mathbf{E}(X_i)}{i} \xrightarrow{\text{a.s.}} Y \quad \text{as} \quad n \to \infty,$$

for some finite random variable Y, and deduce that

$$\frac{1}{n}\sum_{i=1}^{n}(X_i - \mathbf{E}X_i) \xrightarrow{\text{a.s.}} 0 \quad \text{as} \quad n \to \infty.$$

(You may find Kronecker's lemma to be useful: if (a_n) and (b_n) are real sequences with $b_n \to \infty$ and $\sum_i a_i/b_i < \infty$, then $b_n^{-1}\sum_{i=1}^{n} a_i \to 0$ as $n \to \infty$.)

(12) Let S be a martingale with respect to X, such that $\mathbf{E}(S_n^2) < K < \infty$ for some $K \in \mathbb{R}$. Suppose that $\text{var}(S_n) \to 0$ as $n \to \infty$, and prove that $S = \lim_{n\to\infty} S_n$ exists and is constant almost surely.

7.9 Prediction and conditional expectation

Probability theory is not merely an intellectual pursuit, but provides also a framework for estimation and prediction. Practical men and women often need to make guesses about things which are not easily measurable, either because they lie in the future or because of some intrinsic inaccessibility; in doing so they usually make use of some current or feasible observation. Economic examples are commonplace (business trends, inflation rates, and so on); other examples include weather prediction, the climate in prehistoric times, the state of the core of a nuclear reactor, the cause of a disease in an individual or a population, or the paths of celestial bodies. This last problem has the distinction of being amongst the first to be tackled by mathematicians using a modern approach to probability.

At its least complicated, a question of prediction or estimation involves an unknown or unobserved random variable Y, about which we are provided with the value of some (observable) random variable X. The problem is to deduce information about the value of Y from a knowledge of the value of X. Thus we seek a function $h(X)$ which is (in some sense) close to Y; we write $\hat{Y} = h(X)$ and call $\hat{Y}$ an 'estimator' of Y. As we saw in Section 7.1, there are many different ways in which two random variables may be said to be close to one another—pointwise, in rth mean, in probability, and so on. A particular way of especial convenience is to work with the norm given by

$$\|U\|_2 = \{\mathbf{E}(U^2)\}^{\frac{1}{2}}, \tag{1}$$

so that the distance between two random variables U and V is

$$\|U - V\|_2 = [\mathbf{E}\{(U - V)^2\}]^{\frac{1}{2}}; \tag{2}$$

$\|\cdot\|_2$ is often called the L_2 norm, and the corresponding notion of convergence is of course convergence in mean square:

$$\|U_n - U\|_2 \to 0 \quad \text{if and only if} \quad U_n \xrightarrow{\text{m.s.}} U. \tag{3}$$

This norm is a special case of the 'L_p norm' given by $\|X\|_p = \{\mathbf{E}|X^p|\}^{1/p}$ where $p \geqslant 1$.

We recall that $\|\cdot\|_2$ satisfies the *triangle inequality*:

(4) $$\|U + V\|_2 \leqslant \|U\|_2 + \|V\|_2.$$

With this notation, we make the following definition.

(5) **Definition.** Let X and Y be random variables on $(\Omega, \mathscr{F}, \mathbf{P})$ such that $\mathbf{E}(Y^2) < \infty$. The **minimum mean-squared-error predictor** (or **best predictor**) of Y given X is the function $\hat{Y} = h(X)$ of X for which $\|Y - \hat{Y}\|_2$ is a minimum.

We shall commonly use the term 'best predictor' in this context; the word 'best' is only shorthand, and should not be interpreted literally.

Let H be the set of all functions of X having finite second moment:

(6) $$H = \{h(X): h \text{ maps } \mathbb{R} \text{ to } \mathbb{R}, \mathbf{E}(h(X)^2) < \infty\}.$$

The best (or minimum mean-squared-error) predictor of Y is (if it exists) a random variable $\hat{Y}$ belonging to H such that $\mathbf{E}((Y - \hat{Y})^2) \leqslant \mathbf{E}((Y - Z)^2)$ for all $Z \in H$. Does there exist such a $\hat{Y}$? The answer is yes, and furthermore there is (essentially) a unique such $\hat{Y}$ in H. In proving this we shall make use of two properties of H, that it is a linear space, and that it is closed (with respect to the norm $\|\cdot\|_2$); that is to say, for $Z_1, Z_2, \ldots \in H$ and $a_1, a_2, \ldots \in \mathbb{R}$,

(7) $$a_1 Z_1 + a_2 Z_2 + \cdots + a_n Z_n \in H,$$

and

(8) $$\text{if } \|Z_n - Z_m\|_2 \to 0 \text{ as } m, n \to \infty, \text{ there exists } Z \in H \text{ such that } Z_n \xrightarrow{\text{m.s.}} Z.$$

(See Exercise (7.9.34a).) More generally, we call a set H of random variables a *closed linear space* (with respect to $\|\cdot\|_2$) if $\|X\|_2 < \infty$ for all $X \in H$, and H satisfies (7) and (8).

(9) **Theorem.** *Let H be a closed linear space (with respect to $\|\cdot\|_2$) of random variables. Let Y be a random variable on $(\Omega, \mathscr{F}, \mathbf{P})$ with finite variance. There exists a random variable $\hat{Y}$ in H such that*

(10) $$\|Y - \hat{Y}\|_2 \leqslant \|Y - Z\|_2 \quad \textit{for all } Z \in H,$$

and furthermore $\hat{Y}$ is unique in the sense that $\mathbf{P}(\hat{Y} = \bar{Y}) = 1$ for any $\bar{Y} \in H$ with $\|Y - \bar{Y}\|_2 = \|Y - \hat{Y}\|_2$.

Proof. Let $d = \inf\{\|Y - Z\|_2 : Z \in H\}$, and find a sequence $Z_1, Z_2, \ldots$ in H such that $\lim_{n\to\infty} \|Y - Z_n\|_2 = d$. Now, for any $A, B \in H$, the 'parallelogram rule' holds:

(11) $$\|A - B\|_2^2 = 2\big[\|Y - A\|_2^2 - 2\|Y - \tfrac{1}{2}(A + B)\|_2^2 + \|Y - B\|_2^2\big];$$

to show this, just expand the right-hand side as in Exercise (7.9.1b). Note that $\frac{1}{2}(A + B) \in H$ since H is a linear space. Setting $A = Z_n$, $B = Z_m$, we obtain using the definition of d that

$$\|Z_n - Z_m\|_2^2 \leqslant 2(\|Y - Z_n\|_2^2 - 2d + \|Y - Z_m\|_2^2) \to 0 \quad \text{as} \quad n \to \infty.$$

Therefore $\|Z_n - Z_m\|_2 \to 0$, so that there exists $\hat{Y} \in H$ such that $Z_n \xrightarrow{\text{m.s.}} \hat{Y}$; it is here that we use the fact (8) that H is closed. It follows by the triangle inequality (4) that

$$\|Y - \hat{Y}\|_2 \leqslant \|Y - Z_n\|_2 + \|Z_n - \hat{Y}\|_2 \to d \quad \text{as} \quad n \to \infty,$$

so that $\hat{Y}$ satisfies (10).

Finally, suppose that $\bar{Y} \in H$ satisfies $\|Y - \bar{Y}\|_2 = d$. Apply (11) with $A = \bar{Y}$, $B = \hat{Y}$, to obtain

$$\|\bar{Y} - \hat{Y}\|_2^2 = 4[d^2 - \|Y - \tfrac{1}{2}(\bar{Y} + \hat{Y})\|_2^2] \leqslant 4(d^2 - d^2) = 0.$$

Hence $\mathbf{E}((\bar{Y} - \hat{Y})^2) = 0$ and so $\mathbf{P}(\hat{Y} = \bar{Y}) = 1$. ■

(12) **Example.** Let Y have mean μ and variance σ^2. With no information about Y, it is appropriate to ask for the real number h which minimizes $\|Y - h\|_2$. Now

$$\|Y - h\|_2^2 = \mathbf{E}((Y - h)^2) = \sigma^2 + (\mu - h)^2$$

so that μ is the best predictor of Y. The set H of possible estimators is the real line $\mathbb{R}$. ●

(13) **Example.** Let $X_1, X_2, \ldots$ be uncorrelated random variables with zero means and unit variances. It is desired to find the best predictor of Y amongst the class of linear combinations of the X_i. Clearly

$$\begin{aligned}\mathbf{E}\left(\left(Y - \sum_i a_i X_i\right)^2\right) &= \mathbf{E}(Y^2) - 2\sum_i a_i \mathbf{E}(X_i Y) + \sum_i a_i^2 \\ &= \mathbf{E}(Y^2) + \sum_i [a_i - \mathbf{E}(X_i Y)]^2 - \sum_i \mathbf{E}(X_i Y)^2.\end{aligned}$$

This is a minimum when $a_i = \mathbf{E}(X_i Y)$ for all i, so that $\hat{Y} = \sum_i X_i \mathbf{E}(X_i Y)$. (*Exercise*: prove that $\mathbf{E}(\hat{Y}^2) < \infty$.) This is seen best in the following light. Thinking of the X_i as orthogonal (that is, uncorrelated) unit vectors in the space H of linear combinations of the X_i, we have found that $\hat{Y}$ is the weighted average of the X_i, weighted in proportion to the magnitudes of their 'projections' on to Y. ●

The implied geometry of this example is relevant to the next theorem.

(14) **Projection theorem.** *Let H be a closed linear space (with respect to $\|\cdot\|_2$) of random variables, and let Y satisfy $\mathbf{E}(Y^2) < \infty$. Let $M \in H$. The following two*

statements are equivalent:

(15) $$\mathbf{E}((Y-M)Z)=0 \quad \textit{for all } Z \in H,$$

(16) $$\|Y-M\|_2 \leqslant \|Y-Z\|_2 \quad \textit{for all } Z \in H.$$

Here is the geometrical intuition. Let $L_2(\Omega, \mathcal{F}, \mathbf{P})$ be the set of random variables on $(\Omega, \mathcal{F}, \mathbf{P})$ having finite variance. Now H is a linear subspace of $L_2(\Omega, \mathcal{F}, \mathbf{P})$; think of H as a hyperplane in a vector space of very large dimension. If $Y \notin H$ then the shortest route from Y to H is along the perpendicular from Y on to H. Writing $\hat{Y}$ for the foot of this perpendicular, we have that $Y-\hat{Y}$ is perpendicular to any vector in the hyperplane H. Translating this geometrical remark back into the language of random variables, we conclude that $\langle Y-\hat{Y}, Z\rangle = 0$ for all $Z \in H$, where $\langle U, V\rangle$ is the scalar product in $L_2(\Omega, \mathcal{F}, \mathbf{P})$ defined by $\langle U, V\rangle = \mathbf{E}(UV)$. These remarks do not of course constitute a proof of the theorem.

Proof. Suppose first that $M \in H$ satisfies (15). Then, for $M' \in H$,

$$\mathbf{E}((Y-M')^2) = \mathbf{E}((Y-M+M-M')^2) = \mathbf{E}((Y-M)^2) + \mathbf{E}((M-M')^2)$$

by (15), since $M-M' \in H$; therefore $\|Y-M\|_2 \leqslant \|Y-M'\|_2$ for all $M' \in H$.

Conversely, suppose that M satisfies (16), but that there exists $Z \in H$ such that $\mathbf{E}((Y-M)Z) = d > 0$. We may assume without loss of generality that $\mathbf{E}(Z^2)=1$; otherwise replace Z by $Z/\sqrt{(\mathbf{E}(Z^2))}$, noting that $\mathbf{E}(Z^2) \neq 0$ since $\mathbf{P}(Z=0) \neq 1$. Writing $M' = M + dZ$, we have that

$$\begin{aligned}\mathbf{E}((Y-M')^2) &= \mathbf{E}((Y-M+M-M')^2)\\ &= \mathbf{E}((Y-M)^2) - 2d\mathbf{E}((Y-M)Z) + d^2\mathbf{E}(Z^2)\\ &= \mathbf{E}((Y-M)^2) - d^2,\end{aligned}$$

contradicting the minimality of $\mathbf{E}((Y-M)^2)$. ■

It is only a tiny step from the projection theorem (14) to the observation, well known to statisticians, that the best predictor of Y given X is just the conditional expectation $\mathbf{E}(Y \mid X)$. This fact, easily proved directly (*exercise*), follows immediately from the projection theorem.

(17) **Theorem.** *Let X and Y be random variables, and suppose that $\mathbf{E}(Y^2) < \infty$. The best predictor of Y given X is the conditional expectation $\mathbf{E}(Y \mid X)$.*

Proof. Let H be the closed linear space of functions of X having finite second moment. Define $\psi(x) = \mathbf{E}(Y \mid X = x)$. Certainly $\psi(X)$ belongs to H, since

$$\mathbf{E}(\psi(X)^2) = \mathbf{E}(\mathbf{E}(Y \mid X)^2) \leqslant \mathbf{E}(\mathbf{E}(Y^2 \mid X)) = \mathbf{E}(Y^2),$$

where we have used the Cauchy–Schwarz inequality. On the other hand, for

$Z = h(X) \in H$,

$$\begin{aligned}\mathbf{E}([Y - \psi(X)]Z) &= \mathbf{E}(Yh(X)) - \mathbf{E}(\mathbf{E}(Y \mid X)h(X)) \\ &= \mathbf{E}(Yh(X)) - \mathbf{E}(\mathbf{E}(Yh(X) \mid X)) \\ &= \mathbf{E}(Yh(X)) - \mathbf{E}(Yh(X)) = 0,\end{aligned}$$

using the elementary fact that $\mathbf{E}(Yh(X) \mid X) = h(X)\mathbf{E}(Y \mid X)$. Applying the projection theorem, we find that $M = \psi(X)$ $(= \mathbf{E}(Y \mid X))$ minimizes $\| Y - M \|_2$ for $M \in H$, which is the claim of the theorem. ■

Here is an important step. We may take the conclusion of (17) as a *definition* of the conditional expectation $\mathbf{E}(Y \mid X)$: if $\mathbf{E}(Y^2) < \infty$, the *conditional expectation* $\mathbf{E}(Y \mid X)$ of Y given X is defined to be the best predictor of Y given X.

There are two major advantages of defining conditional expectation in this way. First, it is a definition which is valid for all pairs X, Y such that $\mathbf{E}(Y^2) < \infty$, regardless of their types (discrete, continuous, and so on). Secondly, it provides a route to a much more general notion of conditional expectation which is particularly relevant to the martingale theory of Chapter 12.

(18) **Example.** Let $X = \{X_i : i \in I\}$ be a family of random variables, and let H be the space of all functions of the X_i with finite second moments. If $\mathbf{E}(Y^2) < \infty$, the *conditional expectation* $\mathbf{E}(Y \mid X_i, i \in I)$ of Y given the X_i is defined to be the function $M = \psi(X) \in H$ which minimizes the mean squared error $\| Y - M \|_2$ over all M in H. Note that $\psi(X)$ satisfies

(19) $$\mathbf{E}([Y - \psi(X)]Z) = 0 \quad \text{for all } Z \in H,$$

and $\psi(X)$ is unique in the sense that $\mathbf{P}(\psi(X) = N) = 1$ if $\| Y - \psi(X) \|_2 = \| Y - N \|_2$ for any $N \in H$. We note here that, strictly speaking, conditional expectations are not actually *unique*; this causes no difficulty, and we shall therefore continue to speak in terms of *the* conditional expectation. ●

We move on to an important generalization of the idea of conditional expectation, involving 'conditioning on a σ-field'. Let Y be a random variable on $(\Omega, \mathscr{F}, \mathbf{P})$ having finite second moment, and let $\mathscr{G}$ be a sub-σ-field of $\mathscr{F}$. Let H be the space of random variables which are $\mathscr{G}$-measurable and have finite second moment. That is to say, H contains those random variables Z such that $\mathbf{E}(Z^2) < \infty$ and $\{Z \leqslant z\} \in \mathscr{G}$ for all $z \in \mathbb{R}$. It is not difficult to see that H is a closed linear space with respect to $\| \cdot \|_2$. We have from (9) that there exists an element M of H such that $\| Y - M \|_2 \leqslant \| Y - Z \|_2$ for all $Z \in H$, and furthermore M is unique (in the usual way) with this property. We call M the 'conditional expectation of Y given the σ-field $\mathscr{G}$', written $\mathbf{E}(Y \mid \mathscr{G})$.

This is a more general definition of conditional expectation than that obtained by conditioning on a family of random variables (as in the previous example). To see this, take $\mathscr{G}$ to be the smallest σ-field with respect to which every member of the family $X = \{X_i : i \in I\}$ is measurable. It is clear that $\mathbf{E}(Y \mid \mathscr{G}) = \mathbf{E}(Y \mid X_i, i \in I)$, in the sense that they are equal with probability 1.

We arrive at the following definition by use of the projection theorem (14).

(20) **Definition.** Let $(\Omega, \mathscr{F}, \mathbf{P})$ be a probability space, and let Y be a random variable satisfying $\mathbf{E}(Y^2) < \infty$. If $\mathscr{G}$ is a sub-σ-field of $\mathscr{F}$, the **conditional expectation** $\mathbf{E}(Y \mid \mathscr{G})$ is a $\mathscr{G}$-measurable random variable satisfying

(21) $$\mathbf{E}([Y - \mathbf{E}(Y \mid \mathscr{G})]Z) = 0 \quad \text{for all } Z \in H,$$

where H is the collection of all $\mathscr{G}$-measurable random variables with finite second moment.

There are certain members of H with particularly simple form, being the indicator functions of events in $\mathscr{G}$. It may be shown without great difficulty that condition (21) may be replaced by

(22) $$\mathbf{E}([Y - \mathbf{E}(Y \mid \mathscr{G})]I_G) = 0 \quad \text{for all } G \in \mathscr{G}.$$

Setting $G = \Omega$, we deduce the important fact that

(23) $$\mathbf{E}(\mathbf{E}(Y \mid \mathscr{G})) = \mathbf{E}(Y).$$

(24) **Example. Doob's martingale** (though some ascribe this to Lévy). Let Y have finite second moment, and let $X_1, X_2, \ldots$ be a sequence of random variables. Define $Y_n = \mathbf{E}(Y \mid X_1, X_2, \ldots, X_n)$. Then $\{Y_n\}$ is a martingale with respect to $\{X_n\}$. To show this it is necessary to prove that $\mathbf{E}|Y_n| < \infty$ and $\mathbf{E}(Y_{n+1} \mid X_1, X_2, \ldots, X_n) = Y_n$. Certainly $\mathbf{E}|Y_n| < \infty$, since $\mathbf{E}(Y_n^2) < \infty$. For the other part, let H_n be the space of functions of $X_1, X_2, \ldots, X_n$ having finite second moment. We have by (19) that, for $Z \in H_n$,

$$\begin{aligned} 0 = \mathbf{E}((Y - Y_n)Z) &= \mathbf{E}((Y - Y_{n+1} + Y_{n+1} - Y_n)Z) \\ &= \mathbf{E}((Y_{n+1} - Y_n)Z) \quad \text{since} \quad Z \in H_n \subseteq H_{n+1}. \end{aligned}$$

Therefore $Y_n = \mathbf{E}(Y_{n+1} \mid X_1, X_2, \ldots, X_n)$.

Here is a more general formulation. Let Y be a random variable on $(\Omega, \mathscr{F}, \mathbf{P})$ with $\mathbf{E}(Y^2) < \infty$, and let $\{\mathscr{G}_n : n \geqslant 1\}$ be a sequence of σ-fields contained in $\mathscr{F}$ and satisfying $\mathscr{G}_n \subseteq \mathscr{G}_{n+1}$ for all n. Such a sequence $\{\mathscr{G}_n\}$ is called a *filtration*; in the context of the previous paragraph we might take $\mathscr{G}_n$ to be the smallest σ-field with respect to which $X_1, X_2, \ldots, X_n$ are each measurable. We define $Y_n = \mathbf{E}(Y \mid \mathscr{G}_n)$. As before $\{Y_n\}$ satisfies $\mathbf{E}|Y_n| < \infty$ and $\mathbf{E}(Y_{n+1} \mid \mathscr{G}_n) = Y_n$; such a sequence is called a 'martingale with respect to the filtration $\{\mathscr{G}_n\}$'. ●

This new type of conditional expectation has many useful properties. We single out one of these.

(25) **Theorem.** *Let Y have finite second moment and let $\mathcal{G}$ be a sub-σ-field of the σ-field $\mathcal{F}$. Then $\mathbf{E}(XY \mid \mathcal{G}) = X\mathbf{E}(Y \mid \mathcal{G})$ for all $\mathcal{G}$-measurable random variables X with finite second moments.*

Proof. Let X be $\mathcal{G}$-measurable with finite second moment. Clearly $Z = \mathbf{E}(XY \mid \mathcal{G}) - X\mathbf{E}(Y \mid \mathcal{G})$ is $\mathcal{G}$-measurable and satisfies

$$Z = X[Y - \mathbf{E}(Y \mid \mathcal{G})] - [XY - \mathbf{E}(XY \mid \mathcal{G})]$$

so that, for $G \in \mathcal{G}$,

$$\mathbf{E}(ZI_G) = \mathbf{E}([Y - \mathbf{E}(Y \mid \mathcal{G})]XI_G) - \mathbf{E}([XY - \mathbf{E}(XY \mid \mathcal{G})]I_G] = 0,$$

the first term being zero by the fact that XI_G is $\mathcal{G}$-measurable with finite second moment, and the second by the definition of $\mathbf{E}(XY \mid \mathcal{G})$. Any $\mathcal{G}$-measurable random variable Z satisfying $\mathbf{E}(ZI_G) = 0$ for all $G \in \mathcal{G}$ is such that $\mathbf{P}(Z = 0) = 1$ (just set $G_1 = \{Z > 0\}$, $G_2 = \{Z < 0\}$ in turn), and the result follows. ■

In all our calculations so far, we have used the norm $\|\cdot\|_2$, leading to a definition of $\mathbf{E}(Y \mid \mathcal{G})$ for random variables Y with $\mathbf{E}(Y^2) < \infty$. This condition of finite second moment is of course too strong in general, and needs to be replaced by the natural weaker condition that $\mathbf{E}|Y| < \infty$. One way of doing this would be to rework the previous arguments using instead the norm $\|\cdot\|_1$. An easier route is to use the technique of 'truncation' as in the following proof.

(26) **Theorem.** *Let Y be a random variable on $(\Omega, \mathcal{F}, \mathbf{P})$ with $\mathbf{E}|Y| < \infty$, and let $\mathcal{G}$ be a sub-σ-field of $\mathcal{F}$. There exists a random variable Z such that*

(a) *Z is $\mathcal{G}$-measurable*
(b) $\mathbf{E}|Z| < \infty$
(c) $\mathbf{E}((Y - Z)I_G) = 0$ *for all $G \in \mathcal{G}$.*

Z is unique in the sense that, for any Z' satisfying (a), (b), *and* (c), *we have that* $\mathbf{P}(Z = Z') = 1$.

The random variable Z in the theorem is called the 'conditional expectation of Y given $\mathcal{G}$', and is written $\mathbf{E}(Y \mid \mathcal{G})$. It is an *exercise* to prove that

(27) $$\mathbf{E}(XY \mid \mathcal{G}) = X\mathbf{E}(Y \mid \mathcal{G})$$

for all $\mathcal{G}$-measurable X, whenever both sides exist, and also that this definition coincides (almost surely) with the previous one when Y has finite second moment. A meaningful value can be assigned to $\mathbf{E}(Y \mid \mathcal{G})$ under the weaker assumption on Y that either $\mathbf{E}(Y^+) < \infty$ or $\mathbf{E}(Y^-) < \infty$.

Proof. Suppose first that $Y \geqslant 0$ and $\mathbf{E}|Y| < \infty$. Let $Y_n = \min\{Y, n\}$, so that $Y_n \uparrow Y$ as $n \to \infty$. Certainly $\mathbf{E}(Y_n^2) < \infty$, and hence we may use (20) to find the conditional expectation $\mathbf{E}(Y_n \mid \mathscr{G})$, a $\mathscr{G}$-measurable random variable satisfying

(28) $$\mathbf{E}([Y_n - \mathbf{E}(Y_n \mid \mathscr{G})]I_G) = 0 \quad \text{for all } G \in \mathscr{G}.$$

Now $Y_n \leqslant Y_{n+1}$, and so we may take $\mathbf{E}(Y_n \mid \mathscr{G}) \leqslant \mathbf{E}(Y_{n+1} \mid \mathscr{G})$; see Exercise (7.9.32iii). Hence $\lim_{n\to\infty} \mathbf{E}(Y_n \mid \mathscr{G})$ exists, and we write $\mathbf{E}(Y \mid \mathscr{G})$ for this limit, a $\mathscr{G}$-measurable random variable. By monotone convergence (5.6.12) and (23), $\mathbf{E}(Y_n I_G) \uparrow \mathbf{E}(YI_G)$, and

$$\mathbf{E}(\mathbf{E}(Y_n \mid \mathscr{G})I_G) \uparrow \mathbf{E}(\mathbf{E}(Y \mid \mathscr{G})I_G) = \mathbf{E}(\mathbf{E}(YI_G \mid \mathscr{G})) = \mathbf{E}(YI_G),$$

so that, by (28), $\mathbf{E}([Y - \mathbf{E}(Y \mid \mathscr{G})]I_G) = 0$ for all $G \in \mathscr{G}$.

Next we lift in the usual way the restriction that Y be non-negative. We express Y as $Y = Y^+ - Y^-$ where $Y^+ = \max\{Y, 0\}$ and $Y^- = -\min\{Y, 0\}$ are non-negative; we define $\mathbf{E}(Y \mid \mathscr{G}) = \mathbf{E}(Y^+ \mid \mathscr{G}) - \mathbf{E}(Y^- \mid \mathscr{G})$. It is easy to check that $\mathbf{E}(Y \mid \mathscr{G})$ satisfies (a), (b), and (c). To see the uniqueness, suppose that there exist two $\mathscr{G}$-measurable random variables Z_1 and Z_2 satisfying (c). Then $\mathbf{E}((Z_1 - Z_2)I_G) = \mathbf{E}((Y - Y)I_G) = 0$ for all $G \in \mathscr{G}$. Setting $G = \{Z_1 > Z_2\}$ and $G = \{Z_1 < Z_2\}$ in turn, we find that $\mathbf{P}(Z_1 = Z_2) = 1$ as required. ∎

Having defined $\mathbf{E}(Y \mid \mathscr{G})$, we can of course define conditional probabilities also: if $A \in \mathscr{F}$, we define $\mathbf{P}(A \mid \mathscr{G}) = \mathbf{E}(I_A \mid \mathscr{G})$. It may be checked that $\mathbf{P}(\varnothing \mid \mathscr{G}) = 0$, $\mathbf{P}(\Omega \mid \mathscr{G}) = 1$ a.s., and $\mathbf{P}(\bigcup_i A_i \mid \mathscr{G}) = \sum_i \mathbf{P}(A_i \mid \mathscr{G})$ a.s. for any sequence $\{A_i : i \geqslant 1\}$ of disjoint events in $\mathscr{F}$.

It looks as though there should be a way of defining $\mathbf{P}(\cdot \mid \mathscr{G})$ so that it is a probability measure on $(\Omega, \mathscr{F})$. This turns out to be impossible in general, but the details are beyond the scope of this book.

Exercises

(29) Let Y be uniformly distributed on $[-1, 1]$ and let $X = Y^2$.

(a) Find the best predictor of X given Y, and of Y given X.
(b) Find the best linear predictor of X given Y, and of Y given X.

(30) Let the pair (X, Y) have a general bivariate normal distribution. Find $\mathbf{E}(Y \mid X)$.

(31) Let $X_1, X_2, \dots, X_n$ be random variables with zero means and covariance matrix $V = (v_{ij})$, and let Y have finite second moment. Find the linear function h of the X_i which minimizes the mean squared error $\mathbf{E}\{(Y - h(X_1, \dots, X_n))^2\}$.

(32) Verify the following properties of conditional expectation. You may assume that the relevant expectations exist.

(i) $\mathbf{E}\{\mathbf{E}(Y \mid \mathscr{G})\} = \mathbf{E}(Y)$.
(ii) $\mathbf{E}(\alpha Y + \beta Z \mid \mathscr{G}) = \alpha\mathbf{E}(Y \mid \mathscr{G}) + \beta\mathbf{E}(Z \mid \mathscr{G})$ for $\alpha, \beta \in \mathbb{R}$.
(iii) $\mathbf{E}(Y \mid \mathscr{G}) \geqslant 0$ if $Y \geqslant 0$.
(iv) $\mathbf{E}(Y \mid \mathscr{G}) = \mathbf{E}\{\mathbf{E}(Y \mid \mathscr{H}) \mid \mathscr{G}\}$ if $\mathscr{G} \subseteq \mathscr{H}$.
(v) $\mathbf{E}(Y \mid \mathscr{G}) = \mathbf{E}(Y)$ if Y is independent of I_G for every $G \in \mathscr{G}$.
(vi) **Jensen's inequality.** $g\{\mathbf{E}(Y \mid \mathscr{G})\} \leqslant \mathbf{E}\{g(Y) \mid \mathscr{G}\}$ for all convex functions g.
(vii) If $Y_n \xrightarrow{\text{a.s.}} Y$ and $|Y_n| \leqslant Z$ a.s. where $\mathbf{E}(Z) < \infty$, then $\mathbf{E}(Y_n \mid \mathscr{G}) \xrightarrow{\text{a.s.}} \mathbf{E}(Y \mid \mathscr{G})$.

(Statements (ii)–(vi) are of course to be interpreted 'almost surely'.)

(33) Let X and Y have joint mass function $f(x, y) = \{x(x+1)\}^{-1}$ for $x = y = 1, 2, \ldots$. Show that $\mathbf{E}(Y \mid X) < \infty$ while $\mathbf{E}(Y) = \infty$.

(34) Let $(\Omega, \mathscr{F}, \mathbf{P})$ be a probability space and let $\mathscr{G}$ be a sub-σ-field of $\mathscr{F}$. Let H be the space of $\mathscr{G}$-measurable random variables with finite second moment.

(a) Show that H is closed with respect to the norm $\|\cdot\|_2$.
(b) Let Y be a random variable satisfying $\mathbf{E}(Y^2) < \infty$, and show the equivalence of the following two statements for any $M \in H$:

(i) $\mathbf{E}\{(Y - M)Z\} = 0$ for all $Z \in H$,
(ii) $\mathbf{E}\{(Y - M)I_G\} = 0$ for all $G \in \mathscr{G}$.

7.10 Uniform integrability

Suppose that we are presented with a sequence $\{X_n : n \geqslant 1\}$ of random variables, and we are able to prove that $X_n \xrightarrow{P} X$. Convergence in probability tells us little about the behaviour of $\mathbf{E}(X_n)$, as the trite example

$$Y_n = \begin{cases} n & \text{with probability } n^{-1} \\ 0 & \text{otherwise} \end{cases}$$

shows; in this special case, $Y_n \xrightarrow{P} 0$ but $\mathbf{E}(Y_n) = 1$ for all n. Should we wish to prove that $\mathbf{E}(X_n) \to \mathbf{E}(X)$, or further that $X_n \xrightarrow{1} X$ (which is to say that $\mathbf{E}|X_n - X| \to 0$), then an additional condition is required.

We encountered in an earlier exercise (7.2.23) an argument of the kind required. If $X_n \xrightarrow{P} X$ and $|X_n| \leqslant Y$ for some Y such that $\mathbf{E}|Y| < \infty$, then $X_n \xrightarrow{1} X$. This extra condition, that $\{X_n\}$ is dominated *uniformly*, is often too strong in cases of interest. A weaker condition is provided by the following definition. As usual, I_A denotes the indicator function of the event A.

(1) **Definition.** A sequence $X_1, X_2, \ldots$ of random variables is said to be **uniformly integrable** if

(2) $$\sup_n \mathbf{E}(|X_n| I_{\{|X_n| \geqslant a\}}) \to 0 \quad \text{as} \quad a \to \infty.$$

Let us investigate this condition briefly. A random variable Y is called 'integrable' if $\mathbf{E}|Y| < \infty$, which is to say that

$$\mathbf{E}(|Y| I_{\{|Y| \geqslant a\}}) = \int_{|y| \geqslant a} |y| \, dF_Y(y)$$

tends to 0 as $a \to \infty$ (see (5.6.19)). Therefore, a family $\{X_n : n \geqslant 1\}$ is 'integrable' if

$$\mathbf{E}(|X_n| I_{\{X_n \geqslant a\}}) \to 0 \quad \text{as} \quad a \to \infty$$

for all n, and 'uniformly integrable' if the convergence is uniform in n. Roughly speaking, the condition of integrability restricts the amount of probability in the tails of the distribution, and uniform integrability restricts such quantities *uniformly* over the family of random variables in question.

The principal use of uniform integrability is demonstrated by the following theorem.

(3) **Theorem.** *Suppose that* $X_1, X_2, \ldots$ *is a sequence of random variables satisfying* $X_n \xrightarrow{\text{P}} X$. *The following three statements are equivalent to one another.*

(a) $\{X_n : n \geq 1\}$ *is uniformly integrable.*
(b) $\mathbf{E}|X_n| < \infty$ *for all* n, $\mathbf{E}|X| < \infty$, *and* $X_n \xrightarrow{1} X$.
(c) $\mathbf{E}|X_n| < \infty$ *for all* n, *and* $\mathbf{E}|X_n| \to \mathbf{E}|X| < \infty$.

In advance of proving this, we note some sufficient conditions for uniform integrability.

(4) **Example.** Suppose $|X_n| \leq Y$ for all n, where $\mathbf{E}|Y| < \infty$. Then $|X_n| I_{\{|X_n| \geq a\}} \leq |Y| I_{\{|Y| \geq a\}}$, so that

$$\sup_n \mathbf{E}(|X_n| I_{\{|X_n| \geq a\}}) \leq \mathbf{E}(|Y| I_{\{|Y| \geq a\}})$$

which tends to zero as $a \to \infty$, since $\mathbf{E}|Y| < \infty$. ●

(5) **Example.** Suppose that there exist $\delta > 0$ and $K < \infty$ such that $\mathbf{E}(|X_n|^{1+\delta}) \leq K$ for all n. Then

$$\begin{aligned} \mathbf{E}(|X_n| I_{\{|X_n| \geq a\}}) &\leq \frac{1}{a^\delta} \mathbf{E}(|X_n|^{1+\delta} I_{\{|X_n| \geq a\}}) \\ &\leq \frac{1}{a^\delta} \mathbf{E}(|X_n|^{1+\delta}) \leq \frac{K}{a^\delta} \to 0 \end{aligned}$$

as $a \to \infty$, so that the family is uniformly integrable. ●

Turning to the proof of Theorem (3), we note first a preliminary lemma which is of value in its own right.

(6) **Lemma.** *A family* $\{X_n : n \geq 1\}$ *is uniformly integrable if and only if both of the following hold*:

(a) $\sup_n \mathbf{E}|X_n| < \infty$,
(b) *for all* $\varepsilon > 0$, *there exists* $\delta > 0$ *such that, for all* n, $\mathbf{E}(|X_n| I_A) < \varepsilon$ *for any event* A *such that* $\mathbf{P}(A) < \delta$.

The equivalent statement for a single random variable X is the assertion that $\mathbf{E}|X| < \infty$ if and only if

(7) $$\sup_{A:\, \mathbf{P}(A) < \delta} \mathbf{E}(|X| I_A) \to 0 \quad \text{as} \quad \delta \to 0;$$

see Exercise (5.6.19).

Proof of (6). Suppose first that $\{X_n\}$ is uniformly integrable. For any $a > 0$,

$$\mathbf{E}|X_n| = \mathbf{E}(|X_n|I_{\{X_n < a\}}) + \mathbf{E}(|X_n|I_{\{X_n \geqslant a\}}),$$

and therefore

$$\sup_n \mathbf{E}|X_n| \leqslant a + \sup_n \mathbf{E}(|X_n|I_{\{X_n \geqslant a\}}).$$

We use uniform integrability to find that $\sup_n \mathbf{E}|X_n| < \infty$. Next,

(8) $$\mathbf{E}(|X_n|I_A) = \mathbf{E}(|X_n|I_{A \cap B_n(a)}) + \mathbf{E}(|X_n|I_{A \cap B_n(a)^c})$$

where $B_n(a) = \{|X_n| \geqslant a\}$. Now

$$\mathbf{E}(|X_n|I_{A \cap B_n(a)}) \leqslant \mathbf{E}(|X_n|I_{B_n(a)})$$

and

$$\mathbf{E}(|X_n|I_{A \cap B_n(a)^c}) \leqslant a\mathbf{E}(I_A) = a\mathbf{P}(A).$$

Let $\varepsilon > 0$ and pick a such that $\mathbf{E}(|X_n|I_{B_n(a)}) < \frac{1}{2}\varepsilon$ for all n. We have from (8) that $\mathbf{E}(|X_n|I_A) \leqslant \frac{1}{2}\varepsilon + a\mathbf{P}(A)$, which is smaller than ε whenever $\mathbf{P}(A) < \varepsilon/(2a)$.

Secondly, suppose that (a) and (b) hold; let $\varepsilon > 0$ and pick δ according to (b). We have that

$$\mathbf{E}|X_n| \geqslant \mathbf{E}(|X_n|I_{B_n(a)}) \geqslant a\mathbf{P}(B_n(a))$$

(this is Markov's inequality) so that

$$\sup_n \mathbf{P}(B_n(a)) \leqslant \frac{1}{a} \sup_n \mathbf{E}|X_n| < \infty.$$

Pick a such that $a^{-1} \sup_n \mathbf{E}|X_n| < \delta$, implying that $\mathbf{P}(B_n(a)) < \delta$ for all n. It follows from (b) that $\mathbf{E}(|X_n|I_{B_n(a)}) < \varepsilon$ for all n, and hence $\{X_n\}$ is uniformly integrable. ■

Proof of Theorem (3). The main part is the statement that (a) implies (b), and we prove this first. Suppose that the family is uniformly integrable. Certainly each member is integrable, so that $\mathbf{E}|X_n| < \infty$ for all n. Since $X_n \xrightarrow{\mathrm{P}} X$, there exists a subsequence $\{X_{n_k}: k \geqslant 1\}$ such that $X_{n_k} \xrightarrow{\text{a.s.}} X$ (see (7.2.13)). By Fatou's lemma (5.6.13)

(9) $$\mathbf{E}|X| = \mathbf{E}\left(\liminf_{k \to \infty} |X_{n_k}|\right) \leqslant \liminf_{k \to \infty} \mathbf{E}|X_{n_k}| \leqslant \sup_n \mathbf{E}|X_n|,$$

which is finite as a consequence of Lemma (6).

To prove convergence in mean, we write, for $\varepsilon > 0$,

(10) $$\begin{aligned}\mathbf{E}|X_n - X| &= \mathbf{E}(|X_n - X|I_{\{|X_n - X| < \varepsilon\}} + |X_n - X|I_{\{|X_n - X| \geqslant \varepsilon\}}) \\ &\leqslant \varepsilon + \mathbf{E}(|X_n|I_{A_n}) + \mathbf{E}(|X|I_{A_n})\end{aligned}$$

where $A_n = \{|X_n - X| > \varepsilon\}$. Now $\mathbf{P}(A_n) \to 0$ as $n \to \infty$, and hence $\mathbf{E}(|X_n|I_{A_n}) \to 0$ as $n \to \infty$, by Lemma (6). Similarly $\mathbf{E}(|X|I_{A_n}) \to 0$ as $n \to \infty$, by (7), so that $\limsup_{n\to\infty} \mathbf{E}|X_n - X| \leqslant \varepsilon$. Let $\varepsilon \downarrow 0$ to obtain that $X_n \xrightarrow{1} X$.

That (b) implies (c) is immediate from the observation that $|\mathbf{E}|X_n| - \mathbf{E}|X|| \leqslant \mathbf{E}|X_n - X|$, and it remains to prove that (c) implies (a). Suppose then that (c) holds. Clearly

(11) $$\mathbf{E}(|X_n|I_{\{|X_n|\geqslant a\}}) = \mathbf{E}|X_n| - \mathbf{E}(u(X_n))$$

where $u(x) = |x|I_{(-a,a)}(x)$. Now u is a continuous bounded function on $(-a, a)$ and $X_n \xrightarrow{D} X$; hence

$$\mathbf{E}(u(X_n)) \to \mathbf{E}(u(X)) = \mathbf{E}(|X|I_{\{|X|<a\}})$$

if a and $-a$ are points of continuity of the distribution function F_X of X (see Theorem (7.2.19) and the comment thereafter). F_X is monotone, and therefore the set Δ of discontinuities of F_X is at most countable. It follows from (11) that

(12) $$\mathbf{E}(|X_n|I_{\{|X_n|\geqslant a\}}) \to \mathbf{E}|X| - \mathbf{E}(|X|I_{\{|X|<a\}}) = \mathbf{E}(|X|I_{\{|X|\geqslant a\}})$$

if $a \notin \Delta$. For any $\varepsilon > 0$, there exists $b \notin \Delta$ such that $\mathbf{E}(|X|I_{\{|X|\geqslant b\}}) < \varepsilon$; with this choice of b, there exists by (12) an integer N such that $\mathbf{E}(|X_n|I_{\{|X_n|\geqslant b\}}) < 2\varepsilon$ for all $n \geqslant N$. On the other hand, there exists c such that $\mathbf{E}(|X_k|I_{\{|X_k|\geqslant c\}}) < 2\varepsilon$ for all $k < N$, since only finitely many terms are involved. If $a > \max\{b, c\}$, we have that $\mathbf{E}(|X_n|I_{\{|X_n|\geqslant a\}}) < 2\varepsilon$ for all n, and we have proved that $\{X_n\}$ is uniformly integrable. ■

The concept of uniform integrability will be of particular value when we return in Chapter 12 to the theory of martingales. The following example may be seen as an illustration of this.

(13) **Example.** Let Y be a random variable on $(\Omega, \mathcal{F}, \mathbf{P})$ with $\mathbf{E}|Y| < \infty$, and let $\{\mathcal{G}_n : n \geqslant 1\}$ be a filtration, which is to say that $\mathcal{G}_n$ is a sub-σ-field of $\mathcal{F}$, and furthermore $\mathcal{G}_n \subseteq \mathcal{G}_{n+1}$ for all n. Let $X_n = \mathbf{E}(Y \mid \mathcal{G}_n)$. The sequence $\{X_n : n \geqslant 1\}$ is uniformly integrable, as may be seen in the following way.

It is a consequence of Jensen's inequality (7.9.32vi) that $|X_n| = |\mathbf{E}(Y \mid \mathcal{G}_n)| \leqslant \mathbf{E}(|Y| \mid \mathcal{G}_n)$ almost surely, so that $\mathbf{E}(|X_n|I_{\{|X_n|\geqslant a\}}) \leqslant \mathbf{E}(Z_n I_{\{Z_n \geqslant a\}})$ where $Z_n = \mathbf{E}(|Y| \mid \mathcal{G}_n)$. By the definition of conditional expectation, $\mathbf{E}\{(|Y| - Z_n)I_{\{Z_n\geqslant a\}}\} = 0$, so that

(14) $$\mathbf{E}(|X_n|I_{\{|X_n|\geqslant a\}}) \leqslant \mathbf{E}(|Y|I_{\{Z_n\geqslant a\}}).$$

We now repeat an argument used before. By Markov's inequality,

$$\mathbf{P}(Z_n \geqslant a) \leqslant a^{-1}\mathbf{E}(Z_n) = a^{-1}\mathbf{E}|Y|,$$

and therefore $\mathbf{P}(Z_n \geqslant a) \to 0$ as $a \to \infty$, uniformly in n. Using (7), we deduce that $\mathbf{E}(|Y|I_{\{Z_n\geqslant a\}}) \to 0$ as $a \to \infty$, uniformly in n, implying that $\{X_n\}$ is uniformly integrable. ●

We finish this section with an application.

(15) **Example. Convergence of moments.** Suppose that $X_1, X_2, \ldots$ is a sequence satisfying $X_n \xrightarrow{D} X$, and furthermore $\sup_n \mathbf{E}(|X_n|^\alpha) < \infty$ for some $\alpha > 1$. It follows that

(16) $$\mathbf{E}(X_n^\beta) \to \mathbf{E}(X^\beta)$$

for any integer β satisfying $1 \leqslant \beta < \alpha$. This may be proved either directly or via Theorem (3). First, if β is an integer satisfying $1 \leqslant \beta < \alpha$, then $\{X_n^\beta : n \geqslant 1\}$ is uniformly integrable by (5), and furthermore $X_n^\beta \xrightarrow{D} X^\beta$ (easy *exercise*, or use (7.2.18)). If it were the case that $X_n^\beta \xrightarrow{P} X^\beta$ then Theorem (3) would imply the result. In any case, by the Skorokhod representation theorem (7.2.14), there exist random variables $Y, Y_1, Y_2, \ldots$ having the same distributions as $X, X_1, X_2, \ldots$ such that $Y_n^\beta \xrightarrow{P} Y^\beta$. Thus $\mathbf{E}(Y_n^\beta) \to \mathbf{E}(Y^\beta)$ by Theorem (3). However, $\mathbf{E}(Y_n^\beta) = \mathbf{E}(X_n^\beta)$ and $\mathbf{E}(Y^\beta) = \mathbf{E}(X^\beta)$, and the proof is complete. ●

Exercises

(17) Show that the sum $\{X_n + Y_n\}$ of two uniformly integrable sequences $\{X_n\}$ and $\{Y_n\}$ gives a uniformly integrable sequence.

(18) (a) Suppose that $X_n \xrightarrow{r} X$ where $r \geqslant 1$. Show that $\{|X_n|^r : n \geqslant 1\}$ is uniformly integrable, and deduce that $\mathbf{E}(X_n^r) \to \mathbf{E}(X^r)$ if r is an integer.
(b) Conversely, suppose that $\{|X_n|^r : n \geqslant 1\}$ is uniformly integrable where $r \geqslant 1$, and show that $X_n \xrightarrow{r} X$ if $X_n \xrightarrow{P} X$.

(19) Let $g: [0, \infty) \to [0, \infty)$ be an increasing function satisfying $g(x)/x \to \infty$ as $x \to \infty$. Show that the sequence $\{X_n : n \geqslant 1\}$ is uniformly integrable if $\sup_n \mathbf{E}\{g(|X_n|)\} < \infty$.

(20) Let $\{Z_n : n \geqslant 0\}$ be a branching process with $Z_0 = 1$, $\mathbf{E}(Z_1) = 1$, $\operatorname{var}(Z_1) \neq 0$. Show that $\{Z_n : n \geqslant 0\}$ is not uniformly integrable.

(21) **Pratt's lemma**. Suppose that $X_n \leqslant Y_n \leqslant Z_n$ where $X_n \xrightarrow{P} X$, $Y_n \xrightarrow{P} Y$, and $Z_n \xrightarrow{P} Z$. If $\mathbf{E}X_n \to \mathbf{E}X$ and $\mathbf{E}Z_n \to \mathbf{E}Z$, show that $\mathbf{E}Y_n \to \mathbf{E}Y$.

(22) Let $\{X_n : n \geqslant 1\}$ be a sequence of variables satisfying $\mathbf{E}(\sup_n |X_n|) < \infty$. Show that $\{X_n\}$ is uniformly integrable.

7.11 Problems

1. Let X_n have density function

$$f_n(x) = \frac{n}{\pi(1 + n^2 x^2)}, \qquad n \geqslant 1.$$

With respect to which modes of convergence does X_n converge as $n \to \infty$?

2. (i) Suppose that $X_n \xrightarrow{\text{a.s.}} X$ and $Y_n \xrightarrow{\text{a.s.}} Y$, and show that $X_n + Y_n \xrightarrow{\text{a.s.}} X + Y$. Show that the corresponding result holds for convergence in rth mean and in probability, but not in distribution.
(ii) Show that if $X_n \xrightarrow{\text{a.s.}} X$ and $Y_n \xrightarrow{\text{a.s.}} Y$ then $X_n Y_n \xrightarrow{\text{a.s.}} XY$. Does the corresponding result hold for the other modes of convergence?

3. Let $g: \mathbb{R} \to \mathbb{R}$ be continuous. Show that $g(X_n) \xrightarrow{P} g(X)$ if $X_n \xrightarrow{P} X$.

4. Let $Y_1, Y_2, \ldots$ be independent identically distributed variables, each of which can take any value in $\{0, 1, \ldots, 9\}$ with equal probability $\frac{1}{10}$. Let

$$X_n = \sum_{i=1}^{n} Y_i 10^{-i}.$$

Show by the use of characteristic functions that X_n converges in distribution to the uniform distribution on $[0, 1]$. Deduce that $X_n \xrightarrow{\text{a.s.}} Y$ for some Y which is uniformly distributed on $[0, 1]$.

5. Let $N(t)$ be a Poisson process.

(a) Find the covariance of $N(s)$ and $N(t)$.
(b) Show that N is continuous in mean square, which is to say that

$$\mathbf{E}(\{N(t+h) - N(t)\}^2) \to 0 \quad \text{as} \quad h \to 0.$$

(c) Prove that N is continuous in probability, which is to say that

$$\mathbf{P}(|N(t+h) - N(t)| > \varepsilon) \to 0 \quad \text{as} \quad h \to 0, \quad \text{for all } \varepsilon > 0.$$

(d) Show that N is differentiable in probability but not in mean square.

6. Prove that $n^{-1} \sum_{i=1}^{n} X_i \xrightarrow{\text{a.s.}} 0$ whenever the X_i are independent identically distributed variables with zero means and such that $\mathbf{E}(X_1^4) < \infty$.

7. Show that $X_n \xrightarrow{\text{a.s.}} X$ whenever

$$\sum_n \mathbf{E}(|X_n - X|^r) < \infty \quad \text{for some} \quad r > 0.$$

8. Show that if $X_n \xrightarrow{\text{D}} X$ then $aX_n + b \xrightarrow{\text{D}} aX + b$ for any real a and b.

9. If X has zero mean and variance σ^2, show that

$$\mathbf{P}(X \geqslant t) \leqslant \frac{\sigma^2}{\sigma^2 + t^2}, \quad \text{for} \quad t > 0.$$

10. Show that $X_n \xrightarrow{\text{P}} 0$ if and only if

$$\mathbf{E}\left(\frac{|X_n|}{1 + |X_n|}\right) \to 0 \quad \text{as} \quad n \to \infty.$$

11. The sequence $\{X_n\}$ of random variables is called *mean-square Cauchy-convergent* if $\mathbf{E}\{(X_n - X_m)^2\} \to 0$ as $m, n \to \infty$. Show that $\{X_n\}$ converges in mean square to some limit X if and only if it is mean-square Cauchy-convergent. Does the corresponding result hold for the other modes of convergence?

12. Suppose that $\{X_n\}$ is a sequence of uncorrelated variables with zero means and uniformly bounded variances. Show that $n^{-1} \sum_{i=1}^{n} X_i \xrightarrow{\text{m.s.}} 0$.

13. Let $X_1, X_2, \ldots$ be independent identically distributed random variables with the common distribution function $F(x)$, and suppose that $F(x) < 1$ for all x. Let $M_n = \max\{X_1, X_2, \ldots, X_n\}$ and suppose that there exists a strictly increasing unbounded positive sequence $a_1, a_2, \ldots$ such that $\mathbf{P}(M_n/a_n \leqslant x) \to H(x)$ for some distribution function H. Let us assume that H is continuous with $0 < H(1) < 1$; substantially weaker conditions suffice but introduce extra difficulties.

(a) Show that $n(1 - F(a_n x)) \to -\log H(x)$ as $n \to \infty$ and deduce that

$$\frac{1 - F(a_n x)}{1 - F(a_n)} \to \frac{\log H(x)}{\log H(1)} \quad \text{if} \quad x > 0.$$

(b) Deduce that if $x > 0$

$$\frac{1 - F(tx)}{1 - F(t)} \to \frac{\log H(x)}{\log H(1)} \quad \text{as} \quad t \to \infty.$$

(c) Set $x = x_1 x_2$ and make the substitution

$$g(x) = \frac{\log H(\mathrm{e}^x)}{\log H(1)}$$

to find that $g(x + y) = g(x)g(y)$, and deduce that

$$H(x) = \begin{cases} \exp(-\alpha x^{-\beta}) & \text{if} \quad x \geqslant 0 \\ 0 & \text{if} \quad x < 0 \end{cases}$$

for some non-negative constants α and β.

You have shown that H is the distribution function of Y^{-1}, where Y has a Weibull distribution.

14. Let $X_1, X_2, \ldots, X_n$ be independent and identically distributed random variables with the Cauchy distribution. Show that $M_n = \max\{X_1, X_2, \ldots, X_n\}$ is such that $\pi M_n/n$ converges in distribution, the limiting distribution function being given by $H(x) = \mathrm{e}^{-1/x}$ if $x \geqslant 0$.

15. Let $X_1, X_2, \ldots$ be independent and identically distributed random variables whose common characteristic function ϕ satisfies $\phi'(0) = \mathrm{i}\mu$. Show that $n^{-1} \sum_{j=1}^{n} X_j \xrightarrow{\mathrm{P}} \mu$.

16. The *total variation distance* $d(F, G)$ between two distribution functions F and G is defined by

$$d(F, G) = \sup_{u:\, \|u\| = 1} |\mathbf{E}u(X) - \mathbf{E}u(Y)|$$

where X and Y have distribution functions F and G, and the supremum is over all (measurable) functions u: $\mathbb{R} \to \mathbb{R}$ such that $\|u\| = \sup_x |u(x)|$ satisfies $\|u\| = 1$.

(a) If F and G represent discrete distributions which put masses f_n and g_n at the points x_n, show that

$$d(F, G) = \sum_n |f_n - g_n|.$$

(b) If F and G have density functions f and g, show that

$$d(F, G) = \int_{-\infty}^{\infty} |f(x) - g(x)| \, \mathrm{d}x.$$

(c) Show that $d(F_n, F) \to 0$ implies that $F_n \to F$ (in the sense that $F_n(x) \to F(x)$ for all x at which F is continuous), but that the converse is false.

17. Let g: $\mathbb{R} \to \mathbb{R}$ be bounded and continuous. Show that

$$\sum_{k=0}^{\infty} g(k/n) \frac{(n\lambda)^k}{k!} \mathrm{e}^{-n\lambda} \to g(\lambda) \quad \text{as} \quad n \to \infty.$$

18. Let X_n and Y_m be independent random variables having the Poisson distribution with parameters n and m, respectively. Show that

$$\frac{(X_n - n) - (Y_m - m)}{\sqrt{(X_n + Y_m)}} \xrightarrow{\text{D}} N(0,1) \quad \text{as} \quad m, n \to \infty.$$

19. (a) Suppose that $X_1, X_2, \ldots$ is a sequence of random variables, each having a normal distribution, and such that $X_n \xrightarrow{\text{D}} X$. Show that X has a normal distribution, possibly degenerate.

(b) For each n ($\geqslant 1$), let (X_n, Y_n) be a pair of random variables having a bivariate normal distribution. Suppose that $X_n \xrightarrow{\text{P}} X$ and $Y_n \xrightarrow{\text{P}} Y$, and show that the pair (X, Y) has a bivariate normal distribution.

20. Let $X_1, X_2, \ldots$ be random variables satisfying $\text{var}(X_n) < c$ for all n and some constant c. Show that the sequence obeys the weak law, in the sense that $n^{-1}\sum_1^n (X_i - \mathbf{E}X_i)$ converges in probability to 0, if the correlation coefficients satisfy either of the following:

(i) $\rho(X_i, X_j) \leqslant 0$ for all $i \neq j$,
(ii) $\rho(X_i, X_j) \to 0$ as $|i - j| \to \infty$.

21. Let $X_1, X_2, \ldots$ be independent random variables with common density function

$$f(x) = \begin{cases} 0 & \text{if } \ |x| \leqslant 2 \\ \dfrac{c}{x^2 \log|x|} & \text{if } \ |x| > 2 \end{cases}$$

where c is a constant. Show that the X_i have no mean, but $n^{-1}\sum_{i=1}^n X_i \xrightarrow{\text{P}} 0$ as $n \to \infty$. Show that convergence does not take place almost surely.

22. Let X_n be the Euclidean distance between two points chosen independently and uniformly from the n-dimensional unit cube. Show that $\mathbf{E}(X_n)/\sqrt{n} \to 1/\sqrt{6}$ as $n \to \infty$.

23. Let $X_1, X_2, \ldots$ be independent random variables having the uniform distribution on $[-1, 1]$. Show that

$$\mathbf{P}\left(\left|\sum_{i=1}^n X_i^{-1}\right| > \tfrac{1}{2}n\pi\right) \to \tfrac{1}{2} \quad \text{as} \quad n \to \infty.$$

24. Let $X_1, X_2, \ldots$ be independent random variables, each X_k having mass function given by

$$\mathbf{P}(X_k = k) = \mathbf{P}(X_k = -k) = \frac{1}{2k^2},$$

$$\mathbf{P}(X_k = 1) = \mathbf{P}(X_k = -1) = \frac{1}{2}\left(1 - \frac{1}{k^2}\right) \quad \text{if} \quad k > 1.$$

Show that $U_n = \sum_1^n X_i$ satisfies $n^{-\frac{1}{2}}U_n \xrightarrow{\text{D}} N(0,1)$ but $\text{var}(n^{-\frac{1}{2}}U_n) \to 2$ as $n \to \infty$.

25. Let $X_1, X_2, \ldots$ be random variables, and let $N_1, N_2, \ldots$ be random variables taking values in the positive integers such that $N_k \xrightarrow{P} \infty$ as $k \to \infty$. Show that

(i) if $X_n \xrightarrow{D} X$ and the X_i are independent of the N_j, then $X_{N_k} \xrightarrow{D} X$ as $k \to \infty$;
(ii) if $X_n \xrightarrow{\text{a.s.}} X$ then $X_{N_k} \xrightarrow{P} X$ as $k \to \infty$.

26. **Stirling's formula.**

(a) Let $a(k, n) = n^k/(k-1)!$ for $1 \leqslant k \leqslant n+1$. Use the fact that $1 - x \leqslant e^{-x}$ if $x \geqslant 0$ to show that

$$\frac{a(n-k, n)}{a(n+1, n)} \leqslant e^{-k^2/(2n)} \quad \text{if} \quad k \geqslant 0.$$

(b) Let $X_1, X_2, \ldots$ be independent Poisson variables with parameter 1, and let $S_n = X_1 + \cdots + X_n$. Define the function $g\colon \mathbb{R} \to \mathbb{R}$ by

$$g(x) = \begin{cases} -x & \text{if} \quad 0 \geqslant x \geqslant -M, \\ 0 & \text{otherwise}, \end{cases}$$

where M is large and positive. Show that, for large n,

$$\mathbf{E}\left(g\left\{\frac{S_n - n}{\sqrt{n}}\right\}\right) = \frac{e^{-n}}{\sqrt{n}}\{a(n+1, n) - a(n-k, n)\}$$

where $k = \lfloor Mn^{\frac{1}{2}} \rfloor$. Now use the central limit theorem and (a) above to deduce Stirling's formula:

$$\frac{n!\, e^n}{n^{n+\frac{1}{2}}\sqrt{(2\pi)}} \to 1 \quad \text{as} \quad n \to \infty.$$

27. A bag contains red and green balls. A ball is drawn from the bag, its colour noted, and then it is returned to the bag together with a new ball of the same colour. Initially the bag contained one ball of each colour. If R_n denotes the number of red balls in the bag after n additions, show that $S_n = R_n/(n+2)$ is a martingale. Deduce that the ratio of red to green balls converges almost surely to some limit as $n \to \infty$.

28. **Anscombe's theorem.** Let $\{X_i\colon i \geqslant 1\}$ be independent identically distributed random variables with zero mean and finite positive variance σ^2. Let $S_n = \sum_1^n X_i$. Suppose that the integer-valued random process $M(t)$ satisfies $t^{-1}M(t) \xrightarrow{P} \theta$ as $t \to \infty$, where θ is a positive constant. Show that

$$\frac{S_{M(t)}}{\sigma\sqrt{(\theta t)}} \xrightarrow{D} N(0,1) \quad \text{as} \quad t \to \infty,$$

and

$$\frac{S_{M(t)}}{\sigma\sqrt{M(t)}} \xrightarrow{D} N(0,1) \quad \text{as} \quad t \to \infty.$$

You should not assume that the process M is independent of the X_i.

29. **Kolmogorov's inequality.** Let $X_1, X_2, \ldots$ be independent random variables with zero means, and $S_n = X_1 + X_2 + \cdots + X_n$. Let $M_n = \max_{1 \leqslant k \leqslant n} |S_k|$ and show that $\mathbf{E}(S_n^2 I_{A_k}) > c^2 \mathbf{P}(A_k)$ where $A_k = \{M_{k-1} \leqslant c < M_k\}$ and $c > 0$. Deduce Kolmogorov's inequality:

$$\mathbf{P}\left(\max_{1 \leqslant k \leqslant n} |S_k| > c\right) \leqslant \frac{\mathbf{E}(S_n^2)}{c^2}, \qquad c > 0.$$

30. Let $X_1, X_2, \ldots$ be independent random variables with zero means, and let $S_n = X_1 + X_2 + \cdots + X_n$. Using Kolmogorov's inequality or the martingale convergence theorem, show that

(i) $\sum_1^\infty X_i$ converges almost surely if $\sum_1^\infty \mathbf{E}(X_k^2) < \infty$,
(ii) if there exists an increasing real sequence (b_n) such that $b_n \to \infty$, and satisfying $\sum_1^\infty \mathbf{E}(X_k^2)/b_k^2 < \infty$, then $b_n^{-1} \sum_1^\infty X_k \xrightarrow{\text{a.s.}} 0$ as $n \to \infty$.

8 Random processes

8.1 Introduction

Recall that a 'random process' X is a family $\{X_t: t \in T\}$ of random variables which map the sample space Ω into some set S. There are many possible choices for the index set T and the state space S; the characteristics of the process depend strongly upon these choices. For example, in Chapter 6 we studied discrete-time ($T = \{0, 1, 2, \ldots\}$) and continuous-time ($T = [0, \infty)$) Markov chains which took values in some countable set S. Other possible choices for T include $\mathbb{R}^n$ and $\mathbb{Z}^n$, whilst S might be an uncountable set, such as $\mathbb{R}$. The mathematical analysis of a random process varies greatly depending on whether S and T are countable or uncountable, just as discrete random variables are distinguishable from continuous variables. The main differences are indicated by those cases in which

(a) $T = \{0, 1, 2, \ldots\}$ or $T = [0, \infty)$,
(b) $S = \mathbb{Z}$ or $S = \mathbb{R}$.

There are two levels at which we can observe the evolution of a random process X.

(a) Each X_t is a function which maps Ω into S. For any fixed $\omega \in \Omega$, there is a corresponding collection $\{X_t(\omega): t \in T\}$ of members of S; this is called the *realization* or *sample path* of X at ω. We can study properties of sample paths.

(b) The X_t are not independent in general. If $S \subseteq \mathbb{R}$ and $\mathbf{t} = (t_1, t_2, \ldots, t_n)$ is a vector of members of T, then the vector $(X_{t_1}, X_{t_2}, \ldots, X_{t_n})$ has joint distribution function $F_{\mathbf{t}}: \mathbb{R}^n \to [0, 1]$ given by

$$F_{\mathbf{t}}(\mathbf{x}) = \mathbf{P}(X_{t_1} \leqslant x_1, \ldots, X_{t_n} \leqslant x_n).$$

The collection $\{F_{\mathbf{t}}\}$, as $\mathbf{t}$ ranges over all vectors of members of T of any finite length, is called the collection of *finite-dimensional distributions* (abbreviated to *fdds*) of X, and it contains all the information which is available about X from the distributions of its constituent variables X_t. We can study the distributional properties of X by using its fdds.

It is not generally the case that these two approaches yield the same information about the process in question, since knowledge of its fdds does not yield complete information about the properties of its sample paths. We shall see an example of this in the final section of this chapter.

We are not concerned here with the general theory of random processes, but prefer to study certain specific collections of processes which are characterized by one or more special properties. This is not a new approach for us. In Chapter 6 we devoted our attention to processes which satisfy the

Markov property, whilst large parts of Chapter 7 were devoted to sequences $\{S_n\}$ which were either martingales or the partial sums of independent sequences. In this short chapter we introduce certain other types of process and their characteristic properties. These can be divided broadly under four headings, covering 'stationary processes', 'renewal processes', 'queues', and 'diffusions'; their detailed analysis is left for Chapters 9, 10, 11, and 13 respectively.

We shall only be concerned with the cases when T is one of the sets $\mathbb{Z}$, $\{0, 1, 2, \dots\}$, $\mathbb{R}$, or $[0, \infty)$ here. If T is an uncountable subset of $\mathbb{R}$, representing continuous time say, then we shall write $X(t)$ rather than X_t for ease of notation. Evaluation of $X(t)$ at some $\omega \in \Omega$ yields a point in S, which we shall denote by $X(t; \omega)$.

The final section contains a technical discussion about the construction of a process with specified fdds; it may be omitted without prejudicing your understanding of the rest of the book.

8.2 Stationary processes

Many important processes have the property that their finite-dimensional distributions are invariant under time shifts (or space shifts if T is a subset of some Euclidean space $\mathbb{R}^n$, say).

(1) **Definition.** The process $X = \{X(t): t \geqslant 0\}$, taking values in $\mathbb{R}$, is called **strongly stationary** if the families

$$\{X(t_1), X(t_2), \dots, X(t_n)\} \quad \text{and} \quad \{X(t_1 + h), X(t_2 + h), \dots, X(t_n + h)\}$$

have the same joint distribution for all $t_1, t_2, \dots, t_n$ and $h > 0$.

Note that, if X is strongly stationary, then the distribution of $X(t)$ is the same for all times t.

We saw in Section 3.6 that the covariance of two random variables X and Y contains some information, albeit incomplete, about their joint distribution. With this in mind we formulate another stationarity property for processes with $\text{var}(X(t)) < \infty$; for processes with finite variance, it is weaker than (1).

(2) **Definition.** $X = \{X(t): t \geqslant 0\}$ is **weakly** (or **second-order** or **covariance**) **stationary** if

$$\mathbf{E}(X(t_1)) = \mathbf{E}(X(t_2))$$

and

$$\text{cov}(X(t_1), X(t_2)) = \text{cov}(X(t_1 + h), X(t_2 + h))$$

for all t_1, t_2, and $h > 0$.

Thus, X is weakly stationary if and only if it has constant means and the *autocovariance function*

(3) $$c(t, t+h) = \text{cov}(X(t), X(t+h))$$

satisfies

$$c(t, t+h) = c(0, h) \quad \text{for all } t, h \geqslant 0.$$

We emphasize that the autocovariance function $c(s, t)$ of a weakly stationary process is a function of $t - s$ only.

Definitions similar to (1) and (2) hold for processes with $T = \mathbb{R}$ and for discrete-time processes $X = \{X_n: n \geqslant 0\}$; the autocovariance function of a weakly stationary discrete-time process X is just a sequence $\{c(0, m): m \geqslant 0\}$ of real numbers.

Weak stationarity interests us more than strong stationarity for two reasons. First, the condition of strong stationarity is often too restrictive for certain applications; secondly, many substantial and useful properties of stationary processes are derivable from weak stationarity alone. Thus, the assertion that X is *stationary* should be interpreted to mean that X is *weakly stationary*. Of course, there exist processes which are stationary but not strongly stationary (see Example (5)), and conversely processes without finite second moments may be strongly stationary but not weakly stationary.

(4) **Example. Markov chains.** Let $X = \{X(t): t \geqslant 0\}$ be an irreducible Markov chain taking values in some countable subset S of $\mathbb{R}$ and with a unique stationary distribution $\boldsymbol{\pi}$. Then (see (6.9.21))

$$\mathbf{P}(X(t) = j \mid X(0) = i) \to \pi_j \quad \text{as} \quad t \to \infty$$

for all $i, j \in S$. The fdds of X depend on the initial distribution $\boldsymbol{\mu}^{(0)}$ of $X(0)$, and it is not generally true that X is stationary (in either sense). Suppose, however, that $\boldsymbol{\mu}^{(0)} = \boldsymbol{\pi}$. Then the distribution $\boldsymbol{\mu}^{(t)}$ of $X(t)$ satisfies

$$\boldsymbol{\mu}^{(t)} = \boldsymbol{\pi} \mathbf{P}_t = \boldsymbol{\pi}$$

from (6.9.19), where $\{\mathbf{P}_t\}$ is the transition semigroup of the chain. Thus $X(t)$ has distribution $\boldsymbol{\pi}$ for all t. Furthermore, if $0 < s < s + t$ and $h > 0$, then the pairs

$$\{X(s), X(s+t)\} \quad \text{and} \quad \{X(s+h), X(s+t+h)\}$$

have the same joint distribution since

(a) $X(s)$ and $X(s+h)$ are identically distributed,
(b) the distribution of $X(s+h)$ (respectively $X(s+t+h)$) depends only on the distribution of $X(s)$ (respectively $X(s+t)$) and on the transition matrix $\mathbf{P}_h$.

A similar argument holds for collections of the $X(u)$ which contain more than two elements, and we have shown that X is strongly stationary. ●

(5) **Example.** Let A and B be uncorrelated (but not necessarily independent) random variables, each of which has mean 0 and variance 1. Fix a number $\lambda \in [0, \pi]$ and define

(6)
$$X_n = A\cos(\lambda n) + B\sin(\lambda n).$$

Then $\mathbf{E}X_n = 0$ for all n and $X = \{X_n\}$ has autocovariance function

$$\begin{aligned} c(n, n+m) &= \mathbf{E}(X_n X_{n+m}) \\ &= \mathbf{E}([A\cos(\lambda n) + B\sin(\lambda n)]\{A\cos[\lambda(n+m)] + B\sin[\lambda(n+m)]\}) \\ &= \mathbf{E}(A^2\cos(\lambda n)\cos[\lambda(n+m)] + B^2\sin(\lambda n)\sin[\lambda(n+m)]) \\ &= \cos(\lambda m) \end{aligned}$$

since $\mathbf{E}(AB) = 0$. Thus $c(n, n+m)$ depends on m alone and so X is stationary. In general X is not strongly stationary unless extra conditions are imposed on the joint distribution of A and B; to see this for the case $\lambda = \frac{1}{2}\pi$, simply calculate that

$$\{X_0, X_1, X_2, X_3, \ldots\} = \{A, B, -A, -B, \ldots\}$$

which is strongly stationary if and only if the pairs (A, B), $(B, -A)$, and $(-A, -B)$ have the same joint distributions. It can be shown that X is strongly stationary for any λ if A and B are $N(0, 1)$ variables. The reason for this lies in (4.5.9), where we saw that normal variables are independent whenever they are uncorrelated. ●

Two major results in the theory of stationary processes are the 'spectral theorem' and the 'ergodic theorem'; we close this section with a short discussion of these. First, recall the theory of Fourier analysis. Any function $f: \mathbb{R} \to \mathbb{R}$ which

(a) is periodic with period 2π (that is, $f(x + 2\pi) = f(x)$ for all x),
(b) is continuous, and
(c) has bounded variation

has a unique Fourier expansion

$$f(x) = \tfrac{1}{2}a_0 + \sum_{n=1}^{\infty} [a_n\cos(nx) + b_n\sin(nx)]$$

which expresses f as the sum of varying proportions of regular oscillations. In some sense to be specified, a stationary process X is similar to a periodic function since its autocovariances are invariant under time shifts. The spectral theorem asserts that, subject to certain conditions, stationary processes can be decomposed in terms of regular underlying oscillations whose magnitudes are random variables; the set of frequencies of oscillations which contribute to this combination is called the 'spectrum' of the process. For example, the process X in (5) is specified precisely in these terms by (6). In spectral theory it is convenient to allow the processes in question to take values in the complex plane. In this case (6) can be rewritten as

(7) $$X_n = \mathrm{Re}(Y_n) \quad \text{where} \quad Y_n = C\,\mathrm{e}^{\mathrm{i}\lambda n};$$

here C is a complex-valued random variable and $\mathrm{i} = \sqrt{(-1)}$. $Y = \{Y_n\}$ is stationary also whenever

$$\mathbf{E}(C) = 0 \quad \text{and} \quad \mathbf{E}(C\bar{C}) < \infty$$

where $\bar{C}$ is the complex conjugate of C (but see (9.1.1)).

The ergodic theorem deals with the partial sums of a stationary sequence $X = \{X_n : n \geqslant 0\}$. Consider first the following two extreme examples of stationarity.

(8) **Example. Independent sequences.** Let $X = \{X_n : n \geqslant 0\}$ be a sequence of independent identically distributed variables with zero means and unit variances. Certainly X is stationary, and its autocovariance function is given by

$$c(n, n+m) = \mathbf{E}(X_n X_{n+m}) = \begin{cases} 1 & \text{if} \quad m = 0 \\ 0 & \text{if} \quad m \neq 0. \end{cases}$$

The strong law of large numbers asserts that

$$\frac{1}{n}\sum_{j=1}^{n} X_j \xrightarrow{\text{a.s.}} 0.$$

●

(9) **Example. Identical sequences.** Let Y be a random variable with zero mean and unit variance, and let $X = \{X_n : n \geqslant 0\}$ be the stationary sequence given by

$$X_n = Y \quad \text{for all } n.$$

X has autocovariance function

$$c(n, n+m) = \mathbf{E}(X_n X_{n+m}) = 1 \quad \text{for all } m.$$

It is clear that

$$\frac{1}{n}\sum_{j=1}^{n} X_j \xrightarrow{\text{a.s.}} Y$$

since each term in the sum is Y itself. ●

These two examples are, in some sense, extreme examples of stationarity since the first deals with independent variables and the second deals with identical variables. In both examples, however, the averages $n^{-1}\sum_{j=1}^{n} X_j$ converge as $n \to \infty$. In the first case the limit is constant, whilst in the second the limit is a random variable. This indicates a shared property of nice stationary processes, and we shall see that any stationary sequence

$X = \{X_n: n \geqslant 0\}$ satisfies

$$\frac{1}{n}\sum_{j=1}^{n} X_j \xrightarrow{\text{a.s.}} Y$$

for some random variable Y. This result is called the ergodic theorem for stationary sequences. A similar result holds for continuous-time stationary processes.

The theory of stationary processes is important and useful in statistics. Many sequences $\{x_n: 0 \leqslant n \leqslant N\}$ of observations, indexed by the time at which they were taken, are suitably modelled by random processes, and statistical problems such as the estimation of unknown parameters and the prediction of the future values of the sequence are often studied in this context. Such sequences are called 'time series' and they include many examples which are well known to us already, such as the successive values of the Financial Times Share Index, or the frequencies of sunspots in successive years. Statisticians and politicians often seek to find some underlying structure in such sequences, and to this end they may study 'moving average' processes Y, which are smoothed versions of a stationary sequence X,

$$Y_n = \sum_{i=0}^{r} \alpha_i X_{n-i},$$

where $\alpha_0, \alpha_1, \ldots, \alpha_r$ are constants. Alternatively, they may try to fit a model to their observations, and may typically consider 'autoregressive schemes' Y, being sequences which satisfy

$$Y_n = \sum_{i=1}^{r} \alpha_i Y_{n-i} + Z_n$$

where $\{Z_n\}$ is a sequence of uncorrelated variables with zero means and constant finite variance.

An introduction to the theory of stationary processes is given in Chapter 9.

8.3 Renewal processes

We are often interested in the successive occurrences of events such as the emission of radioactive particles, the failures of light bulbs, or the incidences of earthquakes.

(1) **Example. Light bulb failures.** This is the archetype of renewal processes. A room is lit by a single light bulb. When this bulb fails it is replaced immediately by an apparently identical copy. Let X_i be the (random) lifetime of the ith bulb, and suppose that the first bulb is installed at time $t = 0$. Then

$$T_n = X_1 + \cdots + X_n$$

is the time until the nth failure (where, by convention, we set $T_0 = 0$), and

$$N(t) = \max\{n: T_n \leqslant t\}$$

is the number of bulbs which have failed by time t. It is natural to assume that the X_i are independent and identically distributed random variables. ●

(2) **Example. Markov chains.** Let $\{Y_n: n \geqslant 0\}$ be a Markov chain, and choose some state i. We are interested in the time epochs at which the chain is in the state i. The times $0 < T_1 < T_2 < \cdots$ of successive visits to i are given by

$$T_1 = \min\{n \geqslant 1: Y_n = i\}$$

$$T_{m+1} = \min\{n > T_m: Y_n = i\} \quad \text{for} \quad m \geqslant 1;$$

they may be defective unless the chain is irreducible and persistent. Let $\{X_m: m \geqslant 1\}$ be given by

$$X_m = T_m - T_{m-1} \quad \text{for} \quad m \geqslant 1,$$

where we set $T_0 = 0$ by convention. It is clear that the X_m are independent, and that $X_2, X_3, \ldots$ are identically distributed since each is the elapsed time between two successive visits to i. On the other hand, X_1 does *not* have this shared distribution in general, unless the chain began in the state $Y_0 = i$. The number of visits to i which have occurred by time t is given by

$$N(t) = \max\{n: T_n \leqslant t\}.$$ ●

Both these examples contain a continuous-time random process $N = \{N(t): t \geqslant 0\}$, where $N(t)$ represents the number of occurrences of some event in the time interval $[0, t]$. Such a process N is called a 'renewal' or 'counting' process for obvious reasons; the Poisson process of Section 6.8 provides another example of a renewal process.

(3) **Definition.** A **renewal process** $N = \{N(t): t \geqslant 0\}$ is a process for which

$$N(t) = \max\{n: T_n \leqslant t\}$$

where

$$T_0 = 0, \qquad T_n = X_1 + \cdots + X_n \quad \text{for} \quad n \geqslant 1,$$

and the X_m are independent identically distributed non-negative random variables.

This definition describes N in terms of an underlying sequence $\{X_n\}$. In the absence of knowledge about this sequence we can construct it from N; just define

(4) $$T_n = \inf\{t: N(t) = n\}, \qquad X_n = T_n - T_{n-1}.$$

Note that the finite-dimensional distributions of a renewal process N are specified by the distribution of the X_m. For example, if the X_m are exponentially distributed then N is a Poisson process. We shall try to use the notation of (3) consistently in Chapter 10, in the sense that $\{N(t)\}$, $\{T_n\}$, and $\{X_n\}$ will always denote variables satisfying (4).

It is sometimes appropriate to allow X_1 to have a different distribution from the shared distribution of $X_2, X_3, \ldots$; in this case N is called a *delayed* (or *modified*) renewal process. The process N in (2) is a delayed renewal process whatever the initial Y_0; if $Y_0 = i$ then N is an ordinary renewal process.

Those readers who paid attention to (6.9.13) will be able to prove the following little result, which relates renewal processes to Markov chains.

(5) **Theorem.** *Poisson processes are the only renewal processes which are Markov chains.*

If you like, think of renewal processes as a generalization of Poisson processes in which we have dropped the condition that interarrival times be exponentially distributed.

There are two principal areas of interest concerning renewal processes. First, suppose that we interrupt a renewal process N at some specified time s. By this time $N(s)$ occurrences have already taken place and we are waiting for the $(N(s) + 1)$th. That is, s belongs to the random interval

$$I_s = [T_{N(s)}, T_{N(s)+1}).$$

Here are three random variables of interest.

(6) The **excess** (or **residual**) **lifetime** of I_s: $E(s) = T_{N(s)+1} - s$.

(7) The **current lifetime** (or **age**) of I_s: $C(s) = s - T_{N(s)}$.

(8) The **total lifetime** of I_s: $D(s) = E(s) + C(s)$.

We shall be interested in the distributions of these random variables; they are illustrated in Figure 8.1.

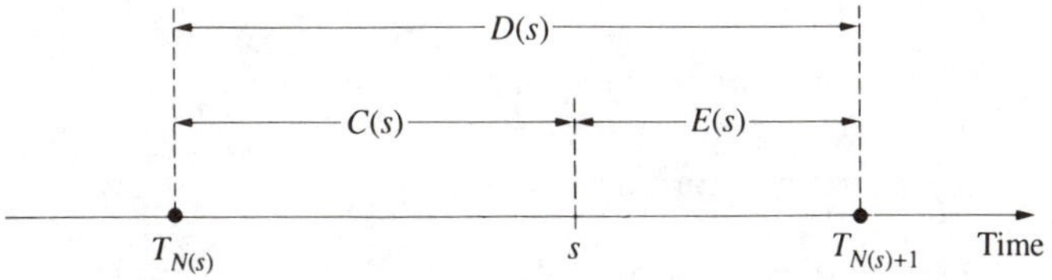

Fig. 8.1 Excess, current, and total lifetimes at time s.

It will come as no surprise to the reader to learn that the other principal topic concerns the asymptotic behaviour of a renewal process N as $t \to \infty$. Here we turn our attention to the *renewal function* $m(t)$ given by

(9) $$m(t) = \mathbf{E}(N(t)).$$

For a Poisson process N with intensity λ, (6.8.2) shows that

$$m(t) = \lambda t.$$

In general m is *not* a linear function of t; however, it is not too difficult to show that m is asymptotically linear, in that

$$\frac{1}{t} m(t) \to \frac{1}{\mu} \quad \text{as} \quad t \to \infty, \quad \text{where} \quad \mu = \mathbf{E}(X_1).$$

The 'renewal theorem' is a refinement of this result and asserts that

$$m(t+h) - m(t) \to \frac{h}{\mu} \quad \text{as} \quad t \to \infty$$

subject to a certain condition on X_1.

An introduction to the theory of renewal processes is given in Chapter 10.

8.4 Queues

The theory of queues is attractive and popular for two main reasons. First, queueing models are easily described and draw strongly from our intuitions about activities such as shopping or dialling the telephone operator. Secondly, even the solutions to the simplest models use much of the apparatus which we have developed in this book. Queues are, in general, non-Markovian, non-stationary, and quite difficult to study. Subject to certain conditions, however, their analysis uses ideas related to imbedded Markov chains, convergence of sequences of random variables, martingales, stationary processes, and renewal processes. We present a broad account of their theory in Chapter 11.

Customers arrive at a service point or counter at which a number of servers are stationed. An arriving customer may have to wait until one of these servers becomes available. Then he moves to the head of the queue and is served; he leaves the system on the completion of his service. We must specify a number of details about this queueing system before we are able to model it adequately. For example,

(a) in what manner do customers enter the system?
(b) in what order are they served?
(c) how long are their service times?

For the moment we shall suppose that the answers to these questions are as follows.

(a) The number $N(t)$ of customers who have entered by time t is a renewal process. That is, if T_n is the time of arrival of the nth customer (with the convention that $T_0 = 0$) then the *interarrival times*

$$X_n = T_n - T_{n-1}$$

are independent and identically distributed.

(b) Arriving customers join the end of a single line of people who receive attention on a 'first come, first served' basis. There are a certain number of servers. When a server becomes free, he turns his attention to the customer at the head of the waiting line. We shall usually suppose that the queue has a single server only.

(c) Service times are independent identically distributed random variables. That is, if S_n is the service time of the nth customer to arrive, then $\{S_n\}$ is a sequence of independent identically distributed non-negative random variables which do not depend on the arriving stream N of customers.

It requires only a little imagination to think of various other systems. Here are some examples.

(1) *Queues with baulking.* If the line of waiting customers is long then an arriving customer may, with a certain probability, decide not to join it.

(2) *Continental queueing.* In the absence of queue discipline, unoccupied servers pick a customer at random from the waiting mêlée.

(3) *Post Office queues.* The waiting customers divide into several lines, one for each server. The servers themselves enter and leave the system at random, causing the attendant customers to change lines as necessary.

(4) *Last come, first served.* No explanation is necessary.

(5) *Group service.* Waiting customers are served in batches. This is appropriate for lift queues and bus queues.

(6) *Student discipline.* Arriving customers jump the queue, joining it where a friend is standing.

Specific examples of some of these occur in the problems at the end of Chapter 11. Henceforth we shall consider only single-server queues described by (a), (b), and (c) above. Such queues are specified by the distribution of a typical interarrival time and the distribution of a typical service time; the method of analysis depends partly upon how much information we have about these quantities.

The state of the queue at time t is described by the number $Q(t)$ of waiting customers ($Q(t)$ *includes* customers who are in the process of being served at this time). It would be unfortunate if $Q(t) \to \infty$ as $t \to \infty$, and we devote special attention to finding out when this occurs. We call a queue *stable* if the distribution of $Q(t)$ settles down as $t \to \infty$ in some well-behaved way; otherwise we call it *unstable*. We choose not to define stability more precisely at this stage, wishing only to distinguish between such extremes as

(a) queues which either grow beyond all bounds or enjoy large wild fluctuations in length,

(b) queues whose lengths, say, converge in distribution, as $t \to \infty$, to some 'equilibrium distribution'.

Let S and X be a typical service time and a typical interarrival time, respectively; the ratio

$$\rho = \frac{\mathbf{E}(S)}{\mathbf{E}(X)}$$

is called the *traffic density*.

(7) **Theorem.** *Let Q be a queue with a single server and traffic density ρ.*

(a) *If $\rho < 1$ then Q is stable.*
(b) *If $\rho > 1$ then Q is unstable.*
(c) *If $\rho = 1$ and at least one of S and X has strictly positive variance then Q is unstable.*

The conclusions of this theorem are intuitively very attractive. Why?

A more satisfactory account of this theorem is given in Section 11.5.

8.5 The Wiener process

Most of the random processes considered so far are 'discrete' in the sense that they take values in the integers or in some other countable set. Perhaps the simplest example is simple random walk $\{S_n\}$, a process which jumps one unit to the left or to the right at each step. This random walk $\{S_n\}$ has two interesting and basic properties:

(a) *time-homogeneity*, in that, for all non-negative m and n, S_m and $S_{m+n} - S_n$ have the same distribution (we assume $S_0 = 0$); and
(b) *independent increments*, in that the increments $S_{n_i} - S_{m_i}$ $(i \geqslant 1)$ are independent whenever the intervals $(m_i, n_i]$ are disjoint.

What is the 'continuous' analogue of this random walk? It is reasonable to require that such a 'continuous' random process has the two properties above, and it turns out that, subject to some extra assumptions about means and variances, there is essentially only one such process called the *Wiener process*. This is a process $\{W(t): t \geqslant 0\}$, indexed by continuous time and taking values in the real line $\mathbb{R}$, which is time-homogeneous with independent increments, and with the vital extra property that $W(t)$ has the normal distribution with mean 0 and variance $\sigma^2 t$ for some constant σ^2. This process is sometimes called *Brownian motion*, and is a cornerstone of the modern theory of random processes. Think about it as a model for a particle which diffuses randomly along a line. There is no difficulty in constructing Wiener processes in higher dimensions, leading to models for such processes as the diffusion of a gas molecule in a container.

What are the finite-dimensional distributions of the Wiener process W? These are easily calculated as follows.

(1) **Lemma.** *The random variables* $W(t_1), W(t_2), \ldots, W(t_n)$ *have the multivariate normal distribution with zero means and covariance matrix* (v_{ij}) *where* $v_{ij} = \sigma^2 \min\{t_i, t_j\}$.

Proof. By assumption, $W(t_i)$ has the normal distribution with zero mean and variance $\sigma^2 t_i$. It therefore suffices to prove that $\operatorname{cov}(W(s), W(t)) = \sigma^2 \min\{s, t\}$. Now, if $s < t$, then

$$\mathbf{E}(W(s)W(t)) = \mathbf{E}\big(W(s)^2 + W(s)[W(t) - W(s)]\big) = \mathbf{E}(W(s)^2) + 0,$$

since W has independent increments and $\mathbf{E}(W(s)) = 0$. Hence

$$\operatorname{cov}(W(s), W(t)) = \operatorname{var}(W(s)) = \sigma^2 s$$

as required. ■

8.6 What is in a name?

In our discussions of the properties of random variables, only scanty reference has been made to the underlying probability space $(\Omega, \mathcal{F}, \mathbf{P})$; indeed we have felt some satisfaction and relief from this omission. We have often made assumptions about hypothetical random variables without even checking that such variables exist. For example, we are in the habit of making statements such as 'let $X_1, X_2, \ldots$ be independent variables with common distribution function F', but we have made no effort to show that there exists some probability space on which such variables can be constructed. The foundations of such statements require examination. It is the purpose of this section to indicate that our assumptions are fully justifiable. Move immediately to the next chapter if you are prepared to take our word for this and require no further insight.

First, suppose that $(\Omega, \mathcal{F}, \mathbf{P})$ is a probability space and that $X = \{X_t : t \in T\}$ is some collection of random variables mapping Ω into $\mathbb{R}$. We saw in Section 8.1 that to any vector $\boldsymbol{t} = (t_1, t_2, \ldots, t_n)$ containing members of T and of finite length there corresponds a joint distribution function $F_{\boldsymbol{t}}$; the collection of such functions $F_{\boldsymbol{t}}$, as $\boldsymbol{t}$ ranges over all possible vectors of any length, is called the set of *fdds* of X. It is clear that these distribution functions satisfy the two *Kolmogorov consistency conditions*:

(1) $$F_{(t_1,\ldots,t_n,t_{n+1})}(x_1, \ldots, x_n, x_{n+1}) \to F_{(t_1,\ldots,t_n)}(x_1, \ldots, x_n) \quad \text{as} \quad x_{n+1} \to \infty;$$

(2) if π is a permutation of $(1, 2, \ldots, n)$ and $\pi\boldsymbol{y}$ denotes the vector

$$\pi\boldsymbol{y} = (y_{\pi(1)}, \ldots, y_{\pi(n)})$$

for any n-vector $\boldsymbol{y}$, then

$$F_{\pi\boldsymbol{t}}(\pi\boldsymbol{x}) = F_{\boldsymbol{t}}(\boldsymbol{x}) \quad \text{for all } \boldsymbol{x}, \boldsymbol{t}, \pi, \text{ and } n.$$

Condition (1) is just a higher-dimensional form of (2.1.6a), and condition (2) says that the operation of permuting the X_t has the obvious corresponding effect on their joint distributions. So fdds always satisfy (1) and (2); furthermore (1) and (2) characterize fdds.

(3) **Theorem.** *Let T be any set, and suppose that to each vector $\boldsymbol{t} = (t_1, \ldots, t_n)$, containing members of T and of finite length, there corresponds a joint distribution function $F_{\boldsymbol{t}}$. If the collection $\{F_{\boldsymbol{t}}\}$ satisfies the Kolmogorov consistency conditions then there exists a probability space $(\Omega, \mathscr{F}, \mathbf{P})$ and a collection $X = \{X_t : t \in T\}$ of random variables on this space such that $\{F_{\boldsymbol{t}}\}$ is the set of fdds of X.*

The proof of this result lies in the heart of measure theory, as the following sketch indicates.

Sketch proof. Let $\Omega = \mathbb{R}^T$, the product of T copies of $\mathbb{R}$; the points of Ω are collections $\boldsymbol{y} = \{y_t : t \in T\}$ of real numbers. Let $\mathscr{F} = \mathscr{B}^T$, the σ-field generated by subsets of the form $\prod_{t \in T} B_t$ for Borel sets B_t all but finitely many of which equal $\mathbb{R}$. It is a fundamental result in measure theory that there exists a probability measure $\mathbf{P}$ on $(\Omega, \mathscr{F})$ such that

$$\mathbf{P}(\{\boldsymbol{y} \in \Omega : y_{t_1} \leqslant x_1, y_{t_2} \leqslant x_2, \ldots, y_{t_n} \leqslant x_n\}) = F_{\boldsymbol{t}}(\boldsymbol{x})$$

for all $\boldsymbol{t}$ and $\boldsymbol{x}$; this follows by an extension of the argument of Section 1.6. Then $(\Omega, \mathscr{F}, \mathbf{P})$ is the required space. Define $X_t : \Omega \to \mathbb{R}$ by

$$X_t(\boldsymbol{y}) = y_t$$

to obtain the required family $\{X_t\}$. ■

We have seen that fdds are characterized by the consistency conditions (1) and (2). But how much do they tell us about the sample paths of the corresponding process X? A simple example is enough to indicate some of the dangers here.

(4) **Example.** Let U be a random variable which is uniformly distributed on $[0, 1]$. Define two processes $X = \{X_t : 0 \leqslant t \leqslant 1\}$ and $Y = \{Y_t : 0 \leqslant t \leqslant 1\}$ by

$$X_t = 0 \quad \text{for all } t, \qquad Y_t = \begin{cases} 1 & \text{if} \quad U = t \\ 0 & \text{otherwise.} \end{cases}$$

Clearly X and Y have the same fdds, since

$$\mathbf{P}(U = t) = 0 \quad \text{for all } t.$$

But X and Y are different processes. In particular

$$\mathbf{P}(X_t = 0 \text{ for all } t) = 1$$
$$\mathbf{P}(Y_t = 0 \text{ for all } t) = 0.$$ ●

One may easily construct less trivial examples of different processes having the same fdds; such processes are called *versions* of one another. This complication should not be overlooked with a casual wave of the hand; it is central to any theory which attempts to study properties of sample paths, such as first-passage times. As the above example illustrates, such properties are not generally specified by the fdds, and their validity may therefore depend on which version of the process is under study.

For the random process $\{X(t): t \in T\}$, where $T = [0, \infty)$ say, knowledge of the fdds amounts to being given a probability space of the form $(\mathbb{R}^T, \mathcal{B}^T, \mathbf{P})$, as in the sketch proof of (3) above. Many properties of sample paths do not correspond to events in $\mathcal{B}^T$. For example, the subset of Ω given by $A = \{\omega \in \Omega: X(t) = 0 \text{ for all } t \in T\}$ is an *uncountable* intersection of events $A = \bigcap_{t \in T} \{X(t) = 0\}$, and may not itself be an event. Such difficulties would be avoided if all sample paths of X were continuous, since then A is the intersection of $\{X(t) = 0\}$ over all *rational* $t \in T$; this is a *countable* intersection.

(5) **Example.** Let W be the Wiener process of Section 8.5, and let T be the time of the first passage of W to the point 1, $S = \inf\{t: W(t) = 1\}$. Then

$$\{S > t\} = \bigcap_{0 \leqslant s \leqslant t} \{W(s) \neq 1\}$$

is a set of configurations which does not belong to the Borel σ-field $\mathcal{B}^{[0, \infty)}$. If all sample paths of W were continuous, one might write

$$\{S > t\} = \bigcap_{\substack{0 \leqslant s \leqslant t \\ s \text{ rational}}} \{W(s) \neq 1\},$$

the countable intersection of events. As the construction of Example (4) indicates, there are versions of the Wiener process which have discontinuous sample paths. One of the central results of Chapter 13 is that there exists a version with continuous sample paths, and it is with this version that one normally works. ●

It is too restrictive to require continuity of sample paths in general; after all, processes such as the Poisson process most definitely do not have continuous sample paths. The most which can be required is continuity from either the left or the right. Following a convention, we go for the latter here. Under what conditions may one assume that there exists a version with right-continuous sample paths? An answer is provided by the next theorem; see Breiman (1968, p. 300) for a proof.

(6) **Theorem.** *Let $\{X(t): t \geqslant 0\}$ be a real-valued random process. Let D be a subset of $[0, \infty)$ which is dense in $[0, \infty)$. If*

(i) *X is continuous in probability from the right, that is, $X(t + h) \xrightarrow{\mathrm{P}} X(t)$ as $h \downarrow 0$, for all t, and*

(ii) *at any accumulation point a of D, X has finite right and left limits with probability 1, that is, $\lim_{h \downarrow 0} X(a + h)$ and $\lim_{h \uparrow 0} X(a + h)$ exist, a.s.,*

then there exists a version Y of X such that

(a) *the sample paths of Y are right-continuous,*

(b) *Y has left limits, in that $\lim_{h \uparrow 0} Y(t + h)$ exists for all t.*

In other words, if (i) and (ii) hold, then there exists a probability space and a process Y defined on this space, such that Y has the same fdds as X in addition to properties (a) and (b). A process which is right-continuous with left limits is called càdlàg by some (largely French speakers), and a Skorokhod map or R-process by others.

8.7 Problems

1. Let $\{Z_n\}$ be a sequence of uncorrelated real-valued variables with zero means and unit variances, and define the 'moving average'

$$Y_n = \sum_{i=0}^{r} \alpha_i Z_{n-i},$$

for constants $\alpha_0, \alpha_1, \dots, \alpha_r$. Show that Y is stationary and find its autocovariance function.

2. Let $\{Z_n\}$ be a sequence of uncorrelated real-valued variables with zero means and unit variances. Suppose that $\{Y_n\}$ is an 'autoregressive' stationary sequence in that it satisfies $Y_n = \alpha Y_{n-1} + Z_n$, $-\infty < n < \infty$, for some real α satisfying $|\alpha| < 1$. Show that Y has autocovariance function $c(m) = \alpha^{|m|}/(1 - \alpha^2)$.

3. Let $\{X_n\}$ be a sequence of independent identically distributed Bernoulli variables, each taking values 0 and 1 with probabilities $1 - p$ and p respectively. Find the mass function of the renewal process $N(t)$ with inter-arrival times $\{X_n\}$.

4. Customers arrive in a shop in the manner of a Poisson process with parameter λ. There are infinitely many servers, and each service time is exponentially distributed with parameter μ. Show that the number $Q(t)$ of waiting customers at time t constitutes a birth–death process. Find its stationary distribution.

5. Let $X(t) = Y\cos(\theta t) + Z\sin(\theta t)$ where Y and Z are independent $N(0,1)$ random variables, and let $\tilde{X}(t) = R\cos(\theta t + \Psi)$ where R and Ψ are independent. Find distributions for R and Ψ such that the processes X and $\tilde{X}$ have the same fdds.

6. In a Prague teashop (U Myšáka) customers queue at the entrance for the blank bill. In the shop there are separate counters for coffee, sweetcakes, pretzels, milk, drinks, and ice cream, and queues form at each of these. At each service point the customers' bills are marked appropriately. There is a restricted number N of seats, and departing customers have to queue in order to pay their bills. If interarrival times and service times are exponentially distributed and the process is in equilibrium, find how much longer a greedy customer must wait if he insists on sitting down. Answers on a postcard to the authors, please.

9 Stationary processes

9.1 Introduction

Recall that a process X is *strongly stationary* whenever its *finite-dimensional distributions* are invariant under time shifts; it is (*weakly*) *stationary* whenever it has constant means and its *autocovariance function* is invariant under time shifts. Section 8.2 contains various examples of such processes. Next, we shall explore the consequences of stationarity and see how they include the spectral theorem and the ergodic theorem†.

A special class of random processes comprises those processes whose joint distributions are multivariate normal; these are called 'Gaussian processes'. Section 9.6 contains a brief account of some of the properties of such processes. In general, a Gaussian process is not stationary, but it is easy to characterize those which are.

We shall be interested mostly in continuous-time processes $X = \{X(t): -\infty < t < \infty\}$, indexed by the whole real line, and will indicate any necessary variations for processes with other index sets, such as discrete-time processes. It is convenient to suppose that X takes values in the complex plane $\mathbb{C}$. This entails few extra complications and provides the natural setting for the theory. No conceptual difficulty is introduced by this generalization, since any complex-valued process X can be decomposed as

$$X = X_1 + \mathrm{i}X_2$$

where X_1 and X_2 are real-valued processes. However, we must take care when discussing the fdds of X since the distribution function of a complex-valued random variable $C = R + \mathrm{i}I$ is no longer a function of a single real variable. Thus, our definition of strong stationarity requires revision; we leave this to the reader. The concept of weak stationarity concerns covariances; we must note an important amendment to the real-valued theory in this context. As before, the expectation operator $\mathbf{E}$ is well defined by

$$\mathbf{E}(R + \mathrm{i}I) = \mathbf{E}(R) + \mathrm{i}\mathbf{E}(I).$$

(1) **Definition.** The **covariance** of two complex-valued random variables C_1 and C_2 is defined to be

$$\operatorname{cov}(C_1, C_2) = \mathbf{E}\big((C_1 - \mathbf{E}C_1)\overline{(C_2 - \mathbf{E}C_2)}\big)$$

where $\bar{z}$ denotes the complex conjugate of z.

† The word 'ergodic' has several meanings, and probabilists tend to use it rather carelessly. We conform to this custom here.

This reduces to the usual definition (3.6.7) when C_1 and C_2 are real. Note that the operator 'cov' is not symmetrical in its arguments, since

$$\operatorname{cov}(C_2, C_1) = \overline{\operatorname{cov}(C_1, C_2)}.$$

Variances are defined as follows.

(2) **Definition.** The **variance** of a complex-valued random variable C is defined to be

$$\operatorname{var}(C) = \operatorname{cov}(C, C).$$

Decompose C into its real and imaginary parts

$$C = R + \mathrm{i}I$$

and apply (2) to obtain

$$\operatorname{var}(C) = \operatorname{var}(R) + \operatorname{var}(I).$$

We can write

$$\operatorname{var}(C) = \mathbf{E}(|C - \mathbf{E}C|^2).$$

We do not generally speak of complex random variables as being 'uncorrelated', preferring to use a word which emphasizes the geometrical properties of the complex plane.

(3) **Definition.** Complex-valued random variables C_1 and C_2 are called **orthogonal** if $\operatorname{cov}(C_1, C_2) = 0$.

If $X = X_1 + \mathrm{i}X_2$ is a complex-valued process with real part X_1 and imaginary part X_2 then $\bar{X}$ denotes the complex conjugate process of X:

$$\bar{X} = X_1 - \mathrm{i}X_2.$$

(4) **Example. Functions of the Poisson process.** Let N be a Poisson process with intensity λ. Let α be a positive number, and define $X(t) = N(t + \alpha) - N(t)$, for $t \geqslant 0$. It is easily seen (*exercise*) from the definition of a Poisson process that X is a strongly stationary process with mean $\mathbf{E}(X(t)) = \lambda\alpha$ and autocovariance function

$$c(t, t+h) = \mathbf{E}(X(t)X(t+h)) - (\lambda\alpha)^2 = \begin{cases} 0 & \text{if } h \geqslant \alpha \\ \lambda(\alpha - h) & \text{if } h < \alpha, \end{cases}$$

where $t, h \geqslant 0$.

Here is a second example based on the Poisson process. Let $\beta = \mathrm{e}^{2\pi \mathrm{i}/m}$ be a complex mth root of unity, where $m \geqslant 2$, and define $Y(t) = \beta^{Z+N(t)}$ where Z is a random variable that is independent of N with mass function $\mathbf{P}(Z = j) = 1/m$, for $1 \leqslant j \leqslant m$. Once again, it is left as an *exercise* to show that Y is a strictly stationary (complex-valued) process with mean $\mathbf{E}(Y(t)) = 0$.

Its autocovariance function is given by the following calculation:

$$\begin{aligned}\mathbf{E}(Y(t)\overline{Y(t+h)}) &= \mathbf{E}(\beta^{N(t)}\bar{\beta}^{N(t+h)}) = \mathbf{E}((\beta\bar{\beta})^{N(t)}\bar{\beta}^{N(t+h)-N(t)})\\ &= \mathbf{E}(\bar{\beta}^{N(h)}) \qquad \text{since} \quad \beta\bar{\beta} = 1\\ &= \exp[\lambda h(\bar{\beta} - 1)] \quad \text{for} \quad t, h \geqslant 0,\end{aligned}$$

where we have used elementary properties of the Poisson process. ●

Exercises

(5) Let $\dots, Z_{-1}, Z_0, Z_1, Z_2, \dots$ be independent real random variables with means 0 and variances 1, and let $\alpha, \beta \in \mathbb{R}$. Show that there exists a (weakly) stationary sequence $\{W_n\}$ satisfying $W_n = \alpha W_{n-1} + \beta W_{n-2} + Z_n$, $n = \dots, -1, 0, 1, \dots$, if the (possibly complex) zeros of the quadratic equation $z^2 - \alpha z - \beta = 0$ are smaller than 1 in absolute value.

(6) Let U be uniformly distributed on $[0, 1]$ with binary expansion

$$U = \sum_{i=1}^{\infty} X_i 2^{-i}.$$

Show that the sequence

$$V_n = \sum_{i=1}^{\infty} X_{i+n} 2^{-i}, \qquad n \geqslant 0,$$

is strongly stationary, and calculate its autocovariance function.

(7) Let $\{X_n: n = \dots, -1, 0, 1, \dots\}$ be a stationary real sequence with mean 0 and autocovariance function c.

(i) Show that the infinite series $\sum_0^\infty a_n X_n$ converges almost surely, and in mean square, whenever $\sum_0^\infty |a_n| < \infty$.

(ii) Let

$$Y_n = \sum_{k=0}^{\infty} a_k X_{n-k}, \qquad n = \dots, -1, 0, 1, \dots$$

where $\sum_{k=0}^\infty |a_k| < \infty$. Find an expression for the autocovariance function c_Y of Y, and show that

$$\sum_{m=-\infty}^{\infty} |c_Y(m)| < \infty.$$

(8) Let $X = \{X_n: n \geqslant 0\}$ be a discrete-time Markov chain with countable state space S and stationary distribution π, and suppose that X_0 has distribution π. Show that the sequence $\{f(X_n): n \geqslant 0\}$ is strongly stationary for any function $f: S \to \mathbb{R}$.

9.2 Linear prediction

Statisticians painstakingly observe and record processes which evolve in time, not merely for the benefit of historians but also in the belief that it is an advantage to know the past when attempting to predict the future. Most scientific schemes (and many non-scientific schemes) for prediction are 'model' based, in that they make some specific assumptions about the process, and then use past data to extrapolate into the future. For example, in the statistical theory of 'time series', one often assumes that the process is some combination of general trend, periodic fluctuations, and random noise, and it is common to suppose that the noise component is a stationary process having an autocovariance function of a certain form.

Suppose that we are observing a sequence $\{x_n\}$ of numbers, the number x_n being revealed to us at time n, and that we are prepared to accept that these numbers are the outcomes of a stationary sequence $\{X_n\}$ with known mean $\mathbf{E}X_n = \mu$ and autocovariance function $c(m) = \text{cov}(X_n, X_{n+m})$. We may be required to estimate the value of X_{r+k} (where $k \geqslant 1$), given the values $X_r, X_{r-1}, \ldots, X_{r-s}$. We saw in Section 7.9 that the 'best' (that is, the minimum mean-squared-error) predictor of X_{r+k} given $X_r, X_{r-1}, \ldots, X_{r-s}$ is the conditional mean $M = \mathbf{E}(X_{r+k} \mid X_r, X_{r-1}, \ldots, X_{r-s})$; that is to say, the mean squared error $\mathbf{E}((Y - X_{r+k})^2)$ is minimized over all choices of functions Y of $X_r, X_{r-1}, \ldots, X_{r-s}$ by the choice $Y = M$. The calculation of such quantities requires a knowledge of the fdds of X which we do not generally possess. For various reasons, it is not realistic to attempt to estimate the fdds in order to *estimate* the conditional mean. The problem becomes more tractable, and its solution more elegant, if we restrict our attention to *linear* predictors of X_{r+k}, which is to say that we seek the best predictor of X_{r+k} amongst the class of linear functions of $X_r, X_{r-1}, \ldots, X_{r-s}$.

(1) **Theorem.** *Let X be a real stationary sequence with zero mean and autocovariance function c. Amongst the class of linear functions $h(X_r, X_{r-1}, \ldots, X_{r-s})$ of $X_r, X_{r-1}, \ldots, X_{r-s}$, the best predictor of X_{r+k} (where $k \geqslant 1$) is*

(2)
$$\hat{X}_{r+k} = \sum_{i=0}^{s} a_i X_{r-i}$$

where the a_i satisfy the equations

(3)
$$\sum_{i=0}^{s} a_i c(|i-j|) = c(k+j) \quad \textit{for} \quad 0 \leqslant j \leqslant s.$$

Proof. Let H be the closed linear space of linear functions of $X_r, X_{r-1}, \ldots, X_{r-s}$. We have from the projection theorem (7.9.14) that the element M of H for which $\mathbf{E}((X_{r+k} - M)^2)$ is a minimum is the (a.s.) unique M such that

(4)
$$\mathbf{E}((X_{r+k} - M)Z) = 0 \quad \text{for all } Z \in H.$$

Certainly $X_{r-j} \in H$ for $0 \leqslant j \leqslant s$. Writing $M = \sum_{i=0}^{s} a_i X_{r-i}$ and substituting $Z = X_{r-j}$ in (4), we obtain

$$\mathbf{E}(X_{r+k}X_{r-j}) = \mathbf{E}(MX_{r-j}) = \sum_{i=0}^{s} a_i \mathbf{E}(X_{r-i}X_{r-j}),$$

whence (3) follows by the assumption of zero mean. ∎

Therefore, if we know the autocovariance function c, then equation (3) tells us how to find the best linear predictor of future values of the stationary sequence X. In practice we may not know c, and may instead have to estimate it. Rather than digress further in this direction, the reader is referred to the time series literature, for example Chatfield (1980).

(5) **Example. Autoregressive scheme.** Let $\{Z_n\}$ be a sequence of independent variables with zero means and unit variances, and let $\{Y_n\}$ satisfy

(6) $$Y_n = \alpha Y_{n-1} + Z_n, \qquad -\infty < n < \infty,$$

where α is a real number satisfying $|\alpha| < 1$. We have from Problem (8.7.2) that Y is stationary with zero mean and autocovariance function $c(m) = \mathbf{E}(Y_n Y_{n+m})$ given by

(7) $$c(m) = \frac{\alpha^{|m|}}{1-\alpha^2}, \qquad -\infty < m < \infty.$$

Suppose we wish to estimate Y_{r+k} (where $k \geq 1$) from a knowledge of $Y_r, Y_{r-1}, \dots, Y_{r-s}$. The best linear predictor is $\hat{Y}_{r+k} = \sum_{i=0}^{s} a_i Y_{r-i}$ where the a_i satisfy equations (3):

$$\sum_{i=0}^{s} a_i \alpha^{|i-j|} = \alpha^{k+j}, \qquad 0 \leq j \leq s.$$

A solution is $a_0 = \alpha^k$, $a_i = 0$ for $i \geq 1$, so that the best linear predictor is $\hat{Y}_{r+k} = \alpha^k Y_r$. The mean squared error of prediction is

$$\begin{aligned}\mathbf{E}((Y_{r+k} - \hat{Y}_{r+k})^2) &= \operatorname{var}(Y_{r+k} - \alpha^k Y_r)\\ &= \operatorname{var}(Y_{r+k}) - 2\alpha^k \operatorname{cov}(Y_{r+k}, Y_r) + \alpha^{2k}\operatorname{var}(Y_r)\\ &= c(0) - 2\alpha^k c(k) + \alpha^{2k} c(0) = \frac{1-\alpha^{2k}}{1-\alpha^2}\end{aligned}$$

by (7). ●

(8) **Example.** Let $X_n = (-1)^n X_0$ where X_0 is equally likely to take each of the values -1 and $+1$. It is easily checked that X is stationary with zero mean and autocovariance function $c(m) = (-1)^m \mathbf{E}(X_0^2) = (-1)^m$, $-\infty < m < \infty$. The best linear predictor of X_{r+k} (where $k \geq 1$) based on $X_r, X_{r-1}, \dots, X_{r-s}$ is obtained by solving the equations

$$\sum_{i=0}^{s} a_i (-1)^{|i-j|} = (-1)^{k+j}, \qquad 0 \leq j \leq s.$$

A solution is $a_0 = (-1)^j$, $a_i = 0$ for $i \geq 1$, so that $\hat{X}_{r+k} = (-1)^k X_r$, and the mean squared error of prediction is zero. ●

Exercises

(9) Let X be a (weakly) stationary sequence with zero mean and autocovariance function c.

(i) Find the best linear predictor $\hat{X}_{n+1}$ of X_{n+1} given X_n.
(ii) Find the best linear predictor $\tilde{X}_{n+1}$ of X_{n+1} given X_n and X_{n-1}.
(iii) Find an expression for $D = \mathbf{E}\{(X_{n+1} - \hat{X}_{n+1})^2\} - \mathbf{E}\{(X_{n+1} - \tilde{X}_{n+1})^2\}$, and evaluate this expression when

(a) $X_n = \cos(nU)$ where U is uniform on $[-\pi, \pi]$,
(b) X is an autoregressive scheme with $c(k) = \alpha^{|k|}$ where $|\alpha| < 1$.

(10) Does there exist a (weakly) stationary sequence $\{X_n: -\infty < n < \infty\}$ with zero means and autocovariance function

$$c(k) = \begin{cases} 1 & \text{if} \quad k = 0 \\ \dfrac{a}{1 + a^2} & \text{if} \quad |k| = 1 \\ 0 & \text{if} \quad |k| > 1 \end{cases}$$

where $|a| < 1$?

Assuming that such a sequence exists, find the best linear predictor $\hat{X}_n$ of X_n given $X_{n-1}, X_{n-2}, \ldots$, and show that the mean squared error of prediction is $(1 + a^2)^{-1}$. Verify that $\{\tilde{X}_n\}$ is (weakly) stationary.

9.3 Autocovariances and spectra

Let $X = \{X(t): -\infty < t < \infty\}$ be a (weakly) stationary process which takes values in $\mathbb{C}$. It has autocovariance function c given by

$$c(s, s + t) = \operatorname{cov}(X(s), X(s + t)) \quad \text{for} \quad s, t \in \mathbb{R}$$

where $c(s, s + t)$ depends on t alone. We think of c as a complex-valued function of the single variable t, and abbreviate it to

$$c(t) = c(s, s + t) \quad \text{for any } s.$$

Notice that the variance of $X(t)$ is constant for all t since

(1) $$\operatorname{var}(X(t)) = \operatorname{cov}(X(t), X(t)) = c(0).$$

We shall sometimes assume that the mean value $\mathbf{E}(X(t))$ of X equals zero; if this is not true, then define $X'(t) = X(t) - \mathbf{E}(X(t))$ to obtain another stationary process with zero means and the same autocovariance function c.

Autocovariances have the following properties.

(2) **Theorem.**

(a) $c(-t) = \overline{c(t)}$

(b) *c is a non-negative definite function, which is to say that*

$$\sum_{j,k} c(t_k - t_j) z_j \bar{z}_k \geq 0$$

for all real $t_1, \ldots, t_n$ *and all complex* $z_1, \ldots, z_n$.

Proof.

(a)
$$\begin{aligned} c(-t) &= \operatorname{cov}(X(t), X(0)) \\ &= \overline{\operatorname{cov}(X(0), X(t))} = \overline{c(t)}. \end{aligned}$$

(b) This is like the proof of (5.7.3c). Just write

$$\begin{aligned} \sum_{j,k} c(t_k - t_j) z_j \bar{z}_k &= \sum_{j,k} \operatorname{cov}(z_j X(t_j), z_k X(t_k)) \\ &= \operatorname{cov}(Z, Z) \geq 0 \end{aligned}$$

where

$$Z = \sum_j z_j X(t_j).$$ ■

Of more interest than the autocovariance function is the 'autocorrelation function' (see (3.6.7)).

(3) **Definition.** The **autocorrelation function** of a weakly stationary process X with autocovariance function c is defined by

$$\rho(t) = \frac{\operatorname{cov}(X(0), X(t))}{[\operatorname{var}(X(0)) \operatorname{var}(X(t))]^{\frac{1}{2}}} = \frac{c(t)}{c(0)}$$

whenever $c(0) = \operatorname{var}(X(t)) > 0$.

Of course, $\rho(t)$ is just the correlation between $X(s)$ and $X(s+t)$, for any s.

Following the discussion in Section 8.2, we seek to assess the incidence of certain regular oscillations within the random fluctuation of X. For a weakly stationary process this is often a matter of studying regular oscillations in its autocorrelation function.

(4) **Theorem. Spectral theorem for autocorrelation functions.** *The autocorrelation function* $\rho(t)$ *of a weakly stationary process* X *with strictly positive variance is the characteristic function of some distribution function* F *whenever* $\rho(t)$ *is continuous at* $t = 0$. *That is to say,*

(5)
$$\rho(t) = \int_{-\infty}^{\infty} e^{it\lambda} \, dF(\lambda).$$

Proof. This follows immediately from the discussion after (5.7.3), and is a simple application of Bochner's theorem. Following (2), we need only show that ρ is uniformly continuous. Without loss of generality we can suppose that $\mathbf{E}(X(t)) = 0$ for all t. Let c be the autocovariance function of X, and use the Cauchy–Schwarz inequality (3.6.9) to obtain

$$\begin{aligned}|c(t+h) - c(t)| &= \big|\mathbf{E}\big(X(0)[X(t+h) - X(t)]\big)\big| \\ &\leqslant \mathbf{E}(|X(0)|\,|X(t+h) - X(t)|) \\ &\leqslant [\mathbf{E}(|X(0)|^2)\mathbf{E}(|X(t+h) - X(t)|^2)]^{\frac{1}{2}} \\ &= \{c(0)[2c(0) - c(h) - c(-h)]\}^{\frac{1}{2}}\end{aligned}$$

whenever $c(h)$ is continuous at $h = 0$. Thus ρ is uniformly continuous, and the result follows. ■

Think of equation (5) as follows. With any real λ we may associate a complex-valued oscillating function g_λ which has period $2\pi/|\lambda|$ and some non-negative amplitude f_λ, say:

$$g_\lambda(t) = f_\lambda\, e^{it\lambda};$$

in the less general real-valued theory we might consider oscillations such as

$$g'_\lambda(t) = f_\lambda \cos(t\lambda)$$

(see (8.2.6) and (8.2.7)). With any collection $\lambda_1, \lambda_2, \ldots$ of frequencies we can associate a mixture

(6) $$g_\lambda(t) = \sum_j f_j \exp(it\lambda_j)$$

of pure oscillations, where the f_j indicate the relative strengths of the various components. As the number of component frequencies in (6) grows, the summation may approach an integral

(7) $$g(t) = \int_{-\infty}^{\infty} f(\lambda)\, e^{it\lambda}\, d\lambda$$

where f is some non-negative function which assigns weights to the λ. The progression from (6) to (7) is akin to the construction of the abstract integral (see Section 5.6). We have seen many expressions which are similar to (7), but in which f is the density function of some continuous random variable. Just as continuous variables are only a special subclass of the larger family of all random variables, so (7) is not the most general limiting form for (6); the general form is

(8) $$g(t) = \int_{-\infty}^{\infty} e^{it\lambda}\, dF(\lambda)$$

where F is a function which maps $\mathbb{R}$ into $[0, \infty)$ and which is right-continuous, non-decreasing, and such that $F(-\infty) = 0$; we omit the details of this, which are very much the same as in part B of Section 5.6. It is easy to see that F is a distribution function if and only if $g(0) = 1$. Theorem (4) asserts that ρ enjoys a decomposition in the form of (8), as a mixture of pure oscillations.

There is an alternative view of (5) which differs slightly from this. If Λ is a random variable with distribution function F, then $g_\Lambda(t) = e^{it\Lambda}$ is a pure oscillation with a random frequency. Theorem (4) asserts that ρ is the mean value of this random oscillation for some special distribution F. Of course, by the uniqueness theorem (5.9.3) there is a unique distribution function F such that (5) holds.

(9) **Definition.** If the autocorrelation function ρ satisfies

$$\rho(t) = \int_{-\infty}^{\infty} e^{it\lambda}\, dF(\lambda)$$

then F is called the **spectral distribution function** of the process. The **spectral density function** is the density function which corresponds to the distribution function F whenever this density exists.

For a given autocorrelation function ρ, we can find the spectral distribution function by the inversion techniques of Section 5.9.

In general, there may be certain frequency bands which make no contribution to (5). For example, if the spectral distribution function F satisfies

$$F(\lambda) = 0 \quad \text{for all } \lambda \leqslant 0,$$

then only positive frequencies make non-trivial contributions. If the frequency band $(\lambda - \varepsilon, \lambda + \varepsilon)$ makes a non-trivial contribution to (5) for all $\varepsilon > 0$, then we say that λ belongs to the 'spectrum' of the process.

(10) **Definition.** The **spectrum** of X is the set of all real numbers λ with the property that

$$F(\lambda + \varepsilon) - F(\lambda - \varepsilon) > 0 \quad \text{for all } \varepsilon > 0$$

where F is the spectral distribution function.

If X is a discrete-time process then the above account is inadequate, since the autocorrelation function ρ now maps $\mathbb{Z}$ into $\mathbb{C}$ and cannot be a characteristic function unless its domain is extended. Theorem (4) remains broadly true, but asserts now that ρ has a representation

(11) $$\rho(n) = \int_{-\infty}^{\infty} e^{in\lambda}\, dF(\lambda)$$

for some distribution function F and all integral n. No condition of continuity is appropriate here. This representation (11) is not unique because the integrand $g_\lambda(n) = e^{in\lambda}$ is periodic in λ:

$$g_{\lambda+2\pi}(n) = g_\lambda(n) \quad \text{for all } n.$$

In this case it is customary to rewrite (11) as

$$\rho(n) = \sum_{k=-\infty}^{\infty} \int_{((2k-1)\pi,(2k+1)\pi]} e^{in\lambda}\, dF(\lambda),$$

yielding the usual statement of the spectral theorem for discrete-time processes

(12) $$\rho(n) = \int_{(-\pi,\pi]} e^{in\lambda}\, d\widetilde{F}(\lambda)$$

for some appropriate distribution function $\widetilde{F}$ obtained from F and satisfying

$$\widetilde{F}(-\pi) = 0, \qquad \widetilde{F}(\pi) = 1.$$

A further simplification is possible if X is real valued, since then $\rho(n) = \rho(-n)$, so that

(13) $$\rho(n) = \tfrac{1}{2}[\rho(n) + \rho(-n)] = \int_{(-\pi,\pi]} \tfrac{1}{2}(e^{in\lambda} + e^{-in\lambda})\, d\widetilde{F}(\lambda) \quad \text{by (12)}$$
$$= \int_{(-\pi,\pi]} \cos(n\lambda)\, d\widetilde{F}(\lambda).$$

Furthermore $\cos(n\lambda) = \cos(-n\lambda)$, and it follows that ρ may be expressed as

(14) $$\rho(n) = \int_{[-\pi,\pi]} \cos(n\lambda)\, dG(\lambda)$$

for some distribution function G of a symmetric distribution on $[-\pi, \pi]$. We note that the validity of (14) for some such G is both necessary and sufficient for ρ to be the autocorrelation function of a real-valued stationary sequence. The necessity of (14) has been shown. For its sufficiency, we shall see at the beginning of Section 9.6 that all symmetric, non-negative definite functions ρ with $\rho(0) = 1$ are autocorrelation functions of stationary sequences whose fdds are multivariate normal.

Equations (12)–(14) express ρ as the Fourier transform of some distribution function. Fourier transforms may be inverted in the usual way to obtain an expression for the spectral distribution in terms of ρ. One such expression is the following.

(15) **Theorem.** *Let ρ be the autocorrelation function of a stationary sequence. If the function $\widetilde{F}$ in (12) is differentiable with derivative f, then*

(16) $$f(\lambda) = \frac{1}{2\pi} \sum_{n=-\infty}^{\infty} e^{-in\lambda} \rho(n)$$

at every point λ at which f is differentiable.

For real-valued sequences, (16) may be written as

(17)
$$f(\lambda) = \frac{1}{2\pi} \sum_{n=-\infty}^{\infty} \rho(n) \cos(n\lambda), \qquad -\pi \leqslant \lambda \leqslant \pi.$$

As in the discussion after (5.9.1) of characteristic functions, a sufficient (but not necessary) condition for the existence of the spectral density function f is

(18)
$$\sum_{n=-\infty}^{\infty} |\rho(n)| < \infty.$$

(19) **Example. Independent sequences.** Let $X = \{X_n: n \geqslant 0\}$ be a sequence of independent variables with zero means and unit variances. In (8.2.8) we found that the autocorrelation function is given by

$$\rho(n) = \begin{cases} 1 & \text{if} \quad n = 0 \\ 0 & \text{if} \quad n \neq 0. \end{cases}$$

To find the spectral density function, either use (15) or recognize that

$$\rho(n) = \frac{1}{2\pi} \int_{-\pi}^{\pi} e^{in\lambda} \, d\lambda$$

to see that the spectral density function is the uniform density function on $[-\pi, \pi]$. The spectrum of X is $[-\pi, \pi]$. ●

(20) **Example. Identical sequences.** Let Y be a random variable with zero mean and unit variance, and let $X = \{X_n: n \geqslant 0\}$ be the stationary sequence given by

$$X_n = Y \quad \text{for all } Y.$$

In (8.2.9) we calculated the autocorrelation function as

$$\rho(n) = 1 \quad \text{for all } n$$

and we recognize this as the characteristic function of a distribution which is concentrated at 0. The spectrum of X is the set $\{0\}$. ●

(21) **Example. Two-state Markov chains.** Let $X = \{X(t): t \geqslant 0\}$ be a Markov chain with state space $S = \{1, 2\}$. Suppose, as in Example (6.9.15), that the times spent in states 1 and 2 are exponentially distributed with parameters α and β respectively where $\alpha\beta > 0$. That is to say, X has generator $\boldsymbol{G}$ given by

$$\boldsymbol{G} = \begin{pmatrix} -\alpha & \alpha \\ \beta & -\beta \end{pmatrix}.$$

In our solution to (6.9.15) we wrote down the Kolmogorov forward equations and found that the transition probabilities

$$p_{ij}(t) = \mathbf{P}(X(t) = j \mid X(0) = i), \qquad 1 \leqslant i, j \leqslant 2$$

are given by

$$p_{11}(t) = 1 - p_{12}(t) = \frac{\beta}{\alpha+\beta} + \frac{\alpha}{\alpha+\beta}\,\mathrm{e}^{-t(\alpha+\beta)}$$

$$p_{22}(t) = 1 - p_{21}(t) = \frac{\alpha}{\alpha+\beta} + \frac{\beta}{\alpha+\beta}\,\mathrm{e}^{-t(\alpha+\beta)}$$

in agreement with (6.10.12). Let $t \to \infty$ to find that the chain has a stationary distribution $\boldsymbol{\pi}$ given by

$$\pi_1 = \frac{\beta}{\alpha+\beta}, \qquad \pi_2 = \frac{\alpha}{\alpha+\beta}.$$

Suppose now that $X(0)$ has distribution $\boldsymbol{\pi}$. As in (8.2.4), X is a strongly stationary process. We are going to find its spectral representation. First, find the autocovariance function. If $t \geqslant 0$, then a short calculation yields

$$\begin{aligned}\mathbf{E}(X(0)X(t)) &= \sum_i i\,\mathbf{E}(X(t) \mid X(0) = i)\pi_i \\ &= \sum_{i,j} ij p_{ij}(t)\pi_i \\ &= \frac{(2\alpha+\beta)^2}{(\alpha+\beta)^2} + \frac{\alpha\beta}{(\alpha+\beta)^2}\,\mathrm{e}^{-t(\alpha+\beta)}\end{aligned}$$

and so the autocovariance function c is given by

$$\begin{aligned}c(t) &= \mathbf{E}(X(0)X(t)) - \mathbf{E}(X(0))\mathbf{E}(X(t)) \\ &= \frac{\alpha\beta}{(\alpha+\beta)^2}\,\mathrm{e}^{-t(\alpha+\beta)} \quad \text{if} \quad t \geqslant 0.\end{aligned}$$

Hence

$$c(0) = \frac{\alpha\beta}{(\alpha+\beta)^2}$$

and the autocorrelation function ρ is given by

$$\rho(t) = \frac{c(t)}{c(0)} = \mathrm{e}^{-t(\alpha+\beta)} \quad \text{if} \quad t \geqslant 0.$$

X is real and so ρ is symmetric; thus

(22) $$\rho(t) = \mathrm{e}^{-|t|(\alpha+\beta)}.$$

The spectral theorem asserts that ρ is the characteristic function of some distribution. We may use the inversion theorem (5.9.2) to find this distribution; however, this method is long and complicated and we prefer to rely on our experience. Compare (22) with the result of (5.8.4), where we saw that if Y is a random variable with the Cauchy density function

$$f(\lambda) = \frac{1}{\pi(1+\lambda^2)}, \qquad -\infty < \lambda < \infty$$

then Y has characteristic function

$$\phi(t) = e^{-|t|}.$$

Thus

$$\rho(t) = \phi(t(\alpha + \beta))$$

and ρ is the characteristic function of $(\alpha + \beta)Y$ (see (5.7.6)). By (4.7.2) the density function of $\Lambda = (\alpha + \beta)Y$ is

$$f_\Lambda(\lambda) = \frac{1}{\alpha + \beta} f_Y\left(\frac{\lambda}{\alpha + \beta}\right)$$

$$= \frac{\alpha + \beta}{\pi[(\alpha + \beta)^2 + \lambda^2]}, \qquad -\infty < \lambda < \infty$$

and this is the spectral density function of X. The spectrum of X is the whole real line $\mathbb{R}$. ●

(23) **Example. Autoregressive scheme.** Let $\{Z_n\}$ be uncorrelated random variables with zero means and unit variances, and suppose that

$$X_n = \alpha X_{n-1} + Z_n, \qquad -\infty < n < \infty$$

where α is real and satisfies $|\alpha| < 1$. We saw in Problem (8.7.2) that X has autocorrelation function

$$\rho(n) = \alpha^{|n|}, \qquad -\infty < n < \infty.$$

Use (16) to find the spectral density function f_X of X:

$$f_X(\lambda) = \frac{1}{2\pi} \sum_{n=-\infty}^{\infty} e^{-in\lambda} \alpha^{|n|}$$

$$= \frac{1 - \alpha^2}{2\pi|1 - \alpha e^{i\lambda}|^2} = \frac{1 - \alpha^2}{2\pi(1 - 2\alpha \cos \lambda + \alpha^2)}, \qquad -\pi \leqslant \lambda \leqslant \pi.$$

More generally, suppose that the process Y satisfies

$$Y_n = \sum_{j=1}^{r} \alpha_j Y_{n-j} + Z_n, \qquad -\infty < n < \infty$$

where $\alpha_1, \dots, \alpha_r$ are constants. The same techniques can be applied, though with some difficulty, to find that Y is stationary if the complex roots $\theta_1, \dots, \theta_r$ of the polynomial

$$A(z) = z^r - \alpha_1 z^{r-1} - \cdots - \alpha_r = 0$$

satisfy $|\theta_j| < 1$. If this holds then the spectral density function f_Y of Y is given by

$$f_Y(\lambda) = \frac{1}{2\pi\sigma^2 |A(e^{-i\lambda})|^2}, \qquad -\pi \leqslant \lambda \leqslant \pi$$

where $\sigma^2 = \text{var}(Y_0)$. ●

Exercises

(24) Let $X_n = A\cos(n\lambda) + B\sin(n\lambda)$ where A and B are uncorrelated random variables with zero means and unit variances. Show that X is stationary with a spectrum containing exactly one point.

(25) Let U be uniformly distributed on $(-\pi, \pi)$, and let V be independent of U with distribution function F. Show that $X_n = e^{i(U-Vn)}$ defines a stationary (complex) sequence with spectral distribution function F.

(26) Find the autocorrelation function of the stationary process $\{X(t): -\infty < t < \infty\}$ whose spectral density function is

(i) $N(0, 1)$, (ii) $f(x) = \frac{1}{2}e^{-|x|}$, $-\infty < x < \infty$.

(27) Let $X_1, X_2, \ldots$ be a real stationary sequence with zero means and autocovariance function c. Show that

$$\operatorname{var}\left(\frac{1}{n}\sum_{j=1}^{n} X_j\right) = c(0)\int_{(-\pi,\pi]}\left(\frac{\sin(n\lambda/2)}{n\sin(\lambda/2)}\right)^2 dF(\lambda)$$

where F is the spectral distribution function. Deduce that $n^{-1}\sum_{j=1}^{n} X_j \xrightarrow{\text{m.s.}} 0$ if and only if $F(0) - F(0-) = 0$, and show that

$$c(0)\{F(0) - F(0-)\} = \lim_{n\to\infty}\frac{1}{n}\sum_{j=0}^{n-1} c(j).$$

9.4 Stochastic integration and the spectral representation

Let $X = \{X(t): -\infty < t < \infty\}$ be a stationary process which takes values in $\mathbb{C}$, as before. In the last section we saw that the autocorrelation function ρ enjoys the representation

(1) $$\rho(t) = \int_{-\infty}^{\infty} e^{it\lambda}\, dF(\lambda)$$

as the characteristic function of some distribution function F whenever ρ is continuous at $t = 0$. This spectral representation is very useful in many contexts, including for example statistical analyses of sequences of data, but it is not the full story. Equation (1) is an analytical result with limited probabilistic content; of more interest to us is the process X, and (1) leads us to ask whether X itself enjoys a similar representation. The answer to this is in the affirmative, but the statement of the result is complicated and draws deeply from abstract theory.

Without much loss of generality we can suppose that $X(t)$ has mean 0 and variance 1 for all t. With each such stationary process X we can associate another process S called the 'spectral process' of X, in much the same way as the spectral distribution function F is associated with ρ.

(2) **Spectral theorem.** *If X is a stationary process with zero mean, unit variance, continuous autocorrelation function, and spectral distribution function F, then*

there exists a complex-valued process $S = \{S(\lambda): -\infty < \lambda < \infty\}$ *such that*

(3)
$$X(t) = \int_{-\infty}^{\infty} e^{it\lambda}\, dS(\lambda).$$

Furthermore S has orthogonal increments in the sense that

$$\mathbf{E}([S(v) - S(u)][\bar{S}(t) - \bar{S}(s)]) = 0 \quad \text{if} \quad u \leqslant v \leqslant s \leqslant t,$$

and in addition $\mathbf{E}(|S(v) - S(u)|^2) = F(v) - F(u)$ *if* $u \leqslant v$.

The discrete-time stationary process $X = \{X_n: -\infty < n < \infty\}$ has a spectral representation also. The only significant difference is that the domain of the spectral process may be taken to be $(-\pi, \pi]$.

(4) **Spectral theorem.** *If X is a discrete-time stationary process with zero mean, unit variance, and spectral distribution function F, then there exists a complex-valued process* $S = \{S(\lambda): -\pi < \lambda \leqslant \pi\}$ *such that*

(5)
$$X_n = \int_{(-\pi, \pi]} e^{in\lambda}\, dS(\lambda).$$

Furthermore S has orthogonal increments, and

(6)
$$\mathbf{E}(|S(v) - S(u)|^2) = F(v) - F(u) \quad \text{for} \quad u \leqslant v.$$

A proof of (4) is presented later in this section. The proof of (2) is very similar, Fourier sums being replaced by Fourier integrals; this proof is therefore omitted. The process S in (3) and (5) is called the *spectral process* of X.

Before proving the above spectral representation, we embark upon an exploration of the 'stochastic integral', of which (3) and (5) are examples. The theory of stochastic integration is of major importance in modern probability theory, particularly in the study of diffusion processes.

As amply exemplified by the material in this book, probabilists are very often concerned with partial sums $\sum_{i=1}^{n} X_i$ and weighted sums $\sum_{i=1}^{n} a_i X_i$ of sequences of random variables. If X is a continuous-time process rather than a discrete-time sequence, the corresponding objects are integrals of the form $\int_{\alpha}^{\beta} a(u)\, dX(u)$; how should such an integral be defined? It is not an easy matter to discuss the 'stochastic integral' before an audience some of whom have seen only little beyond the Riemann integral. There follows such an attempt.

Let $S = \{S(t): t \in \mathbb{R}\}$ be a complex-valued continuous-time random process on the probability space $(\Omega, \mathscr{F}, \mathbf{P})$, and suppose that S has the following properties:

(7)
$$\mathbf{E}(|S(t)|^2) < \infty \quad \text{for all } t,$$

(8)
$$\mathbf{E}(|S(t+h) - S(t)|^2) \to 0 \quad \text{as} \quad h \downarrow 0, \quad \text{for all } t,$$

(9) S has *orthogonal increments* in that

$$\mathbf{E}([S(v) - S(u)][\bar{S}(t) - \bar{S}(s)]) = 0 \quad \text{whenever} \quad u \leqslant v \leqslant s \leqslant t.$$

Condition (7) is helpful, since we shall work with random variables with finite second moments, and with mean-square convergence. Condition (8) is a continuity assumption which will be useful for technical reasons. Condition (9) will be of central importance in demonstrating the existence of limits necessary for the definition of the stochastic integral.

Let $G(t)$ be defined by

$$G(t) = \begin{cases} \mathbf{E}(|S(t) - S(0)|^2) & \text{if} \quad t \geqslant 0 \\ -\mathbf{E}(|S(t) - S(0)|^2) & \text{if} \quad t < 0. \end{cases} \tag{10}$$

It is an elementary calculation that

$$\mathbf{E}(|S(t) - S(s)|^2) = G(t) - G(s), \quad \text{for} \quad s \leqslant t. \tag{11}$$

To see that this holds when $0 \leqslant s \leqslant t$, for example, we argue as follows.

$$\begin{aligned} G(t) &= \mathbf{E}(|[S(t) - S(s)] + [S(s) - S(0)]|^2) \\ &= \mathbf{E}(|S(t) - S(s)|^2) + \mathbf{E}(|S(s) - S(0)|^2) \\ &\quad + \mathbf{E}([S(t) - S(s)][\bar{S}(s) - \bar{S}(0)] + [\bar{S}(t) - \bar{S}(s)][S(s) - S(0)]) \\ &= \mathbf{E}(|S(t) - S(s)|^2) + G(s) \end{aligned}$$

by the assumption of orthogonal increments. It follows from (11) that G is monotonic non-decreasing, and is right-continuous in that

$$G(t + h) \to G(t) \quad \text{as} \quad h \downarrow 0. \tag{12}$$

The function G is central to the analysis which follows.

Let $a_1 < a_2 < \cdots < a_n$, and let $c_1, c_2, \ldots, c_{n-1}$ be complex numbers. Define the step function ϕ on $\mathbb{R}$ by

$$\phi(t) = \begin{cases} 0 & \text{if} \quad t < a_1 \text{ or } t \geqslant a_n \\ c_j & \text{if} \quad a_j \leqslant t < a_{j+1}, \end{cases}$$

and define the integral $I(\phi)$ of ϕ with respect to S by

$$I(\phi) = \int_{-\infty}^{\infty} \phi(t)\, \mathrm{d}S(t) = \sum_{j=1}^{n-1} c_j[S(a_{j+1}) - S(a_j)]; \tag{13}$$

this is a finite sum, and therefore there is no problem concerning its existence.

Suppose that ϕ_1 and ϕ_2 are step functions of the type given above. We may assume, by a suitable 'refinement' argument, that ϕ_1 and ϕ_2 are of the form

$$\phi_1(t) = \phi_2(t) = 0 \quad \text{if} \quad t < a_1 \text{ or } t \geqslant a_n,$$

$$\phi_2(t) = c_j, \quad \phi_2(t) = d_j \quad \text{if} \quad a_j \leqslant t < a_{j+1},$$

for some $a_1 < a_2 < \cdots < a_n$. Then, using the assumption of orthogonal increments,

$$\mathbf{E}(I(\phi_1)\overline{I(\phi_2)}) = \sum_{j,k} c_j \overline{d_k} \mathbf{E}([S(a_{j+1}) - S(a_j)][\bar{S}(a_{k+1}) - \bar{S}(a_k)])$$
$$= \sum_j c_j \overline{d_j} \mathbf{E}(|S(a_{j+1}) - S(a_j)|^2)$$
$$= \sum_j c_j \overline{d_j} [G(a_{j+1}) - G(a_j)] \quad \text{by (11)},$$

which may be written as

(14) $$\mathbf{E}(I(\phi_1)\overline{I(\phi_2)}) = \int_{-\infty}^{\infty} \phi_1(t)\overline{\phi_2(t)}\, \mathrm{d}G(t).$$

It is now immediate by expansion of the squares that

(15) $$\mathbf{E}(|I(\phi_1) - I(\phi_2)|^2) = \int_{-\infty}^{\infty} |\phi_1(t) - \phi_2(t)|^2\, \mathrm{d}G(t),$$

which is to say that 'integration is distance preserving' in the sense that

(16) $$\|I(\phi_1) - I(\phi_2)\|_2 = \|\phi_1 - \phi_2\|,$$

where the first norm is given by

(17) $$\|U - V\|_2 = [\mathbf{E}(|U - V|^2)]^{\frac{1}{2}} \quad \text{for random variables } U, V,$$

and the second by

(18) $$\|f - g\| = \left(\int_{-\infty}^{\infty} |f(t) - g(t)|^2\, \mathrm{d}G(t)\right)^{\frac{1}{2}} \quad \text{for suitable } f, g\colon \mathbb{R} \to \mathbb{C}.$$

We are ready to take limits. Let $\psi\colon \mathbb{R} \to \mathbb{C}$ and let $\{\phi_n\}$ be a sequence of step functions such that $\|\phi_n - \psi\| \to 0$ as $n \to \infty$. Then

$$\|\phi_n - \phi_m\| \leqslant \|\phi_n - \psi\| + \|\phi_m - \psi\| \to 0 \quad \text{as} \quad m, n \to \infty,$$

whence it follows from (16) that the sequence $\{I(\phi_n)\}$ is mean-square Cauchy-convergent, and hence convergent in mean square (see (7.11.11)). That is, there exists a random variable $I(\psi)$ such that $I(\phi_n) \xrightarrow{\text{m.s.}} I(\psi)$; we call $I(\psi)$ the integral of ψ with respect to S, writing

(19) $$I(\psi) = \int_{-\infty}^{\infty} \psi(t)\, \mathrm{d}S(t).$$

Note that the integral is not defined uniquely, but only as any mean-square limit of $I(\phi_n)$; any two such limits I_1 and I_2 are such that $\mathbf{P}(I_1 = I_2) = 1$.

For which functions ψ do there exist approximating sequences $\{\phi_n\}$ of step functions? The answer is those (measurable) functions for which

(20) $$\int_{-\infty}^{\infty} |\psi(t)|^2\, \mathrm{d}G(t) < \infty.$$

To recap, for any given function $\psi: \mathbb{R} \to \mathbb{C}$ satisfying (20), there exists a random variable

$$I(\psi) = \int_{-\infty}^{\infty} \psi(t)\, \mathrm{d}S(t) \tag{21}$$

defined as above. Such integrals have many of the usual properties of integrals, for example

(a) the integral of the zero function is zero,
(b) $I(\alpha\psi_1 + \beta\psi_2) = \alpha I(\psi_1) + \beta I(\psi_2)$ for $\alpha, \beta \in \mathbb{C}$,

and so on. Such statements should be qualified by the phrase 'almost surely', since integrals are not defined uniquely; we shall omit this qualification here.

Integrals may be defined on *bounded* intervals just as on the whole of the real line. For example, if $\psi: \mathbb{R} \to \mathbb{C}$ and (a, b) is a bounded interval, we define

$$\int_{(a,b)} \psi(t)\, \mathrm{d}S(t) = \int_{-\infty}^{\infty} \psi_{ab}(t)\, \mathrm{d}S(t)$$

where $\psi_{ab}(t) = \psi(t) I_{(a,b)}(t)$.

The above exposition is directed at integrals $\int \psi(t)\, \mathrm{d}S(t)$ where ψ is a given function from $\mathbb{R}$ to $\mathbb{C}$. It is possible to extend this definition to the situation where ψ is itself a random process. Such an integral may be constructed very much as above, but at the expense of adding certain extra assumptions concerning the pair (ψ, S).

Proof of Theorem (4). Let H_X be the set of all linear combinations of the X_j, so that H_X is the set of all random variables of the form $\sum_{j=1}^{n} a_j X_{m(j)}$ for $a_1, a_2, \dots, a_n \in \mathbb{C}$ and integers $n, m(1), m(2), \dots, m(n)$. The space H_X is a vector space over $\mathbb{C}$ with a natural inner product given by

$$\langle U, V \rangle_2 = \mathbf{E}(U\overline{V}). \tag{22}$$

The *closure* $\overline{H}_X$ of H_X is defined to be the space H_X together with all limits of mean-square Cauchy-convergent sequences in H_X.

Similarly, we let H_F be the set of all linear combinations of the functions $e_n: \mathbb{R} \to \mathbb{C}$ defined by $e_n(x) = \mathrm{e}^{inx}$ for $-\infty < x < \infty$. We impose an inner product on H_F by

$$\langle u, v \rangle = \int_{(-\pi,\pi]} u(\lambda)\overline{v(\lambda)}\, \mathrm{d}F(\lambda) \quad \text{for} \quad u, v \in H_F, \tag{23}$$

and we write $\overline{H}_F$ for the closure of H_F, being the space H_F together with all Cauchy-convergent sequences in H_F (a sequence $\{u_n\}$ is Cauchy-convergent if $\langle u_n - u_m, u_n - u_m \rangle \to 0$ as $m, n \to \infty$).

The two spaces $\overline{H}_X$ and $\overline{H}_F$ are Hilbert spaces, and we place them in one–one correspondence in the following way. Define the linear mapping

μ: $H_F \to H_X$ by $\mu(e_j) = X_j$, so that

$$\mu\left(\sum_{j=1}^{n} a_j e_j\right) = \sum_{j=1}^{n} a_j X_j;$$

it is seen easily that μ is one–one, in a formal sense. Furthermore

$$\langle \mu(e_n), \mu(e_m) \rangle_2 = \langle X_n, X_m \rangle_2 = \int_{(-\pi,\pi]} e^{i(n-m)\lambda}\, dF(\lambda) = \langle e_n, e_m \rangle$$

by (9.3.12) and (23); therefore, by linearity, $\langle \mu(u), \mu(v) \rangle_2 = \langle u, v \rangle$ for $u, v \in H_F$, so that μ is 'distance preserving' on H_F. The domain of μ may be extended to $\bar{H}_F$ in the natural way: if $u \in \bar{H}_F$, $u = \lim_{n\to\infty} u_n$ where $u_n \in H_F$, we define $\mu(u) = \lim_{n\to\infty} \mu(u_n)$ where the latter limit is taken in the usual sense for $\bar{H}_X$. The new mapping μ from $\bar{H}_F$ to $\bar{H}_X$ is not quite one–one, since mean-square limits are not defined uniquely, but this difficulty is easily avoided (μ is one–one when viewed as a mapping from equivalence classes of functions to equivalence classes of random variables). Furthermore it may easily be checked that μ is distance preserving on $\bar{H}_F$, and linear in that

$$\mu\left(\sum_{j=1}^{n} a_j u_j\right) = \sum_{j=1}^{n} a_j \mu(u_j)$$

for $a_1, a_2, \ldots, a_n \in \mathbb{C}$, $u_1, u_2, \ldots, u_n \in \bar{H}_F$.

The mapping μ is sometimes called an *isometric isomorphism*. We now define the process $S = \{S(\lambda): -\pi < \lambda \leqslant \pi\}$ by

(24) $$S(\lambda) = \mu(I_\lambda) \quad \text{for} \quad -\pi < \lambda \leqslant \pi,$$

where I_λ: $\mathbb{R} \to \{0, 1\}$ is the indicator function of the interval $(-\pi, \lambda]$. It is a standard result of Fourier analysis that $I_\lambda \in \bar{H}_F$, so that $\mu(I_\lambda)$ is well defined. We introduce one more piece of notation, defining $J_{\alpha\beta}$ to be the indicator function of the interval $(\alpha, \beta]$; thus $J_{\alpha\beta} = I_\beta - I_\alpha$.

We need to show that X and S are related (almost surely) by (5). To this end, we check first that S satisfies conditions (7)–(9). Certainly $\mathbf{E}(|S(\lambda)|^2) < \infty$ since $S(\lambda) \in \bar{H}_X$. Secondly,

$$\begin{aligned}\mathbf{E}(|S(\lambda+h) - S(\lambda)|^2) &= \langle S(\lambda+h) - S(\lambda), S(\lambda+h) - S(\lambda) \rangle_2 \\ &= \langle J_{\lambda,\lambda+h}, J_{\lambda,\lambda+h} \rangle\end{aligned}$$

by the linearity and the isometry of μ. Now $\langle J_{\lambda,\lambda+h}, J_{\lambda,\lambda+h} \rangle \to 0$ as $h \downarrow 0$, and (8) has been verified. Thirdly, if $u \leqslant v \leqslant s \leqslant t$, then

$$\langle S(v) - S(u), S(t) - S(s) \rangle_2 = \langle J_{uv}, J_{st} \rangle = 0$$

since $J_{uv}(x) J_{st}(x) = 0$ for all x. Thus S has orthogonal increments. Furthermore, by (23),

$$\begin{aligned}\mathbf{E}(|S(v) - S(u)|^2) = \langle J_{uv}, J_{uv} \rangle &= \int_{(u,v]} dF(\lambda) \\ &= F(v) - F(u)\end{aligned}$$

since F is right-continuous; this confirms (6), and it remains to check that (5) holds.

The process S satisfies conditions (7)–(9), and it follows that the stochastic integral

$$I(\psi) = \int_{(-\pi,\pi]} \psi(\lambda)\,\mathrm{d}S(\lambda)$$

is defined for a broad class of functions $\psi: (-\pi, \pi] \to \mathbb{C}$. We claim that

(25) $$I(\psi) = \mu(\psi) \quad \text{(almost surely)} \quad \text{for} \quad \psi \in \bar{H}_F.$$

The result of the theorem will follow immediately by the choice $\psi = e_n$, for which (25) implies that (almost surely) $I(e_n) = \mu(e_n) = X_n$, which is to say that

$$\int_{(-\pi,\pi]} \mathrm{e}^{in\lambda}\,\mathrm{d}S(\lambda) = X_n$$

as required.

It remains to prove (25), which we do by systematic approximation. Suppose first that ψ is a step function,

(26) $$\psi(x) = \begin{cases} 0 & \text{if} \quad x < a_1 \text{ or } x \geqslant a_n \\ c_j & \text{if} \quad a_j \leqslant x < a_{j+1} \end{cases}$$

where $-\pi < a_1 < a_2 < \cdots < a_n \leqslant \pi$ and $c_1, c_2, \ldots, c_n \in \mathbb{C}$. Then

$$\begin{aligned} I(\psi) &= \sum_{j=1}^{n} c_j[S(a_{j+1}) - S(a_j)] = \sum_{j=1}^{n} c_j \mu(J_{a_j, a_{j+1}}) \quad \text{by (24)} \\ &= \mu\left(\sum_{j=1}^{n} c_j J_{a_j, a_{j+1}}\right) = \mu(\psi) \quad \text{by (26).} \end{aligned}$$

Hence $I(\psi) = \mu(\psi)$ for all step functions ψ. More generally, if $\psi \in \bar{H}_F$ and $\{\psi_n\}$ is a sequence of step functions converging to ψ, then $\mu(\psi_n) \to \mu(\psi)$. By the definition of the stochastic integral, it is the case that $I(\psi_n) \to I(\psi)$, and it follows that $I(\psi) = \mu(\psi)$, which proves (25). ■

Exercises

(27) Let S be the spectral process of a stationary process X with zero mean and unit variance. Show that the increments of S have zero means.

(28) **Moving average representation.** Let X be a discrete-time stationary process having zero means, continuous strictly positive spectral density function f, and with spectral process S. Let

$$Y_n = \int_{(-\pi,\pi]} \frac{\mathrm{e}^{in\lambda}}{\sqrt{\{2\pi f(\lambda)\}}}\,\mathrm{d}S(\lambda).$$

Show that $\ldots, Y_{-1}, Y_0, Y_1, \ldots$ is a sequence of uncorrelated random variables with zero means and unit variances.

Show that X_n may be represented as a moving average of the Y_j,

$$X_n = \sum_{j=-\infty}^{\infty} a_j Y_{n-j}$$

where the a_j are constants satisfying

$$\sqrt{2\pi f(\lambda)} = \sum_{j=-\infty}^{\infty} a_j e^{-ij\lambda} \quad \text{for} \quad \lambda \in (-\pi, \pi].$$

(29) **Gaussian process.** Let X be a discrete-time stationary sequence with zero mean and unit variance, and whose fdds are of the multivariate–normal type. Show that the spectral process of X has independent increments having normal distributions.

9.5 The ergodic theorem

The law of large numbers asserts that

(1) $$\frac{1}{n}\sum_{j=1}^{n} X_j \to \mu$$

whenever $\{X_j\}$ is an independent identically distributed sequence with mean μ; the convergence takes place almost surely. This section is devoted to a complete generalization of the law of large numbers, the assumption that the X_j be independent being replaced by the assumption that they form a stationary process. This generalization is called the 'ergodic theorem' and it has more than one form depending on the type of stationarity—weak or strong—and the required mode of convergence; recall the various corresponding forms of the law of large numbers.

It is usual to state the ergodic theorem for discrete-time processes, and we conform to this habit here. Similar results hold for continuous-time processes, sums like $\sum_1^n X_j$ being replaced by integrals like $\int_0^n X(t)\,dt$.

Here is the usual form of the ergodic theorem.

(2) **Theorem. Ergodic theorem for strongly stationary processes.** *Let* $X = \{X_n: n \geq 1\}$ *be a strongly stationary process such that* $\mathbf{E}|X_1| < \infty$. *Then there exists a random variable* Y *with the same mean as the* X_n *such that*

$$\frac{1}{n}\sum_{j=1}^{n} X_j \to Y \quad \textit{a.s. and in mean.}$$

The proof of this is difficult, as befits a complete generalization of the strong law of large numbers (see Problem (9.7.10)). The following result is considerably more elementary.

(3) **Theorem. Ergodic theorem for weakly stationary processes.** *If* $X = \{X_n: n \geq 1\}$ *is a (weakly) stationary process then there exists a random variable* Y *such that* $\mathbf{E}Y = \mathbf{E}X_1$ *and*

$$\frac{1}{n}\sum_{j=1}^{n} X_j \xrightarrow{\text{m.s.}} Y.$$

We prove the latter theorem first. The normal proof of the 'strong ergodic theorem' (2) is considerably more difficult, and makes use of harder ideas than those required for the 'weak ergodic theorem' (3). The second part of this section is devoted to a discussion of the strong ergodic theorem, together with a relatively straightforward proof.

Theorems (2) and (3) generalize the laws of large numbers. There are similar generalizations of the central limit theorem and the law of the iterated logarithm, although such results hold only for stationary processes which satisfy certain extra conditions. We give no details of this here, save for pointing out that these extra conditions take the form 'X_m and X_n are "nearly independent" when $|m - n|$ is large'.

We give two proofs of (3). Proof A is conceptually easy but has some technical difficulties; we show that $n^{-1}\sum_1^n X_j$ is a mean-square Cauchy-convergent sequence (see Problem (7.11.11)). Proof B uses the spectral representation of X; we sketch this here and show that it yields an explicit form for the limit Y as the contribution made towards X by 'oscillations of zero frequency'.

Proof A. Recall from (7.11.11) that a sequence $\{Y_n\}$ converges in mean square to some limit if and only if $\{Y_n\}$ is *mean-square Cauchy-convergent*, which is to say that

(4) $$\mathbf{E}(|Y_n - Y_m|^2) \to 0 \quad \text{as} \quad m, n \to \infty.$$

A similar result holds for complex-valued sequences. We shall show that the sequence $\{n^{-1}\sum_1^n X_j\}$ satisfies (4) whenever X is stationary. This is easy in concept, since it involves expressions involving the autocovariance function of X alone; the proof of the mean-square version of the law of large numbers was easy for the same reason. Unfortunately, the verification of (4) is not a trivial calculation.

For any complex-valued random variable Z, define

$$\|Z\| = (\mathbf{E}(|Z|^2))^{\frac{1}{2}};$$

the function $\|\cdot\|$ is a norm (see Section 7.2). We wish to show that

(5) $$\|\langle X\rangle_n - \langle X\rangle_m\| \to 0 \quad \text{as} \quad n, m \to \infty$$

where

$$\langle X\rangle_n = \frac{1}{n}\sum_{j=1}^{n} X_j;$$

physicists often use the notation $\langle \cdot \rangle$ to denote expectation. Set

$$\mu_N = \inf_{\lambda} \|\lambda_1 X_1 + \lambda_2 X_2 + \cdots + \lambda_N X_N\|$$

where the infimum is calculated over all vectors $\boldsymbol{\lambda} = (\lambda_1, \ldots, \lambda_N)$ with $\lambda_i \geqslant 0$ and $\sum_1^N \lambda_i = 1$. Clearly $\mu_N \geqslant \mu_{N+1}$ and so

$$\mu = \lim_{N \to \infty} \mu_N = \inf_N \mu_N$$

exists. If $m < n$ then

$$\|\langle X \rangle_n + \langle X \rangle_m\| = 2\left\|\sum_1^n \lambda_j X_j\right\|$$

where

$$\lambda_j = \begin{cases} \dfrac{1}{2}\left(\dfrac{1}{m} + \dfrac{1}{n}\right) & \text{if} \quad 1 \leqslant j \leqslant m \\[2ex] \dfrac{1}{2n} & \text{if} \quad m < j \leqslant n, \end{cases}$$

and so

$$\|\langle X \rangle_n + \langle X \rangle_m\| \geqslant 2\mu.$$

It is not difficult to deduce (see Exercise (7.1.9) for the first line here) that

$$\begin{aligned} \|\langle X \rangle_n - \langle X \rangle_m\|^2 &= 2\|\langle X \rangle_n\|^2 + 2\|\langle X \rangle_m\|^2 - \|\langle X \rangle_n + \langle X \rangle_m\|^2 \\ &\leqslant 2\|\langle X \rangle_n\|^2 + 2\|\langle X \rangle_m\|^2 - 4\mu^2 \\ &= 2\big|\,\|\langle X \rangle_n\|^2 - \mu^2\big| + 2\big|\,\|\langle X \rangle_m\|^2 - \mu^2\big| \end{aligned}$$

and (5) follows as soon as we can show that

(6) $$\|\langle X \rangle_n\| \to \mu \quad \text{as} \quad n \to \infty.$$

The remaining part of the proof is devoted to demonstrating (6).

Choose any $\varepsilon > 0$ and pick N and $\boldsymbol{\lambda}$ such that

$$\|\lambda_1 X_1 + \cdots + \lambda_N X_N\| \leqslant \mu + \varepsilon$$

where $\lambda_i \geqslant 0$ and $\sum_1^N \lambda_i = 1$. Define the moving average

$$Y_k = \lambda_1 X_k + \lambda_2 X_{k+1} + \cdots + \lambda_N X_{k+N-1};$$

it is not difficult to see that $Y = \{Y_k\}$ is a stationary process (see Problem (8.7.1)). We shall show that

(7) $$\|\langle Y \rangle_n - \langle X \rangle_n\| \to 0 \quad \text{as} \quad n \to \infty$$

where

$$\langle Y \rangle_n = \frac{1}{n} \sum_{j=1}^n Y_j.$$

Note first that, by the triangle inequality (7.1.5),

(8) $$\|\langle Y\rangle_n\| \leqslant \|Y_1\| \leqslant \mu + \varepsilon \quad \text{for all } n$$

since $\|Y_n\| = \|Y_1\|$ for all n. Now

$$\langle Y\rangle_n = \lambda_1\langle X\rangle_{1,n} + \lambda_2\langle X\rangle_{2,n} + \cdots + \lambda_N\langle X\rangle_{N,n}$$

where

$$\langle X\rangle_{k,n} = \frac{1}{n}\sum_{j=k}^{k+n-1} X_j;$$

now use the facts that $\langle X\rangle_{1,n} = \langle X\rangle_n$,

$$1 - \lambda_1 = \lambda_2 + \cdots + \lambda_N,$$

and the triangle inequality to deduce that

$$\|\langle Y\rangle_n - \langle X\rangle_n\| \leqslant \sum_{j=2}^{N} \lambda_j \|\langle X\rangle_{j,n} - \langle X\rangle_{1,n}\|.$$

But, by the triangle inequality again,

$$\begin{aligned}
\|\langle X\rangle_{j,n} - \langle X\rangle_{1,n}\| &= \frac{1}{n}\|(X_j + \cdots + X_{j+n-1}) - (X_1 + \cdots + X_n)\| \\
&= \frac{1}{n}\|(X_{n+1} + \cdots + X_{j+n-1}) - (X_1 + \cdots + X_{j-1})\| \\
&\leqslant \frac{2j}{n}\|X_1\|
\end{aligned}$$

since $\|X_n\| = \|X_1\|$ for all n, and so

$$\|\langle Y\rangle_n - \langle X\rangle_n\| \leqslant \sum_{j=2}^{N} \lambda_j \frac{2j}{n}\|X_1\| \leqslant \frac{2N}{n}\|X_1\|;$$

let $n \to \infty$ to deduce that (7) holds. Use (8) to obtain

$$\begin{aligned}
\mu \leqslant \|\langle X\rangle_n\| &\leqslant \|\langle X\rangle_n - \langle Y\rangle_n\| + \|\langle Y\rangle_n\| \\
&\leqslant \|\langle X\rangle_n - \langle Y\rangle_n\| + \mu + \varepsilon \\
&\to \mu + \varepsilon \quad \text{as} \quad n \to \infty.
\end{aligned}$$

But ε was arbitrary; let $\varepsilon \downarrow 0$ to see that (6) holds.

Since $\langle X\rangle_n \xrightarrow{\text{m.s.}} Y$, we have that $\langle X\rangle_n \xrightarrow{1} Y$, which implies that $\mathbf{E}\langle X\rangle_n \to \mathbf{E}Y$. However, $\mathbf{E}\langle X\rangle_n = \mathbf{E}X_1$, whence $\mathbf{E}Y = \mathbf{E}X_1$. ■

Sketch proof B. Suppose that $\mathbf{E}(X_n) = 0$ for all n. X has a spectral representation

$$X_n = \int_{(-\pi,\pi]} e^{in\lambda}\, dS(\lambda).$$

Now,

(9)
$$\langle X\rangle_n = \frac{1}{n}\sum_{j=1}^{n} X_j = \int_{(-\pi,\pi]} \frac{1}{n}\sum_{j=1}^{n} e^{ij\lambda}\, dS(\lambda)$$
$$= \int_{(-\pi,\pi]} g_n(\lambda)\, dS(\lambda)$$

where

(10)
$$g_n(\lambda) = \begin{cases} 1 & \text{if } \lambda = 0 \\ \dfrac{e^{i\lambda}}{n}\dfrac{1-e^{in\lambda}}{1-e^{i\lambda}} & \text{if } \lambda \neq 0. \end{cases}$$

Now,

$$|g_n(\lambda)| \leqslant 1 \quad \text{for all } n \text{ and } \lambda,$$

and, as $n \to \infty$,

(11)
$$g_n(\lambda) \to g(\lambda) = \begin{cases} 1 & \text{if } \lambda = 0 \\ 0 & \text{if } \lambda \neq 0. \end{cases}$$

It can be shown that

$$\int_{(-\pi,\pi]} g_n(\lambda)\, dS(\lambda) \xrightarrow{\text{m.s.}} \int_{(-\pi,\pi]} g(\lambda)\, dS(\lambda) \quad \text{as} \quad n \to \infty,$$

implying that

$$\langle X\rangle_n \xrightarrow{\text{m.s.}} \int_{(-\pi,\pi]} g(\lambda)\, dS(\lambda) = S(0) - S(0-),$$

by the right-continuity of S, where

$$S(0-) = \lim_{y \uparrow 0} S(y).$$

This shows that $\langle X\rangle_n$ converges in mean square to the random magnitude of the discontinuity of $S(\lambda)$ at $\lambda = 0$ (this quantity may be zero); in other words, $\langle X\rangle_n$ converges to the 'zero frequency' or 'infinite wavelength' contribution of the spectrum of X. This conclusion is natural and memorable, since the average of any oscillation with non-zero frequency is zero. ∎

The second proof of Theorem (3) is particularly useful in that it provides an explicit representation for the limit in terms of the spectral process of X. It is easy to calculate the first two moments of this limit.

(12) **Lemma.** *If X is a stationary process with zero means and autocovariance function c then the limit variable*

$$Y = \lim_{n\to\infty} \frac{1}{n}\sum_{j=1}^{n} X_j$$

satisfies

$$\mathbf{E}(Y) = 0, \qquad \mathbf{E}(|Y|^2) = \lim_{n\to\infty} \frac{1}{n}\sum_{j=1}^{n} c(j).$$

A similar result holds for processes with non-zero means.

Proof. $\langle X \rangle_n \xrightarrow{\text{m.s.}} Y$, and so $\langle X \rangle_n \xrightarrow{1} Y$ by (7.2.3). The result of Exercise (7.2.22) implies that as $n \to \infty$,

$$\mathbf{E}(\langle X \rangle_n) \to \mathbf{E}(Y);$$

but $\mathbf{E}(\langle X \rangle_n) = \mathbf{E}(X_1) = 0$ for all n.

To prove the second part, either use (7.2.22) again and expand $\mathbf{E}(\langle X \rangle_n^2)$ in terms of c (see Exercise (37)), or use the method of Proof B of (3). We use the latter method. The autocovariance function c satisfies

$$\begin{aligned}\frac{1}{n}\sum_{j=1}^{n} c(j) &= c(0)\int_{(-\pi,\pi]} g_n(\lambda)\,\mathrm{d}F(\lambda)\\ &\to c(0)\int_{(-\pi,\pi]} g(\lambda)\,\mathrm{d}F(\lambda) \quad \text{as} \quad n\to\infty\\ &= c(0)[F(0) - F(0-)]\end{aligned}$$

where g_n and g are given by (10) and (11), F is the spectral distribution function, and

$$F(0-) = \lim_{y \uparrow 0} F(y)$$

as usual. We can now use (9.4.6) and the continuity properties of S to show that

$$c(0)[F(0) - F(0-)] = \mathbf{E}(|S(0) - S(0-)|^2) = \mathbf{E}(|Y|^2). \qquad \blacksquare$$

We turn now to the strong ergodic theorem (2), which we shall first rephrase slightly. Here is some terminology and general discussion.

A vector $\boldsymbol{X} = (X_1, X_2, \ldots)$ of real-valued random variables takes values in the set of real vectors of the form $\boldsymbol{x} = (x_1, x_2, \ldots)$. We write $\mathbb{R}^T$ for the set of all such real sequences, where T denotes the set $\{1, 2, \ldots\}$ of positive integers. The natural σ-field for $\mathbb{R}^T$ is the product $\mathcal{B}^T$ of the appropriate number of copies of the Borel σ-field $\mathcal{B}$ of subsets of $\mathbb{R}$. Let $\mathbf{Q}$ be a probability measure on the pair $(\mathbb{R}^T, \mathcal{B}^T)$. The triple $(\mathbb{R}^T, \mathcal{B}^T, \mathbf{Q})$ is our basic probability space, and we make the following crucial definitions.

There is a natural 'shift operator' τ mapping $\mathbb{R}^T$ onto itself, defined by $\tau(\boldsymbol{x}) = \boldsymbol{x}'$ where $\boldsymbol{x}' = (x_2, x_3, \ldots)$; that is, the vector $\boldsymbol{x} = (x_1, x_2, \ldots)$ is mapped to the vector $(x_2, x_3, \ldots)$. The measure $\mathbf{Q}$ is called **stationary** if and only if $\mathbf{Q}(A) = \mathbf{Q}(\tau^{-1}A)$ for all $A \in \mathcal{B}^T$ (remember that $\tau^{-1}A = \{\boldsymbol{x} \in \mathbb{R}^T : \tau(\boldsymbol{x}) \in A\}$). If $\mathbf{Q}$ is stationary, we call the shift τ 'measure preserving'. Stationary measures correspond to strongly stationary sequences of random variables, as the following example indicates.

(13) **Example.** Let $X = (X_1, X_2, \ldots)$ be a strongly stationary sequence on the probability space $(\Omega, \mathscr{F}, \mathbf{P})$. Define the probability measure $\mathbf{Q}$ on $(\mathbb{R}^T, \mathscr{B}^T)$ by $\mathbf{Q}(A) = \mathbf{P}(X \in A)$ for $A \in \mathscr{B}^T$. Now X and $\tau(X)$ have the same fdds, and therefore $\mathbf{Q}(\tau^{-1}A) = \mathbf{P}(\tau(X) \in A) = \mathbf{P}(X \in A) = \mathbf{Q}(A)$ for all (measurable) subsets A of $\mathbb{R}^T$.

We have seen that every strongly stationary sequence generates a stationary measure on $(\mathbb{R}^T, \mathscr{B}^T)$. The converse is true also. Let $\mathbf{Q}$ be a stationary measure on $(\mathbb{R}^T, \mathscr{B}^T)$, and define the sequence $Y = (Y_1, Y_2, \ldots)$ of random variables by $Y_n(\mathbf{x}) = x_n$, the nth component of the real vector $\mathbf{x}$. We have from the stationarity of $\mathbf{Q}$ that, for $A \in \mathscr{B}^T$,

$$\mathbf{Q}(Y \in A) = \mathbf{Q}(A) = \mathbf{Q}(\tau^{-1}A) = \mathbf{Q}(\tau(Y) \in A)$$

so that Y and $\tau(Y)$ have the same fdds. Hence Y is a strongly stationary sequence. ●

There is a certain special class of events in $\mathscr{B}^T$ called *invariant* events.

(14) **Definition.** An event A in $\mathscr{B}^T$ is called **invariant** if $A = \tau^{-1}A$.

An event A is invariant if

(15) $$\mathbf{x} \in A \quad \text{if and only if} \quad \tau(\mathbf{x}) \in A,$$

for any $\mathbf{x} \in \mathbb{R}^T$. Now (15) is equivalent to the statement '$\mathbf{x} \in A$ if and only if $\tau^n(\mathbf{x}) \in A$ for all $n \geqslant 0$'; remembering that $\tau^n(\mathbf{x}) = (x_{n+1}, x_{n+2}, \ldots)$, we see therefore that the membership by $\mathbf{x}$ of an invariant event A does not depend on any finite collection of the components of $\mathbf{x}$. Here are some examples of invariant events:

$$A_1 = \left\{\mathbf{x}: \limsup_{n \to \infty} x_n \leqslant 3\right\},$$

$$A_2 = \{\mathbf{x}: \text{the sequence } n^{-1}x_n \text{ converges}\},$$

$$A_3 = \{\mathbf{x}: x_n = 0 \text{ for all large } n\}.$$

We denote by $\mathscr{I}$ the set of all invariant events. It is not difficult to see (Exercise (36)) that $\mathscr{I}$ is a σ-field, and therefore $\mathscr{I}$ is a sub-σ-field of $\mathscr{B}^T$, called the *invariant σ-field.*

Finally, we need the idea of conditional expectation. Let U be a random variable on $(\mathbb{R}^T, \mathscr{B}^T, \mathbf{Q})$ with finite mean $\mathbf{E}(U)$; here, $\mathbf{E}$ denotes expectation with respect to the measure $\mathbf{Q}$. We saw in Theorem (7.9.26) that there exists an $\mathscr{I}$-measurable random variable Z such that $\mathbf{E}|Z| < \infty$ and $\mathbf{E}((U - Z)I_G) = 0$ for all $G \in \mathscr{I}$; Z is usually denoted by $Z = \mathbf{E}(U \mid \mathscr{I})$ and is called the conditional expectation of U given $\mathscr{I}$.

We are now ready to restate the strong ergodic theorem (2) in the following way.

(16) **Ergodic theorem.** *Let* $\mathbf{Q}$ *be a stationary probability measure on* $(\mathbb{R}^T, \mathscr{B}^T)$, *and let* Y *be a real-valued random variable on the space* $(\mathbb{R}^T, \mathscr{B}^T, \mathbf{Q})$. *Let* $Y_1, Y_2, \ldots$ *be the sequence of random variables defined by*

(17)
$$Y_i(\boldsymbol{x}) = Y(\tau^{i-1}(\boldsymbol{x})) \quad \text{for} \quad \boldsymbol{x} \in \mathbb{R}^T.$$

If Y *has finite mean, then*

(18)
$$\frac{1}{n}\sum_{i=1}^{n} Y_i \to \mathbf{E}(Y \mid \mathscr{I}) \quad \text{a.s. and in mean.}$$

The sequence $\boldsymbol{Y} = (Y_1, Y_2, \ldots)$ is of course strongly stationary: since $\mathbf{Q}$ is stationary,

$$\begin{aligned}\mathbf{Q}((Y_2, Y_3, \ldots) \in A) &= \mathbf{Q}(\tau(\boldsymbol{Y}) \in A)\\ &= \mathbf{Q}(\boldsymbol{Y} \in \tau^{-1}A) = \mathbf{Q}(\boldsymbol{Y} \in A) \quad \text{for} \quad A \in \mathscr{B}^T.\end{aligned}$$

The above theorem asserts that the average of the first n values of $\boldsymbol{Y}$ converges as $n \to \infty$, the limit being the conditional mean of Y given $\mathscr{I}$; this is a conclusion very similar to that of the strong law of large numbers (7.5.1).

To understand the relationship between Theorems (2) and (16), consider the situation treated by (2). Let $X_1, X_2, \ldots$ be a strongly stationary sequence on $(\Omega, \mathscr{F}, \mathbf{P})$, and let $\mathbf{Q}$ be the stationary measure on $(\mathbb{R}^T, \mathscr{B}^T)$ defined by $\mathbf{Q}(A) = \mathbf{P}(X \in A)$ for $A \in \mathscr{B}^T$. We define $Y: \mathbb{R}^T \to \mathbb{R}$ by $Y(\boldsymbol{x}) = x_1$ for $\boldsymbol{x} = (x_1, x_2, \ldots) \in \mathbb{R}^T$, so that Y_i in (17) is given by $Y_i(\boldsymbol{x}) = x_i$. It is clear that the sequences $\{X_n : n \geq 1\}$ and $\{Y_n : n \geq 1\}$ have the same joint distributions, and it follows that the convergence of $n^{-1}\sum_1^n Y_i$ entails the convergence of $n^{-1}\sum_1^n X_i$.

(19) **Definition.** The stationary measure $\mathbf{Q}$ on $(\mathbb{R}^T, \mathscr{B}^T)$ is called **ergodic** if each invariant event has probability either 0 or 1, which is to say that $\mathbf{Q}(A) = 0$ or 1 for all $A \in \mathscr{I}$.

Ergodic stationary measures are of particular importance. The simplest example of such a measure is product measure.

(20) **Example. Independent sequences.** Let $\mathbf{S}$ be a probability measure on $(\mathbb{R}, \mathscr{B})$, and let $\mathbf{Q} = \mathbf{S}^T$, the appropriate product measure on $(\mathbb{R}^T, \mathscr{B}^T)$. Product measures arise in the context of independent random variables, as follows. Let $X_1, X_2, \ldots$ be a sequence of independent identically distributed random variables on a probability space $(\Omega, \mathscr{F}, \mathbf{P})$, and let $\mathbf{S}(A) = \mathbf{P}(X_1 \in A)$ for $A \in \mathscr{B}$. Then $\mathbf{S}$ is a probability measure on $(\mathbb{R}, \mathscr{B})$. The probability space $(\mathbb{R}^T, \mathscr{B}^T, \mathbf{S}^T)$ is the natural space for the vector $\boldsymbol{X} = (X_1, X_2, \ldots)$; that is, $\mathbf{S}^T(A) = \mathbf{P}(\boldsymbol{X} \in A)$ for $A \in \mathscr{B}^T$.

Suppose that A $(\in \mathscr{B}^T)$ is invariant. Then A belongs to the σ-field generated by $(X_n, X_{n+1}, \ldots)$ for all n, and hence A belongs to the tail σ-field of the X. By Kolmogorov's zero–one law (7.3.15), the latter σ-field is trivial, in that all events therein have probability either 0 or 1. Hence all invariant events have probability either 0 or 1, and therefore the measure $\mathbf{S}^T$ is ergodic. ●

The conclusion (18) of the ergodic theorem takes on a particularly simple form when the measure $\mathbf{Q}$ is ergodic as well as stationary. In this case, the random variable $\mathbf{E}(Y \mid \mathscr{I})$ is (a.s.) constant, as the following argument demonstrates. The conditional expectation $\mathbf{E}(Y \mid \mathscr{I})$ is $\mathscr{I}$-measurable, and therefore the event $A_y = \{\mathbf{E}(Y \mid \mathscr{I}) \leqslant y\}$ belongs to $\mathscr{I}$ for all y. However, $\mathscr{I}$ is trivial, in that it contains only events having probability 0 or 1. Hence $\mathbf{E}(Y \mid \mathscr{I})$ takes almost surely the value $\sup\{y\colon \mathbf{Q}(A_y) = 0\}$. Taking expectations, we find that this value is $\mathbf{E}(Y)$, so that the conclusion (18) becomes

(21) $$\frac{1}{n}\sum_{i=1}^{n} Y_i \to \mathbf{E}(Y) \quad \text{a.s. and in mean}$$

in the ergodic case.

Proof of ergodic theorem (16). We give full details of this for the case when $\mathbf{Q}$ is ergodic, and finish the proof with brief notes describing how to adapt the argument to the general case.

Assume then that $\mathbf{Q}$ is ergodic, so that $\mathbf{E}(Y \mid \mathscr{I}) = \mathbf{E}(Y)$. First we prove almost-sure convergence, which is to say that

(22) $$\frac{1}{n}\sum_{i=1}^{n} Y_i \to \mathbf{E}(Y) \quad \text{a.s.}$$

It suffices to prove that

(23) $$\text{if} \quad \mathbf{E}(Y) < 0 \quad \text{then} \quad \limsup_{n\to\infty}\left\{\frac{1}{n}\sum_{i=1}^{n} Y_i\right\} \leqslant 0 \quad \text{a.s.}$$

To see that (23) suffices, we argue as follows. Suppose that (23) holds, and that Z is a (measurable) function on $(\mathbb{R}^T, \mathscr{B}^T, \mathbf{Q})$ with finite mean, and let $\varepsilon > 0$. Then $Y' = Z - \mathbf{E}(Z) - \varepsilon$ and $Y'' = -Z + \mathbf{E}(Z) - \varepsilon$ have negative means. Applying (23) to Y' and Y'' we obtain

$$\limsup_{n\to\infty}\left\{\frac{1}{n}\sum_{i=1}^{n} Z_i\right\} \leqslant \mathbf{E}(Z) + \varepsilon \quad \text{a.s., and}$$

$$\liminf_{n\to\infty}\left\{\frac{1}{n}\sum_{i=1}^{n} Z_i\right\} \geqslant \mathbf{E}(Z) - \varepsilon \quad \text{a.s.}$$

where Z_i is the random variable given by $Z_i(\boldsymbol{x}) = Z(\tau^{i-1}(\boldsymbol{x}))$. These inequalities hold for all $\varepsilon > 0$, and therefore

$$\limsup_{n\to\infty}\left\{\frac{1}{n}\sum_{i=1}^{n} Z_i\right\} = \liminf_{n\to\infty}\left\{\frac{1}{n}\sum_{i=1}^{n} Z_i\right\} = \mathbf{E}(Z) \quad \text{a.s.}$$

as required for almost-sure convergence.

Turning to the proof of (23), suppose that $\mathbf{E}(Y) < 0$, and introduce the notation $S_n = \sum_{i=1}^{n} Y_i$. Now $S_n \leqslant M_n$ where $M_n = \max\{0, S_1, S_2, \dots, S_n\}$

satisfies $M_n \leqslant M_{n+1}$. Hence $S_n \leqslant M_\infty$ where $M_\infty = \lim_{n\to\infty} M_n$. Therefore

(24) $$\limsup_{n\to\infty}\left\{\frac{1}{n}S_n\right\} \leqslant \limsup_{n\to\infty}\left\{\frac{1}{n}M_\infty\right\},$$

and (23) will be proved once we know that $M_\infty < \infty$ a.s. It is easily seen that the event $\{M_\infty < \infty\}$ is an invariant event, and hence has probability either 0 or 1; it is here that we use the hypothesis that $\mathbf{Q}$ is ergodic. We must show that $\mathbf{Q}(M_\infty < \infty) = 1$, and to this end we assume the contrary, that $\mathbf{Q}(M_\infty = \infty) = 1$.

Now

(25) $$\begin{aligned} M_{n+1} &= \max\{0, S_1, S_2, \dots, S_{n+1}\} \\ &= \max\{0, S_1 + \max\{0, S_2 - S_1, \dots, S_{n+1} - S_1\}\} \\ &= \max\{0, S_1 + M_n'\} \end{aligned}$$

where $M_n' = \max\{0, S_1', S_2', \dots, S_n'\}$, and $S_j' = \sum_{i=1}^{j} Y_{i+1}$. It follows from (25) that $M_{n+1} = M_n' + \max\{-M_n', Y\}$, since $S_1 = Y$. Taking expectations and using the fact that $\mathbf{E}(M_n') = \mathbf{E}(M_n)$, we find that

(26) $$0 \leqslant \mathbf{E}(M_{n+1}) - \mathbf{E}(M_n) = \mathbf{E}(\max\{-M_n', Y\}).$$

If $M_n \uparrow \infty$ a.s. then $M_n' \uparrow \infty$ a.s., implying that $\max\{-M_n', Y\} \downarrow Y$ a.s. It follows by (26) (and dominated convergence) that $0 \leqslant \mathbf{E}(Y)$ in contradiction of the assumption that $\mathbf{E}(Y) < 0$. Our initial hypothesis was therefore false, which is to say that $\mathbf{Q}(M_\infty < \infty) = 1$, and (23) is proved.

Having proved almost-sure convergence, convergence in mean will follow by Theorem (7.10.3) once we have proved that the family $\{n^{-1}S_n : n \geqslant 1\}$ is uniformly integrable. The random variables $Y_1, Y_2, \dots$ are identically distributed with finite mean; hence (see Exercise (5.6.19)) for any $\varepsilon > 0$, there exists $\delta > 0$ such that, for all i,

(27) $$\mathbf{E}(|Y_i| I_A) < \varepsilon \quad \text{for all } A \text{ satisfying } \mathbf{Q}(A) < \delta.$$

Hence, for all n,

$$\mathbf{E}(|n^{-1}S_n| I_A) \leqslant \frac{1}{n}\sum_{i=1}^{n} \mathbf{E}(|Y_i| I_A) < \varepsilon$$

whenever $\mathbf{Q}(A) < \delta$. We deduce by an appeal to (7.10.6) that $\{n^{-1}S_n : n \geqslant 1\}$ is a uniformly integrable family as required.

This completes the proof in the ergodic case. The proof is only slightly more complicated in the general case, and here is a sketch of the additional steps required.

1. Use the definition of $\mathscr{I}$ to show that $\mathbf{E}(Y \mid \mathscr{I}) = \mathbf{E}(Y_i \mid \mathscr{I})$ for all i.
2. Replace (23) by the following statement: on the event $\{\mathbf{E}(Y \mid \mathscr{I}) < 0\}$, we have that

$$\limsup_{n\to\infty}\left\{\frac{1}{n}\sum_{i=1}^{n} Y_i\right\} \leqslant 0$$

except possibly for an event of probability 0. Check that this is sufficient for the required result by applying it to the random variables $Y' = Z - \mathbf{E}(Z \mid \mathscr{I}) - \varepsilon$, $Y'' = -Z + \mathbf{E}(Z \mid \mathscr{I}) - \varepsilon$, where $\varepsilon > 0$.

3. Moving to (26), prove that $\mathbf{E}(M'_n \mid \mathscr{I}) = \mathbf{E}(M_n \mid \mathscr{I})$, and deduce the inequality $\mathbf{E}(\max\{-M'_n, Y\} \mid \mathscr{I}) \geqslant 0$.

4. Continuing from (26), show that $\{M_n \to \infty\} = \{M'_n \to \infty\}$, and deduce that $\mathbf{E}(Y \mid \mathscr{I}) \geqslant 0$ on the event $\{M_n \to \infty\}$. This leads us to the same contradiction as in the ergodic case, and we conclude the proof as before. ■

Here are some applications of the ergodic theorem.

(28) **Example. Markov chains.** Let $X = \{X_n\}$ be an irreducible ergodic Markov chain with countable state space S, and let $\boldsymbol{\pi}$ be the unique stationary distribution of the chain. Suppose that $X(0)$ has distribution $\boldsymbol{\pi}$; then the argument of (8.2.4) shows that X is strongly stationary. Choose some state k and define the collection $I = \{I_n : n \geqslant 0\}$ of indicator functions by

$$I_n = \begin{cases} 1 & \text{if} \quad X_n = k \\ 0 & \text{otherwise.} \end{cases}$$

Clearly I is strongly stationary. It has autocovariance function

$$c(n, n+m) = \text{cov}(I_n, I_{n+m}) = \pi_k[p_{kk}(m) - \pi_k], \qquad m \geqslant 0,$$

where $p_{kk}(m) = \mathbf{P}(X_m = k \mid X_0 = k)$. The partial sum

$$S_n = \sum_{j=0}^{n-1} I_j$$

is the number of visits to the state k before the nth jump, and a short calculation gives

$$\frac{1}{n}\mathbf{E}(S_n) = \pi_k \quad \text{for all } n.$$

It is a consequence of the ergodic theorem (2) that

$$\frac{1}{n} S_n \xrightarrow{\text{a.s.}} S \quad \text{as} \quad n \to \infty,$$

where S is a random variable with mean

$$\mathbf{E}(S) = \mathbf{E}(I_0) = \pi_k.$$

Actually S is constant,

$$\mathbf{P}(S = \pi_k) = 1;$$

just note that

$$c(n, n+m) \to 0 \quad \text{as} \quad m \to \infty$$

and use the result of Problem (9.7.9). ●

(29) **Example.** Let X be uniformly distributed on $[0, 1]$. X has a binary expansion

$$X = 0 \cdot X_1 X_2 \cdots = \sum_{j=1}^{\infty} X_j 2^{-j}$$

where $X_1, X_2, \ldots$ is a sequence of independent identically distributed random variables, each taking one of the values 0 or 1 with probability $\frac{1}{2}$ (see (7.11.4)). Define

(30) $$Y_n = 0 \cdot X_n X_{n+1} \cdots \quad \text{for} \quad n \geqslant 1$$

and check for yourself that $Y = \{Y_n : n \geqslant 1\}$ is strongly stationary. Use (2) to see that

$$\frac{1}{n} \sum_{j=1}^{n} Y_j \xrightarrow{\text{a.s.}} \frac{1}{2} \quad \text{as} \quad n \to \infty.$$

Generalize this example as follows. Let $g: \mathbb{R} \to \mathbb{R}$ be such that

(a) g has period 1, so that $g(x + 1) = g(x)$ for all x
(b) g is uniformly continuous and integrable over $[0, 1]$,

and define $Z = \{Z_n : n \geqslant 1\}$ by

$$Z_n = g(2^{n-1} X)$$

where X is uniform on $[0, 1]$ as before. The process Y, above, may be constructed in this way by choosing

$$g(x) = x \quad \text{modulo } 1.$$

Check for yourself that Z is strongly stationary, and deduce that

$$\frac{1}{n} \sum_{j=1}^{n} g(2^{j-1} X) \xrightarrow{\text{a.s.}} \int_0^1 g(x)\, dx \quad \text{as} \quad n \to \infty.$$

Can you adapt this example to show that

$$\frac{1}{n} \sum_{j=1}^{n} g(X + (j-1)\pi) \xrightarrow{\text{a.s.}} \int_0^1 g(x)\, dx \quad \text{as} \quad n \to \infty$$

for any fixed positive irrational number π? ●

(31) **Example. Range of random walk.** Let $X_1, X_2, \ldots$ be independent identically distributed random variables taking integer values, and let $S_n = X_1 + X_2 + \cdots + X_n$; think of S_n as being the position of a random walk after n steps. Let R_n be the *range* of the walk up to time n, which is to say that R_n is the number of distinct values taken by the sequence $S_1, S_2, \ldots, S_n$. It was proved by elementary means in Problem (3.11.27) that

(32) $$\frac{1}{n} \mathbf{E}(R_n) \to \mathbf{P}(\text{no return}) \quad \text{as} \quad n \to \infty$$

where the event {no return} is the event that the walk never revisits its starting point $S_0 = 0$; that is, $\{\text{no return}\} = \{S_k \neq 0 \text{ for all } k \geqslant 1\}$.

Of more interest than (32) is the fact that

(33) $$\frac{1}{n} R_n \xrightarrow{\text{a.s.}} \mathbf{P}(\text{no return}),$$

and we shall prove this with the aid of the ergodic theorem (16).

First, let N be a positive integer, and let Z_k be the number of distinct points visited by $S_{(k-1)N+1}, S_{(k-1)N+2}, \ldots, S_{kN}$; clearly $Z_1, Z_2, \ldots$ are independent identically distributed variables. Now, if $KN \leqslant n < (K+1)N$, then $|R_n - R_{KN}| \leqslant N$ and $R_{KN} \leqslant Z_1 + Z_2 + \cdots + Z_K$. Therefore

$$\frac{1}{n} R_n \leqslant \frac{1}{KN}(R_{KN} + N) \leqslant \frac{1}{KN}(Z_1 + Z_2 + \cdots + Z_K) + \frac{1}{K}$$

$$\xrightarrow{\text{a.s.}} \frac{1}{N} \mathbf{E}(Z_1) \quad \text{as} \quad K \to \infty$$

by the strong law of large numbers. It is easily seen that $Z_1 = R_N$, and therefore, almost surely,

(34) $$\limsup_{n \to \infty} \left\{ \frac{1}{n} R_n \right\} \leqslant \frac{1}{N} \mathbf{E}(R_N) \to \mathbf{P}(\text{no return})$$

as $N \to \infty$, by (32). This is the required upper bound.

For the lower bound, we must work a little harder. Let V_k be the indicator function of the event that the position of the walk at time k is not revisited subsequently; that is,

$$V_k = \begin{cases} 1 & \text{if} \quad S_j \neq S_k \quad \text{for all } j > k, \\ 0 & \text{otherwise.} \end{cases}$$

The collection of points S_k for which $V_k = 1$ is a collection of distinct points, and it follows that

(35) $$R_n \geqslant V_1 + V_2 + \cdots + V_n.$$

On the other hand, V_k may be represented as $Y(X_{k+1}, X_{k+2}, \ldots)$ where $Y: \mathbb{R}^T \to \{0, 1\}$ is defined by

$$Y(x_1, x_2, \ldots) = \begin{cases} 1 & \text{if} \quad x_1 + \cdots + x_l \neq 0 \quad \text{for all } l \geqslant 1, \\ 0 & \text{otherwise.} \end{cases}$$

The X_j are independent and identically distributed, and therefore Theorem (16) may be applied to deduce that

$$\frac{1}{n}(V_1 + V_2 + \cdots + V_n) \xrightarrow{\text{a.s.}} \mathbf{E}(V_1);$$

note that $\mathbf{E}(V_1) = \mathbf{P}(\text{no return})$.

It follows from (35) that

$$\liminf_{n\to\infty}\left\{\frac{1}{n}R_n\right\} \geqslant \mathbf{P}(\text{no return}) \quad \text{a.s.,}$$

which may be combined with (34) to obtain the claimed result (33). ●

Exercises

(36) Let $T = \{1, 2, \ldots\}$ and let $\mathscr{I}$ be the set of invariant events of $(\mathbb{R}^T, \mathscr{B}^T)$. Show that $\mathscr{I}$ is a σ-field.

(37) Assume that $X_1, X_2, \ldots$ is a stationary sequence with autocovariance function c. Show that

$$\operatorname{var}\left(\frac{1}{n}\sum_{i=1}^{n} X_i\right) = \frac{2}{n^2}\sum_{j=1}^{n}\sum_{i=0}^{j-1} c(i) - \frac{c(0)}{n}.$$

Assuming that $j^{-1}\sum_{i=0}^{j-1} c(i) \to \sigma^2$ as $j \to \infty$, show that

$$\operatorname{var}\left(\frac{1}{n}\sum_{i=1}^{n} X_i\right) \to \sigma^2 \quad \text{as} \quad n \to \infty.$$

(38) Let $X_1, X_2, \ldots$ be independent identically distributed random variables with zero mean and unit variance. Let

$$Y_n = \sum_{i=0}^{\infty} \alpha_i X_{n+i} \quad \text{for} \quad n \geqslant 1$$

where the α_i are constants satisfying $\sum_i \alpha_i^2 < \infty$. Use the martingale convergence theorem to show that the above summation converges almost surely and in mean square. Prove that $n^{-1}\sum_{i=1}^{n} Y_i \to 0$ a.s. and in mean, as $n \to \infty$.

9.6 Gaussian processes

Let $X = \{X(t): -\infty < t < \infty\}$ be a real-valued stationary process with autocovariance function c; in line with (9.3.2), c is a real-valued function which satisfies

(a) $c(-t) = c(t)$
(b) c is a non-negative definite function.

It is not difficult to see that a function $c: \mathbb{R} \to \mathbb{R}$ is the autocovariance function of some real-valued stationary process if and only if c satisfies (a) and (b). Subject to these conditions on c, there is an explicit construction of a corresponding stationary process.

(1) **Theorem.** *If $c: \mathbb{R} \to \mathbb{R}$ and c satisfies* (a) *and* (b) *above then there exists a real-valued strongly stationary process X with autocovariance function c.*

Proof. We shall construct X by defining its finite-dimensional distributions and then use the Kolmogorov consistency conditions (8.6.3). For any vector $\boldsymbol{t} = (t_1, \ldots, t_n)$ of real numbers with some finite length n, let $F_{\boldsymbol{t}}$ be the multivariate normal distribution function with zero means and covariance matrix $\mathbf{V} = (v_{jk})$ with entries $v_{jk} = c(t_k - t_j)$ (see Section 4.9).

The family $\{F_{\boldsymbol{t}}: \boldsymbol{t} \in \mathbb{R}^n, n = 1, 2, \ldots\}$ satisfies the Kolmogorov consistency conditions (8.6.3) and so there exists a process X with this family of fdds. It is clear that X is strongly stationary with autocovariance function c. ■

A result similar to (1) holds for complex-valued functions $c: \mathbb{R} \to \mathbb{C}$, (a) being replaced by

(2) (a′) $c(-t) = \overline{c(t)}$.

We do not explore this here, but choose to consider real-valued processes only. The process X which we have constructed in the foregoing proof is an example of a (real-valued) 'Gaussian process'.

(3) **Definition.** A real-valued continuous-time process X is called **Gaussian** if each finite-dimensional vector $(X(t_1), \ldots, X(t_n))$ has the multivariate normal distribution $N(\boldsymbol{\mu}(\boldsymbol{t}), \mathbf{V}(\boldsymbol{t}))$ for some mean vector $\boldsymbol{\mu}$ and some covariance matrix $\mathbf{V}$ which may depend on $\boldsymbol{t} = (t_1, \ldots, t_n)$.

The $X(t_j)$ may have a singular multivariate normal distribution. We shall often restrict our attention to Gaussian processes with $\mathbf{E}(X(t)) = 0$ for all t; as before, similar results are easily found when this fails to hold.

A Gaussian process is not necessarily stationary.

(4) **Theorem.** *The Gaussian process X is stationary if and only if $\mathbf{E}(X(t))$ is constant for all t and the covariance matrix $\mathbf{V}(\boldsymbol{t})$ in Definition* (3) *satisfies*

$$\mathbf{V}(\boldsymbol{t}) = \mathbf{V}(\boldsymbol{t} + h)$$

for all $\boldsymbol{t}$ and $h > 0$, where $\boldsymbol{t} + h = (t_1 + h, \ldots, t_n + h)$.

Proof. This is an easy *exercise*. ■

It is clear that a Gaussian process is strongly stationary if and only if it is weakly stationary.

Can a Gaussian process be a Markov process? The answer is in the affirmative. First, we must rephrase the Markov property (6.1.1) to deal with processes which take values in the real line.

(5) **Definition.** The continuous-time process X, taking values in $\mathbb{R}$, is called a **Markov process** if

(6)
$$\mathbf{P}\big(X(t_n) \leqslant x \mid X(t_1) = x_1, \dots, X(t_{n-1}) = x_{n-1}\big) = \mathbf{P}\big(X(t_n) \leqslant x \mid X(t_{n-1}) = x_{n-1}\big)$$

for all $x, x_1, x_2, \dots, x_{n-1}$, and all increasing sequences $t_1 < t_2 < \cdots < t_n$ of times.

(7) **Theorem.** *The Gaussian process X is a Markov process if and only if*

(8)
$$\mathbf{E}\big(X(t_n) \mid X(t_1) = x_1, \dots, X(t_{n-1}) = x_{n-1}\big) = \mathbf{E}\big(X(t_n) \mid X(t_{n-1}) = x_{n-1}\big)$$

for all $x_1, x_2, \dots, x_{n-1}$ and all increasing sequences $t_1 < \cdots < t_n$ of times.

Proof. This is not difficult. It is clear from (5) that (8) holds whenever X is Markov. Conversely, suppose that X is Gaussian and satisfies (8). Both the left- and right-hand sides of (6) are normal distribution functions. But any normal distribution is specified by its mean and variance, and so we need only show that the left- and right-hand sides of (6) have equal first two moments. The equality of the first moments is trivial, since this is simply the assertion of (8). Also, if $1 \leqslant r < n$, then

$$\mathbf{E}(YX_r) = 0$$

where

(9)
$$Y = X_n - \mathbf{E}(X_n \mid X_1, \dots, X_{n-1}) = X_n - \mathbf{E}(X_n \mid X_{n-1})$$

and we have written $X_r = X(t_r)$ for ease of notation; to see this, write

$$\begin{aligned}\mathbf{E}(YX_r) &= \mathbf{E}\big(X_n X_r - \mathbf{E}(X_n X_r \mid X_1, \dots, X_{n-1})\big) \\ &= \mathbf{E}(X_n X_r) - \mathbf{E}(X_n X_r) = 0.\end{aligned}$$

However, Y and X_r are normally distributed, and furthermore $\mathbf{E}(Y) = 0$; as in (4.5.9), Y and X_r are independent. It follows that Y is independent of the collection $X_1, X_2, \dots, X_{n-1}$, using properties of the multivariate normal distribution.

Write $A_r = \{X_r = x_r\}$ and $A = A_1 \cap A_2 \cap \cdots \cap A_{n-1}$. By the proven independence, $\mathbf{E}(Y^2 \mid A) = \mathbf{E}(Y^2 \mid A_{n-1})$, which may be written as $\operatorname{var}(X_n \mid A) = \operatorname{var}(X_n \mid A_{n-1})$, by (9). Thus the left- and right-hand sides of (6) have the same second moment also, and the result is proved. ■

(10) **Example. A stationary Gaussian Markov process.** Suppose X is stationary, Gaussian and Markov, and has zero means. Use the result of (4.11.13) to obtain that

$$c(0)\mathbf{E}[X(s+t) \mid X(s)] = c(t)X(s) \quad \text{whenever} \quad t \geqslant 0,$$

where c is the autocovariance function of X. Thus, if $0 \leqslant s \leqslant s+t$ then

$$\begin{aligned} c(0)\mathbf{E}[X(0)X(s+t)] &= c(0)\mathbf{E}\big(\mathbf{E}(X(0)X(s+t) \mid X(0), X(s))\big) \\ &= c(0)\mathbf{E}[X(0)\mathbf{E}(X(s+t) \mid X(s))] \\ &= c(t)\mathbf{E}(X(0)X(s)) \end{aligned}$$

by (7.7.10). Thus

(11) $$c(0)c(s+t) = c(s)c(t) \quad \text{for} \quad s, t \geqslant 0.$$

This is satisfied whenever

(12) $$c(t) = c(0)\,\mathrm{e}^{-\alpha|t|}.$$

Following (4.11.5) we can see that (12) is the general solution to (11) subject to some condition of regularity such as that c be continuous. We shall see later (see (13.8.4)) that such a process is called a stationary *Ornstein–Uhlenbeck process.* ●

(13) **Example. The Wiener process.** Suppose that $\sigma^2 > 0$ and define

(14) $$c(s, t) = \sigma^2 \min\{s, t\} \quad \text{whenever} \quad s, t \geqslant 0.$$

We claim that there exists a Gaussian process $W = \{W(t): t \geqslant 0\}$ with zero means such that $W(0) = 0$ and

$$\operatorname{cov}(W(s), W(t)) = c(s, t).$$

By the argument in the proof of (1), it is sufficient to show that the matrix $\boldsymbol{V}(\boldsymbol{t})$ with entries (v_{jk}), where $v_{jk} = c(t_k, t_j)$, is positive definite for all $\boldsymbol{t} = (t_1, t_2, \ldots, t_n)$. To see this let $z_1, z_2, \ldots, z_n$ be complex numbers and suppose that $0 = t_0 < t_1 < \cdots < t_n$. Then it is not difficult to check that

$$\sum_{j,k=1}^{n} c(t_k, t_j) z_j \bar{z}_k = \sigma^2 \sum_{j=1}^{n} (t_j - t_{j-1}) \left| \sum_{k=j}^{n} z_k \right|^2 > 0$$

whenever one of the z_j is non-zero; this guarantees the existence of W. It is called the *Wiener process*; we explore its properties in more detail in Chapter 13, noting only two facts here.

(15) **Lemma.** *The Wiener process W satisfies*

$$\mathbf{E}(W(t)^2) = \sigma^2 t \quad \textit{for all } t \geqslant 0.$$

Proof. $$\mathbf{E}(W(t)^2) = \operatorname{cov}(W(t), W(t)) = c(t, t) = \sigma^2 t.$$ ■

(16) **Lemma.** *The Wiener process W has stationary independent increments, which is to say that*

(a) *the distribution of $W(t) - W(s)$ depends on $t - s$ alone*
(b) *the variables $W(t_j) - W(s_j)$, $1 \leqslant j \leqslant n$, are independent whenever the intervals $(s_j, t_j]$ are disjoint.*

Proof. The increments of W are jointly normally distributed; their independence follows as soon as we have shown that they are uncorrelated. However, if $u \leqslant v \leqslant s \leqslant t$,

$$\mathbf{E}([W(v) - W(u)][W(t) - W(s)]) = c(v, t) - c(v, s) + c(u, s) - c(u, t)$$
$$= \sigma^2(v - v + u - u) = 0$$

by (14).

Finally, $W(t) - W(s)$ is normally distributed with zero mean, and with variance

$$\mathbf{E}([W(t) - W(s)]^2) = \mathbf{E}(W(t)^2) - 2c(s, t) + \mathbf{E}(W(s)^2)$$
$$= \sigma^2(t - s) \quad \text{if} \quad s \leqslant t. \qquad \blacksquare\bullet$$

Exercises

(17) Show that the function $c(s, t) = \min\{s, t\}$ is positive definite. That is, show that

$$\sum_{j,k=1}^{n} c(t_k, t_j) z_j \bar{z}_k > 0$$

for all $0 \leqslant t_1 < t_2 < \cdots < t_n$ and all complex numbers $z_1, z_2, \ldots, z_n$ at least one of which is non-zero.

(18) Let $X_1, X_2, \ldots$ be a stationary Gaussian sequence with zero means and unit variances which satisfies the Markov property. Find the spectral density function of the sequence in terms of the constant $\rho = \text{cov}(X_1, X_2)$.

(19) Show that a Gaussian process is strongly stationary if and only if it is weakly stationary.

(20) Let X be a stationary Gaussian process with zero mean, unit variance, and autocovariance function c. Find the autocovariance functions of the processes $X^2 = \{X(t)^2\colon -\infty < t < \infty\}$ and $X^3 = \{X(t)^3\colon -\infty < t < \infty\}$.

9.7 Problems

1. Let $\ldots, X_{-1}, X_0, X_1, \ldots$ be uncorrelated random variables with zero means and unit variances, and define

$$Y_n = X_n + \alpha \sum_{i=1}^{\infty} \beta^{i-1} X_{n-i} \quad \text{for} \quad -\infty < n < \infty,$$

where α and β are constants satisfying $|\beta| < 1$, $|\beta - \alpha| < 1$. Find the best linear predictor of Y_{n+1} given the entire past $Y_n, Y_{n-1}, \ldots$.

2. Let $\{Y_k\colon -\infty < k < \infty\}$ be a stationary sequence with variance σ_Y^2, and let

$$X_n = \sum_{k=0}^{r} a_k Y_{n-k}, \qquad -\infty < n < \infty,$$

where $a_1, a_2, \ldots, a_r$ are constants. Show that X has spectral density function $f_X(\lambda) = (\sigma_Y^2/\sigma_X^2) f_Y(\lambda)|G_a(e^{i\lambda})|^2$ where f_Y is the spectral density function of Y, $\sigma_X^2 = \text{var}(X_1)$, and $G_a(z) = \sum_{k=0}^{r} a_k z^k$.

Calculate this spectral density explicitly in the case of 'exponential smoothing', when $r = \infty$, $a_k = \mu^k(1 - \mu)$, and $0 < \mu < 1$.

3. Suppose that $\widehat{Y}_{n+1} = \alpha Y_n + \beta Y_{n-1}$ is the best linear predictor of Y_{n+1} given the entire past $Y_n, Y_{n-1}, \ldots$ of the stationary sequence $\{Y_k: -\infty < k < \infty\}$. Find the spectral density function of the sequence.

4. **Recurrent events (5.2.15).** Meteorites fall from the sky at integer times $T_1, T_2, \ldots$ where $T_n = X_1 + X_2 + \cdots + X_n$. We assume that the X_j are independent, $X_2, X_3, \ldots$ are identically distributed, and the distribution of X_1 is such that the probability that a meteorite falls at time n is constant for all n. Let Y_n be the indicator function of the event that a meteorite falls at time n. Show that $\{Y_n\}$ is stationary and find its spectral density function in terms of the characteristic function of X_2.

5. Let $X = \{X_n: n \geqslant 1\}$ be given by $X_n = \cos(nU)$ where U is uniformly distributed on $[-\pi, \pi]$. Show that X is stationary but not strongly stationary. Find the autocorrelation function of X and its spectral density function.

6. (a) Let N be a Poisson process with intensity λ, and let $\alpha > 0$. Define $X(t) = N(t+\alpha) - N(t)$ for $t \geqslant 0$. Show that X is strongly stationary, and find its spectral density function.
   (b) Let W be a Wiener process and define $X = \{X(t): t \geqslant 1\}$ by $X(t) = W(t) - W(t-1)$. Show that X is strongly stationary and find its autocovariance function. Find the spectral density function of X.

7. Let $Z_1, Z_2, \ldots$ be uncorrelated variables, each with zero mean and unit variance.
   (a) Define the moving average process X by $X_n = Z_n + \alpha Z_{n-1}$ where α is a constant. Find the spectral density function of X.
   (b) More generally, let

$$Y_n = \sum_{i=0}^{r} \alpha_i Z_{n-i},$$

   where $\alpha_0 = 1$ and $\alpha_1, \ldots, \alpha_r$ are constants. Find the spectral density function of Y.

8. Show that the complex-valued stationary process $X = \{X(t): -\infty < t < \infty\}$ has a spectral density function which is bounded and uniformly continuous whenever its autocorrelation function ρ is continuous and satisfies $\int_0^\infty |\rho(t)|\, dt < \infty$.

9. Let $X = \{X_n: n \geqslant 1\}$ be stationary with constant mean $\mu = \mathbf{E}(X_n)$ for all n, and such that $\text{cov}(X_0, X_n) \to 0$ as $n \to \infty$. Show that $n^{-1}\sum_{j=1}^{n} X_j \xrightarrow{\text{m.s.}} \mu$.

10. Deduce the strong law of large numbers from an appropriate ergodic theorem.

11. Let $\mathbf{Q}$ be a stationary measure on $(\mathbb{R}^T, \mathscr{B}^T)$ where $T = \{1, 2, \ldots\}$. Show that $\mathbf{Q}$ is ergodic if and only if

$$\frac{1}{n}\sum_{i=1}^{n} Y_i \to \mathbf{E}(Y) \quad \text{a.s. and in mean}$$

for all $Y: \mathbb{R}^T \to \mathbb{R}$ for which $\mathbf{E}(Y)$ exists, where $Y_i: \mathbb{R}^T \to \mathbb{R}$ is given by $Y_i(\mathbf{x}) = Y(\tau^{i-1}(\mathbf{x}))$. As usual, τ is the natural shift operator on $\mathbb{R}^T$.

12. The stationary measure $\mathbf{Q}$ on $(\mathbb{R}^T, \mathscr{B}^T)$ is called *strongly mixing* if $\mathbf{Q}(A \cap \tau^{-n}B) \to \mathbf{Q}(A)\mathbf{Q}(B)$ as $n \to \infty$, for all $A, B \in \mathscr{B}^T$; as usual, $T = \{1, 2, \ldots\}$ and τ is the shift operator on $\mathbb{R}^T$. Show that every strongly mixing measure is ergodic.

13. **Ergodic theorem.** Let $(\Omega, \mathscr{F}, \mathbf{P})$ be a probability space, and let $T: \Omega \to \Omega$ be measurable and measure-preserving (i.e., $\mathbf{P}(T^{-1}A) = \mathbf{P}(A)$ for all $A \in \mathscr{F}$). Let $X: \Omega \to \mathbb{R}$ be a random variable, and let X_i be given by $X_i(\omega) = X(T^{i-1}(\omega))$. Show that

$$\frac{1}{n}\sum_{i=1}^{n} X_i \to \mathbf{E}(X \mid \mathscr{I}) \quad \text{a.s. and in mean}$$

where $\mathscr{I}$ is the σ-field of invariant events of T.

If T is ergodic (in that $\mathbf{P}(A)$ equals 0 or 1 whenever A is invariant), prove that $\mathbf{E}(X \mid \mathscr{I}) = \mathbf{E}(X)$ a.s.

14. Consider the probability space $(\Omega, \mathscr{F}, \mathbf{P})$ where $\Omega = [0, 1)$, $\mathscr{F}$ is the set of Borel subsets, and $\mathbf{P}$ is Lebesgue measure. Show that the shift $T: \Omega \to \Omega$ defined by $T(x) = 2x \pmod 1$ is measurable, measure-preserving, and ergodic (in that $\mathbf{P}(A)$ equals 0 or 1 if $A = T^{-1}A$).

Let $X: \Omega \to \mathbb{R}$ be the random variable given by the identity mapping $X(\omega) = \omega$. Show that the proportion of ones, in the expansion of X to base 2, equals $\frac{1}{2}$ a.s.; this is sometimes called 'Borel's normal number theorem'.

15. Let $g: \mathbb{R} \to \mathbb{R}$ be periodic with period 1, and uniformly continuous and integrable over $[0, 1]$. Define $Z_n = g(X + (n-1)\alpha)$, $n \geqslant 1$, where X is uniform on $[0, 1]$ and α is irrational. Show that, as $n \to \infty$,

$$\frac{1}{n}\sum_{j=1}^{n} Z_j \to \int_0^1 g(u)\, du \quad \text{a.s.}$$

16. Let $X = \{X(t): t \geqslant 0\}$ be a non-decreasing random process such that

(a) $X(0) = 0$, X takes values in the non-negative integers,
(b) X has stationary independent increments,
(c) the sample paths $\{X(t, \omega): t \geqslant 0\}$ have only jump discontinuities of unit magnitude.

Show that X is a Poisson process.

17. Let X be a continuous-time process. Show that

(a) if X has stationary increments and $m(t) = \mathbf{E}(X(t))$ is a continuous function of t, then there exist α and β such that $m(t) = \alpha + \beta t$,
(b) if X has stationary independent increments and $v(t) = \text{var}(X(t) - X(0))$ is a continuous function of t then there exists σ^2 such that $\text{var}(X(s+t) - X(s)) = \sigma^2 t$ for all s.

18. A Wiener process is called *standard* if $W(1)$ has unit variance. Let W be a standard Wiener process, and let α be a positive constant. Show that

(a) $\alpha W(t/\alpha^2)$ is a standard Wiener process,
(b) $W(t + \alpha) - W(\alpha)$ is a standard Wiener process,
(c) the process V, given by $V(t) = tW(1/t)$ for $t > 0$, $V(0) = 0$, is a standard Wiener process,
(d) the process $W(1) - W(1 - t)$ is a standard Wiener process on $[0, 1]$.

19. Let W be a standard Wiener process. Show that the stochastic integrals

$$X(t) = \int_0^t \mathrm{d}W(u), \qquad Y(t) = \int_0^t e^{-(t-u)}\,\mathrm{d}W(u), \qquad t \geqslant 0,$$

are well defined, and prove that $X(t) = W(t)$, and that Y has autocovariance function $\operatorname{cov}(Y(s), Y(t)) = \frac{1}{2}(e^{-|s-t|} - e^{-s-t})$, $s < t$.

20. Let W be a standard Wiener process. Find the means of the following processes, and the autocovariance functions in cases (b) and (c):

(a) $X(t) = |W(t)|$,
(b) $Y(t) = e^{W(t)}$,
(c) $Z(t) = \int_0^t W(u)\,\mathrm{d}u$.

Which of these are Gaussian processes? Which of these are Markov processes?

21. Let W be a standard Wiener process. Find the conditional joint density function of $W(t_2)$ and $W(t_3)$ given that $W(t_1) = W(t_4) = 0$, where $t_1 < t_2 < t_3 < t_4$.
Show that the conditional correlation of $W(t_2)$ and $W(t_3)$ is

$$\rho = \sqrt{\frac{(t_4 - t_3)(t_2 - t_1)}{(t_4 - t_2)(t_3 - t_1)}}.$$

22. **Empirical distribution function.** Let $U_1, U_2, \ldots$ be independent random variables with the uniform distribution on $[0, 1]$. Let $I_j(x)$ be the indicator function of the event $\{U_j \leqslant x\}$, and define

$$F_n(x) = \frac{1}{n}\sum_{j=1}^{n} I_j(x), \qquad 0 \leqslant x \leqslant 1.$$

The function F_n is called the 'empirical distribution function' of the U_j.

(a) Find the mean and variance of $F_n(x)$, and prove that $(F_n(x) - x)\sqrt{n} \xrightarrow{\mathrm{D}} Y(x)$ as $n \to \infty$, where $Y(x)$ is normally distributed.
(b) What is the (multivariate) limit distribution of a collection of random variables of the form $\{(F_n(x_i) - x_i)\sqrt{n}: 1 \leqslant i \leqslant k\}$, where $0 \leqslant x_1 < x_2 < \cdots < x_k \leqslant 1$?
(c) Show that the autocovariance function of the asymptotic finite-dimensional distributions of $(F_n(x) - x)\sqrt{n}$, in the limit as $n \to \infty$, is the same as that of the process $Z(t) = W(t) - tW(1)$, $0 \leqslant t \leqslant 1$, where W is a standard Wiener process. The process Z is called a 'Brownian bridge' or 'tied-down Brownian motion'.

10 Renewals

10.1 The renewal equation

We saw in Section 8.3 that renewal processes provide attractive models for many natural phenomena. Recall their definition.

(1) **Definition.** A **renewal process** $N = \{N(t): t \geq 0\}$ is a process such that

$$N(t) = \max\{n: T_n \leq t\}$$

where

$$T_0 = 0, \qquad T_n = X_1 + \cdots + X_n \quad \text{for} \quad n \geq 1,$$

and $\{X_i\}$ is a sequence of independent identically distributed non-negative† random variables.

We commonly think of a renewal process $N(t)$ as representing the number of occurrences of some event in the time interval $[0, t]$; the event in question might be the arrival of a person or particle, or the failure of a light bulb. With this in mind, we shall speak of T_n as the 'time of the nth arrival' and X_n as the 'nth interarrival time'. We shall try to use the notation of (1) consistently throughout, denoting by X and T a typical interarrival time and a typical arrival time of the process N.

When is N an honest process (see (6.8.18) for the definition of honesty)?

(2) **Theorem.** $\mathbf{P}(N(t) < \infty) = 1$ *for all t if and only if* $\mathbf{E}(X_1) > 0$.

This amounts to saying that N is honest if and only if the interarrival times are not concentrated at zero. The proof is simple and relies upon the following important observation:

(3) $$N(t) \geq n \quad \textit{if and only if} \quad T_n \leq t.$$

We shall make repeated use of (3). It provides a link between $N(t)$ and the sum T_n of independent variables; we know a lot about such sums already.

Proof of (2). Clearly, if $\mathbf{E}(X_1) = 0$ then

$$\mathbf{P}(X_i = 0) = 1 \quad \text{for all } i$$

since the X_i are non-negative, and so

$$\mathbf{P}(N(t) = \infty) = 1 \quad \text{for all } t > 0.$$

† But soon we will impose the stronger condition that the X_i be *strictly* positive.

Conversely, suppose that $\mathbf{E}(X_1) > 0$. There exists $\varepsilon > 0$ such that $\mathbf{P}(X_1 > \varepsilon) = \delta > 0$. Writing A_i for the event $\{X_i > \varepsilon\}$, we see that the event $A = \{X_i > \varepsilon \text{ i.o.}\} = \limsup A_i$, that infinitely many of the X_i exceed ε, occurs with probability 1, since

$$\mathbf{P}(A^c) = \mathbf{P}(X_n \leqslant \varepsilon \text{ for all large } n) = \lim_{m \to \infty} (1-\delta)^m = 0;$$

remember (1.8.16). Therefore, by (3),

$$\mathbf{P}(N(t) = \infty) = \mathbf{P}(T_n \leqslant t \text{ for all } n) \leqslant \mathbf{P}(A^c) = 0. \qquad \blacksquare$$

Thus N is honest if and only if X_1 is *not* concentrated at 0. Henceforth we shall assume not only that $\mathbf{P}(X_1 = 0) < 1$, but also impose the stronger condition that $\mathbf{P}(X_1 = 0) = 0$. That is, *we consider only the case when the X_i are strictly positive.*

It is easy in principle to find the distribution of $N(t)$ in terms of the distribution of a typical interarrival time. Let F be the distribution function of X_1, and let F_k be the distribution function of T_k.

(4) **Lemma.**† *$F_1 = F$ and $F_{k+1}(x) = \int_0^x F_k(x-y)\,\mathrm{d}F(y)$ for $k \geqslant 1$.*

Proof. Clearly $F_1 = F$. Also

$$T_{k+1} = T_k + X_{k+1},$$

and (4.8.1) gives the result when suitably rewritten for independent variables of general type. ■

(5) **Lemma.** $\mathbf{P}(N(t) = k) = F_k(t) - F_{k+1}(t)$.

Proof. $\{N(t) = k\} = \{N(t) \geqslant k\} \backslash \{N(t) \geqslant k+1\}$. Now use (3). ■

We shall be interested largely in the expected value of $N(t)$.

(6) **Definition.** The **renewal function** m is given by $m(t) = \mathbf{E}(N(t))$.

Again, it is easy to find m in terms of the F_k.

(7) **Lemma.** $m(t) = \sum_{k=1}^{\infty} F_k(t)$.

† Readers of Section 5.6 may notice that the statement of this lemma violates our notation for the domain of an integral. We adopt the convention that expressions like $\int_a^b g(y)\,\mathrm{d}F(y)$ denote integrals over the half-open interval $(a, b]$, with the left endpoint excluded.

Proof. Define the indicator variables

$$I_k = \begin{cases} 1 & \text{if} \quad T_k \leqslant t \\ 0 & \text{otherwise.} \end{cases}$$

Then

$$N(t) = \sum_{k=1}^{\infty} I_k$$

and so

$$m(t) = \mathbf{E}\left(\sum_{k=1}^{\infty} I_k \right) = \sum_{k=1}^{\infty} \mathbf{E}(I_k) = \sum_{k=1}^{\infty} F_k(t). \qquad \blacksquare$$

An alternative approach to the renewal function is by way of conditional expectations and the 'renewal equation'. First note that m is the solution of a certain integral equation.

(8) **Lemma.** *The renewal function m satisfies the* ***renewal equation****,*

(9) $$m(t) = F(t) + \int_0^t m(t - x)\, \mathrm{d}F(x).$$

Proof. Use conditional expectation to obtain

$$m(t) = \mathbf{E}(N(t)) = \mathbf{E}(\mathbf{E}[N(t) \mid X_1]);$$

but,

$$\mathbf{E}(N(t) \mid X_1 = x) = 0 \quad \text{if} \quad t < x$$

since the first arrival occurs after time t. On the other hand,

$$\mathbf{E}(N(t) \mid X_1 = x) = 1 + \mathbf{E}(N(t - x)) \quad \text{if} \quad t \geqslant x$$

since the process of arrivals, starting from the epoch of the first arrival, is a copy of N itself. Thus

$$m(t) = \int_0^{\infty} \mathbf{E}(N(t) \mid X_1 = x)\, \mathrm{d}F(x) = \int_0^t [1 + m(t - x)]\, \mathrm{d}F(x)$$

as required. ■

We know from (7) that

$$m(t) = \sum_{k=1}^{\infty} F_k(t)$$

is a solution to the renewal equation (9). Actually, it is the unique solution to (9) which is bounded on finite intervals. This is a consequence of the next lemma. We shall encounter a more general form of (9) later, and it is appropriate to anticipate this now. The more general case involves solutions μ to the *renewal-type equation*

(10) $$\mu(t) = H(t) + \int_0^t \mu(t-x)\,\mathrm{d}F(x), \qquad t \geqslant 0$$

where H is a uniformly bounded function.

(11) **Theorem.** *The function* μ, *given by*

$$\mu(t) = H(t) + \int_0^t H(t-x)\,\mathrm{d}m(x),$$

is a solution of the renewal-type equation (10). *If H is bounded on finite intervals then* μ *is bounded on finite intervals and is the unique solution of* (10) *with this property*†.

We shall make repeated use of this result, the proof of which is simple.

Proof. If $h: [0, \infty) \to \mathbb{R}$, define the functions $h * m$ and $h * F$ by

$$(h * m)(t) = \int_0^t h(t-x)\,\mathrm{d}m(x)$$

$$(h * F)(t) = \int_0^t h(t-x)\,\mathrm{d}F(x)$$

whenever these integrals exist. The operation $*$ is a type of convolution; do not confuse it with the related but different convolution operator of Sections 3.8 and 4.8. It can be shown that

$$(h * m) * F = h * (m * F),$$

and so we write $h * m * F$ for this double convolution. Note also that

(12) $$m = F + m * F \quad \text{by (9)}$$

(13) $$F_{k+1} = F_k * F = F * F_k \quad \text{by (4)}.$$

Using this notation, μ can be written as

$$\mu = H + H * m.$$

† Think of the integral in (11) as $\int H(t-x)m'(x)\,\mathrm{d}x$ if you are unhappy about its present form.

Convolve with F and use (12) to find that

$$\begin{aligned}\mu * F &= H * F + H * m * F\\ &= H * F + H * (m - F)\\ &= H * m = \mu - H,\end{aligned}$$

and so μ satisfies (10).

If H is bounded on finite intervals then

$$\begin{aligned}\sup_{0\leqslant t\leqslant T} |\mu(t)| &\leqslant \sup_{0\leqslant t\leqslant T} |H(t)| + \sup_{0\leqslant t\leqslant T} \left|\int_0^t H(t-x)\,\mathrm{d}m(x)\right|\\ &\leqslant [1 + m(T)] \sup_{0\leqslant t\leqslant T} |H(t)| < \infty,\end{aligned}$$

and so μ is indeed bounded on finite intervals; we have used the finiteness of m here (see Problem (10.5.1b)). To show that μ is the unique such solution of (10), suppose that μ_1 is another bounded solution and write $\delta(t) = \mu(t) - \mu_1(t)$; δ is a bounded function. Also

$$\delta = \delta * F \text{ by (10).}$$

Iterate this equation and use (13) to find that

$$\delta = \delta * F_k \quad \text{for all } k \geqslant 1$$

which implies that

$$|\delta(t)| \leqslant F_k(t) \sup_{0\leqslant u\leqslant t} |\delta(u)| \quad \text{for all } k \geqslant 1.$$

Let $k \to \infty$ to find that $|\delta(t)| = 0$ for all t, since

$$F_k(t) = \mathbf{P}(N(t) \geqslant k) \to 0 \quad \text{as} \quad k \to \infty$$

by (2). The proof is complete. ■

The method of Laplace–Stieltjes transforms is often useful in renewal theory (see Definition (15) of Appendix I). For example, we can transform (10) to obtain the formula

$$\mu^*(\theta) = \frac{H^*(\theta)}{1 - F^*(\theta)} \quad \text{for} \quad \theta \neq 0,$$

an equation which links the Laplace–Stieltjes transforms of μ, H, and F. In particular, setting $H = F$, we find from (8) that

(14) $$m^*(\theta) = \frac{F^*(\theta)}{1 - F^*(\theta)},$$

a formula which is directly derivable from (7) and (13). Hence there is a one–one correspondence between the renewal functions m, and the distribution functions F of the interarrival times.

(15) **Example. Poisson process.** This is the only Markovian renewal process, and has exponentially distributed interarrival times with some parameter λ. The epoch T_k of the kth arrival is distributed as $\Gamma(\lambda, k)$; Lemma (7) gives that

$$m(t) = \sum_{k=1}^{\infty} \int_0^t \frac{\lambda(\lambda s)^{k-1} e^{-\lambda s}}{(k-1)!} \, ds = \int_0^t \lambda \, ds = \lambda t.$$

Alternatively, just remember that $N(t)$ is Poisson with parameter λt to obtain the same result. ●

Exercises

(16) Prove that $\mathbf{E}(e^{\theta N(t)}) < \infty$ for some strictly positive θ whenever $\mathbf{E}(X_1) > 0$. [Hint: consider the renewal process with interarrival times $X'_k = \varepsilon I_{\{X_k \geqslant \varepsilon\}}$ for some suitable ε.]

(17) Let N be a renewal process and let W be the waiting time until the length of some interarrival time has exceeded s. That is, $W = \inf\{t: C(t) > s\}$, where $C(t)$ is the time which has elapsed (at time t) since the last arrival. Show that

$$F_W(x) = \begin{cases} 0 & \text{if } x < s \\ 1 - F(s) + \int_0^s F_W(x-u)\, dF(u) & \text{if } x \geqslant s \end{cases}$$

where F is the distribution function of an interarrival time. If N is a Poisson process with intensity λ, show that

$$\mathbf{E}(e^{\theta W}) = \frac{\lambda - \theta}{\lambda - \theta\, e^{(\lambda-\theta)s}} \quad \text{for} \quad \theta < \lambda$$

and $\mathbf{E}(W) = (e^{\lambda s} - 1)/\lambda$. You may find it useful to rewrite the above integral equation in the form of a renewal-type equation.

10.2 Limit theorems

Next, we study the asymptotic behaviour of $N(t)$ and its renewal function $m(t)$ for large values of t. There are four main results here, two for each of N and m. For the renewal process N itself there is a law of large numbers and a central limit theorem; these rely upon the relation (10.1.3), which links N to the partial sums of independent variables. The two results for m deal also with first- and second-order properties. The first asserts that $m(t)$ is approximately linear in t; the second asserts that the gradient of m is asymptotically constant.

How does $N(t)$ behave when t is large? Let $\mu = \mathbf{E}(X_1)$ be the mean of a typical interarrival time. Henceforth we assume that $\mu < \infty$.

(1) **Theorem.** $\dfrac{1}{t}N(t) \xrightarrow{\text{a.s.}} \dfrac{1}{\mu}$ *as* $t \to \infty$.

(2) **Theorem.** *If* $\sigma^2 = \text{var}(X_1)$ *and* $0 < \sigma < \infty$, *then*

$$\frac{N(t) - t/\mu}{(t\sigma^2/\mu^3)^{\frac{1}{2}}} \xrightarrow{\text{D}} N(0, 1) \quad as \quad t \to \infty.$$

It is not quite so easy to find the asymptotic behaviour of the renewal function.

(3) **Elementary renewal theorem.** $\dfrac{1}{t}m(t) \to \dfrac{1}{\mu}$ *as* $t \to \infty$.

The second-order properties of m are hard to find, and require a preliminary definition.

(4) **Definition.** Call a random variable X and its distribution F_X **arithmetic with span** λ (>0) if X takes values in the set $\{m\lambda\colon m = 0, \pm 1, \dots\}$ with probability 1, and λ is maximal with this property.

If the interarrival times of N are arithmetic, with span λ say, then so is T_k for each k. In this case $m(t)$ may be discontinuous at values of t which are multiples of λ, and this affects the second-order properties of m.

(5) **Renewal theorem.** *If* X_1 *is not arithmetic then*

(6) $$m(t + h) - m(t) \to \frac{h}{\mu} \quad as \quad t \to \infty \quad for\ all\ h.$$

If X_1 *is arithmetic with span* λ, *then* (6) *holds whenever* h *is a multiple of* λ.

It is appropriate to make some remarks about these theorems before we set to their proofs. Theorems (1) and (2) are straightforward, and use the law of large numbers and the central limit theorem for partial sums of independent sequences. It is perhaps surprising that (3) is harder to demonstrate than (1) since it concerns only the mean value of $N(t)$; it has a suitably probabilistic proof which uses the method of truncation, a technique which proved useful in the proof of the strong law (7.5.1). On the other hand, the proof of (5) is difficult. The usual method of proof is largely an exercise in solving integral equations; it is not appropriate to include this here (see Feller 1971, p. 360). There is an alternative proof which is short, beautiful, and probabilistic, and uses 'coupling' arguments related to those in the proof of the ergodic theorem for discrete-time Markov chains. This method requires some results which appear later in this chapter, and so we defer a sketch of the argument until (10.4.21). In the case of arithmetic interarrival times, (5) is essentially the same as Theorem (5.2.24), a result about *integer-valued* random variables. There is an apparently more general form of (5) which is deducible from (5). It is called the 'key renewal theorem' because of its many applications.

In the rest of this chapter we shall commonly assume that the interarrival times are *not* arithmetic. Similar results often hold in the arithmetic case, but they are usually more complicated to state.

(7) **Key renewal theorem.** *If* $g\colon [0, \infty) \to [0, \infty)$ *is such that*

(a) $g(t) \geqslant 0$ *for all* t
(b) $\int_0^\infty g(t)\,\mathrm{d}t < \infty$
(c) g *is monotone non-increasing*

then

$$\int_0^t g(t-x)\,\mathrm{d}m(x) \to \frac{1}{\mu}\int_0^\infty g(x)\,\mathrm{d}x \quad \text{as} \quad t \to \infty$$

whenever X_1 *is not arithmetic.*

In order to deduce this theorem from the renewal theorem (5), first prove it for indicator functions of intervals, then for step functions, and finally for limits of increasing sequences of step functions. We omit the details.

Proof of (1). This is easy. Just note that

(8) $$T_{N(t)} \leqslant t < T_{N(t)+1} \quad \text{for all } t.$$

Therefore, if $N(t) > 0$,

$$\frac{T_{N(t)}}{N(t)} \leqslant \frac{t}{N(t)} < \frac{T_{N(t)+1}}{N(t)+1}\left(1 + \frac{1}{N(t)}\right).$$

As $t \to \infty$, $N(t) \xrightarrow{\text{a.s.}} \infty$, and the strong law of large numbers gives

$$\mu \leqslant \lim_{t\to\infty}\left(\frac{t}{N(t)}\right) \leqslant \mu \quad \text{almost surely.}$$ ■

Proof of (2). This is Problem (10.5.4). ■

In preparation for the proof of (3), we recall an important definition. Let M be a random variable taking values in $\{1, 2, \ldots\}$. We call the random variable M a *stopping time* with respect to the sequence X of interarrival times if, for all $t \geqslant 1$, the event $\{M \leqslant t\}$ belongs to the σ-field of events generated by

$X_1, X_2, \ldots, X_t$. Note that $M = N(t) + 1$ is a stopping time for X, since

$$\{M \leqslant t\} = \{N(t) \leqslant t - 1\} = \left\{\sum_{i=1}^{t} X_i > t\right\},$$

an event defined in terms of $X_1, X_2, \ldots, X_t$. The random variable $N(t)$ is *not* a stopping time.

(9) **Lemma. Wald's equation.** *Let $X_1, X_2, \ldots$ be independent identically distributed random variables with finite mean, and let M be a stopping time with respect to the X_i satisfying $\mathbf{E}(M) < \infty$. Then*

$$\mathbf{E}\left(\sum_{i=1}^{M} X_i\right) = \mathbf{E}(X_1)\mathbf{E}(M).$$

Applying Wald's equation to the sequence of interarrival times, with $M = N(t) + 1$, we obtain

(10) $$\mathbf{E}(T_{N(t)+1}) = \mu[m(t) + 1].$$

Wald's equation may seem trite, but this is far from being the case. For example, it is not generally true that $\mathbf{E}(T_{N(t)}) = \mu m(t)$; the forthcoming Example (10.3.2) is an example of some of the dangers here.

Proof of Wald's equation (9). The basic calculation is elementary. Just note that

$$\sum_{i=1}^{M} X_i = \sum_{i=1}^{\infty} X_i I_{\{M \geqslant i\}},$$

so that (using dominated convergence or (5.6.16))

$$\mathbf{E}\left(\sum_{i=1}^{M} X_i\right) = \sum_{i=1}^{\infty} \mathbf{E}(X_i I_{\{M \geqslant i\}}) = \sum_{i=1}^{\infty} \mathbf{E}(X_i)\mathbf{P}(M \geqslant i) \quad \text{by independence,}$$

since $\{M \geqslant i\} = \{M \leqslant i-1\}^c$, an event definable in terms of $X_1, X_2, \ldots, X_{i-1}$ and therefore independent of X_i. The final sum equals

$$\mathbf{E}(X_1)\sum_{i=1}^{\infty} \mathbf{P}(M \geqslant i) = \mathbf{E}(X_1)\mathbf{E}(M). \qquad \blacksquare$$

Proof of (3). Half of this is easy. From (8),

$$t < T_{N(t)+1};$$

take expectations of this and use (10) to obtain

$$\frac{m(t)}{t} > \frac{1}{\mu} - \frac{1}{t}.$$

Letting $t \to \infty$, we obtain

(11) $$\liminf_{t\to\infty} \frac{1}{t} m(t) \geqslant \frac{1}{\mu}.$$

We may be tempted to proceed as follows in order to bound $m(t)$ above. From (8)

$$T_{N(t)} \leqslant t$$

and so

(12) $$\begin{aligned} t \geqslant \mathbf{E}(T_{N(t)}) &= \mathbf{E}(T_{N(t)+1} - X_{N(t)+1}) \\ &= \mu[m(t)+1] - \mathbf{E}(X_{N(t)+1}). \end{aligned}$$

The problem is that $X_{N(t)+1}$ depends on $N(t)$, and so $\mathbf{E}(X_{N(t)+1}) \neq \mu$ in general. To cope with this, truncate the X_i at some $c > 0$ to obtain a new sequence

$$X_j^c = \begin{cases} X_j & \text{if} \quad X_j < c \\ c & \text{if} \quad X_j \geqslant c. \end{cases}$$

Now consider the renewal process N^c with associated interarrival times $\{X_j^c\}$. Apply (12) to N^c, noting that $\mu^c = \mathbf{E}(X_j^c) \leqslant c$, to obtain

(13) $$t \geqslant \mu^c[\mathbf{E}(N^c(t)) + 1] - c.$$

However, $X_j^c \leqslant X_j$ for all j, and so $N^c(t) \geqslant N(t)$ for all t. Therefore

$$\mathbf{E}(N^c(t)) \geqslant \mathbf{E}(N(t)) = m(t)$$

and (13) becomes

$$\frac{m(t)}{t} \leqslant \frac{1}{\mu^c} + \frac{c - \mu^c}{\mu^c t}.$$

Let $t \to \infty$ to obtain

$$\limsup_{t\to\infty} \frac{1}{t} m(t) \leqslant \frac{1}{\mu^c};$$

now let $c \to \infty$ and use monotone convergence (5.6.12) to find that $\mu^c \to \mu$, and so

$$\limsup_{t\to\infty} \frac{1}{t} m(t) \leqslant \frac{1}{\mu}.$$

Combine this with (11) to obtain the result. ■

Exercises

(14) Planes land at Heathrow airport at the times of a renewal process with interarrival time distribution function F. Each plane contains a random number of people with

a given common distribution and finite mean. Assuming as much independence as usual, find an expression for the rate of arrival of passengers over a long time period.

(15) Let $Z_1, Z_2, \ldots$ be independent identically distributed random variables with mean 0 and finite variance σ^2, and let $T_n = \sum_{i=1}^n Z_i$. Let M be a finite stopping time with respect to the Z_i such that $\mathbf{E}(M) < \infty$. Show that $\text{var}(T_M) = \mathbf{E}(M)\sigma^2$.

(16) Show that $\mathbf{E}(T_{N(t)+k}) = \mu(m(t) + k)$ for all $k \geq 1$, but that it is not generally true that $\mathbf{E}(T_{N(t)}) = \mu m(t)$.

(17) Show that, using the usual notation, the family $\{N(t)/t: 0 \leq t < \infty\}$ is uniformly integrable. How might one make use of this observation?

10.3 Excess life

Suppose that we begin to observe a renewal process N at some epoch t of time. A certain number $N(t)$ of arrivals have occurred by then, and the next arrival will be the $(N(t) + 1)$th. That is to say, we have begun our observation at a point in the random interval

$$I_t = [T_{N(t)}, T_{N(t)+1}),$$

the endpoints of which are arrival times.

(1) **Definition.**

(a) The **excess lifetime** at t is $E(t) = T_{N(t)+1} - t$.
(b) The **current lifetime** (or **age**) at t is $C(t) = t - T_{N(t)}$.
(c) The **total lifetime** at t is $D(t) = E(t) + C(t) = X_{N(t)+1}$.

$E(t)$ is the time which elapses before the next arrival, $C(t)$ is the elapsed time since the last arrival (with the convention that the zeroth arrival occurs at time 0), and $D(t)$ is the length of the interarrival time which contains t (see Figure 8.1 for a diagram of these random variables).

(2) **Example. Waiting time paradox.** Suppose that N is a Poisson process with parameter λ. How big is $\mathbf{E}(E(t))$? Consider the two following lines of reasoning.

(A) N is a Markov chain, and so the distribution of $E(t)$ does not depend on the arrivals prior to time t. Thus $E(t)$ has the same mean as $E(0) = X_1$, and so $\mathbf{E}(E(t)) = \lambda^{-1}$.
(B) If t is fairly large, then on average it lies near the midpoint of the interarrival interval I_t which contains it. That is

$$\mathbf{E}(E(t)) \simeq \tfrac{1}{2}\mathbf{E}(T_{N(t)+1} - T_{N(t)}) = \tfrac{1}{2}\mathbf{E}(X_{N(t)+1}) = (2\lambda)^{-1}.$$

These arguments cannot both be correct. The reasoning of (B) is false, in that $X_{N(t)+1}$ does *not* have mean λ^{-1}; we have already observed this after (10.2.12). In fact, $X_{N(t)+1}$ is a very special interarrival time; longer intervals have a higher chance of catching t in their interiors than small intervals. In

Problem (10.5.5) we shall see that

$$\mathbf{E}(X_{N(t)+1}) = \frac{1}{\lambda}(2 - e^{-\lambda t}).$$

For this process, $E(t)$ and $C(t)$ are independent for any t; this property holds for no other renewal process with non-arithmetic interarrival times. ●

Now we find the distribution of $E(t)$.

(3) **Theorem.** *The distribution function of the excess life $E(t)$ is given by*

$$\mathbf{P}(E(t) \leqslant y) = F(t + y) - \int_0^t [1 - F(t + y - x)]\,dm(x).$$

Proof. Condition on X_1 in the usual way to obtain

$$\mathbf{P}(E(t) > y) = \mathbf{E}[\mathbf{P}(E(t) > y \mid X_1)].$$

However, you will see after a little thought that

$$\mathbf{P}(E(t) > y \mid X_1 = x) = \begin{cases} \mathbf{P}(E(t - x) > y) & \text{if } x \leqslant t \\ 0 & \text{if } t < x \leqslant t + y \\ 1 & \text{if } x > t + y \end{cases}$$

since $E(t) > y$ if and only if no arrivals occur in $(t, t + y]$. Thus

$$\begin{aligned} \mathbf{P}(E(t) > y) &= \int_0^\infty \mathbf{P}(E(t) > y \mid X_1 = x)\,dF(x) \\ &= \int_0^t \mathbf{P}(E(t - x) > y)\,dF(x) + \int_{t+y}^\infty dF(x). \end{aligned}$$

So

$$\mu(t) = \mathbf{P}(E(t) > y)$$

satisfies (10.1.10) with $H(t) = 1 - F(t + y)$; use (10.1.11) to see that

$$\mu(t) = 1 - F(t + y) + \int_0^t [1 - F(t + y - x)]\,dm(x)$$

as required. ■

(4) **Corollary.** *The distribution of the current life $C(t)$ is given by*

$$\mathbf{P}(C(t) \geqslant y) = \begin{cases} 0 & \textit{if } \; y > t \\ 1 - F(t) + \int_0^{t-y} [1 - F(t - x)]\,dm(x) & \textit{if } \; y \leqslant t. \end{cases}$$

Proof. $C(t) \geqslant y$ if and only if there are no arrivals in $(t-y, t]$. Thus

$$\mathbf{P}(C(t) \geqslant y) = \mathbf{P}(E(t-y) > y) \quad \text{if} \quad y \leqslant t$$

and the result follows from (3). ■

Might the renewal process N have stationary increments, in that the distribution of $N(t+s) - N(t)$ depends on s alone when $s \geqslant 0$? This is true for the Poisson process but fails in general. The reason is simple: generally speaking, the process of arrivals after time t depends on the age t of the process to date. When t is very large, however, it is plausible that the process may forget the date of its inception, thereby settling down into a stationary existence. This turns out to be the case. To show this asymptotic stationarity we need to demonstrate that the distribution of $N(t+s) - N(t)$ converges as $t \to \infty$. It is not difficult to see that this is equivalent to the assertion that the distribution of the excess life $E(t)$ settles down as $t \to \infty$, an easy consequence of the key renewal theorem (10.2.7) and (4.3.4).

(5) **Theorem.** *If X_1 is not arithmetic and $\mu = \mathbf{E}(X_1) < \infty$ then*

$$\mathbf{P}(E(t) \leqslant y) \to \frac{1}{\mu}\int_0^y [1 - F(x)]\,\mathrm{d}x \quad \textit{as} \quad t \to \infty.$$

Some difficulties arise if X_1 is arithmetic. For example, if the X_j are concentrated at the value 1 then, as $n \to \infty$,

$$\mathbf{P}(E(n+c) \leqslant \tfrac{1}{2}) \to \begin{cases} 1 & \text{if} \quad c = \frac{1}{2} \\ 0 & \text{if} \quad c = \frac{1}{4}. \end{cases}$$

Exercises

(6) Suppose that the distribution of the excess lifetime $E(t)$ does not depend on t. Show that the renewal process is a Poisson process.

(7) Show that the current and excess lifetime processes, $C(t)$ and $E(t)$, are Markov processes.

(8) Suppose that X_1 is non-arithmetic with finite mean μ.

(a) Show that $E(t)$ converges in distribution as $t \to \infty$, the limit distribution function being

$$\alpha(x) = \int_0^x \frac{1}{\mu}(1 - F(y))\,\mathrm{d}y.$$

(b) Show that the rth moment of this limit distribution is given by

$$\int_0^\infty x^r\,\mathrm{d}\alpha(x) = \frac{\mathbf{E}(X_1^{r+1})}{\mu(r+1)},$$

assuming that this is finite.

(c) Show that

$$\mathbf{E}(E(t)^r) = \mathbf{E}(\{(X_1 - t)^+\}^r) + \int_0^t h(t-x)\,dm(x)$$

for some suitable function h to be found, and deduce by the key renewal theorem that $\mathbf{E}(E(t)^r) \to \mathbf{E}(X_1^{r+1})/\{\mu(r+1)\}$ as $t \to \infty$, assuming this limit is finite.

(9) Find an expression for the mean value of the excess lifetime $E(t)$ conditional on the event that the current lifetime $C(t)$ equals x.

(10) Let $M(t) = N(t) + 1$, and suppose that X_1 has finite non-zero variance σ^2.

(a) Show that $\text{var}(T_{M(t)} - \mu M(t)) = \sigma^2(m(t) + 1)$.
(b) In the non-arithmetic case, show that $\text{var}(M(t))/t \to \sigma^2/\mu^3$ as $t \to \infty$.

10.4 Applications

Here are some examples of the ways in which renewal theory can be applied.

(1) **Example. Counters, and their dead periods.** In Section 6.8 we used an idealized Geiger counter which was able to register radioactive particles, irrespective of the rate of their arrival. In practice, after the detection of a particle such counters require a certain interval of time in order to complete its registration. These intervals are called 'dead periods'; during its dead periods the counter is locked and fails to register arriving particles. There are two common types of counter.

Type 1. Each detected arrival locks the counter for a period of time, possibly of random length, during which it ignores all arrivals.

Type 2. Each arrival locks the counter for a period of time, possibly of random length, irrespective of whether the counter is already locked or not. The counter registers only those arrivals that occur whilst it is unlocked.

Genuine Geiger counters are of Type 1; this case might also be used to model the process (8.3.1) of replacement of light bulbs in rented property when the landlord is either mean or lazy. We consider Type 1 counters briefly; Type 2 counters are harder to analyse, and so are left to the reader.

Suppose that arrivals occur as a renewal process N with renewal function m and interarrival times $X_1, X_2, \dots$ with distribution function F. Let L_n be the length of the dead period induced by the nth detected arrival. It is customary and convenient to suppose that an additional dead period, of length L_0, begins at time $t = 0$; the reason for this will soon be clear. We suppose that $\{L_n\}$ is a family of independent variables with the common distribution function F_L, where $F_L(0) = 0$. Let $\tilde{N}(t)$ be the number of arrivals detected by the Type 1 counter by time t. Then $\tilde{N}$ is a stochastic process

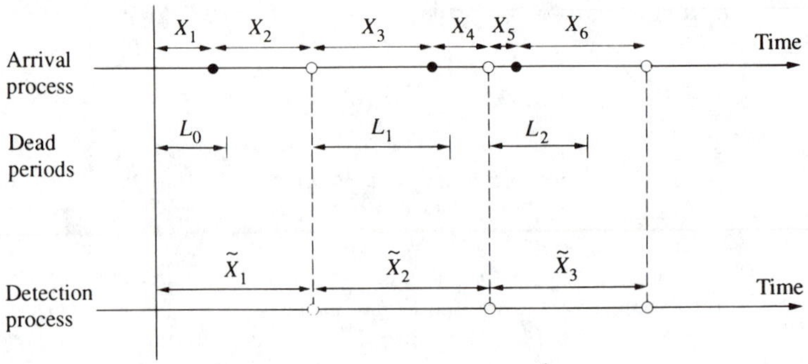

Fig. 10.1 Arrivals and detections by a Type I counter; ● indicates an undetected arrival, and ○ indicates a detected arrival.

with interarrival times $\tilde{X}_1, \tilde{X}_2, \ldots$ where

$$\tilde{X}_{n+1} = L_n + E_n$$

and E_n is the excess life of N at the end of the nth dead period (see Figure 10.1). $\tilde{N}$ is *not* in general a renewal process, because the $\tilde{X}$ need be neither independent nor identically distributed. In the very special case when N is a Poisson process the E_n are independent exponential variables and $\tilde{N}$ is a renewal process; it is easy to construct other examples for which this conclusion fails.

It is not difficult to find the elapsed time $\tilde{X}_1$ until the first detection. Condition on L_0 to obtain

$$\begin{aligned}\mathbf{P}(\tilde{X}_1 \leqslant x) &= \mathbf{E}(\mathbf{P}(\tilde{X}_1 \leqslant x \mid L_0)) \\ &= \int_0^x \mathbf{P}(L_0 + E_0 \leqslant x \mid L_0 = l)\, \mathrm{d}F_L(l).\end{aligned}$$

However, $E_0 = E(L_0)$, the excess lifetime of N at L_0, and so

(2) $$\mathbf{P}(\tilde{X}_1 \leqslant x) = \int_0^x \mathbf{P}(E(l) \leqslant x - l)\, \mathrm{d}F_L(l).$$

Now use (10.3.3) and the integral representation

$$m(t) = F(t) + \int_0^t F(t-x)\, \mathrm{d}m(x),$$

which follows from (10.1.11), to find that

(3) $$\begin{aligned}\mathbf{P}(\tilde{X}_1 \leqslant x) &= \int_0^x \left(\int_l^x [1 - F(x-y)]\, \mathrm{d}m(y) \right) \mathrm{d}F_L(l) \\ &= \int_0^x [1 - F(x-y)] F_L(y)\, \mathrm{d}m(y).\end{aligned}$$

If N is a Poisson process with intensity λ, then (2) becomes

$$\mathbf{P}(\tilde{X}_1 \leqslant x) = \int_0^x (1 - e^{-\lambda(x-l)})\, dF_L(l).$$

$\tilde{N}$ is now a renewal process, and this equation describes the common distribution of the interarrival times.

If the counter is registering the arrival of radioactive particles, then we may seek an estimate $\hat{\lambda}$ of the unknown emission rate λ of the source based upon our knowledge of the mean length $\mathbf{E}(L)$ of a dead period and the counter reading $\tilde{N}(t)$. Assume that the particles arrive like a Poisson process. Let

$$\gamma = \frac{\tilde{N}(t)}{t}$$

be the density of observed particles. Then

$$\gamma \simeq \frac{1}{\mathbf{E}(\tilde{X}_1)} = \frac{1}{\mathbf{E}(L) + \lambda^{-1}} \quad \text{for large } t,$$

and so

$$\lambda \simeq \frac{\gamma}{1 - \gamma\mathbf{E}(L)}.$$

We may estimate λ by $\hat{\lambda}$, given by

$$\hat{\lambda} = \frac{\gamma}{1 - \gamma\mathbf{E}(L)}.$$

●

(4) **Example. Alternating renewal process.** A machine breaks down repeatedly. After the nth breakdown the repairman takes a period of time, length Y_n, to repair it; subsequently the machine runs for a period of length Z_n before it breaks down for the next time. We assume that the Y_n and the Z_n are independent of each other, the Y_n having common distribution function F_Y and the Z_n having common distribution function F_Z. Suppose that the machine was installed at time $t = 0$. Let $N(t)$ be the number of completed repairs by time t (see Figure 10.2). N is a renewal process with interarrival times $X_1, X_2, \ldots$ given by

$$X_n = Z_{n-1} + Y_n$$

with distribution function

$$F(x) = \int_0^x F_Y(x - y)\, dF_Z(y).$$

Let $p(t)$ be the probability that the machine is working at time t.

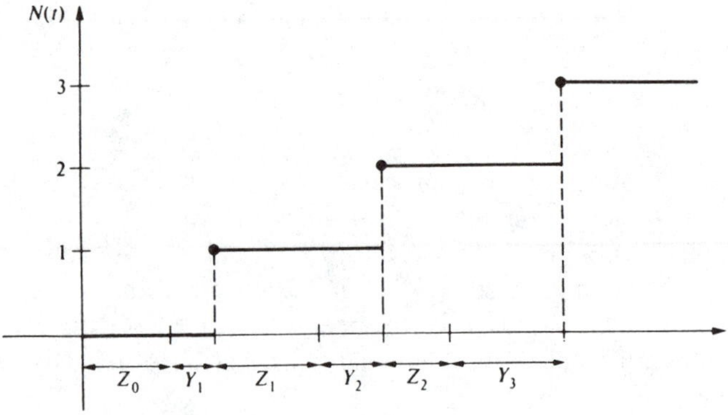

Fig. 10.2 An alternating renewal process.

(5) **Lemma.**

$$p(t) = 1 - F_Z(t) + \int_0^t p(t-x)\,\mathrm{d}F(x)$$

and hence

$$p(t) = 1 - F_Z(t) + \int_0^t [1 - F_Z(t-x)]\,\mathrm{d}m(x)$$

where m is the renewal function of N.

Proof. The probability that the machine is on at time t satisfies

$$\begin{aligned} p(t) &= \mathbf{P}(\text{on at } t) \\ &= \mathbf{P}(Z_0 > t) + \mathbf{P}(\text{on at } t,\ Z_0 \leqslant t) \\ &= \mathbf{P}(Z_0 > t) + \mathbf{E}[\mathbf{P}(\text{on at } t,\ Z_0 \leqslant t \mid X_1)] \\ &= \mathbf{P}(Z_0 > t) + \int_0^t \mathbf{P}(\text{on at } t \mid X_1 = x)\,\mathrm{d}F(x) \\ &\qquad\qquad \text{since } \mathbf{P}(\text{on at } t,\ Z_0 \leqslant t \mid X_1 > t) = 0 \\ &= \mathbf{P}(Z_0 > t) + \int_0^t p(t-x)\,\mathrm{d}F(x). \end{aligned}$$

Now use (10.1.11). ■

(6) **Corollary.** *If X_1 is not arithmetic then*

$$p(t) \to \frac{1}{1+\rho} \quad \text{as} \quad t \to \infty$$

where $\rho = \mathbf{E}(Y)/\mathbf{E}(Z)$ is the ratio of the mean lengths of a typical repair period and a typical working period.

Proof. Use the key renewal theorem (10.2.7). ■●

(7) **Example. Superposition of renewal processes.** Suppose that a room is illuminated by two lights, the bulbs of which fail independently of each other. On failure, they are replaced immediately. Let N_1 and N_2 be the renewal processes describing the occurrences of bulb failures in the first and second lights respectively, and suppose that these are independent processes with the same interarrival time distribution function F. Let $\tilde{N}$ be the superposition of these two processes; that is, $\tilde{N}(t) = N_1(t) + N_2(t)$ is the total number of failures by time t. In general $\tilde{N}$ is not a renewal process. Let us assume for the sake of simplicity that the interarrival times of N_1 and N_2 are not arithmetic.

(8) **Theorem.** *$\tilde{N}$ is a renewal process if and only if N_1 and N_2 are Poisson processes.*

Proof. It is easy to see that $\tilde{N}$ is a Poisson process with intensity 2λ whenever N_1 and N_2 are Poisson processes with intensity λ. Conversely, suppose that $\tilde{N}$ is a renewal process, and write $\{X_n(1)\}$, $\{X_n(2)\}$, and $\{\tilde{X}_n\}$ for the interarrival times of N_1, N_2, and $\tilde{N}$ respectively. Clearly

$$\tilde{X}_1 = \min\{X_1(1), X_1(2)\}$$

and so the distribution function $\tilde{F}$ of $\tilde{X}_1$ satisfies

(9)
$$1 - \tilde{F}(y) = [1 - F(y)]^2.$$

Let $E_1(t)$, $E_2(t)$, and $\tilde{E}(t)$ denote the excess lifetimes of N_1, N_2, and $\tilde{N}$ respectively at time t. Clearly,

$$\tilde{E}(t) = \min\{E_1(t), E_2(t)\},$$

and so

$$\mathbf{P}(\tilde{E}(t) > y) = \mathbf{P}(E_1(t) > y)^2.$$

Let $t \to \infty$ and use (10.3.5) to obtain

(10)
$$\frac{1}{\tilde{\mu}}\int_y^\infty [1 - \tilde{F}(x)]\,dx = \frac{1}{\mu^2}\left(\int_y^\infty [1 - F(x)]\,dx\right)^2$$

where $\tilde{\mu} = \mathbf{E}(\tilde{X}_1)$ and $\mu = \mathbf{E}(X_1(1))$. Differentiate (10) and use (9) to obtain

$$\begin{aligned}\frac{1}{\tilde{\mu}}[1 - \tilde{F}(y)] &= \frac{2}{\mu^2}[1 - F(y)]\int_y^\infty [1 - F(x)]\,dx \\ &= \frac{1}{\tilde{\mu}}[1 - F(y)]^2\end{aligned}$$

(this step needs further justification if F is not continuous). Thus

$$1 - F(y) = \frac{2\tilde{\mu}}{\mu^2} \int_y^\infty [1 - F(x)]\,\mathrm{d}x$$

which is an integral equation with solution

$$F(y) = 1 - \exp\left(-\frac{2\tilde{\mu}}{\mu^2} y\right).$$

■●

(11) **Example. Delayed renewal processes.** The Markov chain example (8.3.2) indicates that it is sometimes appropriate to allow the first interarrival time X_1 to have a distribution which differs from the shared distribution of $X_2, X_3, \ldots$.

(12) **Definition.** Let $X_1, X_2, \ldots$ be independent positive variables such that $X_2, X_3, \ldots$ have the same distribution. Let

$$T_0 = 0, \qquad T_n = \sum_1^n X_i, \qquad N^D(t) = \max\{n: T_n \leqslant t\}.$$

Then N^D is called a **delayed** (or **modified**) **renewal process**.

Another example of a delayed renewal process is provided by a variation of the Type 1 counter of (1) when particles arrive like Poisson process. It was convenient there to assume that the life of the counter began with a dead period in order that the process $\tilde{N}$ of detections be a renewal process. In the absence of this assumption $\tilde{N}$ is a delayed renewal process. The theory of delayed renewal processes is very similar to that of ordinary renewal processes and we do not explore it in detail. The renewal equation (10.1.9) becomes

$$m^D(t) = F^D(t) + \int_0^t m(t-x)\,\mathrm{d}F^D(x)$$

where F^D is the distribution function of X_1 and m is the renewal function of an ordinary renewal process N whose interarrival times are $X_2, X_3, \ldots$. It is left to the reader to check that

(13) $$m^D(t) = F^D(t) + \int_0^t m^D(t-x)\,\mathrm{d}F(x)$$

and

(14) $$m^D(t) = \sum_{k=1}^\infty F_k^D(t)$$

where F_k^D is the distribution function of $T_k = X_1 + \cdots + X_k$ and F is the shared distribution function of $X_2, X_3, \ldots$.

With our knowledge of the properties of m, it is not too hard to show that m^D satisfies the renewal theorems. Write μ for $\mathbf{E}(X_2)$.

(15) **Theorem.** $\dfrac{1}{t}m^D(t) \to \dfrac{1}{\mu}$ *as* $t \to \infty$.

If X_2 *is not arithmetic then*

(16) $$m^D(t+h) - m^D(t) \to \frac{h}{\mu} \quad \textit{as} \quad t \to \infty \quad \textit{for any} \quad h.$$

If X_2 *is arithmetic with span* λ *then* (16) *remains true whenever* h *is a multiple of* λ.

There is an important special case for the distribution function F^D.

(17) **Theorem.** N^D *has stationary increments if and only if*

(18) $$F^D(y) = \frac{1}{\mu}\int_0^y [1 - F(x)]\,\mathrm{d}x.$$

If F^D is given by (18), then N^D is called a *stationary* (or *equilibrium*) *renewal process*. We should recognize (18) as the asymptotic distribution (10.3.5) of the excess lifetime of the ordinary renewal process N. So the result of (17) is no surprise since N^D starts off with this 'equilibrium' distribution. We shall see that in this case $m^D(t) = t/\mu$ for all $t \geqslant 0$.

Proof of (17). Suppose that N^D has stationary increments. Then

$$\begin{aligned} m^D(s+t) &= \mathbf{E}([N^D(s+t) - N^D(s)] + N^D(s)) \\ &= \mathbf{E}(N^D(t)) + \mathbf{E}(N^D(s)) \\ &= m^D(t) + m^D(s). \end{aligned}$$

By monotonicity,

$$m^D(t) = ct$$

for some $c > 0$. Substitute into (13) to obtain

$$F^D(t) = c\int_0^t [1 - F(x)]\,\mathrm{d}x$$

and let $t \to \infty$ to obtain $c = 1/\mu$.

Conversely, suppose that F^D is given by (18). Substitute (18) into (13) and use the method of Laplace–Stieltjes transforms to deduce that

(19) $$m^D(t) = \frac{t}{\mu}.$$

Now, N^D has stationary increments if and only if the distribution of $E^D(t)$, the excess lifetime of N^D at t, does not depend on t. But

$$\begin{aligned}\mathbf{P}(E^D(t) > y) &= \sum_{k=0}^{\infty} \mathbf{P}(E^D(t) > y,\, N^D(t) = k) \\ &= \mathbf{P}(E^D(t) > y,\, N^D(t) = 0) \\ &\quad + \sum_{k=1}^{\infty} \int_0^t \mathbf{P}(E^D(t) > y,\, N^D(t) = k \mid T_k = x)\, \mathrm{d}F_k^D(x) \\ &= 1 - F^D(t+y) + \int_0^t [1 - F(t+y-x)]\, \mathrm{d}\left(\sum_{k=1}^{\infty} F_k^D(x)\right) \\ &= 1 - F^D(t+y) + \int_0^t [1 - F(t+y-x)]\, \mathrm{d}m^D(x)\end{aligned}$$

from (14). Now substitute (18) and (19) into this equation to obtain the result. ■●

(20) **Example. Markov chains.** Let $Y = \{Y_n : n \geq 0\}$ be a discrete-time Markov chain with countable state space S. At last we are able to prove the ergodic theorem (6.4.21) for Y, as a consequence of the renewal theorem (16). Suppose that $Y_0 = i$ and let j be an aperiodic state. We can suppose that j is persistent, since the result follows from (6.2.5) if j is transient. Observe the sequence of visits of Y to the state j. That is, let

$$T_0 = 0, \qquad T_{n+1} = \min\{k > T_n : Y_k = j\} \quad \text{for} \quad n \geq 0.$$

T_1 may equal $+\infty$; actually $\mathbf{P}(T_1 < \infty) = f_{ij}$. Conditional on $\{T_1 < \infty\}$, the inter-visit times

$$X_n = T_n - T_{n-1} \quad \text{for} \quad n \geq 2$$

are independent and identically distributed; following (8.3.2),

$$N^D(t) = \max\{n : T_n \leq t\}$$

defines a delayed renewal process with a renewal function

$$m^D(t) = \sum_{n=1}^{t} p_{ij}(n) \quad \text{for integral } t.$$

Now, adapt (16) to deal with the possibility that the first interarrival time $X_1 = T_1$ equals infinity, to obtain

$$p_{ij}(n) = m^D(n) - m^D(n-1) \to \frac{f_{ij}}{\mu_j} \quad \text{as} \quad n \to \infty$$

where μ_j is the mean recurrence time of j. ●

(21) **Example. Sketch proof of the renewal theorem.** There is an elegant proof of the renewal theorem (10.2.5) which proceeds by coupling the renewal process

N to an independent delayed renewal process N^D; here is a sketch of the method. Let N be a renewal process with interarrival times $\{X_n\}$ and interarrival time distribution function F with mean μ. We suppose that F is non-arithmetic; the proof in the arithmetic case is easier. Let N^D be a stationary renewal process (see (17)) with interarrival times $\{Y_n\}$, where Y_1 has distribution function

$$F^D(y) = \frac{1}{\mu}\int_0^y [1 - F(x)]\,\mathrm{d}x$$

and $Y_2, Y_3, \ldots$ have distribution function F; suppose further that the X_i are independent of the Y_i. The idea of the proof is as follows.

(a) For any $\delta > 0$, there must exist an arrival time $T_a = \sum_1^a X_i$ of N and an arrival time $T_b^D = \sum_1^b Y_i$ of N^D such that

$$|T_a - T_b^D| < \delta.$$

(b) If we replace $X_{a+1}, X_{a+2}, \ldots$ by $Y_{b+1}, Y_{b+2}, \ldots$ in the construction of N, then the distributional properties of N are unchanged since all these variables are identically distributed.

(c) But the Y_i are the interarrival times of a stationary renewal process, for which (19) holds; this implies that $m^D(t+h) - m^D(t) = h/\mu$ for all t, h. However, $m(t)$ and $m^D(t)$ are nearly the same for large t, by the previous remarks, and so

$$m(t+h) - m(t) \simeq h/\mu \quad \text{for large } t.$$

The details of the proof are slightly too difficult for inclusion here (see Lindvall 1977). ●

(22) **Example. Age-dependent branching process.** Consider the branching process $Z(t)$ of Section 5.5 in which each individual lives for a random length of time before splitting into its offspring. We have seen that the expected number

$$m(t) = \mathbf{E}(Z(t))$$

of individuals alive at time t satisfies the integral equation (5.5.4):

(23)
$$m(t) = \nu\int_0^t m(t-x)\,\mathrm{d}F_T(x) + \int_t^\infty \mathrm{d}F_T(x)$$

where F_T is the distribution function of a typical lifetime and ν is the mean family size; we assume that F_T is continuous for simplicity. We have changed some of the notation of (5.5.4) for obvious reasons. Equation (23) reminds us of the renewal-type equation (10.1.10) but the factor ν must be assimilated before the solution can be found using the method of (10.1.11). This presents few difficulties in the supercritical case. If $\nu > 1$ then there is a unique $\beta > 0$ such that

$$F_T^*(\beta) = \int_0^\infty e^{-\beta x}\, dF_T(x) = \frac{1}{\nu};$$

this holds because the Laplace–Stieltjes transform $F_T^*(\theta)$ is a strictly decreasing continuous function of θ with

$$F_T^*(0) = 1, \qquad F_T^*(\theta) \to 0 \quad \text{as} \quad \theta \to \infty.$$

Now, with this choice of β, define

$$\tilde{F}(t) = \nu \int_0^t e^{-\beta x}\, dF_T(x)$$

$$g(t) = e^{-\beta t} m(t).$$

Multiply throughout (23) by $e^{-\beta t}$ to obtain

(24) $$g(t) = h(t) + \int_0^t g(t - x)\, d\tilde{F}(x)$$

where

$$h(t) = e^{-\beta t}[1 - F_T(t)];$$

(24) has the same general form as (10.1.10), since our choice for β ensures that $\tilde{F}$ is the distribution function of a positive random variable. ●

10.5 Problems

In the absence of indications to the contrary, $\{X_n : n \geq 1\}$ denotes the interarrival times of either a renewal process N or a delayed renewal process N^D. In either case F^D and F are the distribution functions of X_1 and X_2 respectively, though $F^D \neq F$ only if the renewal process is delayed. We write $\mu = \mathbf{E}(X_2)$, and shall usually assume that $0 < \mu < \infty$. The quantities m and m^D denote the renewal functions of N and N^D. We write $T_n = \sum_{i=1}^n X_i$, the time of the nth arrival.

1. (a) Show that $\mathbf{P}(N(t) \to \infty \text{ as } t \to \infty) = 1$.
(b) Show that $m(t) < \infty$ if $\mu \neq 0$.
(c) More generally show that, for all $k > 0$, $\mathbf{E}(N(t)^k) < \infty$ if $\mu \neq 0$.

2. Let $v(t) = \mathbf{E}(N(t)^2)$. Show that

$$v(t) = m(t) + 2\int_0^t m(t - s)\, dm(s).$$

Find $v(t)$ when N is a Poisson process.

3. Suppose that $\sigma^2 = \text{var}(X_1) > 0$. Show that the renewal process N satisfies

$$\frac{N(t) - (t/\mu)}{\sqrt{(t\sigma^2/\mu^3)}} \xrightarrow{D} N(0, 1), \quad \text{as} \quad t \to \infty.$$

4. Find the asymptotic distribution of the current life $C(t)$ of N as $t \to \infty$ when X_1 is not arithmetic.

5. Let N be a Poisson process with intensity λ. Show that the total life $D(t)$ at time t has distribution function $\mathbf{P}(D(t) \leqslant x) = 1 - (1 + \lambda \min\{t, x\}) e^{-\lambda x}$ for $x \geqslant 0$. Deduce that $\mathbf{E}(D(t)) = (2 - e^{-\lambda t})/\lambda$. This is thc 'inspection paradox'.

6. A Type 1 counter records the arrivals of radioactive particles. Suppose that the arrival process is Poisson with intensity λ, and that the counter is locked for a dead period of fixed length T after each detected arrival. Show that the detection process $\tilde{N}$ is a renewal process with interarrival time distribution $\tilde{F}(x) = 1 - e^{-\lambda(x-T)}$ if $x \geqslant T$. Find an expression for $\mathbf{P}(\tilde{N}(t) \geqslant k)$.

7. Particles arrive at a Type 1 counter in the manner of a renewal process N; each detected arrival locks the counter for a dead period of random positive length. Show that

$$\mathbf{P}(\tilde{X}_1 \leqslant x) = \int_0^x \{1 - F(x-y)\} F_L(y)\, dm(y)$$

where F_L is the distribution function of a typical dead period.

8. (a) Show that $m(t) = \frac{1}{2}\lambda t - \frac{1}{4}(1 - e^{-2\lambda t})$ if the interarrival times have the gamma distribution $\Gamma(\lambda, 2)$.
(b) Radioactive particles arrive like a Poisson process, intensity λ, at a counter. The counter fails to register the nth arrival whenever n is odd but suffers no dead periods. Find the renewal function $\tilde{m}$ of the detection process $\tilde{N}$.

9. Show that Poisson processes are the only renewal processes with non-arithmetic interarrival times having the property that the excess lifetime $E(t)$ and the current lifetime $C(t)$ are independent for each choice of t.

10. Let N_1 be a Poisson process, and let N_2 be a renewal process which is independent of N_1 with non-arithmetic interarrival times having finite mean. Show that $N(t) = N_1(t) + N_2(t)$ is a renewal process if and only if N_2 is a Poisson process.

11. Let N be a renewal process, and suppose that F is non-arithmetic and that $\sigma^2 = \text{var}(X_1) < \infty$. Use the properties of the moment generating function $F^*(-\theta)$ of X_1 to deduce the formal expansion

$$m^*(\theta) = \frac{1}{\theta\mu} + \frac{\sigma^2 - \mu^2}{2\mu^2} + o(1) \quad \text{as} \quad \theta \to 0.$$

Invert this Laplace–Stieltjes transform formally to obtain

$$m(t) = \frac{t}{\mu} + \frac{\sigma^2 - \mu^2}{2\mu^2} + o(1) \quad \text{as} \quad t \to \infty.$$

Prove this rigorously by showing that

$$m(t) = \frac{t}{\mu} - F_E(t) + \int_0^t \{1 - F_E(t-x)\}\, dm(x),$$

where F_E is the asymptotic distribution function of the excess lifetime (see Exercise (10.3.8)), and applying the key renewal theorem. Compare the result with the renewal theorems.

12. Show that the renewal function m^D of a delayed renewal process satisfies

$$m^D(t) = F^D(t) + \int_0^t m^D(t-x)\,dF(x).$$

Show that $v^D(t) = \mathbf{E}(N^D(t)^2)$ satisfies

$$v^D(t) = m^D(t) + 2\int_0^t m^D(t-x)\,dm(x)$$

where m is the renewal function of the renewal process with interarrival times $X_2, X_3, \dots$.

13. Let $m(t)$ be the mean number of living individuals at time t in an age-dependent branching process with exponential lifetimes, parameter λ, and mean family size ν (>1). Prove that $m(t) = I\,e^{(\nu-1)\lambda t}$ where I is the number of initial members.

14. **Alternating renewal process.** The interarrival times of this process are $Z_0, Y_1, Z_1, Y_2, \dots$, where the Y_i and Z_i are independent with respective common moment generating functions M_Y and M_Z. Let $p(t)$ be the probability that the epoch t of time lies in an interval of type Z. Show that the Laplace–Stieltjes transform p^* of p satisfies

$$p^*(\theta) = \frac{1 - M_Z(-\theta)}{1 - M_Y(-\theta)M_Z(-\theta)}.$$

15. **Type 2 counters.** Particles are detected by a Type 2 counter of the following sort. The incoming particles constitute a Poisson process with intensity λ. The jth particle locks the counter for a length Y_j of time, and annuls any after-effect of its predecessors. Suppose that $Y_1, Y_2, \dots$ are independent of each other and of the Poisson process, each having distribution function G. The counter is unlocked at time 0.

Let L be the (maximal) length of the first interval of time during which the counter is locked. Show that $H(t) = \mathbf{P}(L > t)$ satisfies

$$H(t) = e^{-\lambda t}(1 - G(t)) + \int_0^t H(t-x)(1 - G(x))\lambda e^{-\lambda x}\,dx.$$

Solve for H in terms of G, and evaluate the ensuing expression in the case $G(x) = 1 - e^{-\mu x}$ where $\mu > 0$.

16. **Thinning.** Consider a renewal process N, and suppose that each arrival is 'overlooked' with probability q, independently of all other arrivals. Let $M(t)$ be the number of arrivals which are detected up to time t/p where $p = 1 - q$.

(a) Show that M is a renewal process whose interarrival time distribution function F_p is given by $F_p(x) = \sum_{r=1}^{\infty} pq^{r-1}F_r(x/p)$, where F_n is the distribution function of the time of the nth arrival in the original process N.

(b) Find the characteristic function of F_p in terms of that of F, and use the continuity theorem to show that, as $p \downarrow 0$, $F_p(s) \to 1 - e^{-s/\mu}$ for $s > 0$, so long as the interarrival times in the original process have finite mean μ. Interpret!

(c) Suppose that $p < 1$, and M and N are processes with the same fdds. Show that N is a Poisson process.

17. (a) A word processor has 100 different keys and a monkey is tapping them (uniformly) at random. Assuming no power failure, use the elementary renewal theorem to find the expected number of keys tapped until the first appearance of the sequence 'W. Shakespeare'.
Answer the same question for the sequence 'omo'.
(b) A coin comes up heads with probability p on each toss. Find the mean numbers of tosses until the first appearances of the sequences (i) HHH, and (ii) HTH.

18. Let N be a stationary renewal process. Let s be a fixed positive real number, and define $X(t) = N(s + t) - N(t)$ for $t \geqslant 0$. Show that X is a strongly stationary process.

11 Queues

11.1 Single-server queues

We consider only the simpler queueing systems which are described in Section 8.4. With each queue we can associate two sequences $\{X_n: n \geqslant 1\}$ and $\{S_n: n \geqslant 1\}$ of independent positive random variables, the X_n being interarrival times with common distribution function F_X and the S_n being service times with common distribution function F_S. We assume that customers arrive like a renewal process with interarrival times $\{X_n\}$, the nth customer arriving at

$$T_n = X_1 + \cdots + X_n.$$

Each arriving customer joins the line of customers who are waiting for the attention of the *single* server. When the nth customer reaches the head of this line he is served for a period of length S_n, after which he leaves the system. Let $Q(t)$ be the number of waiting customers at time t (including any customer whose service is in progress at t); clearly $Q(O) = 0$. Then $Q = \{Q(t): t \geqslant 0\}$ is a random process whose fdds are specified by the distribution functions F_X and F_S. We seek information about Q. For example, we may ask:

(a) When is Q a Markov chain, or when does Q contain an imbedded Markov chain?
(b) When is Q asymptotically stationary, in the sense that the distribution of $Q(t)$ settles down as $t \to \infty$?
(c) When does the queue length grow beyond all bounds, in that the server is not able to cope with the high density of arrivals?

The answers to these and other similar questions take the form of applying conditions to F_X and F_S; the style of the analysis may depend on the types of these distribution functions. With this in mind, it is convenient to use a notation for the queue system which incorporates information about F_X and F_S. The most common notation scheme describes each system by a triple $A/B/s$, where A describes F_X, B describes F_S, and s is the number of servers ($s = 1$ always, for us). Typically, A and B may each be one of the following:

$D(d) \equiv$ almost surely concentrated at the value d (D for 'deterministic')
$M(\lambda) \equiv$ exponential, parameter λ (M for 'Markovian')
$\Gamma(\lambda, k) \equiv$ gamma, parameters λ and k
$G \equiv$ some general distribution, fixed but unspecified.

(1) **Example.** $\boldsymbol{M(\lambda)/M(\mu)/1}$. Interarrival times are exponential with parameter λ and service times are exponential with parameter μ. Thus, customers arrive like a Poisson process with intensity λ. $Q = \{Q(t)\}$ is a continuous-time Markov chain with state space $\{0, 1, 2, \ldots\}$; this follows from the lack-of-memory property of the exponential distribution. Furthermore, such systems are the *only* systems whose queue lengths are homogeneous Markov chains. Why is this? ●

(2) **Example.** $\boldsymbol{M(\lambda)/D(1)/1}$. Customers arrive like a Poisson process, and each requires a service time of constant length 1. Q is not a Markov chain, but we shall see later that there exists an imbedded discrete-time Markov chain $\{Q_n: n \geqslant 0\}$ whose properties provide information about Q. ●

(3) **Example.** $\boldsymbol{G/G/1}$. In this case we have no special information about F_X or F_S. Some authors denote this system by $GI/G/1$, reserving the title $G/G/1$ to denote a more complicated system in which the interarrival times may not be independent. ●

The notation $M(\lambda)$ is sometimes abbreviated to M alone. Thus Example (1) becomes $M/M/1$; this rather unfortunate abbreviation does *not* imply that F_X and F_S are the same. A similar remark holds for systems described as $G/G/1$.

Broadly speaking, there are two types of statement to be made about Q:

(a) 'time-dependent' statements, which contain information about the queue for finite values of t;

(b) 'limiting' results, which discuss the asymptotic properties of the queue as $t \to \infty$. These include conditions for the queue length to grow beyond all bounds.

Statements of the former type are most easily made about $M(\lambda)/M(\mu)/1$, since this is the only Markovian system; such conclusions are more elusive for more general systems, and we shall generally content ourselves with the asymptotic properties of such queues.

In the subsequent sections we explore the systems $M/M/1$, $M/G/1$, $G/M/1$, and $G/G/1$, in that order. We present these cases roughly in order of increasing difficulty for the readers' convenience. This is not really satisfactory, since we are progressing from the specific to the general, so we should like to stress that queues with Markovian characteristics are very special systems and that their properties do not always indicate features of more general systems.

Here is a final piece of notation.

(4) **Definition.** The **traffic density** ρ of a queue is defined as

$$\rho = \frac{\mathbf{E}(S)}{\mathbf{E}(X)},$$

the ratio of the mean of a typical service time to the mean of a typical interarrival time.

We assume throughout that neither $\mathbf{E}(S)$ nor $\mathbf{E}(X)$ takes the value zero or infinity.

We shall see that queues behave in qualitatively different manners depending on whether $\rho < 1$ or $\rho > 1$. In the latter case, service times exceed interarrival times, on average, and the queue length grows beyond all bounds with probability 1; in the former case, the queue attains an equilibrium as $t \to \infty$. It is a remarkable conclusion that the threshold between instability and stability depends on the mean values of F_X and F_S alone.

11.2 *M*/*M*/1

The queue $M(\lambda)/M(\mu)/1$ is very special in that Q is a continuous-time Markov chain. Furthermore, reference to (6.11.1) reminds us that Q is a birth–death process with birth and death rates given by

$$\lambda_n = \lambda \quad \text{for all } n$$

$$\mu_n = \begin{cases} \mu & \text{if } \; n \geqslant 1 \\ 0 & \text{if } \; n = 0. \end{cases}$$

The probabilities $p_n(t) = \mathbf{P}(Q(t) = n)$ satisfy the Kolmogorov forward equations in the usual way:

(1) $$\frac{dp_n}{dt} = \lambda p_{n-1}(t) - (\lambda + \mu)p_n(t) + \mu p_{n+1}(t) \quad \text{for} \quad n \geqslant 1$$

(2) $$\frac{dp_0}{dt} = -\lambda p_0(t) + \mu p_1(t)$$

subject to the boundary conditions $p_n(0) = \delta_{0n}$, the Kronecker delta. It is a bit tricky to solve these equations, but routine methods provide the answer after some manipulation. There are at least two possible routes: either use generating functions or use Laplace transforms with respect to t. We proceed in the latter way here. Let

$$\hat{p}_n(\theta) = \int_0^\infty e^{-\theta t} p_n(t)\, dt$$

be the Laplace transform† of p_n.

(3) **Theorem.** $\hat{p}_n(\theta) = \dfrac{1}{\theta}[1 - \alpha(\theta)][\alpha(\theta)]^n$ *where*

(4) $$\alpha(\theta) = \frac{(\lambda + \mu + \theta) - [(\lambda + \mu + \theta)^2 - 4\lambda\mu]^{\frac{1}{2}}}{2\mu}.$$

† Do not confuse $\hat{p}_n$ with the Laplace–Stieltjes transform $p_n^*(\theta) = \int_0^\infty e^{-\theta t}\, dp_n(t)$.

The actual probabilities $p_n(t)$ can be deduced in terms of Bessel functions. It turns out that

$$p_n(t) = J_n(t) - J_{n+1}(t)$$

where

$$J_n(t) = \int_0^t (\lambda/\mu)^{\frac{1}{2}n} n s^{-1} \, \mathrm{e}^{-s(\lambda+\mu)} I_n(2s(\lambda\mu)^{\frac{1}{2}}) \, \mathrm{d}s$$

and $I_n(x)$ is a modified Bessel function (see Feller 1971, p. 482), defined to be the coefficient of z^n in the power series expansion of $\exp[\frac{1}{2}x(z + z^{-1})]$. See Exercise (15) for another representation of $p_n(t)$.

Proof. Transform (1) and (2) to obtain

(5) $$\mu \hat{p}_{n+1} - (\lambda + \mu + \theta)\hat{p}_n + \lambda \hat{p}_{n-1} = 0 \quad \text{for} \quad n \geqslant 1$$

(6) $$\mu \hat{p}_1 - (\lambda + \theta)\hat{p}_0 = -1$$

where we have used the fact (see equation (14) of Appendix I) that

$$\int_0^\infty \mathrm{e}^{-\theta t} \frac{\mathrm{d}p_n}{\mathrm{d}t} \, \mathrm{d}t = \theta \hat{p}_n - \delta_{0n}, \quad \text{for all } n.$$

Equation (5) is an ordinary difference equation, and standard techniques (see Appendix I) show that it has a unique solution which is bounded as $\theta \to \infty$ and which is given by

(7) $$\hat{p}_n(\theta) = \hat{p}_0(\theta)[\alpha(\theta)]^n$$

where α is given by (4). Substitute (7) into (6) to deduce that

$$\hat{p}_0(\theta) = \frac{1}{\theta}[1 - \alpha(\theta)]$$

and the proof is complete. Alternatively, $\hat{p}_0(\theta)$ may be calculated from the fact that $\sum_n p_n(t) = 1$, implying that $\sum_n \hat{p}_n(\theta) = \theta^{-1}$. ■

The asymptotic behaviour of $Q(t)$ as $t \to \infty$ is deducible from (3), but more direct methods yield the answer more quickly. Remember that Q is a Markov chain.

(8) **Theorem.** *Let $\rho = \lambda/\mu$ be the traffic density.*

(a) *If $\rho < 1$ then*

$$\mathbf{P}(Q(t) = n) \to (1 - \rho)\rho^n = \pi_n, \quad \textit{for} \quad n \geqslant 0$$

where $\boldsymbol{\pi}$ is the unique stationary distribution.

(b) *If $\rho \geqslant 1$ then there is no stationary distribution, and*

$$\mathbf{P}(Q(t) = n) \to 0 \quad \textit{for all } n.$$

This result is very natural. It asserts that the queue settles down into equilibrium if and only if interarrival times exceed service times on average. We shall see later that if $\rho > 1$ then $\mathbf{P}(Q(t) \to \infty \text{ as } t \to \infty) = 1$, whilst if $\rho = 1$ then the queue length experiences wild oscillations with no reasonable bound on their magnitudes.

Proof. Q is an irreducible chain. Let us try to find a stationary distribution. Let $t \to \infty$ in (1) and (2) to find that the mass function $\boldsymbol{\pi}$ is a stationary distribution if and only if

(9) $$\pi_{n+1} - (1+\rho)\pi_n + \rho\pi_{n-1} = 0 \quad \text{for} \quad n \geqslant 1$$
$$\pi_1 - \rho\pi_0 = 0.$$

(The operation of taking limits is justifiable by (6.9.20) and the uniformity of Q.) The general solution to (9) is

$$\pi_n = \begin{cases} A + B\rho^n & \text{if} \quad \rho \neq 1 \\ A + Bn & \text{if} \quad \rho = 1 \end{cases}$$

where A and B are arbitrary constants. Thus the only bounded solution to (9) with bounded sum is

$$\pi_n = \begin{cases} B\rho^n & \text{if} \quad \rho < 1 \\ 0 & \text{if} \quad \rho \geqslant 1. \end{cases}$$

Hence, if $\rho < 1$,

$$\pi_n = (1-\rho)\rho^n$$

is a stationary distribution, whilst if $\rho \geqslant 1$ then there exists no stationary distribution. By Theorem (6.9.21), the proof is complete. ■

There is an alternative derivation of the asymptotic behaviour (8) of Q, which has other consequences also. Let U_n be the epoch of time at which the nth change in Q occurs. That is to say

$$U_0 = 0, \qquad U_{n+1} = \inf\{t > U_n : Q(t) \neq Q(U_n+)\}.$$

Now let $Q_n = Q(U_n+)$ be the number of waiting customers immediately after the nth change in Q. Clearly $\{Q_n : n \geqslant 0\}$ is a random walk on the non-negative integers, with

$$Q_{n+1} = \begin{cases} Q_n + 1 & \text{with probability} \quad \dfrac{\lambda}{\lambda+\mu} = \dfrac{\rho}{1+\rho} \\[2ex] Q_n - 1 & \text{with probability} \quad \dfrac{\mu}{\lambda+\mu} = \dfrac{1}{1+\rho} \end{cases}$$

whenever $Q_n \geqslant 1$ (see paragraph A after (6.11.12) for a similar result for another birth–death process). When $Q_n = 0$ we have that

$$\mathbf{P}(Q_{n+1} = 1 \mid Q_n = 0) = 1,$$

so that the walk leaves 0 immediately after arriving there; it is only in this regard that the walk differs from the random walk (6.4.15) with a retaining barrier. Look for stationary distributions of the walk in the usual way to find (*exercise*) that there exists such a distribution if and only if $\rho < 1$, and it is given by

(10) $$\pi_0 = \tfrac{1}{2}(1-\rho), \qquad \pi_n = \tfrac{1}{2}(1-\rho^2)\rho^{n-1} \quad \text{for} \quad n \geqslant 1.$$

Follow the argument of (6.4.15) to find that

$$\{Q_n\} \text{ is } \begin{cases} \text{non-null persistent} & \text{if} \quad \rho < 1 \\ \text{null persistent} & \text{if} \quad \rho = 1 \\ \text{transient} & \text{if} \quad \rho > 1. \end{cases}$$

Equation (10) differs from the result of (8) because the walk $\{Q_n\}$ and the process Q behave differently at the state 0. It is possible to deduce (8) from (10) by taking account of the times which elapse between the jumps of the walk (see Ross 1970, p. 105, for details). It is clear now that $Q_n \to \infty$ almost surely as $n \to \infty$ if $\rho > 1$, whilst $\{Q_n\}$ experiences large fluctuations in the symmetric case $\rho = 1$.

Exercises

(11) Consider a random walk on the non-negative integers with a reflecting barrier at 0, and which moves rightwards or leftwards with respective probabilities $\rho/(1+\rho)$ and $1/(1+\rho)$; when at 0, the particle moves to 1 at the next step. Show that the walk has a stationary distribution if and only if $\rho < 1$, and in this case the unique such distribution $\boldsymbol{\pi}$ is given by $\pi_0 = \frac{1}{2}(1-\rho)$, $\pi_n = \frac{1}{2}(1-\rho^2)\rho^{n-1}$ for $n \geqslant 1$.

(12) Suppose now that the random walker of Exercise (11) delays its steps in the following way. When at the point n, it waits a random length of time having the exponential distribution with parameter θ_n before moving to its next position; different 'holding times' are independent of each other and of further information concerning the steps of the walk. Show that, subject to reasonable assumptions on the θ_n, the ensuing continuous-time process settles into an equilibrium distribution $\boldsymbol{\nu}$ given by $\nu_n = C\pi_n/\theta_n$ for some appropriate constant C.

By applying this result to the case when $\theta_0 = \lambda$, $\theta_n = \lambda + \mu$ for $n \geqslant 1$, deduce that the equilibrium distribution of the $M(\lambda)/M(\mu)/1$ queue is $\nu_n = (1-\rho)\rho^n$, $n \geqslant 0$, where $\rho = \lambda/\mu < 1$.

(13) **Waiting time.** Consider an $M(\lambda)/M(\mu)/1$ queue with $\rho = \lambda/\mu$ satisfying $\rho < 1$, and suppose that the number $Q(0)$ of people in the queue at time 0 has the stationary distribution $\pi_n = (1-\rho)\rho^n$, $n \geqslant 0$. Let W be the time spent by a typical new arrival before he begins his service. Show that the distribution of W is given by $\mathbf{P}(W \leqslant x) = 1 - \rho\,\mathrm{e}^{-x(\mu-\lambda)}$ for $x \geqslant 0$, and note that $\mathbf{P}(W = 0) = 1 - \rho$.

(14) A box contains i red balls and j lemon balls, and they are drawn at random without replacement. Each time a red (respectively lemon) ball is drawn, a particle doing a walk on $\{0, 1, 2, \ldots\}$ moves one step to the right (respectively left); the origin is a retaining barrier, so that leftward steps from the origin are suppressed. Let $\pi(n; i, j)$ be the probability that the particle ends at position n, having started at the origin.

Write down a set of difference equations, and deduce that

$$\pi(n; i, j) = A(n; i, j) - A(n+1; i, j) \quad \text{for} \quad i \leqslant j + n$$

where $A(n; i, j) = \binom{i}{n} \bigg/ \binom{j+n}{n}$.

(15) Let Q be an $M(\lambda)/M(\mu)/1$ queue with $Q(0) = 0$. Show that $p_n(t) = \mathbf{P}(Q(t) = n)$ satisfies

$$p_n(t) = \sum_{i, j \geqslant 0} \pi(n; i, j) \left(\frac{(\lambda t)^i \, e^{-\lambda t}}{i!} \right) \left(\frac{(\mu t)^j \, e^{-\mu t}}{j!} \right)$$

where the $\pi(n; i, j)$ are given in the previous exercise.

11.3 $M/G/1$

$M/M/1$ is the only queue which is a Markov chain; the analysis of other queueing systems requires greater ingenuity. If either interarrival times or service times are exponentially distributed then the general theory of Markov chains still provides a method for studying the queue. For, in these two cases we may find a discrete-time Markov chain which is imbedded in the continuous-time process Q. We consider $M/G/1$ in this section, which is divided into three parts dealing with equilibrium theory, the 'waiting time' of a typical customer, and the length of a typical 'busy period' during which the server is continuously occupied.

(A) Asymptotic queue length. Consider $M(\lambda)/G/1$. Customers arrive like a Poisson process with intensity λ. Let D_n be the time of departure of the nth customer from the system, and let $Q(D_n)$ be the number of customers which he leaves behind him in the system on his departure (really, we should write $Q(D_n+)$ instead of $Q(D_n)$ to make clear that the departing customer is not included). Then $Q(D) = \{Q(D_n): n \geqslant 1\}$ is a sequence of random variables. What can we say about a typical increment $Q(D_{n+1}) - Q(D_n)$? If $Q(D_n) > 0$, then the $(n+1)$th customer begins his service time immediately at D_n; during this service time S_{n+1}, a random number, U_n say, of customers arrive and join the waiting line. Therefore the $(n+1)$th customer leaves $U_n + Q(D_n) - 1$ customers behind him as he departs. That is,

(1) $$Q(D_{n+1}) = U_n + Q(D_n) - 1 \quad \text{if} \quad Q(D_n) > 0.$$

If $Q(D_n) = 0$ then the server must wait for the $(n+1)$th arrival before he sets to work again. When this service is complete, the $(n+1)$th customer leaves exactly U_n customers behind him where U_n is the number of arrivals during his service time, as before. That is

(2) $$Q(D_{n+1}) = U_n \quad \text{if} \quad Q(D_n) = 0.$$

Combine (1) and (2) to obtain

(3) $$Q(D_{n+1}) = U_n + Q(D_n) - h(Q(D_n))$$

where h is defined by

$$h(x) = \begin{cases} 1 & \text{if } x > 0 \\ 0 & \text{if } x \leqslant 0. \end{cases}$$

Equation (3) holds for any queue. However, in the case of $M(\lambda)/G/1$ the random variable U_n depends *only* on the length of time S_{n+1}, and is independent of $Q(D_n)$ because of the special properties of the Poisson process of arrivals. We conclude from (3) that $Q(D)$ is a Markov chain.

(4) **Theorem.** *$Q(D)$ is a Markov chain with transition matrix*

$$\boldsymbol{P}_D = \begin{pmatrix} \delta_0 & \delta_1 & \delta_2 & \cdots \\ \delta_0 & \delta_1 & \delta_2 & \cdots \\ 0 & \delta_0 & \delta_1 & \cdots \\ 0 & 0 & \delta_0 & \cdots \\ \vdots & \vdots & \vdots & \end{pmatrix}$$

where

$$\delta_j = \mathbf{E}\left(\frac{(\lambda S)^j}{j!} e^{-\lambda S}\right)$$

and S is a typical service time.

Of course, δ_j is just the probability that exactly j customers join the queue during a typical service time.

Proof. We need only show that $\boldsymbol{P}_D$ is the correct transition matrix. In the notation of Chapter 6,

$$\begin{aligned} p_{0j} &= \mathbf{P}(Q(D_{n+1}) = j \mid Q(D_n) = 0) \\ &= \mathbf{E}(\mathbf{P}(U_n = j \mid S)) \end{aligned}$$

where $S = S_{n+1}$ is the service time of the $(n + 1)$th customer. Thus

$$p_{0j} = \mathbf{E}\left(\frac{(\lambda S)^j}{j!} e^{-\lambda S}\right) = \delta_j$$

as required, since, conditional on S, U_n has the Poisson distribution with parameter λS. Likewise, if $i \geqslant 1$ then

$$\begin{aligned} p_{ij} &= \mathbf{E}(\mathbf{P}(U_n = j - i + 1 \mid S)) \\ &= \begin{cases} \delta_{j-i+1} & \text{if } j - i + 1 \geqslant 0 \\ 0 & \text{if } j - i + 1 < 0. \end{cases} \end{aligned}$$

■

This result enables us to observe the behaviour of the process $Q = \{Q(t)\}$ by evaluating it at the time epochs $D_1, D_2, \ldots$ and using the theory of Markov chains. It is important to note that this course of action provides reliable information about the asymptotic behaviour of Q only because $D_n \to \infty$ almost surely as $n \to \infty$. The asymptotic behaviour of $Q(D)$ is described by the next theorem.

(5) **Theorem.** *Let $\rho = \lambda \mathbf{E}(S)$ be the traffic density.*

(a) *If $\rho < 1$ then $Q(D)$ is ergodic with a unique stationary distribution π, with generation function*

$$G(s) = \sum_j \pi_j s^j$$

$$= (1-\rho)(s-1)\frac{M_S(\lambda(s-1))}{s - M_S(\lambda(s-1))},$$

where M_S is the moment generating function of a typical service time.

(b) *If $\rho > 1$ then $Q(D)$ is transient.*

(c) *If $\rho = 1$ then $Q(D)$ is null persistent.*

Here are two consequences of this.

(6) **Busy period.** A *busy period* is a period of time during which the server is continuously occupied. The length B of a typical busy period behaves similarly to the time B' between successive visits of the chain $Q(D)$ to the state 0. Thus

$$\begin{aligned} &\text{if} \quad \rho < 1 \quad \text{then} \quad \mathbf{E}(B) < \infty \\ &\text{if} \quad \rho = 1 \quad \text{then} \quad \mathbf{E}(B) = \infty, \qquad \mathbf{P}(B = \infty) = 0 \\ &\text{if} \quad \rho > 1 \quad \text{then} \quad \mathbf{P}(B = \infty) > 0. \end{aligned}$$

See the forthcoming Theorems (17) and (18) for more details about B.

(7) **Stationarity of Q.** It is an immediate consequence of (5) and (6.4.17) that $Q(D)$ is asymptotically stationary whenever $\rho < 1$. In this case it can be shown that Q is asymptotically stationary also, in that

$$\mathbf{P}(Q(t) = n) \to \pi_n \quad \text{as} \quad t \to \infty.$$

Roughly speaking, this is because $Q(t)$ forgets more and more about its origins as t becomes larger.

Proof of (5). $Q(D)$ is irreducible and aperiodic. We proceed by applying (6.4.3), (6.4.10), and (6.4.13).

(a) Look for a root of the equation $\boldsymbol{\pi} = \boldsymbol{\pi} \boldsymbol{P}_D$. Any such $\boldsymbol{\pi}$ satisfies

(8) $$\pi_j = \pi_0 \delta_j + \sum_{i=1}^{j+1} \pi_i \delta_{j-i+1}, \quad \text{for} \quad j \geq 0.$$

First, note that if π_0 (≥ 0) is given, then (8) has a unique solution $\boldsymbol{\pi}$.

Furthermore, this solution has non-negative entries. To see this, add equations (8) for $j = 0, 1, \dots, n$ and solve for π_{n+1} to obtain

(9) $$\pi_{n+1}\delta_0 = \pi_0\varepsilon_n + \sum_{i=1}^{n} \pi_i \varepsilon_{n-i+1} \quad \text{for} \quad n \geqslant 0$$

where

$$\varepsilon_n = 1 - \delta_0 - \delta_1 - \cdots - \delta_n > 0 \quad \text{because} \quad \sum_j \delta_j = 1.$$

From (9), $\pi_{n+1} \geqslant 0$ whenever $\pi_i \geqslant 0$ for all $i \leqslant n$, and so

(10) $$\pi_n \geqslant 0 \quad \text{for all} \quad n$$

if $\pi_0 \geqslant 0$, by induction. Return to (8) to see that the generating functions

$$G(s) = \sum_j \pi_j s^j, \qquad \Delta(s) = \sum_j \delta_j s^j$$

satisfy

$$G(s) = \pi_0 \Delta(s) + \frac{1}{s}[G(s) - \pi_0]\Delta(s)$$

and so

(11) $$G(s) = \frac{\pi_0(s-1)\Delta(s)}{s - \Delta(s)}.$$

$\boldsymbol{\pi}$ is a stationary distribution if and only if $\pi_0 > 0$ and $\lim_{s \uparrow 1} G(s) = 1$. Apply L'Hôpital's rule to (11) to discover that

$$\pi_0 = 1 - \Delta'(1) > 0$$

is a necessary and sufficient condition for this to occur, and thus there exists a stationary distribution if and only if

(12) $$\Delta'(1) < 1.$$

However,

$$\begin{aligned}\Delta(s) &= \sum_j s^j \mathbf{E}\left(\frac{(\lambda S)^j}{j!} e^{-\lambda S}\right) \\ &= \mathbf{E}\left(e^{-\lambda S} \sum_j \frac{(\lambda s S)^j}{j!}\right) \\ &= \mathbf{E}(\exp[\lambda S(s-1)]) = M_S(\lambda(s-1))\end{aligned}$$

where M_S is the moment generating function of S. Thus

(13) $$\Delta'(1) = \lambda M_S'(0) = \lambda \mathbf{E}(S) = \rho$$

and condition (12) becomes

$$\rho < 1.$$

Thus $Q(D)$ is non-null persistent if and only if $\rho < 1$. In this case, $G(s)$ takes the form given in (5a).

(b) Recall from (6.4.10) that $Q(D)$ is transient if and only if there is a bounded non-zero solution $\{y_j : j \geqslant 1\}$ to the equations

(14) $$y_1 = \sum_{i=1}^{\infty} \delta_i y_i$$

(15) $$y_j = \sum_{i=0}^{\infty} \delta_i y_{j+i-1} \quad \text{for} \quad j \geqslant 2.$$

If $\rho > 1$ then $\Delta(s)$ satisfies

$$0 < \Delta(0) < 1, \qquad \Delta(1) = 1, \qquad \Delta'(1) > 1$$

from (13). Draw a picture to see that there exists a number $b \in (0, 1)$ such that

$$\Delta(b) = b.$$

By inspection,

$$y_j = 1 - b^j$$

solves (14) and (15), and (b) is shown.

(c) $Q(D)$ is transient if $\rho > 1$ and non-null persistent if and only if $\rho < 1$. We need only show that $Q(D)$ is persistent if $\rho = 1$. But it is not difficult to see that $\{y_j : j \neq 0\}$ solves equation (6.4.14), when y_j is given by

$$y_j = j \quad \text{for} \quad j \geqslant 1,$$

and the result follows. ■

(B) Waiting time. When $\rho < 1$ the queue length settles down into an equilibrium distribution π. Suppose that a customer joins the queue after some large time t has elapsed. He will wait a period W of time before his service begins; W is called his *waiting time* (this definition is at odds with that used by some authors who include the customer's service time in W). The distribution of W will not vary much with t since the system is 'nearly' in equilibrium.

(16) **Theorem.** *The waiting time W has moment generating function*

$$M_W(s) = \frac{(1-\rho)s}{\lambda + s - \lambda M_S(s)}$$

when the queue is in equilibrium.

Proof. The condition that the queue be in equilibrium amounts to the supposition that the length $Q(D)$ of the queue on the departure of a customer is distributed according to the stationary distribution π. Suppose that a customer waits for a period of length W and then is served for a period of length S. On departure he leaves behind him all those customers who have arrived during the period, length $W + S$, during which he was in the system.

The number Q of such customers is Poisson with parameter $\lambda(W+S)$, and so

$$\begin{aligned}\mathbf{E}(s^Q) &= \mathbf{E}(\mathbf{E}(s^Q \mid W, S))\\ &= \mathbf{E}(\exp[\lambda(W+S)(s-1)])\\ &= \mathbf{E}(e^{\lambda W(s-1)})\mathbf{E}(e^{\lambda S(s-1)}) \quad \text{by independence}\\ &= M_W(\lambda(s-1))M_S(\lambda(s-1)).\end{aligned}$$

However, Q has distribution $\boldsymbol{\pi}$ given by (5a) and the result follows. ■

(C) Busy period: a branching process. Finally, put yourself in the server's shoes. He is not so interested in the waiting times of his customers as he is in the frequency of his tea breaks. Recall from (6) that a *busy period* is a period of time during which he is continuously occupied, and let B be the length of a typical busy period. That is, if the first customer arrives at time T_1 then

$$B = \inf\{t > 0: Q(t + T_1) = 0\};$$

B is well defined whether or not $Q(D)$ is ergodic, though it may equal $+\infty$.

(17) **Theorem.** *The moment generating function M_B of B satisfies the functional equation*

$$M_B(s) = M_S(s - \lambda + \lambda M_B(s)).$$

It can be shown that this functional equation has a unique solution which is the moment generating function of a (possibly infinite) random variable (see Feller 1971, pp. 441, 473). The server may wish to calculate the probability

$$\mathbf{P}(B < \infty) = \lim_{x \to \infty} \mathbf{P}(B \leqslant x)$$

that he is eventually free. It is no surprise to find the following, in agreement with (6).

(18) **Theorem.**

$$\mathbf{P}(B < \infty) \begin{cases} = 1 & \text{if } \rho \leqslant 1 \\ < 1 & \text{if } \rho > 1. \end{cases}$$

This may remind you of a similar result for the extinction probability of a branching process. This is no coincidence; we prove (17) and (18) by methods first encountered in the study of branching processes.

Proof of (17) and (18). Here is an imbedded branching process. Call customer C_2 an 'offspring' of customer C_1 if C_2 joins the queue while C_1 is being served. Let $\{G_n: n \geqslant 0\}$ be disjoint sets of customers given as follows. G_0 contains the first customer only; G_{n+1} contains the set of offspring of customers in G_n. Let Z_n be the size of G_n. Then $Z = \{Z_n: n \geqslant 0\}$ is a branching process; this assertion relies upon the fact that customers arrive like a Poisson process so that the numbers of arrivals during different service times are independent and identically distributed. The process Z is ultimately extinct if and only if the queue is empty at some time later than the first arrival. That is,

$$\mathbf{P}(B < \infty) = \mathbf{P}(Z_n = 0 \text{ for some } n).$$

But, by (5.4.5),

$$\eta = \mathbf{P}(Z_n = 0 \text{ for some } n)$$

satisfies

$$\eta = 1 \quad \text{if and only if} \quad \mu \leqslant 1$$

where μ is the mean number of offspring of the first arrival. However, in the notation of the proof of (5),

$$\mu = \Delta'(1) = \rho$$

by (13), and (18) is proved.

Each individual in this branching process has a service time; B is the sum of these service times. Thus

(19) $$B = S + \sum_{j=1}^{Z_1} B_j$$

where S is the service time of the first customer and B_j is the sum of the service times of the jth member of G_1 together with all his descendants (this is similar to the argument of Problem (5.12.11)). The two terms on the right-hand side of (19) are *not* independent of each other; after all, if S is large then Z_1 is likely to be large as well. However, condition on S to obtain

$$M_B(s) = \mathbf{E}\left(\mathbf{E}\left(\exp\left[s\left(S + \sum_{j=1}^{Z_1} B_j\right)\right] \,\middle|\, S\right)\right)$$

and remember that, conditional on Z_1, the random variables $B_1, \ldots, B_{Z_1}$ are independent with the same distribution as B to obtain

$$M_B(s) = \mathbf{E}(e^{sS} G_P\{M_B(s)\})$$

where G_P is the probability generating function of the Poisson distribution with parameter λS. Thus

$$M_B(s) = \mathbf{E}(\exp[S(s - \lambda + \lambda M_B(s))])$$

as required. ■

Exercises

(20) Consider $M(\lambda)/D(d)/1$ where $\rho = \lambda d < 1$. Show that the mean queue length at moments of departure in equilibrium is $\frac{1}{2}\rho(2 - \rho)/(1 - \rho)$.

(21) Consider $M(\lambda)/M(\mu)/1$, and show that the moment generating function of a typical busy period is given by

$$M_B(s) = \frac{(\lambda + \mu - s) - \sqrt{(\lambda + \mu - s)^2 - 4\lambda\mu}}{2\lambda}$$

for all sufficiently small but positive values of s.

(22) Show that, for an $M/G/1$ queue, the sequence of times at which the server passes from being busy to being free constitutes a renewal process.

11.4 $G/M/1$

The system $G/M(\mu)/1$ contains an imbedded discrete-time Markov chain also, and this chain provides information about the properties of $Q(t)$ for large t. This section is divided into two parts, dealing with the asymptotic behaviour of $Q(t)$ and the waiting time distribution.

(A) Asymptotic queue length. This time, consider the epoch of time at which the nth customer *joins* the queue, and let $Q(A_n)$ be the number of individuals who are ahead of him in the system at the moment of his arrival. $Q(A_n)$ includes any customer whose service is in progress; more specifically, $Q(A_n) = Q(T_n-)$ where T_n is the instant of the nth arrival. The argument of the last section shows that

(1) $$Q(A_{n+1}) = Q(A_n) + 1 - V_n$$

where V_n is the number of departures from the system during the interval $[T_n, T_{n+1})$ between the nth and $(n+1)$th arrival. This time, V_n depends on $Q(A_n)$ since not more than $Q(A_n) + 1$ individuals may depart during this interval. However, service times are exponentially distributed, and so, conditional upon $Q(A_n)$ and $X_{n+1} = T_{n+1} - T_n$, V_n has a truncated Poisson distribution

(2) $$\mathbf{P}(V_n = d \mid Q(A_n) = q, X_{n+1} = x) = \begin{cases} \dfrac{(\mu x)^d}{d!} e^{-\mu x} & \text{if } d \leq q \\ \displaystyle\sum_{m>q} \frac{(\mu x)^m}{m!} e^{-\mu x} & \text{if } d = q + 1. \end{cases}$$

Anyway, given $Q(A_n)$, V_n is independent of $Q(A_1), \ldots, Q(A_{n-1})$, and so $Q(A) = \{Q(A_n): n \geq 1\}$ is a Markov chain.

(3) **Theorem.** *$Q(A)$ is a Markov chain with transition matrix*

$$\mathbf{P}_A = \begin{pmatrix} 1 - \alpha_0 & \alpha_0 & 0 & 0 & \cdots \\ 1 - \alpha_0 - \alpha_1 & \alpha_1 & \alpha_0 & 0 & \cdots \\ 1 - \alpha_0 - \alpha_1 - \alpha_2 & \alpha_2 & \alpha_1 & \alpha_0 & \cdots \\ \vdots & \vdots & \vdots & \vdots & \end{pmatrix}$$

where

$$\alpha_j = \mathbf{E}\left(\frac{(\mu X)^j}{j!} e^{-\mu X}\right)$$

and X is a typical interarrival time.

Of course, α_j is just the probability that exactly j events of a Poisson process occur during a typical interarrival time.

Proof. This proceeds as for (11.3.4). ■

(4) **Theorem.** *Let $\rho = \{\mu \mathbf{E}(X)\}^{-1}$ be the traffic density.*

(a) *If $\rho < 1$ then $Q(A)$ is ergodic with a unique stationary distribution $\boldsymbol{\pi}$ given by*

$$\pi_j = (1-\eta)\eta^j \quad \textit{for} \quad j \geq 0$$

where η is the smallest positive root of

$$\eta = M_X(\mu(\eta - 1))$$

and M_X is the moment generating function of X.

(b) *If $\rho > 1$ then $Q(A)$ is transient.*

(c) *If $\rho = 1$ then $Q(A)$ is null persistent.*

If $\rho < 1$ then $Q(A)$ is asymptotically stationary. Unlike the case of $M/G/1$, however, the stationary distribution $\boldsymbol{\pi}$ given by (4a) need *not* be the limiting distribution of Q itself; to see an example of this, just consider $D(1)/M/1$.

Proof. Let Q_d be an $M(\mu)/G/1$ queue whose service times have the same distribution as the interarrival times of Q (Q_d is called the *dual* of Q, but more about that later). The traffic density ρ_d of Q_d satisfies

(5) $$\rho\rho_d = 1.$$

From the results of Section 11.3, Q_d has an imbedded Markov chain $Q_d(D)$, obtained from the values of Q_d at the epochs of time at which customers depart. We shall see that $Q(A)$ is non-null persistent (respectively transient) if and only if the imbedded chain $Q_d(D)$ of Q_d is transient (respectively non-null persistent) and the results will follow immediately from (11.3.5) and its proof.

(a) Look for non-negative solutions $\boldsymbol{\pi}$ to the equation

(6) $$\boldsymbol{\pi} = \boldsymbol{\pi}\mathbf{P}_A$$

which have sum $\boldsymbol{\pi}\mathbf{1}' = 1$. Expand (6), set

(7) $$y_j = \pi_0 + \cdots + \pi_{j-1} \quad \text{for} \quad j \geq 1,$$

and remember that $\sum_j \alpha_j = 1$ to obtain

(8) $$y_1 = \sum_{i=1}^{\infty} \alpha_i y_i$$

(9) $$y_j = \sum_{i=0}^{\infty} \alpha_i y_{j+i-1} \quad \text{for} \quad j \geqslant 2.$$

These are the same equations as (11.3.14) and (11.3.15) for Q_d. As in the proof of (11.3.5), it is easy to check that

(10) $$y_j = 1 - \eta^j$$

solves (8) and (9) whenever

$$A(s) = \sum_{j=0}^{\infty} \alpha_j s^j$$

satisfies

$$A'(1) > 1$$

where η is the unique root in the interval (0, 1) of the equation

$$A(s) = s.$$

However, write A in terms of M_X, as before, to find that

$$A(s) = M_X(\mu(s-1))$$

giving

$$A'(1) = \rho_d = \rho^{-1}.$$

Combine (7) and (10) to find the stationary distribution for the case $\rho < 1$. If $\rho \geqslant 1$ then $\rho_d \leqslant 1$ by (5), and so (8) and (9) have no bounded non-zero solution since otherwise $Q_d(D)$ would be transient, contradicting (11.3.5). Thus $Q(A)$ is non-null persistent if and only if $\rho < 1$.

(b) To prove transience, we seek bounded non-zero solutions $\{y_j : j \geqslant 1\}$ to the equations

(11) $$y_j = \sum_{i=1}^{j+1} y_i \alpha_{j-i+1} \quad \text{for} \quad j \geqslant 1.$$

Suppose that $\{y_j\}$ satisfies (11), and that $y_1 \geqslant 0$. Define $\boldsymbol{\pi} = \{\pi_j : j \geqslant 0\}$ as follows:

$$\pi_0 = y_1 \alpha_0, \qquad \pi_1 = y_1(1-\alpha_0), \qquad \pi_j = y_j - y_{j-1} \quad \text{for} \quad j \geqslant 2.$$

It is an easy exercise to show that $\boldsymbol{\pi}$ satisfies (11.3.8) with the δ_j replaced by the α_j throughout. But (11.3.8) possesses a non-zero solution with bounded sum if and only if $\rho_d < 1$, which is to say that $Q(A)$ is transient if and only if $\rho = (\rho_d)^{-1} > 1$.

(c) $Q(A)$ is transient if and only if $\rho > 1$, and is non-null persistent if and only if $\rho < 1$. If $\rho = 1$ then $Q(A)$ has no choice but null persistence. ■

(B) Waiting time. An arriving customer waits for just as long as the server needs to complete the service period in which he is currently engaged and to serve the other waiting customers. That is, the nth customer waits for a length W_n of time:

$$W_n = Z_1^* + Z_2 + Z_3 + \cdots + Z_{Q(A_n)} \quad \text{if} \quad Q(A_n) > 0$$

where Z_1^* is the *excess* (or *residual*) *service time* of the customer at the head of the queue, and $Z_2, \ldots, Z_{Q(A_n)}$ are the service times of the others. Given $Q(A_n)$, the Z_i are independent, but Z_1^* does not in general have the same distribution as $Z_2, Z_3, \ldots$. In the case of $G/M(\mu)/1$, however, the lack-of-memory property helps us around this difficulty.

(12) **Theorem.** *The waiting time W of an arriving customer has distribution*

$$\mathbf{P}(W \leqslant x) = \begin{cases} 0 & \text{if } x < 0 \\ 1 - \eta\, e^{-\mu(1-\eta)x} & \text{if } x \geqslant 0 \end{cases}$$

where η is given in (4a), *when the queue is in equilibrium.*

Note that W has an atom of size $1 - \eta$ at the origin.

Proof. By the lack-of-memory property, W_n is the sum of $Q(A_n)$ independent exponential variables. Use the equilibrium distribution of $Q(A)$ to find that

$$M_W(s) = (1 - \eta) + \eta \frac{\mu(1 - \eta)}{\mu(1 - \eta) - s}$$

which we recognize as the moment generating function of a random variable which either equals zero with probability $1 - \eta$, or is exponentially distributed with parameter $\mu(1 - \eta)$ with probability η. ■

Finally, here is a word of caution. There is another quantity called *virtual* waiting time, which must not be confused with *actual* waiting time. The latter is the actual time spent by a customer after his arrival; the former is the time which a customer *would* spend if he were to arrive at some particular instant. The equilibrium distributions of these waiting times may differ whenever the stationary distribution of Q differs from the stationary distribution of the imbedded Markov chain $Q(A)$.

Exercises

(13) Consider $G/M(\mu)/1$, and let $\alpha_j = \mathbf{E}((\mu X)^j e^{-\mu X}/j!)$ where X is a typical interarrival time. Suppose the traffic density ρ is less than 1. Show that the equilibrium distribution π of the imbedded chain at moments of arrivals satisfies

$$\pi_n = \sum_{i=0}^{\infty} \alpha_i \pi_{n+i-1} \quad \text{for} \quad n \geqslant 1.$$

Look for a solution of the form $\pi_n = \theta^n$ for some θ, and deduce that the unique stationary distribution is given by $\pi_j = (1 - \eta)\eta^j$ for $j \geq 0$, where η is the smallest positive root of the equation $s = M_X(\mu(s - 1))$.

(14) Consider a $G/M(\mu)/1$ queue in equilibrium. Let η be the smallest positive root of the equation $x = M_X(\mu(x - 1))$ where M_X is the moment generating function of an interarrival time. Show that the mean number of customers ahead of a new arrival is $\eta(1 - \eta)^{-1}$, and the mean waiting time is $\eta\{\mu(1 - \eta)\}^{-1}$.

(15) Consider $D(1)/M(\mu)/1$ where $\mu > 1$. Show that the continuous-time queue length $Q(t)$ does not converge in distribution as $t \to \infty$, even though the imbedded chain at the times of arrivals is ergodic.

11.5 $G/G/1$

If neither interarrival times nor service times are exponentially distributed then the methods of the last three sections fail. This apparent setback leads us to the remarkable discovery that queueing problems are intimately related to random walk problems. This section is divided into two parts, one dealing with the equilibrium theory of $G/G/1$ and the other dealing with the imbedded random walk.

(A) Asymptotic waiting time. Let W_n be the waiting time of the nth customer. There is a useful simple relationship between W_n and W_{n+1} in terms of the service time S_n of the nth customer and the length X_{n+1} of time between the nth and the $(n + 1)$th arrivals.

(1) **Theorem. Lindley's equation.**

$$W_{n+1} = \max\{0, W_n + S_n - X_{n+1}\}.$$

Proof. The nth customer is in the system for a length $W_n + S_n$ of time. If $X_{n+1} > W_n + S_n$ then the queue is empty at the $(n + 1)$th arrival, and so $W_{n+1} = 0$. If $X_{n+1} \leq W_n + S_n$ then the $(n + 1)$th customer arrives while the nth is still present, but only waits for a period of length $W_n + S_n - X_{n+1}$ before the previous customer leaves. ∎

We shall see that Lindley's equation implies that the distribution functions

$$F_n(x) = \mathbf{P}(W_n \leq x)$$

of the W_n converge as $n \to \infty$ to some limit function $F(x)$. Of course, F need not be a proper distribution function; indeed, it is intuitively clear that the queue settles down into equilibrium if and only if F is a distribution function which is not defective.

(2) **Theorem.** *Let* $F_n(x) = \mathbf{P}(W_n \leq x)$. *Then*

$$F_{n+1}(x) = \begin{cases} 0 & \text{if } x < 0 \\ \displaystyle\int_{-\infty}^{x} F_n(x - y)\, dG(y) & \text{if } x \geq 0 \end{cases}$$

where G is the distribution function of $U_n = S_n - X_{n+1}$. Thus

$$F(x) = \lim_{n \to \infty} F_n(x)$$

exists.

Note that $\{U_n : n \geqslant 1\}$ is a collection of independent identically distributed random variables.

Proof. If $x \geqslant 0$ then

$$\begin{aligned}\mathbf{P}(W_{n+1} \leqslant x) &= \int_{-\infty}^{\infty} \mathbf{P}(W_n + U_n \leqslant x \mid U_n = y)\, \mathrm{d}G(y) \\ &= \int_{-\infty}^{x} \mathbf{P}(W_n \leqslant x - y)\, \mathrm{d}G(y) \quad \text{by independence,}\end{aligned}$$

and the first part is proved. We claim that

(3) $$F_{n+1}(x) \leqslant F_n(x) \quad \text{for all } x \text{ and } n.$$

If (3) holds then the second result follows immediately; we prove (3) by induction. Trivially,

$$F_2(x) \leqslant F_1(x)$$

because $F_1(x) = 1$ for all $x \geqslant 0$. Suppose that (3) holds for $n = k - 1$, say. Then, for $x \geqslant 0$,

$$F_{k+1}(x) - F_k(x) = \int_{-\infty}^{x} [F_k(x - y) - F_{k-1}(x - y)]\, \mathrm{d}G(y) \leqslant 0$$

by the induction hypothesis. The proof is complete. ■

So the distribution functions of $\{W_n\}$ converge as $n \to \infty$. It is clear, by monotone convergence, that the limit $F(x)$ satisfies the Wiener–Hopf equation

$$F(x) = \int_{-\infty}^{x} F(x - y)\, \mathrm{d}G(y) \quad \text{for} \quad x \geqslant 0;$$

this is not easily solved for F in terms of G. However, it is not too difficult to find a criterion for F to be a proper distribution function.

(4) **Theorem.** *Let $\rho = \mathbf{E}(S)/\mathbf{E}(X)$ be the traffic density.*

(a) *If $\rho < 1$ then F is a non-defective distribution function.*
(b) *If $\rho > 1$ then $F(x) = 0$ for all x.*
(c) *If $\rho = 1$ and $\operatorname{var}(U) > 0$ then $F(x) = 0$ for all x.*

An explicit formula for the moment generating function of F when $\rho < 1$

is given in Theorem (14) below. Theorem (4) classifies the stability of $G/G/1$ in terms of the sign of $1 - \rho$; note that this information is obtainable from the distribution function G since

(5) $$\rho < 1 \quad \Leftrightarrow \quad \mathbf{E}(S) < \mathbf{E}(X) \quad \Leftrightarrow \quad \mathbf{E}(U) = \int_{-\infty}^{\infty} u \, \mathrm{d}G(u) < 0$$

where U is a typical member of the U_i.

The crucial step in the proof of (4) is very important in its own right. Use Lindley's equation (1) to see that

$$W_1 = 0$$
$$W_2 = \max\{0, W_1 + U_1\} = \max\{0, U_1\}$$
$$W_3 = \max\{0, W_2 + U_2\} = \max\{0, U_2, U_2 + U_1\}$$

and in general

(6) $$W_{n+1} = \max\{0, U_n, U_n + U_{n-1}, \ldots, U_n + \cdots + U_1\}$$

which expresses W_{n+1} in terms of the partial sums of a sequence of independent identically distributed variables. It is difficult to derive asymptotic properties of W_{n+1} directly from (6) since every non-zero term changes its value as n increases from the value k, say, to the value $k + 1$. The following theorem is the crucial observation.

(7) **Theorem.** W_{n+1} *has the same distribution as*

$$W'_{n+1} = \max\{0, U_1, U_1 + U_2, \ldots, U_1 + \cdots + U_n\}.$$

Proof. $(U_1, \ldots, U_n)$ and $(U_n, \ldots, U_1)$ are sequences with the same joint distribution. Replace each U_i in (6) by U_{n+1-i}. ■

That is to say, W_{n+1} and W'_{n+1} are *different* random variables but they have the *same* distribution. Thus

$$F(x) = \lim_{n\to\infty} \mathbf{P}(W_n \leqslant x) = \lim_{n\to\infty} \mathbf{P}(W'_n \leqslant x).$$

Furthermore,

(8) $$W'_n \leqslant W'_{n+1} \quad \text{for all } n \geqslant 1,$$

a monotonicity property which is not shared by $\{W_n\}$. This property provides another method for deriving the existence of F in (2).

Proof of (4). From (8)

$$W' = \lim_{n\to\infty} W'_n$$

exists almost surely (and, in fact, pointwise) but may be $+\infty$. Furthermore,

(9) $$W' = \max\{0, \Sigma_1, \Sigma_2, \dots\}$$

where

$$\Sigma_n = \sum_{j=1}^{n} U_j$$

and $F(x) = \mathbf{P}(W' \leqslant x)$. Thus

$$F(x) = \mathbf{P}(\Sigma_n \leqslant x \text{ for all } n) \quad \text{if} \quad x \geqslant 0,$$

and the proof proceeds by using properties of the sequence $\{\Sigma_n\}$ of partial sums, such as the strong law (7.5.1):

(10) $$\frac{1}{n}\Sigma_n \xrightarrow{\text{a.s.}} \mathbf{E}(U) \quad \text{as} \quad n \to \infty.$$

Suppose first that $\mathbf{E}(U) < 0$. Then

$$\begin{aligned}\mathbf{P}(\Sigma_n > 0 \text{ for infinitely many } n) &= \mathbf{P}\left(\frac{1}{n}\Sigma_n - \mathbf{E}(U) > |\mathbf{E}(U)| \text{ i.o.}\right)\\ &= 0\end{aligned}$$

by (10). Thus, from (9), W$'$ is almost surely the maximum of only finitely many terms, and so

$$\mathbf{P}(W' < \infty) = 1,$$

implying that F is a non-defective distribution function.

Next suppose that $\mathbf{E}(U) > 0$. Pick any $x > 0$ and choose N such that

$$N \geqslant \frac{2x}{\mathbf{E}(U)}.$$

Then, if $n \geqslant N$,

$$\begin{aligned}\mathbf{P}(\Sigma_n \geqslant x) &= \mathbf{P}\left(\frac{1}{n}\Sigma_n - \mathbf{E}(U) \geqslant \frac{x}{n} - \mathbf{E}(U)\right)\\ &\geqslant \mathbf{P}\left(\frac{1}{n}\Sigma_n - \mathbf{E}(U) \geqslant -\tfrac{1}{2}\mathbf{E}(U)\right)\end{aligned}$$

Let $n \to \infty$ and use the weak law to find that

$$\mathbf{P}(W' \geqslant x) \geqslant \mathbf{P}(\Sigma_n \geqslant x) \to 1 \quad \text{for all } x.$$

Therefore W' almost surely exceeds any finite number, and so

$$\mathbf{P}(W' < \infty) = 0$$

as required.

In the case when $\mathbf{E}(U) = 0$ these crude arguments do not work and we need a more precise measure of the fluctuations of Σ_n; one way of doing this is by way of the law of the iterated logarithm (7.6.1). If $\text{var}(U) > 0$ and

$\mathbf{E}(U_1^2) < \infty$, then $\{\Sigma_n\}$ enjoys fluctuations of order $O((n \log \log n)^{\frac{1}{2}})$ in both positive and negative directions with probability 1, and so

$$\mathbf{P}(\Sigma_n \geqslant x \text{ for some } n) = 1 \quad \text{for all } x.$$

There are other arguments which yield the same result. ■

(B) Imbedded random walk. The sequence $\Sigma = \{\Sigma_n : n \geqslant 0\}$ given by

(11) $$\Sigma_0 = 0, \qquad \Sigma_n = \sum_{j=1}^{n} U_j \quad \text{for} \quad n \geqslant 1$$

describes the path of a particle which performs a random walk on $\mathbb{R}$, jumping by an amount U_n at the nth step. This simple observation leads to a wealth of conclusions about queueing systems. For example, we have just seen that the waiting time W_n of the nth customer has the same distribution as the maximum W'_n of the first n positions of the walking particle. If $\mathbf{E}(U) < 0$ then the waiting time distributions converge as $n \to \infty$, which is to say that the maximum displacement $W' = \lim W'_n$ is almost surely finite. Other properties also can be expressed in terms of this random walk, and the techniques of reflection and reversal which we discussed in Section 3.10 are useful here.

The limiting waiting time distribution is the same as the distribution of the maximum

$$W' = \max\{0, \Sigma_1, \Sigma_2, \ldots\},$$

and so it is appropriate to study the so-called 'ladder points' of Σ. Define an increasing sequence $L(0), L(1), \ldots$ of random variables by

$$L(0) = 0, \qquad L(n+1) = \min\{m > L(n) : \Sigma_m > \Sigma_{L(n)}\};$$

that is, $L(n+1)$ is the earliest epoch m of time at which Σ_m exceeds the walk's previous maximum $\Sigma_{L(n)}$. The L_n are called *ladder points*; *negative ladder points* of Σ are defined similarly as the epochs at which Σ attains new minimum values. The result of (4) amounts to the assertion that

$$\mathbf{P}(\text{there exist infinitely many ladder points}) = \begin{cases} 0 & \text{if } \mathbf{E}(U) < 0 \\ 1 & \text{if } \mathbf{E}(U) > 0. \end{cases}$$

The total number of ladder points is given by the next lemma.

(12) **Lemma.** *Let $\eta = \mathbf{P}(\Sigma_n > 0$ for some $n \geqslant 1)$ be the probability that at least one ladder point exists. The total number Λ of ladder points has mass function*

$$\mathbf{P}(\Lambda = l) = (1 - \eta)\eta^l \quad \text{for} \quad l \geqslant 0.$$

Proof. Σ is a discrete-time Markov process. Thus

$$\mathbf{P}(\Lambda \geqslant l + 1 \mid \Lambda \geqslant l) = \eta$$

since the path of the walk after the lth ladder point is a copy of Σ itself. ■

Thus the queue is stable if $\eta < 1$, in which case the maximum W' of Σ is related to the height of a typical ladder point. Let

$$Y_j = \Sigma_{L(j)} - \Sigma_{L(j-1)}$$

be the difference in the displacements of the walk at the $(j-1)$th and jth ladder points. Conditional on the value of Λ, $\{Y_j\colon 1 \leqslant j \leqslant \Lambda\}$ is a collection of independent identically distributed variables, by the Markov property. Furthermore,

(13) $$W' = \Sigma_{L(\Lambda)} = \sum_{j=1}^{\Lambda} Y_j;$$

this leads to the next lemma, relating waiting time distribution to the distribution of a typical Y_j.

(14) **Lemma.** *If Q is stable then its equilibrium waiting time distribution has moment generating function*

$$M_W(s) = \frac{1-\eta}{1-\eta M_Y(s)}$$

where M_Y is the moment generating function of Y.

Proof. Q is stable if and only if $\eta < 1$. Use (13) and (5.1.25) to find that

$$M_W(s) = G_\Lambda(M_Y(s)).$$

Now use the result of (12). ■

Lemma (14) describes the waiting time distribution in terms of the distribution of Y. Analytical properties of Y are a little tricky to obtain, and we restrict ourselves here to an elegant description of Y which provides a curious link between pairs of 'dual' queueing systems.

The server of the queue enjoys busy periods during which he works continuously; in between busy periods he has *idle periods* during which he drinks tea. Let I be the length of his first idle period.

(15) **Lemma.** *Let $L = \min\{m > 0\colon \Sigma_m < 0\}$ be the first negative ladder point of Σ. Then $I = -\Sigma_L$.*

That is, I equals the absolute value of the depth of the first negative ladder point. It is of course possible that Σ has *no* negative ladder points.

Proof. Call a customer *lucky* if he finds the queue empty as he arrives (customers who arrive at exactly the same time as the previous customer departs are deemed to be unlucky). We claim that the $(L+1)$th customer is the first lucky customer after the very first arrival. If this holds then (15) follows immediately since I is the elapsed time between the Lth departure and the $(L+1)$th arrival:

$$I = \sum_{j=1}^{L} X_{j+1} - \sum_{j=1}^{L} S_j = -\Sigma_L.$$

To verify the claim remember that

(16) $$W_n = \max\{0, V_n\} \quad \text{where} \quad V_n = \max\{U_{n-1}, U_{n-1} + U_{n-2}, \ldots, \Sigma_{n-1}\}$$

and note that the nth customer is lucky if and only if $V_n < 0$. Now

$$V_n \geqslant \Sigma_{n-1} \geqslant 0 \quad \text{for} \quad 2 \leqslant n \leqslant L,$$

and it remains to show that

$$V_{L+1} < 0.$$

To see this, note that

$$U_L + U_{L-1} + \cdots + U_{L-k} = \Sigma_L - \Sigma_{L-k-1} \leqslant \Sigma_L < 0$$

whenever $0 \leqslant k < L$. Now use (16) to obtain the result. ■

Now we are ready to extract a remarkable identity which relates 'dual pairs' of queueing systems.

(17) **Definition.** If Q is a queueing process with interarrival time distribution F_X and service time distribution F_S, then the **dual process** Q_d of Q is a queueing process with interarrival time distribution F_S and service time distribution F_X.

For example, the dual of $M(\lambda)/G/1$ is $G/M(\lambda)/1$, and vice versa; we made use of this fact in the proof of (11.4.4). The traffic densities ρ and ρ_d of Q and Q_d satisfy

$$\rho\rho_d = 1;$$

Q and Q_d cannot both be stable except in pathological instances when all their interarrival and service times almost surely take the same constant value.

(18) **Theorem.** *Let Σ and Σ_d be the random walks associated with the queue Q and its dual Q_d. Then $-\Sigma$ and Σ_d are identically distributed random walks.*

Proof. Let Q have interarrival times $\{X_n\}$ and service times $\{S_n\}$; Σ has jumps of size $U_n = S_n - X_{n+1}$ $(n \geqslant 1)$. The reflected walk $-\Sigma$, which is obtained by reflecting Σ in the x-axis, has jumps of size $-U_n = X_{n+1} - S_n$ $(n \geqslant 1)$ (see Section 3.10 for more details of the reflection principle). Write $\{S'_n\}$ and $\{X'_n\}$ for the interarrival and service times of Q_d; Σ_d has jumps of size $U'_n = X'_n - S'_{n+1}$ $(n \geqslant 1)$, which have the same distribution as the jumps of $-\Sigma$. ■

This leads to a corollary.

(19) **Theorem.** *The height Y of the first ladder point of Σ has the same distribution as the length I_d of a typical idle period in the dual queue.*

Proof. From (15), $-I_d$ is the height of the first ladder point of $-\Sigma_d$, which by (18) is distributed like the height Y of the first ladder point of Σ. ■

Here is an example of an application of these facts.

(20) **Theorem.** *Let Q be a stable queueing process with dual process Q_d. Let W be a typical equilibrium waiting time of Q and I_d a typical idle period of Q_d. Their moment generating functions are related by*

$$M_W(s) = \frac{1-\eta}{1-\eta M_{I_d}(s)}$$

where $\eta = \mathbf{P}(W > 0)$.

Proof. Use (14) and (19). ■

An application of this result is given in Exercise (22). Another application is a second derivation of the equilibrium waiting time distribution (11.4.12) of $G/M/1$; just remark that the dual of $G/M/1$ is $M/G/1$, and that idle periods of $M/G/1$ are exponentially distributed (though, of course, the server does not have many such periods if the queue is unstable).

Exercises

(21) Show that, for a $G/G/1$ queue, the starting times of the busy periods of the server constitute a renewal process.

(22) Consider a $G/M(\mu)/1$ queue in equilibrium, together with the dual (unstable) $M(\mu)/G/1$ queue. Show that the idle periods of the latter queue are exponentially distributed. Use the theory of duality of queues to deduce for the former queue that: (a) the waiting-time distribution is a mixture of an exponential distribution and an atom at zero, and (b) the equilibrium queue length is geometric.

(23) Consider $G/M(\mu)/1$, and let G be the distribution function of $S - X$ where S and X are typical (independent) service and interarrival times. Show that the *Wiener–Hopf equation*

$$F(x) = \int_{-\infty}^{x} F(x-y)\,\mathrm{d}G(y), \qquad x \geqslant 0,$$

for the limiting waiting-time distribution F is satisfied by $F(x) = 1 - \eta\,\mathrm{e}^{-\mu(1-\eta)x}$, $x \geqslant 0$. Here, η is the smallest positive root of the equation $x = M_X(\mu(x-1))$, where M_X is the moment generating function of X.

11.6 Heavy traffic

A queue settles into equilibrium if its traffic density ρ is less than 1; it is unstable if $\rho > 1$. It is our shared personal experience that many queues (such as in doctors' waiting rooms and at airport check-in desks) have a tendency to become unstable. The reason is simple: employers do not like to see their employees idle, and so they provide only just as many servers as are necessary to cope with the arriving customers. That is, they design the queueing system so that ρ is only slightly smaller than 1; the ensuing queue is long but stable, and the server experiences 'heavy traffic'. As $\rho \uparrow 1$ the equilibrium queue length Q_ρ becomes longer and longer, and it is interesting to ask for the rate at which Q_ρ approaches infinity. Often it turns out that a suitably scaled form of Q_ρ is asymptotically exponentially distributed. We describe this here for the $M/D/1$ system, leaving it to the readers to amuse themselves by finding corresponding results for other queues. In this special case, $Q_\rho \simeq Z/(1-\rho)$ as $\rho \uparrow 1$ where Z is an exponential variable.

(1) **Theorem.** *Let $\rho = \lambda d$ be the traffic density of the $M(\lambda)/D(d)/1$ queue, and let Q_ρ be a random variable with the equilibrium queue length distribution. Then $(1-\rho)Q_\rho$ converges in distribution as $\rho \uparrow 1$ to the exponential distribution with parameter $\frac{1}{2}$.*

Proof. Use (11.3.5) to see that Q_ρ has moment generating function

(2)
$$M_\rho(s) = \frac{(1-\rho)(e^s - 1)}{\exp[s - \rho(e^s - 1)] - 1} \quad \text{if} \quad \rho < 1.$$

The moment generating function of $(1-\rho)Q_\rho$ is $M_\rho((1-\rho)s)$; make the appropriate substitution in (2). Now let $\rho \uparrow 1$ and use L'Hôpital's rule to deduce that

$$M_\rho((1-\rho)s) \to \frac{1}{1-2s}. \qquad \blacksquare$$

Exercise

(3) Consider the $M(\lambda)/M(\mu)/1$ queue with $\rho = \lambda/\mu < 1$. Let Q_ρ be a random variable with the equilibrium queue length distribution, and show that $(1-\rho)Q_\rho$ converges in distribution as $\rho \uparrow 1$, the limit distribution being exponential with parameter 1.

11.7 Problems

1. **Finite waiting room.** Consider $M(\lambda)/M(\mu)/1$ with the constraint that arriving customers who see N customers in the line ahead of them leave and never return. Find the stationary distribution of queue length.

2. **Baulking.** Consider $M(\lambda)/M(\mu)/1$ with the constraint that if an arriving customer sees n customers in the line ahead of him, he joins the queue with probability $p(n)$ and otherwise leaves in disgust.

    (a) Find the stationary distribution of queue length if $p(n) = (n+1)^{-1}$.
    (b) Find the stationary distribution π of queue length if $p(n) = 2^{-n}$, and show that the probability that an arriving customer joins the queue (in equilibrium) is $\mu(1-\pi_0)/\lambda$.

3. **Series.** In a Moscow supermarket customers queue at the cash desk to pay for the

goods they want; then they proceed to a second line where they wait for the goods in question. If customers arrive in the shop like a Poisson process with parameter λ and all service times are independent and exponentially distributed, parameter μ_1 at the first desk and μ_2 at the second, find the stationary distributions of queue lengths, when they exist, and show that, at any given time, the two queue lengths are independent in equilibrium.

4. **Batch (or bulk) service.** Consider $M/G/1$, with the modification that the server may serve up to m customers simultaneously. If the queue length is less than m at the beginning of a service period then he serves everybody waiting at that time. Find a formula which is satisfied by the probability generating function of the stationary distribution of queue length at the times of departures, and evaluate this generating function explicitly in the case when $m = 2$ and service times are exponentially distributed.

5. Consider $M(\lambda)/M(\mu)/1$ where $\lambda < \mu$. Find the moment generating function of the length B of a typical busy period, and show that $\mathbf{E}(B) = (\mu - \lambda)^{-1}$ and $\operatorname{var}(B) = (\lambda + \mu)/(\mu - \lambda)^3$. Show that the density function of B is

$$f_B(x) = \frac{\sqrt{\mu/\lambda}}{x} e^{-(\lambda+\mu)x} I_1(2x\sqrt{\lambda\mu}) \quad \text{for} \quad x > 0$$

where I_1 is a modified Bessel function.

6. Consider $M(\lambda)/G/1$ in equilibrium. Obtain an expression for the mean queue length at departure times. Show that the mean waiting time in equilibrium of an arriving customer is $\frac{1}{2}\lambda\mathbf{E}(S^2)/(1 - \rho)$ where S is a typical service time and $\rho = \lambda\mathbf{E}(S)$.

Amongst all possible service-time distributions with given mean, find the one for which the mean waiting time is a minimum.

7. Let W_t be the time which a customer would have to wait in an $M(\lambda)/G/1$ queue if he were to arrive at time t. Show that the distribution function $F(x; t) = \mathbf{P}(W_t \leqslant x)$ satisfies

$$\frac{\partial F}{\partial t} = \frac{\partial F}{\partial x} - \lambda F + \lambda\mathbf{P}(W_t + S \leqslant x)$$

where S is a typical service time, independent of W_t.

Suppose that $F(x, t) \to H(x)$ for all x as $t \to \infty$, where H is a distribution function satisfying $0 = h - \lambda H + \lambda\mathbf{P}(U + S \leqslant x)$ for $x > 0$, where U is independent of S with distribution function H, and h is the density function of H on $(0, \infty)$. Show that the moment generating function M_U of U satisfies

$$M_U(\theta) = \frac{(1 - \rho)\theta}{\lambda + \theta - \lambda M_S(\theta)}$$

where ρ is the traffic density. You may assume that $\mathbf{P}(S = 0) = 0$.

8. Consider a $G/G/1$ queue in which the service times are constantly equal to 2, whilst the interarrival times take either of the values 1 and 4 with equal probability $\frac{1}{2}$. Find the limiting waiting time distribution.

9. Consider an extremely idealized model of a telephone exchange having infinitely many channels available. Calls arrive in the manner of a Poisson process with intensity λ, and each requires one channel for a length of time having the exponential distribution with parameter μ, independently of the arrival process and of the

durations of other calls. Let $Q(t)$ be the number of calls being handled at time t, and suppose that $Q(0) = I$.

Determine the probability generating function of $Q(t)$, and deduce $\mathbf{E}(Q(t))$, $\mathbf{P}(Q(t) = 0)$, and the limiting distribution of $Q(t)$ as $t \to \infty$.

In the case when $\lambda < \mu$ and the queue is in equilibrium, find the proportion of time that no channels are occupied, and the mean length of an idle period. Deduce that the mean length of a busy period is $(e^{\lambda/\mu} - 1)/\lambda$.

10. Customers arrive in a shop in the manner of a Poisson process with intensity λ, where $0 < \lambda < 1$. They are served one by one in the order of their arrival, and each requires a service time of unit length. Let $Q(t)$ be the number in the queue at time t. By comparing $Q(t)$ with $Q(t + 1)$, determine the limiting distribution of $Q(t)$ as $t \to \infty$ (you may assume that the quantities in question converge). Hence show that the mean queue length in equilibrium is $\lambda(1 - \frac{1}{2}\lambda)/(1 - \lambda)$.

Let W be the waiting time of a newly arrived customer when the queue is in equilibrium. Deduce from the results above that $\mathbf{E}(W) = \frac{1}{2}\lambda/(1 - \lambda)$.

11. Consider $M(\lambda)/D(1)/1$, and suppose that the queue is empty at time 0. Let T be the earliest time at which a customer departs leaving the queue empty. Show that the moment generating function M_T of T satisfies

$$\log\left(1 - \frac{s}{\lambda}\right) + \log M_T(s) = (s - \lambda)(1 - M_T(s)),$$

and deduce the mean value of T, distinguishing between the cases $\lambda < 1$ and $\lambda \geqslant 1$.

12. Suppose $\lambda < \mu$, and consider an $M(\lambda)/M(\mu)/1$ queue Q in equilibrium.

(a) Show that Q is a time-reversible Markov chain.
(b) Deduce the equilibrium distributions of queue length and waiting time.
(c) Show that the times of departures of customers form a Poisson process, and that $Q(t)$ is independent of the times of departures prior to t.
(d) Consider a sequence of K single-server queues such that customers arrive at the first in the manner of a Poisson process, and (for each j) on completing service in the jth queue each customer moves to the $(j + 1)$th. Service times in the jth queue are exponentially distributed with parameter μ_j, with as much independence as usual. Determine the (joint) equilibrium distribution of the queue lengths, when $\lambda < \mu_j$ for all j.

13. Consider the queue $M(\lambda)/M(\mu)/s$, where $s \geqslant 1$. Show that a stationary distribution π exists if and only if $\lambda < s\mu$, and calculate it in this case.

Suppose that the cost of operating this system in equilibrium is

$$As + B\sum_{n=s}^{\infty}(n - s + 1)\pi_n,$$

the positive constants A and B representing respectively the costs of employing a server and of the dissatisfaction of delayed customers.

Show that, for fixed μ, there is a unique value λ^* in the interval $(0, \mu)$ such that it is cheaper to have $s = 1$ than $s = 2$ if and only if $\lambda < \lambda^*$.

14. Customers arrive in a shop in the manner of a Poisson process with intensity λ. They form a single queue. There are two servers, labelled 1 and 2, server i requiring an exponentially distributed time with parameter μ_i to serve any given customer. The customer at the head of the queue is served by the first idle server; when both are

idle, an arriving customer is equally likely to choose either.

(a) Show that the queue length settles into equilibrium if and only if $\lambda < \mu_1 + \mu_2$.
(b) Show that, when in equilibrium, the queue length is a time-reversible Markov chain.
(c) Deduce the equilibrium distribution of queue length.
(d) Generalize your conclusions to queues with many servers.

15. Consider an $M(\lambda)/M(\mu)/2$ queue with the constraint that potential customers who see N customers ahead of them leave the system forever. Find the stationary distribution of queue length.

16. Consider the $D(1)/M(\mu)/1$ queue where $\mu > 1$, and let Q_n be the number of people in the queue just before the nth arrival. Let Q_μ be a random variable having as distribution the stationary distribution of the Markov chain $\{Q_n\}$. Show that $(1 - \mu^{-1})Q_\mu$ converges in distribution as $\mu \downarrow 1$, the limit distribution being exponential with parameter 2.

17. Taxis arrive at a stand in the manner of a Poisson process with intensity τ, and passengers arrive in the manner of an (independent) Poisson process with intensity π. If there are no waiting passengers, the taxis wait until passengers arrive, and then move off with the passengers, one to each taxi. If there is no taxi, passengers wait until they arrive. Suppose that initially there are neither taxis nor passengers at the stand. Show that the probability that n passengers are waiting at time t is $(\pi/\tau)^{\frac{1}{2}n} e^{-(\pi+\tau)t} I_n(2t\sqrt{\pi\tau})$, where $I_n(x)$ is the modified Bessel function, i.e., the coefficient of z^n in the power series expansion of $\exp\{\frac{1}{2}x(z + z^{-1})\}$.

18. Machines arrive for repair as a Poisson process with intensity λ. Each repair involves two stages, the ith machine to arrive being under repair for a time $X_i + Y_i$, where the pairs (X_i, Y_i), $i = 1, 2, \ldots$, are independent with a common joint distribution. Let $U(t)$ and $V(t)$ be the numbers of machines in the X-stage and Y-stage of repair at time t. Show that $U(t)$ and $V(t)$ are independent Poisson random variables.

12 Martingales

12.1 Introduction

Random processes come in many forms, and their analysis depends heavily on the assumptions that one is prepared to make about them. There are certain broad classes of processes whose general properties enable one to build attractive theories. Two such classes are Markov processes and stationary processes. A third is the class of martingales.

(1) **Definition.** A sequence $Y = \{Y_n : n \geqslant 0\}$ is a **martingale** with respect to the sequence $X = \{X_n : n \geqslant 0\}$ if, for all $n \geqslant 0$,

(a) $\mathbf{E}|Y_n| < \infty$,
(b) $\mathbf{E}(Y_{n+1} \mid X_0, X_1, \ldots, X_n) = Y_n$.

A warning note: conditional expectations are ubiquitous in this chapter. Remember that they are random variables, and that formulae of the form $\mathbf{E}(A \mid B) = C$ generally hold only 'almost surely'. We shall omit the term 'almost surely' throughout the chapter.

Here are some examples of martingales; further examples may be found in Section 7.7.

(2) **Example. Simple random walk.** A particle jumps either one step to the right or one step to the left, with corresponding probabilities p and q $(=1-p)$. Assuming the usual independence of different moves, it is clear that the position $S_n = X_1 + X_2 + \cdots + X_n$ of the particle after n steps satisfies $\mathbf{E}|S_n| \leqslant n$ and

$$\mathbf{E}(S_{n+1} \mid X_1, X_2, \ldots, X_n) = S_n + (p - q),$$

whence it is easily seen that $Y_n = S_n - n(p-q)$ defines a martingale with respect to X. ●

(3) **Example. The martingale.** The following gambling strategy is called a martingale. A gambler has a large fortune. He wagers £1 on an evens bet. If he loses then he wagers £2 on the next bet. If he loses the first n plays, then he bets £2^n on the $(n+1)$th. He is bound to win sooner or later, say on the Tth bet, at which point he ceases to play, and leaves with his profit of $2^T - (1 + 2 + 4 + \cdots + 2^{T-1})$. Thus, following this strategy, he is assured an ultimate profit. This sounds like a good policy.

Writing Y_n for the accumulated gain of the gambler after the nth play (losses count negative), we have that $Y_0 = 0$ and $|Y_n| \leqslant 1 + 2 + \cdots + 2^{n-1} = 2^n - 1$. Furthermore $Y_{n+1} = Y_n$ if the gambler has stopped by time $n + 1$, and

$$Y_{n+1} = \begin{cases} Y_n - 2^n & \text{with probability } \frac{1}{2} \\ Y_n + 2^n & \text{with probability } \frac{1}{2} \end{cases}$$

otherwise, implying that $\mathbf{E}(Y_{n+1} \mid Y_1, Y_2, \ldots, Y_n) = Y_n$. Therefore Y is a martingale (with respect to itself).

As remarked in (7.7.1), this martingale possesses a particularly disturbing feature. The random time T has a geometric distribution, $\mathbf{P}(T = n) = (\frac{1}{2})^n$ for $n \geqslant 1$, so that the mean loss of the gambler just before his ultimate win is

$$\sum_{n=1}^{\infty} (\tfrac{1}{2})^n(1 + 2 + \cdots + 2^{n-2})$$

which equals infinity. Do not follow this strategy unless your initial capital is considerably greater than that of the casino. ●

(4) **Example. De Moivre's martingale.** About a century before the martingale was fashionable amongst Paris gamblers, Abraham de Moivre made use of a (mathematical) martingale to answer the following 'gambler's ruin' question. A simple random walk on $\{0, 1, 2, \ldots, N\}$ stops when it first hits either of the absorbing barriers at 0 and at N; what is the probability that it stops at 0?

Write $X_1, X_2, \ldots$ for the steps of the walk, and S_n for the position after n steps, where $S_0 = k$. Define $Y_n = (q/p)^{S_n}$ where $p = \mathbf{P}(X_i = 1)$, $p + q = 1$, and $0 < p < 1$. We claim that

(5) $$\mathbf{E}(Y_{n+1} \mid X_1, X_2, \ldots, X_n) = Y_n \quad \text{for all } n.$$

If S_n equals 0 or N then the process has stopped by time n, implying that $S_{n+1} = S_n$ and therefore $Y_{n+1} = Y_n$. If on the other hand $S_n \neq 0, N$, then

$$\begin{aligned} \mathbf{E}(Y_{n+1} \mid X_1, X_2, \ldots, X_n) &= \mathbf{E}((q/p)^{S_n + X_{n+1}} \mid X_1, X_2, \ldots, X_n) \\ &= (q/p)^{S_n}[p(q/p) + q(q/p)^{-1}] = Y_n, \end{aligned}$$

and (5) is proved. It follows, by taking expectations of (5), that $\mathbf{E}(Y_{n+1}) = \mathbf{E}(Y_n)$ for all n, and hence $\mathbf{E}|Y_n| = \mathbf{E}|Y_0| = (q/p)^k$ for all n. In particular Y is a martingale (with respect to X).

Let T be the number of steps before the absorption of the particle at either 0 or N. De Moivre argued as follows: $\mathbf{E}(Y_n) = (q/p)^k$ for all n, and therefore $\mathbf{E}(Y_T) = (q/p)^k$. If you are willing to accept this remark, then the answer to the original question is a simple consequence, as follows. Expanding $\mathbf{E}(Y_T)$, we have that

$$\mathbf{E}(Y_T) = (q/p)^0 p_k + (q/p)^N(1 - p_k)$$

where $p_k = \mathbf{P}(\text{absorbed at } 0 \mid S_0 = k)$. However, $\mathbf{E}(Y_T) = (q/p)^k$ by assumption, and therefore

$$p_k = \frac{\rho^k - \rho^N}{1 - \rho^N} \quad \text{where} \quad \rho = q/p$$

(so long as $\rho \neq 1$), in agreement with thc calculation of (3.9.6).

This is a very attractive method, which relies on the statement that $\mathbf{E}(Y_T) = \mathbf{E}(Y_0)$ for a certain random variable T. A major part of our investigation of martingales will be to determine conditions on such random variables T which ensure that such statements are true. ●

(6) **Example. Markov chains.** Let X be a discrete-time Markov chain taking values in the countable state space S with transition matrix $\boldsymbol{P}$. Suppose that $\psi: S \rightarrow S$ is bounded and *harmonic*, which is to say that

$$\sum_{j \in S} p_{ij}\psi(j) = \psi(i) \quad \text{for all } i \in S.$$

It is easily seen that $Y = \{\psi(X_n): n \geqslant 0\}$ is a martingale with respect to X: simply use the Markov property in order to perform the calculation:

$$\mathbf{E}(\psi(X_{n+1})|X_1, X_2, \ldots, X_n) = \mathbf{E}(\psi(X_{n+1})|X_n) = \sum_{j \in S} p_{X_n, j}\psi(j) = \psi(X_n).$$

More generally, suppose that ψ is a right eigenvector of $\boldsymbol{P}$, which is to say that

$$\sum_{j \in S} p_{ij}\psi(j) = \lambda\psi(i)$$

for some λ ($\neq 0$) and all $i \in S$. Then

$$\mathbf{E}(\psi(X_{n+1})|X_1, X_2, \ldots, X_n) = \lambda\psi(X_n),$$

implying that $\lambda^{-n}\psi(X_n)$ defines a martingale so long as $\mathbf{E}|\psi(X_n)| < \infty$ for all n. ●

Central to the definition of a martingale is the idea of conditional expectation, a subject developed to some extent in Chapter 7. As described there, the most general form of conditional expectation is of the following nature. Let Y be a random variable on the probability space $(\Omega, \mathscr{F}, \mathbf{P})$ having finite mean, and let $\mathscr{G}$ be a sub-σ-field of $\mathscr{F}$. The conditional expectation of Y given $\mathscr{G}$, written $\mathbf{E}(Y \mid \mathscr{G})$, is a $\mathscr{G}$-measurable random variable satisfying

(7) $$\mathbf{E}([Y - \mathbf{E}(Y \mid \mathscr{G})]I_G) = 0 \quad \text{for all } G \in \mathscr{G},$$

where I_G is the indicator function of G. There is a corresponding general definition of a martingale. In preparation for this, we introduce the following terminology. Suppose that $\mathscr{F} = \{\mathscr{F}_0, \mathscr{F}_1, \ldots\}$ is a sequence of sub-σ-fields of $\mathscr{F}$; we call $\mathscr{F}$ a *filtration* if $\mathscr{F}_n \subseteq \mathscr{F}_{n+1}$ for all n. A sequence $Y = \{Y_n: n \geqslant 0\}$ is said to be *adapted* to the filtration $\mathscr{F}$ if Y_n is $\mathscr{F}_n$-measurable for all n. Given a filtration $\mathscr{F}$, we normally write $\mathscr{F}_\infty = \lim_{n \to \infty} \mathscr{F}_n$ for the smallest σ-field containing $\mathscr{F}_n$ for all n.

(8) **Definition.** Let $\mathscr{F}$ be a filtration of the probability space $(\Omega, \mathscr{F}, \mathbf{P})$, and let Y be a sequence of random variables which is adapted to $\mathscr{F}$. We call the pair $(Y, \mathscr{F}) = \{(Y_n, \mathscr{F}_n): n \geqslant 0\}$ a **martingale** if, for all $n \geqslant 0$,

(a) $\mathbf{E}|Y_n| < \infty$,
(b) $\mathbf{E}(Y_{n+1} \mid \mathscr{F}_n) = Y_n$.

The former definition (1) is retrieved by choosing $\mathscr{F}_n = \sigma(X_0, X_1, \ldots, X_n)$, the smallest σ-field with respect to which each of the variables $X_0, X_1, \ldots, X_n$ is measurable. We shall sometimes suppress reference to the filtration $\mathscr{F}$, speaking only of a martingale Y.

Note that, if Y is a martingale with respect to $\mathscr{F}$, then it is also a martingale with respect to $\mathscr{G}$ where $\mathscr{G}_n = \sigma(Y_0, Y_1, \ldots, Y_n)$. A further minor point is that martingales need not be infinite in extent: a finite sequence

$$\{(Y_n, \mathscr{F}_n): 0 \leqslant n \leqslant N\}$$

satisfying the above definition is also termed a martingale.

There are many cases of interest in which the martingale condition $\mathbf{E}(Y_{n+1} \mid \mathscr{F}_n) = Y_n$ does not hold, being replaced instead by an inequality: $\mathbf{E}(Y_{n+1} \mid \mathscr{F}_n) \geqslant Y_n$ for all n, or $\mathbf{E}(Y_{n+1} \mid \mathscr{F}_n) \leqslant Y_n$ for all n. Sequences satisfying such inequalities have many of the properties of martingales, and we have special names for them.

(9) **Definition.** Let $\mathscr{F}$ be a filtration of the probability space $(\Omega, \mathscr{F}, \mathbf{P})$, and let Y be a sequence of random variables which is adapted to $\mathscr{F}$. We call the pair $(Y, \mathscr{F})$ a **submartingale** if, for all $n \geqslant 0$,

(a) $\mathbf{E}(Y_n^+) < \infty$,
(b) $\mathbf{E}(Y_{n+1} \mid \mathscr{F}_n) \geqslant Y_n$,

or a **supermartingale** if, for all $n \geqslant 0$,

(c) $\mathbf{E}(Y_n^-) < \infty$,
(d) $\mathbf{E}(Y_{n+1} \mid \mathscr{F}_n) \leqslant Y_n$.

Remember that $X^+ = \max\{0, X\}$ and $X^- = -\min\{0, X\}$, so that $X = X^+ - X^-$ and $|X| = X^+ + X^-$. The moment conditions (a) and (c) are weaker than the condition that $\mathbf{E}|Y_n| < \infty$. Note that Y is a martingale if and only if it is both a submartingale and a supermartingale. Also, Y is a submartingale if and only if $-Y$ is a supermartingale.

Sometimes we shall write that $(Y_n, \mathscr{F}_n)$ is a (sub/super)martingale in cases where we mean the corresponding statement for $(Y, \mathscr{F})$.

It can be somewhat tiresome to deal with sub(/super)martingales and martingales separately, keeping track of their various properties. The general picture is somewhat simplified by the following result, which expresses a submartingale as the sum of a martingale and an increasing 'predictable' process. We shall not make use of this decomposition in the rest of the chapter. Here is a piece of notation. We call the pair $(S, \mathscr{F}) = \{(S_n, \mathscr{F}_n): n \geqslant 0\}$ *predictable* if S_n is $\mathscr{F}_{n-1}$-measurable for all $n \geqslant 1$. We call a predictable process $(S, \mathscr{F})$ *increasing* if $S_0 = 0$ and $\mathbf{P}(S_n \leqslant S_{n+1}) = 1$ for all n.

(10) **Theorem. Doob decomposition.** *A submartingale* $(Y, \mathscr{F})$ *with finite means may be expressed in the form*

(11) $$Y_n = M_n + S_n$$

where $(M, \mathscr{F})$ *is a martingale, and* $(S, \mathscr{F})$ *is an increasing predictable process. This decomposition is unique.*

The process $(S, \mathscr{F})$ in (11) is called the *compensator* of the submartingale $(Y, \mathscr{F})$. Note that compensators have finite mean, since $0 \leqslant S_n \leqslant Y_n^+ - M_n$, implying that

(12) $$\mathbf{E}|S_n| \leqslant \mathbf{E}(Y_n^+) + \mathbf{E}|M_n|.$$

Proof. We define M and S explicitly as follows: $M_0 = Y_0$, $S_0 = 0$,

$$M_{n+1} - M_n = Y_{n+1} - \mathbf{E}(Y_{n+1} \mid \mathscr{F}_n), \qquad S_{n+1} - S_n = \mathbf{E}(Y_{n+1} \mid \mathscr{F}_n) - Y_n,$$

for $n \geqslant 0$. It is easy to check (*exercise*) that $(M, \mathscr{F})$ and $(S, \mathscr{F})$ satisfy the statement of the theorem. To see uniqueness, suppose that $Y_n = M'_n + S'_n$ is another such decomposition. Then

$$\begin{aligned} Y_{n+1} - Y_n &= (M'_{n+1} - M'_n) + (S'_{n+1} - S'_n) \\ &= (M_{n+1} - M_n) + (S_{n+1} - S_n). \end{aligned}$$

Take conditional expectations given $\mathscr{F}_n$ to obtain $S'_{n+1} - S'_n = S_{n+1} - S_n$, $n \geqslant 0$. However, $S'_0 = S_0 = 0$, and therefore $S'_n = S_n$, implying that $M'_n = M_n$. (Most of the last few statements should be qualified by 'almost surely'.) ■

Exercises

(13) (i) If $(Y, \mathscr{F})$ is a martingale, show that $\mathbf{E}(Y_n) = \mathbf{E}(Y_0)$ for all n.
(ii) If $(Y, \mathscr{F})$ is a submartingale (respectively supermartingale) with finite means, show that $\mathbf{E}(Y_n) \geqslant \mathbf{E}(Y_0)$ (respectively $\mathbf{E}(Y_n) \leqslant \mathbf{E}(Y_0)$).

(14) Let $(Y, \mathscr{F})$ be a martingale, and show that $\mathbf{E}(Y_{n+m} \mid \mathscr{F}_n) = Y_n$ for all $n, m \geqslant 0$.

(15) Let Z_n be the size of the nth generation of a branching process with $Z_0 = 1$, having mean family-size μ and extinction probability η. Show that $Z_n \mu^{-n}$ and η^{Z_n} define martingales.

(16) Let $\{S_n : n \geqslant 0\}$ be a simple symmetric random walk on the integers with $S_0 = k$. Show that S_n and $S_n^2 - n$ are martingales. Making assumptions similar to those of de Moivre (see Example (4)), find the probability of ruin and the expected duration of the game for the gambler's ruin problem.

(17) Let $(Y, \mathscr{F})$ be a martingale such that $\mathbf{E}(Y_n^2) < \infty$ for all n. Show that, for $i \leqslant j \leqslant k$, $\mathbf{E}\{(Y_k - Y_j)Y_i\} = 0$, and $\mathbf{E}\{(Y_k - Y_j)^2 \mid \mathscr{F}_i\} = \mathbf{E}(Y_k^2 \mid \mathscr{F}_i) - \mathbf{E}(Y_j^2 \mid \mathscr{F}_i)$. Suppose there exists K such that $\mathbf{E}(Y_n^2) \leqslant K$ for all n. Show that the sequence $\{Y_n\}$ converges in mean square as $n \to \infty$.

(18) Let Y be a martingale and let u be a convex function mapping $\mathbb{R}$ to $\mathbb{R}$. Show that $\{u(Y_n) : n \geqslant 0\}$ is a submartingale provided that $\mathbf{E}(u(Y_n)^+) < \infty$ for all n.
Show that $|Y_n|$, Y_n^2, and Y_n^+ constitute submartingales whenever the appropriate moment conditions are satisfied.

(19) Let Y be a submartingale and let u be a convex non-decreasing function mapping $\mathbb{R}$ to $\mathbb{R}$. Show that $\{u(Y_n): n \geq 0\}$ is a submartingale provided that $\mathbf{E}(u(Y_n)^+) < \infty$ for all n.

Show that (subject to a moment condition) Y_n^+ constitutes a submartingale, but that $|Y_n|$ and Y_n^2 need not constitute submartingales.

(20) Let X be a discrete-time Markov chain with countable state space S and transition matrix $\boldsymbol{P}$. Suppose that $\psi: S \to \mathbb{R}$ is bounded and satisfies $\sum_{j \in S} p_{ij}\psi(j) \leq \lambda\psi(i)$ for some $\lambda > 0$ and all $i \in S$. Show that $\lambda^{-n}\psi(X_n)$ constitutes a supermartingale.

12.2 Martingale differences and Hoeffding's inequality

Much of the theory of martingales is concerned with their behaviour as $n \to \infty$, and particularly with their properties of convergence. Of supreme importance is the martingale convergence theorem, a general result of great power and with many applications. Before giving an account of that theorem (in the next section), we describe a bound on the degree of fluctuation of a martingale. This bound is straightforward to derive and has many important applications.

Let $(Y, \mathcal{F})$ be a martingale. The sequence of *martingale differences* is the sequence $D = \{D_n: n \geq 1\}$ defined by $D_n = Y_n - Y_{n-1}$, so that

(1) $$Y_n = Y_0 + \sum_{i=1}^{n} D_i.$$

Note that the sequence D is such that D_n is $\mathcal{F}_n$-measurable, $\mathbf{E}|D_n| < \infty$, and

(2) $$\mathbf{E}(D_{n+1} \mid \mathcal{F}_n) = 0 \quad \text{for all } n.$$

(3) **Theorem. Hoeffding's inequality.** *Let $(Y, \mathcal{F})$ be a martingale, and suppose that there exists a sequence $K_1, K_2, \ldots$ of real numbers such that $\mathbf{P}(|Y_n - Y_{n-1}| \leq K_n) = 1$ for all n. Then*

$$\mathbf{P}(|Y_n - Y_0| \geq x) \leq 2\exp\left(-\tfrac{1}{2}x^2 \Big/ \sum_{i=1}^{n} K_i^2\right), \qquad x > 0.$$

That is to say, if the martingale differences are bounded (almost surely) then there is only a small chance of a large deviation of Y_n from its initial value Y_0.

Proof. We begin with an elementary inequality. If $\psi > 0$, then the function $g(d) = e^{\psi d}$ is convex, whence it follows that

(4) $$e^{\psi d} \leq \tfrac{1}{2}(1-d)\,e^{-\psi} + \tfrac{1}{2}(1+d)\,e^{\psi} \quad \text{if } |d| \leq 1.$$

Applying this to a random variable D having mean 0 and satisfying $\mathbf{P}(|D| \leq 1) = 1$, we obtain

(5) $$\mathbf{E}(e^{\psi D}) \leq \tfrac{1}{2}(e^{-\psi} + e^{\psi}) < e^{\psi^2/2},$$

by a comparison of the coefficients of ψ^{2n} for $n \geq 0$.

Moving to the proof proper, it is a consequence of Markov's inequality (7.3.1) that

(6) $$\mathbf{P}(Y_n - Y_0 \geqslant x) \leqslant e^{-\theta x}\mathbf{E}(e^{\theta(Y_n - Y_0)})$$

for $\theta > 0$. Writing $D_n = Y_n - Y_{n-1}$, we have that

$$\mathbf{E}(e^{\theta(Y_n - Y_0)}) = \mathbf{E}(e^{\theta(Y_{n-1} - Y_0)}\, e^{\theta D_n}).$$

By conditioning on $\mathscr{F}_{n-1}$, we obtain

(7) $$\begin{aligned}\mathbf{E}(e^{\theta(Y_n - Y_0)} \mid \mathscr{F}_{n-1}) &= e^{\theta(Y_{n-1} - Y_0)}\mathbf{E}(e^{\theta D_n} \mid \mathscr{F}_{n-1}) \\ &\leqslant e^{\theta(Y_{n-1} - Y_0)}\, e^{\theta^2 K_n^2/2},\end{aligned}$$

where we have used the fact that $Y_{n-1} - Y_0$ is $\mathscr{F}_{n-1}$-measurable, in addition to (5) applied to the random variable D_n/K_n. We take expectations of (7) and iterate to find that

$$\mathbf{E}(e^{\theta(Y_n - Y_0)}) \leqslant \mathbf{E}(\exp[\theta(Y_{n-1} - Y_0)])\exp(\theta^2 K_n^2/2) \leqslant \exp\left(\tfrac{1}{2}\theta^2 \sum_{i=1}^{n} K_i^2\right).$$

Therefore, by (6),

$$\mathbf{P}(Y_n - Y_0 \geqslant x) \leqslant \exp\left(-\theta x + \tfrac{1}{2}\theta^2 \sum_{i=1}^{n} K_i^2\right)$$

for all $\theta > 0$. Suppose $x > 0$, and set $\theta = x/\sum_{i=1}^{n} K_i^2$ (this is the value which minimizes the exponent); we obtain

$$\mathbf{P}(Y_n - Y_0 \geqslant x) \leqslant \exp\left(-\tfrac{1}{2}x^2 \Big/ \sum_{i=1}^{n} K_i^2\right), \qquad x > 0.$$

The same argument is valid with $Y_n - Y_0$ replaced by $Y_0 - Y_n$, and the claim of the theorem follows by adding the two (identical) bounds together. ■

(8) **Example. Large deviations.** Let $X_1, X_2, \ldots$ be independent random variables, X_i having the Bernoulli distribution with parameter p. We set

$$S_n = X_1 + X_2 + \cdots + X_n$$

and $Y_n = S_n - np$ to obtain a martingale Y. It is a consequence of Hoeffding's inequality that

$$\mathbf{P}(|S_n - np| \geqslant x\sqrt{n}) \leqslant 2\exp[-x^2/(2\mu)] \quad \text{for} \quad x > 0,$$

where $\mu = \max\{p, 1 - p\}$. This is an inequality of a type encountered already as Bernstein's inequality (2.2.4), and explored in greater depth in Section 5.11. ●

(9) **Example. Bin packing.** The bin packing problem is a basic problem of operations research. Given n objects with sizes $x_1, x_2, \ldots, x_n$, and an unlimited collection of bins each of size 1, what is the minimum number of

bins required in order to pack the objects? In the randomized version of this problem, we suppose that the objects have independent random sizes $X_1, X_2, \ldots$ having some common distribution on $[0, 1]$. Let B_n be the (random) number of bins required in order to pack $X_1, X_2, \ldots, X_n$ efficiently. It may be shown that B_n grows approximately linearly in n, in that there exists a positive constant β such that $n^{-1}B_n \to \beta$ a.s. and in mean square as $n \to \infty$. We shall not prove this here, but note its consequence:

(10) $$\frac{1}{n}\mathbf{E}(B_n) \to \beta \quad \text{as} \quad n \to \infty.$$

The next question might be to ask how close B_n is to its mean value $\mathbf{E}(B_n)$, and Hoeffding's inequality may be brought to bear here. For $i \leqslant n$, let $Y_i = \mathbf{E}(B_n \mid \mathcal{F}_i)$, where $\mathcal{F}_i$ is the σ-field generated by $X_1, X_2, \ldots, X_i$. It is easily seen that $(Y, \mathcal{F})$ is a martingale, albeit one of finite length. Furthermore $Y_n = B_n$, and $Y_0 = \mathbf{E}(B_n)$ since $\mathcal{F}_0$ is the trivial σ-field $\{\varnothing, \Omega\}$.

Now, let $B_n(i)$ be the minimal number of bins required in order to pack all the objects *except* the ith. Since the objects are packed efficiently, we must have $B_n(i) \leqslant B_n \leqslant B_n(i) + 1$. Taking conditional expectations given $\mathcal{F}_{i-1}$ and $\mathcal{F}_i$, we obtain

(11) $$\begin{aligned} \mathbf{E}(B_n(i) \mid \mathcal{F}_{i-1}) \leqslant Y_{i-1} &\leqslant \mathbf{E}(B_n(i) \mid \mathcal{F}_{i-1}) + 1, \\ \mathbf{E}(B_n(i) \mid \mathcal{F}_i) \leqslant Y_i \quad &\leqslant \mathbf{E}(B_n(i) \mid \mathcal{F}_i) + 1. \end{aligned}$$

However, $\mathbf{E}(B_n(i) \mid \mathcal{F}_{i-1}) = \mathbf{E}(B_n(i) \mid \mathcal{F}_i)$, since we are not required to pack the ith object, and hence knowledge of X_i is irrelevant. It follows from (11) that $|Y_i - Y_{i-1}| \leqslant 1$. We may now apply Hoeffding's inequality (3) to find that

(12) $$\mathbf{P}(|B_n - \mathbf{E}(B_n)| \geqslant x) \leqslant 2\exp(-\tfrac{1}{2}x^2/n), \qquad x > 0.$$

For example, setting $x = \varepsilon n$, we see that the chance that B_n deviates from its mean by εn (or more) decays exponentially in n as $n \to \infty$. Using (10) we have also that, as $n \to \infty$,

(13) $$\mathbf{P}(|B_n - \beta n| \geqslant \varepsilon n) \leqslant 2\exp\{-\tfrac{1}{2}\varepsilon^2 n[1 + \mathrm{o}(1)]\}. \qquad \bullet$$

(14) **Example. Travelling salesman problem.** A travelling salesman is required to visit n towns but may choose his route. How does he find the shortest possible route, and how long is it? Here is a randomized version of the problem. Let $P_1 = (U_1, V_1), P_2 = (U_2, V_2), \ldots, P_n = (U_n, V_n)$ be independent and uniformly distributed points in the unit square $[0, 1]^2$; that is, suppose that $U_1, U_2, \ldots, U_n, V_1, V_2, \ldots, V_n$ are independent random variables each having the uniform distribution on $[0, 1]$. It is required to tour these points using an airplane. If we tour them in the order $P_{\pi(1)}, P_{\pi(2)}, \ldots, P_{\pi(n)}$, for some permutation π of $\{1, 2, \ldots, n\}$, the total length of the journey is

$$d(\pi) = \sum_{i=1}^{n-1} |P_{\pi(i+1)} - P_{\pi(i)}| + |P_{\pi(n)} - P_{\pi(1)}|$$

where $|\cdot|$ denotes Euclidean distance. The shortest tour has length $D_n = \min_\pi d(\pi)$. It turns out that the asymptotic behaviour of D_n for large n is given as follows: there exists a positive constant τ such that $n^{-\frac{1}{2}}D_n \to \tau$ a.s. and in mean square. We shall not prove this, but note the consequence that

(15) $$\frac{1}{\sqrt{n}}\mathbf{E}(D_n) \to \tau \quad \text{as} \quad n \to \infty.$$

How close is D_n to its mean? As in the case of bin packing, this question may be answered in part with the aid of Hoeffding's inequality. Once again, we set $Y_i = \mathbf{E}(D_n \mid \mathscr{F}_i)$ for $i \leqslant n$, where $\mathscr{F}_i$ is the σ-field generated by $P_1, P_2, \dots, P_i$. As before, $(Y, \mathscr{F})$ is a martingale, and $Y_n = D_n$, $Y_0 = \mathbf{E}(D_n)$.

Let $D_n(i)$ be the minimal tour-length through the points $P_1, P_2, \dots, P_{i-1}, P_{i+1}, \dots, P_n$, and note that $\mathbf{E}(D_n(i) \mid \mathscr{F}_i) = \mathbf{E}(D_n(i) \mid \mathscr{F}_{i-1})$. The vital inequality is

(16) $$D_n(i) \leqslant D_n \leqslant D_n(i) + 2Z_i, \qquad i \leqslant n-1,$$

where Z_i is the shortest distance from P_i to one of the points $P_{i+1}, P_{i+2}, \dots, P_n$. It is obvious that $D_n \geqslant D_n(i)$ since every tour of all n points includes a tour of $P_1, \dots, P_{i-1}, P_{i+1}, \dots, P_n$. To obtain the second inequality of (16), we argue as follows. Suppose that P_j is the closest point to P_i amongst the set $\{P_{i+1}, P_{i+2}, \dots, P_n\}$. One way of visiting all n points is to follow the optimal tour of $P_1, \dots, P_{i-1}, P_{i+1}, \dots, P_n$, and on arriving at P_j we make a return trip to P_i. The resulting trajectory is not quite a tour, but it can be turned into a tour by not landing at P_j on the return but going directly to the next point; the resulting tour has length no greater than $D_n(i) + 2Z_i$.

We take conditional expectations of (16) to obtain

$$\begin{aligned}\mathbf{E}(D_n(i) \mid \mathscr{F}_{i-1}) \leqslant Y_{i-1} &\leqslant \mathbf{E}(D_n(i) \mid \mathscr{F}_{i-1}) + 2\mathbf{E}(Z_i \mid \mathscr{F}_{i-1})\\ \mathbf{E}(D_n(i) \mid \mathscr{F}_i) \leqslant Y_i \quad &\leqslant \mathbf{E}(D_n(i) \mid \mathscr{F}_i) + 2\mathbf{E}(Z_i \mid \mathscr{F}_i)\end{aligned}$$

and hence

(17) $$|Y_i - Y_{i-1}| \leqslant 2\max\{\mathbf{E}(Z_i \mid \mathscr{F}_i), \mathbf{E}(Z_i \mid \mathscr{F}_{i-1})\}, \qquad i \leqslant n-1.$$

In order to estimate the right-hand side here, let $Q \in [0,1]^2$, and let $Z_i(Q)$ be the shortest distance from Q to the closest of a collection of $n-i$ points chosen uniformly at random from the unit square. If $Z_i(Q) > x$ then no point lies within the circle $C(x, Q)$ having radius x and centre at Q. Note that $\sqrt{2}$ is the largest possible distance between two points in the square. Now, there exists c such that, for all $x \in (0, \sqrt{2}]$, the intersection of $C(x, Q)$ with the unit square has area at least cx^2, uniformly in Q. Therefore

(18) $$\mathbf{P}(Z_i(Q) > x) \leqslant (1 - cx^2)^{n-i}, \qquad 0 < x \leqslant \sqrt{2}.$$

Integrating over x, we find that

$$\mathbf{E}(Z_i(Q)) \leqslant \int_0^{\sqrt{2}} (1 - cx^2)^{n-i}\,dx \leqslant \int_0^{\sqrt{2}} \exp[-cx^2(n-i)]\,dx < \frac{C}{\sqrt{(n-i)}}$$

for some constant C; *exercise*. Returning to (17), we deduce that $\mathbf{E}(Z_i \mid \mathcal{F}_i)$ and $\mathbf{E}(Z_i \mid \mathcal{F}_{i-1})$ are smaller than $C(n-i)^{-\frac{1}{2}}$, whence $|Y_i - Y_{i-1}| \leqslant 2C(n-i)^{-\frac{1}{2}}$, $i \leqslant n-1$. For the case $i = n$, we use the trivial bound $|Y_n - Y_{n-1}| \leqslant 2\sqrt{2}$, being twice the length of the diagonal of the square.

Applying Hoeffding's inequality, we obtain

(19) $$\mathbf{P}(|D_n - \mathbf{E}D_n| \geqslant x) \leqslant 2\exp\left(-\frac{x^2}{2(8 + \sum_{i=1}^{n-1} 4C^2/i)}\right)$$
$$\leqslant 2\exp(-Ax^2/\log n), \qquad x > 0,$$

for some positive constant A. Combining this with (15), we find that

$$\mathbf{P}(|D_n - \tau\sqrt{n}| \geqslant \varepsilon\sqrt{n}) \leqslant 2\exp(-B\varepsilon^2 n/\log n), \qquad \varepsilon > 0,$$

for some positive constant B and all large n. ●

(20) **Example. Markov chains.** Let $X = \{X_n : n \geqslant 0\}$ be an irreducible aperiodic Markov chain on the finite state space S with transition matrix $\boldsymbol{P}$. Denote by $\boldsymbol{\pi}$ the stationary distribution of X, and suppose that X_0 has distribution $\boldsymbol{\pi}$, so that X is stationary. Fix a state $s \in S$, and let $N(n)$ be the number of visits of $X_1, X_2, \ldots, X_n$ to s. The sequence N is a delayed renewal process, and therefore $n^{-1}N(n) \xrightarrow{\text{a.s.}} \pi_s$ as $n \to \infty$. The convergence is rather fast, as the following (somewhat overcomplicated) argument indicates.

Let $\mathcal{F}_0 = \{\varnothing, \Omega\}$ and, for $0 < m \leqslant n$, let $\mathcal{F}_m = \sigma(X_1, X_2, \ldots, X_m)$. Set $Y_m = \mathbf{E}(N(n) \mid \mathcal{F}_m)$ for $m \geqslant 0$, so that $(Y_m, \mathcal{F}_m)$ is a martingale. Note that $Y_n = N(n)$ and $Y_0 = \mathbf{E}(N(n)) = n\pi_s$ by stationarity.

We write $N(m, n) = N(n) - N(m)$, $0 \leqslant m \leqslant n$, the number of visits to s by the subsequence $X_{m+1}, \ldots, X_n$. Now

$$Y_m = \mathbf{E}(N(m) \mid \mathcal{F}_m) + \mathbf{E}(N(m,n) \mid \mathcal{F}_m) = N(m) + \mathbf{E}(N(m,n) \mid X_m)$$

by the Markov property. Therefore, if $m \geqslant 1$,

$$\begin{aligned} Y_m - Y_{m-1} &= [N(m) - N(m-1)] \\ &\quad + [\mathbf{E}(N(m,n) \mid X_m) - \mathbf{E}(N(m-1,n) \mid X_{m-1})] \\ &= \mathbf{E}(N(m-1,n) \mid X_m) - \mathbf{E}(N(m-1,n) \mid X_{m-1}) \end{aligned}$$

since $N(m) - N(m-1) = \delta_{X_m, s}$, the Kronecker delta. It follows that

$$\begin{aligned} |Y_m - Y_{m-1}| &\leqslant \max_{t, u \in S} |\mathbf{E}(N(m-1,n) \mid X_m = t) - \mathbf{E}(N(m-1,n) \mid X_{m-1} = u)| \\ &= \max_{t, u \in S} |D_m(t,u)| \end{aligned}$$

where, by the time-homogeneity of the process,

(21) $$D_m(t,u) = \mathbf{E}(N(n-m+1) \mid X_1 = t) - \mathbf{E}(N(n-m+1) \mid X_0 = u).$$

It is easily seen that

$$\mathbf{E}(N(n-m+1) \mid X_1 = t) \leqslant \delta_{ts} + \mathbf{E}(T_{tu}) + \mathbf{E}(N(n-m+1) \mid X_0 = u),$$

where $\mathbf{E}(T_{xy})$ is the mean first-passage time from state x to state y; just wait for the first passage to u, counting one for each moment which elapses. Similarly

$$\mathbf{E}(N(n-m+1) \mid X_0 = u) \leqslant \mathbf{E}(T_{ut}) + \mathbf{E}(N(n-m+1) \mid X_1 = t).$$

Hence, by (21), $|D_m(t, u)| \leqslant 1 + \max\{\mathbf{E}(T_{tu}), \mathbf{E}(T_{ut})\}$, implying that

(22) $$|Y_m - Y_{m-1}| \leqslant 1 + \mu$$

where $\mu = \max\{\mathbf{E}(T_{xy}) : x, y \in S\}$; note that $\mu < \infty$ since S is finite. Applying Hoeffding's inequality, we deduce that

$$\mathbf{P}(|N(n) - n\pi_s| \geqslant x) \leqslant 2\exp\left(-\frac{x^2}{2n(\mu+1)}\right), \qquad x > 0.$$

Setting $x = n\varepsilon$, we obtain

(23) $$\mathbf{P}\left(\left|\frac{1}{n}N(n) - \pi_s\right| \geqslant \varepsilon\right) \leqslant 2\exp\left(-\frac{n\varepsilon^2}{2(\mu+1)}\right), \qquad \varepsilon > 0,$$

a large-deviation estimate which decays exponentially fast as $n \to \infty$. Similar inequalities may be established by other means, more elementary than those used above. ●

Exercises

(24) **Knapsack problem.** It is required to pack a knapsack to maximum benefit. Suppose you have n objects, the ith object having volume V_i and worth W_i, where $V_1, V_2, \dots, V_n, W_1, W_2, \dots, W_n$ are independent non-negative random variables with finite means, and $W_i \leqslant M$ for all i and some fixed M. Your knapsack has volume c, and you wish to maximize the total worth of the objects packed in it. That is, you wish to find the vector $z_1, z_2, \dots, z_n$ of zeros and ones such that $\sum_1^n z_i V_i \leqslant c$ and which maximizes $\sum_1^n z_i W_i$. Let Z be the maximal possible worth of the knapsack's contents, and show that $\mathbf{P}(|Z - \mathbf{E}Z| \geqslant x) \leqslant 2\exp\{-x^2/(2nM^2)\}$ for $x > 0$.

(25) **Graph colouring.** Given n vertices $v_1, v_2, \dots, v_n$, for each $1 \leqslant i < j \leqslant n$ we place an edge between v_i and v_j with probability p; different pairs are joined independently of each other. We call v_i and v_j *neighbours* if they are joined by an edge. The *chromatic number* χ of the ensuing graph is the minimal number of pencils of different colours which are required in order that each vertex may be coloured differently from each of its neighbours. Show that $\mathbf{P}(|\chi - \mathbf{E}\chi| \geqslant x) \leqslant 2\exp\{-\frac{1}{2}x^2/n\}$ for $x > 0$.

12.3 Crossings and convergence

Martingales are of immense value in proving convergence theorems, and the following famous result has many applications.

(1) **Martingale convergence theorem.** *Let $(Y, \mathcal{F})$ be a submartingale and suppose that $\mathbf{E}(Y_n^+) \leqslant M$ for some M and all n. There exists a random variable Y_∞ such that $Y_n \xrightarrow{\text{a.s.}} Y_\infty$ as $n \to \infty$. We have in addition that* (i) *Y_∞ has finite mean if $\mathbf{E}|Y_0| < \infty$, and* (ii) *$Y_n \xrightarrow{1} Y_\infty$ if $\{Y_n: n \geqslant 0\}$ is uniformly integrable.*

It follows of course that any submartingale or supermartingale $(Y, \mathcal{F})$ converges almost surely if it satisfies $\mathbf{E}|Y_n| \leqslant M$.

The key step in the classical proof of this theorem is 'Snell's upcrossings-inequality'. Suppose that $y = \{y_n: n \geqslant 0\}$ is a real sequence, and $[a, b]$ is a real interval. An upcrossing of $[a, b]$ is defined to be a crossing by y of $[a, b]$ in the upwards direction. More precisely, we define $T_1 = \min\{n: y_n \leqslant a\}$, the first time that y hits the interval $(-\infty, a]$, and $T_2 = \min\{n > T_1: y_n \geqslant b\}$, the first subsequent time when y hits $[b, \infty)$; we call $[T_1, T_2]$ an *upcrossing* of $[a, b]$. In addition, let

$$T_{2k-1} = \min\{n > T_{2k-2}: y_n \leqslant a\}, \qquad T_{2k} = \min\{n > T_{2k-1}: y_n \geqslant b\},$$

for $k \geqslant 2$, so that the upcrossings of $[a, b]$ are the intervals $[T_{2k-1}, T_{2k}]$ for $k \geqslant 1$. Let $U_n(a, b; y)$ be the number of upcrossings of $[a, b]$ by the subsequence $y_0, y_1, \ldots, y_n$, and let $U(a, b; y) = \lim_{n\to\infty} U_n(a, b; y)$ be the total number of such upcrossings by y.

(2) **Lemma.** *If $U(a, b; y) < \infty$ for all rationals a and b satisfying $a < b$, then $\lim_{n\to\infty} y_n$ exists (but may be infinite).*

Proof. If $\lambda = \liminf_{n\to\infty} y_n$ and $\mu = \limsup_{n\to\infty} y_n$ satisfy $\lambda < \mu$ then there exist rationals a, b such that $\lambda < a < b < \mu$. Now $y_n \leqslant a$ for infinitely many n, and $y_n \geqslant b$ similarly, implying that $U(a, b; y) = \infty$, a contradiction. Therefore $\lambda = \mu$. ■

Suppose now that $(Y, \mathcal{F})$ is a submartingale, and let $U_n(a, b; Y)$ be the number of upcrossings of $[a, b]$ by Y up to time n.

(3) **Theorem. Upcrossings inequality.** *If $a < b$ then*

$$\mathbf{E}U_n(a, b; Y) \leqslant \frac{\mathbf{E}((Y_n - a)^+)}{b - a}.$$

Proof. Setting $Z_n = (Y_n - a)^+$, we have by Exercise (12.1.19) that $(Z, \mathcal{F})$ is a non-negative submartingale. Upcrossings by Y of $[a, b]$ correspond to upcrossings by Z of $[0, b - a]$, so that $U_n(a, b; Y) = U_n(0, b - a; Z)$.

Let $[T_{2k-1}, T_{2k}]$, $k \geqslant 1$, be the upcrossings by Z of $[0, b - a]$, and define the indicator functions

$$I_i = \begin{cases} 1 & \text{if } i \in (T_{2k-1}, T_{2k}] \text{ for some } k, \\ 0 & \text{otherwise.} \end{cases}$$

Note that I_i is $\mathcal{F}_{i-1}$-measurable, since

$$\{I_i = 1\} = \bigcup_k \left(\{T_{2k-1} \leqslant i-1\}\setminus\{T_{2k} \leqslant i-1\}\right),$$

an event which depends on $Y_0, Y_1, \ldots, Y_{i-1}$ only. Now

(4) $$(b-a)U_n(0, b-a; Z) \leqslant \mathbf{E}\left(\sum_{i=1}^{n} (Z_i - Z_{i-1})I_i\right),$$

since each upcrossing of $[0, b-a]$ contributes an amount of at least $b-a$ to the summation. However

(5) $$\begin{aligned}\mathbf{E}((Z_i - Z_{i-1})I_i) &= \mathbf{E}(\mathbf{E}[(Z_i - Z_{i-1})I_i \mid \mathcal{F}_{i-1}]) = \mathbf{E}(I_i[\mathbf{E}(Z_i \mid \mathcal{F}_{i-1}) - Z_{i-1}]) \\ &\leqslant \mathbf{E}[\mathbf{E}(Z_i \mid \mathcal{F}_{i-1}) - Z_{i-1}] = \mathbf{E}(Z_i) - \mathbf{E}(Z_{i-1})\end{aligned}$$

where we have used the fact that Z is a submartingale to obtain the inequality. Summing over i, we obtain from (4) that

$$(b-a)U_n(0, b-a; Z) \leqslant \mathbf{E}(Z_n) - \mathbf{E}(Z_0) \leqslant \mathbf{E}(Z_n)$$

and the lemma is proved. ■

Proof of Theorem (1). Suppose $(Y, \mathcal{F})$ is a submartingale and $\mathbf{E}(Y_n^+) \leqslant M$ for all n. We have from the upcrossings inequality that, if $a < b$,

$$\mathbf{E}U_n(a, b; Y) \leqslant \frac{\mathbf{E}(Y_n^+) + |a|}{b-a}$$

so that $U(a, b; Y) = \lim_{n\to\infty} U_n(a, b; Y)$ satisfies

$$\mathbf{E}U(a, b; Y) = \lim_{n\to\infty} \mathbf{E}U_n(a, b; Y) \leqslant \frac{M + |a|}{b-a}$$

for all $a < b$. Therefore $U(a, b; Y) < \infty$ a.s. for all $a < b$. Since there are only countably many rationals, it follows that, with probability 1, $U(a, b; Y) < \infty$ for all rational a and b. By Lemma (2), the sequence Y_n converges a.s. to some limit Y_∞. We argue as follows to show that $|Y_\infty| < \infty$ a.s. Since $|Y_n| = 2Y_n^+ - Y_n$ and $\mathbf{E}(Y_n \mid \mathcal{F}_0) \geqslant Y_0$, we have that

$$\mathbf{E}(|Y_n| \mid \mathcal{F}_0) = 2\mathbf{E}(Y_n^+ \mid \mathcal{F}_0) - \mathbf{E}(Y_n \mid \mathcal{F}_0) \leqslant 2\mathbf{E}(Y_n^+ \mid \mathcal{F}_0) - Y_0.$$

By Fatou's lemma,

(6) $$\mathbf{E}(|Y_\infty| \mid \mathcal{F}_0) = \mathbf{E}\left(\liminf_{n\to\infty} |Y_n| \mid \mathcal{F}_0\right) \leqslant \liminf_{n\to\infty} \mathbf{E}(|Y_n| \mid \mathcal{F}_0) \leqslant 2Z - Y_0$$

where $Z = \liminf_{n\to\infty} \mathbf{E}(Y_n^+ \mid \mathcal{F}_0)$. However $\mathbf{E}(Z) \leqslant M$ by Fatou's lemma, so that $Z < \infty$ a.s., implying that $\mathbf{E}(|Y_\infty| \mid \mathcal{F}_0) < \infty$ a.s. Hence $\mathbf{P}(|Y_\infty| < \infty \mid \mathcal{F}_0) = 1$, and therefore $\mathbf{P}(|Y_\infty| < \infty) = \mathbf{E}(\mathbf{P}(|Y_\infty| < \infty \mid \mathcal{F}_0)) = 1$.

If $\mathbf{E}|Y_0| < \infty$, we may take expectations of (6) to obtain $\mathbf{E}|Y_\infty| \leqslant 2M - \mathbf{E}(Y_0) < \infty$. That uniform integrability is enough to ensure convergence in mean is a consequence of Theorem (7.10.3). ■

The following is an immediate corollary of the martingale convergence theorem.

(7) **Theorem.** *If $(Y, \mathcal{F})$ is either a non-negative supermartingale or a non-positive submartingale, then $Y_\infty = \lim_{n\to\infty} Y_n$ exists almost surely.*

Proof. If Y is a non-positive submartingale then $\mathbf{E}(Y_n^+) = 0$, whence the result follows from (1). For a non-negative supermartingale Y, apply the same argument to $-Y$. ■

(8) **Example. Random walk.** Consider de Moivre's martingale of (12.1.4), $Y_n = (q/p)^{S_n}$ where S_n is the position after n steps of the usual simple random walk. The sequence $\{Y_n\}$ is a non-negative martingale, and hence converges almost surely to some finite limit Y as $n \to \infty$. This is not of much interest if $p = q$, since $Y_n = 1$ for all n in this case. Suppose then that $p \neq q$. The random variable Y_n takes values in the set $\{\rho^k : k = 0, \pm 1, \ldots\}$ where $\rho = q/p$. Certainly Y_n cannot converge to any given (possibly random) member of this set, since this would necessarily entail that S_n converges to a finite limit (which is obviously false). Therefore Y_n converges to a limit point of the set, not lying within the set. The only such limit point which is finite is 0, and therefore $Y_n \to 0$ a.s. Hence, $S_n \to -\infty$ a.s. if $p < q$, and $S_n \to \infty$ a.s. if $p > q$. Note that Y_n does not converge in mean, since $\mathbf{E}(Y_n) = \mathbf{E}(Y_0) \neq 0$ for all n. ●

(9) **Example. Doob's martingale** (though some ascribe the construction to Lévy). Let Z be a random variable on $(\Omega, \mathcal{F}, \mathbf{P})$ such that $\mathbf{E}|Z| < \infty$. Suppose that $\mathcal{F} = \{\mathcal{F}_0, \mathcal{F}_1, \ldots\}$ is a filtration, and write $\mathcal{F}_\infty = \lim_{n\to\infty} \mathcal{F}_n$ for the smallest σ-field containing $\mathcal{F}_n$ for all n. Now define $Y_n = \mathbf{E}(Z \mid \mathcal{F}_n)$. It is easily seen that $(Y, \mathcal{F})$ is a martingale. First, by Jensen's inequality,

$$\mathbf{E}|Y_n| = \mathbf{E}|\mathbf{E}(Z \mid \mathcal{F}_n)| \leq \mathbf{E}(\mathbf{E}(|Z| \mid \mathcal{F}_n)) = \mathbf{E}|Z| < \infty,$$

and secondly

$$\mathbf{E}(Y_{n+1} \mid \mathcal{F}_n) = \mathbf{E}[\mathbf{E}(Z \mid \mathcal{F}_{n+1}) \mid \mathcal{F}_n] = \mathbf{E}(Z \mid \mathcal{F}_n)$$

since $\mathcal{F}_n \subseteq \mathcal{F}_{n+1}$. Furthermore $\{Y_n\}$ is a uniformly integrable sequence, as shown in (7.10.13). It follows by the martingale convergence theorem that $Y_\infty = \lim_{n\to\infty} Y_n$ exists almost surely and in mean.

It is actually the case that $Y_\infty = \mathbf{E}(Z \mid \mathcal{F}_\infty)$, so that

(10) $$\mathbf{E}(Z \mid \mathcal{F}_n) \to \mathbf{E}(Z \mid \mathcal{F}_\infty) \quad \text{a.s. and in mean.}$$

To see this, one argues as follows. Let N be a positive integer. First, $Y_n I_A \to Y_\infty I_A$ a.s. for all $A \in \mathcal{F}_N$. Now $\{Y_n I_A : n \geq N\}$ is uniformly integrable, and therefore $\mathbf{E}(Y_n I_A) \to \mathbf{E}(Y_\infty I_A)$ for all $A \in \mathcal{F}_N$. On the other hand $\mathbf{E}(Y_n I_A) = \mathbf{E}(Y_N I_A) = \mathbf{E}(Z I_A)$ for all $n \geq N$ and all $A \in \mathcal{F}_N$, by the definition of conditional expectation. Hence $\mathbf{E}(ZI_A) = \mathbf{E}(Y_\infty I_A)$ for all $A \in \mathscr{F}_N$. Letting $N \to \infty$ and using a standard result of measure theory, we find that $\mathbf{E}((Z - Y_\infty)I_A) = 0$ for all $A \in \mathscr{F}_\infty$, whence $Y_\infty = \mathbf{E}(Z \mid \mathscr{F}_\infty)$.

There is an important converse to these results.

(11) **Lemma.** *Let $(Y, \mathscr{F})$ be a martingale. Then Y_n converges in mean if and only if there exists a random variable Z with finite mean such that $Y_n = \mathbf{E}(Z \mid \mathscr{F}_n)$. If $Y_n \xrightarrow{1} Y_\infty$, then $Y_n = \mathbf{E}(Y_\infty \mid \mathscr{F}_n)$.*

If such a random variable Z exists, we say that the martingale $(Y, \mathscr{F})$ is *closed*.

Proof. In the light of the previous discussion, it suffices to prove that, if $(Y, \mathscr{F})$ is a martingale which converges in mean to Y_∞, then $Y_n = \mathbf{E}(Y_\infty \mid \mathscr{F}_n)$. For any positive integer N and $A \in \mathscr{F}_N$, it is the case that $\mathbf{E}(Y_n I_A) \to \mathbf{E}(Y_\infty I_A)$; just note that $Y_n I_A \xrightarrow{1} Y_\infty I_A$ since

$$\mathbf{E}|(Y_n - Y_\infty)I_A| \leqslant \mathbf{E}|Y_n - Y_\infty| \to 0 \quad \text{as} \quad n \to \infty.$$

On the other hand, $\mathbf{E}(Y_n I_A) = \mathbf{E}(Y_N I_A)$ for $n \geqslant N$ and $A \in \mathscr{F}_N$, by the martingale property, and therefore $\mathbf{E}(Y_\infty I_A) = \mathbf{E}(Y_N I_A)$ for all $A \in \mathscr{F}_N$, whence $Y_N = \mathbf{E}(Y_\infty \mid \mathscr{F}_N)$ as required. ■●

(12) **Example. Zero–one law (7.3.12).** Let $X_0, X_1, \ldots$ be independent random variables, and let $\mathscr{T}$ be their tail σ-field; that is to say, $\mathscr{T} = \bigcap_n \mathscr{H}_n$ where $\mathscr{H}_n = \sigma(X_n, X_{n+1}, \ldots)$. Here is a proof that, for all $A \in \mathscr{T}$, either $\mathbf{P}(A) = 0$ or $\mathbf{P}(A) = 1$.

Let $A \in \mathscr{T}$ and define $Y_n = \mathbf{E}(I_A \mid \mathscr{F}_n)$ where $\mathscr{F}_n = \sigma(X_1, X_2, \ldots, X_n)$. Now $A \in \mathscr{T} \subseteq \mathscr{F}_\infty = \lim_{n\to\infty} \mathscr{F}_n$, and therefore $Y_n \to \mathbf{E}(I_A \mid \mathscr{F}_\infty) = I_A$ a.s. and in mean, by (11). On the other hand $Y_n = \mathbf{E}(I_A \mid \mathscr{F}_n) = \mathbf{P}(A)$, since A ($\in \mathscr{T}$) is independent of all events in $\mathscr{F}_n$. Hence $\mathbf{P}(A) = I_A$ almost surely, which is to say that I_A is almost surely constant. However, I_A takes values 0 and 1 only, and therefore either $\mathbf{P}(A) = 0$ or $\mathbf{P}(A) = 1$. ●

This completes the main content of this section. We terminate it with one further result of interest, being a bound related to the upcrossings inequality. For a certain type of process, one may obtain rather tight bounds on the tail of the number of upcrossings.

(13) **Theorem.** *Let $(Y, \mathscr{F})$ be a non-negative supermartingale. Then*

(14) $$\mathbf{P}\{U_n(a, b; Y) \geqslant j\} \leqslant \left(\frac{a}{b}\right)^j \mathbf{E}(\min\{1, Y_0/a\})$$

for $0 < a < b$ and $j \geqslant 0$.

Summing (14) over j, we find that

(15) $$\mathbf{E}U_n(a, b; Y) \leqslant \frac{a}{b-a}\mathbf{E}(\min\{1, Y_0/a\}),$$

an inequality which may be compared with the upcrossings inequality (3).

Proof. This is achieved by an adaptation of the proof of the upcrossings inequality (3), and we use the notation of that proof. Fix a positive integer j. We replace the indicator function I_i by the random variable

$$J_i = \begin{cases} a^{-1}(b/a)^{k-1} & \text{if } i \in (T_{2k-1}, T_{2k}] \text{ for some } k \leqslant j, \\ 0 & \text{otherwise.} \end{cases}$$

Next we let $X_0, X_1, \dots$ be given by $X_0 = \min\{1, Y_0/a\}$,

(16) $$X_n = X_0 + \sum_{i=1}^{n} J_i(Y_i - Y_{i-1}), \qquad n \geqslant 1.$$

If $T_{2j} \leqslant n$, then

$$X_n \geqslant X_0 + \sum_{k=1}^{j} a^{-1}(b/a)^{k-1}(Y_{T_{2k}} - Y_{T_{2k-1}}).$$

However, $Y_{T_{2k}} \geqslant b$ and $Y_{T_{2k+1}} \leqslant a$, so that

(17) $$Y_{T_{2k}} - \frac{b}{a} Y_{T_{2k+1}} \geqslant 0,$$

implying that

$$X_n \geqslant X_0 + a^{-1}(b/a)^{j-1} Y_{T_{2j}} - a^{-1} Y_{T_1}, \quad \text{if} \quad T_{2j} \leqslant n.$$

If $Y_0 \leqslant a$ then $T_1 = 0$ and $X_0 - a^{-1}Y_{T_1} = 0$; on the other hand, if $Y_0 > a$ then $X_0 - a^{-1}Y_{T_1} = 1 - a^{-1}Y_{T_1} > 0$. In either case it follows that

(18) $$X_n \geqslant (b/a)^j \quad \text{if} \quad T_{2j} \leqslant n.$$

Now Y is a non-negative sequence, and hence $X_n \geqslant X_0 - a^{-1}Y_{T_1} \geqslant 0$ by (16) and (17). Take expectations of (18) to obtain

(19) $$\mathbf{E}(X_n) \geqslant (b/a)^j \mathbf{P}(U_n(a, b; Y) \geqslant j),$$

and it remains to bound $\mathbf{E}(X_n)$ above. Arguing as in (5) and using the supermartingale property, we arrive at

$$\mathbf{E}(X_n) = \mathbf{E}(X_0) + \sum_{i=1}^{n} \mathbf{E}(J_i(Y_i - Y_{i-1})) \leqslant \mathbf{E}(X_0).$$

The conclusion of the theorem follows from (19). ■

(20) **Example. Simple random walk.** Consider de Moivre's martingale $Y_n = (q/p)^{S_n}$ of (12.1.4) and (8), with $p < q$. By (13), $\mathbf{P}(U_n(a, b; Y) \geqslant j) \leqslant (a/b)^j$. An upcrossing of $[a, b]$ by Y corresponds to an upcrossing of $[\log a, \log b]$ by S (with logarithms to the base q/p). Hence

$$\mathbf{P}(U_n(0, r; S) \geqslant j) = \mathbf{P}\{U_n(1, (q/p)^r; Y) \geqslant j\} \leqslant (p/q)^{rj}, \qquad j \geqslant 0.$$

Actually equality holds here in the limit as $n \to \infty$: $\mathbf{P}(U(0, r; S) \geqslant j) = (p/q)^{rj}$ for positive integers r; see (5.3.25). ●

Exercises

(21) Give a reasonable definition of a *downcrossing* of the interval $[a, b]$ by the random sequence $Y_0, Y_1, \ldots$.

(a) Show that the number of downcrossings differs from the number of upcrossings by at most 1.
(b) If $(Y, \mathscr{F})$ is a submartingale, show that the number $D_n(a, b; Y)$ of downcrossings of $[a, b]$ by Y up to time n satisfies

$$\mathbf{E}D_n(a, b; Y) \leqslant \frac{\mathbf{E}\{(Y_n - b)^+\}}{b - a}.$$

(22) If $(Y, \mathscr{F})$ is a supermartingale with finite means, show that

$$\mathbf{E}U_n(a, b; Y) \leqslant \frac{\mathbf{E}\{(Y_n - a)^-\}}{b - a}.$$

Deduce that $\mathbf{E}U_n(a, b; Y) \leqslant a/(b - a)$ if Y is non-negative and $a \geqslant 0$.

(23) Let X be a Markov chain with countable state space S and transition matrix $\boldsymbol{P}$. Suppose that X is irreducible and persistent, and that $\psi: S \to S$ is a bounded function satisfying $\sum_{j \in S} p_{ij}\psi(j) \leqslant \psi(i)$ for $i \in S$. Show that ψ is a constant function.

(24) Let $Z_1, Z_2, \ldots$ be independent random variables such that

$$Z_n = \begin{cases} a_n & \text{with probability } \frac{1}{2}n^{-2} \\ 0 & \text{with probability } 1 - n^{-2} \\ -a_n & \text{with probability } \frac{1}{2}n^{-2} \end{cases}$$

where $a_1 = 2$ and $a_n = 4\sum_{j=1}^{n-1} a_j$. Show that $Y_n = \sum_{j=1}^{n} Z_j$ defines a martingale. Show that $Y = \lim Y_n$ exists a.s., but that there exists no M such that $\mathbf{E}|Y_n| \leqslant M$ for all n.

12.4 Stopping times

We are all called upon on occasion to take an action whose nature is fixed but whose timing is optional. Commonly occurring examples include getting married or divorced, employing a secretary, having a baby, and buying a house. An important feature of such actions is that they are taken in the light of the past and present, and they may not depend on the future. Other important examples arise in considering money markets. The management of portfolios is affected by such rules as: (a) sell a currency if it weakens to a predetermined threshold, (b) buy government bonds if the exchange index falls below a given level, and so on. (Such rules are often sufficiently simple to be left to computers to implement, with occasionally spectacular consequences.)

A more mathematical example is provided by the gambling analogy. A gambler pursues a strategy which we may assume to be based upon his experience rather than his clairvoyance. That is to say, his decisions to vary his stake (or to stop gambling altogether) depend on the outcomes of the game up to the time of the decision, and no further. A gambler is able to follow the rule 'stop when ahead' but cannot be expected to follow a rule such as 'stop just before a loss'.

Such actions have the common feature that, at any time, we have sufficient information to decide whether or not to take the action *at that time*. The usual way of expressing this property in mathematical terms is as follows. Let $(\Omega, \mathscr{F}, \mathbf{P})$ be a probability space, and let $\mathscr{F} = \{\mathscr{F}_0, \mathscr{F}_1, \dots\}$ be a filtration. We think of $\mathscr{F}_n$ as representing the information which is available at time n, or more precisely the smallest σ-field with respect to which all observations up to and including time n are measurable.

(1) **Definition.** A random variable T taking values in $\{0, 1, 2, \dots\} \cup \{\infty\}$ is called a **stopping time** (with respect to the filtration $\mathscr{F}$) if $\{T = n\} \in \mathscr{F}_n$ for all $n \geqslant 0$.

Note that stopping times T satisfy

(2) $$\{T > n\} = \{T \leqslant n\}^c \in \mathscr{F}_n \quad \text{for all } n,$$

since $\mathscr{F}$ is a filtration. They are not required to be finite, but may take the value ∞. Stopping times are sometimes called *Markov times*. They were discussed in Section 6.8 in the context of birth processes.

Given a filtration $\mathscr{F}$ and a stopping time T, it is useful to introduce some notation to represent information gained up to the random time T. We denote by $\mathscr{F}_T$ the collection of all events A such that $A \cap \{T \leqslant n\} \in \mathscr{F}_n$ for all n. It is easily seen that $\mathscr{F}_T$ is a σ-field, and we think of $\mathscr{F}_T$ as the set of events whose occurrence or non-occurrence is known by time T.

(3) **Example. The martingale (12.1.3).** A fair coin is tossed repeatedly; let T be the time of the first head. Writing X_i for the number of heads on the ith toss, we have that

$$\{T = n\} = \{X_n = 1, X_j = 0 \text{ for } 1 \leqslant j < n\} \in \mathscr{F}_n$$

where $\mathscr{F}_n = \sigma(X_1, X_2, \dots, X_n)$. Therefore T is a stopping time. In this case T is finite almost surely. ●

(4) **Example. First passage times.** Let $\mathscr{F}$ be a filtration and let the random sequence X be adapted to $\mathscr{F}$, so that X_n is $\mathscr{F}_n$-measurable. For each (nice)

subset B of $\mathbb{R}$ define the *first passage time* of X to B by $T_B = \min\{n: X_n \in B\}$ with $T_B = \infty$ if $X_n \notin B$ for all n. It is easily seen that T_B is a stopping time. ●

Stopping times play an important role in the theory of martingales, as illustrated in the following examples. First, a martingale which is stopped at a random time T remains a martingale, so long as T is a stopping time.

(5) **Theorem.** *Let $(Y, \mathcal{F})$ be a submartingale and let T be a stopping time (with respect to $\mathcal{F}$). Then $(Z, \mathcal{F})$, defined by $Z_n = Y_{T \wedge n}$, is a submartingale.*

Here, as usual, we use the notation $x \wedge y = \min\{x, y\}$. If $(Y, \mathcal{F})$ is a martingale, then it is both a submartingale and a supermartingale, whence $Y_{T \wedge n}$ constitutes a martingale, by (5).

Proof. We may write

(6) $$Z_n = \sum_{t=0}^{n-1} Y_t I_{\{T=t\}} + Y_n I_{\{T \geqslant n\}},$$

whence Z_n is $\mathcal{F}_n$-measurable (using (2)) and

$$\mathbf{E}(Z_n^+) \leqslant \sum_{t=0}^{n} \mathbf{E}(Y_t^+) < \infty.$$

Also, from (6), $Z_{n+1} - Z_n = (Y_{n+1} - Y_n)I_{\{T>n\}}$, whence, using (2) and the submartingale property,

$$\mathbf{E}(Z_{n+1} - Z_n \mid \mathcal{F}_n) = \mathbf{E}(Y_{n+1} - Y_n \mid \mathcal{F}_n)I_{\{T>n\}} \geqslant 0. \qquad \blacksquare$$

One strategy open to a gambler in a casino is to change the game (think of the gambler as an investor in stocks, if you wish). If he is fortunate enough to be playing fair games, then he should not gain or lose (on average) at such a change. More formally, let $(X, \mathcal{F})$ and $(Y, \mathcal{F})$ be two martingales with respect to the filtration $\mathcal{F}$. Let T be a stopping time with respect to $\mathcal{F}$; T is the switching time from X to Y, and X_T is the 'capital' which is carried forward.

(7) **Theorem. Optional switching.** *Suppose that $X_T = Y_T$ on the event $\{T < \infty\}$. Then*

$$Z_n = \begin{cases} X_n & \text{if} \quad n < T \\ Y_n & \text{if} \quad n \geqslant T \end{cases}$$

defines a martingale with respect to $\mathcal{F}$.

Proof. We have that

(8) $$Z_n = X_n I_{\{n<T\}} + Y_n I_{\{n \geqslant T\}};$$

each summand is $\mathscr{F}_n$-measurable, and hence Z_n is $\mathscr{F}_n$-measurable. Also $\mathbf{E}|Z_n| \leqslant \mathbf{E}|X_n| + \mathbf{E}|Y_n| < \infty$. By the martingale property of X and Y,

(9)
$$\begin{aligned} Z_n &= \mathbf{E}(X_{n+1} \mid \mathscr{F}_n)I_{\{n<T\}} + \mathbf{E}(Y_{n+1} \mid \mathscr{F}_n)I_{\{n \geqslant T\}} \\ &= \mathbf{E}\big(X_{n+1}I_{\{n<T\}} + Y_{n+1}I_{\{n\geqslant T\}} \mid \mathscr{F}_n\big), \end{aligned}$$

since T is a stopping time. Now

(10)
$$\begin{aligned} X_{n+1}I_{\{n<T\}} + Y_{n+1}I_{\{n\geqslant T\}} &= Z_{n+1} + X_{n+1}I_{\{n+1=T\}} - Y_{n+1}I_{\{n+1=T\}} \\ &= Z_{n+1} + (X_T - Y_T)I_{\{n+1=T\}} \end{aligned}$$

whence, by (9) and the assumption that $X_T = Y_T$ on $\{T < \infty\}$, we have that $Z_n = \mathbf{E}(Z_{n+1} \mid \mathscr{F}_n)$, so that $(Z, \mathscr{F})$ is a martingale. ■

'Optional switching' does not disturb the martingale property. 'Optional sampling' is somewhat more problematical. Let $(Y, \mathscr{F})$ be a martingale and let $T_1, T_2, \dots$ be a sequence of stopping times satisfying $T_1 \leqslant T_2 \leqslant \cdots < \infty$. Let $Z_0 = Y_0$ and $Z_n = Y_{T_n}$, so that the sequence Z is obtained by 'sampling' the sequence Y at the stopping times T_j. It is natural to set $\mathscr{H}_n = \mathscr{F}_{T_n}$, and to ask whether $(Z, \mathscr{H})$ is a martingale. The answer in general is no. To see this, use the simple example when Y_n is the excess of heads over tails in n tosses of a fair coin, with $T_1 = \min\{n: Y_n = 1\}$; for this example $\mathbf{E}Y_0 = 0$ but $\mathbf{E}Y_{T_1} = 1$. The answer is, however, affirmative if the T_j are bounded.

(11) **Theorem. Optional sampling.** *Let $(Y, \mathscr{F})$ be a submartingale.*

(a) *If T is a stopping time and $\mathbf{P}(T \leqslant N) = 1$ for some deterministic N, then $\mathbf{E}(Y_T^+) < \infty$ and $\mathbf{E}(Y_T \mid \mathscr{F}_0) \geqslant Y_0$.*
(b) *If $T_1 \leqslant T_2 \leqslant \cdots$ is a sequence of stopping times such that $\mathbf{P}(T_j \leqslant N_j) = 1$ for some deterministic sequence N_j, then $(Z, \mathscr{H})$, defined by $(Z_0, \mathscr{H}_0) = (Y_0, \mathscr{F}_0)$, $(Z_j, \mathscr{H}_j) = (Y_{T_j}, \mathscr{F}_{T_j})$, is a submartingale.*

If $(Y, \mathscr{F})$ is a martingale, then it is both a submartingale and a supermartingale; Theorem (11) then implies that $\mathbf{E}(Y_T \mid \mathscr{F}_0) = Y_0$ for any bounded stopping time T, and furthermore $(Y_{T_j}, \mathscr{F}_{T_j})$ is a martingale for any increasing sequence $T_1, T_2, \dots$ of bounded stopping times.

Proof. Part (b) may be obtained without great difficulty by repeated application of part (a), and we therefore confine ourselves to proving (a). Suppose $\mathbf{P}(T \leqslant N) = 1$. Let $Z_n = Y_{T\wedge n}$, so that $(Z, \mathscr{F})$ is a submartingale, by (5). Therefore $\mathbf{E}(Z_N^+) < \infty$ and

(12)
$$\mathbf{E}(Z_N \mid \mathscr{F}_0) \geqslant Z_0 = Y_0,$$

and the proof is finished by observing that $Z_N = Y_{T\wedge N} = Y_T$ a.s. ■

Certain inequalities are of great value when studying the asymptotic properties of martingales. The following simple but powerful 'maximal inequality' is an easy consequence of the optional sampling theorem.

(13) **Theorem.** *Let $(Y, \mathscr{F})$ be a martingale. Then, for $x > 0$,*

(14)
$$\mathbf{P}\Big(\max_{0\leqslant m\leqslant n} Y_m \geqslant x\Big) \leqslant \frac{\mathbf{E}(Y_n^+)}{x} \quad \textit{and} \quad \mathbf{P}\Big(\max_{0\leqslant m\leqslant n} |Y_m| \geqslant x\Big) \leqslant \frac{\mathbf{E}|Y_n|}{x}.$$

Proof. Let $x > 0$, and let $T = \min\{m\colon Y_m \geqslant x\}$ be the first passage time of Y above the level x. Then $T \wedge n$ is a bounded stopping time, and therefore $\mathbf{E}(Y_0) = \mathbf{E}(Y_{T\wedge n}) = \mathbf{E}(Y_n)$ by Theorem (11a) and the martingale property. Now $\mathbf{E}(Y_{T\wedge n}) = \mathbf{E}(Y_T I_{\{T\leqslant n\}} + Y_n I_{\{T>n\}})$. However,

$$\mathbf{E}(Y_T I_{\{T\leqslant n\}}) \geqslant x\mathbf{E}(I_{\{T\leqslant n\}}) = x\mathbf{P}(T \leqslant n)$$

since $Y_T \geqslant x$, and therefore

(15)
$$\mathbf{E}(Y_n) = \mathbf{E}(Y_{T\wedge n}) \geqslant x\mathbf{P}(T \leqslant n) + \mathbf{E}(Y_n I_{\{T>n\}}),$$

whence $x\mathbf{P}(T \leqslant n) \leqslant \mathbf{E}(Y_n I_{\{T\leqslant n\}}) \leqslant \mathbf{E}(Y_n^+)$ as required for the first part of (14). As for the second part, just note that $(-Y, \mathscr{F})$ is a martingale, so that

$$\mathbf{P}\Big(\max_{0\leqslant m\leqslant n} \{-Y_m\} \geqslant x\Big) \leqslant \frac{\mathbf{E}(Y_n^-)}{x} \quad \text{for} \quad x > 0,$$

which may be added to the first part. ∎

We shall explore maximal inequalities for submartingales and supermartingales in Section 12.6.

Exercises

(16) If T_1 and T_2 are stopping times with respect to a filtration $\mathscr{F}$, show that $T_1 + T_2$, $\max\{T_1, T_2\}$, and $\min\{T_1, T_2\}$ are stopping times also.

(17) Let $X_1, X_2, \ldots$ be a sequence of non-negative independent random variables and let $N(t) = \max\{n\colon X_1 + X_2 + \cdots + X_n \leqslant t\}$. Show that $N(t) + 1$ is a stopping time with respect to a suitable filtration to be specified.

(18) Let $(Y, \mathscr{F})$ be a submartingale and $x > 0$. Show that

$$\mathbf{P}\Big(\max_{0\leqslant m\leqslant n} Y_m \geqslant x\Big) \leqslant \frac{\mathbf{E}(Y_n^+)}{x}.$$

(19) Let $(Y, \mathscr{F})$ be a non-negative supermartingale and $x > 0$. Show that

$$\mathbf{P}\Big(\max_{0\leqslant m\leqslant n} Y_m \geqslant x\Big) \leqslant \frac{\mathbf{E}(Y_0)}{x}.$$

(20) Let $(Y, \mathscr{F})$ be a submartingale and let S and T be stopping times satisfying $0 \leqslant S \leqslant T \leqslant N$ for some deterministic N. Show that $\mathbf{E}Y_0 \leqslant \mathbf{E}Y_S \leqslant \mathbf{E}Y_T \leqslant \mathbf{E}Y_N$.

(21) Let $\{S_n\}$ be a simple random walk with $S_0 = 0$ such that $0 < p = \mathbf{P}(S_1 = 1) < \frac{1}{2}$. Use de Moivre's martingale to show that $\mathbf{E}(\sup_m S_m) \leqslant p/(1-2p)$. Show further that this inequality may be replaced by an equality.

(22) Let $\mathscr{F}$ be a filtration. For any stopping time T with respect to $\mathscr{F}$, denote by $\mathscr{F}_T$ the collection of all events A such that, for all n, $A \cap \{T \leqslant n\} \in \mathscr{F}_n$. Let S and T be stopping times.

(a) Show that $\mathscr{F}_T$ is a σ-field, and that T is measurable with respect to this σ-field.
(b) If $A \in \mathscr{F}_S$, show that $A \cap \{S \leqslant T\} \in \mathscr{F}_T$.
(c) Let S and T satisfy $S \leqslant T$. Show that $\mathscr{F}_S \subseteq \mathscr{F}_T$.

12.5 Optional stopping

If you stop a martingale $(Y, \mathscr{F})$ at a fixed time n, the mean value $\mathbf{E}(Y_n)$ satisfies $\mathbf{E}(Y_n) = \mathbf{E}(Y_0)$. Under what conditions is this true if you stop after a *random* time T; that is, when is it the case that $\mathbf{E}(Y_T) = \mathbf{E}(Y_0)$? The answer to this question is very valuable in studying first-passage properties of martingales (see (12.1.4) for example). It would be unreasonable to expect such a result to hold generally unless T is required to be a stopping time.

Let T be a stopping time which is finite (in that $\mathbf{P}(T < \infty) = 1$), and let $(Y, \mathscr{F})$ be a martingale. Then $T \wedge n \to T$ as $n \to \infty$, so that $Y_{T \wedge n} \to Y_T$ a.s. It follows (as in Theorem (7.10.3)) that $\mathbf{E}(Y_0) = \mathbf{E}(Y_{T\wedge n}) \to \mathbf{E}(Y_T)$ so long as the family $\{Y_{T \wedge n} : n \geqslant 0\}$ is uniformly integrable.

The following two theorems provide useful conditions which are sufficient for the conclusion $\mathbf{E}(Y_0) = \mathbf{E}(Y_T)$.

(1) **Optional stopping theorem.** *Let $(Y, \mathscr{F})$ be a martingale and let T be a stopping time. Then $\mathbf{E}(Y_T) = \mathbf{E}(Y_0)$ if*

(a) $\mathbf{P}(T < \infty) = 1$,
(b) $\mathbf{E}|Y_T| < \infty$, *and*
(c) $\mathbf{E}(Y_n I_{\{T > n\}}) \to 0$ *as* $n \to \infty$.

(2) **Theorem.** *Let $(Y, \mathscr{F})$ be a martingale and let T be a stopping time. If Y is uniformly integrable and $\mathbf{P}(T < \infty) = 1$ then $Y_T = \mathbf{E}(Y_\infty \mid \mathscr{F}_T)$ and $Y_0 = \mathbf{E}(Y_T \mid \mathscr{F}_0)$. In particular $\mathbf{E}(Y_0) = \mathbf{E}(Y_T)$.*

Proof of (1). It is easily seen that $Y_T = Y_{T\wedge n} + (Y_T - Y_n) I_{\{T > n\}}$. Taking expectations and using the fact that $\mathbf{E}(Y_{T \wedge n}) = \mathbf{E}(Y_0)$ (see (12.4.11)), we find that

(3)
$$\mathbf{E}(Y_T) = \mathbf{E}(Y_0) + \mathbf{E}(Y_T I_{\{T>n\}}) - \mathbf{E}(Y_n I_{\{T>n\}}).$$

The last term tends to zero as $n \to \infty$, by assumption (c). As for the penultimate term,

$$\mathbf{E}(Y_T I_{\{T>n\}}) = \sum_{k=n+1}^{\infty} \mathbf{E}(Y_T I_{\{T=k\}})$$

is, by assumption (b), the tail of the convergent series $\mathbf{E}(Y_T) = \sum_k \mathbf{E}(Y_T I_{\{T=k\}})$; therefore $\mathbf{E}(Y_T I_{\{T>n\}}) \to 0$ as $n \to \infty$, and (3) yields $\mathbf{E}(Y_T) = \mathbf{E}(Y_0)$ in the limit as $n \to \infty$. ■

Proof of (2). Since $(Y, \mathscr{F})$ is uniformly integrable, we have by (12.3.1) and (12.3.11) that $Y_\infty = \lim_{n\to\infty} Y_n$ exists a.s. and $Y_n = \mathbf{E}(Y_\infty \mid \mathscr{F}_n)$. It follows from the definition (12.1.7) of conditional expectation that

(4) $$\mathbf{E}(Y_n I_A) = \mathbf{E}(Y_\infty I_A) \quad \text{for all } A \in \mathscr{F}_n.$$

Now, if $A \in \mathscr{F}_T$ then $A \cap \{T = n\} \in \mathscr{F}_n$, so that

$$\mathbf{E}(Y_T I_A) = \sum_n \mathbf{E}(Y_n I_{A\cap\{T=n\}}) = \sum_n \mathbf{E}(Y_\infty I_{A\cap\{T=n\}}) = \mathbf{E}(Y_\infty I_A),$$

whence $Y_T = \mathbf{E}(Y_\infty \mid \mathscr{F}_T)$. Secondly, since $\mathscr{F}_0 \subseteq \mathscr{F}_T$,

$$\mathbf{E}(Y_T \mid \mathscr{F}_0) = \mathbf{E}(\mathbf{E}(Y_\infty \mid \mathscr{F}_T) \mid \mathscr{F}_0) = \mathbf{E}(Y_\infty \mid \mathscr{F}_0) = Y_0.$$ ■

(5) **Example. Markov chains.** Let X be an irreducible persistent Markov chain with countable state space S and transition matrix $\boldsymbol{P}$, and let $\psi: S \to \mathbb{R}$ be a bounded function satisfying

$$\sum_{j\in S} p_{ij}\psi(j) = \psi(i) \quad \text{for all } i \in S.$$

Then $\psi(X_n)$ constitutes a martingale. Let T_i be the first passage time of X to the state i, $T_i = \min\{n: X_n = i\}$; it is easily seen that T_i is a stopping time and is (a.s.) finite. Furthermore $\{\psi(X_n)\}$ is bounded and therefore uniformly integrable. Applying Theorem (2), we obtain $\mathbf{E}(\psi(X_T)) = \mathbf{E}(\psi(X_0))$, whence $\mathbf{E}(\psi(X_0)) = \psi(i)$ for all i. Therefore ψ is a constant function. ●

(6) **Example. Symmetric simple random walk.** Let S_n be the position of the particle after n steps and suppose that $S_0 = 0$. Then $S_n = \sum_{i=1}^n X_i$ where $X_1, X_2, \ldots$ are independent and equally likely to take each of the values $+1$ and -1. It is easy to see as in (12.1.2) that $\{S_n\}$ is a martingale. Let a and b be positive integers and let $T = \min\{n: S_n = -a \text{ or } S_n = b\}$ be the earliest time at which the walk visits either $-a$ or b. Certainly T is a stopping time and satisfies the conditions of (1). Let p_a be the probability that the particle visits $-a$ before it visits b. By the optional stopping theorem,

(7) $$\mathbf{E}(S_T) = (-a)p_a + b(1 - p_a), \qquad \mathbf{E}(S_0) = 0;$$

therefore $p_a = b/(a + b)$, which agrees with the earlier result (1.7.7) when the notation is translated suitably. The sequence $\{S_n\}$ is not the only martingale available. Let $\{Y_n\}$ be given by $Y_n = S_n^2 - n$; then $\{Y_n\}$ is a martingale also. Apply (1) with T given as before to obtain $\mathbf{E}(T) = ab$. ●

(8) **Example. De Moivre's martingale (12.1.4).** Consider now a simple random walk $\{S_n\}$ with $0 < S_0 < N$, for which each step is rightwards with probability p where $0 < p = 1 - q < 1$. We have seen that $Y_n = (q/p)^{S_n}$ defines a

martingale, and furthermore the first passage time T of the walk to the set $\{0, N\}$ is a stopping time. It is easily checked that conditions (1a)–(1c) of the optional stopping theorem are satisfied, and hence $\mathbf{E}((q/p)^T) = \mathbf{E}((q/p)^{S_0})$. Therefore $p_k = \mathbf{P}(S_T = 0 \mid S_0 = k)$ satisfies $p_k + (q/p)^N(1 - p_k) = (q/p)^k$, whence p_k may be calculated as in (12.1.4). ●

When applying the optional stopping theorem it is sometimes convenient to use a more restrictive set of conditions.

(9) **Theorem.** *Let $(Y, \mathcal{F})$ be a martingale, and let T be a stopping time. Then* $\mathbf{E}(Y_T) = \mathbf{E}(Y_0)$ *if*

(a) $\mathbf{P}(T < \infty) = 1$, $\mathbf{E}T < \infty$, *and*
(b) *there exists a constant c such that*

$$\mathbf{E}(|Y_{n+1} - Y_n| \mid \mathcal{F}_n) \leqslant c \quad \textit{for all } n < T.$$

Proof. By the discussion prior to (1), it suffices to show that $\{Y_{T \wedge n} : n \geqslant 0\}$ is uniformly integrable. Let $Z_n = |Y_n - Y_{n-1}|$ for $n \geqslant 1$, and $W = Z_1 + Z_2 + \cdots + Z_T$. Certainly $|Y_{T \wedge n}| \leqslant |Y_0| + W$ for all n, and it is enough (by (7.10.4)) to show that $\mathbf{E}(W) < \infty$. We have that

(10) $$W = \sum_{i=1}^{\infty} Z_i I_{\{T \geqslant i\}}.$$

Now

$$\mathbf{E}(Z_i I_{\{T \geqslant i\}} \mid \mathcal{F}_{i-1}) = I_{\{T \geqslant i\}} \mathbf{E}(Z_i \mid \mathcal{F}_{i-1}) \leqslant c I_{\{T \geqslant i\}},$$

since $\{T \geqslant i\} = \{T \leqslant i - 1\}^c \in \mathcal{F}_{i-1}$. Therefore $\mathbf{E}(Z_i I_{\{T \geqslant i\}}) \leqslant c\mathbf{P}(T \geqslant i)$, giving by (10) that

(11) $$\mathbf{E}(W) \leqslant c \sum_{i=1}^{\infty} \mathbf{P}(T \geqslant i) = c\mathbf{E}(T) < \infty.$$ ■

(12) **Example. Wald's equation (10.2.9).** Let $X_1, X_2, \ldots$ be independent identically distributed random variables with finite mean μ, and let $S_n = \sum_{i=1}^n X_i$. It is easy to see that $Y_n = S_n - n\mu$ constitutes a martingale with respect to the filtration $\{\mathcal{F}_n\}$ where $\mathcal{F}_n = \sigma(Y_1, Y_2, \ldots, Y_n)$. Now

$$\mathbf{E}(|Y_{n+1} - Y_n| \mid \mathcal{F}_n) = \mathbf{E}|X_{n+1} - \mu| = \mathbf{E}|X_1 - \mu| < \infty.$$

We deduce from (9) that $\mathbf{E}(Y_T) = \mathbf{E}(Y_0) = 0$ for any stopping time T with finite mean, implying that

(13) $$\mathbf{E}(S_T) = \mu\mathbf{E}(T),$$

a result derived earlier in the context of renewal theory (10.2.9).

If the X_i have finite variance σ^2, it is also the case that

(14) $$\operatorname{var}(Y_T) = \sigma^2 \mathbf{E}(T),$$

if $\mathbf{E}(T) < \infty$. It is possible to prove this by applying the optional stopping theorem to the martingale $Z_n = Y_n^2 - n\sigma^2$, but this is not a simple application of (9). It may also be proved by exploiting Wald's identity (15), or more simply by the method of Exercise (10.2.15). ●

(15) **Example. Wald's identity.** This time, let $X_1, X_2, \dots$ be independent identically distributed random variables with common moment generating function $M(t) = \mathbf{E}(e^{tX})$; suppose that there exists at least one value of t ($\neq 0$) such that $1 \leqslant M(t) < \infty$, and fix t accordingly. Let $S_n = X_1 + X_2 + \cdots + X_n$, define

(16) $$Y_0 = 1, \qquad Y_n = \exp(tS_n)/M(t)^n \quad \text{for} \quad n \geqslant 1,$$

and let $\mathscr{F}_n = \sigma(X_1, X_2, \dots, X_n)$. It is clear that $(Y, \mathscr{F})$ is a martingale. When are the conditions of (9) valid? Let T be a stopping time with finite mean, and note that

(17) $$\mathbf{E}(|Y_{n+1} - Y_n| \mid \mathscr{F}_n) = Y_n \mathbf{E}\left(\left|\frac{e^{tX}}{M(t)} - 1\right|\right) \leqslant \frac{Y_n}{M(t)} \mathbf{E}(e^{tX} + M(t)) = 2Y_n.$$

Suppose that T is such that

(18) $$|S_n| \leqslant C \quad \text{for} \quad n < T,$$

where C is a constant. Now $M(t) \geqslant 1$, and

$$Y_n = \frac{e^{tS_n}}{M(t)^n} \leqslant \frac{e^{|t|C}}{M(t)^n} \leqslant e^{|t|C} \quad \text{for} \quad n < T,$$

giving by (17) that condition (9b) holds. In summary, if T is a stopping time with finite mean such that (18) holds, then

(19) $$\mathbf{E}\{\exp(tS_T)M(t)^{-T}\} = 1 \quad \text{whenever} \quad M(t) \geqslant 1,$$

an equation usually called Wald's identity.

Here is an application of (19). Suppose the X_i are not almost-surely constant, and let $T = \min\{n : S_n \leqslant -a \text{ or } S_n \geqslant b\}$ where $a, b > 0$; T is the 'first exit time' from the interval $(-a, b)$. Certainly $|S_n| \leqslant \max\{a, b\}$ if $n < T$. Furthermore $\mathbf{E}T < \infty$, which may be seen as follows. By the non-degeneracy of the X_i, there exist M and $\varepsilon > 0$ such that $\mathbf{P}(S_M > a+b) > \varepsilon$. If any of the partial sums $S_M, S_{2M} - S_M, \dots, S_{kM} - S_{(k-1)M}$ exceed $a + b$ then the process must have exited $(-a, b)$ by time kM. Therefore $\mathbf{P}(T \geqslant kM) \leqslant (1 - \varepsilon)^k$, implying that

$$\mathbf{E}(T) = \sum_{i=1}^{\infty} \mathbf{P}(T \geqslant i) \leqslant M \sum_{k=0}^{\infty} \mathbf{P}(T \geqslant kM) < \infty.$$

We conclude that (19) is valid. In many concrete cases of interest, there exists θ ($\neq 0$) such that $M(\theta) = 1$. Applying (19) with $t = \theta$, we obtain $\mathbf{E}(\exp(\theta S_T)) = 1$, or

$$\eta_a \mathbf{P}(S_T \leqslant -a) + \eta_b \mathbf{P}(S_T \geqslant b) = 1$$

where

$$\eta_a = \mathbf{E}(\exp(\theta S_T) \mid S_T \leqslant -a), \qquad \eta_b = \mathbf{E}(\exp(\theta S_T) \mid S_T \geqslant b),$$

and therefore

(20) $$\mathbf{P}(S_T \leqslant -a) = \frac{\eta_b - 1}{\eta_b - \eta_a}, \qquad \mathbf{P}(S_T \geqslant b) = \frac{1 - \eta_a}{\eta_b - \eta_a}.$$

If a and b are large, it is reasonable to suppose that $\eta_a \simeq e^{-\theta a}$ and $\eta_b \simeq e^{\theta b}$, giving the approximations

(21) $$\mathbf{P}(S_T \leqslant -a) \simeq \frac{e^{\theta b} - 1}{e^{\theta b} - e^{-\theta a}}, \qquad \mathbf{P}(S_T \geqslant b) \simeq \frac{1 - e^{-\theta a}}{e^{\theta b} - e^{-\theta a}}.$$

These approximations are of course exact if S is a simple random walk and a and b are positive integers. ●

(22) **Example. Simple random walk.** Suppose that $\{S_n\}$ is a simple random walk whose steps $\{X_i\}$ take the values 1 and -1 with respective probabilites p and q ($=1-p$). For positive integers a and b, we have from Wald's identity (19) that

(23) $$e^{-at}\mathbf{E}(M(t)^{-T} I_{\{S_T = -a\}}) + e^{bt}\mathbf{E}(M(t)^{-T} I_{\{S_T = b\}}) = 1 \quad \text{if} \quad M(t) \geqslant 1$$

where T is the first exit time of $(-a, b)$ as before, and $M(t) = p\,e^t + q\,e^{-t}$.

Setting $M(t) = s^{-1}$, we obtain a quadratic for e^t, and hence $e^t = \lambda_1(s)$ or $e^t = \lambda_2(s)$ where

$$\lambda_1(s) = \frac{1 + (1 - 4pqs^2)^{\frac{1}{2}}}{2ps}, \qquad \lambda_2(s) = \frac{1 - (1 - 4pqs^2)^{\frac{1}{2}}}{2ps}.$$

Substituting these into (23), we obtain two linear equations in the quantities

(24) $$P_1(s) = \mathbf{E}(s^T I_{\{S_T = -a\}}), \qquad P_2(s) = \mathbf{E}(s^T I_{\{S_T = b\}}),$$

with solutions

$$P_1(s) = \frac{\lambda_1^a \lambda_2^a (\lambda_1^b - \lambda_2^b)}{\lambda_1^{a+b} - \lambda_2^{a+b}}, \qquad P_2(s) = \frac{\lambda_1^a - \lambda_2^a}{\lambda_1^{a+b} - \lambda_2^{a+b}},$$

which we add to obtain the probability generating function of T,

(25) $$\mathbf{E}(s^T) = P_1(s) + P_2(s), \qquad 0 < s \leqslant 1.$$

Suppose we let $a \to \infty$, so that T becomes the time until the first passage to the point b. From (24), $P_1(s) \to 0$ as $a \to \infty$ if $0 < s < 1$, and a quick calculation gives $P_2(s) \to F_b(s)$ where

$$F_b(s) = \left(\frac{1 - (1 - 4pqs^2)^{\frac{1}{2}}}{2qs}\right)^b$$

in agreement with (5.3.5). Notice that $F_b(1) = (\min\{1, p/q\})^b$. ●

Exercises

(26) Let $(Y, \mathscr{F})$ be a martingale and T a stopping time such that $\mathbf{P}(T < \infty) = 1$. Show that $\mathbf{E}(Y_T) = \mathbf{E}(Y_0)$ if either of the following holds:

(a) $\mathbf{E}(\sup_n |Y_{T \wedge n}|) < \infty$, (b) $\mathbf{E}(|Y_{T \wedge n}|^{1+\delta}) \leqslant c$ for some $c, \delta > 0$ and all n.

(27) Let $(Y, \mathscr{F})$ be a martingale. Show that $(Y_{T \wedge n}, \mathscr{F}_n)$ is a uniformly integrable martingale for any finite stopping time T such that either

(a) $\mathbf{E}|Y_T| < \infty$ and $\mathbf{E}(|Y_n| I_{\{T > n\}}) \to 0$ as $n \to \infty$, or
(b) $\{Y_n\}$ is uniformly integrable.

(28) Let $(Y, \mathscr{F})$ be a uniformly integrable martingale, and let S and T be finite stopping times satisfying $S \leqslant T$. Prove that $Y_T = \mathbf{E}(Y_\infty \mid \mathscr{F}_T)$ and that $Y_S = \mathbf{E}(Y_T \mid \mathscr{F}_S)$, where Y_∞ is the a.s. limit as $n \to \infty$ of Y_n.

(29) Let $\{S_n : n \geqslant 0\}$ be a simple symmetric random walk with $0 < S_0 < N$ and with absorbing barriers at 0 and N. Use the optional stopping theorem to show that the mean time until absorption is $\mathbf{E}\{S_0(N - S_0)\}$.

(30) Let $\{S_n : n \geqslant 0\}$ be a simple symmetric random walk with $S_0 = 0$. Show that

$$Y_n = \frac{\cos\{\lambda[S_n - \frac{1}{2}(b - a)]\}}{(\cos \lambda)^n}$$

constitutes a martingale if $\cos \lambda \neq 0$.

Let a and b be positive integers. Show that the time T until absorption at one of two absorbing barriers at $-a$ and b satisfies

$$\mathbf{E}(\{\cos \lambda\}^{-T}) = \frac{\cos\{\frac{1}{2}\lambda(b - a)\}}{\cos\{\frac{1}{2}\lambda(b + a)\}}, \qquad 0 < \lambda < \frac{\pi}{b + a}.$$

(31) Let $\{S_n : n \geqslant 0\}$ be a simple symmetric random walk on the positive and negative integers, with $S_0 = 0$. For each of the three following random variables, determine whether or not it is a stopping time and find its mean:

$$U = \min\{n \geqslant 5 : S_n = S_{n-5} + 5\}, \quad V = U - 5, \quad W = \min\{n : S_n = 1\}.$$

12.6 The maximal inequality

In proving the convergence of a sequence $X_1, X_2, \ldots$ of random variables, it is often useful to establish an inequality of the form

$$\mathbf{P}(\max\{X_1, X_2, \ldots, X_n\} \geqslant x) \leqslant A_n(x),$$

and such an inequality is sometimes called a maximal inequality. The bound

$A_n(x)$ usually involves an expectation. Examples of such inequalities include Kolmogorov's inequality in the proof of the strong law of large numbers, and the Doob–Kolmogorov inequality (7.8.2) in the proof of the convergence of martingales with bounded second moments. Both these inequalities are special cases of the following maximal inequality for submartingales. In order to simplify the notation of this section, we shall write X_n^* for the maximum of the first $n+1$ members of a sequence $X_0, X_1, \dots$, so that $X_n^* = \max\{X_i : 0 \leqslant i \leqslant n\}$.

(1) **Theorem. Maximal inequality.**

(a) *If $(Y, \mathcal{F})$ is a submartingale, then*

$$\mathbf{P}(Y_n^* \geqslant x) \leqslant \frac{\mathbf{E}(Y_n^+)}{x} \quad \textit{for} \quad x > 0.$$

(b) *If $(Y, \mathcal{F})$ is a supermartingale and $\mathbf{E}|Y_0| < \infty$, then*

$$\mathbf{P}(Y_n^* \geqslant x) \leqslant \frac{\mathbf{E}(Y_0) + \mathbf{E}(Y_n^-)}{x} \quad \textit{for} \quad x > 0.$$

These inequalities may be improved somewhat. For example, a closer look at the proof in case (a) leads to the inequality

$$\mathbf{P}(Y_n^* \geqslant x) \leqslant \frac{1}{x}\mathbf{E}(Y_n^+ I_{\{Y_n^* \geqslant x\}}) \quad \text{for} \quad x > 0. \tag{2}$$

Proof. This is very similar to that of (12.4.13). Let $T = \min\{n: Y_n \geqslant x\}$ where $x > 0$, and suppose first that $(Y, \mathcal{F})$ is a submartingale. Then $(Y^+, \mathcal{F})$ is a non-negative submartingale with finite means by (12.1.19), and $T = \min\{n: Y_n^+ \geqslant x\}$ since $x > 0$. Applying the optional sampling theorem (12.4.11b) with stopping times $T_1 = T \wedge n$, $T_2 = n$, we obtain $\mathbf{E}(Y_{T\wedge n}^+) \leqslant \mathbf{E}(Y_n^+)$. However,

$$\mathbf{E}(Y_{T\wedge n}^+) = \mathbf{E}(Y_T^+ I_{\{T \leqslant n\}}) + \mathbf{E}(Y_n^+ I_{\{T > n\}}) \geqslant x\mathbf{P}(T \leqslant n) + \mathbf{E}(Y_n^+ I_{\{T > n\}})$$

whence

$$x\mathbf{P}(T \leqslant n) \leqslant \mathbf{E}(Y_n^+(1 - I_{\{T > n\}})) = \mathbf{E}(Y_n^+ I_{\{T \leqslant n\}}) \leqslant \mathbf{E}(Y_n^+) \tag{3}$$

as required.

Suppose next that $(Y, \mathcal{F})$ is a supermartingale. By optional sampling, $\mathbf{E}(Y_0) \geqslant \mathbf{E}(Y_{T\wedge n})$. Now

$$\mathbf{E}(Y_{T\wedge n}) = \mathbf{E}(Y_T I_{\{T \leqslant n\}} + Y_n I_{\{T > n\}}) \geqslant x\mathbf{P}(T \leqslant n) - \mathbf{E}(Y_n^-),$$

whence $x\mathbf{P}(T \leqslant n) \leqslant \mathbf{E}(Y_0) + \mathbf{E}(Y_n^-)$. ■

Part (a) of the maximal inequality may be used to handle the maximum of a submartingale, and part (b) may be used as follows to handle its minimum. Suppose that $(Y, \mathscr{F})$ is a submartingale with finite means. Then $(-Y, \mathscr{F})$ is a supermartingale, and therefore

(4) $$\mathbf{P}\left(\min_{0 \leqslant k \leqslant n} Y_k \leqslant -x\right) \leqslant \frac{\mathbf{E}(Y_n^+) - \mathbf{E}(Y_0)}{x} \quad \text{for} \quad x > 0,$$

by (1b). Using (1a) also, we find that

$$\mathbf{P}\left(\max_{0 \leqslant k \leqslant n} |Y_k| \geqslant x\right) \leqslant \frac{2\mathbf{E}(Y_n^+) - \mathbf{E}(Y_0)}{x} \leqslant \frac{3}{x} \sup_k \mathbf{E}|Y_k|;$$

sending n to infinity (and hiding a minor 'continuity' argument), we deduce that

(5) $$\mathbf{P}\left(\sup_k |Y_k| \geqslant x\right) \leqslant \frac{3}{x} \sup_k \mathbf{E}|Y_k|, \quad \text{for} \quad x > 0.$$

A slightly tighter conclusion is valid if $(Y, \mathscr{F})$ is a martingale rather than merely a submartingale. In this case, $(|Y_n|, \mathscr{F}_n)$ is a submartingale, whence (1a) yields

(6) $$\mathbf{P}\left(\sup_k |Y_k| \geqslant x\right) \leqslant \frac{1}{x} \sup_k \mathbf{E}|Y_k|, \quad \text{for} \quad x > 0.$$

(7) **Example. Doob–Kolmogorov inequality (7.8.2).** Let $(Y, \mathscr{F})$ be a martingale such that $\mathbf{E}(Y_n^2) < \infty$ for all n. Then $(Y_n^2, \mathscr{F}_n)$ is a submartingale, whence

(8) $$\mathbf{P}\left(\max_{0 \leqslant k \leqslant n} |Y_k| \geqslant x\right) = \mathbf{P}\left(\max_{0 \leqslant k \leqslant n} Y_k^2 \geqslant x^2\right) \leqslant \frac{\mathbf{E}(Y_n^2)}{x^2}$$

for $x > 0$, in agreement with (7.8.2). This is the major step in the proof of the convergence theorem (7.8.1) for martingales with bounded second moments. ●

(9) **Example. Kolmogorov's inequality.** Let $X_1, X_2, \ldots$ be independent random variables with finite means and variances. Applying the Doob–Kolmogorov inequality (8) to the martingale $Y_n = S_n - \mathbf{E}(S_n)$ where

$$S_n = X_1 + X_2 + \cdots + X_n,$$

we obtain

(10) $$\mathbf{P}\left(\max_{1 \leqslant k \leqslant n} |S_k - \mathbf{E}(S_k)| \geqslant x\right) \leqslant x^{-2} \operatorname{var}(S_n) \quad \text{for} \quad x > 0.$$

This powerful inequality is the principal step in the usual proof of the strong law of large numbers (7.5.1). See Problem (7.11.29) for a simple proof not using martingales. ●

The maximal inequality may be used to address the question of convergence in rth mean of martingales.

(11) **Theorem.** *Let* $(Y, \mathcal{F})$ *be a martingale such that* $\sup_n \mathbf{E}|Y_n^r| < \infty$ *where* $r > 1$. *Then* $Y_n \xrightarrow{r} Y_\infty$ *where* Y_∞ *is the (almost sure) limit of* Y_n.

This is not difficult to prove by way of Fatou's lemma and the theory of uniform integrability. Instead, we shall make use of the following inequality.

(12) **Lemma.** *Let* $(Y, \mathcal{F})$ *be a non-negative submartingale, and suppose that* $\mathbf{E}(Y_n^r) < \infty$ *for all* n, *where* $r > 1$. *Then*

(13)
$$\mathbf{E}(Y_n^r) \leqslant \mathbf{E}((Y_n^*)^r) \leqslant \left(\frac{r}{r-1}\right)^r \mathbf{E}(Y_n^r).$$

Proof. Certainly $Y_n \leqslant Y_n^*$, and therefore the first inequality is trivial. Turning to the second, note first that

$$\mathbf{E}((Y_n^*)^r) \leqslant \mathbf{E}((Y_0 + Y_1 + \cdots + Y_n)^r) < \infty.$$

Now, integrate by parts and use the maximal inequality (2) to obtain

$$\mathbf{E}((Y_n^*)^r) = \int_0^\infty rx^{r-1}\mathbf{P}(Y_n^* \geqslant x)\,dx \leqslant \int_0^\infty rx^{r-2}\mathbf{E}(Y_n I_{\{Y_n^* \geqslant x\}})\,dx$$
$$= \mathbf{E}\left(Y_n \int_0^{Y_n^*} rx^{r-2}\,dx\right) = \frac{r}{r-1}\mathbf{E}(Y_n(Y_n^*)^{r-1}).$$

We have by Hölder's inequality that

$$\mathbf{E}(Y_n(Y_n^*)^{r-1}) \leqslant [\mathbf{E}(Y_n^r)]^{1/r}[\mathbf{E}((Y_n^*)^r)]^{(r-1)/r}.$$

Substituting this, and solving, we obtain

$$[\mathbf{E}((Y_n^*)^r)]^{1/r} \leqslant \frac{r}{r-1}[\mathbf{E}(Y_n^r)]^{1/r}.$$ ■

Proof of Theorem (11). Using the moment condition, $Y_\infty = \lim_{n\to\infty} Y_n$ exists almost surely. Now $(|Y_n|, \mathcal{F}_n)$ is a non-negative submartingale, and hence $\mathbf{E}(\sup_k |Y_k|^r) < \infty$ by (12) and monotone convergence (5.6.12). Hence $\{Y_k^r : k \geqslant 0\}$ is uniformly integrable (Exercise (7.10.22)), implying by Exercise (7.10.18) that $Y_k \xrightarrow{r} Y_\infty$ as required. ■

12.7 Backward martingales and continuous-time martingales

The ideas of martingale theory find expression in several other contexts, of which we consider two in this section. The first of these concerns backward martingales. We call a sequence $\mathcal{G} = \{\mathcal{G}_n : n \geqslant 0\}$ of σ-fields *decreasing* if $\mathcal{G}_n \supseteq \mathcal{G}_{n+1}$ for all n.

(1) **Definition.** Let $\mathcal{G}$ be a decreasing sequence of σ-fields and let Y be a sequence of random variables which is adapted to $\mathcal{G}$. We call $(Y, \mathcal{G})$ a **backward** (or reversed) **martingale** if, for all $n \geq 0$,

(a) $\mathbf{E}|Y_n| < \infty$,
(b) $\mathbf{E}(Y_n \mid \mathcal{G}_{n+1}) = Y_{n+1}$.

Note that $\{(Y_n, \mathcal{G}_n): n = 0, 1, 2, \ldots\}$ is a backward martingale if and only if the reversed sequence $\{(Y_n, \mathcal{G}_n): n = \ldots, 2, 1, 0\}$ is a martingale, an observation which explains the use of the term.

(2) **Example.** Let $X_1, X_2, \ldots$ be independent identically distributed random variables with finite mean. Set $S_n = X_1 + X_2 + \cdots + X_n$ and let $\mathcal{G}_n = \sigma(S_n, S_{n+1}, \ldots)$. Then, using symmetry,

(3) $$\mathbf{E}(S_n \mid \mathcal{G}_{n+1}) = \mathbf{E}(S_n \mid S_{n+1}) = n\mathbf{E}(X_1 \mid S_{n+1}) = n\frac{S_{n+1}}{n+1}$$

since $S_{n+1} = \mathbf{E}(S_{n+1} \mid S_{n+1}) = (n+1)\mathbf{E}(X_1 \mid S_{n+1})$. Therefore $Y_n = S_n/n$ satisfies $\mathbf{E}(Y_n \mid \mathcal{G}_{n+1}) = Y_{n+1}$, whence $(Y, \mathcal{G})$ is a backward martingale. We shall see soon that backward martingales converge almost surely and in mean, and therefore there exists Y_∞ such that $Y_n \to Y_\infty$ a.s. and in mean. By the zero–one law (7.3.15), Y_∞ is almost surely constant, and hence $Y_\infty = \mathbf{E}(X_1)$ a.s. We have proved the strong law of large numbers. ●

(4) **Backward-martingale convergence theorem.** *Let $(Y, \mathcal{G})$ be a backward martingale. Then Y_n converges to a limit Y_∞ almost surely and in mean.*

It is striking that no extra condition is necessary to ensure the convergence of backward martingales.

Proof. Note first that $\ldots, Y_n, Y_{n-1}, \ldots, Y_1, Y_0$ is a martingale with respect to $\ldots, \mathcal{G}_n, \mathcal{G}_{n-1}, \ldots, \mathcal{G}_1, \mathcal{G}_0$, and therefore $Y_n = \mathbf{E}(Y_0 \mid \mathcal{G}_n)$ for all n. However, $\mathbf{E}|Y_0| < \infty$, and therefore $\{Y_n\}$ is uniformly integrable by (7.10.13). It is therefore sufficient to prove that Y_n converges almost surely. The usual way of doing this is via an upcrossings inequality. Applying (12.3.3) to the martingale $Y_n, Y_{n-1}, \ldots, Y_0$, we obtain that

$$\mathbf{E}U_n(a, b; Y) \leq \frac{\mathbf{E}((Y_0 - a)^+)}{b - a}$$

where $U_n(a, b; Y)$ is the number of upcrossings of $[a, b]$ by the sequence $Y_n, Y_{n-1}, \ldots, Y_0$. We let $n \to \infty$, and follow the proof of the martingale convergence theorem (12.3.1) to obtain the required result. ■

Rather than developing the theory of backward martingales in detail, we confine ourselves to one observation and an application. Let $(Y, \mathcal{G})$ be a backward martingale, and let T be a stopping time with respect to $\mathscr{G}$; that is, $\{T = n\} \in \mathscr{G}_n$ for all n. If T is bounded, say $\mathbf{P}(T \leqslant N) = 1$ for some fixed N, then the sequence $Z_N, Z_{N-1}, \dots, Z_0$ defined by $Z_n = Y_{T \vee n}$ is a martingale with respect to the appropriate sequence of σ-fields (remember that $x \vee y = \max\{x, y\}$). Hence, by (12.4.11a),

(5) $$\mathbf{E}(Y_T \mid \mathscr{G}_N) = Y_N.$$

(6) **Example. Ballot theorem (3.10.6).** Let $X_1, X_2, \dots$ be independent identically distributed random variables taking values in $\{0, 1, 2, \dots\}$, and let $S_n = X_1 + X_2 + \cdots + X_n$. We claim that

(7) $$\mathbf{P}(S_k \geqslant k \text{ for some } 1 \leqslant k \leqslant N \mid S_N = b) = \min\{1, b/N\},$$

whenever b is such that $\mathbf{P}(S_N = b) > 0$. It is not immediately clear that this implies the ballot theorem, but look at it this way. In a ballot, each of N voters has two votes; he or she allocates both votes either to candidate A or to candidate B. Let us write X_i for the number of votes allocated to A by the ith voter, so that X_i equals either 0 or 2; assume that the X_i are independent. Now $S_k \geqslant k$ for some $1 \leqslant k \leqslant N$ if and only if B is not always in the lead. Equation (7) implies

(8) $$\mathbf{P}(B \text{ always leads} \mid A \text{ receives a total of } 2a \text{ votes})$$
$$= 1 - \mathbf{P}(S_k \geqslant k \text{ for some } 1 \leqslant k \leqslant N \mid S_N = 2a) = 1 - \frac{2a}{N} = \frac{p-q}{p+q}$$

if $0 \leqslant a < \frac{1}{2}N$, where $p = 2N - 2a$ is the number of votes received by B, and $q = 2a$ is the number received by A. This is the famous ballot theorem discussed after (3.10.6).

In order to prove (7), let $\mathscr{G}_n = \sigma(S_n, S_{n+1}, \dots)$, and recall that $(S_n/n, \mathscr{G}_n)$ is backward martingale. Fix N, and let

$$T = \begin{cases} \max\{k: S_k \geqslant k \text{ and } 1 \leqslant k \leqslant N\} & \text{if this exists} \\ 1 & \text{otherwise.} \end{cases}$$

This does not look like a stopping time, but it is. After all, for $1 < n \leqslant N$, $\{T = n\} = \{S_n \geqslant n,\ S_k < k \text{ for } n < k \leqslant N\}$, an event defined in terms of $S_n, S_{n+1}, \dots$ and therefore lying in the σ-field $\mathscr{G}_n$ generated by these random variables. By a similar argument, $\{T = 1\} \in \mathscr{G}_1$.

We may assume that $S_N = b < N$, since (7) is obvious if $b \geqslant N$. Let $A = \{S_k \geqslant k \text{ for some } 1 \leqslant k \leqslant N\}$. We have that $S_N < N$; therefore, if A occurs, it must be the case that $S_T \geqslant T$ and $S_{T+1} < T + 1$. In this case $X_{T+1} = S_{T+1} - S_T < 1$, so that $X_{T+1} = 0$ and therefore $S_T/T = 1$. On the other hand, if A does not occur then $T = 1$, and also $S_T = S_1 = 0$, implying that $S_T/T = 0$. It follows that $S_T/T = I_A$ if $S_N < N$, where I_A is the indicator function of A. Taking expectations, we obtain

$$\mathbf{E}\left(\frac{1}{T} S_T \,\middle|\, S_N = b\right) = \mathbf{P}(A \mid S_N = b) \quad \text{if} \quad b < N.$$

Finally, we apply (5) to the backward martingale $(n^{-1}S_n, \mathcal{G}_n)$ to obtain

$$\mathbf{E}\left(\frac{1}{T}S_T \,\middle|\, S_N = b\right) = \mathbf{E}\left(\frac{1}{N}S_N \,\middle|\, S_N = b\right) = \frac{b}{N}.$$

The last two equations may be combined to give (7). ●

In contrast to the theory of backward martingales, the theory of continuous-time martingales is hedged about with technical considerations. Let $(\Omega, \mathcal{F}, \mathbf{P})$ be a probability space. A *filtration* is a family $\mathcal{F} = \{\mathcal{F}_t : t \geqslant 0\}$ of sub-σ-fields of $\mathcal{F}$ satisfying $\mathcal{F}_s \subseteq \mathcal{F}_t$ whenever $s \leqslant t$. As before, we say that the (continuous-time) process $Y = \{Y(t): t \geqslant 0\}$ is adapted to $\mathcal{F}$ if $Y(t)$ is $\mathcal{F}_t$-measurable for all t. If Y is adapted to $\mathcal{F}$, we call $(Y, \mathcal{F})$ a martingale if $\mathbf{E}|Y(t)| < \infty$ for all t, and $\mathbf{E}(Y(t) \mid \mathcal{F}_s) = Y(s)$ whenever $s \leqslant t$. A random variable T taking values in $[0, \infty]$ is called a stopping time (with respect to the filtration $\mathcal{F}$) if $\{T \leqslant t\} \in \mathcal{F}_t$ for all $t \geqslant 0$.

Possibly the most important type of stopping time is the first passage time $T(A) = \inf\{t: Y(t) \in A\}$ for a suitable subset A of $\mathbb{R}$. Unfortunately $T(A)$ is not necessarily a stopping time. No problems arise if A is closed and the sample paths $\Pi(\omega) = \{(t, Y(t; \omega)): t \geqslant 0\}$ of Y are continuous, but these conditions are over-restrictive. They may be relaxed at the price of making extra assumptions about the process Y and the filtration $\mathcal{F}$. It is usual to assume in addition that

(a) $(\Omega, \mathcal{F}, \mathbf{P})$ is complete,
(b) $\mathcal{F}_0$ contains all events A of $\mathcal{F}$ satisfying $\mathbf{P}(A) = 0$,
(c) $\mathcal{F}$ is right-continuous in that $\mathcal{F}_t = \mathcal{F}_{t+}$ for all $t \geqslant 0$, where $\mathcal{F}_{t+} = \bigcap_{\varepsilon > 0} \mathcal{F}_{t+\varepsilon}$.

We shall refer to these conditions as the 'usual conditions'. Conditions (a) and (b) pose little difficulty, since an incomplete probability space may be completed (1.6.2), and the null events may be added to $\mathcal{F}_0$. Condition (c) is not of great importance if the process Y has right-continuous sample paths, since then $Y(t) = \lim_{\varepsilon \downarrow 0} Y(t + \varepsilon)$ is $\mathcal{F}_{t+}$-measurable.

Here are some examples of continuous-time martingales.

(9) **Example. Poisson process.** Let $\{N(t): t \geqslant 0\}$ be a Poisson process with intensity λ, and let $\mathcal{F}_t$ be the σ-field generated by $\{N(u): 0 \leqslant u \leqslant t\}$. It is easily seen that

$$U(t) = N(t) - \lambda t, \qquad V(t) = U(t)^2 - \lambda t,$$
$$W(t) = \exp[-\theta N(t) + \lambda t(1 - e^{-\theta})],$$

constitute martingales with respect to $\mathcal{F}$.

There is a converse statement. Suppose $N = \{N(t): t \geqslant 0\}$ is an integer-valued non-decreasing process such that, for all θ,

$$W(t) = \exp[-\theta N(t) + \lambda t(1 - e^{-\theta})]$$

is a martingale. Then, if $s < t$,

$$\mathbf{E}(\exp\{-\theta[N(t) - N(s)]\} \mid \mathscr{F}_s) = \mathbf{E}\left(\frac{W(t)}{W(s)}\exp[-\lambda(t-s)(1-e^{-\theta})] \,\middle|\, \mathscr{F}_s\right)$$
$$= \exp[-\lambda(t-s)(1-e^{-\theta})]$$

by the martingale condition. Hence N has independent increments, $N(t) - N(s)$ having the Poisson distribution with parameter $\lambda(t-s)$. ●

(10) **Example. Wiener process.** Let $\{W(t): t \geqslant 0\}$ be a standard Wiener process with continuous sample paths, and let $\mathscr{F}_t$ be the σ-field generated by $\{W(u): 0 \leqslant u \leqslant t\}$. It is easily seen that $W(t)$, $W(t)^2 - t$, and $\exp[\theta W(t) - \frac{1}{2}\theta^2 t]$ constitute martingales with respect to $\mathscr{F}$. Conversely it may be shown that, if $W(t)$ and $W(t)^2 - t$ are martingales with continuous sample paths, then W is a standard Wiener process (*exercise*). ●

Versions of the convergence and optional stopping theorems are valid in continuous time.

(11) **Convergence theorem.** *Let $(Y, \mathscr{F})$ be a martingale with right-continuous sample paths. If $\mathbf{E}|Y(t)| \leqslant M$ for some M and all t, then $Y_\infty = \lim_{t\to\infty} Y(t)$ exists almost surely. If $(Y, \mathscr{F})$ is uniformly integrable then $Y(t) \xrightarrow{1} Y_\infty$.*

Sketch proof. For each $m \geqslant 1$, the sequence $\{(Y(n2^{-m}), \mathscr{F}_{n2^{-m}}): n \geqslant 0\}$ constitutes a discrete-time martingale. Under the conditions of the theorem, these martingales converge as $n \to \infty$. The right-continuity property of Y may be used to fill in the gaps. ■

(12) **Optional stopping theorem.** *Let $(Y, \mathscr{F})$ be a uniformly integrable martingale with right-continuous sample paths. Suppose that S and T are stopping times such that $S \leqslant T$. Then $\mathbf{E}(Y(T) \mid \mathscr{F}_S) = Y(S)$.*

The idea of the proof is to 'discretize' Y as in the previous proof, use the optional stopping theorem for uniformly integrable discrete-time martingales, and then pass to the continuous limit.

Exercises

(13) Let X be a continuous-time Markov chain with finite state space S and generator $\boldsymbol{G}$. Let $\boldsymbol{\eta} = \{\eta(i): i \in S\}$ be a root of the equation $\boldsymbol{G\eta}' = \mathbf{0}$. Show that $\eta(X(t))$ constitutes a martingale with respect to $\mathscr{F}_t = \sigma(\{X(u): u \leqslant t\})$.

(14) Let N be a Poisson process with parameter λ and $N(0) = 0$, and let $T_a = \min\{t: N(t) = a\}$, where a is a positive integer. Assuming that $\mathbf{E}\{\exp(\psi T_a)\} < \infty$ for sufficiently small positive ψ, use the optional stopping theorem to show that $\mathrm{var}(T_a) = a\lambda^{-2}$.

12.8 Some examples

(1) **Example. Gambling systems.** In practice, gamblers do not invariably follow simple strategies, but they vary their manner of play according to a personal system. One way of expressing this is as follows. For a given game, write $Y_0, Y_1, \ldots$ for the sequence of capitals obtained by wagering one unit on each play; we allow the Y_i to be negative. That is to say, let Y_0 be the initial capital, and let Y_n be the capital after n gambles each involving a unit stake. Take as filtration the sequence $\mathcal{F}$ given by $\mathcal{F}_n = \sigma(Y_0, Y_1, \ldots, Y_n)$. A general betting strategy would allow the gambler to vary his stake. If he bets S_n on the nth play, his profit is $S_n(Y_n - Y_{n-1})$, since $Y_n - Y_{n-1}$ is the profit resulting from a stake of one unit. Hence the gambler's capital Z_n after n plays satisfies

(2)
$$Z_n = Z_{n-1} + S_n(Y_n - Y_{n-1}) = Y_0 + \sum_{i=1}^{n} S_i(Y_i - Y_{i-1}),$$

where Y_0 is the gambler's initial capital. The S_n must have the following special property. The gambler decides the value of S_n in advance of the nth play, which is to say that S_n depends only on $Y_0, Y_1, \ldots, Y_{n-1}$, and therefore S_n is $\mathcal{F}_{n-1}$-measurable. That is, $(S, \mathcal{F})$ must be a predictable process.

The sequence Z given by (2) is called the *transform* of Y by S. If Y is a martingale, we call Z a *martingale transform.*

Suppose $(Y, \mathcal{F})$ is a martingale. The gambler may hope to find a predictable process $(S, \mathcal{F})$ (called a *system*) for which the martingale transform Z (of Y by S) is no longer a martingale. He hopes in vain, since all martingale transforms have the martingale property. Here is a version of that statement.

(3) **Theorem.** *Let $(S, \mathcal{F})$ be a predictable process, and let Z be the transform of Y by S. Then*

(a) *if $(Y, \mathcal{F})$ is a martingale, then $(Z, \mathcal{F})$ is a martingale so long as $\mathbf{E}|Z_n| < \infty$ for all n,*
(b) *if $(Y, \mathcal{F})$ is a submartingale and in addition $S_n \geq 0$ for all n, then $(Z, \mathcal{F})$ is a submartingale so long as $\mathbf{E}(Z_n^+) < \infty$ for all n.*

Proof. From (2),

$$\mathbf{E}(Z_{n+1} \mid \mathcal{F}_n) - Z_n = \mathbf{E}[S_{n+1}(Y_{n+1} - Y_n) \mid \mathcal{F}_n] = S_{n+1}[\mathbf{E}(Y_{n+1} \mid \mathcal{F}_n) - Y_n].$$

The last term is zero if Y is a martingale, and is non-negative if Y is a submartingale and $S_{n+1} \geq 0$. ■

A number of special cases are of value.

(4) *Optional skipping.* At each play, the gambler either wagers a unit stake or skips the round; S_n equals either 0 or 1.

(5) *Optional stopping.* The gambler wagers a unit stake on each play until the (random) time T, when he gambles for the last time. That is,

$$S_n = \begin{cases} 1 & \text{if} \quad n \leqslant T \\ 0 & \text{if} \quad n > T, \end{cases}$$

and $Z_n = Y_{T \wedge n}$. Now $\{T = n\} = \{S_n = 1,\ S_{n+1} = 0\} \in \mathscr{F}_n$, so that T is a stopping time. It is a consequence of (3) that $(Y_{T \wedge n}, \mathscr{F}_n)$ is a martingale whenever Y is a martingale, as established earlier.

(6) *Optional starting.* The gambler does not play until the $(T + 1)$th play, where T is a stopping time. In this case $S_n = 0$ for $n \leqslant T$. ●

(7) **Example. Likelihood ratios.** Let $X_1, X_2, \ldots$ be independent identically distributed random variables with common density function f. Suppose that it is known that $f(\cdot)$ is either $p(\cdot)$ or $q(\cdot)$, where p and q are given (different) densities; the statistical problem is to decide which of the two is the true density. A common approach is to calculate the *likelihood ratio*

$$Y_n = \frac{p(X_1)p(X_2)\cdots p(X_n)}{q(X_1)q(X_2)\cdots q(X_n)}$$

(assume for neatness that $q(x) > 0$ for all x), and to adopt the strategy:

(8) $$\text{decide } p \text{ if } Y_n \geqslant a, \qquad \text{decide } q \text{ if } Y_n < a,$$

where a is some predetermined positive level.

Let $\mathscr{F}_n = \sigma(X_1, X_2, \ldots, X_n)$. If $f = q$, then

$$\mathbf{E}(Y_{n+1} \mid \mathscr{F}_n) = Y_n \mathbf{E}\left(\frac{p(X_{n+1})}{q(X_{n+1})}\right) = Y_n \int_{-\infty}^{\infty} \frac{p(x)}{q(x)} q(x)\, \mathrm{d}x = Y_n$$

since p is a density function. Furthermore

$$\mathbf{E}|Y_n| = \int_{\mathbb{R}^n} \frac{p(x_1)\cdots p(x_n)}{q(x_1)\cdots q(x_n)} q(x_1)\cdots q(x_n)\, \mathrm{d}x_1 \cdots \mathrm{d}x_n = 1.$$

It follows that $(Y, \mathscr{F})$ is a martingale, if q is the common density function of the X_i. By an application of the convergence theorem, the limit $Y_\infty = \lim_{n\to\infty} Y_n$ exists almost surely. We may calculate Y_∞ explicitly as follows:

$$\log Y_n = \sum_{i=1}^{n} \log\left(\frac{p(X_i)}{q(X_i)}\right),$$

the sum of independent identically distributed random variables. The logarithm function is concave, so that

$$\mathbf{E}\left(\log\left(\frac{p(X_1)}{q(X_1)}\right)\right) < \log\left(\mathbf{E}\left(\frac{p(X_1)}{q(X_1)}\right)\right) = 0$$

by Jensen's inequality (5.6.15). Applying the strong law of large numbers (7.5.1), we deduce that $n^{-1}\log Y_n$ converges almost surely to some point in $[-\infty, 0)$, implying that $Y_n \xrightarrow{\text{a.s.}} Y_\infty = 0$. (This is a case when Y_n does not converge to Y_∞ in mean, and $Y_n \neq \mathbf{E}(Y_\infty \mid \mathcal{F}_n)$.)

The fact that $Y_n \xrightarrow{\text{a.s.}} 0$ tells us that $Y_n < a$ for all large n, and hence the decision rule (8) gives the correct answer (that is, that $f = q$) for all large n. Indeed the probability that the outcome of the decision rule is ever in error satisfies

$$\mathbf{P}(Y_n \geqslant a \text{ for any } n \geqslant 1) \leqslant \frac{1}{a}$$

by the maximal inequality (12.6.6). ●

(9) **Example. Epidemics.** A village contains $N + 1$ people, one of whom is suffering from a fatal and infectious illness. Let $S(t)$ be the number of susceptible people at time t (that is, living people who have not yet been infected), let $I(t)$ be the number of infectives (that is, living people with the disease), and let $D(t) = N + 1 - S(t) - I(t)$ be the number of dead people. Assume that $(S(t), I(t), D(t))$ is a (trivariate) Markov chain in continuous time with transition rates

$$(s, i, d) \to \begin{cases} (s-1, i+1, d) & \text{at rate } \lambda si, \\ (s, i-1, d+1) & \text{at rate } \mu i; \end{cases}$$

that is to say, some susceptible becomes infective at rate λsi, and some infective dies at rate μi, where s and i are the numbers of susceptibles and infectives. This is the model of (6.12.4) with the introduction of death. The three variables always add up to $N + 1$, and therefore we may suppress reference to the dead, writing (s, i) for a typical state of the process. Suppose we can find $\boldsymbol{\psi} = \{\psi(s, i): 0 \leqslant s + i \leqslant N + 1\}$ such that $\mathbf{G}\boldsymbol{\psi}' = \mathbf{0}$, where $\mathbf{G}$ is the generator of the chain; think of $\boldsymbol{\psi}$ as a row vector. Then the transition semigroup $\mathbf{P}_t = e^{t\mathbf{G}}$ satisfies

$$\mathbf{P}_t\boldsymbol{\psi}' = \boldsymbol{\psi}' + \sum_{n=1}^{\infty} \frac{1}{n!} t^n \mathbf{G}^n \boldsymbol{\psi}' = \boldsymbol{\psi}',$$

whence it is easily seen (Exercise (12.7.13)) that $Y(t) = \psi(S(t), I(t))$ defines a continuous-time martingale with respect to

$$\mathcal{F}_t = \sigma(\{S(u), I(u): 0 \leqslant u \leqslant t\}).$$

Now $\mathbf{G}\boldsymbol{\psi}' = \mathbf{0}$ if and only if

(10) $$\lambda si\psi(s-1, i+1) - (\lambda si + \mu i)\psi(s, i) + \mu i\psi(s, i-1) = 0$$

for all relevant i and s. If we look for a solution of the form $\psi(s, i) = \alpha(s)\beta(i)$, we obtain

(11) $$\lambda s\alpha(s-1)\beta(i+1) - (\lambda s + \mu)\alpha(s)\beta(i) + \mu\alpha(s)\beta(i-1) = 0.$$

Viewed as a difference equation in the $\beta(i)$, this suggests setting

(12) $$\beta(i) = B^i \quad \text{for some } B.$$

With this choice and a little calculation, one finds that

(13) $$\alpha(s) = \prod_{k=s+1}^{N} \left(\frac{\lambda Bk - \mu(1-B)}{\lambda B^2 k} \right)$$

will do. With such choices for α and β, $\psi(S(t), I(t)) = \alpha(S(t))\beta(I(t))$ defines a martingale.

Two possibilities spring to mind. Either everyone dies ultimately (that is, $S(t) = 0$ before $I(t) = 0$) or the disease dies off before everyone has caught it (that is, $I(t) = 0$ before $S(t) = 0$). Let $T = \inf\{t\colon S(t)I(t) = 0\}$ be the time at which the process terminates. Clearly T is a stopping time, and therefore

$$\mathbf{E}(\psi(S(T), I(T))) = \psi(S(0), I(0)) = \alpha(N)\beta(1) = B,$$

which is to say that

(14) $$\mathbf{E}\left(B^{I(T)} \prod_{k=S(T)+1}^{N} \left(\frac{\lambda Bk - \mu(1-B)}{\lambda B^2 k} \right) \right) = B$$

for all B. From this equation we wish to determine whether $S(T) = 0$ or $I(T) = 0$, corresponding to the two possibilities described above.

We have a free choice of B in (14), and we choose the following values. For $1 \leqslant r \leqslant N$, define $B_r = \mu/(\lambda r + \mu)$, so that $\lambda r B_r - \mu(1 - B_r) = 0$. Substitute $B = B_r$ in (14) to obtain

(15) $$\mathbf{E}\left(B_r^{S(T)-N} \prod_{k=S(T)+1}^{N} \left(\frac{k-r}{k} \right) \right) = B_r$$

(remember that $I(T) = 0$ if $S(T) \neq 0$). Put $r = N$ to get $\mathbf{P}(S(T) = N) = B_N$. More generally, we have from (15) that $p_j = \mathbf{P}(S(T) = j)$ satisfies

(16) $$p_N + \frac{N-r}{NB_r} p_{N-1} + \frac{(N-r)(N-r-1)}{N(N-1)B_r^2} p_{N-2} + \cdots + \frac{(N-r)!r!}{N!B_r^{N-r}} p_r = B_r,$$

for $1 \leqslant r \leqslant N$. From these equations, $p_0 = \mathbf{P}(S(T) = 0)$ may in principle be calculated. ●

(17) **Example.** Our final two examples are relevant to mathematical analysis. Let $f\colon [0, 1] \to \mathbb{R}$ be a (measurable) function such that

(18) $$\int_0^1 |f(x)|\, dx < \infty;$$

that is, f is integrable. We shall show that there exists a sequence $\{f_n\colon n \geqslant 0\}$ of step functions such that $f_n(x) \to f(x)$ as $n \to \infty$, except possibly for an exceptional set of values of x having Lebesgue measure 0.

Let X be uniformly distributed on $[0, 1]$, and define X_n by

(19) $$X_n = k2^{-n} \quad \text{if} \quad k2^{-n} \leqslant X < (k+1)2^{-n}$$

where k and n are non-negative integers. It is easily seen that $X_n \uparrow X$ as $n \to \infty$, and furthermore $2^n(X_n - X_{n-1})$ equals the nth term in the binary expansion of X.

Define $Y = f(X)$ and $Y_n = \mathbf{E}(Y \mid \mathscr{F}_n)$ where $\mathscr{F}_n = \sigma(X_0, X_1, \ldots, X_n)$. Now $\mathbf{E}|f(X)| < \infty$ by (18), and therefore $(Y, \mathscr{F})$ is a uniformly integrable martingale (see (12.3.9)); it follows that

(20) $$Y_n \to Y_\infty = \mathbf{E}(Y \mid \mathscr{F}_\infty) \quad \text{a.s. and in mean,}$$

where $\mathscr{F}_\infty = \sigma(X_0, X_1, X_2, \ldots) = \sigma(X)$. Hence $Y_\infty = \mathbf{E}(f(X) \mid X) = f(X)$, and in addition

(21) $$Y_n = \mathbf{E}(Y \mid \mathscr{F}_n) = \mathbf{E}(Y \mid X_0, X_1, \ldots, X_n) = \int_{X_n}^{X_n + 2^{-n}} f(u)2^n \, du = f_n(X)$$

where $f_n: [0, 1] \to \mathbb{R}$ is the step function defined by

$$f_n(x) = 2^n \int_{x_n}^{x_n + 2^{-n}} f(u) \, du,$$

x_n being the number of the form $k2^{-n}$ satisfying $x_n \leqslant x < x_n + 2^{-n}$. We have from (20) that $f_n(X) \to f(X)$ a.s. and in mean, whence $f_n(x) \to f(x)$ for almost all x, and furthermore

$$\int_0^1 |f_n(x) - f(x)| \, dx \to 0 \quad \text{as} \quad n \to \infty. \qquad \bullet$$

(22) **Example.** This time let $f: [0, 1] \to \mathbb{R}$ be Lipschitz continuous, which is to say that there exists C such that

(23) $$|f(x) - f(y)| \leqslant C|x - y| \quad \text{for all } x, y \in [0, 1].$$

Lipschitz continuity is of course somewhere between continuity and differentiability: Lipschitz-continuous functions are necessarily continuous but need not be differentiable (in the usual sense). We shall see, however, that there must exist a function g such that

$$f(x) - f(0) = \int_0^x g(u) \, du, \quad x \in [0, 1];$$

g is called the *Radon–Nikodým derivative* of f (with respect to Lebesgue measure).

As in the last example, let X be uniformly distributed on $[0, 1]$, define X_n by (19), and let

(24) $$Z_n = 2^n[f(X_n + 2^{-n}) - f(X_n)].$$

It may be seen as follows that $(Z, \mathscr{F})$ is a martingale (remember that $\mathscr{F}_n = \sigma(X_0, X_1, \ldots, X_n)$). First, we check that $\mathbf{E}(Z_{n+1} \mid \mathscr{F}_n) = Z_n$. To this end note that, conditional on $X_0, X_1, \ldots, X_n$, it is the case that X_{n+1} is equally likely to take the value X_n or the value $X_n + 2^{-n-1}$. Therefore

$$\begin{aligned}\mathbf{E}(Z_{n+1} \mid \mathscr{F}_n) &= \tfrac{1}{2}2^{n+1}[f(X_n + 2^{-n-1}) - f(X_n)] \\ &\quad + \tfrac{1}{2}2^{n+1}[f(X_n + 2^{-n}) - f(X_n + 2^{-n-1})] \\ &= 2^n[f(X_n + 2^{-n}) - f(X_n)] = Z_n.\end{aligned}$$

Secondly, by the Lipschitz continuity (23) of f, it is the case that $|Z_n| \leqslant C$, whence $(Z, \mathscr{F})$ is a bounded martingale.

Therefore Z_n converges almost surely and in mean to some limit Z_∞, and furthermore $Z_n = \mathbf{E}(Z_\infty \mid \mathscr{F}_n)$ by (12.3.11). Now Z_∞ is $\mathscr{F}_\infty$-measurable where $\mathscr{F}_\infty = \lim_{n\to\infty} \mathscr{F}_n = \sigma(X_0, X_1, X_2, \ldots) = \sigma(X)$, which implies that Z_∞ is a function of X, say $Z_\infty = g(X)$. As in (21), the relation

$$Z_n = \mathbf{E}(g(X) \mid X_0, X_1, \ldots, X_n)$$

becomes

$$f(X_n + 2^{-n}) - f(X_n) = \int_{X_n}^{X_n + 2^{-n}} g(u)\, du.$$

This is an ('almost sure') identity for X_n, which has positive probability of taking any value of the form $k2^{-n}$ for $0 \leqslant k < 2^n$. Hence

$$f((k+1)2^{-n}) - f(k2^{-n}) = \int_{k2^{-n}}^{(k+1)2^{-n}} g(u)\, du,$$

whence, by summing,

$$f(x) - f(0) = \int_0^x g(u)\, du$$

for all x of the form $k2^{-n}$ for some $n \geqslant 1$ and $0 \leqslant k < 2^n$. The corresponding result for general $x \in [0, 1]$ is obtained by taking a limit along a sequence of such 'dyadic rationals'. ●

12.9 Problems

1. Let Z_n be the size of the nth generation of a branching process with immigration in which the mean family-size is μ ($\neq 1$) and the mean number of immigrants per generation is m. Show that

$$Y_n = \mu^{-n}\left\{Z_n - m\frac{1-\mu^n}{1-\mu}\right\}$$

defines a martingale.

2. In an age-dependent branching process, each individual gives birth to a random number of offspring at random times. At time 0, there exists a single progenitor who

has N children at the subsequent times $B_1 \leqslant B_2 \leqslant \cdots \leqslant B_N$; his family may be described by the vector $(N, B_1, B_2, \ldots, B_N)$. Each subsequent member x of the population has a family described similarly by a vector $(N(x), B_1(x), \ldots, B_{N(x)}(x))$ having the same distribution as $(N, B_1, \ldots, B_N)$ and independent of all other individuals' families. The number $N(x)$ is the number of his offspring, and $B_i(x)$ is the time between the births of the parent and the ith offspring. Let $\{B_{n,r}: r \geqslant 1\}$ be the times of births of individuals in the nth generation. Let $M_n(\theta) = \sum_r e^{-\theta B_{n,r}}$, and show that $Y_n = M_n(\theta)/\mathbf{E}(M_1(\theta))^n$ defines a martingale with respect to

$$\mathscr{F}_n = \sigma(\{B_{m,r}: m \leqslant n, r \geqslant 1\}),$$

for any value of θ such that $\mathbf{E}M_1(\theta) < \infty$.

3. Let $(Y, \mathscr{F})$ be a martingale with $\mathbf{E}Y_n = 0$ and $\mathbf{E}(Y_n^2) < \infty$ for all n. Show that

$$\mathbf{P}\left(\max_{1 \leqslant k \leqslant n} Y_k > x\right) \leqslant \frac{\mathbf{E}(Y_n^2)}{\mathbf{E}(Y_n^2) + x^2}, \qquad x > 0.$$

4. Let $(Y, \mathscr{F})$ be a non-negative submartingale with $Y_0 = 0$, and let $\{c_n\}$ be a non-increasing sequence of positive numbers. Show that

$$\mathbf{P}\left(\max_{1 \leqslant k \leqslant n} c_k Y_k \geqslant x\right) \leqslant \frac{1}{x} \sum_{k=1}^{n} c_k \mathbf{E}(Y_k - Y_{k-1}), \qquad x > 0.$$

Such an inequality is sometimes named after subsets of Hájek, Rényi, and Chow. Deduce Kolmogorov's inequality for the sum of independent random variables. [Hint: Work with the martingale

$$Z_n = c_n Y_n - \sum_{k=1}^{n} c_k \mathbf{E}(X_k \mid \mathscr{F}_{k-1}) + \sum_{k=1}^{n} (c_{k-1} - c_k) Y_{k-1}$$

where $X_k = Y_k - Y_{k-1}$.]

5. Suppose that the sequence $\{X_n: n \geqslant 1\}$ of random variables satisfies

$$\mathbf{E}(X_n \mid X_1, X_2, \ldots, X_{n-1}) = 0$$

for all n, and also $\sum_{k=1}^{\infty} \mathbf{E}(|X_k|^r)/k^r < \infty$ for some $r \in [1, 2]$. Let $S_n = \sum_{i=1}^{n} Z_i$ where $Z_i = X_i/i$, and show that

$$\mathbf{P}\left(\max_{1 \leqslant k \leqslant n} |S_{m+k} - S_m| \geqslant x\right) \leqslant \frac{1}{x^r} \mathbf{E}(|S_{m+n} - S_m|^r), \qquad x > 0.$$

Deduce that S_n converges a.s. as $n \to \infty$, and hence that $n^{-1} \sum_1^n X_k \xrightarrow{\text{a.s.}} 0$. [Hint: In the case $1 < r \leqslant 2$, prove and use the fact that $h(u) = |u|^r$ satisfies

$$h(v) - h(u) \leqslant (v - u)h'(u) + 2h((v - u)/2).$$

Kronecker's lemma is useful for the last part.]

6. Let $X_1, X_2, \ldots$ be independent random variables with

$$X_n = \begin{cases} 1 & \text{with probability } (2n)^{-1} \\ 0 & \text{with probability } 1 - n^{-1} \\ -1 & \text{with probability } (2n)^{-1}. \end{cases}$$

Let $Y_1 = X_1$ and for $n \geqslant 2$

$$Y_n = \begin{cases} X_n & \text{if } Y_{n-1} = 0 \\ nY_{n-1}|X_n| & \text{if } Y_{n-1} \neq 0. \end{cases}$$

Show that Y_n is a martingale with respect to $\mathcal{F}_n = \sigma(Y_1, Y_2, \ldots, Y_n)$. Show that Y_n does not converge almost surely. Does Y_n converge in any way? Why does the martingale convergence theorem not apply?

7. Let $X_1, X_2, \ldots$ be independent identically distributed random variables and suppose that $M(t) = \mathbf{E}(e^{tX_1})$ satisfies $M(t) = 1$ for some $t > 0$. Show that $\mathbf{P}(S_k \geqslant x$ for some $k) \leqslant e^{-tx}$ for $x > 0$ and such a value of t, where $S_k = X_1 + X_2 + \cdots + X_k$.

8. Let Z_n be the size of the nth generation of a branching process with family-size probability generating function $G(s)$, and assume $Z_0 = 1$. Let ξ be the smallest positive root of $G(s) = s$. Use the martingale convergence theorem to show that, if $0 < \xi < 1$, then $\mathbf{P}(Z_n \to 0) = \xi$ and $\mathbf{P}(Z_n \to \infty) = 1 - \xi$.

9. Let $(Y, \mathcal{F})$ be a non-negative martingale, and let $Y_n^* = \max\{Y_k : 0 \leqslant k \leqslant n\}$. Show that

$$\mathbf{E}(Y_n^*) \leqslant \frac{e}{e-1}\{1 + \mathbf{E}(Y_n(\log Y_n)^+)\}.$$

[Hint: $a \log^+ b \leqslant a \log^+ a + b/e$ if $a, b \geqslant 0$, where $\log^+ x = \max\{0, \log x\}$.]

10. Let $X = \{X(t) : t \geqslant 0\}$ be a birth–death process with parameters λ_i, μ_i, where $\lambda_i = 0$ if and only if $i = 0$. Define $h(0) = 0$, $h(1) = 1$, and

$$h(j) = 1 + \sum_{i=1}^{j-1} \frac{\mu_1 \mu_2 \cdots \mu_i}{\lambda_1 \lambda_2 \cdots \lambda_i}, \qquad j \geqslant 2.$$

Show that $h(X(t))$ constitutes a martingale with respect to the filtration

$$\mathcal{F}_t = \sigma(\{X(u) : 0 \leqslant u \leqslant t\}),$$

whenever $\mathbf{E}h(X(t)) < \infty$ for all t. (You may assume that the forward equations are satisfied.)

Fix n, and let $m < n$; let $\pi(m)$ be the probability that the process is absorbed at 0 before it reaches size n, having started at size m. Show that $\pi(m) = 1 - \{h(m)/h(n)\}$.

11. Let $(Y, \mathcal{F})$ be a submartingale such that $\mathbf{E}(Y_n^+) \leqslant M$ for some M and all n.

(a) Show that $M_n = \lim_{m\to\infty} \mathbf{E}(Y_{n+m}^+ \mid \mathcal{F}_n)$ exists (a.s.) and defines a martingale with respect to $\mathcal{F}$.

(b) Show that Y_n may be expressed in the form $Y_n = X_n - Z_n$ where $(X, \mathcal{F})$ is a non-negative martingale, and $(Z, \mathcal{F})$ is a non-negative supermartingale. This representation of Y is sometimes termed the 'Krickeberg decomposition'.

(c) Let $(Y, \mathcal{F})$ be a martingale such that $\mathbf{E}|Y_n| \leqslant M$ for some M and all n. Show that Y may be expressed as the difference of two non-negative martingales.

12. Let £Y_n be the assets of an insurance company after n years of trading. During each year it receives a total (fixed) income of £P in premiums. During the nth year it pays

out a total of £C_n in claims. Thus $Y_{n+1} = Y_n + P - C_{n+1}$. Suppose that Y_0 is fixed, $C_1, C_2, \ldots$ are independent $N(\mu, \sigma^2)$ variables, and show that the probability of ultimate bankruptcy satisfies

$$\mathbf{P}(Y_n \leqslant 0 \text{ for some } n) \leqslant \exp\left\{-\frac{2(P-\mu)Y_0}{\sigma^2}\right\}.$$

13. **Pólya's urn.** A bag contains red and blue balls, with initially r red and b blue where $rb > 0$. A ball is drawn from the bag, its colour noted, and then it is returned to the bag together with a new ball of the same colour. Let R_n be the number of red balls after n such operations.

(a) Show that $Y_n = R_n/(n + r + b)$ is a martingale which converges a.s. and in mean.
(b) Let T be the number of balls drawn until the first blue ball appears, and suppose that $r = b = 1$. Show that $\mathbf{E}\{(T+2)^{-1}\} = \frac{1}{4}$.
(c) Suppose $r = b = 1$, and show that $\mathbf{P}(Y_n \geqslant \frac{3}{4} \text{ for some } n) \leqslant \frac{2}{3}$.

14. Here is a modification of the last problem. Let $\{A_n : n \geqslant 1\}$ be a sequence of random variables, each being a non-negative integer. We are provided with the bag of Problem (13), and we add balls according to the following rules. At each stage a ball is drawn from the bag, and its colour noted; we assume that the distribution of this colour depends only on the current contents of the bag and not on any further information concerning the A_n. We return this ball together with A_n new balls of the same colour. Write R_n and B_n for the numbers of red and blue balls in the urn after n operations, and let $\mathscr{F}_n = \sigma(\{R_k, B_k : 0 \leqslant k \leqslant n\})$. Show that $Y_n = R_n/(R_n + B_n)$ defines a martingale. Suppose $R_0 = B_0 = 1$, let T be the number of balls drawn until the first blue ball appears, and show that

$$\mathbf{E}\left(\frac{1 + A_T}{2 + \sum_{i=1}^{T} A_i}\right) = \frac{1}{2},$$

so long as $\sum_n (2 + \sum_{i=1}^{n} A_i)^{-1} = \infty$ a.s.

15. **Labouchere system.** Here is a gambling system for playing a fair game. Choose a sequence $x_1, x_2, \ldots, x_n$ of positive numbers.

Wager the sum of the first and last numbers on an evens bet. If you win, delete those two numbers; if you lose, append their sum as an extra term x_{n+1} $(= x_1 + x_n)$ at the right-hand end of the sequence.

You play iteratively according to the above rule. If the sequence ever contains one term only, you wager that amount on an evens bet. If you win, you delete the term, and if you lose you append it to the sequence to obtain two terms.

Show that, with probability 1, the game terminates with a profit of $\sum_1^n x_i$, and that the time until termination has finite mean.

This looks like another clever strategy. Show that the mean size of your largest stake before winning is infinite. (When Henry Labouchere was sent down from Trinity College, Cambridge, in 1852, his gambling debts exceeded £6,000.)

16. Here is a martingale approach to the question of determining the mean number of tosses of a coin before the first appearance of the sequence HHH. A large casino contains infinitely many gamblers $G_1, G_2, \ldots$, each with an initial fortune of \$1. A croupier tosses a coin repeatedly. For each n, gambler G_n bets as follows. Just before the nth toss he stakes \$1 on the event that the nth toss shows heads. The game is

assumed fair, so that he receives a total of $\$p^{-1}$ if he wins, where p is the probability of heads. If he wins this gamble, then he *repeatedly* stakes his entire current fortune on heads, at the same odds as his first gamble. At the first subsequent tail he loses his fortune and leaves the casino, penniless. Let S_n be the casino's profit (losses count negative) after the nth toss. Show that S_n is a martingale. Let N be the number of tosses before the first appearance of HHH; show that N is a stopping time and hence find $\mathbf{E}(N)$.

Now adapt this scheme to calculate the mean time to the first appearance of the sequence HTH.

17. Let $\{(X_k, Y_k): k \geqslant 1\}$ be a sequence of independent identically distributed random vectors such that each X_k and Y_k takes values in the set $\{-1, 0, 1, 2, \dots\}$. Suppose that $\mathbf{E}(X_1) = \mathbf{E}(Y_1) = 0$ and $\mathbf{E}(X_1 Y_1) = c$, and furthermore X_1 and Y_1 have finite non-zero variances. Let U_0 and V_0 be positive integers, and define $(U_{n+1}, V_{n+1}) = (U_n, V_n) + (X_{n+1}, Y_{n+1})$ for each $n \geqslant 0$. Let $T = \min\{n: U_n V_n = 0\}$ be the first hitting time by the random walk (U_n, V_n) of the axes of $\mathbb{R}^2$. Show that $\mathbf{E}(T) < \infty$ if and only if $c < 0$, and that $\mathbf{E}(T) = -\mathbf{E}(U_0 V_0)/c$ in this case. [Hint: You might show that $U_n V_n - cn$ is a martingale.]

18. The game 'Red Now' may be played by a single player with a well shuffled conventional pack of 52 playing cards. At times $n = 1, 2, \dots, 52$ the player turns over a new card and observes its colour. Just once in the game he must say, just before exposing a card, 'Red Now'. He wins the game if the next exposed card is red. Let R_n be the number of red cards remaining face down after the nth card has been turned over. Show that $X_n = R_n/(52 - n)$, $0 \leqslant n < 52$, defines a martingale. Show that there is no strategy for the player which results in a probability of winning different from $\frac{1}{2}$.

19. A businessman has a redundant piece of equipment which he advertises for sale, inviting 'offers over £1000'. He anticipates that, each week for the foreseeable future, he will be approached by one prospective purchaser, the offers made in week $0, 1, \dots$ being $£1000X_0$, $£1000X_1, \dots$, where $X_0, X_1, \dots$ are independent random variables with a common density function f and finite mean. Storage of the equipment costs $£1000c$ per week and the prevailing rate of interest is α (> 0) per week. Explain why a sensible strategy for the businessman is to sell in the week T, where T is a stopping time chosen so as to maximize

$$\mu(T) = \mathbf{E}\left\{(1+\alpha)^{-T} X_T - \sum_{n=1}^{T} (1+\alpha)^{-n} c\right\}.$$

Show that this problem is equivalent to maximizing $\mathbf{E}\{(1+\alpha)^{-T} Z_T\}$ where $Z_n = X_n + c/\alpha$.

Show that there exists a unique positive real number γ with the property that $\alpha\gamma = \int_\gamma^\infty \mathbf{P}(Z_n > y)\, dy$, and that, for this value of γ, the sequence

$$V_n = (1+\alpha)^{-n} \max\{Z_n, \gamma\}$$

constitutes a supermartingale. Deduce that the optimal strategy for the businessman is to set a target price τ (which you should specify in terms of γ) and sell the first time he is offered at least this price.

In the case when $f(x) = 2x^{-3}$ for $x \geqslant 1$, and $c = \alpha = \frac{1}{90}$, find his target price and the expected number of weeks he will have to wait before selling.

20. Let Z be a branching process satisfying $Z_0 = 1$, $\mathbf{E}(Z_1) < 1$, and $\mathbf{P}(Z_1 \geqslant 2) > 0$. Show that $\mathbf{E}(\sup_n Z_n) \leqslant \eta/(\eta - 1)$, where η is the largest root of the equation $x = G(x)$ and G is the probability generating function of Z_1.

13 Diffusion processes

13.1 Introduction

Random processes come in many types. For example, they may run in discrete time or continuous time, and their state spaces may also be discrete or continuous. In the main, we have so far considered processes which are *discrete* either in time or space; our purpose in this chapter is to approach the theory of processes indexed by continuous time and taking values in the real line $\mathbb{R}$. Many important examples belong to this category: meteorological data, communication systems with noise, molecular motion, and so on. In other important cases, such random processes provide useful approximations to the physical process in question: processes in population genetics or population evolution, for example.

The archetypal diffusion process is the Wiener process W of (9.6.13), a Gaussian process with stationary independent increments. Think about W as a description of the motion of a particle moving randomly but continuously about $\mathbb{R}$. There are various ways of *defining* the Wiener process, and each such definition has two components. First of all, we require a *distributional* property, such as that the finite-dimensional distributions are Gaussian, and so on. The second component, not explored in Chapter 9, is that the sample paths $\{W(t; \omega): t \geqslant 0\}$, thought of as random functions on the underlying probability space $(\Omega, \mathcal{F}, \mathbf{P})$, are almost surely continuous. This assumption is important and natural, and of particular relevance when studying first passage times of the process.

Similar properties are required of a diffusion process, and we reserve the term 'diffusion' for a process $\{X(t): t \geqslant 0\}$ having the strong Markov property and whose sample paths are almost surely continuous.

13.2 Brownian motion

Let us suppose that we choose to observe a container of water. When seen from a distance the water appears to be motionless, but this is an illusion. If we are able to approach the container so closely as to be able to distinguish individual molecules then we may perceive that each molecule enjoys a motion which is unceasing and without any apparent order. The disorder of this movement arises from the high density of the fluid and the frequent occasions at which the molecule is repulsed by other molecules which are nearby at the time. The first scientific experiments involving this phenomenon were performed by the Dutch scientist A. van Leeuwenhoek (1632–1723), and later by the botanist R. Brown around 1827. Brown studied the motion of pollen particles suspended in water, and lent his name to the type of erratic movement which he observed. It is a classical problem of probability theory to model this movement.

Brownian motion takes place in continuous time and continuous space. Our first attempt to model it might proceed by approximating to it by a discrete process such as a random walk. At any epoch of time the position of an observed particle is constrained to move about the points $\{(a\delta, b\delta, c\delta)$: $a, b, c = 0, \pm 1, \pm 2, \dots\}$ of a three-dimensional 'cubic' lattice in which the distance between neighbouring points is δ; δ is a fixed positive number which is very small. Suppose further that the particle performs a symmetric random walk on this lattice (see Problem (6.13.9) for the case $\delta = 1$) so that its position $\boldsymbol{S}_n$ after n jumps satisfies

$$\mathbf{P}(\boldsymbol{S}_{n+1} = \boldsymbol{S}_n + \delta\boldsymbol{\varepsilon}) = \tfrac{1}{6} \quad \text{if} \quad \boldsymbol{\varepsilon} = (\pm 1, 0, 0), (0, \pm 1, 0), (0, 0, \pm 1).$$

Let us concentrate on the x co-ordinate of the particle, and write

$$\boldsymbol{S}_n = (S_n^1, S_n^2, S_n^3).$$

Then, as in Section 3.9,

$$S_n^1 - S_0^1 = \sum_{i=1}^{n} X_i$$

where $\{X_i\}$ is an independent identically distributed sequence with

$$\mathbf{P}(X_i = k\delta) = \begin{cases} \frac{1}{6} & \text{if} \quad k = -1 \\ \frac{1}{6} & \text{if} \quad k = +1 \\ \frac{2}{3} & \text{if} \quad k = 0. \end{cases}$$

We are interested in the displacement $S_n^1 - S_0^1$ when n is large; the central limit theorem (5.10.4) tells us that the distribution of this displacement is approximately $N(0, \frac{1}{3}n\delta^2)$. Now suppose that the jumps of the random walk take place at time epochs $\tau, 2\tau, 3\tau, \dots$ where $\tau > 0$; τ is the time between jumps and is very small, implying that a very large number of jumps occur in any large time interval. Observe the particle after some time t (>0) has elapsed. By this time it has experienced

$$n = \lfloor t/\tau \rfloor$$

jumps, and so its x co-ordinate $S^1(t)$ is such that $S^1(t) - S^1(0)$ is approximately $N(0, \frac{1}{3}t\delta^2/\tau)$. At this stage in the analysis we let the inter-point distance δ and the inter-jump time τ approach zero; in so doing we hope that the discrete random walk may approach some limit whose properties have something in common with the observed features of Brownian motion. We let $\delta \downarrow 0$ and $\tau \downarrow 0$ in such a way that $\frac{1}{3}\delta^2/\tau$ remains constant, since the variance of the distribution of $S^1(t) - S^1(0)$ fails to settle down to a non-trivial limit otherwise. Set

(1) $$\tfrac{1}{3}\delta^2/\tau = \sigma^2$$

where σ^2 is a positive constant, and pass to the limit to obtain that the distribution of $S^1(t) - S^1(0)$ approaches $N(0, \sigma^2 t)$. We can apply the same argument to the y co-ordinate and the z co-ordinate of the particle to deduce that the particle's position

$$\boldsymbol{S}(t) = (S^1(t), S^2(t), S^3(t))$$

at time t is such that the asymptotic distribution of the co-ordinates of the displacement $\boldsymbol{S}(t) - \boldsymbol{S}(0)$ is multivariate normal whenever $\delta \downarrow 0$, $\tau \downarrow 0$, and (1) holds; furthermore, it is not too hard to see that $S^1(t)$, $S^2(t)$, and $S^3(t)$ are independent of each other.

We may guess from the asymptotic properties of this random walk that an adequate model for Brownian motion will involve a process $\boldsymbol{X} = \{\boldsymbol{X}(t): t \geqslant 0\}$ taking values in $\mathbb{R}^3$ with a co-ordinate representation

$$\boldsymbol{X}(t) = (X^1(t), X^2(t), X^3(t))$$

such that

(a) $\boldsymbol{X}(0) = (0, 0, 0)$, say
(b) X^1, X^2, and X^3 are independent and identically distributed processes
(c) $X^1(s+t) - X^1(s)$ is $N(0, \sigma^2 t)$ for any $s, t \geqslant 0$
(d) X^1 has *independent increments* in that $X^1(v) - X^1(u)$ and $X^1(t) - X^1(s)$ are independent whenever $u \leqslant v \leqslant s \leqslant t$.

We have not yet shown the existence of such a process $\boldsymbol{X}$; the foregoing argument only indicates certain plausible distributional properties without showing that they are attainable. However, properties (c) and (d) are not new to us and remind us of the Wiener process of Example (9.6.13); we deduce that such a process $\boldsymbol{X}$ indeed exists, and is given by

$$\boldsymbol{X}(t) = (W^1(t), W^2(t), W^3(t))$$

where W^1, W^2, and W^3 are independent Wiener processes.

This conclusion is gratifying in that it demonstrates the existence of a random process which seems to enjoy at least some of the features of Brownian motion. A more detailed and technical analysis indicates some weak points of the Wiener model. This is beyond the scope of this text, and we are able only to skim the surface of the main difficulty. For each ω in the sample space Ω, $\{\boldsymbol{X}(t; \omega): t \geqslant 0\}$ is a sample path of the process along which the particle may move. It can be shown that, in some sense to be discussed in the next section,

(a) almost all sample paths are continuous functions of t
(b) almost all sample paths are nowhere differentiable functions of t.

Property (a) is physically necessary, but (b) is a property which *cannot* be shared by the physical phenomenon which we are modelling, since mechanical considerations, such as Newton's laws, imply that only particles with zero mass can move along routes which are nowhere differentiable. So, as a model for the local movement (over a short time interval) of particles, the Wiener process is poor; over longer periods of time the properties of the Wiener process are indeed very similar to experimental results.

A popular improved model for the local behaviour of Brownian paths is the so-called Ornstein–Uhlenbeck process. We close this section with a short account of this. Roughly, it is founded on the assumption that the velocity of the particle (rather than its position) undergoes a random walk; the ensuing motion is damped by the frictional resistance of the fluid. The result is a 'velocity process' with continuous sample paths; their integrals represent the sample paths of the particle itself. Think of the motion in one dimension as before, and write V_n for the velocity of the particle after the nth jump. At the next jump the change $V_{n+1} - V_n$ in the velocity is assumed to have two contributions: the frictional resistance to motion, and some random fluctuation owing to collisions with other particles. We shall assume that the former damping effect is directly proportional to V_n, so that

$$V_{n+1} = V_n + X_{n+1}$$

where

$$\mathbf{E}(X_{n+1} \mid V_n) = -\beta V_n \quad \text{: frictional effect}$$
$$\text{var}(X_{n+1} \mid V_n) = \sigma^2 \quad \text{: collision effect,}$$

where β and σ^2 are constants. $\{V_n\}$ is no longer a random walk on some regular grid of points, but it can be shown that the distributions converge as before, after suitable passage to the limit. Furthermore, there exists a process $V = \{V(t): t \geq 0\}$ with the corresponding distributional properties, and whose sample paths turn out to be almost surely continuous. These sample paths do not represent possible routes of the particle, but rather describe the development of its velocity as time passes. The possible paths of the particle through the space which it inhabits are found by integrating the sample paths of V with respect to time (this is not so easy in practice as it may sound). Of course, the resulting paths are almost surely continuously differentiable functions of time.

13.3 Diffusion processes

We say that a particle is 'diffusing' about a space $\mathbb{R}^n$ whenever it experiences erratic and disordered motion through the space; for example, we may speak of radioactive particles diffusing through the atmosphere, or even of a rumour diffusing through the population. For the moment, we restrict our attention to one-dimensional diffusions, for which the position of the observed particle at any time is a point on the real line; similar arguments will hold for higher dimensions. Our first diffusion model is the Wiener process.

(1) **Definition.** A **Wiener process** $W = \{W(t): t \geq 0\}$, starting from $W(0) = w$, say, is a real-valued Gaussian process such that

(a) W has independent increments (see (9.6.16))
(b) $W(s + t) - W(s)$ is $N(0, \sigma^2 t)$ for all $s, t \geq 0$ where σ^2 is a positive constant
(c) the sample paths of W are almost surely continuous.

Clearly (a) and (b) specify the fdds of a Wiener process W, and the argument of (9.6.1) shows there exists a Gaussian process with these fdds. In agreement with (9.6.14), the autocovariance function of W is given by

$$\begin{aligned} c(s, t) &= \mathbf{E}([W(s) - W(0)][W(t) - W(0)]) \\ &= \mathbf{E}([W(s) - W(0)]^2 + [W(s) - W(0)][W(t) - W(s)]) \\ &= \sigma^2 s + 0 \quad \text{if} \quad 0 \leqslant s \leqslant t \end{aligned}$$

and so

(2) $$c(s, t) = \sigma^2 \min\{s, t\} \quad \text{for all } s, t \geqslant 0.$$

W is called a *standard* Wiener process if $\sigma^2 = 1$. If W is non-standard, then $W_1 = W/\sigma$ is standard. W is said to have 'stationary' independent increments since the distribution of $W(s + t) - W(s)$ depends on t alone. A simple application of (9.6.7) shows that W is a Markov process.

The Wiener process W can be used to model the apparently random displacement of Brownian motion in any chosen direction. For this reason, W is sometimes called 'Brownian motion'; however, *we* reserve the use of this term to describe the motivating physical phenomenon.

Does the Wiener process exist? That is to say, does there exist a probability space $(\Omega, \mathscr{F}, \mathbf{P})$ and a Gaussian process W thereon, satisfying (1a, b, c)? The answer to this non-trivial question is of course in the affirmative, and we defer to the end of this section an explicit construction of such a process.

Roughly speaking, there are two types of statement to be made about diffusion processes in general, and the Wiener process particularly; we made a similar remark in Section 8.1. The first deals with sample path properties, and the second with distributional properties.

Figure 13.1 is a diagram of a typical sample path. Certain distributional properties of continuity are immediate. For example, W is 'continuous in mean square' in that

$$\mathbf{E}([W(s + t) - W(s)]^2) \to 0 \quad \text{as} \quad t \to 0;$$

this follows easily from (2). Consequently, W is 'continuous in probability', in that

$$\mathbf{P}(|W(s + t) - W(s)| > \varepsilon) \to 0 \quad \text{as} \quad t \to 0, \quad \text{for all } \varepsilon > 0.$$

Let us turn our attention to the distributions of a standard Wiener process W. Suppose we are given that $W(s) = x$, say, where $s \geqslant 0$ and $x \in \mathbb{R}$. Conditional on this, $W(t)$ is $N(x, t - s)$ for $t \geqslant s$, which is to say that the conditional distribution function

$$F(y, t \mid x, s) = \mathbf{P}(W(t) \leqslant y \mid W(s) = x)$$

has density function

(3) $$f(y, t \mid x, s) = \frac{\partial}{\partial y} F(y, t \mid x, s)$$

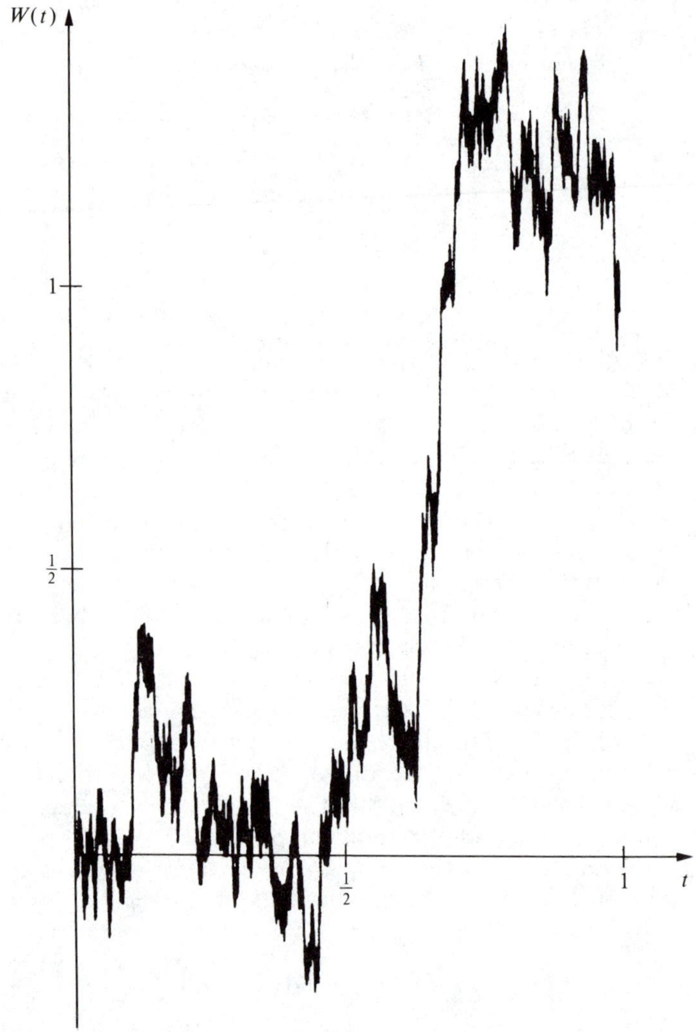

Fig. 13.1 A typical realization of a Wiener process W. This is a scale drawing of a sample path of W over the time interval $[0, 1]$. Note that the path is continuous but very spiky. This picture indicates the general features of the path only; the dense black portions indicate superimposed fluctuations which are too fine for this method of description. Tim Brown and Adrian Bowyer drew the picture, with the aid of a computer, using 100 000 steps of a symmetric random walk and the scaling method of Section 13.2.

which is given by

(4) $$f(y, t \mid x, s) = \frac{1}{[2\pi(t-s)]^{\frac{1}{2}}} \exp\left(-\frac{(y-x)^2}{2(t-s)}\right), \qquad -\infty < y < \infty.$$

This is a function of four variables, but just grit your teeth. It is easy to check that f is the solution of the following differential equations.

(5) **Forward diffusion equation:** $\dfrac{\partial f}{\partial t} = \dfrac{1}{2}\dfrac{\partial^2 f}{\partial y^2}.$

(6) **Backward diffusion equation:** $\dfrac{\partial f}{\partial s} = -\dfrac{1}{2}\dfrac{\partial^2 f}{\partial x^2}.$

We ought to specify the boundary conditions for these equations, but avoid this at the moment. Subject to certain conditions, (4) is the unique density function which solves (5) or (6). There is a good reason why (5) and (6) are called the *forward* and *backward* equations. Remember that W is a Markov process, and use arguments similar to those of Sections 6.8 and 6.9. Equation (5) is obtained by conditioning $W(t+h)$ on the value of $W(t)$ and letting $h \downarrow 0$; (6) is obtained by conditioning $W(t)$ on the value of $W(s+h)$ and letting $h \downarrow 0$. You are treading in Einstein's footprints as you perform these calculations. The derivatives in (5) and (6) have coefficients which do not depend on x, y, s, t; this reflects the fact that the Wiener process is homogeneous in space and time, in that

(a) the increment $W(t) - W(s)$ is independent of $W(s)$ for all $t \geqslant s$
(b) the increments are stationary in time.

Next we turn our attention to diffusion processes which *lack* this homogeneity.

The Wiener process is a Markov process, and the Markov property provides a method for deriving the forward and backward equations. There are other Markov diffusion processes to which this method may be applied in order to obtain similar forward and backward equations; the coefficients in these equations will *not* generally be constant. The existence of such processes can be demonstrated rigorously, but here we explore their distributions only. Let $D = \{D(t): t \geqslant 0\}$ denote a diffusion process. In addition to requiring that D has (almost surely) continuous sample paths, we need to impose some conditions on the transitions of D in order to derive its diffusion equations; these conditions take the form of specifying the mean and variance of increments $D(t+h) - D(t)$ of the process over small time intervals $(t, t+h)$. Suppose that there exist functions $a(x, t)$, $b(x, t)$ such that

$$\mathbf{P}(|D(t+h) - D(t)| > \varepsilon \mid D(t) = x) = \mathrm{o}(h) \quad \text{for all } \varepsilon > 0$$

$$\mathbf{E}(D(t+h) - D(t) \mid D(t) = x) = a(x, t)h + \mathrm{o}(h)$$

$$\mathbf{E}([D(t+h) - D(t)]^2 \mid D(t) = x) = b(x, t)h + \mathrm{o}(h);$$

a and b are called the 'instantaneous mean' and 'instantaneous variance' of D respectively. Subject to certain other technical conditions (see Prabhu 1965*a*, pp. 89, 98), if $s \leqslant t$ then the conditional density function of $D(t)$ given $D(s) = x$,

$$f(y, t \mid x, s) = \frac{\partial}{\partial y}\mathbf{P}(D(t) \leqslant y \mid D(s) = x),$$

satisfies the following partial differential equations:

(7) **forward equation:** $$\frac{\partial f}{\partial t} = -\frac{\partial}{\partial y}[a(y, t)f] + \frac{1}{2}\frac{\partial^2}{\partial y^2}[b(y, t)f]$$

(8) **backward equation:** $$\frac{\partial f}{\partial s} = -a(x, s)\frac{\partial f}{\partial x} - \frac{1}{2}b(x, s)\frac{\partial^2 f}{\partial x^2}.$$

It is an extraordinary fact that the density function f is specified as soon as the instantaneous mean a and variance b are known; we need no further information about the distribution of a typical increment. This is very convenient for many applications, since a and b are often specified in a natural manner by the description of the process.

(9) **Example. The Wiener process.** If increments of any given length have zero means and constant variances then

$$a(x, t) = 0, \qquad b(x, t) = \sigma^2$$

for some $\sigma^2 > 0$, and the diffusion equations (5) and (6) follow from (7) and (8) for processes with variance σ^2. ●

(10) **Example. The Wiener process with drift.** Suppose a particle undergoes a type of Brownian motion in one dimension, in which it experiences a drift at constant rate in some particular direction. That is to say,

$$a(x, t) = m, \qquad b(x, t) = \sigma^2$$

for some drift rate m and constant σ^2. The forward diffusion equation becomes

$$\frac{\partial f}{\partial t} = -m\frac{\partial f}{\partial y} + \frac{1}{2}\sigma^2\frac{\partial^2 f}{\partial y^2}$$

and it follows that the corresponding diffusion process D is such that

$$D(t) = W(t) + mt$$

where W is a Wiener process. ●

(11) **Example. The Ornstein–Uhlenbeck process.** Recall the discussion of this process at the end of Section 13.2. It experiences a drift towards the origin of magnitude proportional to its displacement. So

$$a(x, t) = -\beta x, \qquad b(x, t) = \sigma^2$$

and the forward equation is

$$\frac{\partial f}{\partial t} = \beta\frac{\partial}{\partial y}(yf) + \frac{1}{2}\sigma^2\frac{\partial^2 f}{\partial y^2}.$$

See Problem (13.8.4) for the solution of this equation. ●

(12) **Example. Diffusion approximation to the branching process.** Diffusion models are sometimes useful as continuous approximations to discrete processes. In Section 13.2 we saw that the Wiener process approximates to the random walk under certain circumstances; here is another example of such an approximation. Let $\{Z_n\}$ be the branching process of Section 5.4, with $Z_0 = 1$ and such that

$$\mathbf{E}(Z_1) = \mu, \qquad \text{var}(Z_1) = \sigma^2.$$

A typical increment $Z_{n+1} - Z_n$ has mean and variance given by

$$\mathbf{E}(Z_{n+1} - Z_n \mid Z_n = x) = (\mu - 1)x$$
$$\text{var}(Z_{n+1} - Z_n \mid Z_n = x) = \sigma^2 x;$$

these are directly proportional to the size of Z_n. Now, suppose that the time intervals between successive generations become shorter and shorter, but that the means and variances of the increments retain this proportionality; of course, we need to abandon the condition that the process be integer-valued. This suggests a diffusion model as an approximation to the branching process, with instantaneous mean and variance given by

$$a(x, t) = ax, \qquad b(x, t) = bx,$$

and the forward equation of such a process is

(13)
$$\frac{\partial f}{\partial t} = -a\frac{\partial}{\partial y}(yf) + \frac{1}{2}b\frac{\partial^2}{\partial y^2}(yf).$$

Subject to appropriate boundary conditions, this equation has a unique solution; this may be found by taking Laplace transforms of (13) in order to find the moment generating function of the value of the diffusion process at time t. ●

(14) **Example. A branching diffusion process.** The next example is a modification of the process of (6.12.15) which modelled the distribution in space of the members of a branching process. Read the first paragraph of (6.12.15) again before proceeding with this example. It is often the case that the members of a population move around the space which they inhabit during their lifetimes. With this in mind we introduce a modification into the process of (6.12.15). Suppose a typical individual is born at time s and at position x. We suppose that this individual moves about $\mathbb{R}$ until its lifetime T is finished, at which point it dies and divides, leaving its offspring at the position at which it dies. We suppose further that it moves as a standard Wiener process W, so that it is at position $x + W(t)$ at time $s + t$ whenever $0 \leqslant t \leqslant T$. We assume that each individual moves independently of the positions of all the other individuals. We retain the notation of (6.12.15) whenever it is suitable, writing N for the number of offspring of the initial individual, W for the

process describing its motion, and T for its lifetime. This individual dies at the point $W(T)$.

We no longer seek complete information about the distribution of the individuals around the space, but restrict ourselves to a less demanding task. It is natural to wonder about the rate at which members of the population move away from the place of birth of the founding member. Let $M(t)$ denote the position of the individual who is furthest right from the origin at time t. That is,

$$M(t) = \sup\{x: Z_1(x, t) > 0\}$$

where $Z_1(x, t)$ is the number of living individuals at time t who are positioned at points in the interval $[x, \infty)$. We shall study the distribution function of $M(t)$

$$F(x, t) = \mathbf{P}(M(t) \leqslant x),$$

and proceed roughly as before, noting that

(15) $$F(x, t) = \int_0^\infty \mathbf{P}(M(t) \leqslant x \mid T = s) f_T(s)\, \mathrm{d}s$$

where f_T is the density function of T. However,

$$\mathbf{P}(M(t) \leqslant x \mid T = s) = \mathbf{P}(W(t) \leqslant x) \quad \text{if} \quad s > t,$$

whilst, if $s \leqslant t$, use of conditional probabilities gives

$$\mathbf{P}(M(t) \leqslant x \mid T = s) = \sum_{n=0}^{\infty} \int_{w=-\infty}^{\infty} \mathbf{P}(M(t) \leqslant x \mid T = s, N = n, W(s) = w) \times \mathbf{P}(N = n) f_{W(s)}(w)\, \mathrm{d}w$$

where $f_{W(s)}$ is the density function of $W(s)$. But, if $s \leqslant t$, then

$$\mathbf{P}(M(t) \leqslant x \mid T = s, N = n, W(s) = w) = [\mathbf{P}(M(t - s) \leqslant x - w)]^n,$$

and so (15) becomes

(16) $$F(x, t) = \int_{s=0}^{t} \int_{w=-\infty}^{\infty} G_N[F(x - w, t - s)] f_{W(s)}(w) f_T(s)\, \mathrm{d}w\, \mathrm{d}s + \mathbf{P}(W(t) \leqslant x) \int_{s=t}^{\infty} f_T(s)\, \mathrm{d}s.$$

We consider here only the Markovian case when T is exponentially distributed, so that

$$f_T(s) = \mu\, \mathrm{e}^{-\mu s} \quad \text{for} \quad s \geqslant 0.$$

Multiply throughout (16) by $\mathrm{e}^{\mu t}$, substitute $x - w = v$ and $t - s = u$ within the integral, and differentiate with respect to t to obtain

$$e^{\mu t}\left(\mu F + \frac{\partial F}{\partial t}\right) = \mu \int_{v=-\infty}^{\infty} G_N(F(v, t)) f_{W(0)}(x - v)\, e^{\mu t}\, dt$$

$$+ \mu \int_{u=0}^{t} \int_{v=-\infty}^{\infty} G_N(F(v, u)) \left(\frac{\partial}{\partial t} f_{W(t-u)}(x - v)\right) e^{\mu u}\, dv\, du$$

$$+ \frac{\partial}{\partial t} \mathbf{P}(W(t) \leqslant x).$$

Now differentiate the same equation twice with respect to x, remembering that $f_{W(s)}(w)$ satisfies the diffusion equations and that $\delta(v) = f_{W(0)}(x - v)$ needs to be interpreted as the Dirac δ function at the point $v = x$ to find that

(17) $$\mu F + \frac{\partial F}{\partial t} = \mu G_N(F) + \frac{1}{2}\frac{\partial^2 F}{\partial x^2}.$$

Many eminent mathematicians have studied this equation; for example, Kolmogorov and Fisher were concerned with it in connection with the distribution of gene frequencies. It is difficult to extract precise information from (17). One approach is to look for solutions of the form

$$F(x, t) = \psi(x - ct)$$

for some constant c to obtain the following second-order ordinary differential equation for ψ:

(18) $$\psi'' + 2c\psi' + 2\mu H(\psi) = 0$$

where

$$H(\psi) = G_N(\psi) - \psi.$$

Solutions to (18) yield information about the asymptotic distribution of the so-called 'advancing' wave of the members of the process. ●

Finally in this section, we show that Wiener processes exist. The problem is the requirement that sample paths be continuous. Certainly there exist Gaussian processes with independent normally-distributed increments as required in (1a, b), but there is no reason in general why such a process should have continuous sample paths. We shall show here that one may construct such a Gaussian process with this extra property of continuity.

Let us restrict ourselves for the moment to the time interval $[0, 1]$, and suppose that X is a Gaussian process on $[0, 1]$ with independent increments, such that $X(0) = 0$, and $X(s + t) - X(s)$ is $N(0, t)$ for $s, t \geqslant 0$. We shall concentrate on a certain countable subset Q of $[0, 1]$, namely the set of 'dyadic rationals', being the set of points of the form $m2^{-n}$ for some $n \geqslant 1$ and $0 \leqslant m \leqslant 2^n$. For each $n \geqslant 1$, we define the process $X_n(t)$ by $X_n(t) = X(t)$ if $t = m2^{-n}$ for some integer m, and by linear interpolation otherwise; that is to say,

$$X_n(t) = X(m2^{-n}) + 2^n(t - m2^{-n})[X((m + 1)2^{-n}) - X(m2^{-n})]$$

if $m2^{-n} < t < (m+1)2^{-n}$. Thus X_n is a piecewise-linear and continuous function comprising 2^n line segments. Think of X_{n+1} as being obtained from X_n by repositioning the centres of these line segments by amounts which are independent and normally distributed. It is clear that

(19) $$X_n(t) \to X(t) \quad \text{for} \quad t \in Q,$$

since, if $t \in Q$, then $X_n(t) = X(t)$ for all large n. The first step is to show that the convergence in (19) is (almost surely) uniform on Q, since this will imply that the limit function X is (a.s.) continuous on Q. Now

(20) $$X_n(t) = \sum_{j=1}^{n} Z_j(t)$$

where $Z_j(t) = X_j(t) - X_{j-1}(t)$ and $X_0(t) = 0$. This series representation for X_n converges uniformly on Q if

(21) $$\sum_{j=1}^{\infty} \sup_{t \in Q} |Z_j(t)| < \infty.$$

We note that $Z_j(t) = 0$ for values of t having the form $m2^{-j}$ where m is even. It may be seen by drawing a diagram that

$$\sup_{t \in Q} |Z_j(t)| = \max\{|Z_j(m2^{-j})| : m = 1, 3, \dots, 2^j - 1\}$$

and therefore

(22) $$\mathbf{P}\left(\sup_{t \in Q} |Z_j(t)| > x\right) \leqslant \sum_{m \text{ odd}} \mathbf{P}(|Z_j(m2^{-j})| > x).$$

Now

$$\begin{aligned} Z_j(2^{-j}) &= X(2^{-j}) - \tfrac{1}{2}[X(0) + X(2^{-j+1})] \\ &= \tfrac{1}{2}[X(2^{-j}) - X(0)] - \tfrac{1}{2}[X(2^{-j+1}) - X(2^{-j})], \end{aligned}$$

and therefore $\mathbf{E}Z_j(2^{-j}) = 0$ and, using the independence of increments, $\text{var}(Z_j(2^{-j})) = 2^{-j-1}$; a similar calculation is valid for $Z_j(m2^{-j})$ for $m = 1, 3, \dots, 2^j - 1$. It follows by the bound (4.11.1c) on the tail of the normal distribution that, for all such m,

$$\mathbf{P}(|Z_j(m2^{-j})| > x) \leqslant \frac{1}{x2^{j/2}} e^{-x^2 2^j}, \qquad x > 0.$$

Setting $x = c\sqrt{(j2^{-j} \log 2)}$, we obtain from (22) that

$$\mathbf{P}\left(\sup_{t \in Q} |Z_j(t)| > x\right) \leqslant 2^{j-1} \frac{2^{-c^2 j}}{c\sqrt{(j \log 2)}}.$$

Choosing $c > 1$, the last term is summable in j, implying by the Borel–Cantelli lemma (7.3.10a) that

$$\sup_{t\in Q} |Z_t(t)| > c\sqrt{\frac{j \log 2}{2^j}}$$

for only finitely many values of j (a.s.). Hence

$$\sum_j \sup_{t\in Q} |Z_j(t)| < \infty \quad \text{a.s.,}$$

and the argument prior to (21) yields that X is (a.s.) continuous on Q.

We have proved that X has (a.s.) continuous sample paths on the set of dyadic rationals; a similar argument is valid for other countable dense subsets of $[0, 1]$. It is quite another thing for X to be continuous on the entire interval $[0, 1]$, and actually this need not be the case. We can, however, extend X by continuity from the dyadic rationals to the whole of $[0, 1]$: for $t \in [0, 1]$, define

$$Y(t) = \lim_{\substack{s\to t\\ s\in Q}} X(s),$$

the limit being taken as s approaches t through the dyadic rationals. Such a limit exists for all t since X is continuous on Q. It is not difficult to check that the extended process Y is indeed a Gaussian process with covariance function $\text{cov}(Y(s), Y(t)) = \min\{s, t\}$, and, most important, the sample paths of Y are (a.s.) continuous.

This completes the proof of the existence of a Wiener process on $[0, 1]$. A similar argument can be made to work on $[0, \infty)$, but it is easier either (a) to patch together continuous Wiener processes on $[n, n + 1]$ for $n = 0, 1, \ldots$, or (b) to use the result of Problem (9.7.18c).

Exercises

(23) Let $X = \{X(t): t \geqslant 0\}$ be a simple birth–death process with parameters $\lambda_n = n\lambda$ and $\mu_n = n\mu$. Suggest a diffusion approximation to X.

(24) Let D be a diffusion with instantaneous mean and variance $a(x, t)$ and $b(x, t)$, and let $M(\theta, t) = \mathbf{E}(e^{\theta D(t)})$, the moment generating function of $D(t)$. Use the forward diffusion equation to derive Bartlett's equation:

$$\frac{\partial M}{\partial t} = \theta a\left(\frac{\partial}{\partial\theta}, t\right)M + \tfrac{1}{2}\theta^2 b\left(\frac{\partial}{\partial\theta}, t\right)M$$

where we interpret

$$g\left(\frac{\partial}{\partial\theta}, t\right)M = \sum_n \gamma_n(t)\frac{\partial^n M}{\partial\theta^n}$$

if $g(x, t) = \sum_{n=0}^{\infty} \gamma_n(t)x^n$.

(25) Write down Bartlett's equation in the case of the Wiener process D having drift m and instantaneous variance 1, and solve it subject to the boundary condition $D(0) = 0$.

(26) Write down Bartlett's equation in the case of an Ornstein–Uhlenbeck process D having instantaneous mean $a(x, t) = -x$ and variance $b(x, t) = 1$, and solve it subject to the boundary condition $D(0) = 0$.

13.4 First passage times

We have often been interested in the time which elapses before a Markov chain visits a specified state for the first time, and we continue this chapter with an account of some of the corresponding problems for a diffusion process.

Consider first a standard Wiener process W starting from $W(0) = 0$. W has the property that the process W_1 given by

(1) $$W_1(t) = W(t + T) - W(T), \qquad t \geq 0$$

is a standard Wiener process for any fixed value of T and, conditional on $W(T)$, W_1 is independent of $\{W(s): s < T\}$; the Poisson process enjoys a similar property, which in Section 6.8 we called the 'weak Markov property'. It is a very important and useful fact that this holds even when T is a random variable, so long as T is a stopping time for W. We encountered stopping times in the context of continuous-time martingales in Section 12.7.

(2) **Definition.** Let $\mathscr{F}(t)$ be the smallest σ-field with respect to which $W(s)$ is measurable for each $s \leq t$. T is called a *stopping time* for W if $\{T \leq t\} \in \mathscr{F}(t)$ for all t.

We say that W has the 'strong Markov property' in that this independence holds for all stopping times T. Why not try to prove this?

Here, we make use of the strong Markov property for certain particular stopping times T.

(3) **Definition.** The **first passage time** $T(x)$ to the point $x \in \mathbb{R}$ is given by

$$T(x) = \inf\{t: W(t) = x\}.$$

The continuity of sample paths is essential in order that this definition make sense: a Wiener process cannot jump over the value x, but must pass through it. The proof of the following lemma is omitted.

(4) **Lemma.** *$T(x)$ is a stopping time for W.*

(5) **Theorem.** *$T(x)$ has density function*

$$f_{T(x)}(t) = \frac{|x|}{\sqrt{(2\pi t^3)}} \exp\left(-\frac{x^2}{2t}\right), \qquad t \geq 0.$$

Clearly $T(x)$ and $T(-x)$ are identically distributed. For the case when $x = 1$ we encountered this density function and its moment generating function in Problems (5.12.18) and (5.12.19); it is easy to deduce that $T(x)$ has the same distribution as Z^{-2} where Z is $N(0, x^{-2})$. Before the proof of Theorem (5), here is a result about the size of the maximum of a Wiener process.

(6) **Theorem.** *Let*

$$M(t) = \max\{W(s): 0 \leqslant s \leqslant t\}.$$

Then $M(t)$ has the same distribution as $|W(t)|$; thus $M(t)$ has density function

$$f_{M(t)}(m) = \left(\frac{2}{\pi t}\right)^{\frac{1}{2}} \exp\left(-\frac{m^2}{2t}\right), \qquad m \geqslant 0.$$

You should draw your own diagrams to illustrate the translations and reflections used in the proofs of this section.

Proof of (6). Suppose $m > 0$, and observe that

(7)
$$T(m) \leqslant t \quad \text{if and only if} \quad M(t) \geqslant m.$$

Then

$$\mathbf{P}(M(t) \geqslant m) = \mathbf{P}(M(t) \geqslant m,\, W(t) - m \geqslant 0) + \mathbf{P}(M(t) \geqslant m,\, W(t) - m < 0).$$

However, by (7),

$$\begin{aligned} \mathbf{P}(M(t) &\geqslant m,\, W(t) - m < 0) \\ &= \mathbf{P}(W(t) - W(T(m)) < 0 \mid T(m) \leqslant t)\mathbf{P}(T(m) \leqslant t) \\ &= \mathbf{P}(W(t) - W(T(m)) \geqslant 0 \mid T(m) \leqslant t)\mathbf{P}(T(m) \leqslant t) \\ &= \mathbf{P}(M(t) \geqslant m,\, W(t) - m \geqslant 0) \end{aligned}$$

since $W(t) - W(T(m))$ is symmetric whenever $t \geqslant T(m)$ by the strong Markov property; we have used sample path continuity here, in that we have used the fact that

$$\mathbf{P}\big(W(T(m)) = m\big) = 1.$$

Thus

$$\begin{aligned} \mathbf{P}(M(t) \geqslant m) &= 2\mathbf{P}(M(t) \geqslant m,\, W(t) \geqslant m) \\ &= 2\mathbf{P}(W(t) \geqslant m) \end{aligned}$$

since $W(t) \leqslant M(t)$. Hence

$$\mathbf{P}(M(t) \geqslant m) = \mathbf{P}(|W(t)| \geqslant m)$$

and the theorem is proved on noting that $|W(t)|$ is the absolute value of an $N(0, t)$ variable. ■

Proof of (5). This follows immediately from (7), since if $x > 0$ then

$$\begin{aligned} \mathbf{P}(T(x) \leqslant t) &= \mathbf{P}(M(t) \geqslant x) \\ &= \mathbf{P}(|W(t)| \geqslant x) \\ &= \left(\frac{2}{\pi t}\right)^{\frac{1}{2}} \int_x^\infty \exp\left(-\frac{m^2}{2t}\right) dm \\ &= \int_0^t \frac{|x|}{\sqrt{(2\pi y^3)}} \exp\left(-\frac{x^2}{2y}\right) dy \end{aligned}$$

by the substitution $y = x^2 t/m^2$. ■

We are now in a position to derive some famous results about the times at which W returns to its starting point, the origin. We say that 'W has a zero at time t' if $W(t) = 0$.

(8) **Theorem.** *Suppose* $0 \leqslant t_0 < t_1$. *The probability that a standard Wiener process* W *has a zero in the time interval* (t_0, t_1), *starting from* $W(0) = 0$, *is*

$$\frac{2}{\pi} \cos^{-1}\left[\left(\frac{t_0}{t_1}\right)^{\frac{1}{2}}\right].$$

Here as usual, $\cos^{-1}$ denotes the inverse trigonometric function, sometimes written arc cos.

Proof. If $0 \leqslant u < v$, let $E(u, v)$ denote the event

$$E(u, v) = \{W(t) = 0 \text{ for some } t \in (u, v)\}.$$

Condition on $W(t_0)$ to obtain

$$\begin{aligned} \mathbf{P}(E(t_0, t_1)) &= \int_{-\infty}^{\infty} \mathbf{P}(E(t_0, t_1) \mid W(t_0) = w) f_0(w)\, \mathrm{d}w \\ &= 2 \int_{-\infty}^{0} \mathbf{P}(E(t_0, t_1) \mid W(t_0) = w) f_0(w)\, \mathrm{d}w \end{aligned}$$

by the symmetry of W, where f_0 is the density function of $W(t_0)$. However, if $a > 0$,

$$\mathbf{P}(E(t_0, t_1) \mid W(t_0) = -a) = \mathbf{P}(T(a) < t_1 - t_0 \mid W(0) = 0)$$

by the homogeneity of W in time and space. Use (5) to obtain that

$$\begin{aligned} \mathbf{P}(E(t_0, t_1)) &= 2 \int_{a=0}^{\infty} \int_{t=0}^{t_1 - t_0} f_{T(a)}(t) f_0(-a)\, \mathrm{d}t\, \mathrm{d}a \\ &= \frac{1}{\pi\sqrt{t_0}} \int_{t=0}^{t_1 - t_0} t^{-\frac{3}{2}} \int_{a=0}^{\infty} a \exp\left[-\tfrac{1}{2}a^2\left(\frac{t + t_0}{t t_0}\right)\right] da\, \mathrm{d}t \\ &= \frac{\sqrt{t_0}}{\pi} \int_{t=0}^{t_1 - t_0} \frac{\mathrm{d}t}{(t + t_0)\sqrt{t}} \\ &= \frac{2}{\pi} \tan^{-1}\left[\left(\frac{t_1}{t_0} - 1\right)^{\frac{1}{2}}\right] \quad \text{by the substitution } t = t_0 s^2 \\ &= \frac{2}{\pi} \cos^{-1}\left[\left(\frac{t_0}{t_1}\right)^{\frac{1}{2}}\right] \quad \text{as required.} \end{aligned}$$

■

The result of (8) indicates some remarkable properties of the sample paths of W. Set $t_0 = 0$ to obtain

$$\mathbf{P}(\text{there exists a zero in } (0, t) \mid W(0) = 0) = 1 \quad \text{for all } t > 0,$$

and it follows that

$$T(0) = \inf\{t \neq 0 \colon W(t) = 0\}$$

satisfies $T(0) = 0$ almost surely. A deeper analysis shows that, with probability 1, W has infinitely many zeros in any non-empty time interval $[0, t]$; it is no wonder that W has non-differentiable sample paths! The set

$$Z = \{t \colon W(t) = 0\}$$

of zeros of W is rather a large set; in fact it turns out that Z has Hausdorff dimension $\frac{1}{2}$ (see Mandelbrot (1983) for a charming discussion of fractional dimensionality).

The proofs of (5), (6), and (8) have relied heavily upon certain symmetries of the Wiener process; these are similar to the symmetries of the random walk of Section 3.10. Other diffusions may not have these symmetries, and we may need other techniques for finding out about their first passage times. We illustrate this point by a glance at the Wiener process with drift. Let $D = \{D(t) \colon t \geqslant 0\}$ be a diffusion process with instantaneous mean and variance given by

$$a(x, t) = m, \qquad b(x, t) = 1$$

where m is a constant. It is easy to check that, if $D(0) = 0$, then $D(t)$ is $N(mt, t)$. It is not so easy to find the distributions of the sizes of the maxima of D, and we take this opportunity to display the usefulness of martingales and optional stopping.

(9) **Theorem.** *Let* $U(t) = \exp[-2mD(t)]$. *Then* $U = \{U(t) \colon t \geqslant 0\}$ *is a martingale.*

Our only experience to date of continuous-time martingales is contained in Section 12.7.

Proof. D is a Markov process, and so U is a Markov process also. To check that the continuous martingale condition holds, it suffices to show that

(10) $$\mathbf{E}(U(t+s) \mid U(t)) = U(t) \quad \text{for all } s, t \geqslant 0.$$

However,

(11) $$\begin{aligned} \mathbf{E}(U(t+s) \mid U(t) = e^{-2md}) &= \mathbf{E}(\exp[-2mD(t+s)] \mid D(t) = d) \\ &= \mathbf{E}(\exp\{-2m[D(t+s) - D(t)] - 2md\} \mid D(t) = d) \\ &= e^{-2md}\mathbf{E}(\exp\{-2m[D(t+s) - D(t)]\}) \\ &= e^{-2md}\mathbf{E}(\exp[-2mD(s)]) \end{aligned}$$

because D is Markovian with stationary independent increments. Now,

$$\mathbf{E}(\exp[-2mD(s)]) = M(-2m)$$

where M is the moment generating function of an $N(ms, s)$ variable; M is given by (5.8.5) as

$$M(u) = \exp(msu + \tfrac{1}{2}su^2).$$

Thus

$$\mathbf{E}(\exp[-2mD(s)]) = 1$$

and so (10) follows from (11). ■

We can use this martingale to find the distribution of first passage times, just as we did in (12.5.6) for the random walk. Let $x, y > 0$ and define

$$T(x, -y) = \inf\{t: D(t) = x \text{ or } D(t) = -y\}$$

to be the first passage time of D to the set $\{x, -y\}$. It is easily shown that $T(x, -y)$ is a stopping time which is almost surely finite.

(12) **Theorem.** $\mathbf{E}(U[T(x, -y)]) = 1$ *for all* $x, y > 0$.

Proof. This is just an application of a version of the optional stopping theorem (12.7.12). U is a martingale and $T(x, -y)$ is a stopping time. Thus

$$\mathbf{E}(U[T(x, -y)]) = \mathbf{E}(U(0)) = 1.$$ ■

(13) **Corollary.** *If* $m < 0$ *and* $x > 0$ *then the probability that D ever visits the point x is*

$$\mathbf{P}(D(t) = x \textit{ for some } t) = e^{2mx}.$$

Proof. By (12),

$$1 = e^{-2mx}\mathbf{P}(D[T(x, -y)] = x) + e^{2my}\{1 - \mathbf{P}(D[T(x, -y)] = -y)\}.$$

Let $y \to \infty$ to obtain

$$\mathbf{P}(D[T(x, -y)] = x) \to e^{2mx}$$

so long as $m < 0$. Now complete the proof yourself. ■

The condition of (13), that the drift be negative, is natural; it is clear that if $m > 0$ then D almost surely visits all points on the positive part of the real axis. The result of (13) tells us about the size of the maximum of D also, since if $x > 0$

$$\left\{\max_{t \geq 0} D(t) \geq x\right\} = \{D(t) = x \text{ for some } t\}$$

and the distribution of

$$M = \max\{D(t): t \geq 0\}$$

is easily deduced.

(14) **Corollary.** *If* $m < 0$ *then M is exponentially distributed with parameter* $-2m$.

Exercises

(15) Let W be a standard Wiener process and let $X(t) = \exp\{i\theta W(t) + \frac{1}{2}\theta^2 t\}$. Show that X is a martingale with respect to the filtration given by $\mathcal{F}_t = \sigma(\{W(u): u \leqslant t\})$.

(16) Let T be the (random) time at which a standard Wiener process W (with $W(0) = 0$) hits the 'barrier' in space-time given by $y = at + b$ where $a < 0$, $b \geqslant 0$; that is, $T = \inf\{t: W(t) = at + b\}$. Use the result of Exercise (15) to show that the moment generating function of T is given by $\mathbf{E}(e^{\psi T}) = \exp\{-b(\sqrt{a^2 - 2\psi} + a)\}$ for $\psi < \frac{1}{2}a^2$. You may assume that the conditions of the optional stopping theorem are satisfied.

(17) Let W be a standard Wiener process with $W(0) = 0$, and let T be the time of the last zero of W prior to time t. Show that $\mathbf{P}(T \leqslant u) = (2/\pi)\sin^{-1}\{(u/t)^{\frac{1}{2}}\}$, $0 \leqslant u \leqslant t$.

13.5 Barriers

Diffusing particles are rarely allowed to roam freely, but are often restricted to a given part of space; for example, Brown's pollen particles were suspended in fluid which was confined to a container. What may happen when a particle hits a barrier? As with random walks, two simple types of barrier are the *absorbing* and the *reflecting*, although there are various other types of some complexity.

We begin with the case of the Wiener process. Let W be a standard Wiener process with $W(0) = w$, say, where $w > 0$. The Wiener process W_A *absorbed* at 0 is defined to be the process given by

(1)
$$W_A(t) = \begin{cases} W(t) & \text{if} \quad t < T \\ 0 & \text{if} \quad t \geqslant T \end{cases}$$

where $T = \inf\{t: W(t) = 0\}$ is the hitting time of the position 0. The Wiener process W_R *reflected* at 0 is defined as the process $W_R(t) = |W(t)|$.

Viewing the diffusion equations (13.3.7)–(13.3.8) as forward and backward equations, it is clear that W_A and W_R satisfy these equations so long as they are away from the barrier. That is to say, W_A and W_R are diffusion processes. In order to find their transition density functions, we might solve the diffusion equations subject to suitable boundary conditions. For the special case of the Wiener process, however, it is simpler to argue as follows.

(2) **Theorem.** *Let $f(y, t \mid x, s)$ denote the density function of $W(t)$ conditional on $W(s) = x$, and let W_A and W_R be given as above.*

(a) *The density function of $W_A(t)$ is*

$$f_A(y, t) = f(y, t \mid w, 0) - f(-y, t \mid w, 0), \qquad y > 0.$$

(b) *The density function of $W_R(t)$ is*

$$f_R(y, t) = f(y, t \mid w, 0) + f(-y, t \mid w, 0), \qquad y > 0.$$

Of course,

(3) $$f(y, t \mid w, 0) = \frac{1}{\sqrt{(2\pi t)}} \exp\left(-\frac{(y-w)^2}{2t}\right),$$

the $N(w, t)$ density function.

Proof. Let I be a subinterval of $(0, \infty)$, and let $I^r = \{x \in \mathbb{R}: -x \in I\}$ be the reflection of I in the point 0. Then

$$\begin{aligned}\mathbf{P}(W_A(t) \in I) &= \mathbf{P}(\{W(t) \in I\} \cap \{T > t\}) \\ &= \mathbf{P}(W(t) \in I) - \mathbf{P}(\{W(t) \in I\} \cap \{T \leqslant t\}) \\ &= \mathbf{P}(W(t) \in I) - \mathbf{P}(W(t) \in I^r)\end{aligned}$$

using the reflection principle and the strong Markov property. The result follows.

The result of part (b) is immediate from the fact that $W_R(t) = |W(t)|$. ■

We turn now to the absorption and reflection of a *general* diffusion process. Let $D = \{D(t): t \geqslant 0\}$ be a diffusion process; we write a and b for the instantaneous mean and variance functions of D, and shall suppose that $b(x, t) > 0$ for all x ($\geqslant 0$) and t. We make a further assumption, that D is *regular* in that

(4) $$\mathbf{P}(D(t) = y \text{ for some } t \mid D(0) = x) = 1 \quad \text{for all } x, y \geqslant 0.$$

Suppose that the process starts from $D(0) = d$ say, where $d > 0$. Placing an absorbing barrier at 0 amounts to killing D when it first hits 0. The resulting process D_A is given by

$$D_A(t) = \begin{cases} D(t) & \text{if } T > t \\ 0 & \text{if } T \leqslant t \end{cases}$$

where $T = \inf\{t: D(t) = 0\}$; this formulation requires D to have continuous sample paths.

Viewing the diffusion equations (13.3.7)–(13.3.8) as forward and backward equations, it is clear that they are satisfied away from the barrier. The presence of the absorbing barrier affects the solution to the diffusion equations through the boundary conditions.

Denote by $f_A(y, t)$ the density function of $D_A(t)$; we might write $f_A(y, t) = f_A(y, t \mid d, 0)$ to emphasize the value of $D_A(0)$. The boundary condition appropriate to an absorbing barrier at 0 is

(5) $$f_A(0, t) = 0 \quad \text{for all } t.$$

It is not completely obvious that (5) is the correct condition, but the following rough argument may be made rigorous. The idea is that, if the particle is near to the absorbing barrier, then small local fluctuations, arising from the non-zero instantaneous variance, will carry it to the absorbing barrier extremely quickly. Therefore the chance of it being near to the barrier but unabsorbed is extremely small.

A slightly more rigorous justification for (5) is as follows. Suppose that (5) does not hold, which is to say that there exist $\varepsilon, \eta > 0$ and $0 < u < v$ such that

(6) $$f_{\mathrm{A}}(y, t) > \eta \quad \text{for} \quad 0 < y \leqslant \varepsilon,\ u \leqslant t \leqslant v.$$

There is probability at least $\eta\, \mathrm{d}x$ that $0 < D_{\mathrm{A}}(t) \leqslant \mathrm{d}x$ whenever $u \leqslant t \leqslant v$ and $0 < \mathrm{d}x \leqslant \varepsilon$. Hence the probability of absorption in the time interval $(t, t + \mathrm{d}t)$ is at least

(7) $$\eta\, \mathrm{d}x \mathbf{P}\big(D_{\mathrm{A}}(t + \mathrm{d}t) - D_{\mathrm{A}}(t) < -\mathrm{d}x \mid 0 < D_{\mathrm{A}}(t) \leqslant \mathrm{d}x\big).$$

The instantaneous variance satisfies $b(x, t) \geqslant \beta$ for $0 < x \leqslant \varepsilon$, $u \leqslant t \leqslant v$, for some $\beta > 0$, implying that $D_{\mathrm{A}}(t + \mathrm{d}t) - D_{\mathrm{A}}(t)$ has variance at least $\beta \mathrm{d}t$, under the condition that $0 < D_{\mathrm{A}}(t) \leqslant \mathrm{d}x$. Therefore

$$\mathbf{P}\big(D_{\mathrm{A}}(t + \mathrm{d}t) - D_{\mathrm{A}}(t) < -\gamma\sqrt{(\mathrm{d}t)} \mid 0 < D_{\mathrm{A}}(t) \leqslant \mathrm{d}x\big) \geqslant \delta$$

for some $\gamma, \delta > 0$. Substituting $\mathrm{d}x = \gamma\sqrt{(\mathrm{d}t)}$ in (7), we obtain

$$\mathbf{P}(t < T < t + \mathrm{d}t) \geqslant (\eta\gamma\delta)\sqrt{(\mathrm{d}t)},$$

implying by integration that $\mathbf{P}(u < T < v) = \infty$, which is clearly impossible. Hence (5) holds.

(8) **Example. Wiener process with drift.** Suppose that $a(x, t) = m$ and $b(x, t) = 1$ for all x and t. Put an absorbing barrier at 0 and suppose $D(0) = d > 0$. We wish to find a solution $g(y, t)$ to the forward equation

(9) $$\frac{\partial g}{\partial t} = -m\frac{\partial g}{\partial y} + \frac{1}{2}\frac{\partial^2 g}{\partial y^2}, \qquad y > 0,$$

subject to the boundary conditions

(10) $$g(0, t) = 0, \qquad t \geqslant 0$$

(11) $$g(y, 0) = \delta_d(y), \qquad y \geqslant 0,$$

where δ_d is the Dirac δ function centred at d. We know from (13.3.10), and in any case it is easy to check from first principles, that

(12) $$g(y, t \mid x) = \frac{1}{\sqrt{(2\pi t)}}\exp\left(-\frac{(y - x - mt)^2}{2t}\right)$$

satisfies (9), for all possible 'sources' x. Our target is to find a linear combination of such functions $g(\cdot, \cdot \mid x)$ which satisfies (10) and (11). It turns out that

(13) $$f_{\mathrm{A}}(y, t) = g(y, t \mid d) - \mathrm{e}^{-2md} g(y, t \mid -d), \qquad y > 0,$$

is such a function; assuming the solution is unique (which it is), this is therefore the density function of $D_{\mathrm{A}}(t)$. We may think of it as a mixture of the function $g(\cdot, \cdot \mid d)$ with source d together with a corresponding function from the 'image source' $-d$, being the reflection of d in the barrier at 0.

It is a small step to deduce the density function of the time T until the absorption of the particle. At time t, either the process has been absorbed, or its position has density function given by (13). Hence

$$\mathbf{P}(T \leqslant t) = 1 - \int_0^\infty f_A(y, t)\, dy = 1 - \Phi\left(\frac{mt + d}{\sqrt{t}}\right) + e^{-2md}\Phi\left(\frac{mt - d}{\sqrt{t}}\right)$$

by (12) and (13), where Φ is the $N(0, 1)$ distribution function. Differentiate with respect to t to obtain

(14)
$$f_T(t) = \frac{d}{\sqrt{(2\pi t^3)}} \exp\left(-\frac{(d + mt)^2}{2t}\right), \qquad t > 0.$$

It is easily seen that

$$\mathbf{P}(\text{absorption takes place}) = \mathbf{P}(T < \infty) = \begin{cases} 1 & \text{if } m \leqslant 0 \\ e^{-2md} & \text{if } m > 0. \end{cases}$$ ●

Turning to the matter of a reflecting barrier, suppose once again that D is a regular diffusion process with instantaneous mean a and variance b, starting from $D(0) = d > 0$. A reflecting barrier at the origin has the effect of disallowing infinitesimal negative jumps at the origin and replacing them by positive jumps. A formal definition requires careful treatment of the sample paths, and this is omitted here. Think instead about a reflecting barrier as giving rise to an appropriate boundary condition for the diffusion equations. Let us call the reflected process D_R, and let $f_R(y, t)$ be its density function. The reflected process lives on $[0, \infty)$, and therefore

$$\int_0^\infty f_R(y, t)\, dy = 1 \quad \text{for all } t.$$

Differentiating with respect to t and using the forward diffusion equation, we obtain at the expense of mathematical rigour that

$$\begin{aligned} 0 &= \frac{\partial}{\partial t}\int_0^\infty f_R(y, t)\, dy \\ &= \int_0^\infty \frac{\partial f_R}{\partial t}\, dy = \int_0^\infty \left(-\frac{\partial}{\partial y}(af_R) + \frac{1}{2}\frac{\partial^2}{\partial y^2}(bf_R)\right) dy \\ &= \left[-af_R + \frac{1}{2}\frac{\partial}{\partial y}(bf_R)\right]_0^\infty = \left(af_R - \frac{1}{2}\frac{\partial}{\partial y}(bf_R)\right)\bigg|_{y=0}. \end{aligned}$$

This indicates that the density function $f_R(y, t)$ of $D(t)$ is obtained by solving the forward diffusion equation

$$\frac{\partial g}{\partial t} = -\frac{\partial}{\partial y}(ag) + \frac{1}{2}\frac{\partial^2}{\partial y^2}(bg)$$

subject to the boundary condition

(15) $$\left(ag - \frac{1}{2}\frac{\partial}{\partial y}(bg)\right)\bigg|_{y=0} = 0 \quad \text{for} \quad t \geqslant 0$$

as well as the initial condition

(16) $$g(y, 0) = \delta_d(y) \quad \text{for} \quad y \geqslant 0.$$

(17) **Example. Wiener process with drift.** Once again suppose that $a(x, t) = m$ and $b(x, t) = 1$ for all x, t. This time we seek a linear combination of the functions g given in (12) which satisfies (15) and (16). It turns out that the answer contains an image at $-d$ together with a continuous line of images over the range $(-\infty, -d)$. That is to say, the solution has the form

$$f_R(y, t) = g(y, t \mid d) + Ag(y, t \mid -d) + \int_{-\infty}^{-d} B(x)g(y, t \mid x)\, dx$$

for certain A and $B(x)$. Substituting this into (15), one obtains after some work that

(18) $$A = e^{-2md}, \qquad B(x) = -2m\, e^{2mx}.$$ ●

Exercise

(19) Let D be a standard Wiener process with drift m starting from $D(0) = d > 0$, and suppose that there is a reflecting barrier at the origin. Show that the density function $f_R(y, t)$ of $D(t)$ satisfies $f_R(y, t) \to 0$ as $t \to \infty$ if $m \geqslant 0$, whereas $f_R(y, t) \to 2|m|\, e^{-2|m|y}$ for $y > 0$, as $t \to \infty$ if $m < 0$.

13.6 Excursions, and the Brownian bridge

This section is devoted to properties of the Wiener process conditioned on certain special events. We begin with a question concerning the set of zeros of the process. Let $W = \{W(t): t \geqslant 0\}$ be a standard Wiener process with $W(0) = w$, say. What is the probability that W has no zeros in the time interval $(0, v]$ given that it has none in the smaller interval $(0, u]$? The question is not too interesting if $w \neq 0$, since in this case the probability in question is just the ratio

(1) $$\frac{\mathbf{P}(\text{no zeros in } (0, v] \mid W(0) = w)}{\mathbf{P}(\text{no zeros in } (0, u] \mid W(0) = w)}$$

each term of which is easily calculated from the distribution of maxima (13.4.6). The difficulty arises when $w = 0$, since both numerator and denominator in (1) equal 0. In this case, it may be seen that the required probability is the limit of (1) as $w \to 0$. We have that this limit equals $\lim_{w \to 0}\{g_w(v)/g_w(u)\}$ where $g_w(x)$ is the probability that a Wiener process starting from w fails to reach 0 by time x. Using symmetry and (13.4.6),

$$g_w(x) = \left(\frac{2}{\pi x}\right)^{\frac{1}{2}} \int_0^{|w|} \exp[-m^2/(2x)]\, dm,$$

whence $g_w(v)/g_w(u) \to (u/v)^{\frac{1}{2}}$ as $w \to 0$, which we write as

(2) $$\mathbf{P}\big(W \neq 0 \text{ on } (0, v] \mid W \neq 0 \text{ on } (0, u], W(0) = 0\big) = (u/v)^{\frac{1}{2}}, \qquad 0 < u \leqslant v.$$

A similar argument results in

(3) $$\mathbf{P}\big(W > 0 \text{ on } (0, v] \mid W > 0 \text{ on } (0, u], W(0) = 0\big) = (u/v)^{\frac{1}{2}}, \qquad 0 < u \leqslant v,$$

by the symmetry of the Wiener process.

An 'excursion' of W is a trip taken by W away from 0. That is to say, if $W(u) = W(v) = 0$ and $W(t) \neq 0$ for $u < t < v$, then the trajectory of W during the time interval $[u, v]$ is called an *excursion* of the process; excursions are *positive* if $W > 0$ throughout (u, v), and *negative* otherwise. For any time $t > 0$, let $t - Z(t)$ be the time of the last zero prior to t, which is to say that $Z(t) = \sup\{s\colon W(t - s) = 0\}$; we suppose that $W(0) = 0$. At time t, some excursion is in progress whose current duration is $Z(t)$.

(4) **Theorem.** *Let* $Y(t) = \sqrt{[Z(t)]}\, \text{sign}\{W(t)\}$, *and* $\mathscr{F}_t = \sigma(\{Y(u)\colon 0 \leqslant u \leqslant t\})$. *Then* $(Y, \mathscr{F})$ *is a martingale, called the* ***excursions martingale***.

Proof. Clearly $Z(t) \leqslant t$, so that $\mathbf{E}|Y(t)| \leqslant \sqrt{t}$. It suffices to prove that

(5) $$\mathbf{E}(Y(t) \mid \mathscr{F}_s) = Y(s) \quad \text{for} \quad s < t.$$

Suppose $s < t$, and let A be the event that $W(u) = 0$ for some $u \in [s, t]$. Then, with a slight abuse of notation,

$$\mathbf{E}(Y(t) \mid \mathscr{F}_s) = \mathbf{E}(Y(t) \mid \mathscr{F}_s, A)\mathbf{P}(A \mid \mathscr{F}_s) + \mathbf{E}(Y(t) \mid \mathscr{F}_s, A^c)\mathbf{P}(A^c \mid \mathscr{F}_s).$$

Now

(6) $$\mathbf{E}(Y(t) \mid \mathscr{F}_s, A) = 0$$

since, on A, $Y(t)$ is a symmetric random variable. On the other hand

(7) $$\mathbf{E}(Y(t) \mid \mathscr{F}_s, A^c) = [t - s + Z(s)]^{\frac{1}{2}}\, \text{sign}\{W(s)\}$$

since, given $\mathscr{F}_s$ and A^c, the current duration of the excursion at time t is $(t - s) + Z(s)$, and $\text{sign}\{W(t)\} = \text{sign}\{W(s)\}$. Furthermore $\mathbf{P}(A^c \mid \mathscr{F}_s)$ equals the probability that W has strictly the same sign on $(s - Z(s), t]$ given the corresponding event on $(s - Z(s), s]$, which gives

$$\mathbf{P}(A^c \mid \mathscr{F}_s) = \left(\frac{Z(s)}{t - s + Z(s)}\right)^{\frac{1}{2}} \quad \text{by (3).}$$

Combining this with (6) and (7), we obtain $\mathbf{E}(Y(t) \mid \mathscr{F}_s) = Y(s)$ as required. ■

(8) **Corollary.** *The probability that the standard Wiener process W, starting from 0, has a positive excursion of total duration at least a before it has a negative excursion of total duration at least b is* $\sqrt{b}/(\sqrt{a}+\sqrt{b})$.

Proof. Let $T = \inf\{t: Y(t) \geqslant \sqrt{a} \text{ or } Y(t) \leqslant -\sqrt{b}\}$, the time which elapses before W records a positive excursion of duration at least a or a negative excursion of duration at least b. It may be shown that the optional stopping theorem for continuous-time martingales is applicable, and hence $\mathbf{E}(Y(T)) = \mathbf{E}(Y(0)) = 0$. However, $\mathbf{E}(Y(T)) = \pi\sqrt{a} - (1-\pi)\sqrt{b}$ where π is the required probability. ■

We turn next to the Brownian bridge. Think about the sample path of W on the time interval [0, 1] as the shape of a random string with its left end tied to the origin. What does it look like if you tie down its right end also? That is to say, what sort of process is $\{W(t): 0 \leqslant t \leqslant 1\}$ conditioned on the event that $W(1) = 0$? This new process is called the 'tied-down Wiener process' or the 'Brownian bridge'. There are various ways of studying it, the most obvious of which is perhaps to calculate the fdds of W conditional on $W(1) \in (-\eta, \eta)$, and then take the limit as $\eta \downarrow 0$. This is easily done, and leads to the next theorem.

(9) **Theorem.** *Let* $\mathbb{B} = \{\mathbb{B}(t): 0 \leqslant t \leqslant 1\}$ *be a process with continuous sample paths and the same fdds as* $\{W(t): 0 \leqslant t \leqslant 1\}$ *conditioned on* $W(0) = W(1) = 0$. *Then* $\mathbb{B}$ *is a diffusion process with instantaneous mean a and variance b given by*

(10)
$$a(x, t) = -\frac{x}{1-t}, \qquad b(x, t) = 1, \qquad x \in \mathbb{R},\ 0 \leqslant t \leqslant 1.$$

Note that the Brownian bridge has the same instantaneous variance as W, but its instantaneous mean increases in magnitude as $t \to 1$ and has the effect of guiding the process to its finishing point $\mathbb{B}(1) = 0$.

Proof. We make use of an elementary calculation involving conditional density functions. Let W be a standard Wiener process with $W(0) = w$, and suppose that $0 \leqslant u \leqslant v$. It is left as an *exercise* to prove that, conditional on $\{W(v) = y\}$, the distribution of $W(u)$ is normal with mean $y(u/v) + w(v-u)/v$ and variance $u(v-u)/v$. In particular

(11)
$$\mathbf{E}(W(u) \mid W(0) = w, W(v) = y) = \frac{yu}{v} + \frac{w(v-u)}{v}$$

and

(12)
$$\mathbf{E}(W(u)^2 \mid W(0) = w, W(v) = y) = \left(\frac{yu}{v} + \frac{w(v-u)}{v}\right)^2 + \frac{u(v-u)}{v}.$$

Returning to the Brownian bridge $\mathbb{B}$, after a little reflection one sees that it is Gaussian and Markov, since W has these properties. Furthermore the instantaneous mean is given by

$$\mathbf{E}\big(\mathbb{B}(t+h) - \mathbb{B}(t) \mid \mathbb{B}(t) = x\big) = -\frac{xh}{1-t}$$

by (11) with $w = x$, $y = 0$, $u = h$, $v = 1 - t$; similarly the instantaneous variance is given by the following consequence of (12):

$$\mathbf{E}\big(|\mathbb{B}(t+h) - \mathbb{B}(t)|^2 \mid \mathbb{B}(t) = x\big) = h + o(h). \qquad \blacksquare$$

An elementary calculation based on (11) and (12) shows that

(13) $$\operatorname{cov}(\mathbb{B}(s), \mathbb{B}(t)) = \min\{s, t\} - st, \qquad 0 \leqslant s, t \leqslant 1.$$

Exercises

(14) Let W be a standard Wiener process with $W(0) = 0$. Show that the conditional density function of $W(t)$, given that $W(u) > 0$ for $0 < u < t$, is $g(x) = (x/t)\,e^{-x^2/(2t)}$, $x > 0$.

(15) Show that the autocovariance function of the Brownian bridge is $c(s, t) = \min\{s, t\} - st$, $0 \leqslant s, t \leqslant 1$.

(16) Let W be a standard Wiener process with $W(0) = 0$, and let $\hat{W}(t) = W(t) - tW(1)$. Show that $\{\hat{W}(t): 0 \leqslant t \leqslant 1\}$ is a Brownian bridge.

(17) If W is a Wiener process with $W(0) = 0$, show that $\tilde{W}(t) = (1 - t)W(t/(1 - t))$ for $0 \leqslant t < 1$, $\hat{W}(1) = 0$, defines a Brownian bridge.

13.7 Potential theory

Finally, we return to the three-dimensional Wiener process and show that the study of its sample paths is bound up with the theory of gravitational (or electrostatic) potentials and Laplace's equation.

Recall the theory of scalar potentials. Matter is distributed about regions of $\mathbb{R}^3$. According to the laws of Newtonian attraction, this matter gives rise to a function $V: \mathbb{R}^3 \to \mathbb{R}$ which assigns a potential $V(\boldsymbol{x})$ to each point $\boldsymbol{x} \in \mathbb{R}^3$. In open space V satisfies

(1) **Laplace's equation:** $\nabla^2 V = 0$,

whilst in a region containing matter of density ρ, V satisfies

(2) **Poisson's equation:** $\nabla^2 V = -4\pi\rho$

where

$$\nabla^2 V = \frac{\partial^2 V}{\partial x^2} + \frac{\partial^2 V}{\partial y^2} + \frac{\partial^2 V}{\partial z^2}.$$

The following is an application of Green's theorem. If $\boldsymbol{\Sigma}$ is the surface of an empty sphere with radius a and centre $\boldsymbol{x}$, then the potential $V(\boldsymbol{x})$ at $\boldsymbol{x}$ can be obtained as the integral of V over the surface of $\boldsymbol{\Sigma}$, suitably normalized by the area of this surface:

(3) $$V(\mathbf{x}) = \int_{\mathbf{y}\in\Sigma} \frac{V(\mathbf{y})}{4\pi a^2}\, dS.$$

Furthermore, V satisfies (3) for such spheres $\mathbf{\Sigma}$ if and only if V is a solution to Laplace's equation (1) in the appropriate region.

Now we return to the Wiener process. Let

$$\mathbf{W}(t) = (X(t), Y(t), Z(t))$$

be a three-dimensional Wiener process, so that X, Y, and Z are independent Wiener processes. $\mathbf{W}(t)$ is a triple of random variables with joint density function

(4) $$f_{\mathbf{W}(t)}(\mathbf{r}) = (2\pi\sigma^2 t)^{-\frac{3}{2}} \exp\left(-\frac{1}{2\sigma^2 t}|\mathbf{r} - \mathbf{w}|^2\right), \qquad \mathbf{r} \in \mathbb{R}^3$$

where $\mathbf{W}(0) = \mathbf{w}$ and $|\mathbf{r}|^2 = x^2 + y^2 + z^2$ if $\mathbf{r} = (x, y, z)$.

The particle in question performs a random diffusion about $\mathbb{R}^3$. Now, suppose it starts from $\mathbf{W}(0) = \mathbf{w}$, and ask for the probability that it visits some subset G of $\mathbb{R}^3$ before it visits some other subset H, disjoint from G. A particular case of this might arise as follows. Suppose that $\mathbf{w}$ is a point in the interior of some closed bounded connected domain D of $\mathbb{R}^3$, and suppose that the surface ∂D which bounds D is fairly smooth (if D is a ball then ∂D is the bounding spherical surface, for example). Sooner or later the particle will leave D for the first time. If $\partial D = G \cup H$ for some disjoint sets G and H, then we may ask for the probability that the particle leaves D by way of a point in G rather than a point in H (take D to be the solid ball of radius 1 and centre $\mathbf{w}$ for example, and let G be a hemisphere of D).

(5) **Theorem.** *Let G and H be disjoint nice subsets of $\mathbb{R}^3$, and let $p(\mathbf{w}) =$ $\mathbf{P}(\mathbf{W}$ visits G before it visits $H \mid \mathbf{W}(0) = \mathbf{w})$. Then p satisfies Laplace's equation,*

$$\nabla^2 p(\mathbf{w}) = 0$$

at all points $\mathbf{w} \notin G \cup H$, with the boundary conditions

$$p(\mathbf{w}) = \begin{cases} 0 & \text{if } \mathbf{w} \in H \\ 1 & \text{if } \mathbf{w} \in G. \end{cases}$$

We shall not explain the condition that G and H be nice.

Proof. Let $\mathbf{\Sigma}$ be a sphere with radius a and centre $\mathbf{w} \notin G \cup H$, such that no point of $\mathbf{\Sigma}$ or its interior lies in $G \cup H$; we require that $G \cup H$ be closed in order that such a sphere exists for any given $\mathbf{w}$. Let

$$T = \inf\{t: \mathbf{W}(t) \in \mathbf{\Sigma}\}$$

be the first passage time of $\mathbf{W}$ to the sphere $\mathbf{\Sigma}$, starting from $\mathbf{W}(0) = \mathbf{w}$. T is a stopping time for $\boldsymbol{W}$. It is not difficult to see that

$$\mathbf{P}(T < \infty) = 1.$$

For, let A_i be the event

$$A_i = \{|\boldsymbol{W}(i) - \boldsymbol{W}(i-1)| \leqslant 2a\}$$

and note that

$$\begin{aligned}\mathbf{P}(T > n) &\leqslant \mathbf{P}(A_1 \cap A_2 \cap \cdots \cap A_n)\\ &= \mathbf{P}(A_1)^n \quad \text{by independence}\\ &\to 0 \qquad\qquad \text{as } n \to \infty\end{aligned}$$

because $\mathbf{P}(A_1) < 1$. Now, $\boldsymbol{W}(T) \in \Sigma$, and the spherical symmetry of the density function (4) of $\boldsymbol{W}(t)$ implies that $\boldsymbol{W}(T)$ is uniformly distributed over the surface of Σ. The subsequent path of the process is a copy of $\boldsymbol{W}$, but with the new starting point $\boldsymbol{W}(T)$; this follows from the strong Markov property. Thus, the probability that $\boldsymbol{W}$ visits G before it visits H is just $p(\boldsymbol{W}(T))$, and we are led to the surface integral

(6) $$p(\boldsymbol{w}) = \int_{\boldsymbol{y}\in\Sigma} \mathbf{P}(G \text{ before } H \mid \boldsymbol{W}(T) = \boldsymbol{y}) f(\boldsymbol{y})\,\mathrm{d}S$$

where

$$\{G \text{ before } H\} = \{\boldsymbol{W} \text{ visits } G \text{ before it visits } H\}$$

and $f(\boldsymbol{y})$ is the conditional density function of $\boldsymbol{W}(T)$ given $\boldsymbol{W}(0) = \boldsymbol{w}$. However,

$$f(\boldsymbol{y}) = \frac{1}{4\pi a^2} \quad \text{for all } \boldsymbol{y} \in \Sigma$$

by the uniformity of $\boldsymbol{W}(T)$, and (6) becomes

(7) $$p(\boldsymbol{w}) = \int_{\boldsymbol{y}\in\Sigma} \frac{p(\boldsymbol{y})}{4\pi a^2}\,\mathrm{d}S.$$

This integral equation holds for any sphere Σ with centre $\boldsymbol{w}$ whose contents do not overlap $G \cup H$, and we recognize it as the characteristic property (3) of solutions to Laplace's equation (1). Thus p satisfies Laplace's equation. The boundary conditions are derived easily. ■

This simple observation provides us with an elegant technique for finding the probabilities that $\boldsymbol{W}$ visits certain subsets of $\mathbb{R}^3$. The principles of the method are simple, although some of the ensuing calculations may be lengthy (see Example (14) and Problem (13.8.12), for instance).

(8) **Example.** The foregoing discussion holds in two dimensions also. Start a two-dimensional Wiener process $\boldsymbol{W}$ from a point $\boldsymbol{W}(0) = \boldsymbol{w} \in \mathbb{R}^2$. Let G be a circle, radius ε and centre at the origin, such that $\boldsymbol{w}$ does not lie within the interior of G. What is the probability that $\boldsymbol{W}$ visits G ever?

Solution. We shall need two boundary conditions in order to find the appropriate solution to Laplace's equation. The first arises from the case when $\mathbf{w} \in G$. To find the second, let H be a circle, radius R and centre at the origin, and suppose that R is much bigger than ε. We shall solve Laplace's equation in polar co-coordinates

(9)
$$\frac{1}{r}\frac{\partial}{\partial r}\left(r\frac{\partial p}{\partial r}\right) + \frac{1}{r^2}\frac{\partial^2 p}{\partial \theta^2} = 0$$

in the region $\varepsilon \leqslant r \leqslant R$, and use the boundary conditions

(10)
$$p(\mathbf{w}) = \begin{cases} 0 & \text{if} \quad \mathbf{w} \in H \\ 1 & \text{if} \quad \mathbf{w} \in G \end{cases}$$

to find

$$p(\mathbf{w}) = \mathbf{P}(G \text{ before } H \mid \mathbf{W}(0) = \mathbf{w}).$$

Solutions to (9) with circular symmetry take the form

$$p(\mathbf{w}) = A \log r + B \quad \text{if} \quad \mathbf{w} = (r, \theta)$$

where A and B are arbitrary constants. Use (10) to obtain

$$p(\mathbf{w}) = \frac{\log(r/R)}{\log(\varepsilon/R)}.$$

Now let $R \to \infty$ to obtain that

$$\mathbf{P}(G \text{ before } H \mid \mathbf{W}(0) = (r, \theta)) \to 1.$$

We have shown that $\mathbf{W}$ almost surely visits any ε-neighbourhood of the origin regardless of its starting point. Such a process is called *persistent* (or *recurrent*) since its sample paths pass arbitrarily closely to every point in the plane with probability 1. ●

(11) **Example.** Next consider a three-dimensional version of (8). Let G be the sphere with radius ε and centre at the origin of $\mathbb{R}^3$. Again, start a three-dimensional Wiener process $\mathbf{W}$ from some point $\mathbf{W}(0) = \mathbf{w}$ which does not lie within G. What is the probability that $\mathbf{W}$ visits G?

Solution. As before, let H be a sphere with radius R and centre at the origin, where R is much larger than ε. We seek a solution to Laplace's equation in spherical polar coordinates

(12)
$$\frac{\partial}{\partial r}\left(r^2\frac{\partial p}{\partial r}\right) + \frac{1}{\sin\theta}\frac{\partial}{\partial \theta}\left(\sin\theta\frac{\partial p}{\partial \theta}\right) + \frac{1}{\sin^2\phi}\frac{\partial^2 p}{\partial \phi^2} = 0,$$

subject to the boundary conditions (10), in order to find

$$p(\mathbf{w}) = \mathbf{P}(G \text{ before } H \mid \mathbf{W}(0) = \mathbf{w}).$$

Solutions to (12), which depend on r only, take the form

(13) $$p(\mathbf{w}) = \frac{A}{r} + B \quad \text{if} \quad \mathbf{w} = (r, \theta, \phi).$$

Use (10) to obtain

$$p(\mathbf{w}) = \frac{r^{-1} - R^{-1}}{\varepsilon^{-1} - R^{-1}}.$$

Let $R \to \infty$ to obtain

$$\mathbf{P}(G \text{ before } H \mid \mathbf{W}(0) = (r, \theta, \phi)) \to \frac{\varepsilon}{r}$$

and we have shown that $\mathbf{W}$ ultimately visits G with probability ε/r; is it surprising that the answer is *directly* proportional to ε?

We have shown that the three-dimensional Wiener process is *not* persistent, since its sample paths do not pass through every ε-neighbourhood with probability 1. This mimics the behaviour of symmetric random walks. Recall from Problems (5.12.6) and (6.13.9) that the two-dimensional symmetric random walk is persistent whilst the three-dimensional walk is transient (see (15) also). ●

(14) **Example.** Let $\boldsymbol{\Sigma}$ be the unit sphere in $\mathbb{R}^3$ with centre at the origin, and let

$$G = \{(r, \theta, \phi): r = 1, 0 \leqslant \theta \leqslant \tfrac{1}{2}\pi\}$$

be the upper hemisphere of $\boldsymbol{\Sigma}$. Start a three-dimensional Wiener process $\mathbf{W}$ from a point $\mathbf{W}(0) = \mathbf{w}$ which lies in the *interior* of $\boldsymbol{\Sigma}$. What is the probability that $\mathbf{W}$ visits G before it visits $H = \boldsymbol{\Sigma} \backslash G$, the lower hemisphere of $\boldsymbol{\Sigma}$?

Solution. Let

$$p(\mathbf{w}) = \mathbf{P}(G \text{ before } H \mid \mathbf{W}(0) = \mathbf{w}).$$

p satisfies Laplace's equation (12), subject to the boundary conditions (10). Solutions to (12) which are independent of ϕ are also solutions to the simpler equation

$$\frac{\partial}{\partial r}\left(r^2 \frac{\partial p}{\partial r}\right) + \frac{1}{\sin\theta}\frac{\partial}{\partial \theta}\left(\sin\theta \frac{\partial p}{\partial \theta}\right) = 0.$$

We abandon the calculation at this point, leaving it to the reader to complete. Some knowledge of Legendre polynomials and the method of separation of variables may prove useful. ●

(15) **Example. Random walks revisited.** We may think of a Wiener process as a continuous space-time version of a symmetric random walk. In the light of the discussion of Brownian motion in Section 13.2, it is not surprising that Wiener processes and random walks have many properties in common. In particular, potential theory has applications to random walks; we terminate

this chapter with a brief but powerful demonstration of this. For simplicity we consider a symmetric random walk on the two-dimensional square lattice $\mathbb{Z}^2$, but it will be clear that more general results are obtainable by similar arguments.

A particle moves about the points $\{(m, n): m, n = 0, \pm 1, \pm 2, \ldots\}$ according to the following rule. If after n steps it is at the point (x, y), then it moves at the next step to one of the four neighbouring points $(x \pm 1, y)$, $(x, y \pm 1)$, each such point being chosen with probability $\frac{1}{4}$ independently of all previous choices. We write $\boldsymbol{W}_n$ for the particle's position after n steps. We are concerned with first passage probabilities. Let G and H be disjoint sets of points in $\mathbb{Z}^2$, and write

$$p(\boldsymbol{w}) = \mathbf{P}(G \text{ before } H \mid \boldsymbol{W}_0 = \boldsymbol{w})$$

for the probability that the walk $\boldsymbol{W}$ visits some point in G before it visits any point in H, starting from the point $\boldsymbol{w}$. In place of the integral equation (7), p satisfies the difference equation

(16) $$p(\boldsymbol{w}) = \tfrac{1}{4}[p((w_1 + 1, w_2)) + p((w_1 - 1, w_2)) + p((w_1, w_2 + 1)) + p((w_1, w_2 - 1))],$$

where $\boldsymbol{w} = (w_1, w_2)$, subject to the boundary conditions

(17) $$p(\boldsymbol{w}) = \begin{cases} 0 & \text{if } \boldsymbol{w} \in H \\ 1 & \text{if } \boldsymbol{w} \in G. \end{cases}$$

Equation (16) is easily derived by conditioning on the first step of the walk.

Now think of $\mathbb{Z}^2$ as an electrical network in which each pair of neighbouring points ((0, 0) and (0, 1), for example) is joined by a conducting wire with resistance 1 ohm. Connect a battery into the network in such a way that the points in H are joined to earth and the points in G are raised to the potential 1 volt. It is physically clear that this potential difference induces a potential $V(\boldsymbol{w})$ at each point $\boldsymbol{w}$ of $\mathbb{Z}^2$, together with a current along each wire. These potentials and currents satisfy a well-known collection of equations called Kirchhoff's laws and Ohm's law, and it is an easy consequence of these laws (*exercise*) that V satisfies (16) and (17). This equivalence between first passage probabilities and electrical potentials is the discrete analogue of Theorem (5). As a beautiful application of this equivalence, we show that the walk is persistent; you have seen two proofs of this result already (remember Problem (5.12.6)).

(18) **Theorem.** *The two-dimensional symmetric random walk is persistent.*

Proof. Let G be the origin 0 of $\mathbb{Z}^2$ and let H_n be the set of points $\{(x, \pm n), (\pm n, y): -n \leqslant x, y \leqslant n\}$ on the edge of the square with side length $2n$ and centre at 0. Then

$$p_n(\boldsymbol{w}) = \mathbf{P}(0 \text{ before } H_n \mid \boldsymbol{W}_0 = \boldsymbol{w})$$

satisfies (16) and (17). If $\boldsymbol{W}_0 = 0$ then we may see that the probability ${}_n p_{00}$, that W revisits 0 before hitting H_n, satisfies

(19)
$$_np_{00} = \tfrac{1}{4}[p_n((0, 1)) + p_n((1, 0)) + p_n((0, -1)) + p_n((-1, 0))],$$

by conditioning on the first step. Also 0 is a persistent state if and only if $_np_{00} \to 1$ as $n \to \infty$.

Now think of $\mathbb{Z}^2$ as an electrical network. By (19) and the remarks preceding the theorem, $4(1 - {}_np_{00})$ equals the current which flows into the network at 0 as a result of applying 1 volt at 0 and earthing all the points in H_n. By Ohm's law, the resistance R_n between 0 and H_n satisfies

$$R_n = \frac{1}{4(1 - {}_np_{00})},$$

and it follows that 0 is persistent if and only if $R_n \to \infty$ as $n \to \infty$. To show that this holds, construct a lower bound for R_n in the following way. For each $r \leqslant n$, short out all the points in H_r (draw your own diagram), and use the parallel and series resistance laws to find that

$$R_n \geqslant \frac{1}{4} + \frac{1}{12} + \cdots + \frac{1}{8n - 4}.$$

This shows that $R_n \to \infty$ as $n \to \infty$, and the result is shown. ■

(20) **Theorem.** *The three-dimensional symmetric random walk is transient.*

Proof. This is an *exercise.* You must show that the resistance between 0 and the 'points at infinity' is finite in $\mathbb{Z}^3$. See the solution of Problem (6.13.9) for another method of proof. ■●

Exercises

(21) Let G be the closed sphere with radius ε and centre at the origin of $\mathbb{R}^d$ where $d \geqslant 3$. Let W be a d-dimensional Wiener process starting from $W(0) = w \notin G$. Show that the probability that W visits G is $(\varepsilon/r)^{d-2}$, where $r = |w|$.

(22) Let G be an infinite connected graph (that is, an infinite set of points or 'vertices', some pairs of which are joined by edges, such that any two vertices are joined by some path of edges). A random walk starts at some fixed vertex labelled 0; at each step it moves from its current position to one of its neighbours chosen with equal probability and independently of all earlier steps. Let H_n be the set of vertices x which are distance n from 0 (that is, the shortest path from x to 0 contains n edges), and let N_n be the total number of edges joining pairs x, y of vertices with $x \in H_n$, $y \in H_{n+1}$. Show that the walk is persistent if $\sum_i N_i^{-1} = \infty$.

13.8 Problems

1. Let W be a standard Wiener process, that is, a process with independent increments and continuous sample paths such that $W(s + t) - W(s)$ is $N(0, t)$ for $t > 0$. Let α be a positive constant. Show that

(a) $\alpha W(t/\alpha^2)$ is a standard Wiener process,
(b) $W(t+\alpha) - W(\alpha)$ is a standard Wiener process,
(c) the process V, given by $V(t) = tW(1/t)$ for $t > 0$, $V(0) = 0$, is a standard Wiener process.

2. Let $X = \{X(t): t \geqslant 0\}$ be a Gaussian process with continuous sample paths, zero means, and autocovariance function $c(s, t) = u(s)v(t)$ for $s \leqslant t$ where u and v are continuous functions. Suppose that the ratio $r(t) = u(t)/v(t)$ is continuous and strictly increasing with inverse function r^{-1}. Show that $W(t) = X(r^{-1}(t))/v(r^{-1}(t))$ is a standard Wiener process on a suitable interval of time.
If $c(s, t) = s(1-t)$ for $s \leqslant t < 1$, express X in terms of W.

3. Let $\beta > 0$, and show that $U(t) = \mathrm{e}^{-\beta t} W(\mathrm{e}^{2\beta t} - 1)$ is an Ornstein–Uhlenbeck process if W is a standard Wiener process.

4. Let $V = \{V(t): t \geqslant 0\}$ be an Ornstein–Uhlenbeck process with instantaneous mean $a(x, t) = -\beta x$ where $\beta > 0$, with instantaneous variance $b(x, t) = \sigma^2$, and with $U(0) = u$. Show that $V(t)$ is $N(u\,\mathrm{e}^{-\beta t}, \sigma^2(1 - \mathrm{e}^{-2\beta t})/(2\beta))$. Deduce that $V(t)$ is asymptotically $N(0, \frac{1}{2}\sigma^2/\beta)$ as $t \to \infty$, and show that V is strongly stationary if $V(0)$ is $N(0, \frac{1}{2}\sigma^2/\beta)$.
Show that such a process is the *only* stationary Gaussian Markov process with continuous autocovariance function, and find its spectral density function.

5. Let $D = \{D(t): t \geqslant 0\}$ be a diffusion process with instantaneous mean $a(x, t) = \alpha x$ and instantaneous variance $b(x, t) = \beta x$ where α and β are positive constants. Let $D(0) = d$. Show that the moment generating function of $D(t)$ is

$$M(\theta, t) = \exp\left\{\frac{2\alpha d\theta\,\mathrm{e}^{\alpha t}}{\beta\theta(1 - \mathrm{e}^{\alpha t}) + 2\alpha}\right\}.$$

Find the mean and variance of $D(t)$, and show that $\mathbf{P}(D(t) = 0) \to \mathrm{e}^{-2d\alpha/\beta}$ as $t \to \infty$.

6. Let D be an Ornstein–Uhlenbeck process with $D(0) = 0$, and place reflecting barriers at $-c$ and d where $c, d > 0$. Find the limiting distribution of D as $t \to \infty$.

7. Let $X_0, X_1, \ldots$ be independent $N(0, 1)$ variables, and show that

$$W(t) = \frac{t}{\sqrt{\pi}} X_0 + \sqrt{\frac{2}{\pi}} \sum_{k=1}^{\infty} \frac{\sin(kt)}{k} X_k$$

defines a standard Wiener process on $[0, \pi]$.

8. Let W be a standard Wiener process with $W(0) = 0$. Place absorbing barriers at $-b$ and b, where $b > 0$, and let W_{A} be W absorbed at these barriers. Show that $W_{\mathrm{A}}(t)$ has density function

$$f_{\mathrm{A}}(y, t) = \frac{1}{\sqrt{(2\pi t)}} \sum_{k=-\infty}^{\infty} (-1)^k \exp\left\{-\frac{(y - 2kb)^2}{2t}\right\}, \qquad -b < y < b,$$

which may also be expressed as

$$f_{\mathrm{A}}(y, t) = \sum_{n=1}^{\infty} a_n \mathrm{e}^{-\lambda_n t} \sin\left(\frac{n\pi(y + b)}{2b}\right), \qquad -b < y < b,$$

where $a_n = b^{-1} \sin(\frac{1}{2}n\pi)$ and $\lambda_n = n^2\pi^2/(8b^2)$.
Hence calculate $\mathbf{P}(\sup_{0 \leqslant s \leqslant t} |W(s)| > b)$ for the unrestricted process W.

9. Let D be a Wiener process with drift m, and again suppose that $D(0) = 0$. This time place absorbing barriers at the points $x = -a$ and $x = b$ where a and b are positive real numbers. Show that the probability p_a that the process is absorbed at $-a$ is given by

$$p_a = \frac{e^{2mb} - 1}{e^{2m(a+b)} - 1}.$$

10. Let W be a standard Wiener process and let $F(u, v)$ be the event that W has no zero in the interval (u, v).

(a) If $ab > 0$, show that $\mathbf{P}(F(0, t) \mid W(0) = a, W(t) = b) = 1 - e^{-2ab/t}$.
(b) If $W(0) = 0$ and $0 < t_0 \leqslant t_1 \leqslant t_2$, show that

$$\mathbf{P}(F(t_0, t_2) \mid F(t_0, t_1)) = \frac{\sin^{-1}\{(t_0/t_2)^{\frac{1}{2}}\}}{\sin^{-1}\{(t_0/t_1)^{\frac{1}{2}}\}}.$$

(c) Deduce that, if $W(0) = 0$ and $0 < t_1 \leqslant t_2$, then $\mathbf{P}(F(0, t_2) \mid F(0, t_1)) = (t_1/t_2)^{\frac{1}{2}}$.

11. Let W be a standard Wiener process and suppose that $W(0) = 0$. Show that

$$\mathbf{P}\left(\sup_{0 \leqslant s \leqslant t} |W(s)| \geqslant w\right) \leqslant 2\mathbf{P}(|W(t)| \geqslant w) \leqslant \frac{2t}{w^2} \quad \text{for} \quad w > 0.$$

Set $t = 2^n$ and $w = 2^{2n/3}$ and use the Borel–Cantelli lemma to show that $t^{-1}W(t) \to 0$ a.s. as $t \to \infty$.

12. Let $\mathbf{W}$ be a two-dimensional Wiener process with $\mathbf{W}(0) = \mathbf{w}$, and let F be the unit circle. What is the probability that $\mathbf{W}$ visits the upper semicircle G of F before it visits the lower semicircle H?

13. Let W_1 and W_2 be independent standard Wiener processes with $W_1(0) = W_2(0) = 0$; the pair $\mathbf{W}(t) = (W_1(t), W_2(t))$ represents the position of a particle which is experiencing Brownian motion in the plane. Let l be some straight line in $\mathbb{R}^2$, and let P be the point on l which is closest to the origin O. Draw a diagram. Show that

(a) the particle visits l, with probability one,
(b) if the particle hits l for the first time at the point R, then the distance PR (measured as positive or negative as appropriate) has the Cauchy density function $f(x) = d/\{\pi(d^2 + x^2)\}$, $-\infty < x < \infty$, where d is the distance OP,
(c) the angle $\widehat{\text{POR}}$ is uniformly distributed on $[-\frac{1}{2}\pi, \frac{1}{2}\pi]$.

Appendix I
Foundations and notation

Here is a list of topics with which many readers will already be familiar, and which are used in the body of the text. We begin with some notation.

(A) Basic notation. The end of each example or subsection is indicated by the symbol ●; the end of each proof is indicated by ■.

The largest integer which is not larger than the real number x is denoted by $\lfloor x \rfloor$; $\lceil x \rceil$ denotes the smallest integer not smaller than x.

We use the following symbols:

$$\mathbb{R} \equiv \text{the real numbers } (-\infty, \infty)$$

$$\mathbb{Z} \equiv \text{the integers } (\ldots, -2, -1, 0, 1, 2, \ldots)$$

$$\mathbb{C} \equiv \text{the complex plane } \{x + iy : x, y \in \mathbb{R}\}.$$

The symbol $\Leftrightarrow$ is interchangeable with the expression 'if and only if'.

Here are two 'delta' functions.

Kronecker δ: If i and j belong to some set S, define

$$\delta_{ij} = \begin{cases} 1 & \text{if} \quad i = j \\ 0 & \text{if} \quad i \neq j. \end{cases}$$

Dirac δ function: If $x \in \mathbb{R}$, the symbol δ_x represents a notional function with the properties

(a) $\delta_x(y) = 0 \quad \text{if} \quad y \neq x$

(b) $\displaystyle\int_{-\infty}^{\infty} g(y)\, \delta_x(y)\, dy = g(x)$ for all integrable $g: \mathbb{R} \to \mathbb{R}$.

(B) Sets and counting. In addition to the union and intersection symbols, $\cup$ and $\cap$, we employ the following notation:

set difference: $A \setminus B = \{x \in A : x \notin B\}$

symmetric difference: $A \triangle B = (A \setminus B) \cup (B \setminus A) = \{x \in A \cup B : x \notin A \cap B\}$.

The *cardinality* $|A|$ of a set A is the number of elements which A contains. The *complement* of A is denoted by A^c.

The *binomial coefficient* $\binom{n}{r}$ is the number of distinct combinations of r objects that can be drawn from a set containing n distinguishable objects. The following texts treat this material in more detail: Halmos (1960), Ross (1976), and Rudin (1976).

(C) Vectors and matrices. The symbol $\boldsymbol{x}$ denotes the row vector $(x_1, x_2, \ldots)$ of finite or countably infinite length. The transposes of vectors $\boldsymbol{x}$ and matrices $\boldsymbol{V}$ are denoted by $\boldsymbol{x}'$ and $\boldsymbol{V}'$ respectively. The determinant of a square matrix $\boldsymbol{V}$ is written as $|\boldsymbol{V}|$.

The following books contain information about matrices, their eigenvalues, and their canonical forms: Lipschutz (1974), Rudin (1976), and Cox and Miller (1965).

(D) Convergence.

(1) **Limits inferior and superior.** We often use inferior and superior limits, and so we review their definitions. Given any sequence $\{x_n: n \geqslant 1\}$ of real numbers, define

$$g_m = \inf_{n \geqslant m} x_n, \qquad h_m = \sup_{n \geqslant m} x_n.$$

then

$$g_m \leqslant g_{m+1}, \qquad h_m \geqslant h_{m+1} \quad \text{for all } m$$

and so the sequences $\{g_m\}$ and $\{h_m\}$ converge as $m \to \infty$. Their limits

$$g = \lim_{m \to \infty} g_m, \qquad h = \lim_{m \to \infty} h_m$$

are denoted by '$\liminf_{n \to \infty} x_n$' and '$\limsup_{n \to \infty} x_n$' respectively. Clearly

$$\liminf_{n \to \infty} x_n \leqslant \limsup_{n \to \infty} x_n.$$

The following result is very useful.

(2) **Theorem.** *The sequence $\{x_n\}$ converges if and only if*

$$\liminf_{n \to \infty} x_n = \limsup_{n \to \infty} x_n.$$

(3) **Cauchy convergence.** The criterion of convergence (for all $\varepsilon > 0$, there exists N such that $|x_n - x| < \varepsilon$ if $n \geqslant N$) depends on knowledge of the limit x. In many practical instances it is convenient to use a criterion which does not rely on such knowledge.

(4) **Definition.** The sequence $\{x_n\}$ is called **Cauchy convergent** if, for all $\varepsilon > 0$, there exists N such that

$$|x_m - x_n| < \varepsilon \quad \text{whenever} \quad m, n \geqslant N.$$

(5) **Theorem.** *A real sequence converges if and only if it is Cauchy convergent.*

(6) **Continuity.** Recall that the function $g: \mathbb{R} \to \mathbb{R}$ is *continuous* at the point x if

$$g(x + h) \to g(x) \quad \text{as} \quad h \to 0.$$

We often encounter functions which satisfy only part of this condition.

(7) **Definition.** The function $g: \mathbb{R} \to \mathbb{R}$ is called

(i) **right-continuous** if $g(x+h) \to g(x)$ as $h \downarrow 0$ for all x.
(ii) **left-continuous** if $g(x+h) \to g(x)$ as $h \uparrow 0$ for all x.

So g is continuous if and only if g is both right- and left-continuous.

If g is monotone then it has left and right limits, $\lim_{h\uparrow 0} g(x+h)$, $\lim_{h\downarrow 0} g(x+h)$, at all points x; these may differ from $g(x)$ if g is not continuous at x. We write

$$g(x+) = \lim_{h\downarrow 0} g(x+h), \qquad g(x-) = \lim_{h\uparrow 0} g(x+h).$$

(8) **Infinite products.** We make use of the following result concerning products of real numbers.

(9) **Theorem.** *Let* $p_n = \prod_{i=1}^{n} (1 + x_i)$.

(a) *If* $x_i > 0$ *for all i, then* $p_n \to \infty$ *as* $n \to \infty$ *if and only if*

$$\sum_i x_i = \infty.$$

(b) *If* $-1 < x_i \leqslant 0$ *for all i, then* $p_n \to 0$ *as* $n \to \infty$ *if and only if*

$$\sum_i |x_i| = \infty.$$

(10) **Landau's notation.** Use of the O–o notation is standard.If f and g are two functions of a real variable x, then we say that

$$f(x) = o(g(x)) \quad \text{as} \quad x \to \infty \quad \text{if} \quad \lim_{x\to\infty} \{f(x)/g(x)\} = 0$$

$$f(x) = O(g(x)) \quad \text{as} \quad x \to \infty \quad \text{if} \quad |f(x)/g(x)| < C$$

for all large x and some constant C.

Similar definitions hold as $x \downarrow 0$ and for real sequences $\{f(n)\}, \{g(n)\}$ as $n \to \infty$.

For more details about the topics in this section see Apostol (1957) or Rudin (1976).

(E) Complex analysis. We make use of elementary manipulation of complex numbers, the formula

$$e^{itx} = \cos(tx) + i\sin(tx),$$

and the theory of complex integration. Readers are referred to Phillips (1957) and Nevanlinna and Paatero (1969) for further details.

(F) Transforms. An *integral transform* of the function $g: \mathbb{R} \to \mathbb{R}$ is a function $\tilde{g}$ of the form

$$\tilde{g}(\theta) = \int_{-\infty}^{\infty} K(\theta, x) g(x)\, dx;$$

such transforms are very useful in the theory of differential equations. Perhaps the most useful is the *Laplace transform.*

(11) **Definition.** The **Laplace transform** of g is defined to be

$$\hat{g}(\theta) = \int_{-\infty}^{\infty} e^{-\theta x} g(x)\,dx \quad \text{where} \quad \theta \in \mathbb{C}$$

whenever this integral exists.

As a special case of the Laplace transform of g, set $\theta = i\lambda$ for real λ to obtain the *Fourier transform*

$$G(\lambda) = \hat{g}(i\lambda) = \int_{-\infty}^{\infty} e^{-i\lambda x} g(x)\,dx.$$

Often, we are interested in functions g which are defined on the half-line $[0, \infty)$, with Laplace transform

$$\hat{g}(\theta) = \int_{0}^{\infty} e^{-\theta x} g(x)\,dx;$$

such a transform is called 'one-sided'. We often think of g as a function of a *real* variable θ.

Subject to certain conditions (such as existence and continuity) Laplace transforms have the following important properties.

(12) **Inversion.** *g can be retrieved from knowledge of $\hat{g}$ by the 'inversion formula'.*

(13) **Convolution.** *If $k(x) = \int_{-\infty}^{\infty} g(x-y)h(y)\,dy$ then $\hat{k}(\theta) = \hat{g}(\theta)\hat{h}(\theta)$.*

(14) **Differentiation.** *If $G\colon [0, \infty) \to \mathbb{R}$ and $g = dG/dx$ then*

$$\theta \hat{G}(\theta) = \hat{g}(\theta) + G(0).$$

It is sometimes convenient to use a variant of the Laplace transform.

(14) **Definition.** The **Laplace–Stieltjes transform** of g is defined to be

$$g^*(\theta) = \int_{-\infty}^{\infty} e^{-\theta x}\,dg(x) \quad \text{where} \quad \theta \in \mathbb{C}$$

whenever this integral exists.

We do not wish to discuss the definition of this integral (it is called a 'Lebesgue–Stieltjes' integral and is related to the integrals of Section 5.6). You may think about it in the following way. If g is differentiable then its Laplace–Stieltjes transform g^* is defined to be the Laplace transform of the derivative g' of g, since in this case

$$dg(x) = g'(x)\,dx.$$

Laplace–Stieltjes transforms g^* always receive an asterisk to distinguish them from Laplace transforms. They have properties similar to (12), (13), and (14). For example, (13) becomes

(16) **Convolution.** *If $k(x) = \int_{-\infty}^{\infty} g(x-y)\,dh(y)$ then $k^*(\theta) = g^*(\theta)h^*(\theta)$.*

Fourier–Stieltjes transforms are defined similarly.

More details are provided by Apostol (1957) and Hildebrand (1962).

(G) Difference equations. The sequence $\{u_r: r \geqslant 0\}$ is said to satisfy a difference equation if

(17) $$\sum_{i=0}^{m} a_i u_{n+m-i} = f(n), \qquad n \geqslant 0,$$

for some fixed sequence $a_0, a_1, \ldots, a_m$ and given function f. If $a_0 a_m \neq 0$ then the difference equation is said to be of order m. The general solution of this difference equation is

$$u_n = \sum_{i=1}^{r} \sum_{j=0}^{m_i - 1} c_{ij} n^j \theta_i^n + p_n$$

where $\theta_1, \theta_2, \ldots, \theta_r$ are the distinct roots of the polynomial equation

$$\sum_{i=0}^{m} a_i \theta^{m-i} = 0,$$

m_i being the multiplicity of θ_i, and $\{p_n: n \geqslant 0\}$ is any particular solution to (17). In general there are m arbitrary constants, whose determination requires m boundary conditions.

More details are provided by Hall (1967).

(H) Partial differential equations. Let $a = a(x, y, u)$, $b = b(x, y, u)$, and $c = (x, y, u)$ be 'nice' functions on $\mathbb{R}^3$, and suppose that $u(x, y)$ satisfies the partial differential equation

(18) $$a\frac{\partial u}{\partial x} + b\frac{\partial u}{\partial y} = c.$$

The solution $u = u(x, y)$ may be thought of as a surface $\phi(x, y, u) = 0$ where $\phi(x, y, u) = u - u(x, y)$. The normal to ϕ at the point $(x, y, u(x, y))$ lies in the direction

$$\nabla\phi = \left(-\frac{\partial u}{\partial x}, -\frac{\partial u}{\partial y}, 1\right).$$

Consider a curve $\{x(t), y(t), u(t): t \in \mathbb{R}\}$ in $\mathbb{R}^3$ defined by $\dot{x} = a$, $\dot{y} = b$, $\dot{u} = c$. The direction cosines of this curve are proportional to the vector (a, b, c), whose scalar product with $\nabla\phi$ satisfies

$$\nabla\phi \,.\, (a, b, c) = -a\frac{\partial u}{\partial x} - b\frac{\partial u}{\partial y} + c = 0,$$

so that the curve is perpendicular to the normal vector $\nabla\phi$. Hence any such curve lies in the surface $\phi(x, y, u) = 0$, giving that the family of such curves generates the solution to the differential equation (18).

For more details concerning partial differential equations, see Hildebrand (1962) or Piaggio (1965).

Appendix II
Further reading

This list is neither comprehensive nor canonical. The Bibliography lists books which are useful for mathematical background and further exploration.

Probability theory. Ross (1976), Hoel *et al.* (1971*a*), and Grimmett and Welsh (1986) are excellent elementary texts. More elementary but out of print is Gray (1967); see also Chung (1979). There are many fine advanced texts, including Billingsley (1986), Breiman (1968), Chung (1974), and Shiryayev (1984). The probability section of Kingman and Taylor (1966) provides a clear and concise introduction to the modern theory. Moran (1968) is often useful at our level.

The two volumes of Feller's treatise (Feller 1968, 1971) are essential reading for budding probabilists; the first deals largely in discrete probability, and the second is an idiosyncratic and remarkable encyclopaedia of the continuous theory.

Markov chains. There is no account of discrete-time Markov chains which is wholly satisfactory, though various treatments have attractions. Billingsley (1986) proves the ergodic theorem by the coupling argument, Çinlar (1975) and Karlin and Taylor (1975) contain many examples, and Ross (1970) is clear and to the point; do not overlook Cox and Miller (1965), Feller (1968), and Prabhu (1965*a*). Chung (1960) and Freedman (1971) are much more advanced and include rigorous treatments of the continuous-time theory; these are relatively difficult books. More elementary treatments are by Gray (1967) (very simple), Cox and Miller (1965), Karlin and Taylor (1975) (very clear), and Çinlar (1975).

Other random processes. Any full account of processes with continuous state spaces, and their sample path properties, is far beyond the scope of this reading list. Karlin and Taylor (1975) contains excellent chapters on the Wiener process and stationary processes. You may read Prabhu (1965*a*) and Cox and Miller (1965) for interesting discussions of the general diffusion equations.

Ross (1970) and Karlin and Taylor (1975) deal well with renewal theory. Prabhu (1965*a*) includes a proof of the renewal theorem.

Most authors do not treat queueing theory as a single topic, but prefer to consider the various systems at appropriate points in chapters on Markov chains and renewal. Ross (1970), Çinlar (1975), Billingsley (1986), Prabhu (1965*a*), and Karlin and Taylor (1975) follow this scheme. A more consolidated, but heavily analytical, treatment is provided by Prabhu (1965*b*). Asmussen (1987) gives a good account of recent work in queues and related applied random processes.

Martingale theory was expounded by Doob (1953). The superb recent book of Williams (1991) is one of the best introductions to measure theory (for the probabilist) and to martingales in discrete time. Several books contain passages on martingales, for example Breiman (1968), Karlin and Taylor (1975, 1981), and Shiryayev (1984). For accounts of discrete-parameter martingales see Neveu (1975), and Hall and Heyde (1980); for continuous-parameter martingales, and a lot more besides, see Williams (1979), and Rogers and Williams (1987).

Appendix III
History and varieties of probability

History

Mathematical probability has its origins in games of chance, principally in games with dice and cards, since these were the main tools of gambling in Italy during the fifteenth and sixteenth centuries. A number of Italian mathematicians of this period (including Galileo) gave calculations of the number and proportion of winning outcomes in various fashionable games. One of them (G. Cardano) went so far as to write a book, *On games of chance*, sometime shortly after 1550. However, this was not published until 1663, by which time probability theory had already had its official inauguration elsewhere.

It was around 1654 when B. Pascal and P. de Fermat generated a celebrated correspondence about their solutions of the problem of the points. These were soon widely known, and C. Huygens developed these ideas in a book published in 1657, in Latin. Translations into Dutch (1660) and English (1692) soon followed. The preface by John Arbuthnot to the English version (see Appendix IV) makes it clear that the intuitive notions underlying this work were similar to those commonly in force nowadays.

These first simple ideas were soon extended by Jacob (otherwise known as James) Bernoulli in *Ars conjectandi* (1713) and by A. de Moivre in *Doctrine of chances* (1718, 1738, 1756). These books included simple versions of the weak law of large numbers and the central limit theorem. Methods, results, and ideas were all greatly refined and generalized by P. Laplace in a series of books from 1774 to 1827. Many other eminent mathematicians of this period wrote on probability: Euler, Gauss, Lagrange, Legendre, Poisson, and so on.

However, as ever harder problems were tackled by ever more powerful mathematical techniques during the nineteenth century, the lack of a well-defined axiomatic structure was recognized as a serious handicap. In 1900, D. Hilbert included this as his sixth problem, and in 1933, A. Kolmogorov provided the axioms which today underpin most mathematical probability.

Varieties

It is necessary to have an interpretation of probability, for this is what suggests appropriate axioms and useful applications. The oldest interpretations of probability are as

(a) an indication of relative frequency, and
(b) an expression of symmetry.

These views were natural given the origins of the subject. A well-made die is symmetrical and is equally likely to show any face; an ill-made die is biased and in the long run shows its faces in different relative frequencies. (Recall Ambrose Bierce's definition of 'dice': dice are small polka-dotted cubes of ivory constructed like a lawyer to lie upon any side, commonly the wrong one.)

However, there are many chance events which are neither repeatable nor symmetrical, and from earliest times probabilists have been alert to the fact that applications might be sought in fields other than gambling. G. Leibniz considered the degree to which some statement had been proved, and many later authors concerned themselves with the theory of testimony. This leads to more complicated interpretations of probability such as

(c) to what extent some hypothesis is logically implied by the evidence, and
(d) the degree of belief of an individual that some given event will occur.

This last interpretation is commonly known as 'subjective probability', and the concept is extremely fissiparous. Since different schools of thought choose different criteria for judging possible reasons for belief, a wide variety of axiomatic systems have come into being.

However, by a happy chance, in many cases of importance, the axioms can be reasonably reduced to exactly the axioms (1.3.1) with which we have been concerned. And systems not so reduced have in general proved very intractable to extensive analysis.

Finally we note that (a)–(d) do not exhaust the possible interpretations of probability theory, and that there remain areas where interpretations are as yet unagreed, notably in quantum mechanics. The reader may pursue this in books on physics and philosophy; see Krüger *et al.* (1987).

Appendix IV
John Arbuthnot's Preface to *Of the laws of chance* (1692)

It is thought as necessary to write a Preface before a Book, as it is judg'd civil, when you invite a Friend to Dinner, to proffer him a Glass of Hock beforehand for a Whet: And this being maim'd enough for want of a Dedication, I am resolv'd it shall not want an Epistle to the Reader too. I shall not take upon me to determine, whether it is lawful to play at Dice or not, leaving that to be disputed betwixt the *Fanatick Parsons* and the *Sharpers*; I am sure it is lawful to deal with Dice as with other Epidemic Distempers; and I am confident that the writing a Book about it, will contribute as little towards its Encouragement, as Fluxing and Precipitates do to Whoring.

It will be to little purpose to tell my Reader, of how great Antiquity the playing at Dice is. I will only let him know that by the *Aleæ Ludus*, the Antients comprehended all Games, which were subjected to the determination of mere Chance; this sort of Gaming was strictly forbid by the Emperor *Justinian, Cod. Lib. 3. Tit. 43.* under severe Penalties; and *Phocius Nomocan. Tit. 9. Cap. 27.* acquaints us, that the Use of this was altogether denied the Clergy of that time. *Seneca* says very well, *Aleator quantò in arte est melior tantò est nequior*; That by how much the one is more skilful in Games, by so much he is the more culpable; or we may say of this, as an ingenious Man says of Dancing, That to be extraordinary good at it, is to be excellent in a Fault[1]; therefore I hope no body will imagine I had so mean a Design in this, as to teach the Art of Playing at Dice.

A great part of this Discourse is a Translation from Mons. *Huygen's* Treatise, *De ratiociniis in ludo Aleæ*; one, who in his Improvements of Philosophy, has but one Superior[2], and I think few or no Equals. The whole I undertook for my own Divertisement, next to the Satisfaction of some Friends, who would now and then be wrangling about the Proportions of Hazards in some Cases that are here decided. All it requir'd was a few spare Hours, and but little Work for the Brain; my Design in publishing it, was to make it of general Use, and perhaps persuade a raw Squire, by it, to keep his Money in his Pocket; and if, upon this account, I should incur the Clamours of the Sharpers, I do not much regard it, since they are a sort of People the World is not bound to provide for.

You will find here a very plain and easy Method of the Calculation of the Hazards of Game, which a man may understand, without knowing the Quadratures of *Curves*, the Doctrine of *Series's*, or the Laws of *Concentripetation* of Bodies, or the Periods of the *Satellites* of *Jupiter*; yea, without so much as the Elements of *Euclid*. There is nothing required for the comprehending the whole, but common Sense and practical Arithmetick; saving a few Touches of *Algebra*, as in the first Three Propositions, where the Reader, without suspicion of Popery, may make use of a strong implicit Faith; tho' I must confess, it does not much recommend it self to me in these Purposes; for I had rather he would enquire, and I believe he will find the Speculation not unpleasant.

Every man's Success in any Affair is proportional to his Conduct and Fortune.

[1] An apophthegm of Francis Bacon who attributes it to Diogenes.
[2] Isaac Newton.

Fortune (in the sense of most People) signifies an Event which depends on Chance, agreeing with my Wish; and Misfortune signifies such an one, whose immediate Causes I don't know, and consequently can neither foretel nor produce it (for it is no Heresy to believe, that Providence suffers ordinary matters to run in the Channel of second Causes). Now I suppose, that all a wise Man can do in such a Case is, to lay his Business on such Events, as have the most powerful second Causes, and this is true both in the great Events of the World, and in ordinary Games. It is impossible for a Die, with such determin'd force and direction, not to fall on such a determin'd side, only I don't know the force and direction which makes it fall on such a determin'd side, and therefore I call that Chance, which is nothing but want of Art; that only which is left to me, is to wager where there are the greatest number of Chances, and consequently the greatest probability to gain; and the whole Art of Gaming, where there is any thing of Hazard, will be reduc'd to this at last, *viz.* in dubious Cases to calculate on which side there are most Chances; and tho' this can't be done in the midst of Game precisely to an Unit, yet a Man who knows the Principles, may make such a conjecture, as will be a sufficient direction to him; and tho' it is possible, if there are any Chances against him at all, that he may lose, yet when he chuseth the safest side, he may part with his Money with more content (if there can be any at all) in such a Case.

I will not debate, whether one may engage another in a disadvantageous Wager. Games may be suppos'd to be a tryal of Wit as well as Fortune, and every Man, when he enters the Lists with another, unless out of Complaisance, takes it for granted, his Fortune and Judgment, are, at least, equal to those of his Play-Fellow; but this I am sure of, that false Dice, Tricks of *Leger-de-main*, *&c.* are inexcusable, for the question in Gaming is not, Who is the best Jugler?

The Reader may here observe the Force of Numbers, which can be successfully applied, even to those things, which one would imagine are subject to no Rules. There are very few things which we know, which are not capable of being reduc'd to a Mathematical Reasoning; and when they cannot, it's a sign our Knowledge of them is very small and confus'd; and where a mathematical reasoning can be had, it's as great a folly to make use of any other, as to grope for a thing in the dark, when you have a Candle standing by you. I believe the Calculation of the Quantity of Probability might be improved to a very useful and pleasant Speculation, and applied to a great many Events which are accidental besides those of Games; only these Cases would be infinitely more confus'd, as depending on Chances which the most part of Men are ignorant of; and as I have hinted already, all the Politicks in the World are nothing else but a kind of Analysis of the Quantity of Probability in casual Events, and a good Politician signifies no more, but one who is dextrous at such Calculations; only the Principles which are made use of in the Solution of such Problems, can't be studied in a Closet, but acquir'd by the Observation of Mankind.

There is likewise a Calculation of the Quantity of Probability founded on Experience, to be made use of in Wagers about any thing; it is odds, if a Woman is *with Child*, but it shall be a *Boy*; and if you would know the just odds, you must consider the Proportion in the Bills that the Males bear to the Females: The Yearly Bills of Mortality are observ'd to bear such Proportion to the live People as 1 to 30, or 26; therefore it is an even Wager, that one out of thirteen, dies within a Year (which may be a good reason, tho' not the true, of that foolish piece of Superstition), because, at this rate, if 1 out of 26 dies, you are no loser. It is but 1 to 18 if you meet a *Parson* in the Street, that he proves to be a *Non-Juror*[3], because there is but 1 of 36

[3] A 'Non-Juror' is one who refused to take an oath of allegiance to William and Mary in 1688.

that are such. It is hardly 1 to 10, that a *Woman* of Twenty Years old has her *Maidenhead*[4], and almost the same Wager, that a *Town-Spark* of that Age has not been *clap'd*. I think a Man might venture some odds, that 100 of the *Gens d'arms* beats an equal Number of *Dutch Troopers*; and that an *English Regiment* stands its ground as long as another, making Experience our Guide in all these Cases and others of the like nature.

But there are no casual Events, which are so easily subjected to Numbers, as those of Games; and I believe, there the Speculation might be improved so far, as to bring in the Doctrine of the *Series's* and *Logarithms*. Since Gaming is become a Trade, I think it fit the Adventurers should be upon the Square; and therefore in the Contrivance of Games there ought to be strict Calculation made use of, that they mayn't put one Party in more probability to gain them another; and likewise, if a Man has a considerable Venture; he ought to be allow'd to withdraw his Money when he pleases, paying according to the Circumstances he is then in: and it were easy in most Games to make Tables, by Inspection of which, a Man might know what he was either to pay or receive, in any Circumstances you can imagin, it being convenient to save a part of one's Money, rather than venture the loss of it all.

I shall add no more, but that a Mathematician will easily perceive, it is not put in such a Dress as to be taken notice of by him, there being abundance of Words spent to make the more ordinary sort of People understand it.

[4] Karl Pearson has suggested that this may be a reference to a short-lived Company for the Assurance of Female Chastity.

Bibliography

Apostol, T. M. (1957). *Mathematical analysis.* Addison-Wesley, Reading, MA.

Asmussen, S. (1987). *Applied probability and queues.* Wiley, New York.

Athreya, K. B. and Ney, P. E. (1972). *Branching processes.* Springer, Berlin.

Billingsley, P. (1986). *Probability and measure* (2nd edn). Wiley, New York.

Breiman, L. (1968). *Probability.* Addison-Wesley, Reading, MA.

Casanova de Seingalt (Giacomo Girolamo) (1922). *Memoirs* (trans. A. Machen), Vol. IV. Casanova Society, London.

Chatfield, C. (1980). *The analysis of time series.* Chapman and Hall, London.

Chung, K. L. (1960). *Markov chains with stationary transition probabilities.* Springer, Berlin.

Chung, K. L. (1974). *A course in probability theory.* Academic Press, New York.

Chung, K. L. (1979). *Elementary probability theory and stochastic processes.* Springer, Berlin.

Çinlar, E. (1975). *Introduction to stochastic processes.* Prentice-Hall, Englewood Cliffs, NJ.

Clarke, L. E. (1975). *Random variables.* Longman, London.

Cox, D. R. and Miller, H. D. (1965). *The theory of stochastic processes.* Chapman and Hall, London.

Cox, D. R. and Smith, W. L. (1961). *Queues.* Chapman and Hall, London.

Doob, J. L. (1953). *Stochastic processes.* Wiley, New York.

Dubins, L. and Savage, L. (1965). *How to gamble if you must.* McGraw-Hill, New York.

Dudley, R. M. (1989). *Real analysis and probability.* Wadsworth & Brooks/Cole, Pacific Grove, CA.

Etemadi, N. (1981). *Z. Wahrsch'theorie Geb.*, **55**, 119–22.

Feller, W. (1968). *An introduction to probability theory and its applications*, Vol. 1 (3rd edn). Wiley, New York.

Feller, W. (1971). *An introduction to probability theory and its applications*, Vol. 2 (2nd edn). Wiley, New York.

Freedman, D. (1971). *Markov chains.* Holden-Day, San Francisco.

Gray, J. R. (1967). *Probability.* Oliver and Boyd, Edinburgh.

Grimmett, G. R. and Stirzaker, D. R. (1992). *Probability and random processes: problems and solutions.* Clarendon Press, Oxford.

Grimmett, G. R. and Welsh, D. J. A. (1986). *Probability, an introduction.* Clarendon Press, Oxford.

Hall, M. (1967). *Combinatorial theory.* Blaisdell, Waltham, MA.

Hall, P. and Heyde, C. C. (1980). *Martingale limit theory and its application.* Academic Press, New York.

Halmos, P. R. (1960). *Naive set theory.* Van Nostrand, Princeton, NJ.

Harris, T. E. (1963). *The theory of branching processes.* Springer, Berlin.

Hildebrand, F. B. (1962). *Advanced theory of calculus.* Prentice-Hall, Englewood Cliffs, NJ.

Hoel, P. G., Port, S. C., and Stone, C. J. (1971*a*). *Introduction to probability theory.* Houghton Mifflin, Boston.

Hoel, P. G., Port, S. C., and Stone, C. J. (1971*b*). *Introduction to stochastic processes.* Houghton Mifflin, Boston.

Karlin, S. and Taylor, H. M. (1975). *A first course in stochastic processes* (2nd edn). Academic Press, New York.

Karlin, S. and Taylor, H. M. (1981). *A second course in stochastic processes.* Academic Press, New York.

Kelly, F. P. (1979). *Reversibility and stochastic networks.* Wiley, New York.

Kingman, J. F. C. and Taylor, S. J. (1966). *Introduction to measure and probability.* Cambridge University Press, Cambridge.

Krüger, L. *et al.*, eds (1987). *The probabilistic revolution*, (2 vols). MIT Press, Cambridge, MA.

Laha, R. G. and Rohatgi, V. K. (1979). *Probability theory.* Wiley, New York.

Lindvall, T. (1977). *Ann. Probab.*, **5**, 482–5.

Lipschutz, S. (1974). *Linear algebra*, Schaum Outline Series. McGraw-Hill, New York.

Loève, M. (1977). *Probability theory*, Vol. 1 (4th edn). Springer, Berlin.

Loève, M. (1978). *Probability theory*, Vol. 2 (4th edn). Springer, Berlin.

Lukacs, E. (1970). *Characteristic functions* (2nd edn). Griffin, London.

Mandelbrot, B. (1983). *The fractal geometry of nature.* Freeman, San Francisco.

Moran, P. A. P. (1968). *An introduction to probability theory.* Clarendon Press, Oxford.

Nevanlinna, R. and Paatero, V. (1969). *Introduction to complex analysis.* Addison-Wesley, Reading, MA.

Neveu, J. (1975). *Discrete parameter martingales.* North-Holland, Amsterdam.

Parzen, E. (1962). *Stochastic processes.* Holden-Day, San Francisco.

Phillips, E. G. (1957). *Functions of a complex variable.* Oliver and Boyd, Edinburgh.

Piaggio, H. T. H. (1965). *Differential equations.* Bell, London.

Prabhu, N. U. (1965*a*). *Stochastic processes.* Collier Macmillan, New York.

Prabhu, N. U. (1965*b*). *Queues and inventories.* Wiley, New York.

Rogers, L. C. G. and Williams, D. (1987). *Diffusions, Markov processes, and martingales*, Vol. 2. Wiley, New York.

Ross, S. (1970). *Applied probability models with optimization applications.* Holden-Day, San Francisco.

Ross, S. (1976). *A first course in probability.* Collier Macmillan, New York.

Rudin, W. (1976). *Principles of mathematical analysis* (3rd edn). McGraw-Hill, New York.

Shiryayev, A. N. (1984). *Probability.* Springer, Berlin.

Whittle, P. (1970). *Probability.* Penguin, Harmondsworth, Middlesex.

Williams, D. (1979). *Diffusions, Markov processes, and martingales*, Vol. 1. Wiley, New York.

Williams, D. (1991). *Probability and martingales.* Cambridge University Press, Cambridge.

List of notation

a, b, c, d	constants
$c(n), c(t)$	autocovariances
$\mathrm{cov}(X, Y)$	covariance
f, f_j, f_{ij}	probabilities
$f(x), f_X(\cdot)$	mass or density functions
$f_{Y\mid X}(y \mid x)$	conditional density
$f_{X,Y}(x, y)$	joint density
$g(\cdot), h(\cdot)$	nice functions
i	$\sqrt{(-1)}$
i, j, k, l, m, n, r, s	indices
$m(\cdot), m^D(\cdot)$	mean functions
max, min	maximum, minimum
$p, p_i, p_{ij}, p(t), p_i(t)$	probabilities
$\mathrm{var}(X)$	variance
$B(a, b)$	beta function
$B(n, p)$	binomial distribution
$C(t), D(t), E(t)$	current, total, excess life
$\mathbf{E}$	expectation
$F(r, s)$	F distribution
$F(x), F_X(x)$	distribution functions
$F_{Y\mid X}(y \mid x)$	conditional distribution
$F_{X,Y}(x, y)$	joint distribution
$G(s), G_X(s)$	generating functions
H, T	head, tail
I_A	indicator of the event A
J	Jacobian
$M(t)$	moment generating function
$N(\mu, \sigma^2)$	normal distribution
$N(t)$	Poisson process
$\mathbf{P}(\cdot), \mathbf{Q}(\cdot)$	probability measures
$Q(t)$	queue length
$X, Y, Z, X(\omega)$	random variables
$\boldsymbol{X}, \boldsymbol{Y}, \boldsymbol{W}$	random vectors
$\boldsymbol{V}(\boldsymbol{X})$	covariance matrix
$W(t), \boldsymbol{W}(t)$	Wiener processes
W, W_n	waiting times
$\mathscr{A}, \mathscr{B}, \mathscr{F}, \mathscr{G}, \mathscr{H}, \mathscr{I}$	σ-fields
$\mathscr{B}$	Borel σ-field
δ_{ij}	Kronecker delta
$\delta(t)$	Dirac delta
η	probability of extinction
$\chi^2(\cdot)$	chi-squared distribution
$\phi(t)$	characteristic function
$\psi(X)$	conditional expectation
μ	mean
μ_i	mean recurrence time
$\boldsymbol{\pi}$	stationary distribution
σ	standard deviation
$\rho(n)$	autocorrelation
ω	elementary event
$\rho(X, Y)$	correlation between X and Y
$\Gamma(n)$	gamma function
$\Gamma(\lambda, t)$	gamma distribution
Ω	sample space
$\Phi(x)$	normal distribution
$\mathbb{C}$	complex plane
$\mathbb{R}$	real numbers
$\mathbb{Z}$	integers
$\varnothing$	empty set
$f * g$	convolution
$\hat{g}$	Laplace transform
g^*	Laplace–Stieltjes transform
$\\|\cdot\\|$	norm
$\|A\|$	cardinality of A
$\boldsymbol{A}'$	transpose of $\boldsymbol{A}$
$\|\boldsymbol{V}\|$	determinant of $\boldsymbol{V}$
$f'(t)$	derivative of f
$\boldsymbol{A}'$	matrix with entries $a'_{ij}(t)$
$\lfloor x \rfloor$	integer part of x
$\lceil x \rceil$	least integer not smaller than x

Index

Abbreviations used in this index: c.f. characteristic function; p.g.f. probability generating function; r.v. random variable; r.w. random walk